MULTIVARIATE DATA REDUCTION AND DISCRIMINATION WITH SAS® SOFTWARE

Ravindra Khattree
Dayanand N. Naik

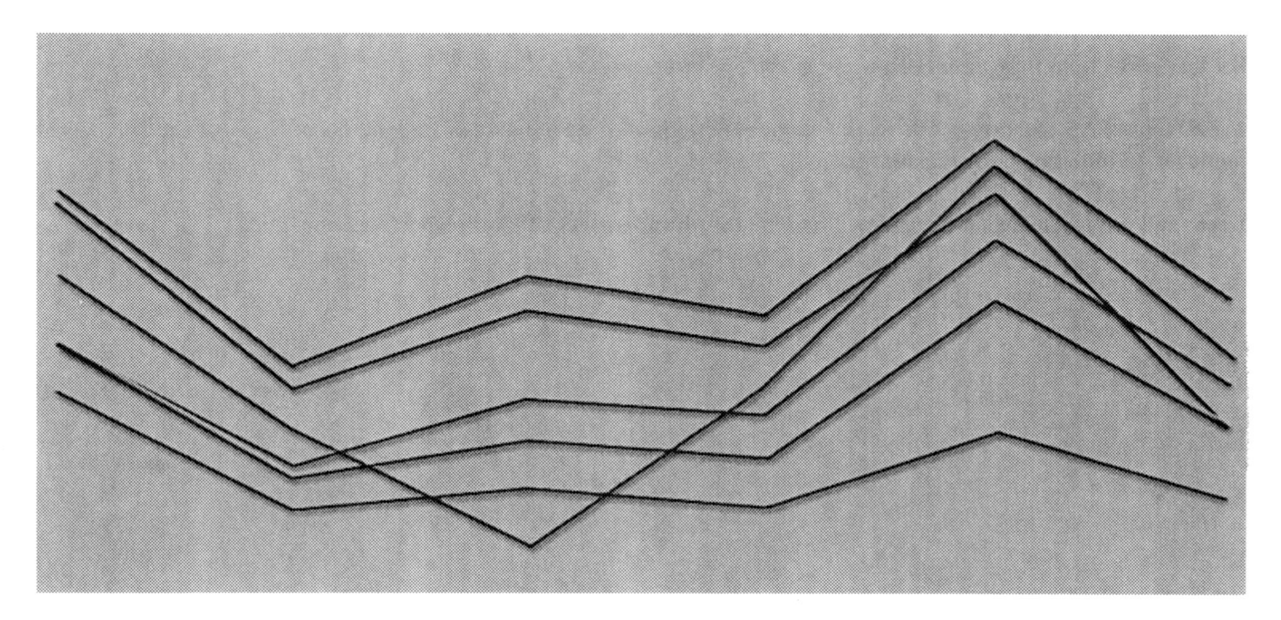

Comments or Questions?

The author assumes complete responsibility for the technical accuracy of the content of this book. If you have any questions about the material in this book, please write to the author at this address:

SAS Institute Inc.
Books By Users
Attn: Ravindra Khattree and Dayanand N. Naik
SAS Campus Drive
Cary, NC 27513

If you prefer, you can send e-mail to sasbbu@sas.com with "comments on Multivariate Data Reduction and Discrimination with SAS Software" as the subject line, or you can fax the Books by Users program at (919) 677-4444.

The correct bibliographic citation for this manual is as follows: Khattree, Ravindra and Dayanand N. Naik, *Multivariate Data Reduction and Discrimination with SAS® Software,* Cary, NC: SAS Institute Inc., 2000.

Multivariate Data Reduction and Discrimination with SAS® Software

SAS Institute Inc. ISBN 1-58025-696-1
John Wiley & Sons, Inc. ISBN 0-471-32300-4

SAS Institute Inc., SAS Campus Drive, Cary, North Carolina 27513.

1st printing, May 2000

Note that text corrections may have been made at each printing.

To the joy of my life, my daughter Vaidehee Gauree (R.K.)

To Baba and all my friends (D.N.N.)

Contents

Preface

It is indeed a pleasure to bring to our audience this companion volume on applied multivariate statistics. Our earlier book, *Applied Multivariate Statistics with SAS Software*, was well received in both the first edition and the extensively revised second edition. Those books largely dealt with those multivariate statistical techniques and analyses for which the assumption of multivariate normality played a pivotal role. The present volume attempts to cover other aspects of multivariate statistics—especially those in which the emphasis is more on data reduction, description, estimation, discrimination, and summarization. In that sense what is presented here can be called multivariate descriptive statistics. In this book, the distributional assumption of either multivariate normality or any other probability distribution, although sometimes made in the text, is not crucial for the most part.

Applied multivariate techniques are routinely used in a variety of disciplines such as agriculture, anthropology, applied statistics, biological science, business, chemistry, econometrics, education, engineering, marketing, medicine, psychology, quality control, and sociology. With such a diverse readership, we thought it essential to present multivariate techniques in a general context while at the same time keeping the mathematical and statistical requirements to a bare minimum. We have sincerely attempted to achieve this objective.

Audience

The book is written both as a handy reference for researchers and practitioners and also as a main or supplementary college text for graduate or senior undergraduate students. Parts of it can also be selected to make an ideal text for a one- or two-day tutorial on descriptive multivariate statistics or for self-study. Many statistical consultants as well as consultants in allied sciences using statistical techniques will find it to be useful. The book can also be conveniently adapted for a statistics service course for students from other disciplines that use multivariate methods.

Another important group that will greatly benefit from this book is those who are interested in data mining. With the current surge of interest in the mining of large data sets, the present book provides an avenue to understand and appreciate the important statistical techniques that come into play in the process. Of special interest are such topics as principal components, discrimination and classification, clustering, and correspondence analysis. These are used first for data reduction and then in the descriptive or predictive data mining processes. While we do not pretend this to be a book solely devoted to data mining, we have deliberately considered the needs of this group of users in writing this book.

Approach

The primary emphasis is on statistical methodology as applied to various scientific disciplines. SAS software is used to carry out various intensive calculations that naturally occur in any typical multivariate analysis application. The general theme of the discussion here is data reduction and its description and estimation from multivariate data.

The concepts, analyses, and programs are integrated in order to impart adequate knowledge of the subject without the reader having to turn to a number of textbooks or software manuals. The mathematical level is kept to a minimum and much of what may be needed is summarized in Chapter 1. We believe that the use of real data reinforces the message about the usefulness of the techniques being applied and often helps the reader to see the overall picture in a practical context. With this in mind, we selected most of the data sets used in the examples from the published literature of a variety of disciplines.

Prerequisites

A course in applied statistics with some emphasis on statistical modeling and some familiarity from matrix algebra provides an adequate preparation to read this book. Much of what is needed in terms of matrix algebra concepts and facts has been compiled in the introductory chapter of the book, and sections from this chapter can be read as the need arises. Some familiarity with SAS is also helpful. Many excellent introductory level books on SAS programming are available from SAS Institute Inc.

Overview of Chapters

Chapter 1 introduces and summarizes the basic concepts. The distinction between the population and sampled data is often overlooked by practitioners from nonstatistical disciplines. Yet not only notationally, but also in terms of interpretations, such distinctions do matter. This distinction has been emphasized here. Along with that, we provide a brief and largely self-sufficient list of matrix theory definitions and results as a separate section. Admittedly, many of our definitions are just the working definitions rather than the formal ones. This is done to keep the mathematical complexity of the discussion to a minimum. Whenever possible or needed, we also point out how the relevant quantities can be computed using the SAS IML procedure. A brief discussion of multivariate normality is also provided since some of the chapters in the book use it.

Chapter 2 is about data reduction using the principal components. Indisputably, apart from the techniques of selection of variables (mostly in regression contexts), principal component analysis is the most popular method used for variable reduction. Chapter 3 deals with canonical correlation analysis. The techniques used here are pretty standard, and we indicate a variety of situations in which this analysis is appropriate. Chapter 4 discusses factor analysis techniques. This is a controversial topic often shunned by classical statisticians who claim that it is ambiguous and dubious and by nonstatisticians who view it as complex and confusing. Being aware of both types of criticism, we have attempted to carefully present a systematic discussion of a number of factor analyses. As a result, it is a large chapter, and we try to propose some order and natural development in what is often termed the *jungle out there*.

Discriminant analysis is discussed in Chapter 5. It can be viewed as a subject in its own right, and thus it deserves a larger number of pages than other chapters in the book. A number (but not all) of the discrimination and classification techniques are systematically presented here. Chapter 6 addresses clustering issues. There is an unending list of various clustering procedures that are available in the literature. Thus, we have confined ourselves to only a few of the leading and most popular ones. While selecting these techniques, we were driven by the practical implementation issues and also by the philosophy that, for most situations, clustering is only an intermediate step of the analysis and not the end of it. Thus, the most sophisticated analyses may not always be called for. All the previous chapters generally assume multivariate data that are quantitative in nature. Chapter 7 covers correspondence analysis. It is a topic dealing with categorical data, and in recent years there has been a surge of interest in correspondence analysis. The chapter attempts to unify much of the relevant and quite recent literature with the available software. Finally, an appendix containing most of the larger data sets used in the book has been included.

We have attempted to cover only some of the main topics in descriptive multivariate analysis. We readily admit that the treatment is not exhaustive. We must also confess that our personal preferences and research interests have played some role in the selection of the material presented here.

In a work of this size, integrating various aspects of statistical methods and data analysis, there are bound to be some errors and gaps that we may have missed. We would greatly appreciate any comments, suggestions, or criticisms that would help us improve this work.

Textual Note

We have used the DATALINES statement and its aliases CARDS and LINES in our programs to indicate that data lines follow. We show variable names in both uppercase and mixed case. The software accepts either form.

Acknowledgments

A number of people have contributed to this project in a number of ways. Our sincere thanks are due to outside peer reviewers and internal reviewers at SAS Institute who critically examined one or several drafts of this book during various stages of writing. These include, Professor Andrezej Galecki (University of Michigan), Professor Robert Ling (Clemson University), Professor Kenneth Portier (University of Florida), Professor David Scott (Rice University), as well as Brent Cohen, Kristin Nauta, Mike Patetta, Donna Sawyer, and David Schlotzhauer of SAS Institute. Their suggestions were most helpful in improving this work. Parts of the book were also read by our student Bonnie Davis, and we sincerely appreciate her comments. Special thanks are also due to our colleague and long-time friend Professor N. Rao Chaganty for all his help during various stages of this project. Parts of the book were typed by Lynnette Folken, Barbara Jeffrey, Kathy Jegla, and Gayle Tarkelsen. We kindly thank them for their assistance. Others at SAS Institute, especially Caroline Brickley, Julie Platt, and Judy Whatley, were also helpful and we appreciate their understanding.

Last, but not least, our sincere thanks go to our wives, Nidhi and Sujatha, and our little ones, Vaidehee and Navin, for allowing us to work during late, odd hours and over the weekends. We thank them for their understanding that this book could not have been completed just by working during regular hours.

R. KHATTREE
Rochester, Michigan

D. N. NAIK
Norfolk, Virginia

Commonly Used Notation

$\mathbf{I}_n$	The n by n identity matrix
$\mathbf{1}_n$	The n by 1 vector of unit elements
$\mathbf{O}$	A matrix of appropriate order with all zero entries
$\lambda_i,\ \hat{\lambda}_i$	The ith largest eigenvalue of the matrix under consideration
$\mathbf{A}'$	The transpose of a matrix $\mathbf{A}$
$diag(\mathbf{A})$	The matrix $\mathbf{A}$ after replacing the nondiagonal entries by zeros
$R(\mathbf{A})$	The rank of a matrix $\mathbf{A}$
$\|\mathbf{A}\|$	The determinant of the square matrix $\mathbf{A}$
$tr(\mathbf{A})$	The trace of the square matrix $\mathbf{A}$
$\mathbf{A}^{-1}$	The inverse of the matrix $\mathbf{A}$
$\mathbf{A}^{1/2}$	The symmetric square root of the matrix $\mathbf{A}$
$\mathbf{A}^{-}$	A generalized inverse of the matrix $\mathbf{A}$
$E(\mathbf{y})$	Expected value of a random variable or vector $\mathbf{y}$
$v(y),\ var(y)$	Variance of a random variable y
$cov(\mathbf{x}, \mathbf{y})$	Covariance *of* a random variable or (vector) $\mathbf{x}$ *with* a random variable (or vector) $\mathbf{y}$
$D(\mathbf{y})$	The variance-covariance or the dispersion matrix of $\mathbf{y}$
$N_p(\boldsymbol{\mu}, \boldsymbol{\Sigma})$	A p-dimensional normal distribution with mean $\boldsymbol{\mu}$ and the variance-covariance matrix $\boldsymbol{\Sigma}$
$\Phi(\cdot)$	The cumulative standard normal probability
$\boldsymbol{\epsilon}$	Error vector
$\boldsymbol{\beta}$	Regression/Design parameter vector
$\mathbf{B}$	Regression/Design parameter matrix
$\boldsymbol{\mu},\ \boldsymbol{\mu}_1,\ \boldsymbol{\mu}_2$	Population mean vectors
$\boldsymbol{\Sigma}$	The dispersion matrix of errors (usually)
$\boldsymbol{\rho},\ \mathbf{R}$	The correlation matrix
df	Degrees of freedom
SS&CP Matrix	Matrix of the sums of squares and crossproducts
$\mathbf{E}$	Error SS&CP matrix
$\mathbf{H}$	Hypothesis SS&CP matrix
$\bar{\mathbf{y}}$	The sample mean vector
$\mathbf{S}$	Sample dispersion matrix (with df as denominator)
$\mathbf{S}_n$	Sample dispersion matrix (with sample size as denominator)
T^2	Hotelling's T^2

$\Lambda, \ \Lambda_p$	Wilks' Lambda
$\beta_{1,p}$	Coefficient of multivariate skewness
$\beta_{2,p}$	Coefficient of multivariate kurtosis
$\otimes$	Kronecker product
AIC	Akaike's information criterion
BIC	Schwarz's Bayesian information criterion
$\mathbf{f}$	Vector of common factors
$\mathbf{L}$	Matrix of factor loadings
ψ_i	The ith specific variance (in the context of factor analysis)
Ψ	The diagonal matrix of specific variances
$\delta(\mathbf{a}, \mathbf{b}), \ d(\mathbf{a}, \mathbf{b})$	Distance between points **a** and **b**
ρ_i	The ith canonical correlation coefficient for the population (usually)
r_i	The ith canonical correlation coefficient for the sample (usually)
$\Pi_1, \ \Pi_2, \ldots$	Populations (in the context of Discriminant Analysis)
$\pi_1, \ \pi_2, \ldots$	Prior probabilities for populations (in the context of Discriminant Analysis)
$P(i \mid \mathbf{x})$	Posterior probability for the ith population given data $\mathbf{x}$
$Logit(\cdot)$	Logit function
$K(\cdot)$	The kernel function
$s(a, b), \ s_{ab}$	Similarity index
D	The distance matrix
$h(B_r, B_s)$	Distance between clusters B_r and B_s

Chapter 1

Basic Concepts for Multivariate Statistics

1.1 Introduction

Data are information. Most crucial scientific, sociological, political, economic, and business decisions are made based on data analyis. Often data are available in abundance, but by themselves they are of little help unless they are summarized and an appropriate interpretation of the summary quantities made. However, such a summary and corresponding interpretation can rarely be made just by looking at the raw data. A careful scientific scrutiny and analysis of these data can usually provide an enormous amount of valuable information. Often such an analysis may not be obtained just by computing simple averages. Admittedly, the more complex the data and their structure, the more involved the data analysis.

The complexity in a data set may exist for a variety of reasons. For example, the data set may contain too many observations that stand out and whose presence in the data cannot be justified by any simple explanation. Such observations are often viewed as influential observations or outliers. Deciding which observation is or is not an influential one is a difficult problem. For a brief review of some graphical and formal approaches to this problem, see Khattree and Naik (1999). A good, detailed discussion of these topics can be found in Belsley, Kuh and Welsch (1980), Belsley (1991), Cook and Weisberg (1982), and Chatterjee and Hadi (1988).

Another situation in which a simple analysis based on averages alone may not suffice occurs when the data on some of the variables are correlated or when there is a trend present in the data. Such a situation often arises when data were collected over time. For example, when the data are collected on a single patient or a group of patients under a given treatment, we are rarely interested in knowing the average response over time. What we are interested in is observing any changes in the values, that is, in observing any patterns or trends.

Many times, data are collected on a number of units, and on each unit not just one, but many variables are measured. For example, in a psychological experiment, many tests are used, and each individual is subjected to all these tests. Since these are measurements on the same unit (an individual), these measurements (or variables) are correlated and, while summarizing the data on all these variables, this set of correlations (or some equivalent quantity) should be an integral part of this summary. Further, when many variables exist, in order to obtain more definite and more easily comprehensible information, this correlation summary (and its structure) should be subjected to further analysis. There are many other possible ways in which a data set can be quite complex for analysis.

However, it is the last situation that is of interest to us in this book. Specifically, we may have n individual units and on each unit we have observed (same) p different characteristics (variables), say $x_1, x_2, \ldots, x_p$. Then these data can be presented as an n by p matrix

$$\mathbf{X} = \begin{bmatrix} x_{11} & x_{12} & \ldots & x_{1p} \\ x_{21} & x_{22} & \ldots & x_{2p} \\ \vdots & & & \vdots \\ x_{n1} & x_{n2} & \ldots & x_{np} \end{bmatrix}.$$

Of course, the measurements in the ith row, namely, $x_{i1}, \ldots, x_{ip}$, which are the measurements on the same unit, are correlated. If we arrange them in a column vector $\mathbf{x}_i$ defined as

$$\mathbf{x}_i = \begin{bmatrix} x_{i1} \\ \vdots \\ x_{ip} \end{bmatrix},$$

then $\mathbf{x}_i$ can be viewed as a multivariate observation. Thus, the n rows of matrix $\mathbf{X}$ correspond to n multivariate observations (written as rows within this matrix), and the measurements within each $\mathbf{x}_i$ are usually correlated. There may or may not be a correlation between columns $\mathbf{x}_1, \ldots, \mathbf{x}_n$. Usually, $\mathbf{x}_1, \ldots, \mathbf{x}_n$ are assumed to be uncorrelated (or statistically independent as a stronger assumption) but this may not always be so. For example, if $\mathbf{x}_i$, $i = 1, \ldots, n$ contains measurements on the height and weight of the ith brother in a family with n brothers, then it is reasonable to assume that some kind of correlation may exist between the rows of $\mathbf{X}$ as well.

For much of what is considered in this book, we will not concern ourselves with the scenario in which rows of the data matrix $\mathbf{X}$ are also correlated. In other words, when rows of $\mathbf{X}$ constitute a sample, such a sample will be assumed to be statistically independent. However, before we elaborate on this, we should briefly comment on sampling issues.

1.2 Population Versus Sample

As we pointed out, the rows in the n by p data matrix $\mathbf{X}$ are viewed as multivariate observations on n units. If the set of these n units constitutes the entire (finite) set of all possible units, then we have data available on the entire reference population. An example of such a situation is the data collected on all cities in the United States that have a population of 1,000,000 or more, and on three variables, namely, cost-of-living, average annual salary, and the quality of health care facilities. Since each U.S. city that qualifies for the definition is included, any summary of these data will be the *true* summary of the population.

However, more often than not, the data are obtained through a survey in which, on each of the units, all p characteristics are measured. Such a situation represents a multivariate sample. A sample (adequately or poorly) represents the underlying population from which it is taken. As the population is now represented through only a few units taken from it, any summary derived from it merely represents the *true* population summary in the sense that we hope that, generally, it will be close to the true summary, although no assurance about an exact match between the two can be given.

How can we measure and ensure that the summary from a sample is a good representative of the population summary? To quantify it, some kinds of indexes based on probabilistic ideas seem appropriate. That requires one to build some kind of probabilistic structure over these units. This is done by artificially and intentionally introducing the probabilistic structure into the sampling scheme. Of course, since we want to ensure that the sample is a good representative of the population, the probabilistic structure should be such that it treats all the population units in an equally fair way. Thus, we require that the sampling is done in such a way that each unit of (finite or infinite) population has an equal chance of being included in the sample. This requirement can be met by a simple random sampling with or without replacement. It may be pointed out that in the case of a finite population and sampling without replacement, observations are *not* independent, although the strength of dependence diminishes as the sample size increases.

Although a probabilistic structure is introduced over different units through random sampling, the same cannot be done for the p different measurements, as there is neither a reference population nor do all p measurements (such as weight, height, etc.) necessarily represent the same thing. However, there is possibly some inherent dependence between these measurements, and this dependence is often assumed and modeled as some joint probability distribution. Thus, we view each row of $\mathbf{X}$ as a multivariate observation from some p-dimensional population that is represented by some p-dimensional multivariate distribution. Thus, the rows of $\mathbf{X}$ often represent a random sample from a p-dimensional population. In much multivariate analysis work, this population is assumed to be infinite and quite frequently it is assumed to have a multivariate normal distribution. We will briefly discuss the multivariate normal distribution and its properties in Section 1.6.

1.3 Elementary Tools for Understanding Multivariate Data

To understand a large data set on several mutually dependent variables, we must somehow summarize it. For univariate data, when there is only one variable under consideration, these are usually summarized by the (population or sample) mean, variance, skewness, and kurtosis. These are the basic quantities used for data description. For multivariate data, their counterparts are defined in a similar way. However, the description is greatly simplified if matrix notations are used. Some of the matrix terminology used here is defined later in Section 1.5.

Let $\mathbf{x}$ be the p by 1 random vector corresponding to the multivariate population under consideration. If we let

$$\mathbf{x} = \begin{bmatrix} x_1 \\ \vdots \\ x_p \end{bmatrix},$$

then each x_i is a random variable, and we assume that $x_1, \ldots, x_p$ are possibly dependent. With $E(\cdot)$ representing the mathematical expectation (interpreted as the long-run average), let $\mu_i = E(x_i)$, and let $\sigma_{ii} = var(x_i)$ be the population variance. Further, let the population covariance between x_i and x_j be $\sigma_{ij} = cov(x_i, x_j)$. Then we define the *population mean vector* $E(\mathbf{x})$ as the vector of term by term expectations. That is,

$$E(\mathbf{x}) = \begin{bmatrix} E(x_1) \\ \vdots \\ E(x_p) \end{bmatrix} = \begin{bmatrix} \mu_1 \\ \vdots \\ \mu_p \end{bmatrix} = \boldsymbol{\mu} \text{ (say)}.$$

Additionally, the concept of population variance is generalized to the matrix with all the population variances and covariances placed appropriately within a variance-covariance matrix. Specifically, if we denote the variance-covariance matrix of $\mathbf{x}$ by $D(\mathbf{x})$, then

$$D(\mathbf{x}) = \begin{bmatrix} var(x_1) & cov(x_1, x_2) & \dots & cov(x_1, x_p) \\ cov(x_2, x_1) & var(x_2) & \dots & cov(x_2, x_p) \\ \vdots & & & \vdots \\ cov(x_p, x_1) & cov(x_p, x_2) & \dots & var(x_p) \end{bmatrix}$$

$$= \begin{bmatrix} \sigma_{11} & \sigma_{12} & \dots & \sigma_{1p} \\ \sigma_{21} & \sigma_{22} & \dots & \sigma_{2p} \\ \vdots & & & \vdots \\ \sigma_{p1} & \sigma_{p2} & \dots & \sigma_{pp} \end{bmatrix} = (\sigma_{ij}) = \mathbf{\Sigma} \text{ (say)}.$$

That is, with the understanding that $cov(x_i, x_i) = var(x_i) = \sigma_{ii}$, the term $cov(x_i, x_j)$ appears as the $(i, j)^{th}$ entry in matrix $\mathbf{\Sigma}$. Thus, the variance of the ith variable appears at the ith diagonal place and all covariances are appropriately placed at the nondiagonal places. Since $cov(x_i, x_j) = cov(x_j, x_i)$, we have $\sigma_{ij} = \sigma_{ji}$ for all i, j. Thus, the matrix $D(\mathbf{x}) = \mathbf{\Sigma}$ is symmetric. The other alternative notations for $D(\mathbf{x})$ are $cov(\mathbf{x})$ and $var(\mathbf{x})$, and it is often also referred to as the dispersion matrix, the variance-covariance matrix, or simply the covariance matrix. We will use the three terms interchangeably.

The quantity $tr(\mathbf{\Sigma})$ (read as trace of $\mathbf{\Sigma}$)$= \sum_{i=1}^{p} \sigma_{ii}$ is called the *total variance* and $|\mathbf{\Sigma}|$ (the determinant of $\mathbf{\Sigma}$) is referred to as the *generalized variance.* The two are often taken as the overall measures of variability of the random vector $\mathbf{x}$. However, sometimes their use can be misleading. Specifically, the total variance $tr(\mathbf{\Sigma})$ completely ignores the nondiagonal terms of $\mathbf{\Sigma}$ that represent the covariances. At the same time, two very different matrices may yield the same value of the generalized variance.

As there exists dependence between $x_1, \dots, x_p$, it is also meaningful to at least measure the degree of linear dependence. It is often measured using the correlations. Specifically, let

$$\rho_{ij} = \frac{cov(x_i, x_j)}{\sqrt{var(x_i)\, var(x_j)}} = \frac{\sigma_{ij}}{\sqrt{\sigma_{ii}\, \sigma_{jj}}}$$

be the Pearson's population correlation coefficient between x_i and x_j. Then we define the population correlation matrix as

$$\boldsymbol{\rho} = (\rho_{ij}) = \begin{bmatrix} \rho_{11} & \rho_{12} & \dots & \rho_{1p} \\ \rho_{21} & \rho_{22} & \dots & \rho_{2p} \\ & & & \\ \rho_{p1} & \rho_{p2} & \dots & \rho_{pp} \end{bmatrix} = \begin{bmatrix} 1 & \rho_{12} & \dots & \rho_{1p} \\ \rho_{21} & 1 & \dots & \rho_{pp} \\ & & & \\ \rho_{p1} & \rho_{p2} & \dots & 1 \end{bmatrix}.$$

As was the case for $\mathbf{\Sigma}$, $\boldsymbol{\rho}$ is also symmetric. Further, $\boldsymbol{\rho}$ can be expressed in terms of $\mathbf{\Sigma}$ as

$$\boldsymbol{\rho} = [\text{diag}(\mathbf{\Sigma})]^{-\frac{1}{2}}\, \mathbf{\Sigma}\, [\text{diag}(\mathbf{\Sigma})]^{-\frac{1}{2}},$$

where diag($\mathbf{\Sigma}$) is the diagonal matrix obtained by retaining the diagonal elements of $\mathbf{\Sigma}$ and by replacing all the nondiagonal elements by zero. Further, the square root of matrix $\mathbf{A}$ denoted by $\mathbf{A}^{\frac{1}{2}}$ is a matrix satisfying $\mathbf{A} = \mathbf{A}^{\frac{1}{2}}\mathbf{A}^{\frac{1}{2}}$. It is defined in Section 1.5. Also, $\mathbf{A}^{-\frac{1}{2}}$ represents the inverse of matrix $\mathbf{A}^{\frac{1}{2}}$.

It may be mentioned that the variance-covariance and the correlation matrices are always nonnegative definite (See Section 1.5 for a discussion). For most of the discussion in this book, these matrices, however, will be assumed to be positive definite. In view of this assumption, these matrices will also admit their respective inverses.

How do we generalize (and measure) the skewness and kurtosis for a multivariate population? Mardia (1970) defines these measures as

$$\text{multivariate skewness: } \beta_{1,p} = E\left[(\mathbf{x} - \boldsymbol{\mu})'\mathbf{\Sigma}^{-1}(\mathbf{y} - \boldsymbol{\mu})\right]^3,$$

where $\mathbf{x}$ and $\mathbf{y}$ are independent but have the same distribution and

$$\text{multivariate kurtosis: } \beta_{2,p} = E\left[(\mathbf{x}-\boldsymbol{\mu})'\boldsymbol{\Sigma}^{-1}(\mathbf{x}-\boldsymbol{\mu})\right]^2.$$

For the univariate case, that is when $p = 1$, $\beta_{1,p}$ reduces to the *square* of the coefficient of skewness, and $\beta_{2,p}$ reduces to the coefficient of kurtosis.

The quantities $\boldsymbol{\mu}$, $\boldsymbol{\Sigma}$, $\boldsymbol{\rho}$, $\beta_{1,p}$ and $\beta_{2,p}$ provide a basic summary of a multivariate population. What about the sample counterparts of these quantities? When we have a p-variate random sample $\mathbf{x}_1, \ldots, \mathbf{x}_n$ of size n, then with the n by p data matrix $\mathbf{X}$ defined as

$$\mathbf{X}_{n\times p} = \begin{bmatrix} \mathbf{x}_1' \\ \vdots \\ \mathbf{x}_n' \end{bmatrix},$$

we define,

$$\text{sample mean vector: } \overline{\mathbf{x}} = n^{-1}\sum_{i=1}^{n}\mathbf{x}_i = n^{-1}\mathbf{X}'\mathbf{1}_n,$$

$$\begin{aligned}\text{sample variance-covariance matrix: } \mathbf{S} &= (n-1)^{-1}\sum_{i=1}^{n}(\mathbf{x}_i-\overline{\mathbf{x}})(\mathbf{x}_i-\overline{\mathbf{x}})' \\ &= (n-1)^{-1}\left\{\sum_{i=1}^{n}\mathbf{x}_i\mathbf{x}_i' - n\,\overline{\mathbf{x}}\,\overline{\mathbf{x}}'\right\} \\ &= (n-1)^{-1}\left\{\mathbf{X}'(\mathbf{I}-n^{-1}\mathbf{1}_n\mathbf{1}_n')\mathbf{X}\right\} \\ &= (n-1)^{-1}\left\{\mathbf{X}'\mathbf{X}-n^{-1}\mathbf{X}'\mathbf{1}_n\mathbf{1}_n'\mathbf{X}\right\} \\ &= (n-1)^{-1}\left\{\mathbf{X}'\mathbf{X}-n\overline{\mathbf{x}}\,\overline{\mathbf{x}}'\right\}.\end{aligned}$$

It may be mentioned that often, instead of the dividing factor of $(n-1)$ in the above expressions, a dividing factor of n is used. Such a sample variance-covariance matrix is denoted by $\mathbf{S}_n$. We also have

$$\begin{aligned}\text{sample correlation matrix: } \hat{\boldsymbol{\rho}} &= \left[\text{diag}(\mathbf{S})\right]^{-\frac{1}{2}}\mathbf{S}\left[\text{diag}(\mathbf{S})\right]^{-\frac{1}{2}} \\ &= \left[\text{diag}(\mathbf{S}_n)\right]^{-\frac{1}{2}}\mathbf{S}_n\left[\text{diag}(\mathbf{S}_n)\right]^{-\frac{1}{2}},\end{aligned}$$

$$\text{sample multivariate skewness: } \hat{\beta}_{1,p} = n^{-2}\sum_{i=1}^{n}\sum_{j=1}^{n}g_{ij}^3,$$

and

$$\text{sample multivariate kurtosis: } \hat{\beta}_{2,p} = n^{-1}\sum_{i=1}^{n}g_{ii}^2.$$

In the above expressions, $\mathbf{1}_n$ denotes an n by 1 vector with all entries 1, $\mathbf{I}_n$ is an n by n identity matrix, and $g_{ij}, i, j = 1, \ldots p$, are defined by $g_{ij} = (\mathbf{x}_i-\overline{\mathbf{x}})'\mathbf{S}_n^{-1}(\mathbf{x}_j-\overline{\mathbf{x}})$. See Khattree and Naik (1999) for details and computational schemes to compute these quantities. In fact, multivariate skewness and multivariate kurtosis are computed later in Chapter 5, Section 5.2 to test the multivariate normality assumption on data. Correlation matrices also play a central role in principal components analysis (Chapter 2, Section 2.2).

1.4 Data Reduction, Description, and Estimation

In the previous section, we presented some of the basic population summary quantities and their sample counterparts commonly referred to as *descriptive statistics*. The basic idea was to summarize the population or sample data through smaller sized matrices or simply numbers. All the quantities (except correlation) defined there were straightforward generalizations of their univariate counterparts. However, the multivariate data do have some of their own unique features and needs, which do not exist in the univariate situation. Even though the idea is still the same, namely that of summarizing or describing the data, such situations call for certain unique ways of handling these, and these unique techniques form the main theme of this book. These can best be described by a few examples.

a. Based on a number of measurements such as average housing prices, cost of living, health care facilities, crime rate, etc., we would like to describe which cities in the country are most livable and also try to observe any unique similarities or differences among cities. There are several variables to be measured, and it is unlikely that attempts to order cities with respect to any one variable will result in the same ordering if another variable were used. For example, a city with a low crime rate (a desirable feature) may have a high cost of living (an undesirable feature), and thus these variables often tend to offset each other. How do we decide which cities are the best to live in? The problem here is that of data reduction. However, this problem can neither be described as that of variable selection (there is no dependent variable and no model) nor can it be viewed as a prediction problem. It is more a problem of attempting to detect and understand the unique features that the data set may contain and then to interpret them. This requires some meaningful approach for data description. The possible analyses for such a data set are principal component analysis (Chapter 2) and cluster analysis (Chapter 6).

b. As another example, suppose we have a set of independent variables which in turn have effects on a large number of dependent variables. Such a situation is quite common in the chemical industry and in economic data, where the two sets can be clearly defined as those containing input and output variables. We are not interested in individual variables, but we want to come up with a few new variables in each group. These may themselves be functions of all variables in the respective groups, so that each new variable from one group can be paired with another new variable in the other group in some meaningful sense, with the hope that these newly defined variables can be appropriately interpreted in the context. We must emphasize that analysis is not being done with any specific purpose of proving or disproving some claims. It is only an attempt to understand the data. As the information is presented in terms of new variables, which are fewer in number, it is easier to observe any striking features or associations in this latter situation. Such problems can be handled using the techniques of canonical correlation (Chapter 3) and in case of qualitative data, using correspondence analysis (Chapter 7).

c. An automobile company wants to know what determines the customer's preference for various cars. A sample of 100 randomly selected individuals were asked to give a score between 1 (low) and 10 (high) on six variables, namely, price, reliability, status symbol related to car, gas mileage, safety in an accident, and average miles driven per week. What kind of analysis can be made for these data? With the assumptions that there are some underlying hypothetical and unobservable variables on which the scores of these six observable variables depend, a natural inquiry would be to identify these hypothetical variables. Intuitively, safety consciousness and economic status of the individual may be two (perhaps of several others) traits that may influence the scores on some of these six observable variables. Thus, some or all of the observed variables can be written as a function of, say, these two unobservable traits. A question in reverse is this: can we quantify the unobservable traits as functions of the observable ones? Such a query can be usually answered by factor analysis techniques (Chapter 4). Note, however, that the analysis provides only the functions and, their interpretations as some meaningful unobservable trait, is left to the analyst. Nonetheless, it is again a problem of data reduction and description in that many measurements are reduced to only a few traits with the objective of providing an appropriate description of the data.

As is clear from these examples, many multivariate problems involve data reduction, description and, in the process of doing so, estimation. These issues form the focus of the next six chapters. As a general theme, most of the situations either require some matrix decomposition and transformations or use a distance-based approach. Distributional assumptions such as multivariate normality are also helpful (usually but not always, in assessing the quality of estimation) but not crucial. With that in mind in the next section we provide a brief review of some important concepts from matrix theory. A review of multivariate normality is presented in Section 1.6.

1.5 Concepts from Matrix Algebra

This section is meant only as a brief review of concepts from matrix algebra. An excellent account of results on matrices with a statistical viewpoint can be found in the recent books by Schott (1996), Harville (1997) and Rao and Rao (1998). We will assume that the reader is already familiar with the addition, multiplication, and transposition of matrices. Also the working knowledge of other elementary concepts such as linear independence of vectors is assumed.

In SAS, matrix computations can be performed using the IML procedure. The first statement is

```
proc iml;
```

Matrix additions, subtractions, and multiplications are performed using the $+$, $-$, and $*$ notations. Thus, if the sum of matrices $\mathbf{A}_1$, $\mathbf{A}_2$ is to be multiplied by the difference of $\mathbf{A}_3$ and $\mathbf{A}_4$, then the final matrix, say $\mathbf{B}$, will be computed using the program

```
proc iml;
b = (a1 + a2) * (a3 - a4);
```

1.5.1 Transpose of a Matrix

For an m by n matrix $\mathbf{A}$, the transpose of $\mathbf{A}$ is obtained by interchanging its rows and columns. It is denoted by $\mathbf{A}'$. Naturally, $\mathbf{A}'$ is of order n by m. For example, if

$$\mathbf{A} = \begin{bmatrix} 1 & 3 & 4 \\ 7 & 0 & 1 \end{bmatrix}$$

then

$$\mathbf{A}' = \begin{bmatrix} 1 & 7 \\ 3 & 0 \\ 4 & 1 \end{bmatrix}.$$

Also for two matrices $\mathbf{A}$ and $\mathbf{B}$ of order m by n and n by r we have, $(\mathbf{AB})' = \mathbf{B}'\mathbf{A}'$.

In PROC IML, the leading quote (‘) is used for transposition. Since some keyboards may not support this key, an alternative way to obtain $\mathbf{A}'$ is to use the function T. Specifically, $\mathbf{B} = \mathbf{A}'$ is obtained as either

```
b = a‘;
```

or as

```
b = t(a);
```

1.5.2 Symmetric Matrices

An n by n matrix $\mathbf{A}$ is said to be symmetric if $\mathbf{A}' = \mathbf{A}$. For example,

$$\mathbf{A} = \begin{bmatrix} 7 & 8 & -3 \\ 8 & 0 & 1 \\ -3 & 1 & 9 \end{bmatrix}$$

is symmetric. Clearly, if a_{ij} is the $(i, j)^{th}$ element in matrix $\mathbf{A}$, then for a symmetric matrix $a_{ij} = a_{ji}$ for all i, j.

1.5.3 Diagonal Matrices

An n by n matrix $\mathbf{A}$ is diagonal if all its nondiagonal entries are zeros. A diagonal matrix is trivially symmetric. For example, $\mathbf{A} = \begin{bmatrix} 3 & 0 & 0 \\ 0 & -1 & 0 \\ 0 & 0 & 0 \end{bmatrix}$ is a diagonal matrix.

We will often use the notation diag($\mathbf{A}$), which stands for a matrix that retains only the diagonal entries of $\mathbf{A}$ and replaces all nondiagonal entries with zeros. Thus, for

$$\mathbf{A} = \begin{bmatrix} 3 & 4 & 5 \\ 1 & 8 & 2 \\ 4 & -1 & 0 \end{bmatrix},$$

the diag($\mathbf{A}$) will be

$$\text{diag}(\mathbf{A}) = \begin{bmatrix} 3 & 0 & 0 \\ 0 & 8 & 0 \\ 0 & 0 & 0 \end{bmatrix}.$$

In PROC IML, the function DIAG(B) requires $\mathbf{B}$ to be a vector or a square matrix. Thus, if $\mathbf{A}$ is an n by n matrix, and we want the n by n matrix $\mathbf{D} = \text{diag}(\mathbf{A})$, then the appropriate SAS statement is

```
d = diag(a);
```

1.5.4 Some Special Matrices

Here are some examples:

- An n by n diagonal matrix with all diagonal entries equal to 1 is called an *identity matrix*. It is denoted by $\mathbf{I}_n$ or simply by $\mathbf{I}$ if there is no confusion.
- An n by 1 column vector with all entries equal to 1 is denoted by $\mathbf{1}_n$ or simply by $\mathbf{1}$.
- An m by n matrix with all elements as zero is called a zero matrix. It is denoted by $\mathbf{0}_{m,n}$ or simply by $\mathbf{0}$.

In PROC IML, the respective functions are $I(n)$, $J(n, 1, 1)$, and $J(m, n, 0)$.

1.5.5 Triangular Matrices

An n by n matrix $\mathbf{A}$ is said to be *upper triangular* if all entries below the main diagonal are zero. The lower triangular matrix is similarly defined as one that has all entries above main diagonal zero. For example,

$$\mathbf{A}_1 = \begin{bmatrix} 1 & 1 & 9 \\ 0 & 3 & 4 \\ 0 & 0 & 4 \end{bmatrix} \text{ and } \mathbf{A}_2 = \begin{bmatrix} 1 & 0 & 0 \\ 0 & 3 & 0 \\ 0 & 3 & 9 \end{bmatrix}$$

are respectively upper and lower triangular.

1.5.6 Linear Independence

A set of nonzero column (or row) vectors is said to be *linearly independent* if none of them can be expressed as a linear combination of some or all of the remaining vectors. If this does not happen, then this set will be called *linearly dependent.* A set containing a zero vector will always be viewed as linearly dependent.

Given a linearly dependent set of vectors, if we discard the zero vector and we continue to discard one by one the vectors that can be expressed as a linear combination of the remaining undiscarded vectors, then we will either end with a subset that is linearly independent or with an empty set. The number of vectors that finally remain is an important concept and is formally defined for a matrix (when viewed as a set of columns or rows) in the next subsection.

1.5.7 Rank of a Matrix

The rank of a matrix $\mathbf{A}$, denoted by $R(\mathbf{A})$, is defined as the number of linearly independent rows (or columns) in the matrix. Since we can either work with only rows or with only columns, it is obvious that $R(\mathbf{A}) = R(\mathbf{A}')$. It can also be established that $R(\mathbf{AB}) \leq \min(R(\mathbf{A}), (R(\mathbf{B}))$. Further, $R(\mathbf{A}'\mathbf{A}) = R(\mathbf{A})$.

1.5.8 Nonsingular and Singular Matrices

An n by n matrix $\mathbf{A}$ is said to be nonsingular if all its rows (or columns) are linearly independent. In other words, $\mathbf{A}$ is nonsingular if $R(\mathbf{A}) = n$. If one or more rows (or columns) of $\mathbf{A}$ can be written as linear combinations of some or all of the remaining rows (or columns) of $\mathbf{A}$, then there exists some linear dependence among the rows (or columns) of $\mathbf{A}$. Consequently, $\mathbf{A}$ is said to be singular in this case. For example,

$$\mathbf{A} = \begin{bmatrix} 1 & 3 \\ 9 & 4 \end{bmatrix}$$

is nonsingular, as neither of the two rows can be linearly expressed in terms of the other. However,

$$\mathbf{B} = \begin{bmatrix} 1 & 3 & 4 \\ 9 & 4 & 3 \\ 11 & 10 & 11 \end{bmatrix}$$

is singular since Row 3 $= 2\times$ Row 1 + Row 2, which indicates that the third row (or any other row, for that matter) can be expressed as the linear combination of the other two.

1.5.9 Inverse of a Square Matrix

An n by n matrix $\mathbf{A}$ admits an inverse if there exists a matrix $\mathbf{B}$ such that $\mathbf{AB} = \mathbf{BA} = \mathbf{I}_n$. The matrix $\mathbf{B}$ is called the inverse of $\mathbf{A}$ and is denoted by $\mathbf{A}^{-1}$. For example, for

$$\mathbf{A} = \begin{bmatrix} 1 & 3 \\ 9 & 4 \end{bmatrix},$$

the $\mathbf{A}^{-1}$ is given by

$$\mathbf{A}^{-1} = \begin{bmatrix} -\frac{4}{23} & \frac{3}{23} \\ \frac{9}{23} & -\frac{1}{23} \end{bmatrix} = \begin{bmatrix} -0.1739 & 0.1304 \\ 0.3913 & -0.0435 \end{bmatrix}.$$

It is obvious that the inverse of $\mathbf{A}^{-1}$, namely, $(\mathbf{A}^{-1})^{-1}$ is $\mathbf{A}$. The inverse is defined only for n by n matrices, that is, when the number of rows and the number of columns are equal. Even for such matrices, it exists if and only if $\mathbf{A}$ is nonsingular. Thus, no inverse exists for matrices that are singular or for which the number of rows is not equal to the number of columns. For such matrices, a weaker concept, known as a *generalized inverse* or simply a g-inverse can be defined. Whenever an inverse of a given matrix exists, it is unique.

In PROC IML, the inverse for a square matrix $\mathbf{A}$ can be computed by the statement

```
a_inv = inv (a);
```

Thus, A_INV is the desired inverse. It is unique.

If two matrices $\mathbf{A}$ and $\mathbf{B}$ are both of order n by n and are nonsingular, then $(\mathbf{AB})^{-1}$ and $(\mathbf{BA})^{-1}$ both exist. However, they are not equal. Specifically,

$$(\mathbf{AB})^{-1} = \mathbf{B}^{-1}\mathbf{A}^{-1}$$

and

$$(\mathbf{BA})^{-1} = \mathbf{A}^{-1}\mathbf{B}^{-1}.$$

Since the product of matrices is not commutative, the right-hand sides of the above two expressions are not equal. This makes it clear why $(\mathbf{AB})^{-1}$ and $(\mathbf{BA})^{-1}$ are not the same.

1.5.10 Generalized Inverses

For an m by n matrix, $\mathbf{B}$, a generalized inverse or simply a g-inverse, say $\mathbf{G}$, is an n by m matrix such that

$$\mathbf{BGB} = \mathbf{B}.$$

In general, the g-inverse always exists. However, it is not necessarily unique. The g-inverse is unique only for nonsingular matrices and in that case, it is the same as the inverse. A g-inverse of $\mathbf{B}$ is denoted by $\mathbf{B}^{-}$.

The matrix

$$\mathbf{B} = \begin{bmatrix} 1 & 3 & 4 \\ 9 & 4 & 3 \\ 11 & 10 & 11 \end{bmatrix}$$

was earlier found to be singular. A g-inverse of $\mathbf{B}$ is given by,

$$\mathbf{B}^- = \begin{bmatrix} -\frac{4}{23} & \frac{3}{23} & 0 \\ \frac{9}{23} & -\frac{1}{23} & 0 \\ 0 & 0 & 0 \end{bmatrix}.$$

Of course, the above choice of $\mathbf{B}^-$ is not unique. In PROC IML, a g-inverse for matrix $\mathbf{B}$ can be computed by the statement

```
b_ginv = ginv(b);
```

Thus, B_GINV is a g-inverse. The specific generalized inverse that SAS computes using the GINV function is the Moore-Penrose g-inverse (which, in fact, has been made unique by additional restrictions (Rao, 1973)).

1.5.11 A System of Linear Equations

Consider a system of n consistent equations in m unknowns, $x_1, \ldots, x_m$, (that is, a system in which no subset of equations violates any of the remaining equations)

$$a_{11}x_1 + a_{12}x_2 + \cdots + a_{1m}x_m = b_1$$

$$\vdots$$

$$a_{n1}x_1 + a_{n2}x_2 + \cdots + a_{nm}x_m = b_n.$$

With $\mathbf{A} = \begin{bmatrix} a_{11} & \ldots & a_{1m} \\ \vdots & & \\ a_{n1} & \ldots & a_{nm} \end{bmatrix}$, $\mathbf{b} = \begin{bmatrix} b_1 \\ \vdots \\ b_n \end{bmatrix}$ and $\mathbf{x} = \begin{bmatrix} x_1 \\ \vdots \\ x_m \end{bmatrix}$, the above system can be written as,

$$\mathbf{Ax} = \mathbf{b}.$$

If $m = n$ and if the matrix $\mathbf{A}$ is nonsingular, then the solution $\mathbf{x}$ is given by $\mathbf{x} = \mathbf{A}^{-1}\mathbf{b}$. If $m = n$ and if $\mathbf{A}$ is singular (in which case some equations may be redundant as they are implied by other equations) or if $m \neq n$, then a solution $\mathbf{x}$ is obtained as $\mathbf{x} = \mathbf{A}^-\mathbf{b}$, where $\mathbf{A}^-$ is a g-inverse of $\mathbf{A}$. Since the g-inverses are not unique, unless $\mathbf{A}$ is nonsingular, in this case there is no unique solution to the above system of linear equations. The reason for this is that changing the choice of g-inverse of $\mathbf{A}$ in the equation above yields another new solution.

In PROC IML, the solution $\mathbf{x}$ can be obtained by using the SOLVE function. Specifically, when $\mathbf{A}$ is nonsingular, we use

```
x=solve(a,b);
```

Alternatively, one can just use the INV function and get the solution by

```
x=inv(a)*b;
```

When $\mathbf{A}$ is singular, there are infinitely many solutions, all of which can be collectively expressed as $\mathbf{x} = \mathbf{A}^-\mathbf{b} + (\mathbf{I} - \mathbf{A}^-\mathbf{A})\mathbf{z}$, where $\mathbf{z}$ is any arbitrary vector and $\mathbf{A}^-$ is a g-inverse of $\mathbf{A}$. Of special interest is the case in which we have a system of consistent linear equations $\mathbf{Ax} = \mathbf{0}$ when $n < m$. In this case, although there are infinitely many solutions, a finite orthonormal (to be defined later) set of solutions can be obtained as a matrix $\mathbf{X}$ by using

```
x=homogen(a);
```

The columns of matrix $\mathbf{X}$ are the orthonormal solutions. The order of $\mathbf{X}$ is determined by the rank of the matrix $\mathbf{A}$.

1.5.12 Euclidean Norm of a Vector

For an n by 1 vector $\mathbf{a}$, the norm (or length) of $\mathbf{a}$ is defined as $\sqrt{\mathbf{a}'\mathbf{a}}$. Clearly, $\mathbf{b}$ defined as $\mathbf{b} = \mathbf{a}/\sqrt{\mathbf{a}'\mathbf{a}}$ has norm 1. In this case, $\mathbf{b}$ is called the normalized version of $\mathbf{a}$.

1.5.13 Euclidean Distance between Two Vectors

Visualizing the n by 1 vectors $\mathbf{a}$ and $\mathbf{b}$ as points in an n-dimensional space, we can define the distance between $\mathbf{a}$ and $\mathbf{b}$ as the norm of the vector $(\mathbf{a}-\mathbf{b})$. That is, the distance $d(\mathbf{a}, \mathbf{b})$ is defined as

$$d(\mathbf{a}, \mathbf{b}) = \sqrt{(\mathbf{a}-\mathbf{b})'(\mathbf{a}-\mathbf{b})}$$

$$= \sqrt{\sum_{i=1}^{n}(a_i - b_i)^2},$$

where a_i and b_i, respectively, are the ith entries of vectors $\mathbf{a}$ and $\mathbf{b}$.

The Euclidean distance is the distance between the points as our eyes see it. However, sometimes distance can be defined after assigning some weights through a positive definite matrix (to be defined later). Specifically, the weighted distance with weight matrix $\mathbf{A}$ is defined as

$$d_{\mathbf{A}}(\mathbf{a}, \mathbf{b}) = \sqrt{(\mathbf{a}-\mathbf{b})'\mathbf{A}(\mathbf{a}-\mathbf{b})},$$

where $\mathbf{A}$ is positive definite. Clearly $d_{\mathbf{I}_n}(\mathbf{a}, \mathbf{b}) = d(\mathbf{a}, \mathbf{b})$. One common weighted distance that we encounter in multivariate analysis is the Mahalanobis distance (Rao, 1973).

In general, a distance function, say, $\delta(a, b)$ can be defined in many other ways. However, a distance function must satisfy the following conditions:

- $\delta(\mathbf{a}, \mathbf{b}) = 0$ if and only if $\mathbf{a} = \mathbf{b}$.
- $\delta(\mathbf{a}, \mathbf{b}) = \delta(\mathbf{b}, \mathbf{a})$.
- $\delta(\mathbf{a}, \mathbf{b}) \geq 0$.
- $\delta(\mathbf{a}, \mathbf{c}) \leq \delta(\mathbf{a}, \mathbf{b}) + \delta(\mathbf{b}, \mathbf{c})$.

Clearly, $d(\mathbf{a}, \mathbf{b})$ and $d_{\mathbf{A}}(\mathbf{a}, \mathbf{b})$ satisfy all of the above conditions. It may be remarked that often in statistics, the *squared distances* are also referred to as distances. This is especially more frequent in case of certain cluster analyses. In this context, we may remark that the distance functions are often used as the measures of dissimilarity between the objects or units. However, various other dissimilarity indexes are also often applied. Many of these are not distance functions in that they do not satisfy all of the above conditions.

1.5.14 Orthogonal Vectors and Matrices

Two n by 1 vectors $\mathbf{a}$ and $\mathbf{b}$ are said to be *orthogonal* to each other if $\mathbf{a}'\mathbf{b} = 0$. Additionally, if $\mathbf{a}$ and $\mathbf{b}$ are normalized (i.e., $\mathbf{a}'\mathbf{a} = 1 = \mathbf{b}'\mathbf{b}$), then they are called *orthonormal*. For example,

$$\mathbf{a} = \begin{bmatrix} 1 \\ 1 \\ 1 \end{bmatrix} \text{ and } \mathbf{b} = \begin{bmatrix} -1 \\ 0 \\ 1 \end{bmatrix}$$

are orthogonal to each other. Their normalized versions, $\mathbf{a}/\sqrt{3}$ and $\mathbf{b}/\sqrt{2}$ are orthonormal to each other.

An n by n matrix $\mathbf{A}$ is said to be an orthogonal matrix if

$$\mathbf{A}'\mathbf{A} = \mathbf{A}\mathbf{A}' = \mathbf{I}_n.$$

This necessarily is equivalent to saying that all rows (or columns) of $\mathbf{A}$ are orthonormal to one another. Since for an orthogonal matrix, $\mathbf{A}'\mathbf{A} = \mathbf{A}\mathbf{A}' = \mathbf{I}_n$, $\mathbf{A}'$ also acts as the inverse of $\mathbf{A}$. Hence, $\mathbf{A}$ is nonsingular as well. Trivially, $\mathbf{A}'$ is also orthogonal.

Let $m < n$ and let $\mathbf{A}$ be of order n by m, such that all m columns of $\mathbf{A}$ are orthonormal to each other. In that case,

$$\mathbf{A}'\mathbf{A} = \mathbf{I}_m,$$

but no such claim can be made for $\mathbf{A}\mathbf{A}'$. In this case the matrix $\mathbf{A}$ is referred to as a *suborthogonal* matrix.

The matrix

$$\mathbf{A} = \begin{bmatrix} \frac{1}{\sqrt{3}} & \frac{1}{\sqrt{2}} & \frac{-1}{\sqrt{6}} \\ \frac{1}{\sqrt{3}} & \frac{-1}{\sqrt{2}} & \frac{-1}{\sqrt{6}} \\ \frac{1}{\sqrt{3}} & 0 & \frac{2}{\sqrt{6}} \end{bmatrix}$$

is orthogonal. However,

$$\mathbf{A}_1 = \begin{bmatrix} \frac{1}{\sqrt{2}} & \frac{-1}{\sqrt{6}} \\ \frac{-1}{\sqrt{2}} & \frac{-1}{\sqrt{6}} \\ 0 & \frac{2}{\sqrt{6}} \end{bmatrix}$$

is suborthogonal because only $\mathbf{A}_1'\mathbf{A}_1 = \mathbf{I}_2$, but $\mathbf{A}_1\mathbf{A}_1'$ is not equal to $\mathbf{I}_3$.

There are many orthogonal matrices, and using PROC IML a variety of suborthogonal matrices can be generated. The premultiplication of a general matrix by an orthogonal matrix amounts to the rotation of the axes. This frequently arises in multivariate contexts such as principal components analysis and factor analysis.

1.5.15 Eigenvalues and Eigenvectors

Let $\mathbf{A}$ be an n by n matrix. The pairs $(\lambda_1, \mathbf{x}_1), \ldots, (\lambda_n, \mathbf{x}_n)$ are said to be pairs of the eigenvalues and corresponding eigenvectors if all $(\lambda_i, \mathbf{x}_i)$ satisfy the matrix equation

$$\mathbf{A}\mathbf{x} = \lambda\mathbf{x}.$$

If $\mathbf{x}_i$ satisfies the above, then a constant multiple of $\mathbf{x}_i$ also satisfies the above. Thus, often we work with the eigenvector $\mathbf{x}_i$ that has norm 1. In general, λ_i as well as elements of $\mathbf{x}_i$ may be complex valued. However, if $\mathbf{A}$ is symmetric, all eigenvalues are necessarily real valued and one can find eigenvectors that are all real valued. If any eigenvalue is zero, then it implies, and is implied by, the fact that the matrix $\mathbf{A}$ is singular.

If $\mathbf{A}$ is nonsingular, then $\mathbf{A}^{-1}$ exists. The eigenvalues of $\mathbf{A}^{-1}$ are $\frac{1}{\lambda_1}, \ldots, \frac{1}{\lambda_n}$, and the corresponding eigenvectors are the same as those of $\mathbf{A}$.

The eigenvalues may be repeated. If an eigenvalue is repeated r times, then we say that it has multiplicity r. If $\mathbf{A}$ is symmetric, then the eigenvectors corresponding to distinct eigenvalues are all orthonormal (provided they all have norm 1). Further, eigenvectors corresponding to an eigenvalue with multiplicity r are not necessarily orthonormal, but one can always find a set of r distinct eigenvectors, corresponding to this eigenvalue, which are orthonormal to each other. Putting all these facts together suggests that we can always find a set of n orthonormal eigenvectors for a symmetric matrix. Thus, in terms of these orthonormal eigenvectors, namely, $\mathbf{x}_1, \ldots, \mathbf{x}_n$, we have n equations

$$\mathbf{A}\mathbf{x}_1 = \lambda_1 \mathbf{x}_1$$
$$\vdots$$
$$\mathbf{A}\mathbf{x}_n = \lambda_n \mathbf{x}_n.$$

Writing these n equations side by side yields the matrix equation,

$$(\mathbf{A}\mathbf{x}_1 : \cdots : \mathbf{A}\mathbf{x}_n) = (\lambda_1 \mathbf{x}_1 : \cdots : \lambda_n \mathbf{x}_n)$$

or

$$\mathbf{A}(\mathbf{x}_1 : \cdots : \mathbf{x}_n) = (\mathbf{x}_1 : \cdots : \mathbf{x}_n) \begin{bmatrix} \lambda_1 & & \\ & \ddots & \\ & & \lambda_n \end{bmatrix}.$$

Let $\mathbf{\Lambda} = diag(\lambda_1, \ldots, \lambda_n)$ and $\mathbf{P} = [\mathbf{x}_1 : \cdots : \mathbf{x}_n]$. Clearly, $\mathbf{\Lambda}$ is diagonal and $\mathbf{P}$ is orthogonal, since all $\mathbf{x}_i$ are orthonormal to each other. Thus, we have

$$\mathbf{AP} = \mathbf{P\Lambda}$$

or

$$\mathbf{A} = \mathbf{P\Lambda P}'.$$

The above fact results in an important decomposition of a symmetric matrix, as stated below.

1.5.16 Spectral Decomposition of a Symmetric Matrix

Let $\mathbf{A}$ be an n by n symmetric matrix. Then $\mathbf{A}$ can be written as

$$\mathbf{A} = \mathbf{P\Lambda P}',$$

for some orthogonal matrix $\mathbf{P}$ and a diagonal matrix $\mathbf{\Lambda}$. Of course, the choices of $\mathbf{P}$ and $\mathbf{\Lambda}$ have been indicated above.

Using PROC IML, the eigenvalues and eigenvectors of a symmetric matrix $\mathbf{A}$ can be found by using the call

```
call eigen(lambda, p, a);
```

The eigenvalues and respective eigenvectors are stored in $\mathbf{\Lambda}$ and $\mathbf{P}$. Columns of $\mathbf{P}$ are the eigenvectors. Of course, this also readily provides a choice for the spectral decomposition matrices. However, the spectral decomposition of $\mathbf{A}$ is not unique.

1.5.17 Generalized Eigenvalues and Eigenvectors

Let $\mathbf{A}$ and $\mathbf{B}$ be two n by n symmetric matrices, and let $\mathbf{B}$ be positive definite. Then $(\delta_1, \mathbf{x}_1), (\delta_2, \mathbf{x}_2), \ldots, (\delta_n, \mathbf{x}_n)$ are the pairs of eigenvalues and eigenvectors of $\mathbf{A}$ *with respect to* $\mathbf{B}$ if they all satisfy the generalized eigenequation

$$\mathbf{Ax} = \delta \mathbf{Bx},$$

for all $i = 1, \ldots, n$. With $\mathbf{Q} = (\mathbf{x}_1 : \mathbf{x}_2 : \cdots : \mathbf{x}_n)$, all of the above n equations (with $\mathbf{x} = \mathbf{x}_i$ and $\delta = \delta_i$) can be written as one matrix equation,

$$\mathbf{AQ} = \mathbf{BQ\Delta},$$

where $\mathbf{\Delta} = diag(\delta_1, \ldots, \delta_n)$.

The generalized eigenvalue problems occur naturally in many statistical contexts. One such context is the construction of the canonical discriminant functions discussed in Chapter 5, Section 5.6. Using PROC IML, and given $\mathbf{A}$ and $\mathbf{B}$, the matrices $\mathbf{Q}$ and $\mathbf{\Delta}$ can be computed by the subroutine call

```
call geneig(d,q,a,b);
```

The vector $\mathbf{d}$ obtained from the above call contains the eigenvalues of $\mathbf{A}$ with respect to $\mathbf{B}$. Thus, $\mathbf{\Delta}$ is computed as $\mathbf{\Delta} = \text{DIAG}(\mathbf{d})$. The columns of $\mathbf{Q}$ are the respective eigenvectors. These eigenvectors are not necessarily orthogonal. It may be remarked that the generalized eigenvalue problem is equivalent to finding the eigenvalues and eigenvectors of a possibly nonsymmetric matrix $\mathbf{B}^{-1}\mathbf{A}$. It is known that these will necessarily be real, even though the particular matrix $\mathbf{B}^{-1}\mathbf{A}$ is possibly asymmetric.

1.5.18 Determinant of a Matrix

For our purpose, we define the determinant of an n by n matrix $\mathbf{A}$ as the product of all eigenvalues $\lambda_1, \ldots, \lambda_n$ of $\mathbf{A}$. Thus, the determinant of $\mathbf{A}$, denoted by $|\mathbf{A}|$, is

$$|\mathbf{A}| = \lambda_1 \ldots \lambda_n.$$

Thus, $|\mathbf{A}| = 0$ if and only if at least one eigenvalue is zero, which occurs if and only if $\mathbf{A}$ is singular. In the IML procedure, the determinant DETER, of a square matrix $\mathbf{A}$ can be computed by using the statement

```
deter = det(a);
```

1.5.19 The Trace of a Matrix

The trace of an n by n matrix $\mathbf{A}$ is defined as the sum of all its eigenvalues. Thus, the trace of $\mathbf{A}$, denoted by $tr(\mathbf{A})$, is

$$tr(\mathbf{A}) = \lambda_1 + \cdots + \lambda_n.$$

It turns out (as already mentioned in Section 1.3) that $tr(\mathbf{A})$ is also equal to $a_{11} + \cdots + a_{nn}$, the sum of all diagonal elements of $\mathbf{A}$. This equivalence is useful in the conceptual development of the theory for principal components. In PROC IML the trace TR of a square matrix $\mathbf{A}$, can be computed by using the TRACE function as follows

```
tr = trace(a);
```

1.5.20 Majorization

Let $\mathbf{a} = \begin{bmatrix} a_1 \\ \vdots \\ a_n \end{bmatrix}$ and $\mathbf{b} = \begin{bmatrix} b_1 \\ \vdots \\ b_n \end{bmatrix}$ be two n by 1 vectors with $a_1 \geq a_2 \geq \cdots \geq a_n$ and $b_1 \geq b_2 \geq \cdots \geq b_n$. Then $\mathbf{a}$ is said to be majorized by $\mathbf{b}$ if

$$\begin{aligned}
a_1 &\le b_1 \\
a_1 + a_2 &\le b_1 + b_2 \\
&\vdots \\
a_1 + \cdots + a_{n-1} &\le b_1 + \cdots + b_{n-1} \\
a_1 + \cdots + a_n &= b_1 + \cdots + b_n.
\end{aligned}$$

One important majorization fact about a symmetric matrix is that the vector of all diagonal elements arranged in increasing order is majorized by the vector of all eigenvalues also arranged in increasing order. This result is useful in principal component analysis, and it justifies why the use of a few principal components may be superior to the use of a few individual variables in certain situations.

For two vectors **a** and **b** as defined, we can verify if **a** is majorized by **b** when we use the following SAS/IML code:

```
if all( cusum(a) <= cusum(b)) then major = 'yes';
else major = 'no';
print major;
```

1.5.21 Quadratic Forms

Let $\mathbf{A} = (a_{ij})$ be an n by n matrix and $\mathbf{x}$ be an n by 1 vector of variables. Then

$$\begin{aligned}
\mathbf{x}'\mathbf{A}\mathbf{x} &= \sum_{i=1}^{n}\sum_{j=1}^{n} a_{ij}x_ix_j \\
&= a_{11}x_1^2 + \cdots + a_{nn}x_n^2 \\
&\qquad + (a_{12} + a_{21})x_1x_2 + \cdots + (a_{n-1,n} + a_{n,n-1})x_{n-1}x_n.
\end{aligned}$$

It is a second degree polynomial in $x_1, \ldots, x_n$, and thus it is referred to as a quadratic form in **x**.

Clearly, $\mathbf{x}'\mathbf{A}\mathbf{x} = \mathbf{x}'\mathbf{A}'\mathbf{x}$ which, by averaging, is also the same as $\mathbf{x}'\left(\frac{\mathbf{A}+\mathbf{A}'}{2}\right)\mathbf{x}$. Since $\frac{\mathbf{A}+\mathbf{A}'}{2}$ is always symmetric, without any loss of generality, the matrix **A** in the above definition of quadratic forms can be taken to be symmetric. In this case, with **A** symmetric, a quadratic form can be expanded into any one of the alternative representations:

$$\begin{aligned}
\mathbf{x}'\mathbf{A}\mathbf{x} &= \sum_{i=1}^{n}\sum_{j=1}^{n} a_{ij}x_ix_j \\
&= \sum_{i=1}^{n} a_{ii}x_i^2 + \sum_{\substack{i=1 \\ i\neq j}}^{n}\sum_{j=1}^{n} a_{ij}x_ix_j \\
&= \sum_{i=1}^{n} a_{ii}x_i^2 + 2\sum_{\substack{i=1 \\ i<j}}^{n}\sum_{j=1}^{n} a_{ij}x_ix_j.
\end{aligned}$$

The equation $\mathbf{x}'\mathbf{A}\mathbf{x} = c$, where c is a constant, represents a quadratic surface in an n-dimensional space. Thus, it may be a paraboloid, a hyperboloid or an ellipsoid (or a hybrid of these). Which it is depends on the elements of matrix **A**. The latter case of ellipsoid is of special interest in statistics, and it occurs if **A** is positive (semi-)definite, which is defined below.

1.5.22 Positive Definite and Semidefinite Matrices

An n by n symmetric matrix $\mathbf{A}$ is said to be *positive definite* if for any vector $\mathbf{x} \neq \mathbf{0}$, the quadratic form $\mathbf{x}'\mathbf{A}\mathbf{x} > 0$. Similarly, it is *positive semidefinite* (also referred to as nonnegative definite) if $\mathbf{x}'\mathbf{A}\mathbf{x} \geq 0$. Of course any positive definite matrix is also positive semidefinite. When $\mathbf{A}$ is positive definite the equation $\mathbf{x}'\mathbf{A}\mathbf{x} = c$, where c is a constant, represents an ellipsoid.

It is known that for a positive definite matrix, all eigenvalues are positive. The converse is also true. Similarly, for a positive semidefinite matrix, these are nonnegative. Since for a positive definite matrix all eigenvalues are positive, so is the determinant, being the product of these. Thus, the determinant is not equal to zero, and hence $\mathbf{A}$ is necessarily nonsingular. Thus, a positive definite matrix $\mathbf{A}$ always admits an inverse.

If $\mathbf{B}$ is an m by n matrix, then $\mathbf{B}\mathbf{B}'$ and $\mathbf{B}'\mathbf{B}$ are positive semidefinite. If $m < n$ and $R(\mathbf{B}) = m$ then $\mathbf{B}\mathbf{B}'$ is also positive definite. However $\mathbf{B}'\mathbf{B}$ is still positive semidefinite only.

1.5.23 Square Root of a Symmetric Positive Semidefinite Matrix

For a symmetric positive semidefinite matrix $\mathbf{A}$, one can find an upper triangular matrix $\mathbf{U}$ such that

$$\mathbf{A} = \mathbf{U}'\mathbf{U}.$$

This is called the Cholesky decomposition. In PROC IML, the statement

```
u=root(a);
```

performs the Cholesky decomposition. The matrix $\mathbf{U}$ in the above is upper triangular and hence not symmetric. A symmetric square root of $\mathbf{A}$ denoted by $\mathbf{A}^{1/2}$ can also be obtained. Specifically, since $\mathbf{A}$ is symmetric, we must have by its spectral decomposition, $\mathbf{A} = \mathbf{P}\boldsymbol{\Lambda}\mathbf{P}' = (\mathbf{P}\boldsymbol{\Lambda}^{1/2}\mathbf{P}')(\mathbf{P}\boldsymbol{\Lambda}^{1/2}\mathbf{P}') = \mathbf{A}^{1/2}\mathbf{A}^{1/2}$, where $\mathbf{P}$ is orthogonal and $\boldsymbol{\Lambda}$ is diagonal. The diagonal matrix $\boldsymbol{\Lambda}$ contains the eigenvalues of $\mathbf{A}$ in the diagonal places, which are nonnegative since the matrix $\mathbf{A}$ is nonnegative definite. Thus, we take $\boldsymbol{\Lambda}^{1/2}$ as just a diagonal matrix with diagonal elements as the positive square roots of the corresponding elements of $\boldsymbol{\Lambda}$. Accordingly, we define $\mathbf{A}^{1/2}$ as $\mathbf{A}^{1/2} = \mathbf{P}\boldsymbol{\Lambda}^{1/2}\mathbf{P}'$. Thus, $\mathbf{A}^{1/2}$ is also symmetric. However, it may not be unique, since the spectral decomposition of $\mathbf{A}$ is not unique.

The SAS statements that obtain $\mathbf{A}^{1/2}$ are

```
proc iml;
a = {
10 3 9,
3 40 8,
9 8 15};
call eigen(lambda,p,a);
lam_half = root(diag(lambda));
a_half = p*lam_half*p';
print a, p, lam_half;
print a_half ;
```

The symmetric square root matrix $\mathbf{A}^{1/2}$ in the above program is denoted by A_HALF. It may be pointed out that $\mathbf{A}^{-1/2}$ may be computed by taking the inverse of $\mathbf{A}^{1/2}$ or by directly computing the symmetric square root of $\mathbf{A}^{-1}$ instead of $\mathbf{A}$ using the program.

1.5.24 Singular Value Decomposition

Any matrix $\mathbf{B}$ of order m by n can be presented as $\mathbf{B} = \mathbf{UQV}'$, where $\mathbf{U}$ and $\mathbf{V}$ are orthogonal or suborthogonal. If m is larger than n, then $\mathbf{U}$ is suborthogonal (only $\mathbf{UU}'$ is equal to the identity matrix but $\mathbf{U}'\mathbf{U}$ is not) and $\mathbf{V}$ is orthogonal. The matrix $\mathbf{Q}$ is a diagonal matrix of order n by n. If m is smaller than n, then after ignoring the last $n - m$ zero columns of $\mathbf{U}$ this reduced matrix, say $\mathbf{U}_*$, and $\mathbf{V}$ are both orthogonal. If $\mathbf{B}$ is square, then both $\mathbf{U}$ and $\mathbf{V}$ are orthogonal. The diagonal places of matrix $\mathbf{Q}$ contain the singular values of $\mathbf{B}$. Denoting $\mathbf{U}$, $\mathbf{Q}$, and $\mathbf{V}$ by LEFT, MID, and RIGHT, the following IML subroutine call results in their computation

```
call svd(left,mid,right,b);
```

Only the diagonal elements of $\mathbf{Q}$—and not the entire matrix—are printed, and hence MID is a column vector, not a square matrix. Thus, for any further calculations involving $\mathbf{Q}$, it should be specified as DIAG(MID).

The singular value decomposition (SVD) is also written in a form when the left and right side matrices of decomposition are orthogonal and not just suborthogonal. In this case the middle matrix $\mathbf{Q}$ is of order m by n. Specifically, when $m = n$, nothing needs to be done as $\mathbf{U}$ and $\mathbf{V}$ are both orthogonal. When $m > n$, we write $\mathbf{B}$ as

$$\mathbf{B} = \left[\mathbf{U}_{m\times n} : \mathbf{Uc}_{m\times(m-n)}\right] \begin{bmatrix} \mathbf{Q}_{n\times n} \\ \cdots \\ \mathbf{0}_{(m-n)\times n} \end{bmatrix} \mathbf{V}'_{n\times n}$$

$$= \mathbf{U}_*\mathbf{Q}_*\mathbf{V}'_*.$$

Here $\mathbf{V}_* = \mathbf{V}$, $\mathbf{Q} = \begin{bmatrix} \mathbf{Q}_{n\times n} \\ \cdots \\ \mathbf{0}_{(m-n)\times n} \end{bmatrix}$ and $\mathbf{U}_* = [\mathbf{U} : \mathbf{U}_c]$. The matrix $\mathbf{U}_c$ is suitably chosen such that $\mathbf{U}_c'\mathbf{U} = \mathbf{0}$. It is called the *orthogonal complement* of $\mathbf{U}$, and one choice of $\mathbf{U}_c$ (as it may not be unique) can be obtained by using the function HOMOGEN. Specifically, in PROC IML, we use the statement

```
uc=homogen(t(u));
```

to obtain the matrix $\mathbf{U}_c$.

When $m < n$, the $m \times n$ matrix $\mathbf{U}$ will necessarily have $(n - m)$ zero columns. The matrix $\mathbf{U}_*$ is obtained by eliminating these columns from $\mathbf{U}$, $\mathbf{V}_*$ is the same as $\mathbf{V}$, and $\mathbf{Q}_* = \mathbf{Q}$. Thus, we again have $\mathbf{B} = \mathbf{U}_*\mathbf{Q}_*\mathbf{V}'_*$.

It may be pointed out that the SVD of $\mathbf{B}'$ is equivalent to the SVD of $\mathbf{B}$. Thus, alternatively, the case of $m < n$ can be derived from the case of $m > n$ and vice versa.

1.5.25 Generalized Singular Value Decomposition

In the singular value decomposition of matrix $\mathbf{B}$ defined above, the matrices $\mathbf{U}_*$ and $\mathbf{V}_*$ were orthogonal. That is, we had

$$\mathbf{U}'_*\mathbf{U}_* = \mathbf{U}_*\mathbf{U}'_* = \mathbf{I}_m$$

and

$$\mathbf{V}'_*\mathbf{V}_* = \mathbf{V}_*\mathbf{V}'_* = \mathbf{I}_n.$$

While dropping the subscripts $*$, we may instead require

$$\mathbf{U}'\mathbf{C}\mathbf{U} = \mathbf{I}_m$$

and

$$\mathbf{V}'\mathbf{D}\mathbf{V} = \mathbf{I}_n,$$

where $\mathbf{C}$ and $\mathbf{D}$ are, respectively, m by m and n by n symmetric positive definite matrices. We can still have the decomposition $\mathbf{B} = \mathbf{U}\mathbf{Q}\mathbf{V}'$, where $\mathbf{U}$ and $\mathbf{V}$ satisfy the latter two requirements instead of the former two. This is known as the *generalized singular value decomposition* of matrix $\mathbf{B}$. Such a decomposition is very useful in correspondence analysis.

The generalized singular value decomposition of $\mathbf{B}$ is closely related to the singular value decomposition of $\mathbf{C}^{\frac{1}{2}}\mathbf{B}\mathbf{D}^{\frac{1}{2}}$. In fact, one can be obtained from the other. Thus, to find the generalized singular value decomposition of $\mathbf{B}$, let us call $\mathbf{C}^{\frac{1}{2}}\mathbf{B}\mathbf{D}^{\frac{1}{2}} = \mathbf{B}_*$. We can perform the singular value decomposition of $\mathbf{B}_*$, using the SVD call as shown in the previous subsection. By calling the corresponding orthogonal matrices as $\mathbf{U}_*$ and $\mathbf{V}_*$, respectively, the matrices $\mathbf{U}$ and $\mathbf{V}$, satisfying the requirements $\mathbf{U}'\mathbf{C}\mathbf{U} = \mathbf{I}_m$ and $\mathbf{V}'\mathbf{D}\mathbf{V} = \mathbf{I}_n$, are obtained as

$$\mathbf{U} = \mathbf{C}^{-\frac{1}{2}}\mathbf{U}_*$$

and

$$\mathbf{V} = \mathbf{D}^{-\frac{1}{2}}\mathbf{V}_*\,.$$

Of course to compute $\mathbf{C}^{\frac{1}{2}}$ and $\mathbf{D}^{\frac{1}{2}}$, the ROOT function can be used.

1.5.26 Kronecker Product

We define the Kronecker product of $\mathbf{C}$ *with* $\mathbf{D}$ (denoted by $\mathbf{C} \otimes \mathbf{D}$) by multiplying every entry of $\mathbf{C}$ by matrix $\mathbf{D}$ and then creating a matrix out of these block matrices. In notations, the Kronecker product is defined as $\mathbf{C} \otimes \mathbf{D} = (c_{ij}\mathbf{D})$. In SAS/IML software, the operator @ does this job. For example, the Kronecker product matrix $KRON\_CD$ is obtained by writing

```
kron_cd = c @ d;
```

With

$$\mathbf{C} = \begin{bmatrix} 1 & 0 & 3 & 4 \\ 0 & 4 & 1 & -1 \\ 1 & 1 & -3 & 2 \end{bmatrix},$$

and

$$\mathbf{D} = \begin{bmatrix} 1 \\ 3 \\ 7 \end{bmatrix},$$

the Kronecker product $\mathbf{C} \otimes \mathbf{D}$ (using SAS syntax **C@D**) is equal to

$$\mathbf{C} \otimes \mathbf{D} = \begin{bmatrix} 1 & 0 & 3 & 4 \\ 3 & 0 & 9 & 12 \\ 7 & 0 & 21 & 28 \\ 0 & 4 & 1 & -1 \\ 0 & 12 & 3 & -3 \\ 0 & 28 & 7 & -7 \\ 1 & 1 & -3 & 2 \\ 3 & 3 & -9 & 6 \\ 7 & 7 & -21 & 14 \end{bmatrix}.$$

The next two subsections cover data manipulation and indicate how to create a matrix from a data set and how to convert a data set into a matrix.

1.5.27 Creating a Matrix from a SAS Data Set

Often, after running a SAS program, we may, for further calculations, need to use PROC IML. That may require converting an input or output data set to a matrix. An example follows.

Suppose we have a data set called MYDATA with three variables X1, X2, and X3 and five data points. From that we want to create a matrix called MYMATRIX. To do so, we use the following SAS statements

```
data mydata;
input x1 x2 x3;
lines;
2 4 8
3 9 1
9 4 8
1 1 1
2 7 8
;
proc iml;
use mydata;
read all into mymatrix;
quit;
print mymatrix;
```

If we want a matrix consisting of only a few variables, say in this case x_3 and x_1 (in that specific order) from the data set, then the appropriate READ statement needs to be slightly more specific:

```
read all var {x3 x1} into mymatrix;
```

1.5.28 Creating a SAS Data Set from a Matrix

Conversely, we can create a SAS data set out of a matrix. An example is presented here. Suppose we have a 5 by 3 matrix titled MYMATRIX that contains five observations from three variables for which we will use the default names COL1, COL2, and COL3. From this, we want to create a data set named NEWDATA. It is done as follows.

```
proc iml;
mymatrix = {
2 4 8,
3 9 1,
```

```
9 4 8,
1 1 1,
2 7 8};
create newdata from mymatrix;
append from mymatrix;
close newdata;
quit;
proc print data = newdata;
```

1.6 Multivariate Normal Distribution

A probability distribution that plays a pivotal role in much of the multivariate analysis is *multivariate normal distribution*. However, as this book concentrates more on the description of multivariate data, we will encounter it only occasionally. With that in mind, we give here only a very brief review of multivariate normal distribution. The material here is adopted from Khattree and Naik (1999). We say that $\mathbf{x}$ has a p-dimensional multivariate normal distribution (with a mean vector $\boldsymbol{\mu}$ and the variance-covariance matrix $\boldsymbol{\Sigma}$) if its probability density is given by

$$f(\mathbf{x}) = \frac{1}{(2\pi)^{p/2}|\boldsymbol{\Sigma}|^{1/2}} \cdot \exp(-\frac{1}{2}(\mathbf{x}-\boldsymbol{\mu})'\boldsymbol{\Sigma}^{-1}(\mathbf{x}-\boldsymbol{\mu})).$$

In notation, we state this fact as $\mathbf{x} \sim N_p(\boldsymbol{\mu}, \boldsymbol{\Sigma})$. Observe that the above density is a straightforward extension of the univariate normal density to which it will reduce when $p = 1$.

Important properties of the multivariate normal distribution include some of the following:

- Let $\mathbf{A}_{r \times p}$ be a fixed matrix, then $\mathbf{Ax} \sim N_r(\mathbf{A}\boldsymbol{\mu}, \mathbf{A}\boldsymbol{\Sigma}\mathbf{A}')(r \leq p)$. It may be added that $\mathbf{Ax}$ will admit the density if $\mathbf{A}\boldsymbol{\Sigma}\mathbf{A}'$ is nonsingular, which will happen if and only if all rows of $\mathbf{A}$ are linearly independent. Further, in principle, r can also be greater than p. However, in that case, the matrix $\mathbf{A}\boldsymbol{\Sigma}\mathbf{A}'$ will not be nonsingular. Consequently, the vector $\mathbf{Ax}$ will not admit a density function.
- Let $\mathbf{G}$ be such that $\boldsymbol{\Sigma}^{-1} = \mathbf{GG}'$, then $\mathbf{G}'\mathbf{x} \sim N_p(\mathbf{G}'\boldsymbol{\mu}, \mathbf{I})$ and $\mathbf{G}'(\mathbf{x}-\boldsymbol{\mu}) \sim N_p(\mathbf{0}, \mathbf{I})$.
- Any fixed linear combination of $x_1, \ldots, x_p$, say, $\mathbf{c}'\mathbf{x}$, $\mathbf{c}_{p\times 1} \neq \mathbf{0}$ is also normally distributed. Specifically, $\mathbf{c}'\mathbf{x} \sim N_1(\mathbf{c}'\boldsymbol{\mu}, \mathbf{c}'\boldsymbol{\Sigma}\mathbf{c})$.
- The subvectors $\mathbf{x}_1$ and $\mathbf{x}_2$ are also normally distributed. Specifically, $\mathbf{x}_1 \sim N_{p_1}(\boldsymbol{\mu}_1, \boldsymbol{\Sigma}_{11})$ and $\mathbf{x}_2 \sim N_{p-p_1}(\boldsymbol{\mu}_2, \boldsymbol{\Sigma}_{22})$, where with appropriate partitioning of $\boldsymbol{\mu}$ and $\boldsymbol{\Sigma}$,

$$\boldsymbol{\mu} = \begin{bmatrix} \boldsymbol{\mu}_1 \\ \boldsymbol{\mu}_2 \end{bmatrix},$$

and

$$\boldsymbol{\Sigma} = \begin{bmatrix} \boldsymbol{\Sigma}_{11} & \boldsymbol{\Sigma}_{12} \\ \boldsymbol{\Sigma}_{21} & \boldsymbol{\Sigma}_{22} \end{bmatrix}.$$

- Individual components $x_1, \ldots, x_p$ are all normally distributed. That is, $x_i \sim N_1(\mu_i, \sigma_{ii})$, $i = 1, \ldots, p$.
- The conditional distribution of $\mathbf{x}_1$ given $\mathbf{x}_2$, written as $\mathbf{x}_1|\mathbf{x}_2$, is also normal. Specifically,

$$\mathbf{x}_1|\mathbf{x}_2 \sim N_{p_1}(\boldsymbol{\mu}_1 + \boldsymbol{\Sigma}_{12}\boldsymbol{\Sigma}_{22}^{-1}(\mathbf{x}_2 - \boldsymbol{\mu}_2), \boldsymbol{\Sigma}_{11} - \boldsymbol{\Sigma}_{12}\boldsymbol{\Sigma}_{22}^{-1}\boldsymbol{\Sigma}_{21}).$$

Let $\boldsymbol{\mu}_1 + \boldsymbol{\Sigma}_{12}\boldsymbol{\Sigma}_{22}^{-1}(\mathbf{x}_2 - \boldsymbol{\mu}_2) = \boldsymbol{\mu}_1 - \boldsymbol{\Sigma}_{12}\boldsymbol{\Sigma}_{22}^{-1}\boldsymbol{\mu}_2 + \boldsymbol{\Sigma}_{12}\boldsymbol{\Sigma}_{22}^{-1}\mathbf{x}_2 = \mathbf{B_0} + \mathbf{B}_1\mathbf{x}_2$, and $\boldsymbol{\Sigma}_{11.2} = \boldsymbol{\Sigma}_{11} - \boldsymbol{\Sigma}_{12}\boldsymbol{\Sigma}_{22}^{-1}\boldsymbol{\Sigma}_{21}$. The conditional expectation of $\mathbf{x}_1$ for given values of $\mathbf{x}_2$ or the regression function of $\mathbf{x}_1$ on $\mathbf{x}_2$ is $\mathbf{B_0} + \mathbf{B}_1\mathbf{x}_2$, which is linear in $\mathbf{x}_2$. This is a key fact for multivariate multiple linear regression modeling. The matrix $\boldsymbol{\Sigma}_{11.2}$ is usually represented by the variance-covariance matrix of error components in these models. An analogous result (and the interpretation) can be stated for the conditional distribution of $\mathbf{x}_2$ given $\mathbf{x}_1$.

- Let $\boldsymbol{\delta}$ be a fixed $p \times 1$ vector, then

$$\mathbf{x} + \boldsymbol{\delta} \sim N_p(\boldsymbol{\mu} + \boldsymbol{\delta}, \boldsymbol{\Sigma}).$$

- The random components $x_1, \ldots, x_p$ are all independent if and only if $\boldsymbol{\Sigma}$ is a diagonal matrix; that is, when all the covariances (or correlations) are zero.
- Let $\mathbf{u}_1$ and $\mathbf{u}_2$ be respectively distributed as $N_p(\boldsymbol{\mu}_{u_1}, \boldsymbol{\Sigma}_{u_1})$ and $N_p(\boldsymbol{\mu}_{u_2}, \boldsymbol{\Sigma}_{u_2})$, then

$$\mathbf{u}_1 \pm \mathbf{u}_2 \sim N_p(\boldsymbol{\mu}_{u_1} \pm \boldsymbol{\mu}_{u_2}, \boldsymbol{\Sigma}_{u_1} + \boldsymbol{\Sigma}_{u_2} \pm (cov(\mathbf{u}_1, \mathbf{u}_2) + cov(\mathbf{u}_2, \mathbf{u}_1))).$$

Note that if $\mathbf{u}_1$ and $\mathbf{u}_2$ were independent, the last two covariance terms would drop out.

There is a vast amount of literature available on the multivariate normal distribution, its properties, and the evaluations of the multivariate normal probabilities. See Anderson (1984), Kshirsagar (1972), Rao (1973), and Tong (1990) for further details.

1.6.1 Random Multivariate Normal Vector Generation

Oftentimes, we may want to generate random observations from a multivariate normal distribution. The following SAS/IML code, illustrated for $n = 10$ random observations from $N_3(\boldsymbol{\mu}, \boldsymbol{\Sigma})$, with

$$\boldsymbol{\mu} = \begin{bmatrix} 1 \\ 2 \\ 3 \end{bmatrix},$$

and

$$\boldsymbol{\Sigma} = \begin{bmatrix} 1 & .7 & .2 \\ .7 & 2 & -.8 \\ .2 & -.8 & 2.5 \end{bmatrix},$$

can be appropriately modified for this purpose. It is necessary to specify the appropriate values of $\boldsymbol{\mu}$ (MU), $\boldsymbol{\Sigma}$ (SIGMA), and the initial seed vector (SEED).

```
proc iml;
start rnorm(mu,sigma,seed);
z=normal(seed);
g=root(sigma);
x=mu+t(g)*z;
return(x);
finish;
do i=1 to 10;
x=rnorm(1,2,3,
1 .7 .2, .7 2 -.8, .2 -.8 2.5,
12345,87948,298765);
matx=matx//x';
end;
print matx;
```

The output, namely the ten vectors from the above trivariate normal population, are stored as a 10 by 3 matrix named MATX. Details about the steps of the generation can be found in Khattree and Naik (1999).

1.7 Concluding Remarks

This chapter is meant to be an introduction in order to prepare readers for what is covered within this book. There are many other concepts, topics, and methods that are not mentioned. However, the sections on matrix results and multivariate normality provide adequate preparation for appreciating and understanding the data analysis approaches discussed in this book. Some of the more advanced concepts are occasionally introduced in other chapters as and when their needs arise. Readers who are interested in the extensive study of matrix theory-related results as they apply in multivariate analysis should see Rao and Rao (1998).

Chapter 2

Principal Component Analysis

2.1 Introduction

One of the most challenging problems in multivariate data analysis is reduction of dimensionality for any large multivariate data set. This is often accomplished by reducing the set of variables to a smaller set that contains fewer variables or new variables. The new variables are functions of the original variables or of those that contain a significant proportion of total information available in the data set. Reducing the dimension of the data is essential in exploratory data analysis (EDA) in which the plots of the data play a major role in conveying the information visually. The use of *principal components*, which are certain linear functions of the variables, is often suggested for such dimensionality reduction.

The statistical procedure of finding the principal components that preserves most of the information in the original data is called *principal component analysis*. The principal components are such that most of the information, measured in terms of total variance, is preserved in only a few of them. Rao (1964) has considered a number of different ways of quantifying information in a set of data and discussed various other optimality properties of the principal components. Principal component analysis can also be viewed as a particular case of *projection pursuit*, which is a general method of selecting low-dimensional projections of multivariate data. The selection of low-dimensional projections is usually done by optimizing some index of "interesting features" in the data over all projection directions. Bolton and Krzanowski (1999) showed that if we take the index of "interesting features" to be the variance of the data, then principal component analysis is simply a projection pursuit technique. Also see Krzanowski and Marriott (1994) for further discussion of projection pursuit.

Principal component analysis (PCA) is one of the most widely used multivariate techniques of exploratory data analysis. It is also one of the oldest multivariate methods, having been introduced by Pearson (1901) and subsequently developed by Hotelling (1933), Rao

(1964), and others. The concept of principal component analysis as applied to a random vector was developed by Hotelling (1933, 1936). Rao (1964) presents principal component analysis as a genuine data analytical technique that can be used in a wide variety of fields. He also gave a solid foundation to PCA by mathematically formalizing the concepts. As in the case with many multivariate techniques, PCA became increasingly popular as an exploratory data analysis technique after high speed computing facilities became easily accessible to most researchers.

The overwhelming surge of interest in the principal component analysis during the last two decades is evident from recent books that, solely or mainly, cover PCA. Especially readable among these with comprehensive coverage are the books by Jolliffe (1986), Flury (1988), Jackson (1991), and Basilevsky (1994). These books have expanded Rao's (1964) work in several directions and present numerous real-life examples. Jackson's book is more data analytic in flavor whereas the others are relatively more theoretical. Jeffers (1967) contains certain early applications of PCA in real-life problems from a variety of disciplines, such as biology and forestry. Some of the other disciplines in which PCA has been applied are business, chemistry, criminology, educational research, industrial quality control, psychology, sociology, and sports. In this chapter we illustrate PCA by considering examples from several fields.

Some of the very closely related fields to principal component analysis are factor analysis, biased regression analysis, and the correspondence analysis. Correspondence analysis is, in fact, one kind of principal component analysis that is performed on categorical data. Factor analysis and correspondence analysis will be dealt with in detail in separate chapters in this book. Principal component regression as a biased regression analysis technique will be discussed in Section 2.7.

2.2 Population Principal Components

In this section we define the principal components for a population. That is, suppose we have a population and a random vector that can be measured on the individuals of the population. Then, given the variance-covariance matrix of this random vector we define the principal components and call these components the *population principal components* of this random vector.

Suppose $\mathbf{\Sigma}$ is the variance-covariance (or simply covariance) matrix of p variables $x_1, \ldots, x_p$. The *total variance* of these variables is defined as $tr\,\mathbf{\Sigma}$ (the *trace* of $\mathbf{\Sigma}$), which is nothing but the sum of all the diagonal elements of the matrix $\mathbf{\Sigma}$. The first principal component of p by 1 vector $\mathbf{x} = (x_1, \ldots, x_p)'$ is a linear combination

$$\mathbf{a}_1'\mathbf{x} = a_{11}x_1 + \cdots + a_{1p}x_p, \quad \text{where } \mathbf{a}_1 = (a_{11}, \ldots, a_{1p})', \quad \text{with } \mathbf{a}_1'\mathbf{a}_1 = 1$$

and such that var $(\mathbf{a}_1'\mathbf{x})$ is the maximum among all linear combinations of $\mathbf{x}$, with the coefficient vector having unit length. Thus the first principal component so obtained accounts for the maximum variation (hence perhaps most of $tr\,\mathbf{\Sigma}$). The second principal component $\mathbf{a}_2'\mathbf{x}$ of $\mathbf{x}$ with $\mathbf{a}_2'\mathbf{a}_2 = 1$ is such that it is uncorrelated with the first and var $(\mathbf{a}_2'\mathbf{x})$ is the highest among all the linear combinations uncorrelated with $\mathbf{a}_1'\mathbf{x}$. In a similar way the third, fourth, ..., and the pth principal components, all uncorrelated with others, are defined. Obviously, the last (p^{th}) principal component $\mathbf{a}_p'\mathbf{x}$ is uncorrelated with all the other $(p-1)$ principal components and accounts for the least of the total variance. In that sense, it is the least informative component.

Let $\lambda_1 \geq \lambda_2 \geq \cdots \geq \lambda_p > 0$ be the eigenvalues and $\mathbf{a}_1, \ldots, \mathbf{a}_p$ be the corresponding eigenvectors of $\mathbf{\Sigma}$, each of them having length 1, that is, $\mathbf{a}_i'\mathbf{a}_i = 1$ for $i = 1, \ldots, p$. Then $y_1 = \mathbf{a}_1'\mathbf{x}$, $y_2 = \mathbf{a}_2'\mathbf{x}, \ldots, y_p = \mathbf{a}_p'\mathbf{x}$ are precisely the first, second, ..., p^{th} principal components of $\mathbf{x}$. Furthermore, var $(y_1) = \lambda_1, \ldots,$ var $(y_p) = \lambda_p$. That is, the eigenvalues of $\mathbf{\Sigma}$ are the variances of the corresponding principal components. Since the total variance represented by $tr\mathbf{\Sigma}$ is also the same as the sum of all the eigenvalues of $\mathbf{\Sigma}$ or $\sum_{i=1}^{p} \lambda_i$, the p principal components cumulatively account for all of the total variance $tr\mathbf{\Sigma}$. Like the eigenvalues, the eigenvectors $\mathbf{a}_1, \ldots, \mathbf{a}_p$, in fact the elements of the eigenvectors, also have some nice interpretations. For example, the covariance between the ith variable x_i and the jth principal component y_j is $\lambda_j a_{ji}$, and hence the correlation coefficient $corr(x_i, y_j)$ between them, is given by

$$corr(x_i, y_j) = a_{ji}\sqrt{\lambda_j / var(x_i)}. \tag{2.1}$$

Thus the variables with coefficients of larger magnitude in a principal component have larger contribution to that component.

The first step in determining the principal components of a vector $\mathbf{x}$ of variables is to obtain the eigenvalues and the eigenvectors of the variance-covariance matrix $\mathbf{\Sigma}$ of $\mathbf{x}$. Thus, the starting point of a principal component analysis is $\mathbf{\Sigma}$. It is also an equally valid, although not equivalent, process to start with a correlation matrix instead of a covariance matrix. It is often advised in the literature that a PCA should be started with a correlation matrix, unless there is an evidence that the measurements on different variables are on the same scale and variances are of similar magnitude. In that case, the use of a covariance matrix is suggested. See Everitt and Dunn (1992) and Jolliffe (1986) for discussion about the choice of correlation matrix over covariance matrix for a PCA. In Section 2.3, we discuss some issues and considerations for and against using a correlation matrix for PCA. However, if the correlation matrix is a starting point then the total variance is simply $\sum_{i=1}^{p} \lambda_i = p$.

It has been pointed out that various eigenvalues of $\mathbf{\Sigma}$ (or of a correlation matrix) represent the variances of various principal components, and the elements of the eigenvectors appear in the correlations between the components and the original variables. In many problems it is also possible to give other meaningful interpretations to the coefficients (eigenvectors) in the principal components. For illustration, consider the correlation matrix $\boldsymbol{\rho}$ of $\mathbf{x}' = (x_1, x_2, x_3, x_4)$ given in Hotelling (1933):

$$\boldsymbol{\rho} = \begin{bmatrix} 1.000 & .698 & .264 & .081 \\ .698 & 1.000 & -.061 & .092 \\ .264 & -.061 & 1.000 & .594 \\ .081 & .092 & .594 & 1.000 \end{bmatrix}.$$

The variables are x_1 = reading speed, x_2 = reading comprehension, x_3 = arithmetic speed, and x_4 = arithmetic comprehension all representing the scores of a seventh-grade child. Using the PRINCOMP procedure, we find that the eigenvalues of $\boldsymbol{\rho}$ are 1.8474, 1.4642, 0.5221, and 0.1663, and the corresponding eigenvectors are

$$\begin{bmatrix} .5985 \\ .5070 \\ .4495 \\ .4273 \end{bmatrix}, \begin{bmatrix} -.3660 \\ -.5156 \\ .5534 \\ .5422 \end{bmatrix}, \begin{bmatrix} -.4025 \\ .3984 \\ -.5208 \\ .6388 \end{bmatrix}, \text{and} \begin{bmatrix} -.5881 \\ .5642 \\ .4695 \\ -.3397 \end{bmatrix}.$$

See Program 2.1 and Output 2.1 .

```
/* Program 2.1 */

title1 "Output 2.1";
options ls=64 ps=45 nodate nonumber;

data score (type=corr);
_type_='corr';
input _name_ $ readsp readpow mathsp mathpow;
datalines;
readsp 1. .698 .264 .081
readpow .698 1. -.061 .092
```

```
mathsp .264 -.061 1. .594
mathpow .081 .092 .594 1.
;
/* Source: Hotelling (1933).  Reprinted by permission of
the American Psychological Association. */

title2 "Population Principal Components";
proc princomp data=score (type=corr);
run;
```

Output 2.1

```
                        Output 2.1
              Population Principal Components

                  The PRINCOMP Procedure

                 Observations        10000
                 Variables               4


            Eigenvalues of the Correlation Matrix

       Eigenvalue    Difference    Proportion    Cumulative

  1    1.84741215    0.38324442        0.4619        0.4619
  2    1.46416773    0.94207560        0.3660        0.8279
  3    0.52209213    0.35576414        0.1305        0.9584
  4    0.16632799                      0.0416        1.0000


                       Eigenvectors

               Prin1         Prin2         Prin3         Prin4

readsp      0.598522      -.365969      -.402499      -.588075
readpow     0.507031      -.515602      0.398421      0.564212
mathsp      0.449516      0.553425      -.520768      0.469528
mathpow     0.427348      0.542165      0.638792      -.339670
```

The first principal component, say $y_1 = \mathbf{a}_1'\mathbf{x} = a_{11}x_1 + a_{12}x_2 + a_{13}x_3 + a_{14}x_4 = 0.5985x_1 + 0.5070x_2 + 0.4495x_3 + 0.4273x_4$ with a variance of 1.8474, which is

$$\frac{1.8474}{tr(\boldsymbol{\rho})}(100) = \frac{1.8474}{4}(100) = 46.1853\%$$

of the total variation, appears to be a weighted average (with all coefficients > 0) of the scores in the four characteristics. The coefficients are of comparable magnitude, and hence it may represent the general ability of the child. Also, using Equation 2.1 and the fact that the correlation matrix is being used for the PCA, we see that $corr(x_i, y_i) = a_{ii}\sqrt{\lambda_1}$ (since $var(x_i) = 1$). Thus $corr(x_1, y_1) = 0.81$, $corr(x_2, y_1) = 0.69$, $corr(x_3, y_1) = 0.61$, and $corr(x_4, y_1) = 0.58$. We see that the first variable is most correlated with the first principal component. This fact is also evident since the largest coefficient of $\mathbf{a}_1$ is $a_{11} = 0.5985$.

The second principal component $-0.3660x_1 - 0.5156x_2 + 0.5534x_3 + 0.5422x_4$ with a variance of 1.4642 (36.604% of the total variation) seems to measure a difference between math (represented by x_3 and x_4) and verbal (represented by x_1 and x_2) ability. These two components together account for about 83% of the total variation. The third principal component $-0.4025x_1 + 0.3984x_2 - 0.5208x_3 + 0.6388x_4$ with a variance of 0.5221 (13.052%) measures a difference between the comprehension and speed. The last principal component does not have any obvious interpretation. However, it explains only about 4% of the total variation. Since in this analysis we try to find the components with the largest variance, this component perhaps can be ignored. If this problem is looked upon as a problem of reduction in dimensionality, then the first three principal components may be used instead of the original four variables, thus reducing the dimension by one, and only at the expense of the 4% loss in the total information. Further, the three principal components chosen here do have meaningful interpretations as the indices of the general ability, difference in mathematical and verbal abilities and a difference in comprehension and speed respectively. They are also uncorrelated with each other, thereby indicating no overlap of information among the three indices.

Principal Components for Structured Covariance Matrices. Suppose the p variables $x_1, \ldots, x_p$ are uncorrelated. Then the covariance matrix of $\mathbf{x} = (x_1, \ldots, x_p)'$ is

$$D(\mathbf{x}) = \mathbf{\Sigma} = \begin{bmatrix} \sigma_{11} & 0 & \cdot & \cdot & 0 \\ 0 & \sigma_{22} & \cdot & \cdot & 0 \\ \cdot & \cdot & \cdot & \cdot & \cdot \\ \cdot & \cdot & \cdot & \cdot & \cdot \\ 0 & 0 & \cdot & \cdot & \sigma_{pp} \end{bmatrix},$$

that is, the correlation matrix of $\mathbf{x}$ is an identity matrix of order p. If we perform a PCA starting with this $\mathbf{\Sigma}$, then the first principal component is the variable with the largest σ_{ii}, the second principal component is that variable with the second largest σ_{ii} and so on. Thus when the original variables are uncorrelated, the principal component analysis does not give any reduction or new insight to the problem.

In many practical problems, the covariance matrix of a set of variables may have a structure. For example, in certain medical and biological problems the covariance matrix of a set of repeated measures is found to have an equicorrelation or intraclass covariance structure given by

$$D(\mathbf{x}) = \mathbf{\Sigma} = \sigma^2 \begin{bmatrix} 1 & \rho & \cdot & \cdot & \rho \\ \rho & 1 & \cdot & \cdot & \rho \\ \cdot & \cdot & \cdot & \cdot & \cdot \\ \cdot & \cdot & \cdot & \cdot & \cdot \\ \rho & \rho & \cdot & \cdot & 1 \end{bmatrix} = \sigma^2[(1-\rho)\mathbf{I}_p + \rho\mathbf{J}_p],$$

where $\mathbf{I}_p$ is the identity matrix of order p and $\mathbf{J}_p$ is a p by p matrix of all ones. The eigenvalues of $\mathbf{\Sigma}$ are $\sigma^2(1 + (p-1)\rho)$, $\sigma^2(1-\rho), \ldots, \sigma^2(1-\rho)$, and to ensure its positive definiteness, it is assumed that $\frac{-1}{p-1} < \rho < 1$. The eigenvector corresponding to the eigenvalue $\sigma^2(1 + (p-1)\rho)$ is $\frac{1}{\sqrt{p}}(1, \ldots, 1)'$ and the other eigenvectors are any $(p-1)$ orthonormal vectors, orthogonal to the vector $(1, \ldots, 1)'$. If ρ is positive then the first principal component is simply $\sqrt{p}\bar{x} = (x_1 + \cdots + x_p)/\sqrt{p}$, which in biological problems often represents the size of an organism. Further, when ρ is large, the first PC explains most of the variation, that is, $(\rho + \frac{1-\rho}{p})100\%$, and hence only one PC may be enough to explain the total variation in the data. On the other hand, if ρ is close to zero then all of the PCs are needed in the analysis.

2.3 Sample Principal Components

Let $\mathbf{x}_1, \ldots, \mathbf{x}_n$ be n data vectors of order p by 1 representing n observations on each of the p variables. Let $\bar{\mathbf{x}}$ be the p by 1 vector of sample means, $\mathbf{S}$ be the p by p matrix of sample variances and covariances, and $\mathbf{R}$ be the sample correlation matrix of order p. Then the sample principal components of each of the observation vector $\mathbf{x}_i$ are obtained in the same way as in the previous section except that the eigenvectors that are used to compute the principal components are those of $\mathbf{S}$ (or $\mathbf{R}$). The eigenvalues $\hat{\lambda}_1, \ldots, \hat{\lambda}_p$ of $\mathbf{S}$ (or $\mathbf{R}$), being the estimates of $\lambda_1, \ldots, \lambda_p$ of the previous section, now play the role of $\lambda_1, \ldots, \lambda_p$.

Let $\boldsymbol{\ell}_1, \ldots, \boldsymbol{\ell}_p$ be the eigenvectors of $\mathbf{S}$ corresponding to the eigenvalues $\hat{\lambda}_1, \ldots, \hat{\lambda}_p$. Then the *scores* of p principal components corresponding to the ith observation vector $\mathbf{x}_i$ are computed as

$$\boldsymbol{\ell}'_1(\mathbf{x}_i - \bar{\mathbf{x}}), \ldots, \boldsymbol{\ell}'_p(\mathbf{x}_i - \bar{\mathbf{x}}), i = 1, \ldots, n.$$

Thus the quantities $\boldsymbol{\ell}'_1(\mathbf{x}_i - \bar{\mathbf{x}})$, $i = 1, \ldots, n$ are the scores on the first principal component. Similarly, $\boldsymbol{\ell}'_j(\mathbf{x}_i - \bar{\mathbf{x}})$, $i = 1, \ldots, n$ are the scores on the jth principal component. On the other hand, if $\boldsymbol{\ell}_1, \ldots, \boldsymbol{\ell}_p$ are the eigenvectors of $\mathbf{R}$, then the scores of p principal components corresponding to the ith observation vector $\mathbf{x}_i$ will be computed as

$$\boldsymbol{\ell}'_1\mathbf{D}^{-1/2}(\mathbf{x}_i - \bar{\mathbf{x}}), \ldots, \boldsymbol{\ell}'_p\mathbf{D}^{-1/2}(\mathbf{x}_i - \bar{\mathbf{x}}), i = 1, \ldots, n,$$

where $\mathbf{D}^{-1/2} = diag\ (1/\sqrt{s_{11}}, \ldots, 1/\sqrt{s_{pp}})$, s_{ii} being the ith diagonal element of $\mathbf{S}$. Since the use of correlation or covariance matrix centers the data, the principal component scores can be positive, negative, or zero. The scores on the first two principal components are often used in the exploratory data analysis. For example, the scatter plots of these scores are used to identify the clusters, groups, or any other patterns in the data set. Also, the scores on the first or second principal components are used to order or rank the observations in some data analysis problems. These aspects will be illustrated in the following example on the national track records of various nations in the 1984 Los Angeles Olympics.

EXAMPLE 1 ***Ranking the Nations, Los Angeles Olympics Track Records Data*** The data on national track records for men and women at various track races are obtained from the *IAAF/ATFS Track and Field Statistics Handbook for the 1984 Los Angeles Olympics*. For details on race events, see Program 2.2. For the first three events (for men as well as women), the data (time to finish a race) are recorded in seconds. For the remaining events, the times are in minutes. Data are available for fifty-five countries.

A problem of interest is to rank these countries according to their overall performance in track events. One way of doing this, of course, is to take the average of the data, after converting all in seconds or minutes, in all the track events for each country and rank them according to this average. However, with $p = 8$ (or 7) the average explains only about $\frac{\mathbf{1}'_8\mathbf{S}\mathbf{1}_8}{8^2}(100)\%$ (or $\frac{\mathbf{1}'_7\mathbf{S}\mathbf{1}_7}{7^2}(100)\%$) of the total variance for men's (or women's) track data, where $\mathbf{1}_n$ represents a n by 1 vector of ones.

Knowing that the first principal component explains maximum variation (measured in terms of total variance), it may be a good idea to use the first principal component for ranking the countries. Dawkins (1989) has done this analysis for both of these data sets. Dawkins argued that if the raw data were to be analyzed using the same time unit, the variable represented by the time taken in the marathon, because of its larger amount of variability, would be weighted excessively in the analysis. Therefore, he first rescaled the variables in each of the two data sets to have mean 0 and standard deviation 1, thereby making them unit-free. This, in turn, amounts to using the sample correlation matrix instead of the sample variance-covariance matrix (of the time taken by the athletes in the events) to obtain the principal components.

For illustration, we select track data for women only. There are seven track events for women. In Program 2.2 we have created a SAS data set named TRACK. It contains raw data (provided in the Appendix as WOMENTRACK.DAT) for the seven variables named M100, M200, M400, M800, M1500, M3000, and MARATHON. The SAS statements

```
proc princomp data = track;
var m100 m200 m400 m800 m1500 m3000 marathon;
run;
```

yield simple statistics; the sample means and the sample standard deviations for each of the seven variables, and the correlation matrix of the variables. By default, the sample correlation matrix **R** is used for the principal component analysis. The ordered eigenvalues (largest to the smallest), difference between the consecutive eigenvalues, proportion of the total variance explained by the individual principal components, the cumulative total of these proportions and the eigenvectors (which provide the coefficients for the individual variables in the respective principal components) corresponding to each of the eigenvalues are also provided in the output. Default names Prin1, ..., Prinp are given to the eigenvectors. We can, however, assign different names by using the PREFIX= option in the proc statement. For example, the statement

```
proc princomp data = track prefix = pc;
```

would assign the names pc1, ..., pc7 to the seven eigenvectors. The output resulting from the execution of Program 2.2 is presented in Output 2.2. Note that some output from this example has been suppressed.

```
/* Program 2.2 */

options ls=64 ps=45 nodate nonumber;
title1 "Output 2.2";
data track;
infile 'womentrack.dat' firstobs=3;
input m100 m200 m400 m800 m1500 m3000 marathon country$;
run;

proc princomp data=track out=pctrack;
var m100 m200 m400 m800 m1500 m3000 marathon;
title2 'PCA of National Track Data: Time';
run;

proc sort data=pctrack;
by prin1;
proc print;
id country;
var prin1;
title2 'Rankings by the First Principal Component: Time';
run;

data labels;
set pctrack;
retain xsys '2' ysys '2';
length text $12 function $8;
text=country;
y=prin2;
x=prin1;
size=1.2;
position='8';
function = 'LABEL';
run;
filename gsasfile "prog22a.graph";
goptions reset=all gaccess=gsasfile autofeed dev=psl;
goptions horigin=1in vorigin=2in;
goptions hsize=6in vsize=8in;
```

```
title1 h=1.2 'Plots of the First Two Principal Components of Time';
title2 j=l 'Output 2.2';
proc gplot data=pctrack;
plot prin2*prin1/annotate=labels;
label prin1='First PC'
      prin2='Second PC';
symbol v=star;
run;

data track;
set track;
title1 'Output 2.2';
title2 'Speeds for the Track Events (Women)';
m800=m800*60;
m1500=m1500*60;
m3000=m3000*60;
marathon=marathon*60;
x1=100/m100;
x2=200/m200;
x3=400/m400;
x4=800/m800;
x5=1500/m1500;
x6=3000/m3000;
x7=42195/marathon;
run;

data speed;
set track;
keep country x1-x7;
proc print data=speed noobs label;
var country x1-x7;
format x1-x7 6.2;
label x1='100m'
x2='200m'
x3='400m'
x4='800m'
x5='1,500m'
x6='3,000m'
x7='marathon';
run;

proc princomp data=track cov out=pctrack;
var x1-x7;
title2 'PCA of National Track Data: Speeds';
run;

proc sort;
by prin1;
run;
proc print;
id country;
var prin1;
title2 'Rankings by the First Principal Component: Speed';
run;

data labels;
set pctrack;
retain xsys '2' ysys '2';
length text $12 function $8;
```

```
text=country;
y=prin2;
x=prin1;
size=1.2;
position='8';
function = 'LABEL';
run;
filename gsasfile "prog22b.graph";
goptions reset=all gaccess=gsasfile autofeed dev=psl;
goptions horigin=1in vorigin=2in;
goptions hsize=6in vsize=8in;
title1 h=1.2 'Plots of the First Two Principal Components of Speed';
title2 j=l 'Output 2.2';
proc gplot data=pctrack;
plot prin2*prin1/annotate=labels;
label prin1='First PC'
      prin2='Second PC';
symbol v=star;
run;
```

Output 2.2

Output 2.2
PCA of National Track Data: Time

The PRINCOMP Procedure

Observations	55
Variables	7

Correlation Matrix

	m100	m200	m400	m800
m100	1.0000	0.9528	0.8347	0.7277
m200	0.9528	1.0000	0.8570	0.7241
m400	0.8347	0.8570	1.0000	0.8984
m800	0.7277	0.7241	0.8984	1.0000
m1500	0.7284	0.6984	0.7878	0.9016
m3000	0.7417	0.7099	0.7776	0.8636
marathon	0.6863	0.6856	0.7054	0.7793

Correlation Matrix

	m1500	m3000	marathon
m100	0.7284	0.7417	0.6863
m200	0.6984	0.7099	0.6856
m400	0.7878	0.7776	0.7054
m800	0.9016	0.8636	0.7793
m1500	1.0000	0.9692	0.8779
m3000	0.9692	1.0000	0.8998
marathon	0.8779	0.8998	1.0000

Output 2.2 (*continued*)

Eigenvalues of the Correlation Matrix

	Eigenvalue	Difference	Proportion	Cumulative
1	5.80568576	5.15204024	0.8294	0.8294
2	0.65364552	0.35376309	0.0934	0.9228
3	0.29988243	0.17440494	0.0428	0.9656
4	0.12547749	0.07166058	0.0179	0.9835
5	0.05381692	0.01476763	0.0077	0.9912
6	0.03904928	0.01660668	0.0056	0.9968
7	0.02244260		0.0032	1.0000

Eigenvectors

	Prin1	Prin2	Prin3	Prin4
m100	0.368356	0.490060	0.286012	-.319386
m200	0.365364	0.536580	0.229819	0.083302
m400	0.381610	0.246538	-.515367	0.347377
m800	0.384559	-.155402	-.584526	0.042076
m1500	0.389104	-.360409	-.012912	-.429539
m3000	0.388866	-.347539	0.152728	-.363120
marathon	0.367004	-.369208	0.484370	0.672497

Rankings by the First Principal Component: Time

country	Prin1
GDR	-3.50602
USSR	-3.46469
USA	-3.33581
Czech	-3.05380
FRG	-2.92578
GB&NI	-2.78316
Poland	-2.67210
Canada	-2.60813
Finland	-2.18184
Italy	-2.13954
.	.
.	.
.	.
Guatemal	3.22730
Guinea	3.98086
Mauritiu	4.23385
Cookis	6.07728
WSamoa	8.33288

Output 2.2
(continued)

Plots of the First Two Principal Components of Time

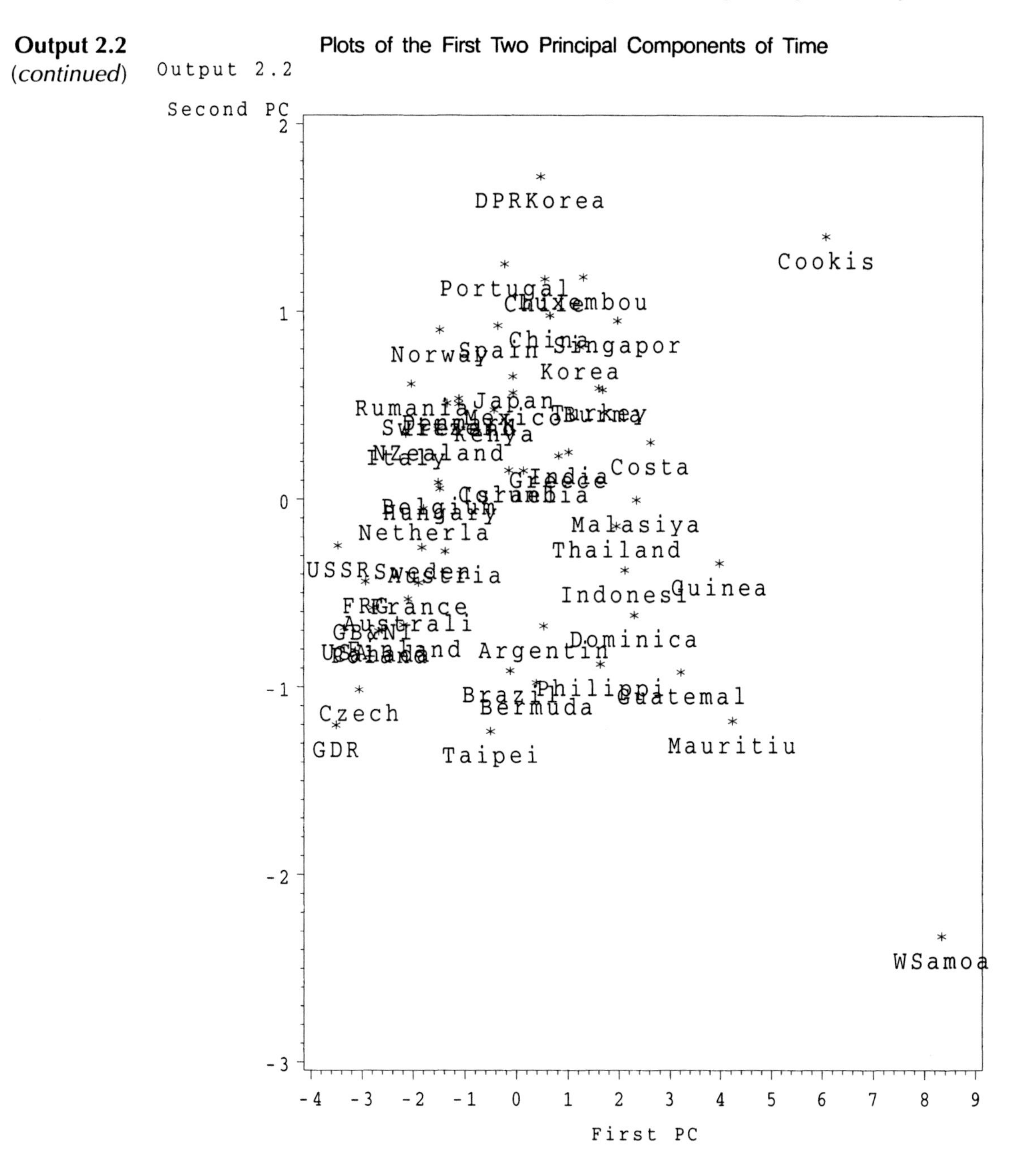

For the early part of Output 2.2, we see that the largest eigenvalue is $\hat{\lambda}_1 = 5.8057$. Thus the first principal component explains about 83% of the total variation in the data. We also observe that the first eigenvector (Prin1 column),

$$(0.3684, 0.3654, 0.3816, 0.3846, 0.3891, 0.3889, 0.3670)'$$

having all positive elements, reveals that the first principal component (like the average) seems to measure the overall performance. Hence it may be used to order the countries based on their overall performance in track records. The following SAS code, a part of Program 2.2, ranks the observations (countries) based on their scores on the first principal component,

```
proc princomp data = track out = pctrack;
var m100 m200 m400 m800 m1500 m3000 marathon;
run;
proc sort data = pctrack;
by prin1;
proc print;
id country;
var prin1;
run;
```

The option OUT=PCTRACK in the PROC PRINCOMP statement creates a SAS data set named PCTRACK. This data set contains all the original data and the principal component scores. However, we have not provided the content of this data set in Output 2.2. The results produced by the above code (selectively provided in Output 2.2) indicate that GDR is ranked as number one followed by USSR, USA, and CZECH (Czechoslovakia).

The second principal component with variance $\hat{\lambda}_2 = 0.6537$ explains about 9% of the total variability, and the first two together explain more than 92% of the variation. The eigenvector (Prin2) corresponding to $\hat{\lambda}_2$ is

$$(0.4901,\ 0.5366,\ 0.2465,\ -0.1554,\ -0.3604,\ -0.3692)'.$$

This indicates that the second principal component represents a difference between the short- and the long-distance races. For example, if for a country the second principal component is close to zero, then it can be interpreted that the performance of this country in both short- and long-distance races is similar.

Since the principal components are uncorrelated, it is a common practice to plot the scores on the first two principal components on the plane to determine the clusters or groups among the observations. As we have mentioned before, the OUT = PCTRACK option in the PROC PRINCOMP statement stores all the original data and principal component scores in a data set named PCTRACK. A scatter plot of the second principal component scores against the first or that of the first against the second can be easily obtained using the PLOTIT macro. For example, the statement

```
%plotit(data=pctrack, labelvar=country, plotvars=prin1 prin2);
```

produces a scatter plot of the scores on the first principal component against the second. The PLOTIT macro is a general macro for plotting the data in many other situations like correspondence analysis, discriminant analysis, and so forth. However, to have more control on our plots in terms of size, fonts, etc., in Program 2.2 we have adopted PROC GPLOT for plotting.

Upon examining the scatter plot reproduced in Output 2.2, we see that there are a few clusters of countries. Further, Western Samoa (WSamoa (W)) and Cook Islands (Cookis (C)) stand out from the rest.

Correlation or Covariance Matrix? The starting point of a principal component analysis is either a covariance matrix or a correlation matrix. One of the problems that a practitioner of principal component analysis faces is whether to begin with the correlation or covariance matrix. Generally it is a difficult question to answer. Because there is no obvious relationship between the corresponding eigenvalues and eigenvectors of the two matrices, the PCA results based on these two matrices can be very different. Also, the number of important principal components that are selected using one matrix may be quite different from the number selected using the other. The interpretations of the principal components can be quite different as well.

Some researchers advocate always using the correlation matrix. This seems like a good idea except for the two important facts outlined here. First, the theoretical development of statistical inference based on the eigenvalues and eigenvectors of a correlation matrix is much more difficult than that based on the eigenvalues and eigenvectors of a covariance matrix. Second, by using the correlation matrix we are forcing all the variables to have equal variance and this in turn may defeat the purpose of identifying those variables that contribute more significantly to the total variability. It is therefore recommended that other practical considerations and issues be taken into account in selecting either of the two matrices as a starting point for any PCA.

For the Olympics track record data, Naik and Khattree (1996) considered the *distance covered per unit time* or *speed* for each track event to bring all the variables to an equal footing instead of standardizing them to have the same variance. This is especially so because of the different nature and running strategies followed by runners in various races. For example, a marathon runner moves slowly in the beginning of a race as compared to a short-distance runner, in order to sustain speed over a longer period of time. Thus, the variable speed succeeds in retaining the possibility of having different degrees of variability in different variables. Hence, instead of using the correlation matrix, we used the variance-covariance matrix of the speed in order to rank the countries using the first principal component. Program 2.2 also contains the code for doing this analysis. First of all, the total times taken are converted to the same unit (in seconds). Then the speeds are computed. For example, speed for 100 meters run is calculated as $X1 = 100/M100$, where M100 is the variable representing total time taken in seconds for running 100 meters. Similarly speeds X2, ..., X7 are calculated. The principal component analysis using the covariance matrix of X1, ..., X7 then can be performed using the code

```
proc princomp data = track cov out = pctrack;
var x1-x7;
run;
```

The COV option is used to utilize a covariance matrix rather than the default choice of correlation matrix as a starting point in the PCA. The first principal component with the variance $\hat{\lambda}_1 = 0.9791$ explains about 83.7% of the variability and the eigenvector corresponding to $\hat{\lambda}_1$ listed under the column Prin1 in Output 2.2 is $(0.2908, 0.3419, 0.3386, 0.3054, 0.3859, 0.3996, 0.5310)'$. The ranks based on the scores on the first principal component indicate that the USA is ranked as number one followed by USSR, GDR, and FRG with the second, third, and fourth ranks respectively. This ranking is different from the ranking that we had obtained using the correlation matrix.

The second principal component with variance $\hat{\lambda}_2 = 0.0986$ explains about 8% of the variance. It represents, as in the case of PCA based on the correlation matrix, a contrast between long- and short-distance races. The scatter plot of the scores on the first and the second principal components yields a pattern similar to what was observed in Dawkins' analysis. See the continuation of Output 2.2.

Output 2.2 *(continued)*

```
Output 2.2
PCA of National Track Data: Speeds

The PRINCOMP Procedure

Observations        55
Variables            7

Covariance Matrix
```

	x1	x2	x3	x4
x1	0.1095994742	0.1237832898	0.1038905488	0.0795460124
x2	0.1237832898	0.1533183476	0.1264920245	0.0939865908

Output 2.2 *(continued)*

x3	0.1038905488	0.1264920245	0.1408367011	0.1111527230
x4	0.0795460124	0.0939865908	0.1111527230	0.1085419120
x5	0.0991434510	0.1136836024	0.1217076143	0.1220487003
x6	0.1031927674	0.1174087274	0.1222224807	0.1199176569
x7	0.1348335245	0.1582593575	0.1518052167	0.1468255042

Covariance Matrix

	x5	x6	x7
x1	0.0991434510	0.1031927674	0.1348335245
x2	0.1136836024	0.1174087274	0.1582593575
x3	0.1217076143	0.1222224807	0.1518052167
x4	0.1220487003	0.1199176569	0.1468255042
x5	0.1624693091	0.1617652421	0.1963723872
x6	0.1617652421	0.1734350606	0.2097175807
x7	0.1963723872	0.2097175807	0.3215983865

Total Variance 1.1697991911

Eigenvalues of the Covariance Matrix

	Eigenvalue	Difference	Proportion	Cumulative
1	0.97910541	0.88053887	0.8370	0.8370
2	0.09856654	0.04551016	0.0843	0.9212
3	0.05305638	0.02981429	0.0454	0.9666
4	0.02324210	0.01603734	0.0199	0.9865
5	0.00720475	0.00198981	0.0062	0.9926
6	0.00521494	0.00180586	0.0045	0.9971
7	0.00340907		0.0029	1.0000

Eigenvectors

	Prin1	Prin2	Prin3	Prin4
x1	0.290843	0.426967	-.250363	0.329132
x2	0.341929	0.558190	-.320174	0.132018
x3	0.338593	0.381781	0.320836	-.537041
x4	0.305423	0.007925	0.475251	-.309251
x5	0.385868	-.197142	0.372461	0.362247
x6	0.399608	-.253973	0.214598	0.474764
x7	0.531023	-.506889	-.566770	-.365473

Rankings by the First Principal Component: Speed

country	Prin1
WSamoa	-3.09597
Cookis	-2.30587
Mauritiu	-1.88921
Guinea	-1.72851
Guatemal	-1.43046
.	.
.	.
.	.
Finland	0.89759
Italy	0.93873
Poland	1.00439

Output 2.2
(continued)

Canada	1.11069
Czech	1.14187
GB&NI	1.18965
FRG	1.24939
GDR	1.35933
USSR	1.45505
USA	1.49962

Plots of the First Two Principal Components of Speed

Output 2.2

```
Second PC
 0.7
 0.6   WSamoa                                   GDR  Czech
 0.5                              Taipei
       Mauritiu
 0.4                                            Poland
               Philippi
 0.3       Guatemal
                    Brazil Bermuda
 0.2              Argentin  Finland  Canada
           *Dominica
       Guinea  Indonesi      Austria GB&NI USSR USA
 0.1                         Austria France FRG
 0.0                                 Sweden
                   India
                   Kenya
       Malasiya Greece  Hungary Netherla
-0.1
          Turkey Columbia  Rumania
          Thailand Burma
-0.2                Israel  Italy
-0.3           Mexico Switzerl NZealand
          Singapor Costa
                    Denmark
-0.4                *Spain Ireland
                   China
                 DPRKorea
       Cookis   Luxembou Korea Japan
-0.5                          Norway
-0.6
                            Portugal
-0.7
     -4     -3     -2     -1     0      1      2
                       First PC
```

2.4 Selection of the Number of Principal Components

The two issues in principal component analysis that have been extensively debated without definite consensus are (a) whether to use a correlation or covariance matrix as the starting point, and (b) the appropriate number of principal components to be selected for study. Some discussion of the first has been provided in the last section. We will discuss the second issue here.

There are three methods that are commonly used. The first one is based on the *cumulative proportion of total variance*. This is also the most commonly used criterion, and it can be adopted whether a covariance or a correlation matrix is used for PCA. An appropriate minimum percentage of total variation desired to be explained by the PCA is prespecified, and the smallest number of principal components that satisfies this criterion is selected. The prespecified percentage is usually taken to be about 90%. If $\lambda_1 \geq \cdots \geq \lambda_p$ are the eigenvalues of a covariance (correlation) matrix, then the cumulative proportion of the first k eigenvalues is

$$\frac{\sum_{i=1}^{k} \lambda_i}{\sum_{i=1}^{p} \lambda_i}, \quad k = 1, \ldots, p.$$

In the case of a correlation matrix, since $\sum_{i=1}^{p} \lambda_i = p$, this cumulative proportion is simply

$$\frac{1}{p} \sum_{i=1}^{k} \lambda_i, \quad k = 1, \ldots, p.$$

These cumulative proportions are printed as the standard output whenever the PRINCOMP procedure is used.

The second method applicable for PCA using the correlation matrix is based on the size (magnitude) of the variances of the principal components. These are nothing but the eigenvalues of the correlation matrix. This method, suggested by Kaiser (1960), is based on the idea that if the original variables are all uncorrelated then the set of principal components is the same as the set of original variables. In the case of a correlation matrix, all the standardized variables have variance one. Hence, any principal component whose variance is much less than one is not selected since it is presumed to contain substantially less information than any of the original variables. Using some simulation studies Jolliffe (1972) suggested that the cutoff point of the eigenvalues for dropping undesirable principal components should be 0.7. Since the eigenvalues of the correlation matrix are printed as the standard output of the PRINCOMP procedure, this method, like the first, is easy to implement. However, the principal component analysis for the example that follows indicates that this method and the suggested cutoff points are arbitrary and unsatisfactory. Further, Kaiser's argument may be questionable on the grounds that the set of first k principal components cumulatively explains more variation than that cumulatively explained by any set of k original variables or their linear combinations. This fact, usually not emphasized in textbooks, follows from the theory of majorization (Marshall and Olkin (1979).

The third method is graphical and uses what is usually called a *scree* diagram. Like the first method, this is applicable to both the correlation and covariance matrices. A scree diagram is a plot of eigenvalues λ_k (of a correlation or covariance matrix) against k, $k = 1, \ldots, p$. Using this plot, the number (k) of principal components to be selected is determined in such a way that the slope of the graph is steep to the left of k but at the same time not steep to the right. The idea behind this plot is that the number of principal components to be selected is such that the differences between consecutive eigenvalues are becoming increasingly smaller. Interpretation of the scree diagram is rather subjective because it involves picking the number of important principal components based on the visual appearance of the plot. Although PROC PRINCOMP does not have the facility to draw a scree diagram, its counterpart the FACTOR procedure can readily produce a scree diagram.

An illustration of these three methods is provided in Example 2. Several other methods are suggested in the literature for selecting the number of principal components. Jolliffe (1986) provides an extensive discussion of these methods. Keep in mind, however, that different methods may result in different conclusions, as we will show at the end of this section.

EXAMPLE 2 ***Selection of PCs, Talent Data*** The data set under consideration is a part of the large data set titled *Talent Data* that appears in Cooley and Lohnes (1971). The variables under consideration are scores on tests in English ($X1$), Reading Comprehension ($X2$), Creativity ($X3$), Mechanical Reasoning ($X4$), Abstract Reasoning ($X5$), and Mathematics ($X6$) for 30 sixth-grade male students who intend to go to college.

We will illustrate various criteria described above for selecting the number of principal components. Both the principal component analysis based on correlation matrix and that based on the covariance matrix will be considered.

Once a SAS data set (TALENT), containing the raw data on the variables $X1$ to $X6$ is created, the following statements use by default the correlation matrix between $X1, \ldots, X6$ for the principal component analysis.

```
proc princomp data = talent;
var x1-x6;
run;
```

If we want to use a covariance matrix instead of a correlation matrix, we use the following statements.

```
proc princomp data = talent cov;
var x1-x6;
run;
```

The detailed program is given in Program 2.3, and the results are shown in Output 2.3.

```
/* Program 2.3 */

options ls=64 ps=45 nodate nonumber;
title1 "Output 2.3";
data talent;
input x1-x6@@;
sub=_n_;
datalines;
82 47 13 14 12 29 94 40 10 15 12 32
76 33 9 12 9 25 99 46 18 20 15 51
79 38 14 18 11 39 85 42 12 17 12 32
82 32 10 18 8 31 81 43 8 10 11 34
92 38 11 14 11 35 81 32 5 14 13 30
92 41 17 17 11 27 86 43 5 11 11 42
92 43 12 15 12 37 92 43 16 19 12 39
78 25 10 15 7 23 97 45 10 16 11 49
76 27 8 10 13 17 99 39 9 17 11 44
96 44 18 15 10 43 83 33 7 15 11 27
89 43 18 17 10 42 104 47 8 13 14 47
84 36 18 16 8 18 85 41 15 19 12 41
86 44 14 20 12 37 94 40 13 15 6 23
99 44 17 20 10 32 71 23 1 5 9 15
89 33 7 16 11 24 106 48 18 13 12 37
;
```

```
/* Source: Cooley and Lohnes (1971). Courtesy of
John Wiley and Sons, Inc. */

proc princomp data=talent;
var x1-x6;
title2 ' PCA Using Correlation Matrix';
title3 ' Methods for Selection of Number of PCs ';
run;

proc factor data=talent scree;
var x1-x6;
*Scree diagram using FACTOR procedure;
run;

proc princomp data=talent cov;
var x1-x6;
title2 ' PCA Using Covariance Matrix';
title3 ' Methods for Selection of Number of PCs ';
run;

proc factor data=talent cov scree;
var x1-x6;
*Scree diagram using FACTOR procedure;
run;
```

Output 2.3

```
                        Output 2.3
              PCA Using Correlation Matrix
          Methods for Selection of Number of PCs

                 The PRINCOMP Procedure

                Observations         30
                Variables             6

                   Correlation Matrix
```

	x1	x2	x3	x4	x5	x6
x1	1.0000	0.7251	0.4709	0.3972	0.3045	0.6313
x2	0.7251	1.0000	0.5760	0.4121	0.4284	0.7335
x3	0.4709	0.5760	1.0000	0.6671	0.0196	0.3140
x4	0.3972	0.4121	0.6671	1.0000	0.1132	0.4335
x5	0.3045	0.4284	0.0196	0.1132	1.0000	0.5076
x6	0.6313	0.7335	0.3140	0.4335	0.5076	1.0000

Eigenvalues of the Correlation Matrix

	Eigenvalue	Difference	Proportion	Cumulative
1	3.33449913	2.11229063	0.5557	0.5557
2	1.22220849	0.62333956	0.2037	0.7595
3	0.59886893	0.19530941	0.0998	0.8593
4	0.40355952	0.10313524	0.0673	0.9265
5	0.30042428	0.15998463	0.0501	0.9766
6	0.14043965		0.0234	1.0000

Output 2.3 *(continued)*

Eigenvectors

	Prin1	Prin2	Prin3	Prin4	Prin5	Prin6
x1	0.449234	0.052976	-.534779	-.077200	0.703830	0.089761
x2	0.490090	0.100148	-.280621	0.253468	-.422287	-.654570
x3	0.380140	-.539171	0.059796	0.543824	-.159450	0.489943
x4	0.368763	-.456507	0.571655	-.395853	0.233644	-.342836
x5	0.268195	0.637653	0.552111	0.396341	0.235757	0.063047
x6	0.453408	0.285410	-.008468	-.565918	-.436694	0.449362

```
Scree Plot of Eigenvalues
      |
    4 +
      |
      |
      |
      |
      |            1
      |
    3 +
      |
E     |
i     |
g     |
e     |
n     |
v   2 +
a     |
l     |
u     |
e     |
s     |                 2
      |
    1 +
      |
      |
      |                       3
      |                             4
      |                                   5
      |                                         6
    0 +
      |
      -----+------+------+------+------+------+------+-----
           0      1      2      3      4      5      6

                            Number
```

In reviewing the output, first consider the principal component analysis based on the correlation matrix. By examining the cumulative proportion of the variation explained by the principal components given in Output 2.3, we see that at least four principal components are needed to account for 90% of the total variability. However, only the first two principal components will be selected if the second approach is used. The reason for this is that, in this case, all principal components having a variance of less than 0.7 will be discarded. If we examine the scree diagram, created by the following statements, we see that perhaps only the first three principal components will be selected.

```
proc factor data = talent scree;
var x1-x6;
run;
```

Obviously, these criteria are not consistent with each other with respect to suggesting the number of important principal components.

If we examine the coefficients of the variables (available as the eigenvectors of the correlation matrix), in these principal components, we see that the first component with all its coefficients positive, that is, 0.4492, 0.4901, 0.3801, 0.3688, 0.2682, and 0.4534 (see Prin1 in Output 2.3) seems to measure the general ability or talent of a student. The second principal component (see the last four values under the column Prin2), with the coefficients -0.5392, -0.4565, 0.6377, and 0.2854, respectively, for the variables X3, X4, X5, and X6, seems to measure a difference between abstract thinking and mechanical thinking. The coefficients for the variables X1 and X2, respectively, are 0.0530 and 0.1001. Hence these two variables can be ignored in the definition of the second principal components. The third component seems to measure a difference between reasoning and language. This is so since the coefficients for X1, X2, X4, and X5 are -0.5348, -0.2806, 0.5717, and 0.5521, respectively. It is not very clear how to interpret the fourth principal component.

Output 2.3
(continued)

Output 2.3
PCA Using Covariance Matrix
Methods for Selection of Number of PCs

The PRINCOMP Procedure

Observations	30
Variables	6

Covariance Matrix

	x1	x2	x3
x1	76.49310345	42.14137931	18.95517241
x2	42.14137931	44.16206897	17.61724138
x3	18.95517241	17.61724138	21.18275862
x4	11.69655172	9.22068966	10.33793103
x5	5.20344828	5.56206897	0.17586207
x6	52.25517241	46.13103448	13.67586207

Covariance Matrix

	x4	x5	x6
x1	11.69655172	5.20344828	52.25517241
x2	9.22068966	5.56206897	46.13103448
x3	10.33793103	0.17586207	13.67586207
x4	11.33793103	0.74482759	13.81379310
x5	0.74482759	3.81724138	9.38620690
x6	13.81379310	9.38620690	89.55862069

Total Variance 246.55172414

Output 2.3 *(continued)*

Eigenvalues of the Covariance Matrix

	Eigenvalue	Difference	Proportion	Cumulative
1	176.375691	143.731554	0.7154	0.7154
2	32.644137	12.417046	0.1324	0.8478
3	20.227092	8.956244	0.0820	0.9298
4	11.270847	7.687289	0.0457	0.9755
5	3.583558	1.133160	0.0145	0.9901
6	2.450398		0.0099	1.0000

Eigenvectors

	Prin1	Prin2	Prin3	Prin4	Prin5	Prin6
x1	0.577696	0.599819	-.521483	0.184627	-.005124	0.020649
x2	0.444875	0.085092	0.305239	-.778526	0.245940	-.187311
x3	0.186443	0.347546	0.688422	0.188339	-.516097	0.262103
x4	0.131490	0.074372	0.397971	0.510614	0.726572	-.173680
x5	0.067421	-.100703	-.047184	-.112062	0.325012	0.929998
x6	0.641684	-.704639	-.018753	0.225950	-.198976	-.027008

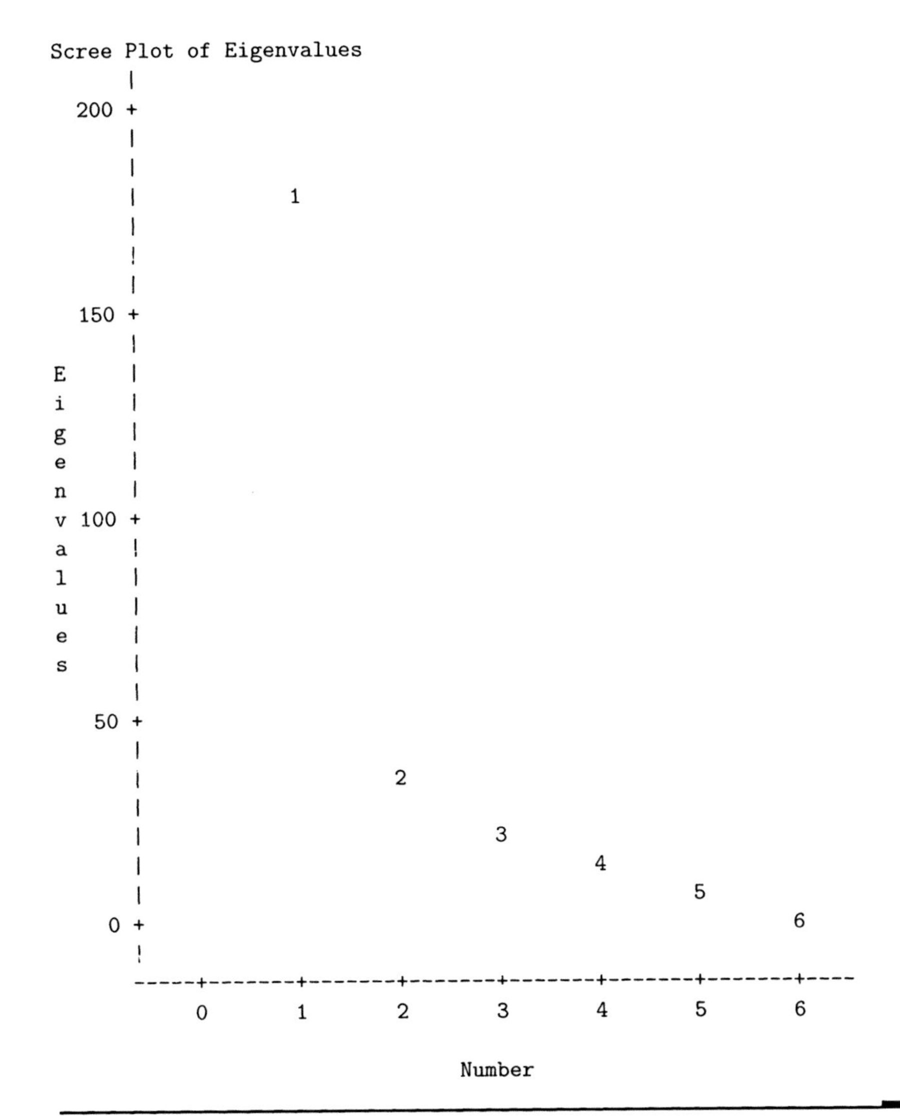

Now consider the principal component analysis based on the covariance matrix. Examining this part of Output 2.3, we see that about 85% of the total variation is explained by the first two principal components, whereas about 93% of the variation is explained by the first three components. Being relevant only in the context of a correlation matrix, the second method of selecting the number of principal components is not applicable in this case. After examining the scree diagram, let us select the first three principal components. The first principal component, as in the previous case, measures the ability of a boy in academic subjects. The second component seems to measure a difference between math ability and English plus creativity. Interpretation of the third component is not clear-cut, although it seems to measure a difference between creativity and competence in English.

Thus, at least in this example, there is very little agreement in the number of components selected using the three methods. Which of the three methods described above to identify the appropriate number of principal components is the best? Our recommendation is to use the first method, that is, to fix a cumulative proportion (percentage) of the variation that needs to be explained by the PCA and then select the minimum number of principal components that satisfy this criterion. An advantage of this approach, as opposed to the second, is that it can be used whether the PCA is done using the correlation matrix or covariance matrix. Also, it is in general less subjective than the approach using a scree diagram. In addition, cumulative proportions are provided as standard output of the PRINCOMP procedure and hence the selection of important principal components is straightforward.

2.5 Some Applications of Principal Component Analysis

In this section we present various applications of principal component analysis in a variety of fields. Through these examples we illustrate many different aspects of the principal components analysis that are specific to the particular context.

2.5.1 Size and Shape Factors

Biologists often use functions of measurements to represent the size and shape of organisms in order to study the differences between the groups. Jolicoeur and Mosimann (1960) suggested the use of the first principal component as a *size* factor, provided that all its coefficients are positive, and other principal components with positive and negative coefficients as *shape* factors. Rao (1964) has suggested some other interesting methods for measuring the size and shape factors of an organism. For a more recent discussion see Bookstein (1989). We will illustrate the application of principal components to extract these factors in the following example.

EXAMPLE 3 ***Size and Shape Based on Body Measurements*** Wright (1954) obtained the correlation matrix of the bone measurements on six variables for a sample of 276 white leghorn fowls. The six measurements are skull length (SL), skull breadth (SB), humerus length (HL), ulna length (UL), femur length (FL), and tibia length (TL). We use the data to determine the size and shape factors by using principal component analysis. The information (data) is available in the form of a correlation matrix. The PRINCOMP procedure accepts the correlation matrix as data when we specify TYPE=CORR. The corresponding statements in the DATA step are

```
data leghorn (type=corr);
infile cards missover;
if _n_=1 then _type_='N';
_type_='corr';
```

These statements create a data set called LEGHORN that contains the 6 by 6 correlation matrix of the six variables defined above. Since the correlation matrix is symmetric, only the lower part of the matrix needs to be input by using the INFILE CARDS MISSOVER statement. The number of observations (276) used in computing the correlation matrix is specified as the first line of data (however, we do not directly use the sample size in this analysis). In the INPUT statement the variable _NAME_ is used to specify the names for the appropriate variables being considered. To perform the principal component analysis, we use the following statement:

```
proc princomp data = datasetname ;
```

Since no raw data are available, the principal component scores cannot be obtained nor can the graphs be plotted. However, using the eigenvectors, we can give some interpretations for the principal components. Consider the following program and its output.

```
/* Program 2.4 */

options ls=64 ps=45 nodate nonumber;
title1 "Output 2.4";
data leghorn (type=corr);
infile cards missover;
_type_='corr';
if _n_=1 then _type_='N';
input _name_ $ sl sb hl ul fl tl;
cards;
N 276
sl 1.0
sb .584 1.0
hl .615 .576 1.0
ul .601 .530 .940 1.0
fl .570 .526 .875 .877 1.0
tl .600 .555 .878 .886 .924 1.0
;
/* Source: Wright (1954).  Reprinted by permission of
the Iowa State University Press. */

proc print;
title2 'Determination of Size and Shape Factors Using PCA';
proc princomp data=leghorn(type=corr);
run;
```

Output 2.4

Output 2.4
Determination of Size and Shape Factors Using PCA

Obs	_type_	_name_	sl	sb	hl	ul	fl	tl
1	N	N	276.000	.	.	.	.	.
2	corr	sl	1.000	.	.	.	.	.
3	corr	sb	0.584	1.000	.	.	.	.
4	corr	hl	0.615	0.576	1.000	.	.	.
5	corr	ul	0.601	0.530	0.940	1.000	.	.
6	corr	fl	0.570	0.526	0.875	0.877	1.000	.
7	corr	tl	0.600	0.555	0.878	0.886	0.924	1

The PRINCOMP Procedure

Observations	276
Variables	6

Output 2.4 *(continued)*

Eigenvalues of the Correlation Matrix

	Eigenvalue	Difference	Proportion	Cumulative
1	4.56757080	3.85344753	0.7613	0.7613
2	0.71412326	0.30199429	0.1190	0.8803
3	0.41212898	0.23894007	0.0687	0.9490
4	0.17318890	0.09733018	0.0289	0.9778
5	0.07585872	0.01872938	0.0126	0.9905
6	0.05712934		0.0095	1.0000

Eigenvectors

	Prin1	Prin2	Prin3	Prin4	Prin5	Prin6
sl	0.347439	0.536974	-.766673	0.049099	0.027212	0.002372
sb	0.326373	0.696467	0.636305	0.002033	0.008044	0.058827
hl	0.443419	-.187301	0.040071	-.524077	0.168397	-.680939
ul	0.439983	-.251382	-.011196	-.488771	-.151153	0.693796
fl	0.434544	-.278168	0.059205	0.514259	0.669483	0.132738
tl	0.440150	-.225698	0.045735	0.468582	-.706953	-.184077

By examining Output 2.4, we conclude that three principal components explain about 95% of the variation. The first principal component has all its coefficients positive, and thus it may represent the size of an individual. The second principal component is a measure of shape factor, which indicates a difference between head (skull) and body (wing and other parts) measurements. The third component seems to measure a difference between head length and head breadth; this is so since the coefficients corresponding to the other variables are very small.

2.5.2 Reduction of Dimensionality

One of the important uses of principal component analysis is its utility in reducing the dimension of a multivariable problem. This can be done by either selecting only a few important variables from a large list of variables, or by determining a few new variables that are certain functions of all the original variables. To illustrate variable selection by using principal components, we consider the following example from Jeffers (1967).

EXAMPLE 4 ***Selection of Variables, Adelges Data*** Measurements on 19 variables; body length (LENGTH), body width (WIDTH), fore-wing length (FORWING), hind-wing length (HINWING), number of spiracles (SPIRAC), length of antennal segment I to V (ANTSEG1 through ANTSEG5), number of antennal spines (ANTSPIN), leg length tarsus III (TARSUS3), leg length tibia III (TIBIA3), leg length femur III (FEMUR3), rostrum (ROSTRUM), ovipositor (OVIPOS), number of ovipositor spines (OVSPIN), anal fold (FOLD), and number of hind-wing hooks (HOOKS) were observed on 40 individual *alate adeleges* (winged aphids) (Jeffers, 1967). The above variables were selected as diagnostic characteristics of aphids. The objective of the study was to determine the number of distinct taxa that were present in the particular habitat from which the samples were collected. Since *adelges* are difficult to identify with certainty by the taxonomical measurements, the principal component method was used to gain information on the number of distinct taxa present in the sample.

The correlation matrix of these variables, presented in the data set ADELGES, is used to compute the principal components. The following code will help us accomplish that.

```
proc princomp data = adelges n=5;
```

The option N= selects a prespecified number of components. Suppose we decide to select at most five components. The complete program for this example is given in Program 2.5. The output from this program is presented in Output 2.5.

```
/* Program 2.5 */

options ls=64 ps=45 nodate nonumber;
title1 "Output 2.5";
title2 'Dimension Reduction and Selection of Variables in PCA';
data adelges(type=corr);
infile cards missover;
_type_='corr';
if _n_=1 then _type_='N';
input length width forwing hinwing spirac antseg1-antseg5 antspin;
input tarsus3 tibia3 femur3 rostrum ovipos ovspin fold hooks;
cards;
40

1.0

.934 1.0

.927 .941 1.0

.909 .944 .933 1.0

.524 .487 .543 .499 1.0

.799 .821 .856 .833 .703 1.0

.854 .865 .886 .889 .719 .923 1.0

.789 .834 .846 .885 .253 .699 .751 1.0

.835 .863 .862 .850 .462 .752 .793 .745 1.0

.845 .878 .863 .881 .567 .836 .913 .787 .805 1.0

-.458 -.496 -.522 -.488 -.174 -.317 -.383 -.497 -.356 -.371 1.0

.917 .942 .940 .945 .516 .846 .907 .861 .848 .902 -.465
1.0
.939 .961 .956 .952 .494 .849 .914 .876 .877 .901 -.447
.981 1.0
.953 .954 .946 .949 .452 .823 .886 .878 .883 .891 -.439
.971 .991 1.0
.895 .899 .882 .908 .551 .831 .891 .794 .818 .848 -.405
.908 .920 .921 1.0
.691 .652 .694 .623 .815 .812 .855 .410 .620 .712 -.198
.725 .714 .676 .720 1.0
.327 .305 .356 .272 .746 .553 .567 .067 .300 .384 -.032
.396 .360 .298 .378 .781 1.0
```

```
-.676 -.712 -.667 -.736 -.233 -.504 -.502 -.758 -.666 -.629 .492
 -.657 -.655 -.687 -.633 -.186 .169 1.0
.702 .729 .746 .777 .285 .499 .592 .793 .671 .668 -.425
 .696 .724 .731 .694 .287 -.026 -.775 1.0
;
/* Source: Jeffers (1967). Reprinted by permission of
the Royal Statistical Society. */

proc princomp data=adelges n=5;
var _numeric_;
run;
```

Output 2.5

Output 2.5
Dimension Reduction and Selection of Variables in PCA

The PRINCOMP Procedure

Observations	40
Variables	19

Eigenvalues of the Correlation Matrix

	Eigenvalue	Difference	Proportion	Cumulative
1	13.8613146	11.4909926	0.7295	0.7295
2	2.3703220	1.6218273	0.1248	0.8543
3	0.7484947	0.2464972	0.0394	0.8937
4	0.5019975	0.2237595	0.0264	0.9201
5	0.2782380		0.0146	0.9348

Eigenvectors

	Prin1	Prin2	Prin3	Prin4	Prin5
length	0.253897	-.031747	0.023746	0.073209	-.167574
width	0.258288	-.066528	0.009997	0.098548	-.131096
forwing	0.259813	-.031480	-.052875	0.067589	0.029241
hinwing	0.259373	-.087559	0.028949	0.001792	0.053512
spirac	0.161724	0.405922	-.190092	-.618137	-.067325
antseg1	0.239357	0.177464	0.039201	-.014380	-.146559
antseg2	0.253350	0.161638	0.003735	0.024644	0.121903
antseg3	0.231543	-.235673	0.053882	0.111762	0.338485
antseg4	0.237975	-.043405	0.164620	0.007061	-.434156
antseg5	0.248516	0.027673	0.104942	-.021418	0.046261
antspin	-.130374	0.203859	0.929154	-.168920	0.027825
tarsus3	0.261273	-.010129	0.031963	0.178445	0.053126
tibia3	0.263479	-.028806	0.082660	0.195039	0.049530
femur3	0.261118	-.065641	0.115012	0.193668	-.034963
rostrum	0.252250	0.010780	0.074815	0.036682	0.033025
ovipos	0.201145	0.396044	-.024095	0.056006	-.056267
ovspin	0.108823	0.546107	-.147398	0.037297	0.239745
fold	-.187971	0.351153	0.041023	0.493503	0.405247
hooks	0.200697	-.282871	0.054814	-.446215	0.609677

The first two principal components account for 85% of the total variation whereas slightly less than 90% of the variation is explained by the first three components. The first four components explain about 92% of the variation. Thus, only four principal components are sufficient to explain most of the variability exhibited by these 19 original variables. This indicates that, although the researchers have collected measurements on 19 variables, only a few basic dimensions of the individuals have actually been measured through these measurements.

Principal component analysis of this data set suggests that we need to use only three or four principal components instead of all the original 19 variables. However, this does not reduce the basic work of collecting the data on all the 19 variables, since every principal component is a linear combination of all of the original variables. Hence, a more challenging problem is to be able to devise a criterion that will select only a few important variables from the long list of original variables so that future researchers in the area need not collect data on all variables but only on the suggested few.

Selection of Variables Using PCA. Selection of variables is a familiar topic in regression analysis. Several methods for selecting a subset of variables using principal components are discussed in Jolliffe (1986). One method is to select as many variables as the number of principal components selected. The variable corresponding to the maximum of the absolute values of the coefficients in the first principal component is selected first. The second variable selected is that variable (if not already selected) whose coefficient in the second principal component has the maximum absolute value. If the variable corresponding to the maximum coefficient was already selected then we choose the variable corresponding to the next maximum. Proceeding this way, we select a subset of required size from the list of original variables.

Returning to Example 4, suppose we wish to select only the first four most important variables. If we apply this procedure (by examining the elements of the eigenvectors in Output 2.5), the variables TIBIA3, OVSPIN, ANTSPIN, and SPIRAC are sequentially selected. Thus in future studies only the following variables—length of tibia III, the number of ovipositor spines, the number of antennal spines, and the number of spiracles—may need to be measured.

Sometimes a substantial reduction in the dimensionality of a set of data may not be possible. Also it may not always be possible to give appropriate interpretations to all the principal components selected by a principal component analysis. We illustrate these points using the example that follows. In this example, several principal components are needed to account for more than 90% of the total variability. Also, it is not easily possible, as it will be seen, to give interpretations to more than two principal components.

EXAMPLE 5 ***Minimal Reduction, Drug Usage Data*** A study conducted by Huba, Wingard, and Bentler (1981), included 1,634 students from the seventh to ninth grades from a school in greater metropolitan Los Angeles. Participants completed a questionnaire about the number of times they had used a particular substance. Frequency-of-drug-use data were collected for cigarettes (CIGARS), beer (BEER), wine (WINE), liquor (LIQUOR), cocaine (COCAINE), tranquilizers (TRANQUIL), drugstore medication to get high (DRUGSTOR), heroin and other opiates (HEROIN), marijuana (MARIJUAN), hashish (HASHISH), inhalents (glue, gasoline) (INHALENT), hallucinogenics (LSD, psilocybin, mescaline) (HALLOCIN), and amphetamine stimulants (AMPHETAM). Responses were recorded on a 5-point scale: never tried (1), only once (2), a few times (3), many times (4), regularly (5). The objective of the study was to see whether a person can be characterized by the pattern of his psychoactive substance use.

The correlation matrix of drug usage rates of the 13 drugs and the program for computing the principal components are presented in Program 2.6. The corresponding output is provided in Output 2.6.

```
/* Program 2.6 */

options ls=64 ps=45 nodate nonumber;
title1 "Output 2.6";
title2 'Minimal Reduction in Dimension Using PCA';
data drug(type=corr);
infile cards missover;
_type_='corr';
if _n_=1 then _type_='N';
input _name_ $ Cigars Beer Wine Liquor Cocaine Tranquil Drugstor
Heroin Marijuan Hashish Inhalent Hallucin Amphetam;
cards;
N 1634
Cigars 1.0
Beer .447 1.0
Wine .422 .619 1.0
Liquor .435 .604 .583 1.0
Cocaine .114 .068 .053 .115 1.0
Tranquil .203 .146 .139 .258 .349 1.0
Drugstor .091 .103 .110 .122 .209 .221 1.0
Heroin .082 .063 .066 .097 .321 .355 .201 1.0
Marijuan .513 .445 .365 .482 .186 .315 .150 .154 1.0
Hashish .304 .318 .240 .368 .303 .377 .163 .219 .534 1.0
Inhalent .245 .203 .183 .255 .272 .323 .310 .288 .301 .302 1.0
Hallucin .101 .088 .074 .139 .279 .367 .232 .320 .204 .368 .340 1.0
Amphetam .245 .199 .184 .293 .278 .545 .232 .314 .394 .467 .392 .511 1.0
;
/* Source: Huba, Wingard, and Bentler (1981).  Reprinted by permission of
the American Psychological Association. */

proc princomp data=drug;
var _numeric_;
run;
```

Output 2.6

```
                         Output 2.6
          Minimal Reduction in Dimension Using PCA

                   The PRINCOMP Procedure

                  Observations        1634
                  Variables             13


            Eigenvalues of the Correlation Matrix

        Eigenvalue    Difference    Proportion    Cumulative

   1    4.38163599    2.33601608        0.3370        0.3370
   2    2.04561991    1.09292004        0.1574        0.4944
   3    0.95269987    0.13522376        0.0733        0.5677
   4    0.81747610    0.05194962        0.0629        0.6306
   5    0.76552648    0.07839085        0.0589        0.6895
   6    0.68713564    0.04360985        0.0529        0.7423
   7    0.64352579    0.02871350        0.0495        0.7918
   8    0.61481229    0.05394684        0.0473        0.8391
   9    0.56086544    0.16210780        0.0431        0.8823
  10    0.39875764    0.00390995        0.0307        0.9129
```

Output 2.6
(continued)

11	0.39484769	0.02061015	0.0304	0.9433
12	0.37423754	0.01137793	0.0288	0.9721
13	0.36285962		0.0279	1.0000

Eigenvectors

	Prin1	Prin2	Prin3	Prin4	Prin5
Cigars	0.278186	-.279931	-.059574	0.016117	0.316082
Beer	0.285892	-.396531	0.129711	0.100628	-.176575
Wine	0.264979	-.391936	0.220090	0.141729	-.308037
Liquor	0.317706	-.324906	0.052843	0.063318	-.177994
Cocaine	0.208150	0.288057	0.052028	0.591037	0.437657
Tranquil	0.292848	0.258752	-.171519	0.085892	-.125470
Drugstor	0.176050	0.189282	0.727048	-.330316	0.247847
Heroin	0.201588	0.315328	0.146530	0.531883	-.323982
Marijuan	0.339180	-.163788	-.236343	-.109348	0.358136
Hashish	0.328735	0.050612	-.354958	-.118052	0.247381
Inhalent	0.276234	0.168891	0.313601	-.195032	0.083960
Hallucin	0.248363	0.329032	-.109434	-.287996	-.357599
Amphetam	0.328497	0.232077	-.232431	-.264454	-.210430

Eigenvectors

	Prin6	Prin7	Prin8	Prin9	Prin10
Cigars	-.462724	-.122966	-.027994	0.620043	0.135044
Beer	0.149582	0.110023	0.054077	-.046802	0.104355
Wine	0.160142	0.062714	-.071930	0.091683	0.422450
Liquor	0.163766	-.002217	-.119963	-.199145	-.622805
Cocaine	0.356458	0.294555	-.225849	0.179525	-.086048
Tranquil	0.054337	-.551141	-.482186	-.077894	0.118907
Drugstor	0.260713	-.338185	0.209300	0.070163	-.006151
Heroin	-.391757	-.162423	0.501834	-.106291	-.021651
Marijuan	-.145539	-.130752	0.253077	-.141456	-.374878
Hashish	0.246872	0.124179	0.375417	-.347645	0.447282
Inhalent	-.505887	0.472230	-.353810	-.368400	0.079265
Hallucin	0.145505	0.401501	0.188533	0.480428	-.185478
Amphetam	0.019101	-.144385	-.178494	0.058121	0.034545

Eigenvectors

	Prin11	Prin12	Prin13
Cigars	0.141779	-.217505	-.201038
Beer	0.071989	0.636735	-.491731
Wine	-.214745	-.159759	0.565242
Liquor	0.159364	-.488567	-.148866
Cocaine	-.157642	0.027442	-.007144
Tranquil	0.430746	0.173798	0.139936
Drugstor	0.004732	-.035025	-.041078
Heroin	-.039863	-.057102	-.047822
Marijuan	-.192712	0.379297	0.467098
Hashish	0.221122	-.290906	-.128449
Inhalent	0.062993	0.019241	0.035955
Hallucin	0.288924	0.107321	0.170682
Amphetam	-.719528	-.084925	-.295657

From Output 2.6, the cumulative proportions of the eigenvalues are 0.3371, 0.4944, 0.5677, 0.6306, 0.6895, 0.7423, 0.7918, 0.8391, 0.8823, 0.9129, 0.9433, 0.9721 and 1.0000. It is clear from these values that 10 principal components out of 13 are needed to account for more than 90% of the total variation in the data. At least in this problem, it is evident that the PCA for reduction of the dimensionality is not very useful except that it gives an indication that the reduction of the dimensionality in this problem is perhaps minimal. This behavior of the PCA is not surprising for this example, since most of the correlations are very weak. For example, the maximum correlation coefficient between any two variables is 0.619.

Low correlation among the variables implies the orthogonality of the variables in the original space. This makes it difficult to reduce the space to a smaller dimension. Thus, the correlation matrix in hand gives us an indication of whether the dimension reduction can be achieved. Performance of a principal component analysis then will enable us to know how much reduction is possible and in the direction of which variable.

Returning to the example in hand, the first eigenvector has all its components positive and close to each other. Hence, the first principal component measures the frequency of overall drug usage of the children. The second principal component seems to measure a difference between the frequency of usage of illegal and legal drugs. Except for marijuana, all the illegal drugs have positive signs, and the legal drugs have negative signs in the coefficients of the second principal component. The second principal component accounts for about 16% of the total variation, which is the main source of variation after the variation in general drug use has been accounted for. All the remaining principal components account for variation in the neighborhood of 5%, except for the last four, which account for about 3% each of the total variation. It also seems difficult to give any clear-cut interpretations to these principal components.

2.5.3 Determination of Typical Points

The principal component analysis thus far has mainly been used for reducing the dimension of the data. In some applications, for example in photographic and optical research, it is important to be able to solve the dual problem of reducing the number of observations to a few *typical points* while keeping the number of dimensions (variables) the same.

Suppose responses of a variable x are available at p time points (or at p levels of an independent variable). For each experimental condition, p values of x form a p by 1 data vector $\mathbf{x}$. Suppose we have a random sample $\mathbf{x}_1, \ldots, \mathbf{x}_n$ of n vectors. Let $\bar{\mathbf{x}}$ be the sample mean vector. It is possible to find a set of eigenvectors which, when added in the proper amounts to the mean vector $\bar{\mathbf{x}}$, will adequately approximate any of the original family of n response vectors. These eigenvectors, together with the mean vector $\bar{\mathbf{x}}$, are called the typical points.

The experiments in the dose-response studies and photographic or optical sciences often yield data of this form. For example, optical densities of photographic images resulting from a series of exposures and spectrophotometric data at various wavelengths are some of the instances. Often these data are presented in the form of curves to aid in comparing the results of several experiments. However, when many curves are being compared simultaneously it may be difficult to draw any definite conclusions using this large crowd of curves. Hence it may be desirable to find an appropriate set of typical profile curves (typical points) to help represent the individual curves.

Let $\mathbf{S}$ be the sample variance-covariance matrix and let $\hat{\lambda}_1 > \cdots > \hat{\lambda}_p$ be the eigenvalues and $\boldsymbol{\ell}_1, \ldots, \boldsymbol{\ell}_p$ be the corresponding eigenvectors of $\mathbf{S}$. Suppose we are interested in a prespecified number $(q+1)$ of typical points. Rao (1964) has determined that

$$\bar{\mathbf{x}}, \sqrt{\hat{\lambda}_1}\boldsymbol{\ell}_1, \ldots, \sqrt{\hat{\lambda}_q}\boldsymbol{\ell}_q \ (q \leq min(p, n))$$

are the $(q+1)$ typical points having the property that the Euclidean norm of the difference between the sample variance-covariance matrix $\mathbf{S}$ and $\mathbf{TT}'$ is minimum when the p by q matrix $\mathbf{T}$ is chosen as $\mathbf{T} = [\sqrt{\hat{\lambda}_1}\boldsymbol{\ell}_1 : \cdots : \sqrt{\hat{\lambda}_q}\boldsymbol{\ell}_q]$. Given $(q+1)$ typical points listed above, any vector, say $\mathbf{x}_i$, is then approximated by

$$\hat{\mathbf{x}}_i = \bar{\mathbf{x}} + b_{1i}\sqrt{\hat{\lambda}_1}\boldsymbol{\ell}_1 + \cdots + b_{qi}\sqrt{\hat{\lambda}_q}\boldsymbol{\ell}_q,$$

where each b_{ij}, the multiplying constants for any vector $\mathbf{x}_i$, is obtained as

$$b_{ij} = \boldsymbol{\ell}'_j(\mathbf{x}_i - \bar{\mathbf{x}})/\sqrt{\hat{\lambda}_j}.$$

EXAMPLE 6 ***Determination of Typical Points*** The following data, with $n = 7$ and $p = 5$, is taken from Simonds (1963). The data vectors $\mathbf{x}_1$ to $\mathbf{x}_7$ are

$$\mathbf{x}_1 = (.12, .16, .36, .68, 1.06)',$$

$$\mathbf{x}_2 = (.11, .12, .29, .74, 1.24)',$$

$$\mathbf{x}_3 = (.14, .18, .40, .87, 1.40)',$$

$$\mathbf{x}_4 = (.12, .14, .33, .82, 1.38)',$$

$$\mathbf{x}_5 = (.16, .20, .46, 1.00, 1.62)',$$

$$\mathbf{x}_6 = (.12, .14, .34, .90, 1.52)',$$

$$\mathbf{x}_7 = (.17, .22, .51, 1.17, 1.90)'.$$

For these data $\bar{\mathbf{x}} = (0.1343, 0.1657, 0.3843, 0.8829, 1.4457)'$.

```
/* Program 2.7 */

options ls=64 ps=45 nodate nonumber;
title1 "Output 2.7";
title2 'Determination of Typical Points';
data simond;
input x1-x5;
cards;
.12 .16 .36 .68 1.06
.11 .12 .29 .74 1.24
.14 .18 .40 .87 1.40
.12 .14 .33 .82 1.38
.16 .20 .46 1.00 1.62
.12 .14 .34 .90 1.52
.17 .22 .51 1.17 1.90
;
/* Source: Simonds (1963).  Reprinted by permission of
the Optical Society of America. */

proc princomp data=simond cov out=b;
var x1-x5;
run;

proc print data=b;
var prin1-prin5;
title3 'Principal Component Scores';
format prin1-prin5 7.4;
run;
```

Output 2.7

```
                    Output 2.7
          Determination of Typical Points

              The PRINCOMP Procedure

             Observations           7
             Variables              5

          Total Variance     0.1081714286


        Eigenvalues of the Covariance Matrix

       Eigenvalue    Difference    Proportion    Cumulative

  1    0.10505441    0.10195969        0.9712        0.9712
  2    0.00309472    0.00308210        0.0286        0.9998
  3    0.00001262    0.00000663        0.0001        0.9999
  4    0.00000599    0.00000229        0.0001        1.0000
  5    0.00000370                      0.0000        1.0000


                      Eigenvectors

         Prin1        Prin2        Prin3        Prin4        Prin5

x1    0.061242     0.201467     0.741837     0.467980     -.431663
x2    0.084110     0.421011     0.171176     0.304595     0.832824
x3    0.197784     0.787063     -.013872     -.542505     -.216587
x4    0.506112     0.186623     -.582127     0.562605     -.231574
x5    0.833011     -.357582     0.285154     -.278174     0.139766

              Principal Component Scores

 Obs       Prin1        Prin2        Prin3        Prin4        Prin5

  1      -0.4301       0.0757      -0.0031      -0.0021      -0.0003
  2      -0.2676      -0.0514      -0.0000       0.0027      -0.0028
  3      -0.0399       0.0335       0.0009       0.0040       0.0026
  4      -0.1003      -0.0447       0.0036      -0.0022       0.0019
  5       0.2239       0.0387       0.0054      -0.0012      -0.0017
  6       0.0588      -0.0719      -0.0032      -0.0015       0.0008
  7       0.5554       0.0201      -0.0036       0.0002      -0.0004
```

In the output (Output 2.7) produced by Program 2.7, we see that the five ordered eigenvalues, $\hat{\lambda}_1 \geq \cdots \geq \hat{\lambda}_5$, of the variance-covariance matrix, respectively are 0.105054, 0.003095, 0.000013, 0.000006, and 0.000004. About 99.98% of the total variation is explained by the first two principal components. The two eigenvectors corresponding to the first two eigenvalues listed in the first part of Output 2.7 under the columns Prin1 and Prin2, respectively, are

$$\ell_1 = (.0612, .0841, .1978, .5061, .8330)',$$

$$\ell_2 = (.2015, .4210, .7871, .1866, -.3576)'.$$

Thus in this example, the three typical points are $\bar{\mathbf{x}}$, $\sqrt{\hat{\lambda}_1}\boldsymbol{\ell}_1$, $\sqrt{\hat{\lambda}_2}\boldsymbol{\ell}_2$. Various linear combinations of these three vectors will approximately reproduce the original data vectors. Using SAS, it is easy to find the coefficients in these linear combinations. First note that using Rao's formula,

$$\hat{\mathbf{x}}_i = \bar{\mathbf{x}} + \boldsymbol{\ell}_1'(\mathbf{x}_i - \bar{\mathbf{x}})\boldsymbol{\ell}_1 + \boldsymbol{\ell}_2'(\mathbf{x}_i - \bar{\mathbf{x}})\boldsymbol{\ell}_2.$$

However, $\boldsymbol{\ell}_j'(\mathbf{x}_i - \bar{\mathbf{x}})$, $j = 1, \ldots, q$ are nothing but the scores on the first q principal components corresponding to vector $\mathbf{x}_i$. Using the OUT= option in the PROC PRINCOMP statement, we can obtain the principal component scores. These scores with the eigenvectors and $\bar{\mathbf{x}}$ can then be used to reconstruct approximate values of the original data. The SCORE procedure can be employed for this purpose. To avoid digression from the main theme we do not discuss the details here.

For our example, the principal component scores named (by default) Prin1 and Prin2 are stored in the SAS data set B, which is obtained by using OUT=B. It may be noted that the eigenvectors are also printed under the variable names Prin1 and Prin2. The principal component scores for each of the seven observations that serve as the coefficients for the first and the second eigenvectors are

$$\text{Prin1} = (-.4301, -.2676, -.0399, -.1003, .2239, .0588, .5554)',$$
$$\text{Prin2} = (.0757, -.0514, .0335, -.0447, .0387, -.0719, .0201)'.$$

Using $\bar{\mathbf{x}}$, $\boldsymbol{\ell}_1$ and $\boldsymbol{\ell}_2$ and using the elements of Prin1 and Prin2 as the multiplying coefficients for $\boldsymbol{\ell}_1$ and $\boldsymbol{\ell}_2$, respectively, we obtain approximate values of $\mathbf{x}_1, \ldots, \mathbf{x}_7$. For example, $\hat{\mathbf{x}}_1 = \bar{\mathbf{x}} + (-0.4301)\boldsymbol{\ell}_1 + (0.0757)\boldsymbol{\ell}_2$.

The approximate values are summarized below:

$$\hat{\mathbf{x}}_1 = (.12319, .16139, .35877, .67928, 1.06035)',$$
$$\hat{\mathbf{x}}_2 = (.10753, .12154, .29086, .73780, 1.24116)',$$
$$\hat{\mathbf{x}}_3 = (.13859, .17645, .40274, .86890, 1.40048)',$$
$$\hat{\mathbf{x}}_4 = (.11914, .13847, .32929, .82374, 1.37811)',$$
$$\hat{\mathbf{x}}_5 = (.15580, .20086, .45907, 1.00341, 1.61837)',$$
$$\hat{\mathbf{x}}_6 = (.12339, .14038, .33930, .89917, 1.52038)',$$
$$\hat{\mathbf{x}}_7 = (.17236, .22090, .50998, 1.16769, 1.90114)'.$$

An examination of the above values indicates that the typical points determine the original observations very closely.

2.6 Principal Component Analysis of Compositional Data

In many natural science research experiments it is common that a sample of chemical compound or geological specimen is analyzed, and percentages (or proportions) of various chemicals or ingredients are recorded. Since by definition the percentages of various ingredients for each sample add up to 100 (or in general to a constant), there is a built-in constraint among the measurements on the variables. Hence, we must take due care in analyzing these data, usually referred to as compositional, mixture, or proportional data.

Suppose $x_1, \ldots, x_p$ are the measurements on (percent amounts of) the p variables, with $\sum_{i=1}^{p} x_i = 100$ (or $= 1$). Then exactly one of the eigenvalues of the covariance matrix of

$\mathbf{x} = (x_1, \ldots, x_p)'$ will be zero due to the constraint. Thus, it appears that there is no theoretical problem in performing the PCA based on the covariance or the correlation matrix of $\mathbf{x}$. However, because of the constraint present in this problem, the usual interpretation of the variances and the covariances is lost. Hence Aitchison (1983, p. 109) suggested that the principal component analysis based on the sample variance-covariance matrix of the p log-contrasts of the original variables, that is,

$$v_j = \log x_j - \frac{1}{p}\sum_{i=1}^{p} \log x_i, \; j = 1, \ldots, p$$

be performed instead of that based on the sample variance-covariance matrix of the original percentages.

EXAMPLE 7 *Geology Data* Many times in geological experiments the interest lies in comparing the proportions of elements in a specimen found in two or more sources. Since only a few principal components can often explain most of the variability, it may be simple to use one or two principal components from each source and use the standard statistical techniques to compare the specimens from different sources.

Fifteen samples of a rhyolite-basalt complex from the Gardiner River, Yellowstone National Park, (Miesch, 1976) were collected and percentages of S_iO_2, Al_2O_3, F_eO, M_gO, C_aO, Na_2O, K_2O, and H_2O were recorded. We are interested in observing any special features that this data set may contain, after the suitable dimensionality reduction through principal component analysis.

The data and the code for performing a PCA on these data are provided in Program 2.8.

```
/* Program 2.8 */

options ls=64 ps=45 nodate nonumber;
title1 "Output 2.8";

data percent;
input x1-x8;
array x{8} x1-x8;
array u{8} u1-u8;
array v{8} v1-v8;
do i=1 to 8;
u{i}=log(x{i});
end;
ub=mean(of u1-u8);
do i=1 to 8;
v{i}=u{i}-ub;
end;
datalines;
51.64 16.25 10.41 7.44 10.53 2.77 .52 .44
54.33 16.06 9.49 6.70 8.98 2.87 1.04 .53
54.49 15.74 9.49 6.75 9.30 2.76 .98 .49
55.07 15.72 9.40 6.27 9.25 2.77 1.13 .40
55.33 15.74 9.40 6.34 8.94 2.61 1.13 .52
58.66 15.31 7.96 5.35 7.28 3.13 1.58 .72
59.81 14.97 7.76 5.09 7.02 2.94 1.97 .45
62.24 14.82 6.79 4.27 6.09 3.27 2.02 .51
64.94 14.11 5.78 3.45 5.15 3.36 2.66 .56
65.92 14.00 5.38 3.19 4.78 3.13 2.98 .61
67.30 13.94 4.99 2.55 4.22 3.22 3.26 .53
68.06 14.20 4.30 1.95 4.16 3.58 3.22 .53
72.23 13.13 3.26 1.02 2.22 3.37 4.16 .61
```

```
75.48 12.71 1.85 .37 1.10 3.58 4.59 .31
75.75 12.70 1.72 .40 .83 3.44 4.80 .37
;
/* Source: Miesch (1976).  Courtesy: U.S. Department
of Interior. */

proc princomp data=percent n=4;
var v1-v8;
title2 'PCA on the Transformed Variables';
run;
```

Output 2.8

```
                    Output 2.8
        PCA on the Transformed Variables

             The PRINCOMP Procedure

             Observations          15
             Variables              8

        Eigenvalues of the Correlation Matrix

     Eigenvalue    Difference    Proportion    Cumulative

1    6.95526905    6.18789694        0.8694        0.8694
2    0.76737211    0.52731532        0.0959        0.9653
3    0.24005679    0.21735750        0.0300        0.9953
4    0.02269929                      0.0028        0.9982


                    Eigenvectors

          Prin1          Prin2          Prin3          Prin4

v1     0.374348       0.174662       -.006147       -.079825
v2     0.305425       0.610383       0.512994       -.261481
v3     -.371041       0.163706       0.279571       -.241410
v4     -.377963       -.030443       0.066662       -.103165
v5     -.376886       0.042543       0.078874       0.478984
v6     0.371392       0.186946       0.028815       0.731665
v7     0.362851       -.172393       -.494657       -.302612
v8     0.272795       -.709028       0.634396       -.000844
```

Using the ARRAY statement and the MEAN function, we first transform the data to v_j and then we use the PRINCOMP procedure to determine the principal components. Results are shown in Output 2.8.

The first two principal components account for more than 96% of the total variation in these data and so must suffice for all practical purposes. The first principal component that explains about 87% of the variability seems to represent a contrast between the compounds S_iO_2, Al_2O_3, Na_2O, K_2O, H_2O and F_eO, M_gO, C_aO. The second principal component that explains about 10% of the variability seems to measure the difference between H_2O and Al_2O_3 in the rhyolite-basalt complex.

2.7 Principal Component Regression

In a typical linear regression problem, a set of n measurements on a dependent (or response) variable y and on p independent (or predictor) variables $x_1, \ldots, x_p$ is given. The problem is to determine the prediction equation by estimating the parameters $\beta_0, \beta_1, \ldots, \beta_p$ in the linear model

$$y = \beta_0 + \beta_1 x_1 + \beta_2 x_2 + \cdots + \beta_p x_p + \epsilon.$$

The random error ϵ for various observations is assumed to be uncorrelated with an expected value of zero and variance σ^2. The estimated regression function, namely $\hat{\beta}_0 + \hat{\beta}_1 x_1 + \hat{\beta}_2 x_2 + \cdots + \hat{\beta}_p x_p$, is used for prediction of a response variable for a given set of values for predictor variables $x_1, \ldots, x_p$.

Given the data $(y_i, x_{1i}, x_{2i}, \ldots, x_{pi}), i = 1, \ldots, n$, define

$$\mathbf{Y} = \begin{bmatrix} y_1 \\ \vdots \\ y_n \end{bmatrix}, \quad \mathbf{X} = \begin{bmatrix} 1 & x_{11} & \cdots & x_{p1} \\ \vdots & \vdots & \cdots & \vdots \\ 1 & x_{1n} & \cdots & x_{pn} \end{bmatrix},$$

and let

$$\boldsymbol{\beta} = \begin{bmatrix} \beta_0 \\ \beta_1 \\ \vdots \\ \beta_p \end{bmatrix}, \quad \boldsymbol{\epsilon} = \begin{bmatrix} \epsilon_1 \\ \vdots \\ \epsilon_n \end{bmatrix}.$$

Then the n equations corresponding to n observations can be written in the matrix form as

$$\mathbf{Y} = \mathbf{X}\boldsymbol{\beta} + \boldsymbol{\epsilon}.$$

The least-squares procedure, which minimizes $\boldsymbol{\epsilon}'\boldsymbol{\epsilon}$ with respect to $\boldsymbol{\beta}$, applied to the above model yields the normal equation

$$\mathbf{X}'\mathbf{X}\boldsymbol{\beta} = \mathbf{X}'\mathbf{Y}.$$

If the rank of $\mathbf{X}$ is $p + 1$ $(< n)$, then the unique solution $\hat{\boldsymbol{\beta}} = (\mathbf{X}'\mathbf{X})^{-1}\mathbf{X}'\mathbf{Y}$ to the above normal equations serves as the best linear unbiased estimator (BLUE) of $\boldsymbol{\beta}$. If the rank of $\mathbf{X}$ is $r < (p+1)$ then the normal equations do not have a unique solution. However, the above system of equations can be solved as $\hat{\boldsymbol{\beta}}^- = (\mathbf{X}'\mathbf{X})^-\mathbf{X}'\mathbf{Y}$, where $(\mathbf{X}'\mathbf{X})^-$ is a generalized inverse of $\mathbf{X}'\mathbf{X}$. Since the solution $\hat{\boldsymbol{\beta}}^-$ is nonunique, it is *not* really an estimator of $\boldsymbol{\beta}$, but it does serve all the purposes of $\hat{\boldsymbol{\beta}}$.

In many practical problems, we may come across a situation in which the rank of $\mathbf{X}$ is theoretically $p + 1$ but one or more eigenvalues of $\mathbf{X}'\mathbf{X}$ are close to zero. This situation creates a problem in the computation of $\hat{\boldsymbol{\beta}}$ since $(\mathbf{X}'\mathbf{X})^{-1}$ then becomes numerically unstable. A full rank matrix with some of its eigenvalues close to zero is termed an ill-conditioned matrix. If $\mathbf{X}'\mathbf{X}$ is ill-conditioned, then it creates a problem of multicollinearity in regression analysis. Under multicollinearity, the estimates $\hat{\beta}_i$ of $\beta_i, i = 0, 1, \ldots, p$, are unstable and have very large standard errors. There are methods, such as ridge-regression and James-Stein estimation, to handle the situation of multicollinearity in regression analysis. Principal component regression is yet another alternative method to approach this problem. All these methods, including principal component regression, yield the biased estimates of $\beta_0, \beta_1, \ldots, \beta_p$. The idea behind proposing these methods is that, by allowing a small bias in the estimation it is possible to find estimators with a smaller mean square error than the usual estimators.

Principal component analysis seems to be an ideal method to use in the case of regression analysis with a multicollinearity problem. The core idea behind principal component analysis is to forego the last few principal components, which explain only a small percentage of the total variability. Hence, in principal component regression, all the principal components of the independent variables $x_1, \ldots, x_p$ are computed and the first few, say r ($< p$), are used as the transformed new independent variables in the model, provided that the percentage of variability explained by the remaining $(p - r)$ principal components is negligible. Thus the corresponding linear regression model will be

$$y_i = \beta_0^* + \beta_1^* PC_{1i} + \cdots + \beta_r^* PC_{ri} + \epsilon_i, i = 1, \ldots, n,$$

where $PC_{1i}, \ldots, PC_{ri}$, $i = 1, \ldots, n$ are, respectively,the principal component scores of the $1^{st}, \ldots, r^{th}$ principal components. Although we are using only the r principal components, all nine original predictor variables are still represented in the model through them. Usually, the predictor variables are standardized for both location and scale (to have zero mean and standard deviation 1 for all independent variables) before the PCA is applied and hence the sample correlation matrix of the independent variables $x_1, \ldots, x_p$ is used to find the eigenvalues and eigenvectors.

Besides the obvious fact that the last few principal components are dropped from the regression equation to handle the multicollinearity problem, it is not entirely clear, unlike the case of the usual PCA, how to select the principal component linear regression model. It has been found that the natural ordering in which the principal components are included in the analysis may not be the best way to select the predictors from the list of p principal components. For example, a model with second and third principal components as the predictor variables may actually have a smaller mean square error of prediction than that with first and second principal components. This is considered to be a serious drawback of principal component regression. However, it is not surprising since in the computation of the principal components, the dependent variable has not played any role. The dependent variable and its observed values nonetheless play a crucial role in selecting an appropriate model. See Jolliffe (1982), Jolliffe (1986), and Hadi and Ling (1998) for a detailed discussion of this and other drawbacks in using principal components for regression. This drawback of principal component regression may, however, be correctable.

Having dropped the last few principal components corresponding to the smallest eigenvalues to overcome the problem of multicollinearity, we may apply a selection of variables' algorithms on the remaining principal components. The set of components giving maximum predictive ability can then be selected as the desired set of variables to be included in the model. Any appropriate selection of variables method may be used for this purpose. Using the SELECTION= option in the REG procedure, this can be easily achieved. This will be illustrated in Example 8. Many times regression equations are developed for the prediction of future observations. This especially is so in the case of growth curves, time series, and other types of forecasting models. A set of principal components of the predictor variables may yield better (in some sense) predictors. A common measure of predictive ability of a prediction formula is the Predicted Residual Error Sum of Squares, or what is popularly known as PRESS. This is the sum of squares of the predicted residuals for the observations. The predicted residual for an observation is defined as the residual for that observation resulting from a linear model constructed from the remaining observations. A smaller value of PRESS corresponds to a better predictive model. The value of PRESS is printed in the output when predicted values are requested in the REG procedure.

EXAMPLE 8 ***A Model for Prediction, Skull Data*** Rao and Shaw (1948) considered the problem of predicting cranial capacity (C) of skulls, which may be broken or damaged (so that the cranial capacity cannot be directly measured). Using the measurements on the capacity and some external variables on well-preserved skulls, they developed an estimated regression function that can be used to predict the capacity of a skull on which the measurements on the same set of external variables are available. The three external variables selected to build the regression function were glabellar-occipital length (L), the maximum parietal breadth (B), and the basio-bregmatic height (HP). Since cranial capacity represents the volume, the regression formula of the form

$$C = \gamma L^{\beta_1} B^{\beta_2} (HP)^{\beta_3}$$

was used. Using the log transformation it can equivalently be written as

$$\log(C) = \beta_0 + \beta_1 \log(L) + \beta_2 \log(B) + \beta_3 \log(HP)$$

and hence we can consider the regression model

$$Y = \beta_0 + \beta_1 X_1 + \beta_2 X_2 + \beta_3 X_3 + \epsilon,$$

for prediction purposes with $X_1 = \log(L)$, $X_2 = \log(B)$, $X_3 = \log(HP)$ and $Y = \log(C)$.

However, if the measurements on several more external variables are available, one may be interested in exploring whether a few linear combinations of these variables (perhaps a few principal components) will serve as better predictor variables. Hooke (1926) has given measurements on $n = 63$ male skulls (from the Farringdon Street series) on the variables C, L, B, HP, and on the six other variables: Flower's ophryo-occipital length (F), glabellar projective length (LP), least forehead breadth (BP), basion to point vertically above it with skull adjusted on craniophor to Frankfurt horizontal (H), craniophor auricular height (OH), and basion to nasion length (LB). These data are provided in the Appendix under the name, SKULL.DAT and are read into Program 2.9 by using the INFILE statement. Note that some of the corresponding output for this program has been suppressed.

```
/* Program 2.9 */

options ls=64 ps=45 nodate nonumber;
title1 "Output 2.9";
title2 'Selection of Principal Components for Prediction';
data skull;
infile 'skull.dat';
input id c f lp l b bp h hp oh lb;
y=log(c);
x1=log(f); x2=log(lp); x3=log(l);
x4=log(b); x5=log(bp); x6=log(h);
x7=log(hp); x8=log(oh); x9=log(lb);

proc reg data=skull;
model y=x3 x4 x7/p;
title2 'Prediction Equation Obtained by Rao and Shaw (1948)';
run;

proc princomp data=skull prefix=pcx out=pc1x;
var x1-x9;
run;

data pc1;
set pc1x;
keep y pcx1-pcx9;
run;

data pc1;
set pc1;
proc reg;
model y=pcx1-pcx9/selection= rsquare cp;
title2 'Selection of Principal Components for Prediction';
run;
```

```
proc reg data=pc1;
title2 'Principal component regression to get PRESS';
model y=pcx1 pcx2 pcx3/p;
model y=pcx1 pcx3 pcx4/p;
model y=pcx1-pcx4/p;
model y=pcx1-pcx5/p;
model y=pcx1-pcx4 pcx8/p;
run;
```

Output 2.9

Output 2.9
Selection of Principal Components for Prediction

The PRINCOMP Procedure

Observations	63
Variables	9

Eigenvalues of the Correlation Matrix

	Eigenvalue	Difference	Proportion	Cumulative
1	4.31909442	2.33907857	0.4799	0.4799
2	1.98001585	0.82848916	0.2200	0.6999
3	1.15152669	0.50913547	0.1279	0.8278
4	0.64239122	0.15834890	0.0714	0.8992
5	0.48404232	0.17053462	0.0538	0.9530
6	0.31350770	0.23004600	0.0348	0.9878
7	0.08346169	0.06885968	0.0093	0.9971
8	0.01460201	0.00324391	0.0016	0.9987
9	0.01135810		0.0013	1.0000

Eigenvectors

	pcx1	pcx2	pcx3	pcx4	pcx5
x1	0.400394	-.320888	0.060056	-.223169	-.181220
x2	0.375088	-.404299	-.152178	-.141631	0.001584
x3	0.370570	-.431206	-.102664	-.159028	-.007284
x4	0.265750	0.164107	0.543521	-.247084	0.731448
x5	0.229682	-.012828	0.673485	0.431966	-.472044
x6	0.339532	0.442497	-.228329	-.108068	-.117489
x7	0.334120	0.440213	-.257998	-.119271	-.130206
x8	0.342832	0.355883	0.078667	0.002933	-.144466
x9	0.306337	-.065543	-.298012	0.794537	0.397032

Using PROC PRINCOMP (see Program 2.9), the eigenvalues and eigenvectors (first five of them) of the correlation matrix of the log transformation of the nine predictor variables F, LP, L, B, BP, H, HP, OH, and LB are obtained in Output 2.9. The first principal component, which is a measure of the size factor (this will be clear by examining the entries in column pcx1), explains about 48% of the total variability. The second principal component, which explains about 22% of the remaining variability, represents a shape factor and seems to measure a difference between the variables corresponding to height and length. The third component, representing a shape factor, seems to measure a difference between the height and breadth variables (column pcx3). The first three principal components explain about 83% of the variability, whereas the first four explain about 90%.

As is discussed earlier, just because the first four principal components explain 90% of the variation does not mean that they form the best subset of predictors of cranial capacity. First we note that the last (smallest) eigenvalue of the standardized (in correlation form) matrix $\mathbf{X}'\mathbf{X}$ is 0.01136. Since this is not close enough to zero, multicollinearity is not a problem in this example. In the following we will attempt to select an appropriate subset of nine principal components that will give a best prediction equation (in terms of a minimum value of the PRESS statistic).

The log transformed variables are called Y and $X_1, \ldots, X_9$ in Program 2.9. Since we used the option PREFIX = PCX in the PROC PRINCOMP statement, the principal components (as well as the scores) are named PCX1, ..., PCX9. The SAS data set PC1X that was created using OUT=PC1X in PROC PRINCOMP contains the principal component scores and the data on the original variables. Using the KEEP statement we have retained only Y and PCX1-PCX9 in the new data set PC1. Then the following SAS statments are used to select the variables using RSQUARE and CP criteria.

```
proc reg data = pc1;
model y=pcx1-pcx9/selection = rsquare cp;
run;
```

Details about the SELECTION option in PROC REG can be found in the *SAS/STAT User's Guide*.

Using the maximum R^2 procedure of selecting variables in regression, we select as the best two variables PCX1 and PCX4. The best subset of size three is PCX1, PCX3, and PCX4. The best four variables model corresponds to PCX1 to PCX4. Since these four variables yield an R^2 value of 0.7656 and since, by adding one more variable, R^2 is increased at most by 1.17%, we may suggest a model with the first four principal components as the independent variables. On the other hand, if C_p criterion is used, the variables PCX1 to PCX4 and PCX8 are selected. The output of the selection of variables procedure (in Program 2.9) is suppressed in the interest of space.

The predictive ability of these selected principal components is summarized below along with the Rao and Shaw (1948) equation using the variables X_3, X_4, and X_7.

TABLE 2.1 Predictive Ability of Selected Principal Components

Procedure	Variables	PRESS
Rao & Shaw	X_3, X_4, X_7	.1381
First 3 *PC*	$PCX1, PCX2, PCX3$	.1784
Best 3 (R^2)	$PCX1, PCX3, PCX4$	.1358
First & Best 4	$PCX1 - PCX4$	.1265
First & Best 5	$PCX1 - PCX5$	.1242
Best C_p	$PCX1 - PCX4, PCX8$	.1311

Examining the PRESS values in the table, we see that the model that has the five (PCX1 through PCX5) principal components is the best. However, as we have argued earlier, the increase in R^2 from the model with four principal components to that with five components is minimal. In any case it is clear that the use of a few selected principal components for prediction of response variable may be beneficial.

Before we end this section, we should mention that using the option PCOMIT=m in conjunction with OUTEST=*datasetname* of the REG procedure can enable us to perform the principal component regression analysis based on *first* $p - m$ principal components, where m is any number less than p. The final model in the output is described in terms of the original variables. However, this option could not be used for performing the analysis similar to above.

2.8 Principal Component Residuals and Detection of Outliers

Pearson (1901) considered the problem of fitting a subspace to the points $\mathbf{x}_1, \ldots, \mathbf{x}_n$ in a p-dimensional space. He defined a best fitting subspace as that for which the sum of squares of the perpendicular distance from the points to the subspace is a minimum. If $\boldsymbol{\ell}_1, \ldots, \boldsymbol{\ell}_q$ are the eigenvectors of the variance-covariance matrix (or correlation matrix) of $\mathbf{x}_1, \ldots, \mathbf{x}_n$ corresponding to the first q eigenvalues whose cumulative variance accounts for most of the variation in the given data, then the best q-dimensional representation of the points $\mathbf{x}_1, \ldots, \mathbf{x}_n$ is given by $\mathbf{Z}_i = \mathbf{L}'\mathbf{x}_i, i = 1, \ldots, n$, where $\mathbf{L} = [\boldsymbol{\ell}_1 : \cdots : \boldsymbol{\ell}_q]$ is a p by q matrix of the eigenvectors, and $\mathbf{x}_i$ is the ith standardized observational vector. That is, $\mathbf{X}_i = \mathbf{x}_i - \bar{\mathbf{x}}$ if a covariance matrix is used and $\mathbf{X}_i = \mathbf{D}^{-1/2}(\mathbf{x}_i - \bar{\mathbf{x}})$, where $\mathbf{D}^{-1/2} = diag(1/\sqrt{s_{11}}, \ldots, 1/\sqrt{s_{qq}})$, if the correlation matrix is used for determining the eigenvalues and eigenvectors. The sum of squares of the perpendicular distance is $tr\mathbf{S} - tr\mathbf{L}'\mathbf{SL}$, where $\mathbf{S}$ is the covariance matrix (or correlation matrix) of $\mathbf{x}_1, \ldots, \mathbf{x}_n$. If $q = p$, that is, all the eigenvectors are used, then the above sum of squares of perpendicular distance is zero and hence the fit is exact. In that case, $\mathbf{Z}_i = \mathbf{L}'\mathbf{X}_i$ is a p by 1 vector and furthermore, $\mathbf{LZ}_i = \mathbf{X}_i$, since $\mathbf{LL}' = \mathbf{L}'\mathbf{L} = \mathbf{I}_p$. When $q < p$, $\mathbf{L}$ is only a suborthogonal matrix, hence $\mathbf{X}_i$ is not exactly but only approximately equal to $\mathbf{LZ}_i$. Then the p by 1 vector $\mathbf{R}_i$ of the difference, $\mathbf{X}_i - \mathbf{LZ}_i$ is the ith *principal component residual.* The best subspace of dimension q minimizing the sum of squares of these residuals:

$$\begin{aligned}\sum_{i=1}^{n} \mathbf{R}_i'\mathbf{R}_i &= \sum_{i=1}^{n} \mathbf{X}_i'\mathbf{X}_i - \sum_{i=1}^{n} \mathbf{Z}_i'\mathbf{L}'\mathbf{LZ}_i \\ &= \sum_{i=1}^{n} \mathbf{X}_i'\mathbf{X}_i - \sum_{i=1}^{n} \mathbf{Z}_i'\mathbf{Z}_i \\ &= \sum_{i=1}^{n} \mathbf{X}_i'\mathbf{X}_i - \sum_{i=1}^{n} \mathbf{X}_i'\mathbf{LL}'\mathbf{X}_i \\ &= \sum_{i=1}^{n} \mathbf{X}_i'\mathbf{MM}'\mathbf{X}_i,\end{aligned}$$

where $\mathbf{M} = [\boldsymbol{\ell}_{q+1} : \cdots : \boldsymbol{\ell}_p]$, is given by the one that is spanned by $\boldsymbol{\ell}_1, \ldots, \boldsymbol{\ell}_q$. Another form of the sum of squares of the residuals given above is

$$\begin{aligned}tr\ \mathbf{S} - \sum_{i=1}^{n} [\boldsymbol{\ell}_1'\mathbf{X}_i + \cdots + \boldsymbol{\ell}_q'\mathbf{X}_i]^2 \\ = \sum_{i=1}^{n} [\boldsymbol{\ell}_{q+1}'\mathbf{X}_i + \cdots + \boldsymbol{\ell}_p'\mathbf{X}_i]^2.\end{aligned}$$

Rao (1964) suggested that the squares of the perpendicular distance

$$d_i^2 = [\boldsymbol{\ell}_{q+1}'\mathbf{X}_i + \cdots + \boldsymbol{\ell}_p'\mathbf{X}_i]^2, \ i = 1, \ldots, n \tag{2.2}$$

from the point $\mathbf{X}_i$ to the best subspace be used for the detection of outliers. An unusually large square distance d_i^2 would indicate that the ith observation vector may possibly be an outlier. Hawkins (1974) suggested that a weighted form

$$h_i^2 = \sum_{j=q+1}^{p} [(\boldsymbol{\ell}_j'\mathbf{X}_i)^2/\lambda_j] \tag{2.3}$$

instead of d_i^2 be used. When the observations $\mathbf{x}_1, \ldots, \mathbf{x}_n$ are from a multivariate normal distribution, the squared distance h_i^2 has a chi-square distribution with $p - q$ degrees of freedom when n is large. Hence a chi-square cutoff point may serve as a yardstick for determining outliers. Gnanadesikan (1997) presents more discussion of other types of residuals and outlier detection methods.

EXAMPLE 9 ***Detection of Outliers, Hematology Data*** These data, obtained from Royston (1983) are a part of data set obtained from a health survey of paint sprayers in a car assembly plant. Six variables—hemoglobin concentration (HAEMO), packed cell volume (PCV), white blood cell count (WBC), lymphocyte count (LYMPHO), neutrophil count (NEUTRO), and serum lead concentration (LEAD)—were measured on 103 workers assigned to paint-spraying jobs. As it is common in these types of data, the three count variables, WBC, LYMPHO, and NEUTRO were transformed using the log transformation. Royston (1983) analyzed these data to determine whether these data possibly came from a multivariate normal distribution. The data, after deleting the three observations (numbered 21, 47 and 52) suspected to be outliers, seemed to have come from a multivariate normal distribution.

Without assuming any known distributional form, we may be able to detect outliers from a set of multivariate data using the distances d_i^2 and h_i^2. The data are provided in the Appendix with the name HAEMATO.DAT. The SAS code for computing d_i^2 and h_i^2 is given in Program 2.10. We have suppressed the outputs corresponding to this code. However, these distances are collected in a SAS data set and are produced in Output 2.10 along with the original observations.

```
/* Program 2.10 */

options ls=64 ps=45 nodate nonumber;
title1 "Output 2.10";

data hemato;
infile 'haemato.dat';
input haemo pcv wbc lympho neutro lead@@;
lgwbc=log(wbc);
lglympho=log(lympho);
lgneutro=log(neutro);

proc princomp data=hemato prefix=pc out=b1;
var haemo pcv lgwbc lglympho lgneutro lead;
title2 "PCA for Detection of Outliers Using Rao's Method";
run;

proc princomp data=hemato std prefix=spc out=b2 noprint;
var haemo pcv lgwbc lglympho lgneutro lead;
title2 "PCA for Detection of Outliers Using Hawkins' Method";
run;

data c;
set b1;
set b2;
disq=uss(of pc5-pc6);
hisq=uss(of spc5-spc6);
title2 'Computation of Outlier Detection Statistics';
run;

data outlier;
set c;
set hemato;
proc print;
var haemo pcv wbc lympho neutro lead disq hisq;
title2 ' Detection of Outliers Using the Distances';
run;
```

Output 2.10

Output 2.10
PCA for Detection of Outliers Using Rao's Method

The PRINCOMP Procedure

Observations	103
Variables	6

Eigenvalues of the Correlation Matrix

	Eigenvalue	Difference	Proportion	Cumulative
1	2.07100190	0.33178433	0.3452	0.3452
2	1.73921757	0.71888140	0.2899	0.6350
3	1.02033617	0.10790294	0.1701	0.8051
4	0.91243323	0.70000538	0.1521	0.9572
5	0.21242785	0.16784457	0.0354	0.9926
6	0.04458328		0.0074	1.0000

Detection of Outliers Using the Distances

Obs	haemo	pcv	wbc	lympho	neutro	lead	disq	hisq
1	13.4	39	4100	14	25	17	0.04252	0.3323
2	14.6	46	5000	15	30	20	0.53470	2.6490
3	13.5	42	4500	19	21	18	0.19436	0.9167
4	15.0	46	4600	23	16	18	0.11755	1.1223
5	14.6	44	5100	17	31	19	0.04437	0.3227
6	14.0	44	4900	20	24	19	0.34512	1.6390
7	16.4	49	4300	21	17	18	0.00358	0.0258
8	14.8	44	4400	16	26	29	0.02887	0.2525
9	15.2	46	4100	27	13	27	0.00550	0.0259
10	15.5	48	8400	34	42	36	0.03916	0.2150
11	15.2	47	5600	26	27	22	0.19539	1.6306
12	16.9	50	5100	28	17	23	0.05037	0.4199
13	14.8	44	4700	24	20	23	0.03426	0.3690
14	16.2	45	5600	26	25	19	0.75295	3.7313
15	14.7	43	4000	23	13	17	0.09459	0.6244
16	14.7	42	3400	9	22	13	0.07632	0.6186
17	16.5	45	5400	18	32	17	0.92077	4.4654
18	15.4	45	6900	28	36	24	0.10260	0.6710
19	15.1	45	4600	17	29	17	0.11081	2.2649
20	14.2	46	4200	14	25	28	0.76878	3.6248
21	15.9	46	5200	8	34	16	0.75625	16.8520
22	16.0	47	4700	25	14	18	0.15459	2.4628
23	17.4	50	8600	37	39	17	0.17096	0.9326
24	14.3	43	5500	20	31	19	0.02233	0.2132
25	14.8	44	4200	15	24	19	0.00793	0.1630
26	14.9	43	4300	9	32	17	0.09328	1.6661
27	15.5	45	5200	16	30	20	0.06848	0.5259
28	14.5	43	3900	18	18	25	0.03952	0.2524
29	14.4	45	6000	17	37	23	0.34276	2.0574
30	14.6	44	4700	23	21	27	0.01077	0.1880
31	15.3	45	7900	43	23	23	0.26671	3.7369
32	14.9	45	3400	17	15	24	0.01932	0.3984
33	15.8	47	6000	23	32	21	0.00942	0.2112
34	14.4	44	7700	31	39	23	0.04177	0.2197
35	14.7	46	3700	11	23	23	0.38258	1.8870

Output 2.10
(continued)

Obs	haemo	pcv	wbc	lympho	neutro	lead	disq	hisq
36	14.8	43	5200	25	19	22	0.22184	1.8533
37	15.4	45	6000	30	25	18	0.07763	0.5479
38	16.2	50	8100	32	38	18	0.26251	1.2367
39	15.0	45	4900	17	26	24	0.01232	0.2473
40	15.1	47	6000	22	33	16	0.41344	2.1753

Obs	haemo	pcv	wbc	lympho	neutro	lead	disq	hisq
41	16.0	46	4600	20	22	22	0.22691	1.2250
42	15.3	48	5500	20	23	23	0.47687	4.7132
43	14.5	41	6200	20	36	21	0.33870	1.6941
44	14.2	41	4900	26	20	20	0.21692	1.1841
45	15.0	45	7200	40	25	25	0.02609	0.2696
46	14.2	46	5800	22	31	22	0.87653	4.1734
47	14.9	45	8400	61	17	17	0.23323	5.1868
48	16.2	48	3100	12	15	18	0.00054	0.0040
49	14.5	45	4000	20	18	20	0.21162	1.6476
50	16.4	49	6900	35	22	24	0.10435	2.1209
51	14.7	44	7800	38	34	16	0.02085	0.4334
52	17.0	52	6300	19	21	16	1.01034	18.8063
53	15.4	47	3400	12	19	18	0.16280	0.8450
54	13.8	40	4500	19	23	21	0.13986	0.8244
55	16.1	47	4600	17	28	20	0.09813	1.6578
56	14.6	45	4700	23	22	27	0.07650	0.8561
57	15.0	44	5800	14	39	21	0.06397	1.3065
58	16.2	47	4100	16	24	18	0.13740	2.1782
59	17.0	51	5700	26	29	20	0.12376	2.4317
60	14.0	44	4100	16	24	18	0.48133	3.3569
61	15.4	46	6200	32	25	16	0.01533	0.3254
62	15.6	46	4700	28	16	16	0.02945	0.2087
63	15.8	48	4500	24	20	23	0.10246	1.8439
64	13.2	38	5300	16	26	20	0.35982	5.4645
65	14.9	47	5000	22	25	15	0.64160	3.9771
66	14.9	47	3900	15	19	16	0.59883	2.8214
67	14.0	45	5200	23	25	17	0.78079	4.0168
68	16.1	47	4300	19	22	22	0.12271	1.5137
69	14.7	46	6800	35	25	18	0.26406	1.2907
70	14.8	45	8900	47	36	17	0.05429	0.5934
71	17.0	51	6300	42	19	15	0.01592	0.2415
72	15.2	45	4600	21	22	18	0.03638	0.7511
73	15.2	43	5600	25	28	17	0.37506	2.5747
74	13.8	41	6300	25	27	15	0.04379	0.9759
75	14.8	43	6400	36	24	18	0.14817	0.8312
76	16.1	47	5200	18	28	25	0.08261	0.4641
77	15.0	43	6300	22	34	17	0.14203	0.6776
78	16.2	46	6000	25	25	24	0.50232	3.0307
79	14.8	44	3900	9	25	14	0.14840	2.5741
80	17.2	44	4100	12	27	18	2.98504	14.1037

Obs	haemo	pcv	wbc	lympho	neutro	lead	disq	hisq
81	17.2	48	5000	25	19	25	0.93027	4.44096
82	14.6	43	5500	22	31	19	0.06118	1.13585
83	14.4	44	4300	20	20	15	0.13881	1.17954
84	15.4	48	5700	29	26	24	0.25906	2.22912
85	16.0	52	4100	21	15	22	1.15214	5.46210
86	15.0	45	5000	27	18	20	0.00180	0.03332
87	14.8	44	5700	29	23	23	0.02790	0.13964

Output 2.10 *(continued)*

88	15.4	43	3300	10	20	19	0.42906	2.08885
89	16.0	47	6100	32	23	26	0.12323	0.59543
90	14.8	43	5100	18	31	19	0.07947	0.81555
91	13.8	41	8100	52	24	17	0.05026	0.43155
92	14.7	43	5200	24	24	17	0.06064	0.58852
93	14.6	44	9899	69	28	18	0.00385	0.04061
94	13.6	42	6100	24	30	15	0.18190	0.85639
95	14.5	44	4800	14	29	15	0.14074	0.80761
96	14.3	39	5000	25	20	19	1.09064	5.14561
97	15.3	45	4000	19	19	16	0.07000	1.38294
98	16.4	49	6000	34	22	17	0.01753	0.38105
99	14.8	44	4500	22	18	25	0.03994	0.26826
100	16.6	48	4700	17	27	20	0.11044	0.92671
101	16.0	49	7000	36	28	18	0.12608	0.82361
102	15.5	46	6600	30	33	13	0.09241	2.00304
103	14.3	46	5700	26	20	21	0.71501	5.28060

The first three principal components account for about 80% of the variability whereas the first four components account for about 96% of the total variability. Then the subspace determined by these four principal components can be taken as the best subspace of dimension 4. Therefore, d_i^2 values given by Equation 2.2 are calculated using the last $p-q=6-4=2$ principal components. In SAS we have used the USS function to calculate d_i^2. Using the STD option of PROC PRINCOMP, we obtain the standardized (scaled by the square root of the corresponding eigenvalues) principal component scores, and then we calculate h_i^2 using Equation 2.3. The criterion based on d_i^2 seems to indicate that observation 80 is the only outlier. For this observation d_i^2, the value is $d_{80}^2 = 2.9850$ whereas the next largest d_i^2 value is $d_{85}^2 = 1.15$. On the other hand, h_i^2 indicates observation 21 with a $h_{21}^2 = 16.8520$, 52 with $h_{52}^2 = 18.8063$, and observation 80 with $h_{80}^2 = 14.1037$ to be outliers, since all three of these observations have distances that exceed the chi-square upper 5% cutoff value of 5.99 for a chi-square distribution with 2 degrees of freedom. According to the analysis done here, observation 47 does not come out to be an outlier even though Royston classified this observation as an outlier in his analysis.

2.9 Principal Component Biplot

Biplots studied by Gabriel (1971) and discussed extensively by Gower and Hand (1996) are useful tools in the exploratory data analysis of multivariate data. The biplot is a graphical representation of a data matrix on a plane by means of two sets of points representing its rows and columns. Some discussion of biplots and plotting them using the Friendly (1991) BIPLOT macro (in SAS) is provided in Chapter 2 of Khattree and Naik (1999).

Let $\mathbf{X}$ be an n by p matrix of rank r. Then it is well-known that (Rao, 1973, p. 19) $\mathbf{X}$ can be written using the rank factorization as

$$\mathbf{X} = \mathbf{G}\mathbf{H}', \tag{2.4}$$

where the n by r matrix $\mathbf{G}$ and the p by r matrix $\mathbf{H}$ are both of rank r. Each x_{ij}, the $(i, j)^{th}$ element of $\mathbf{X}$, is thus expressed as $x_{ij} = \mathbf{g}_i'\mathbf{h}_j$, where $\mathbf{g}_i'$ is the ith row of $\mathbf{G}$ and $\mathbf{h}_j$ is the jth column of $\mathbf{H}'$. Thus, each element x_{ij} of $\mathbf{X}$ is represented by two r-dimensional vectors, $\mathbf{g}_i'$, corresponding to the ith row, and $\mathbf{h}_j$, corresponding to the jth column of the matrix $\mathbf{X}$.

When the rank of $\mathbf{X}$ is $r=2$, the vectors $\mathbf{g}_i$ and $\mathbf{h}_j$ are all of size 2 by 1. Gabriel (1971) suggested that the $n+p$ points, $\mathbf{g}_1, \ldots, \mathbf{g}_n$ and $\mathbf{h}_1, \ldots, \mathbf{h}_p$, be plotted on the plane, giving a representation of the np elements of **X**. He called this a *biplot* of **X**. For a matrix **X** with rank greater than two, suppose an approximation matrix of rank two can be obtained. The biplot of the approximating matrix can be taken as an approximate biplot of **X** itself (Gabriel, 1971).

To get the representation in Equation 2.4 for **X**, Gabriel (1971) suggested the use of singular value decomposition (SVD) (see Chapter 1, Section 1.5.24 or Rao, 1973, p. 42) of the matrix. That is, write **X** as

$$\mathbf{X} = \mathbf{U}\boldsymbol{\Delta}\mathbf{V}' = \sum_{i=1}^{r} \delta_i \mathbf{u}_i \mathbf{v}_i',$$

where $\boldsymbol{\Delta}$ is an r by r diagonal matrix with positive diagonal elements $\delta_1 \geq \delta_2 \geq \cdots \geq \delta_r > 0$, $\mathbf{U}$ is an n by r matrix with columns $\mathbf{u}_1, \ldots, \mathbf{u}_r$ such that $\mathbf{U}\mathbf{U}' = \mathbf{I}$, and $\mathbf{V}$ is a p by r matrix with columns $\mathbf{v}_1, \ldots, \mathbf{v}_r$, such that $\mathbf{V}'\mathbf{V} = \mathbf{I}$. The values $\delta_1, \ldots, \delta_r$ are called the singular values of **X**. Then, an approximation of dimension two for **X** using the method of least squares is

$$\mathbf{X} \approx \delta_1 \mathbf{u}_1 \mathbf{v}_1' + \delta_2 \mathbf{u}_2 \mathbf{v}_2'. \tag{2.5}$$

From Equation 2.5 we get an approximate biplot for **X**, and the corresponding goodness of fit is measured by

$$\eta = \frac{\delta_1^2 + \delta_2^2}{\sum_{i=1}^{r} \delta_i^2}.$$

If the actual dimension of **X** is 2, then $\eta = 1$. If $r \geq 3$, then $\eta < 1$. Thus, if η is near one, the biplot obtained by using the representation in Equation 2.5 will be a good approximation of the biplot of **X**.

From Equation 2.5, several choices for the coordinates of the biplot are possible. The three popular choices (see Gabriel, 1971) are

a. $\mathbf{g}_i' = (\sqrt{\delta_1} u_{1i}, \sqrt{\delta_2} u_{2i}),\ i = 1, \ldots, n$

$\mathbf{h}_j' = (\sqrt{\delta_1} v_{1j}, \sqrt{\delta_2} v_{2j}),\ j = 1, \ldots, p$

b. $\mathbf{g}_i' = (u_{1i}, u_{2i}),\ i = 1, \ldots, n$

$\mathbf{h}_j' = (\delta_1 v_{1j}, \delta_2 v_{2j}),\ j = 1, \ldots, p$

c. $\mathbf{g}_i' = (\delta_1 u_{1i}, \delta_2 u_{2i}),\ i = 1, \ldots, n$

$\mathbf{h}_j' = (v_{1j}, v_{2j}),\ j = 1, \ldots, p$

Although there are no clear-cut guidelines for selecting one over the other, depending on the practical problem in hand and the interest of the practitioner, one or more of these three choices for plotting the biplots can be made. In choice (a), there is an equal division of weights through δ_1 and δ_2, for representing the observations and variables. Hence, a biplot based on this choice is commonly used in practice. The representation in choice (b) leads to some interesting interpretations. For example, the distance between any two $\mathbf{g}_i$'s, say $\mathbf{g}_i$ and $\mathbf{g}_{i'}$ approximates the Mahalanobis distance between the observation vectors (that is, the ith and the i'^{th} rows of the data matrix **X**). Further, the angle (cosine of the angle) between any two $\mathbf{h}_j$'s, say $\mathbf{h}_j$ and $\mathbf{h}_{j'}$, approximates the angle between the corresponding columns of the data matrix **X** (approximates the correlation between the jth and j'^{th} variables). Finally, under representation (c), the usual Euclidean distance between $\mathbf{g}_i$ and $\mathbf{g}_{i'}$ approximates the Euclidean distance between the ith and the i'^{th} rows of the data matrix **X**.

The three choices (a), (b), and (c), approximate biplots of the biplot of **X**, can be obtained from certain quantities in the present context of principal component analysis. Specifically, suppose **X** is the n by p data matrix in the standardized form. That is, it is standardized to have the mean of each column to be zero or to have mean of each column to be zero and a variance of each column to be equal to one. Let $\mathbf{L} = [\boldsymbol{\ell}_1 : \cdots : \boldsymbol{\ell}_p]$ be the p by p matrix of eigenvectors corresponding to the eigenvalues $\lambda_1 > \lambda_2 > \cdots > \lambda_p > 0$ of the sample variance-covariance (or correlation) matrix of the data. Then the matrix of principal component scores is expressed as

$$\mathbf{Z}_{n\times p} = \mathbf{X}_{n\times p}\mathbf{L}_{p\times p}.$$

Since $\mathbf{LL}' = \mathbf{L}'\mathbf{L} = \mathbf{I}_p$ (being the property of eigenvectors) the data matrix $\mathbf{X}$ has the representation

$$\mathbf{X} = \mathbf{ZL}'. \tag{2.6}$$

In practice, when only the first r ($< p$) principal components are used, the data matrix $\mathbf{X}$ has the following approximation to the above representation

$$\mathbf{X}_{n\times p} \approx \mathbf{G}_{n\times r}\mathbf{H}'_{r\times p},$$

where $\mathbf{G}$ is an n by r matrix of scores on the first r principal components (principal component scores) and $\mathbf{H}'$ is an r by p matrix, r rows of which are the first r eigenvectors (principal component weights) $\boldsymbol{\ell}_1, \ldots, \boldsymbol{\ell}_r$. For $r = 2$, the plot of points $\mathbf{g}_1, \ldots, \mathbf{g}_n$ and $\mathbf{h}_1, \ldots, \mathbf{h}_p$, where

$$\mathbf{G}'_{2\times n} = (\mathbf{g}_{1_{2\times 1}}, \ldots, \mathbf{g}_{n_{2\times 1}}) \text{ and } \mathbf{H}'_{2\times p} = (\mathbf{h}_{1_{2\times 1}}, \ldots, \mathbf{h}_{p_{2\times 1}}),$$

on the same graph is an approximate biplot of the data $\mathbf{X}$ with choice (c). From the representation in Equation 2.6 the other choices, that is, (b) and (a), for biplots can also be obtained. Suppose $\delta_i = \sqrt{\lambda_i}$, $i = 1, \ldots, p$ and $\boldsymbol{\Delta} = diag(\delta_1, \ldots, \delta_p)$. Then Equation 2.6 is equivalent to $\mathbf{X} = \mathbf{Z}\boldsymbol{\Delta}^{-1}\boldsymbol{\Delta}\mathbf{L}'$. By selecting the n by 2 matrix $\mathbf{G}$ to be the first two columns of $\mathbf{Z}\boldsymbol{\Delta}^{-1}$ and the 2 by p matrix $\mathbf{H}'$ to be the first two rows of $\boldsymbol{\Delta}\mathbf{L}'$ we get a biplot with choice (b). The choice (a) can be obtained from Equation 2.6 by considering the equivalent form $\mathbf{X} = \mathbf{Z}\boldsymbol{\Delta}^{-1/2}\boldsymbol{\Delta}^{1/2}\mathbf{L}'$. Now by selecting the n by 2 matrix $\mathbf{G}$ to be first two columns of $\mathbf{Z}\boldsymbol{\Delta}^{-1/2}$ and the 2 by p matrix $\mathbf{H}'$ to be first two rows of $\boldsymbol{\Delta}^{1/2}\mathbf{L}'$, we get a biplot with the choice (a).

EXAMPLE 10 ***Biplot, National Track Record Data (continued)*** A biplot for national track data in terms of speed for women described in Example 1 (and provided in the Appendix as WTRACK.DAT) is given in Output 2.11. This biplot is obtained using a SAS macro given by Friendly (1991). The SAS code to use the BIPLOT macro is given in Program 2.11.

```
/* Program 2.11 */

options ls=64 ps=45 nodate nonumber;
title1 'Output 2.11';
data track;
infile 'wtrack.dat';
input country$ m100 m200 m400 m800 m1500 m3000 marathon;
%include biplot;
%biplot( data = track,  var = m100 m200 m400 m800 m1500
m3000 marathon, id = country, factype=JK, std =MEAN);
filename gsasfile "prog211a.graph";
goptions reset=all gaccess=gsasfile autofeed dev=psl;
goptions horigin=.75in vorigin=2in;
goptions hsize=7in vsize=8in;
proc gplot data=biplot;
plot dim2 * dim1/anno=bianno frame
                 href=0 vref=0 lvref=3 lhref=3
                 vaxis=axis2 haxis=axis1
                 vminor=1 hminor=1;
```

```
axis1 length=6 in order=(-3.5 to 1.5 by .5)
        offset=(2)
label = (h=1.3 'Dimension 1');
axis2 length=6 in order =(-.7 to .72 by .2)
        offset=(2)
label=(h=1.3 a=90 r=0  'Dimension 2');
  symbol v=none;
title1 h=1.2 'Biplot of Speeds for Women Track Records Data';
title2 j=l 'Output 2.11';
title3 'The choice (c): FACTYPE=JK';
run;

%biplot( data = track,  var = m100 m200 m400 m800 m1500
m3000 marathon, id = country, factype=GH, std =MEAN);
filename gsasfile "prog211b.graph";
goptions reset=all gaccess=gsasfile autofeed dev=psl;
goptions horigin=.75in vorigin=2in;
goptions hsize=7in vsize=8in;
proc gplot data=biplot;
plot dim2 * dim1/anno=bianno frame
                 href=0 vref=0 lvref=3 lhref=3
                 vaxis=axis2 haxis=axis1
                 vminor=1 hminor=1;
axis1 length=6 in order=(-.5 to 4 by .5)
        offset=(2)
        label = (h=1.3 'Dimension 1');
  axis2 length=6 in order =(-1.5 to 1.5 by .2)
        offset=(2)
        label=(h=1.3 a=90 r=0  'Dimension 2');
  symbol v=none;
title1 h=1.2 'Biplot of Speeds for Women Track Records Data';
title2 j=l 'Output 2.11';
title3 'The choice (b): FACTYPE=GH';
run;
```

Output 2.11

Output 2.11

COORDINATES FOR BIPLOTS WITH THE TWO CHOICES:

Type	Country	jk_dim1	jk_dim2	gh_dim1	gh_dim2
OBS	Argentin	-0.31595	0.26246	-0.04346	0.11390
OBS	Australi	0.86277	0.19188	0.11867	0.08327
OBS	Austria	0.50033	0.15256	0.06882	0.06621
OBS	Belgium	0.61453	-0.03304	0.08453	-0.01434
OBS	Bermuda	-0.21686	0.28973	-0.02983	0.12573
OBS	Brazil	-0.05349	0.29849	-0.00736	0.12954
OBS	Burma	-0.74531	-0.13240	-0.10251	-0.05746
OBS	Canada	1.11711	0.23282	0.15365	0.10103
OBS	Chile	-0.23290	-0.37966	-0.03203	-0.16476
OBS	China	-0.24672	-0.36546	-0.03394	-0.15859
OBS	Columbia	-0.07293	-0.10923	-0.01003	-0.04740
OBS	Cookis	-2.30362	-0.41746	-0.31685	-0.18116
OBS	Costa	-0.96545	-0.28963	-0.13279	-0.12569
OBS	Czech	1.14604	0.58429	0.15763	0.25356
OBS	Denmark	0.55566	-0.32431	0.07643	-0.14074
OBS	Dominica	-1.05736	0.23935	-0.14544	0.10387
OBS	Finland	0.89992	0.25345	0.12378	0.10999
OBS	France	0.76040	0.14660	0.10459	0.06362

Output 2.11 (*continued*)

```
OBS   GDR        1.35767    0.63704    0.18674    0.27645
OBS   FRG        1.24460    0.16174    0.17119    0.07019
OBS   GB&NI      1.19027    0.20472    0.16372    0.08884
OBS   Greece    -0.45687   -0.01448   -0.06284   -0.00628
OBS   Guatemal  -1.43174    0.32532   -0.19693    0.14118
OBS   Hungary    0.61454   -0.02597    0.08453   -0.01127
OBS   India     -0.55150    0.03591   -0.07586    0.01558
OBS   Indonesi  -0.99296    0.18588   -0.13658    0.08066
OBS   Ireland    0.58480   -0.33411    0.08044   -0.14499
OBS   Israel     0.08744   -0.16146    0.01203   -0.07007
OBS   Italy      0.93696   -0.15268    0.12888   -0.06626
OBS   Japan      0.17446   -0.42190    0.02400   -0.18309
OBS   Kenya      0.01564    0.01103    0.00215    0.00479
OBS   Korea     -0.41181   -0.40609   -0.05664   -0.17623
OBS   DPRKorea  -0.25802   -0.38483   -0.03549   -0.16700
OBS   Luxembou  -0.53680   -0.42039   -0.07383   -0.18243
OBS   Malasiya  -0.95285   -0.02310   -0.13106   -0.01002
OBS   Mauritiu  -1.89438    0.46579   -0.26056    0.20213
OBS   Mexico     0.07625   -0.24757    0.01049   -0.10744
OBS   Netherla   0.76922   -0.02569    0.10580   -0.01115
OBS   NZealand   0.77946   -0.26528    0.10721   -0.11512
OBS   Norway     0.82241   -0.48529    0.11312   -0.21060
OBS   Guinea    -1.72607    0.20090   -0.23741    0.08718
OBS   Philippi  -0.84533    0.39557   -0.11627    0.17166
OBS   Poland     1.00242    0.42039    0.13788    0.18244
OBS   Portugal   0.29102   -0.61345    0.04003   -0.26622
OBS   Rumania    0.76525   -0.08348    0.10526   -0.03623
OBS   Singapor  -0.82329   -0.28645   -0.11324   -0.12431
OBS   Spain      0.17917   -0.34723    0.02465   -0.15069
OBS   Sweden     0.76018    0.04885    0.10456    0.02120
OBS   Switzerl   0.62632   -0.26180    0.08615   -0.11361
OBS   Taipei     0.01733    0.52644    0.00238    0.22845
OBS   Thailand  -0.73883   -0.13277   -0.10162   -0.05762
OBS   Turkey    -0.77550   -0.09511   -0.10667   -0.04128
OBS   USA        1.49906    0.17499    0.20619    0.07594
OBS   USSR       1.45272    0.18855    0.19982    0.08183
OBS   WSamoa    -3.09742    0.60562   -0.42604    0.26282
 VAR   M100      0.29092    0.42644    2.11504    0.98266
 VAR   M200      0.34145    0.55810    2.48242    1.28606
 VAR   M400      0.33904    0.38349    2.46491    0.88369
 VAR   M800      0.30532    0.00670    2.21974    0.01543
 VAR   M1500     0.38574   -0.19889    2.80442   -0.45830
 VAR   M3000     0.39993   -0.25414    2.90759   -0.58564
 VAR   MARATHON  0.53093   -0.50539    3.85998   -1.16460
```

Output 2.11
(continued)

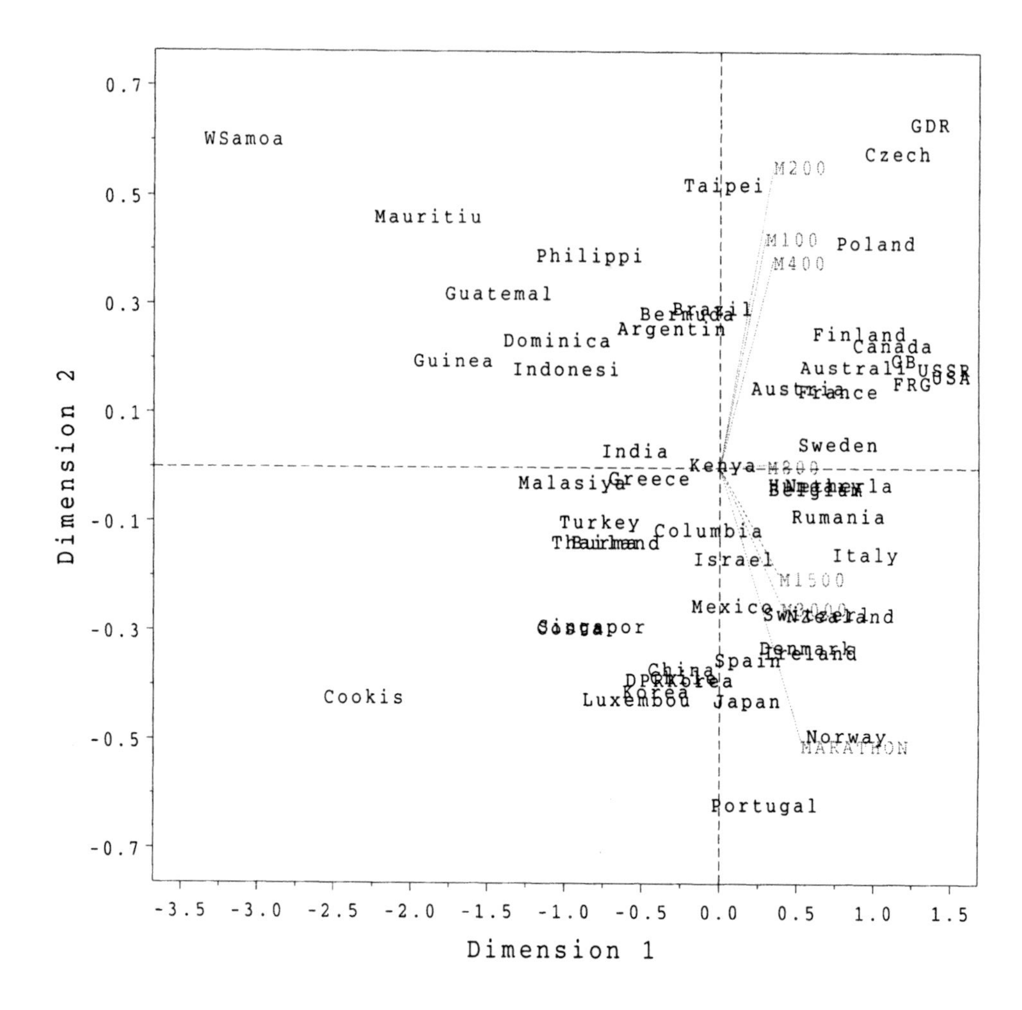

Output 2.11
(continued)

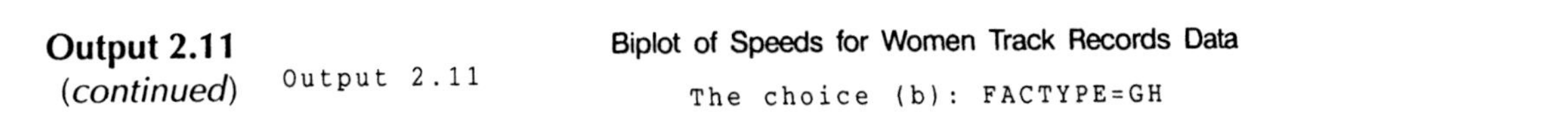

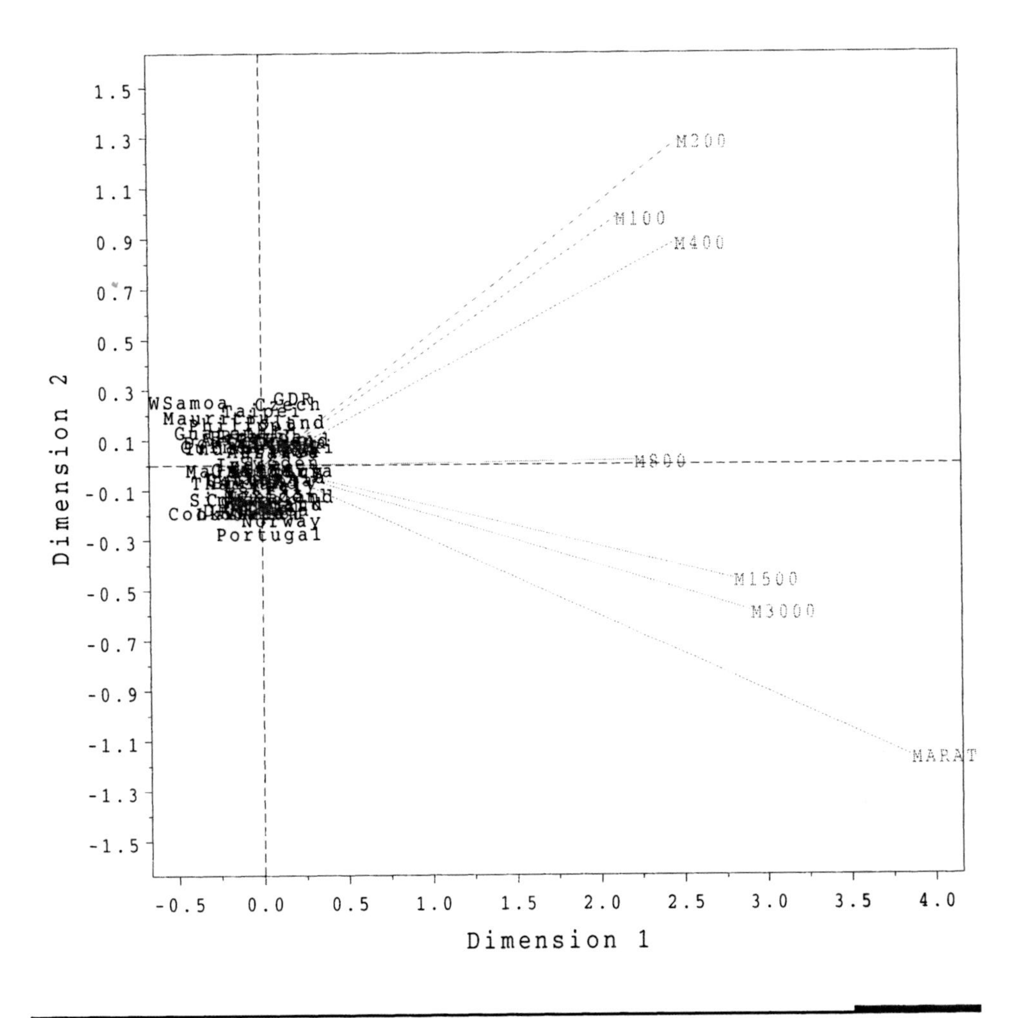

We have selected choice (c), which draws the principal component scores and the weights on the same graph. This is accomplished by choosing the FACTYPE=JK option in the BIPLOT macro. Since the covariance matrix of the speeds is being used for the principal component analysis, we have selected the STD=MEAN option.

The biplot shown in Output 2.11 appears cluttered because of the large number (55) of observations in this data set. However, strong correlation among the speeds for short-distance running events and among the speeds for long-distance running events is clear from the graphs (vectors) of the variables. Further, we clearly see that the correlation between the speeds for long- and short-distance events is small from the large angles between these sets of variables.

In Output 2.11, we have also included the biplot corresponding to choice (b), which uses FACTYPE=GH, as this plot will be later referred to in Chapter 6, Section 6.2.

The two calls of the BIPLOT macro in Program 2.11 provide the coordinates for plotting the biplot for the above two cases. For convenience, in Output 2.11 we have provided the coordinates for both of the biplots in a single display.

2.10 PCA Using SAS/INSIGHT Software

SAS/INSIGHT software is an interactive tool for data exploration and analysis. We can use it to explore data through a variety of interactive graphs and analyses linked across multiple windows. In particular, it is easy to perform principal component analysis by clicking a mouse button. Given a set of data, we first select a correlation or covariance matrix, followed by the number of principal components. The simple summary statistics, the correlation or covariance matrix, the eigenvalues in a descending order, their consecutive differences, the cumulative sums of the eigenvalues are all routinely provided in an output window. Further, the eigenvectors, the principal component scores, and graphs of the first two or three principal component scores are provided by simply clicking and selecting. For more information, see the *SAS/INSIGHT User's Guide*.

2.11 Concluding Remarks

In this chapter, we have provided a fairly detailed account of various features and applications of principal component analysis. Real-life, practical examples are also provided to point out when a PCA may work well and when it may fail in its objective of reducing the dimension.

We want to mention that more general principal component analyses using different weighting schemes have been considered in the literature. For example see Rao (1964), Gnanadesikan (1997), and Jolliffe (1986). Greenacre (1984), in the context of correspondence analysis of two-way tables, uses the weights for both variables and observations. The PRINCOMP procedure has the facility to use weights for the observations but not for the variables. Suppose a data set A contains the observations on the variables $x_1, \ldots, x_p$ and the weights for the observations in a variable named WT. Then the following SAS statements can be used to conduct a PCA using the weights for the observations.

```
proc princomp data = a;
var x1-xp;
weight wt;
run;
```

We have not discussed the use of the PRINQUAL procedure so let us mention it here. PROC PRINQUAL can be used to generalize principal component analysis to handle data that are not necessarily quantitative. For an analysis of data given in the form of a contingency table using the correspondence analysis, see Chapter 7.

Chapter 3

Canonical Correlation Analysis

3.1 Introduction

Canonical correlation analysis is a statistical technique to identify and measure the association between two sets of random variables using a specific matrix function of the variance-covariance matrix of these variables. It is one of the most general types of multivariate analysis in that the multivariate regression analysis can be interpreted as a special case of the canonical correlation analysis. Also, certain aspects of the discriminant analysis and factor analysis can both be shown to be closely related to it, in philosophy as well as in the underlying mathematical ideas. Canonical correlation analysis has been used extensively in a wide variety of disciplines such as psychology, education, sociology, economics, biology, ecology, chemistry, and medicine.

In what situations is canonical correlation analysis applicable? Let us illustrate it by an example. Suppose an investigator wants to explore the possible association between the prices and the consumptions of various food products. She may collect such data from past records. However, since certain food products may sometimes compete with each other (for example, tea and coffee) and others may be often consumed together (for example, wine and cheese), there may be strong associations not only between the price variables and consumption variables but also within the price variables and within the consumption variables themselves. Thus, to explore the association between the two sets of variables, the variance-covariance matrix of *all* the variables, and not just the matrix of covariances *between* the variables in the two sets, should be considered.

Since a large number of weak as well as strong associations may exist between the two sets of variables, canonical correlation analysis is a very convenient data reduction technique. It may be easier to understand and interpret these relationships if the variables in each of the two sets can be transformed to possibly fewer and uncorrelated variables so that there is a one-to-one correspondence between the variables in the two transformed sets.

3.2 Population Canonical Correlations and Canonical Variables

Consider a $(p+q)$ dimensional population and let the $(p+q)$ variables be divided into two sets containing p and q variables respectively. Let the variables in the first set be denoted by a p by 1 vector $\mathbf{x} = (x_1, x_2, \cdots, x_p)'$ and those in the second set by a q by 1 vector $\mathbf{y} = (y_1, y_2, \cdots, y_q)'$. In canonical correlation analysis, we look for vectors $\mathbf{a}$ and $\mathbf{b}$ for which the correlation between $\mathbf{a}'\mathbf{x}$ and $\mathbf{b}'\mathbf{y}$ is maximized. In the multivariate regression context, if $\mathbf{x}$ represents the vector of predictor variables and $\mathbf{y}$ that of the criterion or response variables, then as Mardia, Kent, and Bibby (1979) express it, $\mathbf{a}'\mathbf{x}$ is the best predictor and $\mathbf{b}'\mathbf{y}$ is the *most predictable criterion.* In case $q = 1$, there is only one response variable, and $\mathbf{a}'\mathbf{x}$ is the best linear predictor (possibly apart from a constant) for y.

Let us assume that the variables $\mathbf{x}$ and $\mathbf{y}$ have the respective population variance-covariance matrices $\mathbf{\Sigma}_{xx}$ and $\mathbf{\Sigma}_{yy}$ and $cov(\mathbf{x}, \mathbf{y}) = \mathbf{\Sigma}_{xy} = \mathbf{\Sigma}'_{yx}$. Then, the $p \cdot q$ elements of $\mathbf{\Sigma}_{xy}$ measure the association between $\mathbf{x}$ and $\mathbf{y}$. For convenience, we will assume that $p \leq q$ (and hence vector $\mathbf{x}$ is smaller than or equal in size to vector $\mathbf{y}$). Also in the theoretical presentation, we will assume that the variance-covariance matrix of $\begin{bmatrix} \mathbf{x} \\ \mathbf{y} \end{bmatrix}$, namely

$$D\begin{bmatrix} \mathbf{x} \\ \mathbf{y} \end{bmatrix} = \begin{bmatrix} \mathbf{\Sigma}_{xx} & \mathbf{\Sigma}_{xy} \\ \mathbf{\Sigma}_{yx} & \mathbf{\Sigma}_{yy} \end{bmatrix}$$

is of full rank (viz. $p+q$). The sets of canonical variables, say $\mathbf{v} = (v_1, \ldots, v_p)'$ and $\mathbf{w} = (w_1, \ldots, w_q)'$, are the linearly transformed variables from $\mathbf{x}$ and $\mathbf{y}$, respectively. The first pair of canonical variables (v_1, w_1) is such that the correlation between $v_1 = \mathbf{a}_1'\mathbf{x}$ and $w_1 = \mathbf{b}_1'\mathbf{y}$ is maximum. The correlation between the resulting canonical variables v_1 and w_1 is called the first canonical correlation and will be denoted by ρ_1. The second pair (v_2, w_2) is such that v_2 and w_2 each is uncorrelated with both v_1 and w_1 and the correlation between $v_2 = \mathbf{a}_2'\mathbf{x}$ and $w_2 = \mathbf{b}_2'\mathbf{y}$ is maximized. In general, the ith pair (v_i, w_i) $v_i = \mathbf{a}_i'\mathbf{x}$, $w_i = \mathbf{b}_i'\mathbf{y}$, $i = 1, \ldots, p$ is such that v_i, w_i are both uncorrelated with v_j, w_j, $j = 1, \ldots, i-1$, and the correlation between v_i and w_i is maximized subject to these restrictions. If $\rho_1 \geq \rho_2 \geq \cdots \geq \rho_p \geq 0$ are the corresponding canonical correlations, then their squares $\rho_1^2 \geq \cdots \geq \rho_p^2$ are the p eigenvalues of $\mathbf{\Sigma}_{xy}\mathbf{\Sigma}_{yy}^{-1}\mathbf{\Sigma}_{yx}\mathbf{\Sigma}_{xx}^{-1}$. These are also the nonzero eigenvalues of $\mathbf{\Sigma}_{yx}\mathbf{\Sigma}_{xx}^{-1}\mathbf{\Sigma}_{xy}\mathbf{\Sigma}_{yy}^{-1}$. The vectors $\mathbf{a}_i$ and $\mathbf{b}_i$, such that $(\mathbf{a}_i'\mathbf{x}, \mathbf{b}_i'\mathbf{y})$ is the ith pair of canonical variables, are obtained as the solutions of

$$\mathbf{\Sigma}_{yy}^{-1}\mathbf{\Sigma}_{yx}\mathbf{\Sigma}_{xx}^{-1}\mathbf{\Sigma}_{xy}\mathbf{b}_i = \rho_i^2\mathbf{b}_i$$

and

$$\mathbf{\Sigma}_{xx}^{-1}\mathbf{\Sigma}_{xy}\mathbf{\Sigma}_{yy}^{-1}\mathbf{\Sigma}_{yx}\mathbf{a}_i = \rho_i^2\mathbf{a}_i.$$

It is assumed that the matrices $\mathbf{\Sigma}_{xx}$ and $\mathbf{\Sigma}_{yy}$ both admit respective inverses. If any of them were a singular matrix, a generalized inverse could be used instead. In general, the number of pairs of canonical variables is equal to the number of nonzero eigenvalues of $\mathbf{\Sigma}_{xy}\mathbf{\Sigma}_{yy}^{-}\mathbf{\Sigma}_{yx}\mathbf{\Sigma}_{xx}^{-}$, where $\mathbf{\Sigma}_{yy}^{-}$ and $\mathbf{\Sigma}_{xx}^{-}$ are the two respective choices of the generalized inverses of matrices $\mathbf{\Sigma}_{yy}$ and $\mathbf{\Sigma}_{xx}$.

The canonical correlations and canonical variables have an important and desirable invariance property. Under the nonsingular linear transformations $\mathbf{x}^* = \mathbf{A}\mathbf{x}$ and $\mathbf{y}^* = \mathbf{B}\mathbf{y}$, the canonical correlations remain the same. The ith pair of the canonical variables for sets $\mathbf{x}^*$ and $\mathbf{y}^*$ is $(\mathbf{a}_i'\mathbf{A}^{-1}\mathbf{x}, \mathbf{b}_i'\mathbf{B}^{-1}\mathbf{y})$, where $(\mathbf{a}_i'\mathbf{x}, \mathbf{b}_i'\mathbf{y})$ is the ith pair of canonical variables corresponding to the sets of variables $\mathbf{x}$ and $\mathbf{y}$. This is an important property in that instead of the variance-covariance matrices $\mathbf{\Sigma}_{xx}$, $\mathbf{\Sigma}_{yy}$ and the covariance of $\mathbf{x}$ with $\mathbf{y}$, $\mathbf{\Sigma}_{xy}$, we could use the correlation matrices $\mathbf{R}_{xx}$, $\mathbf{R}_{yy}$, and the matrix of correlations $\mathbf{R}_{xy}$.

The covariance vectors between $\mathbf{x}$ and $\mathbf{a}_i'\mathbf{x}$, $\mathbf{x}$ and $\mathbf{b}_i'\mathbf{y}$ and $\mathbf{y}$ and $\mathbf{b}_i'\mathbf{y}$ are respectively given by $\boldsymbol{\Sigma}_{xx}\mathbf{a}_i$, $\boldsymbol{\Sigma}_{xy}\mathbf{b}_i$, and $\boldsymbol{\Sigma}_{yy}\mathbf{b}_i$.

3.3 Sample Canonical Correlations and Canonical Variables

Usually the matrices $\boldsymbol{\Sigma}_{xx}$, $\boldsymbol{\Sigma}_{yy}$ and $\boldsymbol{\Sigma}_{xy}$ are not known and need to be estimated from a random sample. In this case, the population canonical correlations will be estimated by the sample canonical correlations. Mathematically, the sample canonical correlations and canonical variables are obtained essentially in the same way as those for the population, except that the matrices $\boldsymbol{\Sigma}_{xx}$, $\boldsymbol{\Sigma}_{yy}$ and $\boldsymbol{\Sigma}_{xy}$ are replaced by their sample counterparts $\mathbf{S}_{xx}$, $\mathbf{S}_{yy}$ and $\mathbf{S}_{xy}$. These are obtained using the formula for the sample variance-covariance matrix given in Chapter 1, Section 3 and reproduced below. If $\begin{bmatrix} \mathbf{x}_1 \\ \mathbf{y}_1 \end{bmatrix}, \ldots, \begin{bmatrix} \mathbf{x}_n \\ \mathbf{y}_n \end{bmatrix}$ is a sample of size n from $(p+q)$ dimensional population with mean $\boldsymbol{\mu} = \begin{bmatrix} \boldsymbol{\mu}_x \\ \boldsymbol{\mu}_y \end{bmatrix}$ and a variance-covariance matrix $\boldsymbol{\Sigma} = \begin{bmatrix} \boldsymbol{\Sigma}_{xx} & \boldsymbol{\Sigma}_{xy} \\ \boldsymbol{\Sigma}_{yx} & \boldsymbol{\Sigma}_{yy} \end{bmatrix}$, then we estimate $\boldsymbol{\Sigma}$ by $\mathbf{S} = \begin{bmatrix} \mathbf{S}_{xx} & \mathbf{S}_{xy} \\ \mathbf{S}_{yx} & \mathbf{S}_{yy} \end{bmatrix}$ where

$$\mathbf{S}_{xx} = \frac{1}{n-1}\sum_{j=1}^{n}(\mathbf{x}_j - \bar{\mathbf{x}})(\mathbf{x}_j - \bar{\mathbf{x}})'$$

$$\mathbf{S}_{yy} = \frac{1}{n-1}\sum_{j=1}^{n}(\mathbf{y}_j - \bar{\mathbf{y}})(\mathbf{y}_j - \bar{\mathbf{y}})'$$

$$\mathbf{S}_{xy} = \mathbf{S}_{yx}' = \frac{1}{n-1}\sum_{j=1}^{n}(\mathbf{x}_j - \bar{\mathbf{x}})(\mathbf{y}_j - \bar{\mathbf{y}})'.$$

The ith pair of sample canonical variables $(\hat{v}_i, \hat{w}_i)$ $\hat{v}_i = \hat{\mathbf{a}}_i'\mathbf{x}$, $\hat{w}_i = \hat{\mathbf{b}}_i'\mathbf{y}$ maximizes the sample correlation

$$r_{v_i, w_i} = \frac{\mathbf{a}_i'\mathbf{S}_{xy}\mathbf{b}_i}{\sqrt{\mathbf{a}_i'\mathbf{S}_{xx}\mathbf{a}_i\mathbf{b}_i'\mathbf{S}_{yy}\mathbf{b}_i}}$$

subject to the condition that both v_i and w_i are uncorrelated with previous $(i-1)$ pairs of canonical variables. Computationally, the sample canonical correlations $r_1 \geq r_2 \geq \cdots \geq r_p$ are obtained as the positive square roots of the nonzero eigenvalues of $\mathbf{S}_{yx}\mathbf{S}_{xx}^{-1}\mathbf{S}_{xy}\mathbf{S}_{yy}^{-1}$. The vectors $\hat{\mathbf{a}}_i$, $\hat{\mathbf{b}}_i$ corresponding to the ith pair of canonical variables are obtained as the solutions of

$$\mathbf{S}_{yy}^{-1}\mathbf{S}_{yx}\mathbf{S}_{xx}^{-1}\mathbf{S}_{xy}\mathbf{b}_i = r_i^2\mathbf{b}_i$$

$$\mathbf{S}_{xx}^{-1}\mathbf{S}_{xy}\mathbf{S}_{yy}^{-1}\mathbf{S}_{yx}\mathbf{a}_i = r_i^2\mathbf{a}_i.$$

These computations are similar to those outlined in the previous section. However, since $\mathbf{S}_{xx}$, $\mathbf{S}_{yy}$ and $\mathbf{S}_{xy}$ are all random, hypothesis testing and other relevant inference procedures can be performed in this case. The inference procedures, however, are asymptotic in nature and are appropriate only if the sample size is at least moderately large. Also, the covariances between the original and the canonical variables can be estimated using the formulas given in the previous section, with the population variances and covariances replaced by their sample counterparts.

We may often be interested in performing a canonical analysis on the sets of standardized variables, that is, those that have zero (population) means and unit (population) standard deviations. We may want this because various variables are measured on different scales, and therefore these variables are to be brought to an equal footing. In such a case, the population variance-covariance matrices for the two sets will be same as their respective correlation matrices. However, since the true population means, variances, and covariances may be all unknown, we often use their estimates from the samples for the standardization of the data. A canonical analysis performed on these standardized data will result in different sets of canonical coefficients. These coefficients are often called *standardized canonical coefficients*. They can be used to compute the canonical scores from the standardized input variables. However, the raw coefficients, that is, those corresponding to the unstandardized raw data, are used to compute the canonical variable scores from the centered input variables. It may be remarked that the standardized coefficients can be obtained by multiplying the raw coefficients by the standard deviations of the corresponding variables. We will illustrate the computation of all these quantities through two examples.

EXAMPLE 1 ***Canonical Analysis of Livestock Prices and Meat Consumption*** Waugh (1942) analyzed price and consumption data on meat for the years 1921 through 1940. Beef prices (PRICEB), hog prices (PRICEH), per capita consumption for beef (CONSUMPB), and consumption for pork (CONSUMPH) were available for all 20 years. We ask, "From this data set, what measure can we get of the relation of livestock prices to meat consumption?"

In order to answer this query, we may define prices as one set of variables and consumption as another. With this understanding, for this data set $n = 20$ and $p = q = 2$. Hence, there are only two canonical variables. These can be obtained using the simple SAS statements in the CANCORR procedure:

```
proc cancorr;
var priceb priceh;
with consumpb consumph;
```

where the two sets of variables are separately specified, one listed in the VAR statement and the other in the WITH statement. The convention concerning which list should be given in which statement is arbitrary. In program 3.1, which performs the canonical correlation analysis of this data set, we have used the convention shown above.

The B option in PROC CANCORR can be used to print regression coefficients when the variables listed in either the VAR or the WITH statement are regressed respectively on the variables that are listed in the WITH or the VAR statement. In the present context (see Output 3.1), if we treat the prices as being determined by the amount of consumption, then the corresponding estimated prediction equations are

$$\text{PRICEB} = 26.3500 - 0.1793 * \text{CONSUMPB} - 0.1051 * \text{CONSUMPH}$$

and

$$\text{PRICEH} = 14.2316 + 0.1080 * \text{CONSUMPB} - 0.1819 * \text{CONSUMPH}.$$

These indicate that the price of the particular meat is affected negatively as the corresponding consumption increases. The effect of the consumption of one type of meat on the price of the other is not as clear. The prediction equations include an intercept if it is so indicated by using the INT option (see Program 3.1). The scores on canonical variables can be output as a SAS data set by specifying them in the OUT= option. We have saved these scores in a data set named CSCORES. Similarly, various statistics, canonical correlations, and the coefficients of the **x** and **y** in the respective canonical variables have been output in a SAS data set named CANSTAT by using the OUTSTAT= CANSTAT option. The canonical variables have been respectively named PRISCOR1 and PRISCOR2 (for the set listed in VAR) and CONSCOR1 and CONSCOR2 (for the set listed in WITH) using the options VPREFIX (V for VAR) and WPREFIX (W for WITH).

As $p = q = 2$, there are only two pairs of canonical variables. The first canonical correlation, the correlation between the first pair of canonical variables, namely,

$$\hat{v}_1 = 0.4848 * PRICEB + 0.4095 * PRICEH$$

and

$$\hat{w}_1 = -0.0504 * CONSUMPB - 0.1482 * CONSUMPH$$

is $r_1 = 0.8467$. Similarly, we have,

$$\hat{v}_2 = -0.6467 * PRICEB + 0.4438 * PRICEH,$$
$$\hat{w}_2 = 0.2247 * CONSUMPB - 0.0175 * CONSUMPH$$

and $r_2 = 0.7291$. However, since the prices are in dollars while the consumptions are in pounds, the standard coefficients, that is, when the data on all four variables have been centered to have zero means and scaled to have variances 1, may be more appropriate. It is so since, after such a scaling, all the coefficients are, in some sense, on the same scale.

```
/* Program 3.1 */

options ls=64 ps=45 nodate nonumber;
title1 'Output 3.1';

data livestok;
infile 'livestock.dat' firstobs=5;
input year priceb priceh consumpb consumph ;

proc cancorr data = livestok  b int out = cscores  outstat = canstat
                vprefix =priscor wprefix =conscor ;

var priceb priceh;
with consumpb consumph ;
title2 'Canonical Correlation Analysis: Prices and Meat Consumption' ;

proc print data = cscores;
title2 'Canonical Scores';

proc print data = canstat;
title2 'Canonical Correlation Statistics';
run;
```

Output 3.1

```
                        Output 3.1
  Canonical Correlation Analysis: Prices and Meat Consumption

                  The CANCORR Procedure

 Univariate Multiple Regression Statistics for Predicting the
           VAR Variables from the WITH Variables

               Raw Regression Coefficients

                          priceb            priceh

        consumpb       -0.17928782        0.10803736
        consumph       -0.10514134       -0.18192221
        Intercept      26.34996534       14.23155908
```

Univariate Multiple Regression Statistics for Predicting the WITH Variables from the VAR Variables

Raw Regression Coefficients

	consumpb	consumph
priceb	-2.2534707	-2.0032757
priceh	1.2251962	-2.7563105
Intercept	66.3336021	107.6162849

Canonical Correlation Analysis

	Canonical Correlation	Adjusted Canonical Correlation	Approximate Standard Error	Squared Canonical Correlation
1	0.846663	0.821064	0.064962	0.716838
2	0.729116	.	0.107456	0.531610

Eigenvalues of Inv(E)*H
= CanRsq/(1-CanRsq)

	Eigenvalue	Difference	Proportion	Cumulative
1	2.5315	1.3966	0.6904	0.6904
2	1.1350		0.3096	1.0000

Test of H0: The canonical correlations in the current row and all that follow are zero

	Likelihood Ratio	Approximate F Value	Num DF	Den DF	Pr > F
1	0.13263032	13.97	4	32	<.0001
2	0.46838959	19.29	1	17	0.0004

Multivariate Statistics and F Approximations

S=2 M=-0.5 N=7

Statistic	Value	F Value	Num DF	Den DF
Wilks' Lambda	0.13263032	13.97	4	32
Pillai's Trace	1.24844801	14.12	4	34
Hotelling-Lawley Trace	3.66651723	14.42	4	18.211
Roy's Greatest Root	2.53154237	21.52	2	17

Multivariate Statistics and F Approximations

S=2 M=-0.5 N=7

Statistic	Pr > F
Wilks' Lambda	<.0001
Pillai's Trace	<.0001
Hotelling-Lawley Trace	<.0001
Roy's Greatest Root	<.0001

```
NOTE: F Statistic for Roy's Greatest Root is an upper bound.
      NOTE: F Statistic for Wilks' Lambda is exact.

          Raw Canonical Coefficients for the VAR Variables

                              priscor1           priscor2

          priceb         0.4848242363       -0.646666558
          priceh         0.4095221043       0.4438449077

          Raw Canonical Coefficients for the WITH Variables

                              conscor1           conscor2

          consumpb       -0.050408968       0.2247807312
          consumph       -0.148200986       -0.017492222

                   Canonical Correlation Analysis

      Standardized Canonical Coefficients for the VAR Variables

                              priscor1       priscor2

               priceb           0.6100        -0.8136
               priceh           0.6895         0.7473

      Standardized Canonical Coefficients for the WITH Variables

                              conscor1       conscor2

               consumpb        -0.2200         0.9810
               consumph        -0.9984        -0.1178

                        Canonical Structure

              Correlations Between the VAR Variables
                  and Their Canonical Variables

                              priscor1       priscor2

               priceb           0.7350        -0.6781
               priceh           0.8001         0.5999

              Correlations Between the WITH Variables
                  and Their Canonical Variables

                              conscor1       conscor2

               consumpb        -0.1172         0.9931
               consumph        -0.9758        -0.2188

          Correlations Between the VAR Variables and the
            Canonical Variables of the WITH Variables

                              conscor1       conscor2

               priceb           0.6223        -0.4944
               priceh           0.6774         0.4374
```

```
Correlations Between the WITH Variables and
the Canonical Variables of the VAR Variables
```

	priscor1	priscor2
consumpb	-0.0992	0.7241
consumph	-0.8261	-0.1595

Canonical Scores

Obs	year	priceb	priceh	consumpb	consumph
1	1921	8.41	8.73	55.7	65.0
2	1922	8.68	9.26	59.2	65.9
3	1923	8.22	6.60	59.8	74.5
4	1924	8.24	7.23	59.9	74.7
5	1925	8.64	10.04	60.0	67.3
6	1926	7.63	9.95	60.7	64.6
7	1927	9.50	8.32	54.9	68.2
8	1928	11.41	7.56	49.0	71.3
9	1929	10.49	7.93	49.8	69.8
10	1930	9.84	8.51	48.9	67.0
11	1931	9.20	7.03	48.5	68.3
12	1932	10.58	6.05	46.7	70.6
13	1933	8.92	6.48	51.4	69.9
14	1934	9.52	6.55	55.8	64.2
15	1935	12.76	11.53	53.5	48.6
16	1936	9.47	10.62	58.8	55.6
17	1937	11.37	9.93	55.3	55.9
18	1938	10.10	8.70	54.5	58.3
19	1939	9.88	6.66	54.5	64.4
20	1940	9.91	5.42	55.2	72.5

Obs	priscor1	priscor2	conscor1	conscor2
1	-0.36013	1.04964	0.06781	0.26065
2	-0.01218	1.11028	-0.24200	1.03164
3	-1.32453	0.22712	-1.54678	1.01608
4	-1.05683	0.49381	-1.58146	1.03506
5	0.28785	1.48234	-0.48981	1.18698
6	-0.23868	2.09553	-0.12496	1.39155
7	0.00042	0.16280	-0.36611	0.02485
8	0.61520	-1.40966	-0.52812	-1.35558
9	0.32069	-0.65050	-0.34614	-1.14952
10	0.24307	0.02726	0.11419	-1.30284
11	-0.67331	-0.21576	-0.05831	-1.41549
12	-0.40558	-1.54313	-0.30844	-1.86033
13	-1.03430	-0.27881	-0.44162	-0.79162
14	-0.71473	-0.63574	0.18133	0.29713
15	2.89552	-0.52059	2.60920	0.05301
16	0.92778	1.20304	1.30463	1.12190
17	1.56637	-0.33188	1.43660	0.32992
18	0.44694	-0.05654	1.12125	0.10811
19	-0.49515	-0.81972	0.21722	0.00141
20	-0.98841	-1.38949	-1.01849	0.01707

Canonical Correlation Statistics

Obs	_TYPE_	_NAME_	priceb	priceh	consumpb	consumph	Intercept
1	MEAN		9.6385	8.1550	54.6050	65.8300	.
2	STD		1.2581	1.6838	4.3641	6.7369	.
3	N		20.0000	20.0000	20.0000	20.0000	.
4	CORR	priceb	1.0000	0.1813	-0.5640	-0.4990	.
5	CORR	priceh	0.1813	1.0000	0.3549	-0.7567	.
6	CORR	consumpb	-0.5640	0.3549	1.0000	-0.1029	.
7	CORR	consumph	-0.4990	-0.7567	-0.1029	1.0000	.
8	B	priceb	.	.	-0.1793	-0.1051	26.350
9	B	priceh	.	.	0.1080	-0.1819	14.232
10	B	consumpb	-2.2535	1.2252	.	.	66.334
11	B	consumph	-2.0033	-2.7563	.	.	107.616
12	CANCORR		0.8467	0.7291	.	.	.
13	SCORE	priscor1	0.6100	0.6895	0.0000	0.0000	.
14	SCORE	priscor2	-0.8136	0.7473	0.0000	0.0000	.
15	SCORE	conscor1	0.0000	0.0000	-0.2200	-0.9984	.
16	SCORE	conscor2	0.0000	0.0000	0.9810	-0.1178	.
17	RAWSCORE	priscor1	0.4848	0.4095	0.0000	0.0000	.
18	RAWSCORE	priscor2	-0.6467	0.4438	0.0000	0.0000	.
19	RAWSCORE	conscor1	0.0000	0.0000	-0.0504	-0.1482	.
20	RAWSCORE	conscor2	0.0000	0.0000	0.2248	-0.0175	.
21	STRUCTUR	priscor1	0.7350	0.8001	-0.0992	-0.8261	.
22	STRUCTUR	priscor2	-0.6781	0.5999	0.7241	-0.1595	.
23	STRUCTUR	conscor1	0.6223	0.6774	-0.1172	-0.9758	.
24	STRUCTUR	conscor2	-0.4944	0.4374	0.9931	-0.2188	.

The output also provides the values of $r_1^2 = 0.7186$ and $r_2^2 = 0.5316$ as well as the adjusted canonical correlations $r_{i(adj)}$ defined by Lawley (1959), which asymptotically have slightly smaller bias than raw canonical correlations. However, $r_{i(adj)}$ may be negative and, unlike the raw canonical correlations, may *not* follow the ordering $r_{1(adj)} \geq r_{2(adj)} \geq \ldots$. Whenever this ordering is not satisfied, SAS prints the corresponding $r_{i(adj)}$ as a missing value. Thus, in the output, we have $r_{1(adj)} = 0.8211$, but $r_{2(adj)}$ is reported missing.

Under the assumption that all the variables jointly have multivariate normal distribution, consider the hypothesis testing problem

$$H_0: \ \rho_{s+1} = \cdots = \rho_p = 0,$$

which means that only the first s of the population canonical correlations are nonzero. This hypothesis can be tested by using the chi-square approximation of the likelihood ratio (or Wilks' Λ) test statistic

$$-\{n - (p + q + 3)/2\} \ ln \left[\prod_{i=s+1}^{p} (1 - r_i^2) \right]$$

which approximately follows a $\chi^2_{(p-s)(q-s)}$ under the null hypothesis. However, SAS provides the following F-distribution approximation (called Rao's F) of Wilks' Λ as given by Rao (1973, p. 556),

$$F = \left(\frac{r_{2s}}{r_{1s}}\right)\left(\frac{1-\Lambda_s^{1/m_s}}{\Lambda_s^{1/m_s}}\right),$$

where

$$m_s = \left\{\frac{(p-s)^2(q-s)^2-4}{(p-s)^2+(q-s)^2-5}\right\}^{1/2},$$

$$r_{1s} = (p-s)(q-s),$$

and

$$r_{2s} = m_s\left[n-(p+q+3)/2\right] - \left[(p-s)(q-s)-2\right]/2.$$

Under null hypothesis H_0 and for large samples, Rao's F statistic approximately follows an $F(r_{1s}, r_{2s})$ distribution.

In our example, there are $p = 2$ hypotheses on canonical correlations that we will test in sequence:

$$H_0^{(1)} : \rho_1 = \rho_2 = 0,$$

$$H_0^{(2)} : \rho_2 = 0.$$

The null hypothesis $H_0^{(2)}$ would need to be tested only if $H_0^{(1)}$ is rejected. The first null hypothesis $H_0^{(1)}$ is rejected in view of strong evidence against it (p value < 0.0001 corresponding to an approximate F (4,32) distribution). Subsequently, $H_0^{(2)}$ is also rejected by a similar argument.

PROC CANCORR also provides four popular multivariate tests that are often used to test the linear hypotheses in the multivariate analysis of variance context. These tests are discussed in detail in Khattree and Naik (1999). However, they are used here to test the hypothesis that all population canonical correlations are zero. In general, the test based on Wilks' Λ is equivalent to that described above for H_0 with $s = 0$, using Rao's F approximation. The other multivariate tests may or may not be equivalent to the F test described above. However, in the present example, all multivariate tests reach the same conclusions.

The estimated correlations between the sets of variables and the canonical variables are also provided. There appear to be rather strong correlations between PRISCOR1 and PRICEH (= 0.8001), and between CONSCOR1 and CONSUMPH (= −0.9758). Thus, the hog factor appears to dominate in the first pair of canonical variables and is seen to have been assigned higher weights (as the standardized canonical coefficients) in the equations. However, in the second pair of canonical variables, it is the beef factor that dominates the equations. In terms of standardized canonical coefficients, the canonical variables are

$$\hat{v}_1 = 0.6100 * PRICEB + 0.6895 * PRICEH$$

$$\hat{w}_1 = -0.2200 * CONSUMPB - 0.9984 * CONSUMPH$$

$$\hat{v}_2 = -0.8136 * PRICEB + 0.7473 * PRICEH$$

$$\hat{w}_2 = 0.9810 * CONSUMPB - 0.1178 * CONSUMPH$$

In the equations given above, the variables PRICEB, PRICEH, CONSUMPB, and CONSUMPH are all standardized to have a zero (sample) mean and a (sample) variance 1.

The original data and the corresponding scores of canonical variables have been output in the SAS data set CSSCORE. We could use the PRINT procedure to print all this information. The data set saved by the optional use of OUTSTAT= provides a fairly complete summary of all the computations performed during various analyses.

PROC CANCORR can also perform canonical correlation analysis when, instead of raw data, only the sample variance-covariance or the correlation matrix is provided. Of course, since the raw data are not available in such a situation, we cannot perform regression analysis or compute the canonical variable scores for various observations. However, the coefficients of various canonical variables can still be obtained as well as the appropriate statistical tests performed on the canonical correlations.

EXAMPLE 2 ***Analysis of Canadian Hard Red Spring Wheat Data*** Waugh (1942) presents an example in which two sets of variables respectively represent the wheat and flour characteristics for Canadian hard red spring wheat. The wheat characteristics are kernel texture (X1), test weight (X2), damaged kernels (X3), foreign material (X4), and the presence of crude protein (X5) in wheat. The flour characteristics are wheat per barrel of flour (Y1), ash in flour (Y2), crude protein in flour (Y3), and the gluten quality index (Y4).

Since the quality of wheat should give a good indication of the quality of flour, it is of interest to correlate the two sets of variables to find the set of best predictable quality indexes for flour quality and, correspondingly, their best predictors by using the quality characteristics of the wheat. The data from $n = 138$ samples were available on both the wheat and the flour characteristics. From the samples, the 9 by 9 sample variance-covariance matrix as well as the sample correlation matrix of wheat and flour characteristics were computed. The degrees of freedom for these matrices will be $n - 1 = 137$. The sample correlation matrix is given in Program 3.2.

Since the raw data are not available, we need to indicate in the DATA step that the data being provided are a correlation matrix of the particular collection of random variables. This is done by indicating (_TYPE_ = CORR) in the DATA step. Since the rows of data contain the correlations, we have defined a reference variable _NAME_ to take the values $X1, \ldots, X5, Y1, \ldots, Y4$, respectively. The correlation matrix is specified as the lower triangular matrix. Once the correlation matrix has been provided, we need only to run PROC CANCORR in the usual way except for a few minor changes.

For example, since the raw data are not available, the regression coefficients cannot be computed as they require various sample quantities other than correlations in the computation. Consequently, options such as B and INT are not applicable. Also, no canonical variable scores will be calculated and hence the option OUT = is also not relevant.

On the other hand, since the correlation matrix provided on the data is computed from a sample of finite size, it is only appropriate that the correct degrees of freedom are used in the statistical tests. This is done by using the option EDF =. In the present example, the sample variance-covariance matrix and the correlation matrix have (error) degrees of freedom $(n-1) = 137$. Thus, the error degrees of freedom will be specified as EDF = 137.

The complete program is given in Program 3.2 and the output in Output 3.2.

```
/* Program 3.2 */

options ls=64 ps=45 nodate nonumber;
title1 'Output 3.2';

data wheat(type = corr);
_type_='corr' ;
input _name_  $ x1 x2 x3 x4 x5 y1 y2 y3 y4 ;
lines;
x1 1.0       .  .  .  .  .  .  .  .
x2  .75409 1.0  .  .  .  .  .  .  .
x3  -.69048 -.71235 1.0  .  .  .  .  .  .
```

```
x4  -.44578 -.51483 .32326 1.0  .  .  .  .  .
x5  .69173 .41184 -.44393 -.33439 1.0  .  .  .  .
y1  -.60463 -.72236 .73742 .52744 -.38310 1.0  .  .  .
y2  -.47881 -.41878 .36132 .46092 -.50494 .25056 1.0  .  .
y3  .77978 .54245 -.54624 -.39266 .73666 -.48993 -.43361 1.0  .
y4  -.15205 -.10236 .17224 -.01873 -.14848 .24955 -.07851 -.16276 1.0
;
/*
Data from  Waugh (1942),
Reprinted with  permission from The Econometric Society.
*/

title2 'Wheat and Flour Characteristics in Canadian Wheat';
proc cancorr data = wheat edf = 137 vprefix =flour wprefix =wheat ;
var y1 y2 y3 y4 ;
with x1 x2 x3 x4 x5  ;
run;
```

Output 3.2

Output 3.2
Wheat and Flour Characteristics in Canadian Wheat

The CANCORR Procedure

Canonical Correlation Analysis

	Canonical Correlation	Adjusted Canonical Correlation	Approximate Standard Error	Squared Canonical Correlation
1	0.909388	0.904848	0.014781	0.826987
2	0.636689	0.621293	0.050802	0.405373
3	0.255485	0.215789	0.079859	0.065273
4	0.094466	0.069024	0.084673	0.008924

Eigenvalues of Inv(E)*H
= CanRsq/(1-CanRsq)

	Eigenvalue	Difference	Proportion	Cumulative
1	4.7799	4.0982	0.8627	0.8627
2	0.6817	0.6119	0.1230	0.9858
3	0.0698	0.0608	0.0126	0.9984
4	0.0090		0.0016	1.0000

Test of H0: The canonical correlations in the
current row and all that follow are zero

	Likelihood Ratio	Approximate F Value	Num DF	Den DF	Pr > F
1	0.09530476	22.11	20	428.79	<.0001
2	0.55085361	7.25	12	344.24	<.0001
3	0.92638585	1.70	6	262	0.1208
4	0.99107612	0.59	2	132	0.5534

```
                    Canonical Correlation Analysis

           Multivariate Statistics and F Approximations

                        S=4     M=0     N=63.5

Statistic                        Value   F Value  Num DF  Den DF

Wilks' Lambda               0.09530476     22.11      20  428.79
Pillai's Trace              1.30655725     12.81      20     528
Hotelling-Lawley Trace      5.54048024     35.44      20  276.45
Roy's Greatest Root         4.77991736    126.19       5     132

           Multivariate Statistics and F Approximations

                        S=4     M=0     N=63.5

                Statistic                Pr > F

                Wilks' Lambda            <.0001
                Pillai's Trace           <.0001
                Hotelling-Lawley Trace   <.0001
                Roy's Greatest Root      <.0001

NOTE: F Statistic for Roy's Greatest Root is an upper bound.

          Raw Canonical Coefficients for the VAR Variables

            flour1          flour2          flour3          flour4

y1    -0.536621867    1.0091593137    -0.235648641    -0.130965911
y2    -0.288309403    0.0272598952    0.9686805033     0.504066129
y3    0.4549129597    0.9790822641    0.4318835677    0.4356812442
y4    0.0245656334    -0.179359853    -0.283894992    0.9958509551

          Raw Canonical Coefficients for the WITH Variables

            wheat1          wheat2          wheat3          wheat4

x1    0.2110801172    0.9263730874    0.7421512289    1.5165655059
x2    0.1741269951    -0.583026154    -0.189567334    -0.071705681
x3    -0.331010736    0.6533978723    -0.568361607    0.8026957021
x4    -0.264006717    0.3425979674    0.8497558375    -0.262303405
x5    0.2976949079    0.5497430959    -0.554327371     -1.08185945

      Standardized Canonical Coefficients for the VAR Variables

                 flour1         flour2         flour3         flour4

     y1         -0.5366         1.0092        -0.2356        -0.1310
     y2         -0.2883         0.0273         0.9687         0.5041
     y3          0.4549         0.9791         0.4319         0.4357
     y4          0.0246        -0.1794        -0.2839         0.9959

      Standardized Canonical Coefficients for the WITH Variables

                 wheat1         wheat2         wheat3         wheat4

     x1          0.2111         0.9264         0.7422         1.5166
```

```
x2        0.1741     -0.5830     -0.1896     -0.0717
x3       -0.3310      0.6534     -0.5684      0.8027
x4       -0.2640      0.3426      0.8498     -0.2623
x5        0.2977      0.5497     -0.5543     -1.0819
```

In Output 3.2, the Wilks' test (and corresponding $F = 22.11$ with (20,428) df) as well as other multivariate tests show that there certainly is a strong relationship between the wheat and flour quality characteristics. The p value for all four tests is < 0.0001. This is equivalent to testing $H_0^{(1)}: \rho_1 = \rho_2 = \rho_3 = \rho_4 = 0$, which is obviously rejected (the first line of second table). We would also like to sequentially test

$$H_0^{(2)}: \rho_2 = \rho_3 = \rho_4 = 0,$$
$$H_0^{(3)}: \rho_3 = \rho_4 = 0,$$
$$H_0^{(4)}: \rho_4 = 0.$$

The null hypothesis $H_0^{(2)}$ is also rejected, as there is sufficiently strong evidence against it. The observed value of approximate $F(12{,}344)$ is 7.25, which leads to a p value < 0.0001. The $H_0^{(3)}$, however, is accepted (p value $= 0.1208$); hence, it is not necessary to proceed further and test $H_0^{(4)}$. Thus, there are only two significant canonical correlations, and the pairs of the corresponding canonical variables are given below (with $\hat{v}_1 = FLOUR1$, $\hat{w}_1 = WHEAT1$, $\hat{v}_2 = FLOUR2$, $\hat{w}_2 = WHEAT2$).

$$FLOUR1 = -0.5366 * Y1 - 0.2883 * Y2 + 0.4549 * Y3 + 0.0246 * Y4$$
$$WHEAT1 = 0.2111 * X1 + 0.1174 * X2 - 0.3310 * X3 - 0.2640 * X4 + 0.2977 * X5$$
$$FLOUR2 = 1.0092 * Y1 + 0.0273 * Y2 + 0.9791 * Y3 - 0.1794 * Y4$$
$$WHEAT2 = 0.9264 * X1 - 0.5830 * X2 + 0.6534 * X3 + 0.3426 * X4 + 0.5497 * X5$$

The raw canonical coefficients and the standardized coefficients are the same since the correlation matrix has been used in the computation (and hence all the variables are already implicitly scaled to have variance 1).

We make an important observation about the first pair of canonical variables. From the definition of $X1, \ldots, X5, Y1, \ldots, Y4$, we observe that the variables $X3$, $X4$, and $Y1$, $Y2$ are the "smaller-the-better" type of characteristic while the remaining, namely $X1$, $X2$, $X5$, $Y3$, and $Y4$, are of the "larger-the-better" type. The coefficients for the smaller-the-better type variables are all negative in the first pair, whereas these are positive for the larger-the-better type. In other words, $(\hat{v}_1, \hat{w}_1)$ = (FLOUR1, WHEAT1) represents the pair of quality indices and respectively measures the value of the flour and the value of the wheat. The batches with the high numerical values of these indices are desirable. Also, by looking at the correlation of FLOUR1 with $Y1, \ldots, Y4$ and WHEAT1 with $X1, \ldots, X5$, we see that all the characteristics are important for wheat (the correlations with WHEAT1 are of comparable magnitude), while the prominent flour characteristics are Y1 (wheat per barrel of flour),Y3 (crude protein in flour) and to some extent Y2 (ash in flour). Y4 (the gluten quality index) is not as important to determine the value of the flour. Also observe that the corresponding coefficient of Y4 in FLOUR1 is very small. The second pair of canonical variables $(\hat{v}_2, \hat{w}_2)$ is, however, difficult to interpret.

3.4 Canonical Analysis of Residuals

Sometimes it may be of interest to perform the analysis on sets of residuals rather than on the variables themselves. For example, consider Waugh's livestock data (1942) discussed in Example 1. It is possible that both the prices and consumptions are functions of time, and there may be a trend in the observed values of these variables. In this case, it may be of interest to first remove any trend by fitting a model with time as the independent variable and then analyzing the data on the residuals. Of course, one of the related problems would be to determine if the trend can be represented as a polynomial function (or any other desirable function) of time. These problems can be tackled by using the widely available standard regression model-fitting methods. Once the residuals have been obtained, the degrees of freedom of the sample variance-covariance matrix of residuals may not be $(n - 1)$ and may depend on the type of model under consideration (and the number of parameters estimated). This fact should be given appropriate consideration when testing statistical hypotheses.

EXAMPLE 1 ***Continued*** Suppose there is reason to believe that all four variables are changing linearly with time. To remove this linear trend, we first fit the appropriate linear regression models with YEAR as the independent variable to obtain the data on the residuals. It is done by using the following SAS statements:

```
proc reg;
model priceb priceh consumpb consumph = year;
output out = regout r = y1 y2 x1 x2;
```

Thus, the four linear regression models with YEAR as the independent variable have been fitted. The residuals from the corresponding models are saved in a SAS data set named REGOUT. Note that Y1 and Y2 are the residuals corresponding to PRICEB and PRICEH, and X1, X2 represent the residuals that correspond to CONSUMPB and CONSUMPH. The canonical correlation analysis will be performed on sets {Y1, Y2} and {X1, X2}. It is important to realize that for this analysis, the data set to be used is REGOUT, as the original data set does not contain the corresponding values of the residuals. Also, since the error degree of freedom is $(n - 2) = 18$ and not 19, we must specify $EDF = 18$ in the PROC CANCORR. The four regression models each have the number of independent variables $k = 1$; thus the error degrees of freedom would be $n - (k + 1) = n - 2 = 18$. The applicable SAS statements are

```
proc cancorr data = regout edf = 18 vprefix = v wprefix = w;
var y1 y2;
with x1 x2;
```

The complete program is given in Program 3.3 and selected parts of the output are presented in Output 3.3. The canonical variables are respectively denoted as V1, V2 (for prices) and W1, W2 (for consumptions). It is interesting to observe that the first canonical correlation has improved from 0.8467 (see Output 3.1) to 0.9145, thereby indicating a slightly stronger association. The intensities of the significance of the multivariate tests have also possibly increased slightly (strictly speaking, F or other test statistics are not comparable in the present context because of different denominator degrees of freedom). These are, however, not shown here.

```
/* Program 3.3 */

options ls=64 ps=45 nodate nonumber;
title1 'Output 3.3';
```

```
data livestok;
infile 'livestock.dat' firstobs=5;
input year priceb priceh consumpb consumph;

proc reg data =  livestok;
model priceb priceh consumpb consumph = year;
output out = regout r = y1 y2 x1 x2 ;
title2 'Prices and Meat Consumption: Residuals';

proc cancorr data = regout edf = 18 vprefix =v wprefix =w ;
var y1 y2;
with x1 x2;


proc cancorr data = livestok  b int  outstat = canstat
             vprefix =priscor wprefix =conscor ;
partial year ;
var priceb priceh;
with consumpb consumph ;
title2 'Prices and Meat Consumption: Partialing Out Year' ;
run;
```

Output 3.3

Output 3.3
Prices and Meat Consumption: Residuals

The CANCORR Procedure

Canonical Correlation Analysis

	Canonical Correlation	Adjusted Canonical Correlation	Approximate Standard Error	Squared Canonical Correlation
1	0.914478	0.906705	0.038591	0.836271
2	0.677792	.	0.127420	0.459402

3.5 Partial Canonical Correlations

The analysis done in the previous section can be interpreted as one in which the effect of covariate YEAR has been eliminated or partialed out before computing the canonical correlations or canonical variables. The corresponding canonical correlations and variables can therefore be called partial canonical correlations and partial canonical variables. With the use of covariates such as YEAR, we thus remove the variations in the price and consumption variables that exist because of the individual economic behavior of the particular variable. The elimination of these individual variations is expected to provide a better insight and enables us to see more clearly which variables in one set are more (or less) associated with the variables in the other.

One way to do this analysis in SAS is first to fit the model as we did in the previous section. Alternatively, we could use the PARTIAL statement to specify the variables to be partialed out, after the PROC CANCORR statement. For example, for the livestock data the linear time trend can be eliminated by using the statement

```
partial year;
```

as shown in the later part of the Program 3.3. The resulting output is the same as that previously obtained by directly using the residuals, except for the round-off errors that, in this example, happen to be surprisingly high in some instances. Therefore, we do not present this part of the output.

EXAMPLE 3 ***Analysis of Respiratory Tract Infection Data*** Armitage (1971, p. 255) presents survey data on upper respiratory tract infections. Observations were made on 18 families, each consisting of a father, a mother and three children (CHILD1, CHILD2, and CHILD3) in descending age order. The youngest child, CHILD3, was in each case a preschool child. Out of these 18 families, 6 families were a random sample from those living in overcrowded conditions, another 6 from crowded conditions and the remaining from uncrowded conditions. Thus, the nominal variable CROWDING takes three values, namely HIGH, MEDIUM, and LOW corresponding to the three crowding conditions. The response variable to be analyzed is the number of swabs positive for pneumococcus during a certain period. Its values for father, mother, and the three children are denoted by FATHER, MOTHER, CHILD1, CHILD2 and CHILD3 in the code.

Our interest is in estimating what is commonly called the *familial correlation* or the interclass correlation, which may be defined as the first canonical correlation between the parents and children, and also to obtain the coefficients of the corresponding canonical variables. However, the number of positive swabs may also be dependent on the crowding condition. Hence, the effect of the qualitative variable CROWDING is to be partialed out in the analysis.

The situation here is different from that of the livestock data in that unlike YEAR, the variable CROWDING is not quantitative. Thus, we must define the corresponding dummy variables (or equivalently the columns of the corresponding design matrix), which are to be partialed out. We define three variables OVER, OK, and UNDER. The first of these takes the value 1 for overcrowded families and 0 otherwise; the second takes value 1 for crowded families and 0 otherwise; while the third variable UNDER takes value 1 for undercrowded families and 0 otherwise. This is done in a SAS DATA step by defining

```
over = (crowding eq 'high');
ok = (crowding eq 'medium');
under = (crowding eq 'low');
```

The basic SAS statements to analyze these data by first partialing out the variables OVER, OK, and UNDER are

```
proc cancorr;
partial over ok under;
var child1 child2 child3;
with father mother;
```

Program 3.4 presents this code with certain other options used for further analysis. The selected pieces from the output are presented in Output 3.4. The specified options, B and INT, result in a summary of the regression analysis when all five variables—FATHER, MOTHER, CHILD1, CHILD2, and CHILD3—have been individually regressed on dummy variables OVER, OK, and UNDER. (This regression is equivalent to using the GLM procedure on the CLASS variable CROWDING.) Note that since these models also have an intercepts column that is the sum of the columns corresponding to OVER, OK, and UNDER in the 20 by 4 design matrix, the design matrix is deemed rank deficient and will have rank 3. This rank deficiency is reflected in the regression coefficients (not shown) of the last variable, namely UNDER, in all the five models have been set to zero.

As some of the variables have been partialed out, the output provides the respective blocks of the partial correlation matrix for the five variables. The blocks represent these among the VAR variables, among the WITH variables and between the VAR and the WITH variables.

As the $\min(p, q) = 2$, there are only two partial canonical correlations and correspondingly only two pairs of canonical variables. The respective values of the partial canonical correlations are $r_1 = 0.3029$ and $r_2 = 0.0971$, none of which appears to be particularly strong. This intuitive observation is supported by formal testing of the hypothesis

$$H_0^{(1)} : \rho_1 = \rho_2 = 0,$$

which, in view of the large p value ($= 0.9735$) corresponding to Wilks' Λ, is readily accepted. This particular hypothesis (that all partial canonical correlations are zero) can also be tested using Pillai's Trace, Hotelling-Lawley Trace, and Roy's greatest root tests. In each case, a very high p value results. Consequently, we conclude that after accounting for the crowding condition, the number of positive swabs for parents and children are uncorrelated. In other words, the parent-sib familial correlation is negligible and there may not be any point in attempting to predict one from the other.

It may be useful to do a separate canonical analysis for the three crowding condition groups and explore the possibility that the canonical relationships and their degrees of strength themselves may depend on crowding conditions. Analyses for the separate crowding conditions can be done by using the BY CROWDING statement after the PROC CANCORR statement. However, these analyses may not be particularly reliable and informative in the present context since each of the three groups has a relatively small sample of only six families, while the corresponding correlation matrices themselves are of the size 5 by 5 each. This would not result in sufficient degrees of freedom to take the approximate asymptotic tests on their face values.

```
/* Program 3.4 */

options ls=64 ps=45 nodate nonumber;
title1 'Output 3.4';

data family ;
input crowding $ father mother child1 child2 child3 ;
over =(crowding eq 'high');
ok =(crowding eq 'medium');
under =(crowding eq 'low');
lines;
high  5 7 6 25 19
high  11 8 11 33 35
high  3 12 19 6 21
high  3 19 12 17 17
high  10 9 15 11 17
high  9 0 6 9 5
medium  11 7 7 15 13
medium  10 5 8 13 17
medium  5 4 3 18 10
medium  1 9 4 16 8
medium  5 5 10 16 20
medium 7 3 13 17 18
low 6 3 5 7 3
low 9 6 6 14 10
low 2 2 6 15 8
low 0 2 10 16 21
low 3 2 0 3 14
low 6 2 4 7 20
;
/*
Data from Armitage (1971),
reprinted with  permission from Blackwell Science Ltd.
*/
```

```
title2 'No. of Swabs positive for Pneumococcus';
proc cancorr data = family  b int  outstat = stat
vprefix =chscore wprefix =pscore ;
partial over ok under ;
var child1 child2 child3;
with father mother ;
run;
```

Output 3.4

```
                     Output 3.4
         No. of Swabs positive for Pneumococcus

                 The CANCORR Procedure

   Canonical Correlation Analysis Based on Partial Correlations

                     Adjusted    Approximate      Squared
          Canonical  Canonical      Standard    Canonical
        Correlation  Correlation       Error  Correlation

   1       0.302893    -.008607     0.234511     0.091744
   2       0.097129    -.168833     0.255763     0.009434


        Test of HO: The canonical correlations in the
          current row and all that follow are zero

        Likelihood   Approximate
             Ratio       F Value   Num DF   Den DF   Pr > F

   1   0.89968694          0.20        6       22   0.9735
   2   0.99056587          0.06        2       12   0.9447
```

3.6 Canonical Redundancy Analysis

We may want to know which variables in a given set are most influential in defining the canonical variables. This influence is measured by computing the covariances or correlations between the original variables and the canonical variables. If $\begin{bmatrix} \mathbf{R}_{xx} & \mathbf{R}_{xy} \\ \mathbf{R}_{yx} & \mathbf{R}_{yy} \end{bmatrix}$ is the correlation matrix of $\begin{bmatrix} \mathbf{x} \\ \mathbf{y} \end{bmatrix}$ and if, as earlier, $\mathbf{v}$ and $\mathbf{w}$ are the p by 1 vectors of the respective canonical variables, then this influence can be quantified through vectors

$$\mathbf{s}_j(\mathbf{x}) = \mathbf{R}_{xx}\mathbf{b}_j, \text{ and } \mathbf{s}_j(\mathbf{y}) = \mathbf{R}_{yy}\mathbf{a}_j \qquad j = 1, \ldots, p, \tag{3.1}$$

where the canonical coefficients $\mathbf{a}_j$ and $\mathbf{b}_j$, respectively, correspond to set $\mathbf{x}$ and set $\mathbf{y}$ of variables and are such that $v_j = \mathbf{a}'_j\mathbf{x}$ and $w_j = \mathbf{b}'_j\mathbf{y}$. For the jth canonical variable, the proportion of variation extracted from the set $\mathbf{x}$ is $\mathbf{s}_j(\mathbf{x})'\mathbf{s}_j(\mathbf{x})/p$. Stewart and Love (1968) and Miller (1969) defined the jth redundancy of $\mathbf{x}$ given $\mathbf{y}$ as

$$R_j(\mathbf{x}) = \left(\frac{\mathbf{s}_j(\mathbf{x})'\mathbf{s}_j(\mathbf{x})}{p}\right)\rho_j^2 \qquad j = 1, \ldots, p. \tag{3.2}$$

As Cooley and Lohnes (1971) point out, $R_j(\mathbf{x})$ can be interpreted as the proportion of variance extracted by $\mathbf{x}$ multiplied by the proportion of shared variance between $\mathbf{x}$ and the corresponding canonical variable w_j derived from $\mathbf{y}$. It expresses the amount of actual overlap between the two sets of variables that is packaged in the jth canonical relationship as seen from the side of $\mathbf{x}$ as added to an already available $\mathbf{y}$. The redundancies of $\mathbf{y}$ given $\mathbf{x}$ are similarly defined as

$$R_j(\mathbf{y}) = \left(\frac{\mathbf{s}_j(\mathbf{y})'\mathbf{s}_j(\mathbf{y})}{q}\right)\rho_j^2 \qquad j = 1, \ldots, p \tag{3.3}$$

and can be interpreted similarly. In general $R_j(\mathbf{x})$ and $R_j(\mathbf{y})$ may not be equal.

EXAMPLE 4 ***Redundancy Analysis for Rain Forest Data*** Gittins (1985, p. 305) presents an interesting ecological case study on the dynamic status of a lowland tropical rain forest in Guyana. The objective of this study was to discover whether the field observations were consistent with the view that the forest was likely to retain the same type of vegetation as the trees composing the forest died and were replaced. Thus, the study aimed to establish the tree-seedling relationships. The weaker relationships would imply that the composition was changing with time, while the stronger relationships would indicate stability in the vegetation dynamics. Two sets of six variables each representing the contribution of trees and seedling communities, respectively, were observed at 25 100m × 100m sites chosen by a systematic sampling scheme. These sets are assumed to constitute a random sample. These characteristics are respectively denoted as $X1, \ldots, X6$ (for trees) and $Y1, \ldots, Y6$ (for seedlings). The SAS code is presented in Program 3.5 (we have multiplied Gittins' original data by 100 for this analysis) and selected parts of the output are presented in Output 3.5. As most of the other analyses are standard and have already been discussed through a variety of other examples, we will concentrate on the redundancy analysis.

As $p = q = 6$, there are six canonical correlations and six associated pairs of canonical variables. The first four canonical correlations are all larger than 0.87, while the last one is very small. The fifth canonical correlation ($r_5 = 0.6509$) also appears to be large. A formal sequential hypothesis testing of the hypotheses

$$H_0^{(s+1)} : \rho_{s+1} = \cdots = \rho_6 = 0, \qquad s = 0, 1, \ldots 5,$$

supports this intuition and hence we conclude that the first five canonical correlations are significantly different from zero. However, in view of large values of $p + q$ (= 12), and of the rather moderate sample size, these nonzero canonical correlations are most likely overestimated, and it is somewhat likely that even the fifth canonical correlation is zero. Observe that the associated p value = 0.0460 for the approximate F test is only marginally significant and will possibly be larger, were the estimate not positively biased.

The redundancy analysis is performed by specifying REDUNDANCY as an option in the PROC CANCORR statement. This results in the computation of estimates of ρ_j^2 (that is, r_j^2), $\mathbf{s}_j(\mathbf{x})'\mathbf{s}_j(\mathbf{x})$, $\mathbf{s}_j(\mathbf{y})'\mathbf{s}_j(\mathbf{y})$, $R_j(\mathbf{x})$ and $R_j(\mathbf{y})$, $j = 1, \ldots, p$. In the first four tables of Output 3.5 under this title (Canonical Redundancy Analysis), the quantities in columns 1 and 3 are computed using Equations 3.1 and 3.2 (or 3.3) given above. For example, since the variables $Y1, \ldots, Y6$ are listed in the VAR statements, the first table provides the values of $\hat{\mathbf{s}}_j(\mathbf{y})'\hat{\mathbf{s}}_j(\mathbf{y})/q$ (= 0.3261, ..., 0.0898) in the first column for $j = 1, \ldots, p(= 6)$, which in turn are computed using the second subequation in Equation 3.1 (with parameters replaced by corresponding estimates). In the third column, the values of $r_1^2, \ldots, r_p^2$ (= 0.9766, ..., 0.0032) are listed. The proportions of raw variance of $Y1, \ldots, Y6$ explained by six canonical variables corresponding to other variables $X1, \ldots, X6$ are given in column 4 (= 0.3185, ..., 0.0003) and are obtained using Equation 3.3, that is, by elementwise product of entries in columns 1 and 3.

It may be more meaningful to discuss the redundancies in terms of standardized variances rather than raw variances. The redundancies in the seedling variables ($Y1, \ldots, Y6$) attributable to the opposite canonical variates (of trees, that is of $X1, \ldots, X6$) all fall between 0.25 to 0.0. If the last two canonical variables are ignored, then from the magnitude of redundancies, it appears that each of the four remaining canonical variables provides valuable information to help us understand the tree-seedling relationship. From the third table under this title, we see that the first four canonical variates involving $X1, \ldots, X6$ account for about 75.71% of the total variance of the seedling variables $Y1, \ldots, Y6$. Similar observations can be made to interpret the redundancies corresponding to the other set.

```
/* Program 3.5 */

options ls=64 ps=45 nodate nonumber;
title1 'Output 3.5';

data rainfrst;
input x1 x2 x3 x4 x5 x6 y1 y2 y3 y4 y5 y6 ;
lines;
43 10 22 45 42 59 84 4 20 18 40 8
9 5 49 11 74 23 23 10 4 23 17 88
43 12 37 17 63 46 78 6 10 21 30 48
24 4 86 7 28 25 36 10 5 77 31 15
29 2 71 9 32 48 18 11 7 68 16 33
53 12 28 34 29 55 51 7 48 15 61 20
55 1 6 23 72 6 94 2 13 9 21 7
62 58 14 22 34 1 8 34 7 7 84 28
10 95 7 4 9 6 15 67 3 21 43 23
9 8 91 24 16 -11 6 6 2 87 1 -3
67 8 36 15 42 30 67 6 11 6 63 -4
62 6 20 23 69 0 40 15 8 14 69 2
10 94 4 2 1 2 6 93 3 -4 20 -7
2 97 -1 1 2 2 3 99 2 4 -4 4
64 6 20 37 50 30 40 11 46 5 73 1
67 11 30 20 57 -2 62 18 9 10 51 -4
90 21 8 18 11 -1 16 17 7 10 94 5
2 98 1 1 1 3 5 95 3 1 14 -1
33 8 20 81 39 14 14 7 96 6 18 5
25 2 8 96 9 2 8 0 99 1 9 -1
28 87 5 6 3 2 6 82 4 22 27 16
79 14 24 17 18 38 46 12 9 13 84 7
91 15 4 17 17 16 44 21 5 16 74 12
18 8 28 68 49 38 24 2 96 4 6 3
27 5 32 45 72 21 84 7 31 23 26 19
;
/*
Data from Gittins (1980, p 305),
printed with permission from Springer Verlag
*/

title2 'Redundancy Analysis: Rain Forest Data from Guyana';
proc cancorr data = rainfrst redundancy
vprefix =seedvar wprefix =treevar ;
var y1 y2 y3 y4 y5 y6;
with x1 x2 x3 x4 x5 x6 ;
run;
```

Output 3.5

Output 3.5
Redundancy Analysis: Rain Forest Data from Guyana

The CANCORR Procedure

Canonical Correlation Analysis

	Canonical Correlation	Adjusted Canonical Correlation	Approximate Standard Error	Squared Canonical Correlation
1	0.988238	0.983102	0.004774	0.976614
2	0.956268	0.928425	0.017463	0.914449
3	0.945619	.	0.021597	0.894195
4	0.871496	0.861990	0.049091	0.759506
5	0.650864	0.639713	0.117652	0.423624
6	0.056301	.	0.203477	0.003170

Eigenvalues of Inv(E)*H
= CanRsq/(1-CanRsq)

	Eigenvalue	Difference	Proportion	Cumulative
1	41.7612	31.0722	0.6445	0.6445
2	10.6889	2.2376	0.1650	0.8094
3	8.4514	5.2933	0.1304	0.9399
4	3.1581	2.4231	0.0487	0.9886
5	0.7350	0.7318	0.0113	1.0000
6	0.0032		0.0000	1.0000

Test of H0: The canonical correlations in the
current row and all that follow are zero

	Likelihood Ratio	Approximate F Value	Num DF	Den DF	Pr > F
1	0.00002925	16.25	36	59.848	<.0001
2	0.00125072	10.80	25	53.51	<.0001
3	0.01461964	8.67	16	46.463	<.0001
4	0.13817562	5.45	9	39.09	<.0001
5	0.57454940	2.71	4	34	0.0460
6	0.99683024	0.06	1	18	0.8136

Redundancy Analysis: Rain Forest Data from Guyana

Canonical Redundancy Analysis

Raw Variance of the VAR Variables Explained by
Their Own
Canonical Variables

Canonical Variable Number	Proportion	Cumulative Proportion	Canonical R-Square
1	0.3261	0.3261	0.9766
2	0.1840	0.5101	0.9144
3	0.2237	0.7338	0.8942
4	0.1556	0.8894	0.7595
5	0.0208	0.9102	0.4236
6	0.0898	1.0000	0.0032

Raw Variance of the VAR Variables Explained by The Opposite Canonical Variables

Canonical Variable Number	Proportion	Cumulative Proportion
1	0.3185	0.3185
2	0.1682	0.4867
3	0.2000	0.6867
4	0.1182	0.8049
5	0.0088	0.8138
6	0.0003	0.8141

Raw Variance of the WITH Variables Explained by Their Own Canonical Variables

Canonical Variable Number	Proportion	Cumulative Proportion	Canonical R-Square
1	0.4442	0.4442	0.9766
2	0.1583	0.6025	0.9144
3	0.1637	0.7662	0.8942
4	0.1274	0.8936	0.7595
5	0.0640	0.9576	0.4236
6	0.0424	1.0000	0.0032

Raw Variance of the WITH Variables Explained by The Opposite Canonical Variables

Canonical Variable Number	Proportion	Cumulative Proportion
1	0.4338	0.4338
2	0.1448	0.5786
3	0.1463	0.7249
4	0.0968	0.8217
5	0.0271	0.8488
6	0.0001	0.8489

Standardized Variance of the VAR Variables Explained by Their Own Canonical Variables

Canonical Variable Number	Proportion	Cumulative Proportion	Canonical R-Square
1	0.2541	0.2541	0.9766
2	0.2113	0.4655	0.9144
3	0.2013	0.6668	0.8942
4	0.1785	0.8453	0.7595
5	0.0309	0.8762	0.4236
6	0.1238	1.0000	0.0032

Standardized Variance of the VAR Variables Explained by The Opposite Canonical Variables

Canonical Variable Number	Proportion	Cumulative Proportion
1	0.2482	0.2482
2	0.1932	0.4414
3	0.1800	0.6215
4	0.1356	0.7571
5	0.0131	0.7701
6	0.0004	0.7705

Standardized Variance of the WITH Variables Explained by Their Own Canonical Variables

Canonical Variable Number	Proportion	Cumulative Proportion	Canonical R-Square
1	0.3650	0.3650	0.9766
2	0.1735	0.5385	0.9144
3	0.1574	0.6958	0.8942
4	0.1465	0.8423	0.7595
5	0.1002	0.9425	0.4236
6	0.0575	1.0000	0.0032

Standardized Variance of the WITH Variables Explained by The Opposite Canonical Variables

Canonical Variable Number	Proportion	Cumulative Proportion
1	0.3565	0.3565
2	0.1586	0.5151
3	0.1407	0.6558
4	0.1112	0.7671
5	0.0424	0.8095
6	0.0002	0.8097

Squared Multiple Correlations Between the VAR Variables and the First M Canonical Variables of the WITH Variables

M	1	2	3	4	5	6
y1	0.1649	0.3657	0.4218	0.6664	0.6679	0.6686
y2	0.8850	0.8870	0.9459	0.9636	0.9636	0.9636
y3	0.3989	0.4284	0.7770	0.9015	0.9027	0.9027
y4	0.0383	0.6943	0.7896	0.8215	0.8596	0.8596
y5	0.0014	0.2710	0.7923	0.8543	0.8596	0.8597
y6	0.0007	0.0022	0.0022	0.3351	0.3676	0.3691

```
Squared Multiple Correlations Between the WITH Variables
and the First M Canonical Variables of the VAR Variables
```

M	1	2	3	4	5	6
x1	0.0831	0.3700	0.8469	0.8808	0.8864	0.8864
x2	0.8553	0.8553	0.9174	0.9498	0.9531	0.9531
x3	0.1876	0.6914	0.7640	0.8322	0.8647	0.8647
x4	0.5547	0.5821	0.8095	0.9149	0.9168	0.9168
x5	0.2934	0.3742	0.3797	0.7116	0.7159	0.7164
x6	0.1645	0.2174	0.2174	0.3130	0.5201	0.5206

The output also provides the R^2 values when each of the $Y1, \ldots, Y6$ is regressed on sets of canonical variables $\{\mathbf{a}_1'\mathbf{x}\}$, $\{\mathbf{a}_1'\mathbf{x}, \mathbf{a}_2'\mathbf{x}\} \ldots \{\mathbf{a}_1'\mathbf{x}, \ldots, \mathbf{a}_6'\mathbf{x}\}$, and when each of the $X1, \ldots, X6$ is regressed on sets $\{\mathbf{b}_1'\mathbf{y}\}$, $\{\mathbf{b}_1'\mathbf{y}, \mathbf{b}_2'\mathbf{y}\}, \ldots, \{\mathbf{b}_1'\mathbf{y}, \ldots, \mathbf{b}_6'\mathbf{y}\}$. These values indicate the proportion of variation in a particular variable, explained by a given set of opposite canonical variables. Except for $X6$ and $Y6$, the sets of four opposite canonical variables appear to explain the substantial amount of variation of a given variable.

Canonical redundancy analysis can also be helpful in interpreting the multivariate analysis-of-variance results. SAS course notes *Multivariate Statistical Methods: Practical Applications* (1998) provide an interesting illustration in a designed experiment situation.

3.7 Canonical Correlation Analysis of Qualitative Data

The techniques of canonical correlation analysis can be applied to the qualitative data available by using a two-way frequency table to identify any meaningful associations between the nominal variables that are represented by the rows and columns of the table. Suppose the variables, referred to as, say, A and B, are at levels $1 \ldots, a$ and $1, \ldots, b$ respectively, and let n_{ij} be the frequency corresponding to $(i, j)^{th}$ cell $i = 1, \ldots, a,\ j = 1, \ldots, b$ of the two-way frequency table. Thus, the total sample size is $n = \sum_i \sum_j n_{ij}$, while there are a total of $(a + b)$ dummy variables, each taking values 0 or 1. Specifically, if we define for $i = 1, \ldots, a,\ j = 1, \ldots, b,\ u = 1, \ldots, n,$

$$x_{ui} = \begin{cases} 1 & \text{if the } u^{th} \text{unit belongs to } i^{th} \text{category of } A \\ 0 & \text{otherwise} \end{cases} \tag{3.4}$$

and

$$y_{uj} = \begin{cases} 1 & \text{if the } u^{th} \text{unit belongs to } j^{th} \text{ category of } B \\ 0 & \text{otherwise} \end{cases} \tag{3.5}$$

Then the a dummy variables corresponding to factor A and b dummy variables corresponding to factor B form the two sets of variables. For each of the n units, they take values 0 or 1 showing the absence or presence in the respective categories of A and B. The canonical correlation analysis can be performed on these two data sets. To do so we first need to calculate the matrices $\mathbf{S}_{xx}$, $\mathbf{S}_{yy}$, and $\mathbf{S}_{yx}$ as indicated earlier in the chapter. However, since each unit is in one and only one category of A and in one and only one category of B, the sums $\sum_{i=1}^{a} x_{ui}$ and $\sum_{j=1}^{b} y_{uj}$ both individually add up to 1. Because of this linear dependence and a rank deficiency of 1 in the individual sets of variables, $\mathbf{S}_{xx}$ and $\mathbf{S}_{yy}$ both do not admit their respective inverses. One way to remove this rank deficiency (in fact, it is really not necessary, as a generalized inverse can be used instead of the inverse) is to drop one of the categories from each of the two sets and perform the canonical correlation analysis on the two sets of variables of sizes $p = a - 1$ and $q = b - 1$, respectively. The subsequent calculations are straightforward. However, it may be remembered that individuals in the dropped category are assigned a score of zero for the corresponding canonical variables.

EXAMPLE 5 ***Canonical Analysis for the Status of Mental Health*** Srole et al. (1978, p. 289) presents data on parents' socioeconomic status and the mental health status of the children for a sample of residents of Manhattan. The data have also been illustrated in Agresti (1990). The cross-classified data on cell frequencies for the two factors are presented in Table 3.1.

TABLE 3.1 Cross-Classification of Mental Health Status and Parents' Socioeconomic Status

	Mental Health Status			
Parents' Socioeconomic Status	(Well) 1	(Mild Symptom Formation) 2	(Moderate Symptom Formation) 3	(Impaired) 4
1 (high)	64	94	58	46
2	57	94	54	40
3	57	105	65	60
4	72	141	77	94
5	36	97	54	78
6 (low)	21	71	54	71

In order to perform the canonical correlation analysis, we must first create the dummy variables and assign them the x_{ui} and y_{uj} values defined in Equations 3.4 and 3.5 for each unit (individual) $u = 1, \ldots, n$. For the data set at hand with A representing the socioeconomic status and B the mental health status, we have $a = 6$ and $b = 4$. Thus, $p = a - 1 = 5$, and $q = b - 1 = 3$. As a result, there are three nonzero canonical correlations. Also, $n = \sum_i \sum_j n_{ij} = 1660$. The sample variances and covariances will be computed from this large data set of zeros and ones. The data set is created using a DO loop that is iterated n_{ij} (= FREQ) times. The values of FREQ are supplied as data for $i = 1, \ldots, 6$, and $j = 1, \ldots 4$.

To assign the values x_{ui} and y_{uj} as in Equations 3.4 and 3.5 the following SAS code is used after the INPUT statement and before the LINES statement in the DATA step.

```
proc cancorr;
var priceb priceh;
with consumpb consumph;

do k=1 to freq;
x1 = (i eq '1');
x2 = (i eq '2');
x3 = (i eq '3');
x4 = (i eq '4');
x5 = (i eq '5');
x6 = (i eq '6');
y1 = (j eq '1');
y2 = (j eq '2');
y3 = (j eq '3');
y4 = (j eq '4');
output;
end;
```

The statements given above, for example, create a variable named X1 for which the first (64 + 94 + 58 + 46 =) 262 values are 1 followed by all the remaining (1660 − 262 =) 1398 values zero. The values of Y1 are as follows: the first 64 values are all ones, followed by (94 + 58 + 46 =) 198 zeros, then 57 ones followed by (94 + 54 + 40 =) 188 zeros, and so on. Values for other variables are similarly created.

As mentioned earlier, to eliminate the problem of rank deficiency, one of variables $X1, \ldots, X6$ and one from $Y1, \ldots Y4$ need to be dropped from the analysis and assigned a score of zero. With the higher-the-better convention for the levels of A and B, we will drop the variables corresponding to the lowest levels, namely, the low socioeconomic status (X6) and the impaired mental health status (Y4) from the analysis. This is equivalent to assigning them the coefficients of zeros in the pairs of canonical variables. Thus, the canonical analysis is done for the variables X1 through X5 only with Y1 through Y3 only. All of the analysis discussed here can be done using Program 3.6. Selected parts from the Output are presented in Output 3.6.

The first canonical correlation is 0.1613 and the remaining two nonzero canonical correlations are quite small. The coefficients for the variables in the first pair of canonical vectors are given in Table 3.2

TABLE 3.2 Coefficients in the First Pair of the Canonical Variables

	Levels					
Socioeconomic Status	1	2	3	4	5	6
Raw Can. Coef.	2.9040	2.9292	2.1486	1.7277	0.7579	0
Standardized Can. Coef.	1.0591	1.0393	0.8127	0.7288	0.2777	0
Mental Health Status	1	2	3	4		
Raw Can. Coef.	3.0794	1.6545	1.3830	0		
Standardized Can. Coef.	1.1959	0.7956	0.5713	0		

```
/* Program 3.6 */

options ls=64 ps=45 nodate nonumber;
title1 'Output 3.6';

data mental;
input i $ j $ freq ;
do k  = 1 to freq ;
x1=(i eq '1');
x2=(i eq '2');
x3=(i eq '3');
x4=(i eq '4');
x5=(i eq '5');
x6=(i eq '6');
y1=(j eq '1');
y2=(j eq '2');
y3=(j eq '3');
y4=(j eq '4');
output;
end;
lines;
1 1 64
1 2  94
1 3 58
1 4 46
```

```
2 1 57
2 2  94
2 3  54
2 4 40

3 1 57
3 2  105
3 3 65
3 4 60

4 1 72
4 2 141
4 3 77
4 4 94

5 1 36
5 2 97
5 3 54
5 4 78

6 1 21
6 2 71
6 3 54
6 4 71
;
/*
Data from Srole et. al (1978),
reprinted with permission from New York University Press.
*/

proc cancorr data = mental outstat = canstat
      vprefix =psstatus wprefix =mhstatus ;
var  x1 x2 x3 x4 x5 ;
with  y1 y2 y3 ;
title2 'Mental Health Status vs. Socioeconomic Status of Parents:';
run;
```

Output 3.6

```
                          Output 3.6
   Mental Health Status vs. Socioeconomic Status of Parents:

                    The CANCORR Procedure

              Canonical Correlation Analysis

                         Adjusted    Approximate        Squared
            Canonical    Canonical      Standard      Canonical
          Correlation  Correlation         Error    Correlation

      1      0.161322     0.150213      0.023912       0.026025
      2      0.037137            .      0.024518       0.001379
      3      0.017261            .      0.024544       0.000298
```

Raw Canonical Coefficients for the VAR Variables

	psstatus1	psstatus2	psstatus3
x1	2.9040084847	1.1506859153	-1.314802027
x2	2.929195221	1.3558783856	1.8655812904
x3	2.1486006664	1.0713101314	0.8918829474
x4	1.7277189085	2.8015731473	-0.359121623
x5	0.7578709189	2.8426653485	0.8798868947

Raw Canonical Coefficients for the WITH Variables

	mhstatus1	mhstatus2	mhstatus3
y1	3.079412237	-0.182929581	-0.390445039
y2	1.6544464004	0.1279555964	2.0656790453
y3	1.3830349061	-2.39038447	1.1049574447

Standardized Canonical Coefficients for the VAR Variables

	psstatus1	psstatus2	psstatus3
x1	1.0591	0.4196	-0.4795
x2	1.0393	0.4811	0.6619
x3	0.8127	0.4052	0.3374
x4	0.7288	1.1817	-0.1515
x5	0.2777	1.0415	0.3224

Standardized Canonical Coefficients for the WITH Variables

	mhstatus1	mhstatus2	mhstatus3
y1	1.1959	-0.0710	-0.1516
y2	0.7956	0.0615	0.9934
y3	0.5713	-0.9874	0.4564

These coefficients can be interpreted as the respective scores for various socioeconomic classes of the parents as well as the mental health status of individuals. The standardized scores are in decreasing order as would be expected since the categories for the two factors are ordinally arranged from high to low values. Also, the raw scores for first and second socioeconomic classes for parents are practically indistinguishable, and hence for all practical purposes the same decreasing order is observed in case of raw scores, as well. Thus, there appears to be some positive association between the socioeconomic status of parents and the mental health of their children. However, the association is not very strong, as $r_1 = 0.1613$ is not very large. The magnitudes for the coefficients seem to suggest that the pairs of socioeconomic status groups (1,2), (3,4), and (5,6) each are similar within the pairs, but more distinct from the other two pairs. Also, mental health status groups 2 and 3 are relatively similar but are more distinct from mental health status 1 and 4. Perhaps children of parents in socioeconomic status groups (1,2) are more likely to have mental health status 1, while those whose parents are in socioeconomic status groups (3,4) may most likely belong to mental health groups (2,3), and the children of socioeconomic groups (5,6) may likely fall in mental health group 4.

It may be pointed out that, strictly speaking, dropping X6 and Y4 from the analysis was not necessary. Exactly the same coefficients of the respective canonical variables will be obtained if these were also included in the respective WITH and VAR lists. The reason for this is that SAS sets the coefficients for the last variables in the respective lists to zero because of the rank deficiency of one in each of the two lists. However, the same coefficients would not be obtained if, instead of X6 and Y4, any other variables from the respective lists were dropped. We may further point out that, since the data in the present example are binary, the normal theory-based inference may be inappropriate. Thus we should refrain from attempting to judge the statistical significance of canonical correlations using the F statistics.

In this section, we have presented the canonical analysis of a two-way frequency table. In this situation, both variables (defining two separate sets) were qualitative. We will see later that this analysis is actually equivalent to a correspondence analysis of the same data. Canonical correlation analysis done here is closely related to the techniques of correspondence analysis discussed in Chapter 7 where the above data set is reanalyzed (see Example 2). Another situation, when only one variable (defining one set) is qualitative and the other set consists of a number of quantitative variables, arises in the contexts of experimental designs, analysis of covariance and discriminant analysis. The last topic will be discussed in detail in Chapter 5. The analysis of covariance situation is briefly described in the next section. For a discussion in the framework of experimental designs, see Khattree and Naik (1999).

3.8 'Partial Tests' in Multivariate Regression

The Wilks' Λ test in canonical correlation analysis tests the *model adequacy* hypothesis: whether the variables $\mathbf{x}$ have any effect on $\mathbf{y}$. It is equivalent to testing the null hypothesis H_0: $\rho_1 = \rho_2 = \cdots = \rho_p = 0$. However, suppose we want to test that only a few of the variables in $\mathbf{x}$, say $\mathbf{x}_1$, are important. This is a hypothesis on the regression coefficients corresponding to $\mathbf{x}_2$, where $\mathbf{x} = \begin{bmatrix} \mathbf{x}_1 \\ \mathbf{x}_2 \end{bmatrix}$. An appropriate multivariate test would be the generalization of what is popularly known as the partial F test in the univariate regression analysis.

If $r_1 \ldots, r_p$ are the p nonzero and significant sample canonical correlations, then the Wilks' Λ can be expressed as

$$\Lambda_p = (1 - r_1^2)(1 - r_2^2)\ldots(1 - r_p^2).$$

However, if the variables in $\mathbf{x}_2$ were not important for $\mathbf{y}$, then under this assumption, there may be fewer, say p_1, nonzero and significant sample canonical correlations, denoted as $r_{*1}, \ldots, r_{*p_1}$ and then

$$\Lambda_{p_1} = (1 - r_{*1}^2)(1 - r_{*2}^2)\ldots(1 - r_{*p_1}^2).$$

To test the null hypothesis that the variables in subvector $\mathbf{x}_2$ are unimportant, we could compare the values of Λ_p and Λ_{p_1}. An appropriate test statistic is constructed by taking the ratio of the two,

$$\Lambda_{partial} = \frac{\Lambda_p}{\Lambda_{p_1}} = \frac{(1 - r_1^2)\ldots(1 - r_p^2)}{(1 - r_{*1}^2)\ldots(1 - r_{*p_1}^2)}.$$

A quick way to compute $\Lambda_{partial}$ using SAS is to include the variables corresponding to $\mathbf{x}_1$ in the VAR as well as in the WITH statement. Under such a setup and with p_2 equal to the number of nonzero and non-one sample canonical correlations $R_1, R_2, \ldots, R_{p_2}$, the $\Lambda_{partial}$ given above can be written as (Velu and Wichern, 1985):

$$\Lambda_{partial} = (1 - R_1^2)\ldots(1 - R_{p_2}^2).$$

The SAS output readily provides the values of $R_1^2, \ldots, R_{p_2}^2$ and hence the computation of $\Lambda_{partial}$ given above is straightforward. The appropriate chi-square or F approximations of $\Lambda_{partial}$ can then be used to arrive at the appropriate conclusion about the adequacy of the reduced model.

EXAMPLE 6 ***Partial Tests for Urine Data*** The data are adopted from Smith, Gnanadesikan, and Hughes (1962). Forty-five urine samples from men in four weight categories were taken. On each sample, 11 response variables ($Y1, \ldots, Y11$) and two covariates, namely, urine volume ($Z1$) in ml and scaled specific gravity ($Z2$), were measured. We want to determine if the covariates have any significant effect on the responses.

Note that this is an analysis of covariance problem and that the weight groups define a nominal or class variable to be denoted by GROUP in Program 3.7. We define four dummy variables, $G1, \ldots, G4$, each taking value 1 if the particular individual is in the respective weight group and 0 if otherwise. This is done in the DATA step. As indicated in the discussion of this section earlier, we define the two sets of variables ($Y1, \ldots, Y11$, $G1, \ldots, G4$), and ($Z1$, $Z2$, $G1, \ldots, G4$) and find the canonical correlations between these two sets. As there is a linear dependence between the four dummy variables $G1, \ldots, G4$, there are only five and not six nonzero canonical correlations. Also, $p_2 = 2$, and from Output 3.7, $R_1 = 0.9321$ and $R_2 = 0.6705$. Thus, $\Lambda_{partial} = (1 - R_1^2)(1 - R_2^2) = 0.0723$. The asymptotic chi-square or F approximations for $\Lambda_{partial}$ can be used to compute the appropriate p value. We will compute Rao's approximate F test statistic described earlier. Due to linear dependence between $G1, \ldots, G4$, the values of p and q will respectively be 14 and 5 instead of 15 and 6. Since $p_2 = 2$, under H_0, there are only $5 - 2 = 3$ nonzero canonical correlations. Hence, $s = 3$. Thus, for Rao's approximate F test we have $m_s = 2$, $r_{1s} = 22$, $r_{2s} = 58$. Hence the observed value of the approximate F statistic is $F = \frac{58}{22}\left[\frac{1-(0.0723)^{1/2}}{(0.0723)^{1/2}}\right] = 7.17$, as shown in Output 3.7. Under H_0, the F follows an approximate $F(22, 58)$ distribution and hence the p value corresponding to the observed value of $F = 7.17$ is obtained as < 0.0001. Consequently, we reject the null hypothesis and conclude that the covariates do not have a significant effect on the responses. It may be pointed out that this hypothesis could also be equivalently tested using the MTEST option in PROC REG (Khattree and Naik, 1999, Chapter 3). The SAS statements have been included as part of Program 3.7. The corresponding output is, however, suppressed.

```
/* Program 3.7 */

options ls=64 ps=45 nodate nonumber;
title1 'Output 3.7';

data urisamp;
infile 'urine.dat' firstobs = 7;
input group y1 y2 y3 y4 y5 y6 y7 y8 y9 y10 y11 z1 z2 ;
g1 = group = '1';
g2 = group = '2';
g3 = group = '3';
g4 = group = '4';

proc cancorr data = urisamp short ;
var y1 y2 y3 y4 y5 y6 y7 y8 y9 y10 y11 g1 g2 g3 g4 ;
with g1 g2 g3 g4 z1 z2 ;
title2 'Partial Tests Using Canonical Correlation Analysis';
run;

proc reg data = urisamp ;
model y1 y2 y3 y4 y5 y6 y7 y8 y9 y10 y11 =  z1 z2 g1 g2 g3 g4 ;
mtest z1,z2;
title2 'Partial Tests Using MTEST Option of PROC REG';
run;
```

Output 3.7

```
                         Output 3.7
        Partial Tests Using Canonical Correlation Analysis

                    The CANCORR Procedure

                Canonical Correlation Analysis

NOTE: The correlation matrix for the VAR Variables is less than
full rank. Therefore, some canonical coefficients will be zero.

NOTE: The correlation matrix for the WITH Variables is less
      than full rank.
      Therefore, some canonical correlations and coefficients
      will be zero.
```

	Canonical Correlation	Adjusted Canonical Correlation	Approximate Standard Error	Squared Canonical Correlation
1	1.000000	.	0.000000	1.000000
2	1.000000	.	0.000000	1.000000
3	1.000000	.	0.000000	1.000000
4	0.932047	0.915379	0.019792	0.868711
5	0.670521	0.593391	0.082976	0.449598
6	0.000000	.	0.150756	0.000000

Test of H0: The canonical correlations in the current row and all that follow are zero

	Likelihood Ratio	Approximate F Value	Num DF	Den DF	Pr > F
1	0.00000000	Infty	70	127.87	<.0001
2	0.00000000	Infty	52	106.68	<.0001
3	0.00000000	Infty	36	83.457	<.0001
4	0.07226145	7.17	22	58	<.0001
5	0.55040172	2.45	10	30	0.0282
6	1.00000000	.	.	.	.

3.9 Concluding Remarks

In closing this chapter, we comment that the canonical correlation analysis is basically a descriptive technique with the basic objective to understand the data. Like other descriptive multivariate techniques, no formal objective is defined *a priori* in the sense that we allow the data and the corresponding analyses to reveal any special hidden features that we subsequently try to interpret. Thus, the analysis may not be used to provide formal evidence for or against any hypotheses conceived during the analysis. The technique, at best, enables us to quantify and describe the unique features of the data. Essentially this is the view we have adopted in much of this chapter. Consequently, relatively little emphasis is placed here on hypothesis testing problems.

To comment further on the issues related to hypothesis testing, any significance tests, if performed for canonical correlations, should be viewed with caution. The small sample sizes may make most of the asymptotic tests inappropriate in the particular situations. Also the sequential tests described in this chapter should be regarded only as indicative of the number of important canonical variables pairs. However, their p values and so forth should not be taken so seriously as to draw any formal statistical conclusions. Of course, when the canonical correlation analysis is performed for the data on nominal variables, as in Section 3.7, the statistical tests are completely inappropriate. We may also mention that the canonical correlation analysis of two-way categorical data is essentially equivalent to correspondence analysis, a topic discussed in Chapter 7.

Chapter

4

Factor Analysis

4.1 Introduction

Factor analysis is one of the most widely used multivariate techniques, introduced by Spearman (1904) and developed by Thurstone (1947), Thomson (1951), Lawley (1940, 1941) and others. Started as a controversial and difficult subject, it has emerged as one of the most fascinating and useful multivariate data analysis tools. Thanks mainly to electronic computers and statistical software packages, in the 1970s several books on the subject of factor analysis were written, and its applicability to many diverse fields such as biology, chemistry, ecology, economics, education, political science, psychology, and sociology has been demonstrated. A systematic and detailed treatment of the subject can be found in Harman (1976) and Mulaik (1972). Statistical treatment of the subject is in Lawley and Maxwell (1971). For a more recent treatment of the subject see Flury (1988) and Basilevsky (1994). Brief but excellent discussions of factor analysis are given in several multivariate statistical analysis textbooks. In particular, see Morrison (1976), Mardia, Kent, and Bibby (1979), Seber (1984), and Johnson and Wichern (1998).

In factor analysis, the main concern is identifying the internal relationships between a set of random variables. Spearman (1904), while working with scores obtained in scholastic examinations, noticed a certain systematic pattern in the correlation coefficients matrix of the scores. Specifically, one of the matrices he had considered was the following correlation matrix $\boldsymbol{\rho}$ of the variables, which are the scores on different courses: Classics (C), French (F), English (E), Mathematics (M), Discrimination of pitch (D), and Music (Mu) for 33 boys in a preparatory school.

$$\rho = \begin{array}{c} C \\ F \\ E \\ M \\ D \\ Mu \end{array} \begin{bmatrix} 1.00 & .83 & .78 & .70 & .66 & .63 \\ .83 & 1.00 & .67 & .67 & .65 & .57 \\ .78 & .67 & 1.00 & .64 & .54 & .51 \\ .70 & .67 & .64 & 1.00 & .45 & .51 \\ .66 & .65 & .54 & .45 & 1.00 & .40 \\ .63 & .57 & .51 & .51 & .40 & 1.00 \end{bmatrix}.$$

The variables are arranged to get the present pattern of decreasing correlations within the matrix. It was observed that this matrix has the interesting property that, ignoring the diagonal, any two rows are almost proportional. For example, for rows C and M the ratios,

$$\frac{.83}{.67} \approx \frac{.78}{.64} \approx \frac{.66}{.45} \approx \frac{.63}{.51} \approx 1.2.$$

Spearman argued that this pattern of the correlation matrix can be explained by the model

$$x_i = \mathrm{L_i f} + \epsilon_i, \ i = 1, \ldots, p \ (p = 6), \tag{4.1}$$

where x_i represents the observation on the ith variable, f is a random variable representing general factor (identified with general intelligence) common to all the variables x_i's, ϵ_i is a random variable representing specific factor (identifying specific ability to the particular test) limited to the ith variable x_i, and L_i is an unknown constant. Under the assumptions, $var(f) = \sigma_f^2$, $var(\epsilon_i) = \sigma^2$, $cov(\epsilon_i, \epsilon_j) = 0$ for $i \neq j$, and $cov(\epsilon_i, f) = 0$, the ratios of the correlation coefficients satisfy the constant ratio property. Specifically, for $i \neq j \neq k$, $i = 1, \ldots, p$

$$\frac{cov(x_i, x_j)}{cov(x_i, x_k)} = \frac{L_j}{L_k},$$

is independent of i. Spearman used this observation to justify his theory that "all branches of intellectual activity have in common one fundamental function (or group of functions) whereas the remaining or specific elements of the activity seem in every case to be wholly different from that in all others" (Spearman, 1904, p. 202). The two factors (one common and one specific) were soon found to be inadequate for explaining many practical phenomena. Hence, natural generalizations of this theory to include several common factors and one specific factor were introduced in the literature, leading to what is now commonly known as the factor model.

4.2 Factor Model

Let $\mathbf{x}$ be a p by 1 random vector with mean vector $\boldsymbol{\mu}$ and variance-covariance matrix $\boldsymbol{\Sigma}$. Suppose the interrelationships between the elements of $\mathbf{x}$ can be explained by the *factor model*

$$\mathbf{x} = \boldsymbol{\mu} + \mathbf{Lf} + \boldsymbol{\varepsilon}, \tag{4.2}$$

where $\boldsymbol{\mu}$ is a vector of constants; $\mathbf{f}$ is a random vector of order k by 1 $(k < p)$, with elements $f_1, \ldots, f_k$, which are called the *common factors*; $\mathbf{L}$ is a p by k matrix of unknown constants, called *factor loadings*; and the elements, $\epsilon_1, \ldots, \epsilon_p$ of the p by 1 random vector $\boldsymbol{\varepsilon}$ are called the *specific factors*. It is assumed that the vectors $\mathbf{f}$ and $\boldsymbol{\varepsilon}$ are uncorrelated. Thus the above model implies that a given element of $\mathbf{x}$, say x_i perhaps representing the measurement on certain characteristics, can be viewed as a linear combination of all common factors and one specific factor ϵ_i. Specifically,

$$x_1 = \mu_1 + \ell_{11} f_1 + \cdots + \ell_{1k} f_k + \epsilon_1$$
$$x_2 = \mu_2 + \ell_{21} f_1 + \cdots + \ell_{2k} f_k + \epsilon_2$$
$$\vdots$$
$$x_p = \mu_p + \ell_{p1} f_1 + \cdots + \ell_{pk} f_k + \epsilon_p,$$

where ℓ_{ij}, the $(i, j)^{th}$ element of **L**, is the factor loading of x_i on the jth common factor f_j. If $k = 1$ then the factor model reduces to Spearman's model in Equation 4.1.

In any factor analysis problem an attempt is made to determine the common factors such that the correlations among the components of **x** are completely accounted for by these factors. This amounts to saying, since $cov(\boldsymbol{\varepsilon}, \mathbf{f}) = \mathbf{0}$, that $D(\mathbf{x}) - D(\mathbf{Lf}) = D(\boldsymbol{\varepsilon})$ is a diagonal matrix; that is, the specific factors are uncorrelated.

Under the assumptions

$$E(\mathbf{f}) = \mathbf{0}, E(\boldsymbol{\varepsilon}) = \mathbf{0},$$
$$cov(\mathbf{f}, \boldsymbol{\varepsilon}) = \mathbf{0}, D(\mathbf{f}) = \boldsymbol{\Delta}, \ \boldsymbol{\Delta} \text{ positive definite},$$
$$D(\boldsymbol{\varepsilon}) = \boldsymbol{\Psi} = diag\,(\psi_1, \ldots, \psi_p), \ \psi_i > 0,$$

made on the random quantities in the model given in Equation 4.2, we have,

$$D(\mathbf{x}) = \boldsymbol{\Sigma} = \mathbf{L}\boldsymbol{\Delta}\mathbf{L}' + \boldsymbol{\Psi}.$$

Since **L** and **f** are both unknown, another model equivalent to the model in Equation 4.2 is

$$\mathbf{x} = \boldsymbol{\mu} + \mathbf{L}\boldsymbol{\Delta}^{1/2}\boldsymbol{\Delta}^{-1/2}\mathbf{f} + \boldsymbol{\varepsilon}$$
$$= \boldsymbol{\mu} + \mathbf{L}^*\mathbf{f}^* + \boldsymbol{\varepsilon},$$

where $\mathbf{L}^* = \mathbf{L}\boldsymbol{\Delta}^{1/2}$ and $\mathbf{f}^* = \boldsymbol{\Delta}^{-1/2}\mathbf{f}$. In this form of the model the variance-covariance matrix of **x** is

$$D(\mathbf{x}) = \boldsymbol{\Sigma} = \mathbf{L}^*\mathbf{L}^{*\prime} + \boldsymbol{\Psi}.$$

Hence, without loss of generality, in the model given in Equation 4.2, we can assume that $D(\mathbf{f}) = \mathbf{I}_k$, an identity matrix of order k, which leads to

$$\boldsymbol{\Sigma} = \mathbf{L}\mathbf{L}' + \boldsymbol{\Psi}. \tag{4.3}$$

The model in Equation 4.2 together with the assumption in Equation 4.3 forms the standard factor model. The objective is to determine an **L** and a $\boldsymbol{\Psi}$ such that the assumption specified in Equation 4.3 is satisfied. This determination essentially solves the factor analysis problem.

Note $cov(\mathbf{x}, \mathbf{f}) = \mathbf{L}$, that is, $cov(x_i, f_j) = \ell_{ij}$. This implies that the covariance between the random vector **x** and the vector of common factors **f** is completely determined by the factor loading matrix **L**. Also note that $corr(x_i, f_j) = \ell_{ij}/\sqrt{\sigma_{ii}} = \ell_{ij}$ if $var\,(x_i) = \sigma_{ii} = 1$, that is when $\boldsymbol{\Sigma}$ is in the correlation form. In this case the factor loadings are nothing but the correlation coefficients between the original variables and the common factors.

Suppose k by 1 vectors $\boldsymbol{\ell}_i$ and $\boldsymbol{\ell}_j$, respectively, are the ith and j^{th} rows of **L**. Then for $i \neq j$,

$$\sigma_{ij} = cov(x_i, x_j) = \boldsymbol{\ell}_i'\boldsymbol{\ell}_j = \ell_{i1}\ell_{j1} + \ell_{i2}\ell_{j2} + \cdots + \ell_{ik}\ell_{jk} \text{ and}$$

$$\sigma_{ii} = var(x_i) = \boldsymbol{\ell}_i'\boldsymbol{\ell}_i + \psi_i$$
$$= \ell_{i1}^2 + \ell_{i2}^2 + \cdots + \ell_{ik}^2 + \psi_i$$
$$= h_i^2 + \psi_i,$$

where $h_i^2 = \boldsymbol{\ell}_i'\boldsymbol{\ell}_i$. Thus the variance of x_i is partitioned into two variance components, namely h_i^2 and ψ_i, corresponding to the common factors and specific factor respectively. The quantity ψ_i, the contribution of the specific factor ϵ_i, is called the *uniqueness* or *specific variance*, and the quantity h_i^2, the contribution of common factors, is called *communality of common variance*. Further, l_{i1}^2 is the contribution of the 1st common factor to the common variance, l_{i2}^2 is the contribution of the 2nd common factor to the common variance, and so on.

An illustration. Consider the following correlation matrix

$$\boldsymbol{\rho}_{4\times4} = \begin{bmatrix} 1.00 & .51 & .35 & .20 \\ .51 & 1.00 & .21 & .06 \\ .35 & .21 & 1.00 & .35 \\ .20 & .06 & .35 & 1.00 \end{bmatrix}$$

It may be verified that $\boldsymbol{\rho} = \mathbf{LL}' + \boldsymbol{\Psi}$, with the choices of the factor loading matrix,

$$\mathbf{L}_{4\times2} = \begin{bmatrix} .8 & -.1 \\ .6 & -.3 \\ .5 & .5 \\ .3 & .4 \end{bmatrix}$$

and the specific variance matrix

$$\boldsymbol{\Psi}_{4\times4} = \begin{bmatrix} .35 & 0 & 0 & 0 \\ 0 & .55 & 0 & 0 \\ 0 & 0 & .50 & 0 \\ 0 & 0 & 0 & .75 \end{bmatrix}.$$

Thus $\boldsymbol{\rho}$ has a factor structure produced by $k = 2$ factor model. The communality of x_1 is $h_1^2 = \ell_{11}^2 + \ell_{12}^2 = (.8)^2 + (-.1)^2 = 0.65$. The specific variance for x_1 is $\psi_1 = 0.35$. It is readily seen that $h_1^2 + \psi_1 = 1 = \sigma_{11}$. A similar observation can be made for x_2, x_3, and x_4. Further, $corr(x_1, f_1) = 0.8$, $corr(x_1, f_2) = -0.1$, $corr(x_2, f_1) = 0.6$, and so forth.

In general, the exact representation of the correlation matrix $\boldsymbol{\rho}$ or the variance-covariance matrix $\boldsymbol{\Sigma}$ as $\mathbf{LL}' + \boldsymbol{\Psi}$ may not always be possible. Further, $\boldsymbol{\Sigma} = \mathbf{LL}' + \boldsymbol{\Psi}$ is a useful representation in the context of factor analysis only under certain conditions relating p and k. Specifically, we want the number of unknown parameters in the factor model to be less than that in $\boldsymbol{\Sigma}$. Notice that $\boldsymbol{\Sigma}$ on the left hand side of equation $\boldsymbol{\Sigma} = \mathbf{LL}' + \boldsymbol{\Psi}$ has $p(p+1)/2$ distinct elements, whereas on the right hand side, $\mathbf{L}$ has pk, and $\boldsymbol{\Psi}$ has p distinct elements, giving a total of $pk + p$ parameters. Hence, for a factor model to be useful, it is required that

$$\frac{p(p+1)}{2} - (pk + p) > 0,$$

that is, $p > 2k$.

When $p > 2k$, the exact solution to the equation $\boldsymbol{\Sigma} = \mathbf{LL}' + \boldsymbol{\Psi}$ in terms of $\mathbf{L}$ and $\boldsymbol{\Psi}$ is possible in certain simple cases. For example, suppose $p = 3$, $k = 1$ and

$$x_1 = \ell_1 f + \epsilon_1$$
$$x_2 = \ell_2 f + \epsilon_2$$
$$x_3 = \ell_3 f + \epsilon_3,$$

then

$$\boldsymbol{\Sigma} = \mathbf{LL}' + \boldsymbol{\Psi} = \begin{bmatrix} \ell_1 \\ \ell_2 \\ \ell_3 \end{bmatrix} [\ell_1 \; \ell_2 \; \ell_3] + \begin{bmatrix} \psi_1 & 0 & 0 \\ 0 & \psi_2 & 0 \\ 0 & 0 & \psi_3 \end{bmatrix}.$$

That is,

$$\begin{bmatrix} \sigma_{11} & \sigma_{12} & \sigma_{13} \\ \sigma_{12} & \sigma_{22} & \sigma_{23} \\ \sigma_{13} & \sigma_{23} & \sigma_{33} \end{bmatrix} = \begin{bmatrix} \ell_1^2 & \ell_1\ell_2 & \ell_1\ell_3 \\ \ell_1\ell_2 & \ell_2^2 & \ell_2\ell_3 \\ \ell_1\ell_3 & \ell_2\ell_3 & \ell_3^2 \end{bmatrix} + \begin{bmatrix} \psi_1 & 0 & 0 \\ 0 & \psi_2 & 0 \\ 0 & 0 & \psi_3 \end{bmatrix}.$$

Solving these equations for ℓ_i and ψ_i, $i = 1, 2, 3$, yields

$$\ell_1^2 = \frac{\sigma_{13}\sigma_{12}}{\sigma_{23}}, \quad \ell_2^2 = \frac{\sigma_{23}\sigma_{12}}{\sigma_{13}}, \quad \ell_3^2 = \frac{\sigma_{13}\sigma_{23}}{\sigma_{12}}, \tag{4.4}$$

and

$$\psi_1 = \sigma_{11} - \ell_1^2, \ \psi_2 = \sigma_{22} - \ell_2^2, \ \psi_3 = \sigma_{33} - \ell_3^2. \tag{4.5}$$

Thus at least in this special case the exact solution is easy to obtain. However, in some cases the exact solutions may not be *proper*. For example, consider the correlation matrix

$$\boldsymbol{\rho} = \begin{bmatrix} 1.00 & .75 & .95 \\ .75 & 1.00 & .85 \\ .95 & .85 & 1.00 \end{bmatrix}.$$

Using the formulae given in Equations 4.4 and 4.5 we have $\ell_1^2 = 0.838$, $\ell_2^2 = 0.671$, $\ell_3^2 = 1.077$, $\psi_1 = 1 - \ell_1^2 = 0.162$, $\psi_2 = 0.329$, and $\psi_3 = -0.077$. However ψ_3, being the variance of a random variable (of ϵ_3), cannot be negative. Hence, this solution is not proper. This situation has been called in the literature a case of improper solution or a *Heywood case*. We will discuss the Heywood case in more detail in Section 4.6.

The factor model in Equation 4.2 is not unique in that two different pairs $(\mathbf{L}, \mathbf{f})$ and $(\mathbf{L}^*, \mathbf{f}^*)$ may result in the same covariance structure $\boldsymbol{\Sigma}$. Specifically, suppose $\boldsymbol{\Gamma}$ is any k by k orthogonal matrix. Then model in Equation 4.2 can be rewritten as

$$\begin{aligned} \mathbf{x} &= \boldsymbol{\mu} + \mathbf{L}\boldsymbol{\Gamma}\boldsymbol{\Gamma}'\mathbf{f} + \boldsymbol{\varepsilon} \\ &= \boldsymbol{\mu} + \mathbf{L}^*\mathbf{f}^* + \boldsymbol{\varepsilon}, \end{aligned}$$

where $\mathbf{L}^* = \mathbf{L}\boldsymbol{\Gamma}$ and $\mathbf{f}^* = \boldsymbol{\Gamma}'\mathbf{f}$. Note that $E(\mathbf{f}^*) = E(\boldsymbol{\Gamma}'\mathbf{f}) = \boldsymbol{\Gamma}'E(\mathbf{f}) = \mathbf{0}$, and $D(\mathbf{f}^*) = D(\boldsymbol{\Gamma}'\mathbf{f}) = \boldsymbol{\Gamma}'D(\mathbf{f})\boldsymbol{\Gamma} = \boldsymbol{\Gamma}'\boldsymbol{\Gamma} = \mathbf{I}_k$. Thus any orthogonal transformation of $\mathbf{f}$ will result in a similar covariance structure for $\boldsymbol{\Sigma}$, that is,

$$\begin{aligned} \boldsymbol{\Sigma} &= \mathbf{L}\mathbf{L}' + \boldsymbol{\Psi} \\ &= \mathbf{L}\boldsymbol{\Gamma}\boldsymbol{\Gamma}'\mathbf{L}' + \boldsymbol{\Psi} \\ &= \mathbf{L}^*\mathbf{L}^{*'} + \boldsymbol{\Psi}. \end{aligned}$$

Further, the communalities h_i^2's remain the same, since $\mathbf{L}\mathbf{L}' = \mathbf{L}^*\mathbf{L}^{*'}$. Thus the communalities and more generally the factorization of $\boldsymbol{\Sigma}$ remain invariant of the orthogonal transformation. Hence, the factor loadings are determined only up to an orthogonal transformation of the matrix $\boldsymbol{\Gamma}$. Since any orthogonal transformation is equivalent to rotation of axes, we observe that communalities are not affected by factor rotations.

A uniqueness condition. This problem of nonuniqueness of factor loadings can in fact be to our advantage. Since all factor loadings are related to each other via some orthogonal transformation, we can choose that particular factor loading matrix $\mathbf{L}$ which, in addition to satisfying Equation 4.3, also possesses some other meaningful property. Often we want $\mathbf{L}$ to be a matrix satisfying the property that $\mathbf{L}'\boldsymbol{\Psi}^{-1}\mathbf{L}$ is a diagonal matrix with positive elements. That is,

$$\mathbf{L}'\boldsymbol{\Psi}^{-1}\mathbf{L} = diag(\delta_1, \ldots, \delta_k), \ \text{for some } \delta_i > 0, i = 1, \ldots, k. \tag{4.6}$$

The relevance of this constraint can be justified by the following argument (see Bartholomew (1981) and also Seber (1984, p. 217)). Suppose $\mathbf{f}$ has the k-variate normal distribution with mean vector $\mathbf{0}$ and the covariance matrix $\mathbf{I}_k$. Also suppose that the conditional distribution of $\mathbf{x}$ given $\mathbf{f}$ is a p-variate normal with mean vector $\boldsymbol{\mu}+\mathbf{Lf}$ and the covariance matrix $\boldsymbol{\Psi}$. Then it can be shown that the conditional distribution of $\mathbf{f}$ given $\mathbf{x}$ is k-variate normal with mean vector $\mathbf{L}'\boldsymbol{\Sigma}^{-1}(\mathbf{x}-\boldsymbol{\mu})$ and covariance matrix $[\mathbf{L}'\boldsymbol{\Psi}^{-1}\mathbf{L}+\mathbf{I}_k]^{-1}$. The condition specified by Equation 4.6 ensures that the factor loading matrix $\mathbf{L}$ finally obtained is such that given $\mathbf{x}$, the components of $\mathbf{f}$ are mutually independent.

The constraint given in Equation 4.6 imposes $\frac{k(k-1)}{2}$ linear restrictions (nondiagonal elements of the symmetric matrix $\mathbf{L}'\boldsymbol{\Psi}^{-1}\mathbf{L}$ are all zero) on the number of parameters under the factor model. Hence, a new condition for the factor analysis to be meaningful is that

$$\left\{\frac{p(p+1)}{2}-\left[pk+p-\frac{k(k-1)}{2}\right]\right\}\geq 0. \tag{4.7}$$

For example, from Inequality 4.7 note that for a four-factor model ($k=3$) to be meaningful it is required that $p\geq 6$, whereas for a five-factor model ($k=4$) p is required to be ≥ 8.

4.3 A Difference between PCA and Factor Analysis

Sometimes a clear distinction between principal component analysis and the factor analysis is not made and are both thought to be the same. Although there are some common features between the two techniques a clear distinction exists between the two, as pointed out by Rao (1955).

Principal component analysis is a general technique for the reduction of dimensionality where a random vector $\mathbf{x}$ having the covariance matrix $\boldsymbol{\Sigma}$ is transformed using an orthogonal matrix so that the transformed vector, say $\mathbf{y}$, has uncorrelated components. No structure on the covariance matrix is assumed in the analysis. Reduction in the dimensionality is achieved by dropping some components of $\mathbf{y}$ with negligible variances. If the eigenvalues of $\boldsymbol{\Sigma}$ are distinct, then the orthogonal transformation that is used in principal component analysis can be selected to be unique. For details about the principal component analysis technique and its applications see Chapter 2.

On the other hand, in factor analysis we postulate an underlying structure on a set of measurements ($\mathbf{x}$) in terms of hypothetical variables depending on common factors $\mathbf{f}$ and specific factors $\boldsymbol{\varepsilon}$. Also we make an assumption that there is an underlying linear model that accounts completely for the covariance structure of $\mathbf{x}$. Further, the underlying factors can always be transformed by an orthogonal transformation without changing the form of the model.

More formally, the following distinction between the two techniques exists. Suppose $\mathbf{x}$ is a random vector with covariance matrix $\boldsymbol{\Sigma}$. Let $\boldsymbol{\Gamma}$ be an orthogonal matrix such that for

$$\mathbf{y}=\boldsymbol{\Gamma}'(\mathbf{x}-\boldsymbol{\mu}) \tag{4.8}$$

the variance-covariance matrix of $\mathbf{y}$ is diagonal. Without loss of generality, let this be $\boldsymbol{\Lambda}=\text{diag}(\lambda_1,\ldots,\lambda_p)$, $\lambda_1\geq\cdots\geq\lambda_p$. Then y_i, the ith component of $\mathbf{y}$, is called the ith principal component of $\mathbf{x}$.

Suppose in a principal component analysis, the components $y_1,\ldots,y_k$ are retained, reducing the dimension of the random vector to k from p. Let $\mathbf{y}_1=(y_1,\ldots,y_k)'$ and $\mathbf{y}_2=(y_{k+1},\ldots,y_p)'$. Then $\mathbf{y}=(\mathbf{y}_1',\mathbf{y}_2')'$. Similarly, let $\boldsymbol{\Gamma}=(\boldsymbol{\Gamma}_1,\boldsymbol{\Gamma}_2)$ such that $\mathbf{y}_1=\boldsymbol{\Gamma}_1'(\mathbf{x}-\boldsymbol{\mu})$ and $\mathbf{y}_2=\boldsymbol{\Gamma}_2'(\mathbf{x}-\boldsymbol{\mu})$. Since $\boldsymbol{\Gamma}$ is orthogonal, $\boldsymbol{\Gamma}^{-1}=\boldsymbol{\Gamma}'$. Hence, from Equation 4.8 it follows that

$$\mathbf{x}-\boldsymbol{\mu}=\boldsymbol{\Gamma}\mathbf{y}.$$

That is, we have the following model

$$\begin{aligned}
\mathbf{x} &= \boldsymbol{\mu} + \boldsymbol{\Gamma}\mathbf{y} \\
&= \boldsymbol{\mu} + (\boldsymbol{\Gamma}_1 : \boldsymbol{\Gamma}_2)\begin{pmatrix}\mathbf{y}_1 \\ \mathbf{y}_2\end{pmatrix} \\
&= \boldsymbol{\mu} + \boldsymbol{\Gamma}_1\mathbf{y}_1 + \boldsymbol{\Gamma}_2\mathbf{y}_2 \\
&= \boldsymbol{\mu} + \boldsymbol{\Gamma}_1\boldsymbol{\Lambda}_1^{1/2}\boldsymbol{\Lambda}_1^{-1/2}\mathbf{y}_1 + \boldsymbol{\Gamma}_2\mathbf{y}_2 \\
&= \boldsymbol{\mu} + \mathbf{L}_1\mathbf{f}_1 + \boldsymbol{\eta} \text{ (say)},
\end{aligned} \tag{4.9}$$

where $\boldsymbol{\Lambda}_1 = \text{diag}(\lambda_1, \ldots, \lambda_k)$, $\boldsymbol{\Lambda}_1^{1/2}\boldsymbol{\Lambda}_1^{1/2} = \boldsymbol{\Lambda}_1$, $(\boldsymbol{\Lambda}_1^{1/2})^{-1} = \boldsymbol{\Lambda}_1^{-1/2}$, $\mathbf{L}_1 = \boldsymbol{\Gamma}_1\boldsymbol{\Lambda}_1^{1/2}$, $\mathbf{f}_1 = \boldsymbol{\Lambda}_1^{-1/2}\mathbf{y}_1$, and $\boldsymbol{\eta} = \boldsymbol{\Gamma}_2\mathbf{y}_2$. Note that the model given in Equation 4.9 looks similar to the factor model given in Equation 4.2. Also note that

$$D(\mathbf{f}_1) = D(\boldsymbol{\Lambda}_1^{-1/2}\mathbf{y}_1) = \boldsymbol{\Lambda}_1^{-1/2}D(\mathbf{y}_1)\boldsymbol{\Lambda}_1^{-1/2} = \mathbf{I}_k,$$

since $D(\mathbf{y}_1) = \boldsymbol{\Lambda}_1 = \boldsymbol{\Lambda}_1^{1/2}\boldsymbol{\Lambda}_1^{1/2}$.

Next,

$$cov(\mathbf{f}_1, \boldsymbol{\eta}) = \boldsymbol{\Lambda}_1^{-1/2}cov(\mathbf{y}_1, \mathbf{y}_2)\boldsymbol{\Gamma}_2' = \mathbf{0},$$

since the components of $\mathbf{y}$ are uncorrelated. But unlike the factor model in Equation 4.2, the covariance matrix of specific factor vector $\boldsymbol{\eta}$, that is,

$$D(\boldsymbol{\eta}) = D(\boldsymbol{\Gamma}_2\mathbf{y}_2) = \boldsymbol{\Gamma}_2\boldsymbol{\Lambda}_2\boldsymbol{\Gamma}_2'$$

is not a diagonal matrix. This indicates that the principal component model in Equation 4.9 does not completely explain the covariance structure in $\mathbf{x}$, unlike the factor model of Equation 4.2.

The most important aspect of the factor analysis is to estimate the factor loadings and specific variances. Let $\mathbf{x}_1, \ldots, \mathbf{x}_n$ be a sample of size n from a p variate distribution with $E(\mathbf{x}) = \boldsymbol{\mu}$ and $D(\mathbf{x}) = \boldsymbol{\Sigma}$. The analysis consists of two parts:

- identification of the structure and
- estimation of the parameters.

Once an appropriate factor model is identified and the parameters involved are estimated, it is customary to give practical meaning and interpretations to the estimated factor loadings. Estimated factor loadings are usually rotated (transformed using an orthogonal transformation), in view of nonuniqueness of the solutions, to help interpret them.

Identification of the structure simply amounts to finding an appropriate k ($< p$) such that the factor model in Equation 4.2 with this k holds for $\mathbf{x}$. Most of the time, selection of k is subjective, although in some cases, for example in the maximum likelihood method of estimation, a systematic approach using testing of hypothesis is possible. Some of the ideas for selecting the numbers of principal components in principal component analysis can be used here for selecting the number of common factors, viz. (k).

Solving a factor analysis problem appropriately is a difficult task. This is the reason, we believe, there are many different approaches and viewpoints for solving this problem. For ease of discussion and presentation, we divide the various methods available in the literature into two broad categories (a) noniterative methods, and (b) iterative methods. Most of these methods can be adopted using the METHOD = option in the FACTOR procedure in SAS/STAT software.

As is the case with principal component analysis, the starting point for the factor analysis is the sample variance-covariance matrix or the sample correlation matrix of the data.

All the methods discussed in this chapter can be used either with a variance-covariance matrix or with a correlation matrix. Since many of these methods are scale-invariant in most factor analysis applications, the starting point is often taken as a correlation matrix. However, for the methods that are not scale-invariant, the results of the analysis based on the correlation matrix will not always be the same as those based on the variance-covariance matrix.

4.4 Noniterative Methods of Estimation

By taking the sample correlation matrix $\mathbf{R}$ as a starting point, we will discuss four methods in this category, namely the principal component method, the principal factor method, image analysis, and Harris' noniterative canonical factor analysis.

4.4.1 Principal Component Method

The principal component method for factor analysis is one of the most simple. Suppose $\mathbf{R}$ is the p by p sample correlation matrix. Since $\mathbf{R}$ is symmetric and positive definite, it can be written as

$$\mathbf{R} = \mathbf{\Gamma}\mathbf{\Lambda}\mathbf{\Gamma}',$$

where $\mathbf{\Lambda} = \text{diag}(\lambda_1, \dots, \lambda_p)$, $\lambda_1 \geq, \dots, \geq \lambda_p > 0$ are the eigenvalues, and $\mathbf{\Gamma}\mathbf{\Gamma}' = \mathbf{\Gamma}'\mathbf{\Gamma} = \mathbf{I}_p$; that is, $\mathbf{\Gamma}$ is a p by p orthogonal matrix containing the eigenvectors $\mathbf{\Gamma}_1, \dots, \mathbf{\Gamma}_p$ corresponding to the eigenvalues $\lambda_1, \dots, \lambda_p$. Let k be the number of principal components selected using some meaningful criterion, such as the minimum number of principal components that were able to explain more than a certain percentage of total variability as measured by the total variance. We define a p by k matrix $\hat{\mathbf{L}}$ as

$$\hat{\mathbf{L}} = [\sqrt{\lambda_1}\mathbf{\Gamma}_1 : \cdots : \sqrt{\lambda_k}\mathbf{\Gamma}_k] \tag{4.10}$$

then approximate $\mathbf{R}$ by

$$\hat{\mathbf{L}}\hat{\mathbf{L}}' = \sum_{i=1}^{k} \lambda_i \mathbf{\Gamma}_i \mathbf{\Gamma}_i',$$

where $\mathbf{\Gamma}_i$ is the ith column of $\mathbf{\Gamma}$.

Thus $\hat{\mathbf{L}} = (\hat{\ell}_{ij})$ given in Equation 4.10 serves as an estimate of the factor loading matrix $\mathbf{L}$. The diagonal matrix of specific variance $\mathbf{\Psi}$ is estimated by $\hat{\mathbf{\Psi}}$, the matrix of diagonal elements of $\mathbf{R} - \hat{\mathbf{L}}\hat{\mathbf{L}}'$. Specifically,

$$\hat{\mathbf{\Psi}} = diag(1 - h_1^2, \dots, 1 - h_p^2),$$

where $h_i^2 = \sum_{j=1}^{k} \hat{\ell}_{ij}^2, \quad i = 1, \dots, p$.

Thus we have a k factor model with $\mathbf{L}$ estimated by $\hat{\mathbf{L}}$ and $\mathbf{\Psi}$ estimated by $\hat{\mathbf{\Psi}}$ and thereby giving the approximation of $\mathbf{R}$ as

$$\mathbf{R} \approx \hat{\mathbf{L}}\hat{\mathbf{L}}' + \hat{\mathbf{\Psi}}.$$

In practice, we must also validate the model. Thus, before accepting $\hat{\mathbf{L}}$ and $\hat{\mathbf{\Psi}}$ as the final estimates, the *residual matrix*

$$\mathbf{Res} = \mathbf{R} - (\hat{\mathbf{L}}\hat{\mathbf{L}}' + \hat{\mathbf{\Psi}})$$

is computed, and the magnitude of its elements is examined using certain statistical or intuitive measures.

The residual matrix **Res** always has zeros on the diagonals. In the most ideal case $\mathbf{Res} = \mathbf{0}$. Therefore, intuitively speaking, if the off-diagonal elements are also close to zero then the estimates $\hat{\mathbf{L}}$ and $\hat{\boldsymbol{\Psi}}$ of $\mathbf{L}$ and $\boldsymbol{\Psi}$, respectively, are considered to be reasonably good and acceptable. This motivates the following statistical measure for assessing the quality of estimation.

Formally, if k is selected such that $\sum_{i=1}^{k} \lambda_i$, the total variance explained by the first k eigenvectors is large (close to p), then $\sum_{i=k+1}^{p} \lambda_i$ will be small (close to 0). But we can show that the sum of squares of all the elements of the residual matrix **Res** is always less than or equal to $\sum_{i=k+1}^{p} \lambda_i$ (Rao, 1964). Thus if $\sum_{i=k+1}^{p} \lambda_i$ is small it also ensures that the entries of the residual matrix will be small. Therefore,

$$\frac{\sum_{i=k+1}^{p} \lambda_i}{\sum_{i=1}^{p} \lambda_i} = \frac{\sum_{i=k+1}^{p} \lambda_i}{p}$$

can be taken as a measure of the goodness of fit of the factor model. Smaller values of this measure indicate good fit. It may be mentioned that if the variance-covariance matrix instead of correlation matrix is used then $\sum_{i=1}^{p} \lambda_i$ is not necessarily equal to p.

EXAMPLE 1 *Principal Component Method, Spearman's Data* The previously introduced correlation matrix of the scores on the tests in Classics (C), French (F), English (E), Mathematics (M), Discrimination of pitch (D), and Music (MU) for 33 boys considered by Spearman (1904) is further considered here. This correlation matrix is first stored in a SAS data set named SPEARMAN, as shown in Program 4.1.

```
/* Program 4.1 */

options ls=64 ps=45 nodate nonumber;
title1 'Output 4.1';
title2 "Factor Analysis of Spearman's Data";

data spearman (type=corr);
_type_='corr';
if _n_=1 then _type_='N';
infile cards missover;
input _name_ $ c f e m d mu;
lines;
n 33
c 1.0
f .83 1.0
e .78 .67 1.0
m .70 .67 .64 1.0
d .66 .65 .54 .45 1.0
mu .63 .57 .51 .51 .40 1.0
;
/* Source: Spearman (1904). */

proc factor data=spearman method=prin res scree;
var c f e m d mu;
title3 'Principal Component Method';
run;
proc factor data=spearman method=prin res nfact=2;
title3 'PC Method: Res matrix for nfact=2';
run;
proc factor data=spearman method=principal res nfact=3;
title3 'PC Method: RMS values for nfact=3';
run;
```

```
proc factor data=spearman method=principal res n=4;
title3 'PC Method: RMS values for nfact=4';
run;
proc factor data=spearman method=principal priors=smc res;
title3 'Principal Factor Method: Priors=SMC';
run;
proc factor data=spearman method=principal priors=smc
nfact=3 res;
title3 'Principal Factor Method: Priors=SMC, nfact=3';
run;
proc factor data=spearman method=principal priors=max res;
title3 'Principal Factor Method: Priors=MAX';
run;
proc factor data=spearman method=principal priors=max
nfact=3 res;
title3 'Principal Factor Method: Priors=MAX, nfact=3';
run;
proc factor data=spearman method=principal priors=asmc res;
title3 'Principal Factor Method: Priors=ASMC';
run;
proc factor data=spearman method=principal priors=asmc
nfact=3 res;
title3 'Principal Factor Method: Priors=ASMC, nfact=3';
run;

proc factor data=spearman method=image res;
title3 'Image Analysis: Method=Image';
run;

proc factor data=spearman method=harris res;
title3 'Canonical Factor Analysis: Method=Harris';
run;
proc factor data=spearman method=harris n=3 res;
title3 'Canonical Factor Analysis: Method=Harris, nfact=3';
run;
```

Output 4.1

```
                      Output 4.1
           Factor Analysis of Spearman's Data
               Principal Component Method

                  The FACTOR Procedure
       Initial Factor Method: Principal Components

           Prior Communality Estimates: ONE


  Eigenvalues of the Correlation Matrix: Total = 6  Average = 1

        Eigenvalue    Difference    Proportion    Cumulative

   1    4.10277064    3.48366093        0.6838        0.6838
   2    0.61910971    0.10737750        0.1032        0.7870
   3    0.51173221    0.15468251        0.0853        0.8723
   4    0.35704970    0.08664189        0.0595        0.9318
   5    0.27040780    0.13147786        0.0451        0.9768
   6    0.13892994                      0.0232        1.0000

     1 factor will be retained by the MINEIGEN criterion.
```

Output 4.1
Factor Analysis of Spearman's Data
Principal Component Method

The FACTOR Procedure
Initial Factor Method: Principal Components

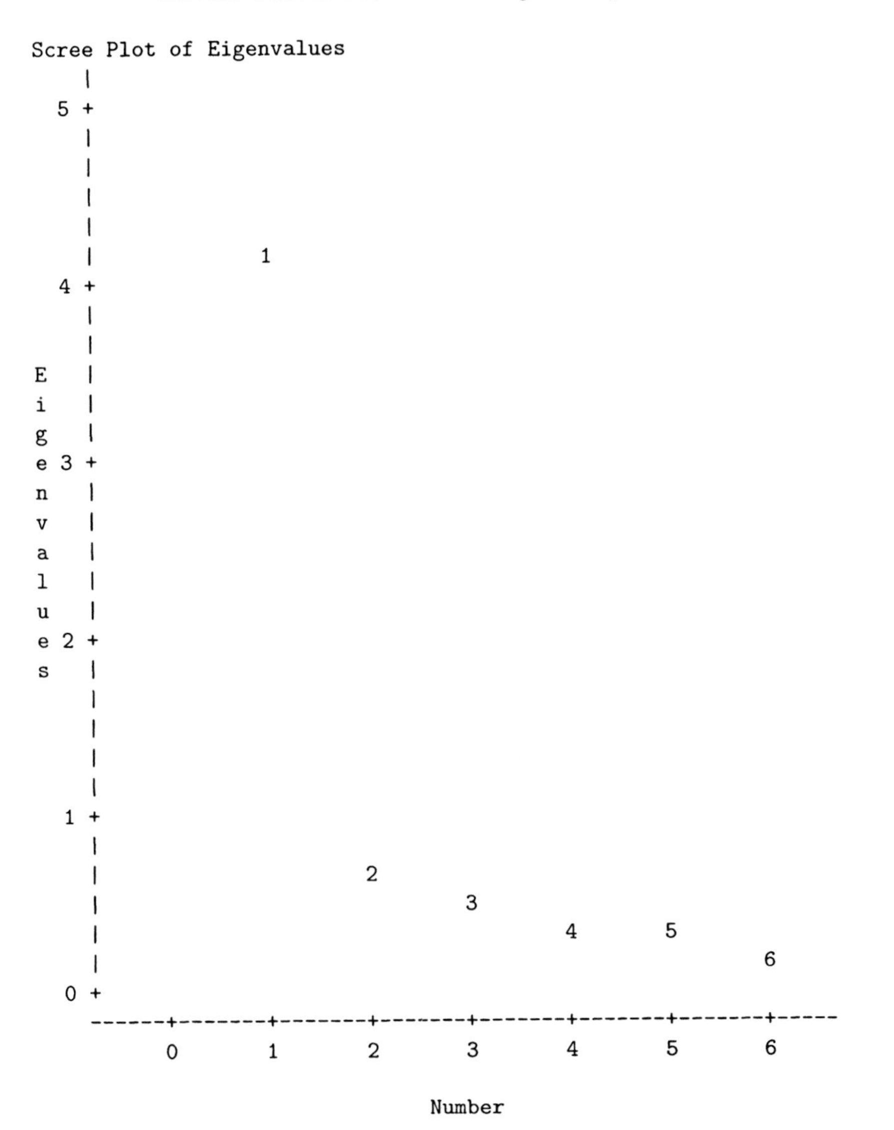

```
                    Output 4.1
         Factor Analysis of Spearman's Data
            Principal Component Method

               The FACTOR Procedure
     Initial Factor Method: Principal Components

                  Factor Pattern

                         Factor1

              c          0.93653
              f          0.89396
              e          0.84204
              m          0.80416
              d          0.74257
              mu         0.72070


          Variance Explained by Each Factor

                     Factor1

                    4.1027706


     Final Communality Estimates: Total = 4.102771

              c              f               e

      0.87708211     0.79915902      0.70902893

              m              d              mu

      0.64668064     0.55141460      0.51940533

 Residual Correlations With Uniqueness on the Diagonal

                     c              f               e

 c             0.12292       -0.00721        -0.00859
 f            -0.00721        0.20084        -0.08275
 e            -0.00859       -0.08275         0.29097
 m            -0.05312       -0.04889        -0.03714
 d            -0.03544       -0.01383        -0.08528
 mu           -0.04495       -0.07427        -0.09686

 Residual Correlations With Uniqueness on the Diagonal

                     m              d              mu

 c            -0.05312       -0.03544        -0.04495
 f            -0.04889       -0.01383        -0.07427
 e            -0.03714       -0.08528        -0.09686
 m             0.35332       -0.14715        -0.06956
 d            -0.14715        0.44859        -0.13517
 mu           -0.06956       -0.13517         0.48059
```

```
Root Mean Square Off-Diagonal Residuals: Overall = 0.07481719

          c              f              e

 0.03528300     0.05476631     0.07064382

Root Mean Square Off-Diagonal Residuals: Overall = 0.07481719

          m              d             mu

 0.08134260     0.09863417     0.08947381


              PC Method: Res matrix for nfact=2

Root Mean Square Off-Diagonal Residuals: Overall = 0.07714068

          c              f              e

 0.03456495     0.05099583     0.07064357

Root Mean Square Off-Diagonal Residuals: Overall = 0.07714068

          m              d             mu

 0.08364814     0.08666589     0.11140151


              PC Method: RMS values for nfact=3

Root Mean Square Off-Diagonal Residuals: Overall = 0.05405025

          c              f              e

 0.03428257     0.05098906     0.07151290

Root Mean Square Off-Diagonal Residuals: Overall = 0.05405025

          m              d             mu

 0.07465358     0.05109302     0.02134595


              PC Method: RMS values for nfact=4

Root Mean Square Off-Diagonal Residuals: Overall = 0.03844670

          c              f              e

 0.03218523     0.05129305     0.02787280

Root Mean Square Off-Diagonal Residuals: Overall = 0.03844670

          m              d             mu

 0.04252327     0.04657046     0.02116835
```

The SAS code

```
proc factor data = spearman method=principal;
var c f e m d mu;
run;
```

can be used to implement the principal component method for determining the factors. The option METHOD = PRINCIPAL (or METHOD = PRIN or simply METHOD = P) in the PROC FACTOR statement is also a default option and so can be omitted.

The FACTOR procedure provides three choices for selecting the number of factors.

(a) Use option NFACTORS = (or NFACT = or simply N =) to request the required number of factors.

(b) Use the option MINEIGEN = (or MIN=) to select all the factors corresponding to the eigenvalues greater than or equal to a specified value. The default value is 1 if a correlation matrix is used. Otherwise it is the average of the eigenvalues.

(c) Use the option PROPORTION = (or P =) to select all the factors needed to account for a certain proportion of common variance. Also if desired, the option PERCENT = instead of the option PROPORTION = can be used. The default values for the PERCENT = option is 100 (it is 1 for PROPORTION = option).

If any two or more of these three options are specified, the number of factors selected by the FACTOR procedure is the minimum of the numbers of factors selected by the criteria specified.

We see from Output 4.1 that for the 6 by 6 correlation matrix used in the data set SPEARMAN, the eigenvalues are 4.1028, 0.6191, 0.5117, 0.3570, 0.2704, and 0.1389. Only one factor ($k = 1$) is selected since the default value for the option MINEIGEN = is 1. With $p = 6$, the vector $\boldsymbol{\ell} = (\ell_1, \ldots, \ell_p)'$ of factor loadings for this factor is given by

$$\boldsymbol{\ell} = (.9365, .8940, .8420, .8042, .7426, .7207)'.$$

Since $k = 1$, the communality estimates $h_i^2 = \ell_i^2$, $i = 1, \ldots, 6$ corresponding to the six variables respectively are 0.8771, 0.7992, 0.7090, 0.6467, 0.5514, and 0.5194. Thus for example, $h_1^2 = \ell_1^2 = (0.9365)^2 = 0.8771$. The estimate of total communality (reported as Final Communality Estimates: Total) is given by

$$\sum_{i=1}^{p} h_i^2 = \sum_{i=1}^{p} \ell_i^2 = 4.1028.$$

This is only about 68.4% of the total variability, which is 6 in our case. The specific variances $\psi_i = 1 - h_i^2, i = 1, \ldots, 6$ for the six variables respectively are

$$.1229 \quad .2008 \quad .2910 \quad .3533 \quad .4486 \quad .4806.$$

All this information has been summarized in the following table.

TABLE 4.1 A Summary of Factor Analysis Results

Variables →	C	F	E	M	D	MU
Factor Loading	.9365	.8940	.8420	.8042	.7426	.7207
Communality	.8771	.7992	.7090	.6467	.5514	.5194
Specific Variance	.1229	.2008	.2910	.3533	.4486	.4806
Root Mean Square	.0353	.0548	.0706	.0813	.0986	.0895

The residual matrix $\mathbf{Res} = \mathbf{R} - (\mathbf{LL}' + \mathbf{\Psi})$ gives an indication of how well the factor model fits the data. This matrix can be obtained using RESIDUALS (or RES) option in the PROC FACTOR statement.

From Output 4.1 the following residual matrix is obtained.

$$\mathbf{Res} = (res_{ij}) = \begin{bmatrix} .0000 & -.0072 & -.0086 & -.0531 & -.0354 & -.0450 \\ -.0072 & .0000 & -.0828 & -.0489 & -.0138 & -.0743 \\ -.0086 & -.0828 & .0000 & -.0371 & -.0853 & -.0969 \\ -.0531 & -.0489 & -.0371 & .0000 & -.1472 & -.0696 \\ -.0354 & -.0138 & -.0853 & -.1472 & .0000 & -.1352 \\ -.0450 & -.0743 & -.0969 & -.0696 & -.1352 & .0000 \end{bmatrix}.$$

Note that the corresponding matrix in Output 4.1 contains nonzero elements on the diagonals. It is so because, as a compact way of conveying information, SAS places the specific variances (ψ_i) on the diagonals of this matrix. The overall root mean square (RMS_overall) is the sum of the off-diagonal entries of the residual matrix **Res**. That is,

$$\text{RMS\_overall} = \sqrt{\frac{1}{p(p-1)} \sum_{i}^{p} \sum_{j \neq i}^{p} res_{ij}^2}. \tag{4.11}$$

The value of RMS_overall is also reported in Output 4.1 and is equal to 0.0748. Although there is no commonly accepted rule for determining the number of common factors based on the value of RMS_overall, we suggest a cutoff value of 0.05 for RMS_overall. That is, a model that produces an RMS_overall value of less than or equal to 0.05 with the smallest number of common factors can be taken as the best model according to this criterion.

Also provided in the same output are the root mean sum of squares of the elements corresponding to each variable. These values corresponding to C, F, E, M, D, and Mu respectively are 0.0353, 0.0548, 0.0706, 0.0813, 0.0986, and 0.0895. The first value, RMS_1 for example, is calculated as

$$\begin{aligned} \text{RMS}_1 &= \sqrt{\frac{1}{p-1} \sum_{j \neq 1}^{p} res_{1j}^2} \\ &= \sqrt{\frac{1}{5}[(-.0072)^2 + (-.0086)^2 + (-.0531)^2 + (-.0354)^2 + (-.0450)^2]} \\ &= .0353. \end{aligned}$$

The RMS values corresponding to various variables are also reported in Table 4.1.

There may be other factors, whose inclusion in the model may increase the percentage of total variability explained by the total communality of the factors. However, this additional inclusion may (or may not) result in smaller root mean squares.

An indication of the number of factors to be included in the factor model may sometimes be obtained using a graph called a *Scree* diagram. It is a graphical display of the eigenvalues $\hat{\lambda}_k$ of the sample correlation matrix $\mathbf{R}$ against k, $k = 1, \ldots, p$. Using this graph, k, the number of factors, is selected such that the slope of the graph is steep to the left of k but at the same time not very steep to its right. The scree diagram was also used in Chapter 2 to select the appropriate number of principal components. In SAS, a scree diagram can be obtained using the SCREE option in the PROC FACTOR statement.

An examination of the scree diagram, provided in Output 4.1, indicates that at least two factors are needed in the factor model. Thus we use the NFACT=2 option in PROC FACTOR statement. The RMS_overall value (Equation 4.11) in this case is found to be 0.0771, indicating that the factor model with two common factors did not fare as well for these data. In fact, this value is much larger than that corresponding to the case $k = 1$. Further, the percent of total variability accounted for by the two factors, namely,

$$100 \times \frac{\lambda_1 + \lambda_2}{\lambda_1 + \cdots + \lambda_6}$$

is still only 78.7%. When we use NFACT=3 and NFACT=4, RMS_overall values for the three and four common factors, respectively, are 0.0541 and 0.0384, which are the considerable improvement from 0.0748 for a factor model with one common factor and 0.0771 for a factor model with two common factors. Moreover, the final total communality estimates are 5.23 and 5.59, respectively, which are 87% and 93% of the total variability explained by the communality part of the factors.

For comparison purposes the following table provides the RMS_overall and RMS_i values for $k = 1, \ldots, 4$.

TABLE 4.2 The Individual and Overall RMS Values

Variables→	C	F	E	M	D	MU	Overall
$k = 1$	.0353	.0548	.0706	.0813	.0986	.0895	.0748
$k = 2$	.0346	.0510	.0706	.0836	.0867	.1114	.0771
$k = 3$	.0343	.0510	.0715	.0747	.0511	.0213	.0541
$k = 4$	.0322	.0513	.0279	.0425	.0466	.0212	.0384

4.4.2 Principal Factor Method

This method is based on the application of the principal component method to the pseudo correlation matrix (or pseudo variance-covariance matrix) with corresponding communalities replacing the diagonal elements.

Suppose the factor structure given in Equation 4.3, $\mathbf{R} = \mathbf{LL}' + \mathbf{\Psi}$, holds for some k, where, as earlier, k is the number of factors. Then the *pseudo correlation matrix* $\mathbf{R} - \mathbf{\Psi} = \mathbf{LL}'$ is a symmetric positive semi-definite matrix with rank k. Our objective is to estimate $\mathbf{L}$. Using the spectral decomposition of $\mathbf{R} - \mathbf{\Psi}$ we write

$$\mathbf{R} - \mathbf{\Psi} = \mathbf{\Gamma \Lambda \Gamma}',$$

where $\mathbf{\Lambda} = diag(\lambda_1, \ldots, \lambda_k, 0, \ldots, 0)$, $\lambda_1 \geq \cdots \geq \lambda_k > 0$, are the eigenvalues and $\mathbf{\Gamma}$ is the matrix of the corresponding eigenvectors of $\mathbf{R} - \mathbf{\Psi}$. Thus the matrix $\mathbf{L}$ can be defined as

$$\mathbf{L} = [\sqrt{\lambda_1}\mathbf{\Gamma}_1 : \cdots : \sqrt{\lambda_k}\mathbf{\Gamma}_k],$$

where $\mathbf{\Gamma}_i$ is the ith column of $\mathbf{\Gamma}$. In practice for the observed $\mathbf{R}$ and a given $\mathbf{\Psi} = \hat{\mathbf{\Psi}}$ (an estimate), $\mathbf{R} - \hat{\mathbf{\Psi}}$ may be of full rank. Thus with $\hat{\mathbf{\Gamma}}$ estimating $\mathbf{\Gamma}$ and $\hat{\mathbf{\Lambda}}$ estimating $\mathbf{\Lambda}$ through the decomposition

$$\mathbf{R}_a = \mathbf{R} - \hat{\mathbf{\Psi}} = \hat{\mathbf{\Gamma}}\hat{\mathbf{\Lambda}}\hat{\mathbf{\Gamma}}',$$

we estimate $\mathbf{L}$ as

$$\hat{\mathbf{L}} = [\sqrt{\hat{\lambda}_1}\hat{\mathbf{\Gamma}}_1 : \cdots : \sqrt{\hat{\lambda}_k}\hat{\mathbf{\Gamma}}_k],$$

where $\hat{\mathbf{\Gamma}}_i$ is the ith column of $\hat{\mathbf{\Gamma}}$. Thus we have estimated the matrix $\mathbf{L}$ by $\hat{\mathbf{L}}$ approximately satisfying the factor structure

$$\mathbf{R} \approx \hat{\mathbf{L}}\hat{\mathbf{L}}' + \hat{\boldsymbol{\Psi}},$$

given $\boldsymbol{\Psi} = \hat{\boldsymbol{\Psi}}$.

Suppose the correlation matrix $\mathbf{R}$ and $\hat{\boldsymbol{\Psi}}$, the diagonal matrix of estimated specific variances, is given. Then using $\hat{h}_i^2 = 1 - \hat{\psi}_i, \quad i = 1, \ldots, p$ we can obtain the spectral decomposition of

$$\mathbf{R}_a = \mathbf{R} - \hat{\boldsymbol{\Psi}} = \begin{bmatrix} \hat{h}_1^2 & r_{12} & \cdots & r_{1p} \\ r_{12} & \hat{h}_2^2 & \cdots & r_{2p} \\ \vdots & & & \\ r_{1p} & r_{2p} & \cdots & \hat{h}_p^2 \end{bmatrix}, \tag{4.12}$$

say, $\mathbf{R}_a = \hat{\boldsymbol{\Gamma}}\hat{\boldsymbol{\Lambda}}\hat{\boldsymbol{\Gamma}}'$. Let k' be the number of nonzero diagonal elements in $\hat{\boldsymbol{\Lambda}}$. Let them be $\hat{\lambda}_1, \ldots, \hat{\lambda}_{k'}$, and $\hat{\boldsymbol{\Gamma}}_1, \ldots, \hat{\boldsymbol{\Gamma}}_{k'}$ be the corresponding eigenvectors of the matrix $\mathbf{R}_a$. Although $\mathbf{R} - \boldsymbol{\Psi}$ is positive semi-definite with rank k, $\mathbf{R}_a = \mathbf{R} - \hat{\boldsymbol{\Psi}}$ can have some negative eigenvalues. We take k' as an estimate of k. In the principal factor method, $\hat{k}$, the estimate of k, may be selected so as to have $\hat{k} < k' < k$ and $\sum_{i=1}^{\hat{k}} \hat{\lambda}_i$ close to the estimated total communality $\sum_{i=1}^{p} h_i^2 = tr\ (\mathbf{R} - \hat{\boldsymbol{\Psi}})$. Accordingly the matrix $\mathbf{L}$ will be estimated by

$$\hat{\mathbf{L}} = [\sqrt{\hat{\lambda}_1}\hat{\boldsymbol{\Gamma}}_1 : \cdots : \sqrt{\hat{\lambda}_{\hat{k}}}\hat{\boldsymbol{\Gamma}}_{\hat{k}}].$$

The estimation of $\mathbf{L}$ described above was performed for a given $\hat{\boldsymbol{\Psi}}$ or equivalently for the given values of communalities h_i^2. In SAS the principal factor method can be adopted by providing the prior values of the estimated communalities, $\hat{h}_i^2 = 1 - \hat{\psi}_i, \ i = 1, \ldots, p$. These values can be stored in a SAS data set and used for the analysis using the PRIORS=INPUT (or PRIORS=I) option in the PROC FACTOR statement. For example, the SAS code

```
data commun (type=factor);
if _n_=1 then _type_='PRIORS';
else if _n_=2 then _type_='N';
else _type_='corr';
infile cards missover;
input _name_ $ c f e m d mu;
datalines;
val .2 .3 .1 .2 .1 .3
n 33
c 1.0
f .83 1.0
e .78 .67 1.0
m .70 .67 .64 1.0
d .66 .65 .54 .45 1.0
mu .63 .57 .51 .51 .40 1.0
;
proc factor data=commun method=prin priors=input;
var c f e m d mu;
run;
```

uses the principal factor method to perform factor analysis starting from a pseudo correlation matrix whose diagonal entries are the communalities provided in the SAS data set COMMUN. The prior communalities must be provided as the first observation using _TYPE_ = 'PRIORS' or _TYPE_ = 'COMMUNAL' in the DATA step, which must be TYPE=FACTOR as shown above.

If so desired, the communalities can also be randomly generated from the uniform distribution on the interval (0,1) and then used to perform the factor analysis. The option PRIORS = RANDOM (or PRIORS = R) in the PROC FACTOR statement can be used for this purpose. If PRIORS=ONE is specified, which is also the default value, then the principal component method is used for the analysis.

In addition to the choices described earlier, there are several methods of estimating ψ_i, the specific variances, or the h_i^2, communalities from the data. Some of these methods are described here.

Estimation using squared multiple correlation (SMC). Suppose $\mathbf{R}^{-1} = (r^{ij})$ is the inverse of the correlation matrix $\mathbf{R}$. A popular approach to estimate the specific variances ψ_i is by taking $\hat{\psi}_i = \frac{1}{r^{ii}}$ $i = 1, \ldots, p$. This is clearly equivalent to estimating the ith communality h_i^2 by $\hat{h}_i^2 = 1 - \hat{\psi}_i = 1 - \frac{1}{r^{ii}} = \text{SMC}_i$ (say), which is equal to the sample squared multiple correlation coefficient between x_i, the ith component of vector of variables $\mathbf{x}$, and the remaining $(p-1)$ components of $\mathbf{x}$. The pseudo correlation matrix $\mathbf{R}_a$ in Equation 4.12, replacing h_i^2 by $\hat{h}_i^2$ on the diagonal, is now used for performing the factor analysis. In SAS, by using the PRIORS=SMC option in PROC FACTOR statement this choice of communality estimates can be implemented whenever the principal factor method is used. The total communality is then estimated by $\sum_{i=1}^{p} SMC_i$.

Estimation using maximum absolute correlation. Another intuitive method of estimating the ith communality h_i^2 is by

$$r_{max}^{(i)} = \max_{i \neq j} |r_{ij}|,$$

the maximum of the absolute value of the correlation coefficients in the off-diagonal elements of the ith row of $\mathbf{R}$. That is,

$$\hat{h}_i^2 = r_{max}^{(i)}, i = 1, \ldots, p.$$

The option PRIORS=MAX (or PRIORS=M) in the PROC FACTOR statement can be used to adopt this estimator.

Estimation using adjusted squared multiple correlation (ASMC). Cureton and D'Agostino (1983) point out that $\sum_{i=1}^{p} r_{max}^{(i)}$ is a better estimator of the total communality $\sum_{i=1}^{p} h_i^2$ than is $\sum_{i=1}^{p} SMC_i$. Hence, they suggest an adjustment to the squared multiple correlation for estimating h_i^2 such that $\sum_{i=1}^{p} h_i^2$ is still estimated by $\sum_{i=1}^{p} r_{max}^{(i)}$. Specifically, in this case we estimate h_i^2 by

$$\hat{h}_i^2 = \text{ASMC}_i = \frac{\sum_{j=1}^{p} r_{max}^{(j)}}{\sum_{j=1}^{p} \text{SMC}_j} \text{SMC}_i, i = 1, \ldots, p.$$

Consequently, the total communality, $\sum_{i=1}^{p} h_i^2$, is estimated by $\sum_{i=1}^{p} r_{max}^{(i)}$.

In SAS, the option PRIORS=ASMC in the PROC FACTOR statement can be specified, to use this estimate in $\mathbf{R}_a$ of Equation 4.12 and to perform the factor analysis using the principal factor method.

EXAMPLE 2 ***Principal Factor Method, Spearman's Data (continued)*** Returning to Spearman's data in the form of a correlation matrix, suppose the interest is to extract factors using the principal factor method with squared multiple correlations (SMC) as the prior communalities. The SAS code

```
proc factor data=speareman method=prin priors=smc res;
var c f e m d mu;
run;
```

(also provided in Program 4.1) performs the intended task.

Output 4.1
(continued)

```
                    Output 4.1
         Factor Analysis of Spearman's Data
         Principal Factor Method: Priors=SMC

                The FACTOR Procedure
        Initial Factor Method: Principal Factors

         Prior Communality Estimates: SMC

              c             f             e

     0.81416371    0.72610719    0.62713210

              m             d            mu

     0.54308031    0.47561114    0.41184072


      Eigenvalues of the Reduced Correlation Matrix:
        Total = 3.59793517  Average = 0.59965586

       Eigenvalue   Difference   Proportion   Cumulative

   1   3.73338880   3.63524819       1.0376       1.0376
   2   0.09814061   0.08780901       0.0273       1.0649
   3   0.01033160   0.04019956       0.0029       1.0678
   4   -.02986796   0.06688801      -0.0083       1.0595
   5   -.09675598   0.02054593      -0.0269       1.0326
   6   -.11730190                    -0.0326       1.0000

   1 factor will be retained by the PROPORTION criterion.

       Final Communality Estimates: Total = 3.733389

              c             f             e

     0.87768219    0.76484345    0.64504099

              m             d            mu

     0.56244212    0.46321485    0.42016520

 Root Mean Square Off-Diagonal Residuals: Overall = 0.03016521


       Principal Factor Method: Priors=SMC, nfact=3

       Final Communality Estimates: Total = 3.841861
```

```
              c              f              e

     0.87785050     0.77435789     0.65565779

              m              d             mu

     0.58503215     0.51350492     0.43545776

Root Mean Square Off-Diagonal Residuals: Overall = 0.02086627


        Principal Factor Method: Priors=MAX

              The FACTOR Procedure
      Initial Factor Method: Principal Factors

        Prior Communality Estimates: MAX

              c              f              e

     0.83000000     0.83000000     0.78000000

              m              d             mu

     0.70000000     0.66000000     0.63000000

      Eigenvalues of the Reduced Correlation Matrix:
           Total = 4.43  Average = 0.73833333

       Eigenvalue   Difference   Proportion   Cumulative

  1    3.85706762   3.57955878       0.8707       0.8707
  2    0.27750884   0.08649482       0.0626       0.9333
  3    0.19101402   0.07474996       0.0431       0.9764
  4    0.11626406   0.08723223       0.0262       1.0027
  5    0.02903183   0.06991820       0.0066       1.0092
  6    -.04088637                   -0.0092       1.0000

  4 factors will be retained by the PROPORTION criterion.

      Final Communality Estimates: Total = 4.441855

              c              f              e

     0.85894965     0.82296494     0.78382648

              m              d             mu

     0.69251509     0.65291425     0.63068413

Root Mean Square Off-Diagonal Residuals: Overall = 0.00708432


      Principal Factor Method: Priors=MAX, nfact=3

      Final Communality Estimates: Total = 4.325590
```

```
              c               f               e

     0.85772232      0.79718589      0.74257915

              m               d              mu

     0.65524622      0.65161940      0.62123750


Root Mean Square Off-Diagonal Residuals: Overall = 0.01882936


            Principal Factor Method: Priors=ASMC

                   The FACTOR Procedure
          Initial Factor Method: Principal Factors

            Prior Communality Estimates: ASMC

              c               f               e

      1.0024486       0.8940280       0.7721638

              m               d              mu

      0.6686740       0.5856018       0.5070837


      Eigenvalues of the Reduced Correlation Matrix:
            Total = 4.43  Average = 0.73833333

       Eigenvalue    Difference    Proportion    Cumulative

  1    3.88066800    3.66005151        0.8760        0.8760
  2    0.22061650    0.07307187        0.0498        0.9258
  3    0.14754463    0.03509880        0.0333        0.9591
  4    0.11244583    0.06423097        0.0254        0.9845
  5    0.04821486    0.02770468        0.0109        0.9954
  6    0.02051018                      0.0046        1.0000

  6 factors will be retained by the PROPORTION criterion.


       Principal Factor Method: Priors=ASMC, nfact=3

        Final Communality Estimates: Total = 4.248829

              c               f               e

     0.94417682      0.87934426      0.75001694

              m               d              mu

     0.64325177      0.57336948      0.45866986

Root Mean Square Off-Diagonal Residuals: Overall = 0.01648017
```

The prior communalities, h_i^2, $i = 1, \ldots, 6$, estimated using the squared multiple correlations are 0.8142, 0.7261, 0.6271, 0.5431, 0.4756, and 0.4118, respectively (see Output 4.1). Since these are the diagonal elements of $\mathbf{R}_a$, the matrix used for performing the factor analysis, the sum of these values, namely 3.5979, is equal to the total variation, $tr\ (\mathbf{R}_a)$. The eigenvalues of this matrix from Output 4.1 are 3.7334, 0.0981, 0.0103, -0.0299, -0.0968, and -0.1173. Thus k', the number of positive eigenvalues, is 3. Note that the proportion of the total variation explained by the first factor is $\hat{\lambda}_1 / \sum_i \hat{\lambda}_i = 3.7334/3.5979$, which already exceeds 1. Hence, $\hat{k}$ $(\leq k' = 3)$ has to be 1. Hence, only one factor will be selected by the FACTOR procedure by default, which is enforced by the PROPORTION criterion. For this analysis the RMS_overall value is 0.0302, which is smaller than what we were able to achieve using the principal component method.

When the PRIORS = option is used, the default value for the option MINEIGEN = that was used to select the number of factors is zero, *not* 1, as in the principal component method. Thus, no factors corresponding to the negative eigenvalues will be selected. Since there are three positive eigenvalues of the matrix $\mathbf{R}_a$, at the most three factors can be extracted using this method. When three factors are used using NFACT=3 option, the final estimates of the communalities are 0.8779, 0.7744, 0.6557, 0.5850, 0.5135, and 0.4535. The RMS_overall value is 0.0209, which is smaller than that for the one common factor model.

Next, using the PRIORS=MAX option we see from the elements of the correlation matrix $\mathbf{R}$ that the prior communalities are estimated by 0.83, 0.83, 0.78, 0.70, 0.66, and 0.63, respectively. Such an ordering of the estimates of the communalities is not surprising since the elements of the Spearman's correlation matrix are also arranged in decreasing magnitude. For this case, the ratio of the sum of the eigenvalues to $tr\ (\mathbf{R}_a)$ exceeds one when four factors are used. Hence, the default choice by the FACTOR procedure is four common factors. This yields the final communality estimates to be 0.8590, 0.8220, 0.7838, 0.6925, 0.6529, and 0.6307 respectively. The RMS_overall value is 0.0071. This is very small compared to any of the RMS_overall values of the previous analyses. If three factors are selected using the NFACT = 3 option, the RMS_overall value turns out to be 0.0188. This value is larger than that for four factors, but smaller than that for any of the previous analyses. It may be preferable in the interest of parsimony and interpretability.

Finally, when PRIORS=ASMC is used, the prior estimates of the communalities are 1.0024, 0.8940, 0.7722, 0.6687, 0.5856, and 0.5071. All six factors are needed to achieve the proportion of total variation to be greater than or equal to one. Hence, the default choice is the trivial model with six factors, which is the number of variables in the problem. Hence, there is really no data reduction. When three factors are selected using the NFACT = 3 option, the RMS_overall value is seen to be 0.0165. Thus, among the three methods discussed here this case indicates the best three common factors model. For four factors (not shown in the output) the final estimates of the communalities are 0.9890, 0.8877, 0.7660, 0.6581, 0.5756, and 0.4850. The RMS_overall value is 0.0077, which is slightly larger than that for four common factors with the use of PRIORS = MAX.

4.4.3 Image Analysis

Although the factor analysis clearly specifies in its model how the common and specific (uniqueness) parts are related, identification of these parts is often done only by trial and error. To resolve this inability of the factor analysis to explicitly define the common and unique parts of a variable, Guttman (1953) introduced what is now popularly known as *image analysis*.

A distinguishing feature of image analysis is its explicit definition of the common part of the variable. Specifically, in Guttman's own words "the common part of a variable is that part which is predictable by linear multiple correlation from all the other variables in the universe of variables." This common part of the variable is called the *image* of the variable based on all the other variables. The specific part of the variable is the remaining part of the variable (residual) that cannot be predicted. This unique part is called the *anti-image* of the variable. Using regression analysis terminology, the image and anti-image parts of the variables can be described as follows.

Suppose $\mathbf{x}$ is a p by 1 random vector in the standardized form so that its variance-covariance matrix is same as the correlation matrix $\mathbf{R}$, which is assumed to be nonsingular. If the regression model

$$x_i = \beta_1 x_1 + \cdots + \beta_{i-1} x_{i-1} + \beta_i x_{i+1} + \cdots + \beta_{p-1} x_p + \epsilon_i$$

is used to predict x_i, then the least squares predictor $\hat{x}_i$ of x_i is

$$\hat{x}_i = \mathbf{R}_{12(i)} \mathbf{R}_{22(i)}^{-1} \mathbf{x}_{2(i)},$$

where $\mathbf{x} = (x_i, \mathbf{x}'_{2(i)})'$, $\mathbf{x}_{2(i)}$ is the vector containing all variables other than x_i. The matrix $\mathbf{R}$ is also partitioned accordingly as

$$\mathbf{R} = \begin{bmatrix} 1 & \mathbf{R}_{12(i)} \\ \mathbf{R}_{21(i)} & \mathbf{R}_{22(i)} \end{bmatrix}.$$

Let the residual $\hat{\epsilon}_i$ be $\hat{\epsilon}_i = x_i - \hat{x}_i$. The x_i can be expressed in terms of common and unique parts, that is, in terms of image and anti-image as

$$x_i = \hat{x}_i + \hat{\epsilon}_i, \quad i = 1, \ldots, p.$$

In a vector form these p equations can be written as

$$\mathbf{x} = \hat{\mathbf{x}} + \hat{\boldsymbol{\varepsilon}}, \tag{4.13}$$

where $\mathbf{x} = (x_1, \ldots, x_p)'$, $\hat{\mathbf{x}} = (\hat{x}_1, \ldots, \hat{x}_p)'$ and $\hat{\boldsymbol{\varepsilon}} = (\hat{\epsilon}_1, \ldots, \hat{\epsilon}_p)'$. Then it can be shown (Mulaik, 1972) that,

$$\hat{\mathbf{x}} = (\mathbf{I} - \mathbf{M}\mathbf{R}^{-1})\mathbf{x},$$

where $\mathbf{M}$ is the diagonal matrix containing the inverse of the diagonal elements of $\mathbf{R}^{-1}$, that is,

$$\mathbf{M} = [diag\,(\mathbf{R}^{-1})]^{-1}.$$

Setting $\mathbf{W} = \mathbf{I} - \mathbf{M}\mathbf{R}^{-1}$, we can express Equation 4.13 as

$$\begin{aligned} \mathbf{x} &= \mathbf{W}\mathbf{x} + (\mathbf{I} - \mathbf{W})\mathbf{x} \\ &= image \; + \; anti\_image. \end{aligned}$$

Note that the variance-covariance matrix of the image,

$$\begin{aligned} D(\mathbf{W}\mathbf{x}) &= \mathbf{W}\mathbf{R}\mathbf{W}' \\ &= (\mathbf{I} - \mathbf{M}\mathbf{R}^{-1})\mathbf{R}(\mathbf{I} - \mathbf{M}\mathbf{R}^{-1})' \\ &= \mathbf{R} + \mathbf{M}\mathbf{R}^{-1}\mathbf{M} - 2\mathbf{M}. \end{aligned} \tag{4.14}$$

It may be noted that a typical diagonal element of this covariance matrix is the squared multiple correlation coefficient of the particular variable with all the other variables. In image factor analysis using the given correlation matrix $\mathbf{R}$, the covariance matrix of the image given in Equation 4.14 is computed first. Using this matrix as a starting point, the principal component method (not the principal factor method) of extracting the factors is used to perform the image factor analysis. All the factors corresponding to the eigenvalues of $\mathbf{W}\mathbf{R}\mathbf{W}'$ greater than or equal to one are usually extracted.

If the interest is to extract factors using image analysis, the following SAS code (see Program 4.1)

```
proc factor data=Spearman method=image res;
var c f e m d mu;
run;
```

will perform the computations.

EXAMPLE 3 ***Image Analysis, Spearman's Data (continued)*** We return to Spearman's data and using the above SAS code (included in Program 4.1), we perform the factor analysis. The resulting output is included as part of Output 4.1.

Output 4.1
(continued)

```
                    Output 4.1
         Factor Analysis of Spearman's Data
           Image Analysis: Method=Image

               The FACTOR Procedure
       Initial Factor Method: Image Components

          Squared Multiple Correlations

              c              f              e

     0.81416371     0.72610719     0.62713210

              m              d             mu

     0.54308031     0.47561114     0.41184072


               Image Coefficients

                    c              f              e

c             0.00000        0.54654        0.61057
f             0.37083        0.00000        0.00260
e             0.30431        0.00191        0.00000
m             0.11788        0.16996        0.18352
d             0.14317        0.18638        0.04839
mu            0.14604        0.06348        0.01091

               Image Coefficients

                    m              d             mu

c             0.28983        0.40400        0.46222
f             0.28353        0.35684        0.13631
e             0.22489        0.06805        0.01721
m             0.00000       -0.09217        0.10713
d            -0.08031        0.00000       -0.05117
mu            0.08323       -0.04562        0.00000
```

Image Covariance Matrix

	c	f	e
c	0.81416	0.72843	0.66653
f	0.72843	0.72611	0.66929
e	0.66653	0.66929	0.62713
m	0.64614	0.59234	0.55614
d	0.58492	0.55226	0.51463
mu	0.54410	0.53267	0.50358

Image Covariance Matrix

	m	d	mu
c	0.64614	0.58492	0.54410
f	0.59234	0.55226	0.53267
e	0.55614	0.51463	0.50358
m	0.54308	0.49212	0.46105
d	0.49212	0.47561	0.42683
mu	0.46105	0.42683	0.41184

Eigenvalues of the Image Covariance Matrix:
Total = 3.59793517 Average = 0.59965586

	Eigenvalue	Difference	Proportion	Cumulative
1	3.46885095	3.39901083	0.9641	0.9641
2	0.06984012	0.02990652	0.0194	0.9835
3	0.03993360	0.02333874	0.0111	0.9946
4	0.01659486	0.01411658	0.0046	0.9992
5	0.00247827	0.00224090	0.0007	0.9999
6	0.00023737		0.0001	1.0000

1 factor will be retained by the MINEIGEN criterion.

Factor Pattern

	Factor1
c	0.97239
f	0.90734
e	0.83290
m	0.77780
d	0.70566
mu	0.67203

Final Communality Estimates: Total = 4.017090

c	f	e
0.94555177	0.82327229	0.69371807

m	d	mu
0.60496646	0.49795747	0.45162358

Root Mean Square Off-Diagonal Residuals: Overall = 0.05072598

The matrix of image coefficients $\mathbf{W}$ and the covariance matrix $\mathbf{WRW}'$ given in Equation 4.14 are provided in the SAS output. Specifically,

$$\mathbf{W} = \begin{bmatrix} .0000 & .5465 & .6106 & .2898 & .4040 & .4622 \\ .3708 & .0000 & .0026 & .2835 & .3568 & .1363 \\ .3043 & .0019 & .0000 & .2249 & .0681 & .0172 \\ .1179 & .1700 & .1835 & .0000 & -.0922 & .1071 \\ .1432 & .1864 & .0484 & -.0803 & .0000 & -.0512 \\ .1460 & .0635 & .0109 & .0832 & -.0456 & .0000 \end{bmatrix}$$

and the image covariance matrix

$$\mathbf{WRW}' = \begin{bmatrix} .8142 & .7284 & .6665 & .6461 & .5841 & .5441 \\ .7284 & .7261 & .6693 & .5923 & .5523 & .5327 \\ .6665 & .6693 & .6271 & .5561 & .5146 & .5036 \\ .6461 & .5923 & .5561 & .5431 & .4921 & .4611 \\ .5849 & .5523 & .5146 & .4921 & .4756 & .4268 \\ .5441 & .5327 & .5036 & .4611 & .4268 & .4118 \end{bmatrix}.$$

The eigenvalues of $\mathbf{WRW}'$ are 3.4698, 0.0698, 0.0399, 0.0166, 0.0025, and 0.0002 respectively. Suppose we use the default option of minimum eigenvalue criterion (MINEIGEN=). As default, MINEIGEN=1 is used. Since only one eigenvalue is greater than one, only one common factor will be selected. The factor loadings matrix (vector) of this single factor corresponding the first eigenvalue is $(0.9724, 0.9073, 0.8329, 0.7778, 0.7057, 0.6720)'$. The RMS_overall value is 0.0507. The final communality estimates are 0.9456, 0.8233, 0.6937, 0.6050, 0.4980, and 0.4516 respectively.

4.4.4 Harris' Noniterative Canonical Factor Analysis

Rao (1955) introduced the canonical factor analysis as an answer to the question "What is that factor variable that is maximally related to the observed variables?" The solution to this problem relates to canonical correlation analysis (see Chapter 3) of the hypothetical factor variables (the common parts) with the observable variables.

Referring to the model given in Equation 4.2, the $p + k$ by $p + k$ covariance matrix of $\mathbf{x}$ and $\mathbf{f}$ is

$$D\left(\begin{bmatrix} \mathbf{x} \\ \mathbf{f} \end{bmatrix}\right) = \begin{bmatrix} \mathbf{\Sigma} & \mathbf{L} \\ \mathbf{L}' & \mathbf{I} \end{bmatrix}.$$

Assume that the variables have been standardized and thus $\mathbf{\Sigma}$ is in the correlation form. Thus, we will use the notation $\boldsymbol{\rho}$ instead of $\mathbf{\Sigma}$. The objective is to estimate $\mathbf{L}$ and $\mathbf{\Psi}$, where $\boldsymbol{\rho} = \mathbf{LL}' + \mathbf{\Psi}$. The squared canonical correlation coefficients between $\mathbf{x}$ and $\mathbf{f}$ are obtained as the roots of the determinantal equation

$$|\mathbf{LL}' - \nu\boldsymbol{\rho}| = 0.$$

Suppose the matrix $\boldsymbol{\rho}$ is estimated by the sample correlation coefficient matrix $\mathbf{R}$ and also suppose that an initial estimate $\hat{\mathbf{\Psi}}$ of $\mathbf{\Psi}$ is available. Then, we expect

$$\mathbf{LL}' \approx \mathbf{R} - \hat{\mathbf{\Psi}}.$$

Thus, with $\hat{\mathbf{L}}$ as an estimate of $\mathbf{L}$ and with the choice for $\hat{\mathbf{L}}\hat{\mathbf{L}}' = \mathbf{R} - \hat{\mathbf{\Psi}}$ the sample version of the above determinantal equation becomes

$$|\mathbf{R} - \hat{\mathbf{\Psi}} - \nu\mathbf{R}| = 0$$

or

$$|\hat{\mathbf{\Psi}}^{-1/2}\mathbf{R}\hat{\mathbf{\Psi}}^{-1/2} - \lambda\mathbf{I}| = 0$$

with $\lambda = \frac{1}{1-\nu}$.

In applications, instead of $\mathbf{R}$, the pseudo correlation matrix $\mathbf{R}_a$ in which the ith diagonal element of $\mathbf{R}$ is replaced by $\hat{h}_i^2$ (see the SMC method described in the earlier section "Estimation using squared multiple correlation (SMC)") is used. Thus the determinantal equation used is

$$|\hat{\mathbf{\Psi}}^{-1/2}\mathbf{R}_a\hat{\mathbf{\Psi}}^{-1/2} - \lambda\mathbf{I}| = 0.$$

The matrix $\mathbf{R}_a$ may not be positive definite. Thus, due to above replacement, some of the eigenvalues λ_i of

$$\hat{\mathbf{\Psi}}^{-1/2}\mathbf{R}\hat{\mathbf{\Psi}}^{-1/2}$$

may be negative. Let $\hat{\lambda}_1 \geq \cdots \geq \hat{\lambda}_k > 0$ be the positive eigenvalues of $\hat{\mathbf{\Psi}}^{-1/2}\mathbf{R}_a\hat{\mathbf{\Psi}}^{-1/2}$ and $\mathbf{\Gamma}_1$ be the p by k matrix of the corresponding eigenvectors. Set $\hat{\mathbf{\Lambda}}_1^{1/2} = \text{diag}\,(\sqrt{\hat{\lambda}_1}, \ldots, \sqrt{\hat{\lambda}_k})$ and take

$$\hat{\mathbf{L}} = \hat{\mathbf{\Psi}}^{1/2}\mathbf{\Gamma}_1\hat{\mathbf{\Lambda}}_1^{1/2},$$

as an estimate of $\mathbf{L}$. If we want, we can obtain a new estimate of $\mathbf{\Psi}$ as $\hat{\mathbf{\Psi}} = \mathbf{R} - \hat{\mathbf{L}}\hat{\mathbf{L}}'$ and iteratively continue the procedure until a stability in the final communality estimates has been achieved. This process of extracting the factors is scale-invariant. Specifically, consider the scale transformation $\mathbf{y} = \mathbf{D}\mathbf{x}$, where the elements of the diagonal matrix $\mathbf{D}$ all have the same sign. The solution $\mathbf{L}$ is not altered if $\mathbf{y}$ instead of $\mathbf{x}$ are used as data. This is because the solution extracted from the matrix

$$(\mathbf{D}\hat{\mathbf{\Psi}}\mathbf{D}')^{-1/2}\mathbf{D}\mathbf{R}_a\mathbf{D}'(\mathbf{D}\hat{\mathbf{\Psi}}\mathbf{D}')^{-1/2}$$

corresponding to $\mathbf{y}$ is the same as that obtained from the matrix $\hat{\mathbf{\Psi}}^{-1/2}\mathbf{R}_a\hat{\mathbf{\Psi}}^{-1/2}$ corresponding to $\mathbf{x}$. Consequently, the use of $\hat{\mathbf{\Sigma}}$ instead of $\mathbf{R}$ in the above procedure will still result in identical solutions.

Harris (1962) has shown an important connection (known as the *Rao-Guttman relationship*) between the Rao's canonical factor analysis and Guttman's image analysis. Making use of a result of Guttman (1956), which states that if k is small compared to p (which is the case in very large dimensional problems) then $[\text{diag}(\mathbf{R}^{-1})]^{-1} \doteq \mathbf{\Psi}$ as $p \to \infty$, he suggested a noniterative algorithm for Rao's canonical factor analysis by extracting the eigenvalues and eigenvectors of the matrix

$$[\text{diag}\,(\mathbf{R}^{-1})]^{1/2}\mathbf{R}_a[\text{diag}\,(\mathbf{R}^{-1})]^{1/2}. \tag{4.15}$$

This noniterative approach is adopted in SAS in the METHOD=HARRIS option in the PROC FACTOR statement. It can be performed using the SAS statements

```
proc factor data=spearman method=harris res;
var c f e m d mu;
run;
```

EXAMPLE 4 ***Harris' Canonical Factor Analysis, Spearman's Data (continued)*** Here we perform the canonical factor analysis using Harris' method for the Spearman's correlation matrix. The relevant lines of SAS code are included in Program 4.1 and the results are shown in Output 4.1.

Output 4.1
(continued)

```
                    Output 4.1
           Factor Analysis of Spearman's Data
        Canonical Factor Analysis: Method=Harris

                 The FACTOR Procedure
        Initial Factor Method: Harris Components

             Squared Multiple Correlations

              c               f                e

        0.81416371      0.72610719       0.62713210

              m               d               mu

        0.54308031      0.47561114       0.41184072


       Eigenvalues of the Weighted Reduced Correlation
      Matrix: Total = 11.5098284  Average = 1.91830474

         Eigenvalue    Difference    Proportion    Cumulative

    1    12.0239645    11.8161974        1.0447        1.0447
    2     0.2077672     0.1839344        0.0181        1.0627
    3     0.0238327     0.0978677        0.0021        1.0648
    4    -0.0740350     0.1989812       -0.0064        1.0584
    5    -0.2730162     0.1256686       -0.0237        1.0346
    6    -0.3986848                      -0.0346        1.0000

    1 factor will be retained by the PROPORTION criterion.

                    Factor Pattern

                             Factor1

                   c         0.92448
                   f         0.87289
                   e         0.80782
                   m         0.75384
                   d         0.69382
                   mu        0.65581


        Final Communality Estimates and Variable Weights
Total Communality: Weighted = 12.023965   Unweighted = 3.748921

        Variable      Communality         Weight

        c              0.85465641     5.38108018
        f              0.76193758     3.65106339
        e              0.65257936     2.68191499
        m              0.56827766     2.18856841
        d              0.48138508     1.90698178
        mu             0.43008443     1.70021971
```

```
Root Mean Square Off-Diagonal Residuals: Overall = 0.03294437

      Canonical Factor Analysis: Method=Harris, nfact=3

Root Mean Square Off-Diagonal Residuals: Overall = 0.02296229
```

As the output shows, the positive eigenvalues of the weighted matrix given in Display 4.15 are 12.0240, 0.2078, and 0.0238. Since there are certain negative eigenvalues, and the total of all the eigenvalues is 11.5000, the proportion of the variation explained by the first eigenvalue already exceeds 1. Hence, only one factor will be selected by the PROPORTION criterion. The factor loadings for this factor are (0.9245, 0.8729, 0.8087, 0.7538, 0.6938, 0.6558). The RMS_overall of Equation 4.11 is 0.0329. The final communality estimates are 0.8547, 0.7619, 0.6526, 0.5683, 0.4814, and 0.4301. If all the three factors corresponding to the positive eigenvalues are used, the RMS_overall value is 0.0230. Since the RMS_overall value is much smaller when three factors are considered as compared to when only one factor is considered, three factors may be recommended to use here.

4.5 Iterative Methods of Estimation

Many iterative methods are often motivated by extracting factors using the maximum likelihood method. As an alternative to this, still other iterative procedures have also been developed, without any distributional assumptions. We discuss the following four iterative methods in this section: the maximum likelihood (ML) method, the unweighted least squares (ULS) method, the iterative principal component method, and alpha factor analysis.

4.5.1 Maximum Likelihood Method

The maximum likelihood (ML) method of factor analysis under multivariate normality was originally introduced by Lawley (1940, 1941) and later algorithmically developed by Jöreskog (1967, 1977). The ML method assumes that the variance-covariance or correlation matrix of all variables is nonsingular. Since the method is scale invariant, we can obtain the solution using either the variance-covariance or correlation matrix. Let $\mathbf{x}_1, \ldots, \mathbf{x}_n$ be a random sample from a p-variate normal distribution with mean vector $\boldsymbol{\mu}$ and variance-covariance matrix $\boldsymbol{\Sigma}$, which is assumed to have the covariance structure $\boldsymbol{\Sigma} = \mathbf{LL}' + \boldsymbol{\Psi}$. The factor analysis problem can be viewed as that of finding the ML estimates of $\mathbf{L}$ and $\boldsymbol{\Psi}$.

Trivially, the sample mean vector $\bar{\mathbf{x}}$ is the maximum likelihood estimator of $\boldsymbol{\mu}$. To find the ML estimates of $\mathbf{L}$ and $\boldsymbol{\Psi}$, we take $\mathbf{S}$, the sample variance-covariance matrix as the initial estimate of $\boldsymbol{\Sigma}$. Since the sample is assumed to be from a multivariate normal distribution, $(n-1)\mathbf{S}$ has a Wishart distribution with $(n-1)$ degrees of freedom and with the expected value of $\mathbf{S}$ equal to $\boldsymbol{\Sigma}$. The probability density function $\mathcal{L}(\mathbf{S})$ of $\mathbf{S}$ is given by

$$\mathcal{L}(\mathbf{S}) = c \cdot |\boldsymbol{\Sigma}|^{-\left(\frac{n-1}{2}\right)} |\mathbf{S}|^{\left(\frac{n-1}{2}\right)-\left(\frac{p+1}{2}\right)} e^{-\left(\frac{n-1}{2}\right) tr(\boldsymbol{\Sigma}^{-1}\mathbf{S})},$$

where c is a constant. Thus, the log-likelihood of $\mathbf{L}$ and $\boldsymbol{\Psi}$, where $\boldsymbol{\Sigma} = \mathbf{LL}' + \boldsymbol{\Psi}$, is

$$ln\ c - \left(\frac{n-1}{2}\right)\left\{tr\ [(\mathbf{LL}' + \boldsymbol{\Psi})^{-1}\mathbf{S}] - ln|(\mathbf{LL}' + \boldsymbol{\Psi})^{-1}\mathbf{S}|\right\}. \tag{4.16}$$

The maximum likelihood estimates of $\mathbf{L}$ and $\mathbf{\Psi}$ are obtained by maximizing the expression in Display 4.16 subject to certain $k(k-1)/2$ uniqueness conditions (Johnson and Wichern, 1998, p. 530).

Determination of number of common factors

(a) Likelihood ratio test. One of the main advantages of using the maximum likelihood method for estimation in factor analysis is that it also provides a way of testing the hypothesis that the k-factor model is appropriate, where k is a known integer. This formalization eliminates to some extent the subjectivity determining the number of common factors. Specifically, using the likelihood ratio test we can test the null hypothesis

$$H_0 : \mathbf{\Sigma} = \mathbf{LL}' + \mathbf{\Psi}, \ \ \text{Rank}(\mathbf{L}) = k \text{ (known)}.$$

Suppose $\hat{\mathbf{L}}$, $\hat{\mathbf{\Psi}}$, and $\hat{\mathbf{\Sigma}} = \hat{\mathbf{L}}\hat{\mathbf{L}}' + \hat{\mathbf{\Psi}}$ are the maximum likelihood estimates of $\mathbf{L}$, $\mathbf{\Psi}$, and $\mathbf{\Sigma}$, respectively, under H_0. The *maximum of the log-likelihood function* in this case is given by

$$ln\mathcal{L}_{H_0} = c^* - \left(\frac{n-1}{2}\right)\left\{tr(\hat{\mathbf{\Sigma}}^{-1}\mathbf{S}) - ln|\hat{\mathbf{\Sigma}}^{-1}\mathbf{S}|\right\}.$$

Under no structure on $\mathbf{\Sigma}$ (i.e., $\mathbf{\Sigma}$ unstructured positive definite), the ML estimate of $\mathbf{\Sigma}$ is $\mathbf{S}$, and thus the maximum of the log-likelihood function is

$$\begin{aligned} ln\mathcal{L} &= c^* - \left(\frac{n-1}{2}\right)\{tr(\mathbf{S}^{-1}\mathbf{S}) - ln|\mathbf{S}^{-1}\mathbf{S}|\} \\ &= c^* - \frac{n-1}{2}p. \end{aligned}$$

Thus under H_0, and for large n the likelihood ratio test statistic

$$\begin{aligned} -2ln\lambda &= -2ln\left(\frac{\mathcal{L}_{H_0}}{\mathcal{L}}\right) \\ &= (n-1)\left\{tr(\hat{\mathbf{\Sigma}}^{-1}\mathbf{S}) - ln|\hat{\mathbf{\Sigma}}^{-1}\mathbf{S}| - p\right\} \\ &= (n-1)\{tr[(\hat{\mathbf{L}}\hat{\mathbf{L}}' + \hat{\mathbf{\Psi}})^{-1}\mathbf{S}] - \ln|(\hat{\mathbf{L}}\hat{\mathbf{L}}' + \hat{\mathbf{\Psi}})^{-1}\mathbf{S}| - p\} \\ &= (n-1)F(\hat{\mathbf{L}}, \hat{\mathbf{\Psi}}) \end{aligned}$$

with $F(\mathbf{L}, \mathbf{\Psi}) = tr[(\mathbf{LL}'+\mathbf{\Psi})^{-1}\mathbf{S}] - ln|(\mathbf{LL}'+\mathbf{\Psi})^{-1}\mathbf{S}| - p$ has an approximate chi-square distribution with $\frac{1}{2}\left\{(p-k)^2 - (p+k)\right\}$ degrees of freedom.

The above degrees of freedom are calculated as the difference between the number of estimated parameters under no structure on $\mathbf{\Sigma}$ and the number of parameters estimated under the null hypothesis. We estimate $p(p+1)/2$ parameters under no restriction on $\mathbf{\Sigma}$ and $(pk+p)$ parameters under the null hypothesis. But the $k(k-1)/2$ restrictions specified reduce the number of parameters (to be estimated) under H_0 by that. Hence, the degrees of freedom for the test is

$$\frac{p(p+1)}{2} - \left\{pk + p - \frac{k(k-1)}{2}\right\} = \frac{1}{2}\left\{(p-k)^2 - (p+k)\right\}.$$

Estimation of L and $\mathbf{\Psi}$. To obtain the maximum likelihood estimates of $\mathbf{L}$ and $\mathbf{\Psi}$, Jöreskog (1977) suggested an iterative two-stage minimization process. First, for a given $\mathbf{\Psi}$, the conditional minimum of

$$F(\mathbf{L}, \mathbf{\Psi}) = tr[(\mathbf{LL}' + \mathbf{\Psi})^{-1}\mathbf{S}] - \ln|(\mathbf{LL}' + \mathbf{\Psi})^{-1}\mathbf{S}| - p$$

with respect to $\mathbf{L}$ is obtained. Suppose $F(\hat{\mathbf{L}}, \boldsymbol{\Psi})$ is such a conditional minimum. Then $F(\hat{\mathbf{L}}, \boldsymbol{\Psi})$ is minimized with respect to $\boldsymbol{\Psi}$. The algorithm iterates between the estimates of $\mathbf{L}$ and $\boldsymbol{\Psi}$ to find the global minimum of $F(\mathbf{L}, \boldsymbol{\Psi})$ with respect to $\mathbf{L}$ and $\boldsymbol{\Psi}$. SAS uses an iterative algorithm that is due to Fuller (1987), which is implemented through the METHOD=ML option in the FACTOR procedure. For this algorithm, the prior estimates of the communalities, h_i^2, also needs to be provided using the PRIOR= option. As we have seen in Section 4.4.2, several choices for prior communalities are available. The default choice is the *squared multiple correlation*, otherwise indicated by the option PRIORS=SMC.

Rao (1955) showed that the factor solutions obtained using the maximum likelihood method under the multivariate normality are identical to the estimates obtained from the Rao's canonical factor analysis (that is, when $\mathbf{R}$ and not $\mathbf{R}_a$ is used in Harris' canonical factor analysis). Hence, for estimating the factor loadings and the specific variances, using the ML method, the multivariate normality is not required. For testing of the hypothesis, however, we do require the assumption of multivariate normality.

(b) Akaike's information criterion. Alternatively, we may want to use Akaike's information criterion (AIC) instead of a significance test for determining the number of factors. AIC was first introduced by Akaike (1973) to estimate the number of parameters in a model and its applications in factor analysis are discussed in Akaike (1987). Under a k factor model, the variance-covariance matrix $\boldsymbol{\Sigma}$ can be written as $\boldsymbol{\Sigma} = \mathbf{L}_k\mathbf{L}_k' + \boldsymbol{\Psi}$, where $\mathbf{L}_k$ is a p by k matrix of factor loadings. Thus the log-likelihood function corresponding to a k factor model, based on a random sample $\mathbf{x}_1, \ldots, \mathbf{x}_n$, from a p-dimensional multivariate normal population, is given by

$$ln\mathcal{L}(k) = c - \frac{n}{2}\left[ln|\mathbf{L}_k\mathbf{L}_k' + \boldsymbol{\Psi}| + tr((\mathbf{L}_k\mathbf{L}_k' + \boldsymbol{\Psi})^{-1}\mathbf{S}_n)\right],$$

where $\mathbf{S}_n = \frac{1}{n}\sum_{i=1}^{n}(\mathbf{x}_i - \bar{\mathbf{x}})(\mathbf{x}_i - \bar{\mathbf{x}})'$. Then, the AIC statistic for a k-factor model is defined as

$$AIC(k) = -2\ln\mathcal{L}(k) + [2p(k+1) - k(k-1)].$$

The k-factor model with k corresponding to the smallest value of $AIC(k)$ is considered best. AIC is available as part of the output provided by the METHOD=ML option in the FACTOR procedure. In addition to AIC, there are several other criteria reported, such as Schwarz's Bayesian Criterion (SBC), and they also can be used to determine the number of factors k.

The SAS code

```
proc factor method = ml nfact = k res;
```

can be used to extract the k-factor model (NFACT = k option) using the maximum likelihood method.

EXAMPLE 5 *Maximum Likelihood Factor Analysis, Ecology Data* Sinha and Lee (1970) considered a study from agricultural ecology where composite samples of wheat, oats, barley, and rye from various locations in the Canadian prairie were taken. Samples were taken from commercial and government terminal elevators at Thunder Bay (Ontario) during the unloading of railway boxcars. The objective of the study was to determine the interrelationships, if any, between arthropod infestation and grain environment. The three grain environmental variables observed for 165 samples were as follows: grade of sample indicating grain quality (1 highest, 6 lowest) (GRADE), percentage of moisture content in grain (MOIST), and dockage, which measures the presence of weed seed, broken kernels, and other foreign matters (DOCK). The counts of (the number found in grain) six types of arthropods considered were Acarus (ACAR), Cheyletus (CHEY), Glycychagus (GLYC), Larsonemus (LARS), Cryptolestes (CRYP), and Psocoptera (PSOC).

The two variables GRADE and DOCK are transformed by the square root transformation and the six arthropod counts were transformed using the logarithm transformation (to the base 10). The correlation matrix of these nine variables (using the same names for the transformed variables as the original variables) is provided in Program 4.2. The correlation matrix, along with the sample size, are stored in a SAS data set named ECOLOGY. Here we would like to first determine the smallest number of factors that can explain the data using the formal statistical tests and the AIC. The null hypothesis for k-factor model is H_0 : *k factors are sufficient.* In Program 4.2, we have used the option NFACT=k for $k = 1, \ldots, 5$.

```
/* Program 4.2 */

options ls=64 ps=45 nodate nonumber;
title1 'Output 4.2';
title2 'Factor Analysis of Ecology Data';

data ecology (type=corr);
infile cards missover;
_type_='corr';
if _n_=1 then _type_='N';
input _name_ $ grade moist dock acar chey glyc lars cryp psoc;
lines;
n 165
grade 1.0
moist .441 1.0
dock .441 .342 1.0
acar .107 .250 .040 1.0
chey .194 .323 .060 .180 1.0
glyc .105 .400 .082 .123 .220 1.0
lars .204 .491 .071 .226 .480 .399 1.0
cryp .197 .158 .051 .019 .138 -.114 .154 1.0
psoc -.236 -.220 -.073 -.199 -.084 -.304 -.134 -.096 1.00
;
/* Source: Sinha and Lee (1970). Reprinted by permission of
the Society for Population Ecology, Kyoto, Japan.*/

proc factor data=ecology method=ml nfact=1 res;
var grade moist dock acar chey glyc lars cryp psoc;
title3 'ML Method, nfact=1';
run;
proc factor data=ecology method=ml nfact=2 res;
var grade moist dock acar chey glyc lars cryp psoc;
title3 'ML Method, nfact=2';
run;
proc factor data=ecology method=ml nfact=3 res;
var grade moist dock acar chey glyc lars cryp psoc;
title3 'ML Method, nfact=3';
run;
proc factor data=ecology method=ml hey nfact=3 res;
var grade moist dock acar chey glyc lars cryp psoc;
title3 'ML Method, nfact=3, HEY option';
run;
proc factor data=ecology method=ml hey nfact=4 res;
var grade moist dock acar chey glyc lars cryp psoc;
title3 'ML Method, nfact=4, HEY option';
run;
```

```
proc factor data=ecology method=ml hey nfact=5 res;
var grade moist dock acar chey glyc lars cryp psoc;
title3 'ML Method, nfact=5, HEY option';
run;

proc factor data=ecology method=uls hey nfact=3 res;
var grade moist dock acar chey glyc lars cryp psoc;
title3 'ULS Method, nfact=3, HEY option';
run;

proc factor data=ecology method=prinit hey nfact=3 res;
var grade moist dock acar chey glyc lars cryp psoc;
title3 'PRINIT Method, nfact=3, HEY option';
run;

proc factor data=ecology method=alpha hey nfact=3 res;
var grade moist dock acar chey glyc lars cryp psoc;
title3 'Alpha Method, nfact=3, HEY option';
run;
```

Output 4.2

```
                         Output 4.2
                Factor Analysis of Ecology Data
                      ML Method, nfact=1

                     The FACTOR Procedure
           Initial Factor Method: Maximum Likelihood

             Prior Community Estimates: SMC

     grade         moist          dock          acar          chey

0.33503944    0.45499985    0.23850409    0.10786071    0.25234462

       glyc            lars            cryp            psoc

 0.31596140      0.41044427      0.11466990      0.17043474


            Preliminary Eigenvalues: Total =
            3.60340214  Average = 0.40037802

        Eigenvalue    Difference    Proportion    Cumulative

    1   3.22852154    2.36126459        0.8960        0.8960
    2   0.86725695    0.43397155        0.2407        1.1366
    3   0.43328540    0.23823513        0.1202        1.2569
    4   0.19505027    0.17827358        0.0541        1.3110
    5   0.01677670    0.12972367        0.0047        1.3157
    6   -.11294697    0.17563764       -0.0313        1.2843
    7   -.28858461    0.03966041       -0.0801        1.2042
    8   -.32824502    0.07946711       -0.0911        1.1131
    9   -.40771213                     -0.1131        1.0000

    1 factor will be retained by the NFACTOR criterion.
```

```
Iteration   Criterion   Ridge   Change        Communalities

    1       0.5206577  0.0000  0.1863  0.23632  0.64126  0.11043
                                       0.09657  0.23524  0.22369
                                       0.40241  0.03601  0.10810
    2       0.5201740  0.0000  0.0138  0.24897  0.64085  0.12110
                                       0.09786  0.22143  0.23471
                                       0.39964  0.03717  0.09793
    3       0.5201440  0.0000  0.0041  0.24921  0.64495  0.12211
                                       0.09728  0.22036  0.23321
                                       0.39671  0.03694  0.09849
    4       0.5201404  0.0000  0.0010  0.25009  0.64566  0.12285
                                       0.09719  0.21935  0.23298
                                       0.39584  0.03696  0.09834
    5       0.5201399  0.0000  0.0004  0.25030  0.64604  0.12307
                                       0.09714  0.21908  0.23284
                                       0.39547  0.03696  0.09834

Convergence criterion satisfied.

          Significance Tests Based on 165 Observations

                                                              Pr >
            Test                          DF    Chi-Square    ChiSq

HO: No common factors                     36      267.4854    <.0001
HA: At least one common factor
HO: 1 Factor is sufficient                27       82.9623    <.0001
HA: More factors are needed

    Chi-Square without Bartlett's Correction        85.302948
    Akaike's Information Criterion                  31.302948
    Schwarz's Bayesian Criterion                   -52.557580
    Tucker and Lewis's Reliability Coefficient       0.677662


         Eigenvalues of the Weighted Reduced Correlation
        Matrix: Total = 3.79263388  Average = 0.42140376

         Eigenvalue   Difference    Proportion    Cumulative

    1    3.79263416   3.08253497        1.0000        1.0000
    2    0.71009920   0.38480689        0.1872        1.1872
    3    0.32529231   0.18462916        0.0858        1.2730
    4    0.14066314   0.14536232        0.0371        1.3101
    5    -.00469918   0.05392672       -0.0012        1.3089
    6    -.05862590   0.23974887       -0.0155        1.2934
    7    -.29837476   0.06806674       -0.0787        1.2147
    8    -.36644151   0.08147208       -0.0966        1.1181
    9    -.44791359                    -0.1181        1.0000

              Variance Explained by Each Factor

              Factor        Weighted    Unweighted

              Factor1     3.79263416    2.09922636
```

Final Communality Estimates and Variable Weights
Total Communality: Weighted = 3.792634 Unweighted = 2.099226

Variable	Communality	Weight
grade	0.25039244	1.33387355
moist	0.64606511	2.82516749
dock	0.12315682	1.14033986
acar	0.09712330	1.10758608
chey	0.21898484	1.28053629
glyc	0.23280633	1.30350495
lars	0.39540049	1.65418359
cryp	0.03696612	1.03838122
psoc	0.09833093	1.10906083

ML Method, nfact=2

The FACTOR Procedure
Initial Factor Method: Maximum Likelihood

Significance Tests Based on 165 Observations

Test	DF	Chi-Square	Pr > ChiSq
H0: No common factors	36	267.4854	<.0001
HA: At least one common factor			
H0: 2 Factors are sufficient	19	36.6356	0.0088
HA: More factors are needed			

Chi-Square without Bartlett's Correction	37.827263
Akaike's Information Criterion	-0.172737
Schwarz's Bayesian Criterion	-59.185701
Tucker and Lewis's Reliability Coefficient	0.855651

Variance Explained by Each Factor

Factor	Weighted	Unweighted
Factor1	4.54076970	2.19366553
Factor2	1.50419212	0.75070237

```
           Final Communality Estimates and Variable Weights
     Total Communality: Weighted = 6.044962   Unweighted = 2.944368

                Variable       Communality        Weight

                grade           0.54385827      2.19247172
                moist           0.57682799      2.36307954
                dock            0.36675950      1.57915275
                acar            0.09677600      1.10716435
                chey            0.30675843      1.44229735
                glyc            0.26033881      1.35209578
                lars            0.64947709      2.85261909
                cryp            0.04873534      1.05126163
                psoc            0.09483647      1.10481867


                        ML Method, nfact=3

                       The FACTOR Procedure
              Initial Factor Method: Maximum Likelihood

Iteration   Criterion   Ridge   Change         Communalities

    1       0.0927983  0.0000  0.4103   0.59084  0.56743  0.32767
                                        0.09224  0.35313  0.72630
                                        0.66655  0.07216  0.10267
    2       0.0780704  0.0000  0.3778   0.56222  0.54115  0.35679
                                        0.08815  0.32337  1.10412
                                        0.71271  0.09334  0.11640

ERROR: Communality greater than 1.0.


                  ML Method, nfact=3, HEY option

                       The FACTOR Procedure
              Initial Factor Method: Maximum Likelihood

       3 factors will be retained by the NFACTOR criterion.


Iteration   Criterion   Ridge   Change         Communalities

    1       0.0927983  0.0000  0.4103   0.59084  0.56743  0.32767
                                        0.09224  0.35313  0.72630
                                        0.66655  0.07216  0.10267
    2       0.0802793  0.0000  0.2737   0.57011  0.54839  0.34877
                                        0.08928  0.33157  1.00000
                                        0.69999  0.08750  0.11262
    3       0.0790542  0.0000  0.0324   0.57308  0.55263  0.34993
                                        0.08899  0.32902  1.00000
                                        0.70562  0.11987  0.14205
    4       0.0790483  0.0000  0.0027   0.57574  0.55199  0.34760
                                        0.08873  0.32863  1.00000
                                        0.70703  0.11993  0.14233
```

```
5        0.0790480  0.0000  0.0004  0.57617  0.55168  0.34731
                                    0.08864  0.32852  1.00000
                                    0.70738  0.11997  0.14256
```

Convergence criterion satisfied.

Significance Tests Based on 165 Observations

Test	DF	Chi-Square	Pr > ChiSq
H0: No common factors	36	267.4854	<.0001
HA: At least one common factor			
H0: 3 Factors are sufficient	12	12.5028	0.4062
HA: More factors are needed			

Chi-Square without Bartlett's Correction	12.963865
Akaike's Information Criterion	-11.036135
Schwarz's Bayesian Criterion	-48.307481
Tucker and Lewis's Reliability Coefficient	0.993484

Variance Explained by Each Factor

Factor	Weighted	Unweighted
Factor1	1.14849087	1.50589100
Factor2	3.68553913	1.64910539
Factor3	1.59463264	0.70725260

Final Communality Estimates and Variable Weights
Total Communality: Weighted = 6.428663 Unweighted = 3.862249

Variable	Communality	Weight
grade	0.57618765	2.35944643
moist	0.55165839	2.23053521
dock	0.34728560	1.53212567
acar	0.08862049	1.09726173
chey	0.32852538	1.48925023
glyc	1.00000000	Infty
lars	0.70738531	3.41744420
cryp	0.11998270	1.13633029
psoc	0.14260347	1.16626838

Residual Correlations With Uniqueness on the Diagonal

	grade	moist	dock
grade	0.42381	-0.01592	0.00372
moist	-0.01592	0.44834	0.03114
dock	0.00372	0.03114	0.65271
acar	-0.02496	0.04848	-0.03305
chey	0.01903	-0.02082	-0.01247
glyc	0.00000	0.00000	0.00000
lars	-0.00005	0.00339	0.00326
cryp	0.01915	0.00290	-0.05780
psoc	-0.03668	0.01724	0.08195

Residual Correlations With Uniqueness on the Diagonal

	acar	chey	glyc
grade	-0.02496	0.01903	0.00000
moist	0.04848	-0.02082	0.00000
dock	-0.03305	-0.01247	0.00000
acar	0.91138	0.01331	0.00000
chey	0.01331	0.67147	0.00000
glyc	0.00000	0.00000	0.00000
lars	-0.01308	0.00216	0.00000
cryp	-0.05271	0.01036	0.00000
psoc	-0.13176	0.01583	0.00000

Residual Correlations With Uniqueness on the Diagonal

	lars	cryp	psoc
grade	-0.00005	0.01915	-0.03668
moist	0.00339	0.00290	0.01724
dock	0.00326	-0.05780	0.08195
acar	-0.01308	-0.05271	-0.13176
chey	0.00216	0.01036	0.01583
glyc	0.00000	0.00000	0.00000
lars	0.29261	-0.00225	0.01790
cryp	-0.00225	0.88002	-0.07955
psoc	0.01790	-0.07955	0.85740

Root Mean Square Off-Diagonal Residuals: Overall = 0.03568955

grade	moist	dock	acar	chey
0.01925157	0.02324905	0.03920881	0.05540073	0.01365107

Root Mean Square Off-Diagonal Residuals: Overall = 0.03568955

glyc	lars	cryp	psoc
0.00000000	0.00808807	0.04021036	0.06385405

ML Method, nfact=4, HEY option

The FACTOR Procedure
Initial Factor Method: Maximum Likelihood

Significance Tests Based on 165 Observations

Test	DF	Chi-Square	Pr > ChiSq
H0: No common factors	36	267.4854	<.0001
HA: At least one common factor			
H0: 4 Factors are sufficient	6	4.0137	0.6748
HA: More factors are needed			

```
Chi-Square without Bartlett's Correction        4.179309
Akaike's Information Criterion                 -7.820691
Schwarz's Bayesian Criterion                  -26.456363
Tucker and Lewis's Reliability Coefficient      1.051485


                ML Method, nfact=5, HEY option

                    The FACTOR Procedure
           Initial Factor Method: Maximum Likelihood

         Significance Tests Based on 165 Observations

                                                             Pr >
        Test                          DF    Chi-Square      ChiSq

HO: No common factors                 36      267.4854     <.0001
HA: At least one common factor
HO: 5 Factors are sufficient           1        0.6152     0.4329
HA: More factors are needed


 Chi-Square without Bartlett's Correction       0.6432655
 Akaike's Information Criterion                -1.3567345
 Schwarz's Bayesian Criterion                  -4.4626800
 Tucker and Lewis's Reliability Coefficient     1.0598500
```

Output 4.2 first provides a test for testing the null hypothesis

H_0 : No common factors (diagonal variance-covariance matrix) *vs*.

H_a : At least one common factor (unstructured variance-covariance matrix)

using the likelihood ratio test. The chi-square test statistic value for our example is 267.485 with the p value based on chi-square with 36 degrees of freedom less than 0.0001. In view of the small p value we reject this null hypothesis. Next, the likelihood ratio test is provided for testing

H_0 : k Factors are sufficient *vs*.

H_a : More factors are needed

A summary of test statistics, p values, AIC statistics, and other information extracted from Output 4.2 is provided in Table 4.3.

TABLE 4.3 Model Selection

k under (H_0)	Chi-squared	d.f.	p-value	AIC
1	82.962	27	.0001	31.3029
2	36.636	19	.0088	-.1727
3*	12.503	12	.4062	-11.0361
4	4.014	6	.6748	-7.8207
5*	0.615	1	.4329	-1.3567

* The HEYWOOD option was used.

Since the p values corresponding to the likelihood ratio test for $k \leq 2$ are very small, the one-factor and two-factor models may not be appropriate for this data set. However for $k = 3$, the p value is large (p value=0.4062). Thus, a three-factor model is declared as sufficient. Although the same is true for $k = 4$ and $k = 5$, our search is for the smallest number of factors that can explain the data. Factor models with at most five common factors can be fit for these data since only the first five eigenvalues (of the matrix that is used) for the analysis are positive. We thus conclude that a three-factor model is sufficient. This fact is also confirmed by the AIC criterion, which has its minimum (= -11.0361) for the three-factor model.

Next, let us examine final communality estimates. The final communality estimates h_i^2 for the one-factor model are 0.2504, 0.6461, 0.1232, 0.0971, 0.2190, 0.2328, 0.3954, 0.0370, 0.0983. Hence, the variance explained by the single factor (provided in the output under the title "Variance Explained by Each Factor" under the column "Unweighted") is $\sum_{i=1}^{9} h_i^2 = 2.0992$, which amounts to $\frac{1}{p}\sum_{i=1}^{9} h_i^2 \times 100 = 23.32\%$ approximately. The weighted sum of communalities is also provided in the output under the column "Weighted." This quantity is

$$\sum_{i=1}^{9} w_i h_i^2 = \sum_{i=1}^{9} \frac{h_i^2}{\psi_i} = \sum_{i=1}^{9} \frac{h_i^2}{1-h_i^2} = 3.7926.$$

The weights $w_i = \frac{1}{1-h_i^2}$ are also printed.

For a two-factor model the final communality estimates are 0.5439, 0.5768, 0.3668, 0.0968, 0.3068, 0.2603, 0.6495, 0.0487, 0.0948. The variance explained by the first and the second factors are 2.1937 and 0.7507, and the corresponding weighted variances are 4.5408 and 1.5042.

For a three-factor model, at the second iteration, the communality for the 6th variable, that is, GLYC, exceeds one (1.1041) causing the future computations to stop. The process stops with the message

```
ERROR: Communality greater than 1.0.
```

There are two ways to overcome this interruption:

(*a*) Set communalities greater than one to one and proceed until the convergence criterion is met. This can be done by using the HEYWOOD (HEY) option in the PROC FACTOR statement.

(*b*) continue the process with the values of communalities greater than one until convergence. This can be done by using the ULTRAHEYWOOD (ULTRA) option. However, the solution may not converge.

It may be pointed out that both of these choices may lead to improper solutions (see Section 4.2). SAS will print warning messages whenever the solution is improper.

When we use the HEYWOOD option for our three-factor model, the solution is obtained in just five iterations. The final communality estimates are 0.5762, 0.5517, 0.3473, 0.0886, 0.3285, 1.0000, 0.7074, 0.1200, 0.1426. Since the weight $w_6 = \frac{1}{1-h_6^2} = \frac{1}{1-1}$ is undefined (that is, infinity), in SAS output the weight corresponding to the 6th variable is printed as "Infty."

When the RES option in the PROC FACTOR statement is used, as seen in the previous section, the matrix of residuals will be printed along with the quantity RMS_overall given in Equation 4.11. This measure can be adopted to discuss the selection of various factor models. However, whenever the HEY option is used to get the ML factor solution, all the entries in one or more rows (and columns) corresponding to the variables where the Heywood case had occurred in the process of iteration will be automatically zero (see Output 4.2). Hence, it may not be appropriate to use the residual measure in this case.

It is advisable to try other prior estimates for the communalities before a solution is finally accepted. This is especially crucial when HEYWOOD or ULTRAHEYWOOD options are used to get the solutions. However, for our data the conclusions of Table 4.3 did not change when other choices for the PRIORS=option were used in this case.

4.5.2 Unweighted Least Squares Method

Jöreskog (1977) has developed an algorithm for estimating $\mathbf{L}$ and $\mathbf{\Psi}$ iteratively, by minimizing the least squares function,

$$F_u(\mathbf{L}, \mathbf{\Psi}) = \frac{1}{2} tr(\mathbf{S} - \mathbf{\Sigma})^2 = \frac{1}{2} tr[(\mathbf{S} - \mathbf{\Sigma})(\mathbf{S} - \mathbf{\Sigma})],$$

with respect to $\mathbf{L}$ and $\mathbf{\Psi}$, where $\mathbf{\Sigma} = \mathbf{LL}' + \mathbf{\Psi}$ and $\mathbf{L}'\mathbf{L}$ is a diagonal matrix of order k by k. The solution obtained using this algorithm is called the *unweighted least squares* (ULS) solution. Although this method is not scale invariant, it has the advantage of not requiring the assumption that the variance-covariance (or correlation) matrix is nonsingular. The algorithm for ULS method is essentially the same as the iterative two-stage algorithm discussed in the previous section.

The ULS solution to factor analysis is also equivalent to the solution obtained using another popular method called the *minimum residual sum of squares* (MINRES) method, developed by Harman (1977). Because of this equivalence we will not discuss the MINRES method further.

EXAMPLE 6 *ULS Method, Ecology Data (continued)* Returning to ecology data, we will use the METHOD=ULS option in the PROC FACTOR statement to use the unweighted least squares method for factor analysis. We use the default choice PRIORS=SMC, and thus we will use the squared multiple correlation for the prior communality estimates.

As in the case of maximum likelihood estimation, during the process of determining the ULS solution, communality estimates of one or more factors during iterations may occasionally exceed 1. This leads to an error and the iterative process for seeking the solution stops. However, as noted earlier, using the HEYWOOD or ULTRAHEYWOOD option in the PROC FACTOR statement we can force the iterative process to continue. In Program 4.2 we have included the SAS statements needed to obtain three common factors using this method. We will suppress the output since the factor analysis results using ULS method for this example (as we will see later in Section 4.7) are similar to those in the previous subsection.

4.5.3 Iterated Principal Factor Method

Hoping to get better estimates of the factor loadings matrix and the matrix of specific variances, we sometimes iterate the principal factor method of Section 4.4.2 until the change in the final communality estimates is smaller than a prespecified value. This is the main rationale of iterated principal factor method.

Suppose $\hat{\mathbf{L}}^{(i)}$ is the estimate of the factor loading matrix at the ith iteration stage. Then, we estimate $\mathbf{\Psi}$ at the next iteration by

$$\hat{\mathbf{\Psi}}^{(i+1)} = diag\,(\mathbf{S} - \hat{\mathbf{L}}^{(i)}\hat{\mathbf{L}}^{(i)'})$$

and extract k principal components of the matrix $\mathbf{S} - \hat{\mathbf{\Psi}}^{(i+1)}$ to obtain $\hat{\mathbf{L}}^{(i+1)}$. The iteration process is continued until a certain convergence criterion is met. A sample correlation matrix $\mathbf{R}$ in place of $\mathbf{S}$ can also be used. However, the results will not be identical.

The iterative procedure described above is equivalent to minimizing

$$tr[(\mathbf{S} - \mathbf{\Sigma})^2] = \sum_i \sum_j (s_{ij} - \sigma_{ij})^2$$

subject to

$$\mathbf{\Sigma} = \mathbf{LL}' + \mathbf{\Psi},$$

in two stages. Hence, the iterated principal factor method is equivalent to the ULS method, except that there is one extra set of restrictions in the ULS method, namely that the $\mathbf{L}'\mathbf{L}$ is a diagonal matrix.

The iterated principal factor analysis can be performed using the METHOD=PRINIT option in the PROC FACTOR statement in SAS. The default choice for priors is PRIORS=ONE. Since the solution of the factor analysis for the ecology data using this method is very similar to that obtained using the ULS method, we will not illustrate this method through an example here. However, we have included in Program 4.2 the appropriate SAS code to do this analysis. In general, the ULS and Iterated Principal Factor methods will not give the same results because of the additional restrictions under which the ULS solutions are obtained.

4.5.4 Alpha Factor Analysis

Kaiser and Caffrey (1965) defined alpha factor analysis as a technique to determine the common factors in a sample of tests or measurements that have the maximum correlation (in some meaningful sense) with corresponding common factors in a universe of tests or measurements from which the sample tests are drawn. A set of common factors satisfies this principle if each of them has the maximum squared correlation with a linear combination of a universe of variables. This squared correlation is called the *coefficient of generalizability* or *Cronbach's coefficient of alpha* (Cronbach, 1951). It is a quantity that measures the extent to which the results obtained by testing on a limited number of sample tests generalize to the population from which these tests were drawn.

The idea of generalizability can be explained explicitly as follows (McDonald, 1970). Suppose the p by 1 data vector $\mathbf{x}$ of tests or measurements is written as a sum of two random vectors,

$$\mathbf{x} = \mathbf{c} + \boldsymbol{\varepsilon},$$

where $\mathbf{c}$ is the common part (representing factors) and $\boldsymbol{\varepsilon}$ is the unique part of the partition. Further, suppose $\mathbf{u}$ is an N by 1 ($N \geq p$) vector of factors from a universe of factors. Let $E(\mathbf{c}) = \mathbf{0}$, $E(\mathbf{u}) = \mathbf{0}$, $cov(\mathbf{c}, \mathbf{u}) = \boldsymbol{\Omega}_{cu}$, $cov(\mathbf{c}, \mathbf{c}) = D(\mathbf{c}) = \boldsymbol{\Omega}_{cc}$, $cov(\mathbf{u}, \mathbf{u}) = D(\mathbf{u}) = \boldsymbol{\Omega}_{uu}$, and $cov(\mathbf{u}, \mathbf{c}) = \boldsymbol{\Omega}_{uc} = \boldsymbol{\Omega}'_{cu}$. Let $f_i = \mathbf{w}'_i\mathbf{c}$, $i = 1, \ldots, k$ and $g = \mathbf{w}'\mathbf{u}$ be thelinear combination of $\mathbf{c}$ and $\mathbf{u}$, respectively. Then the squared correlation coefficient between f_i and g is

$$\rho_i^2 = \frac{(\mathbf{w}'_i\boldsymbol{\Omega}_{cu}\mathbf{w})^2}{(\mathbf{w}'_i\boldsymbol{\Omega}_{cc}\mathbf{w}_i)(\mathbf{w}'\boldsymbol{\Omega}_{uu}\mathbf{w})}.$$

Kaiser and Caffrey (1965) showed that under certain conditions (see McDonald (1970) for an explication of these conditions) that as $N \to \infty$, ρ_i^2 converges to

$$\alpha_i = \frac{p}{p-1}\left\{1 - \frac{\mathbf{w}'_i\mathbf{H}\mathbf{w}_i}{\mathbf{w}'_i\boldsymbol{\Omega}_{cc}\mathbf{w}_i}\right\},$$

where $\mathbf{H} = diag\ (\boldsymbol{\Omega}_{cc})$. These quantities, α_i, $i = 1, \ldots, k$ $(< p)$, are called coefficients of generalizability or Cronbach's coefficients of alpha.

In the context of alpha factor analysis, the quantities α_i are

$$\alpha_i = \frac{p}{p-1}\left\{1 - \frac{\mathbf{w}'_i\mathbf{H}\mathbf{w}_i}{\mathbf{w}'_i(\boldsymbol{\Sigma} - \boldsymbol{\Psi})\mathbf{w}_i}\right\} = \frac{p}{p-1}\left\{1 - \frac{1}{\lambda_i}\right\}, i = 1, \ldots, k,$$

where $\mathbf{H} = diag(h_1^2, \ldots, h_p^2) = diag(\boldsymbol{\Sigma} - \boldsymbol{\Psi})$ is the diagonal matrix of communalities,

Ψ is the diagonal matrix of specific variances, Σ is the variance-covariance matrix of $\mathbf{x}$, and

$$\lambda_i = \frac{\mathbf{w}_i'(\Sigma - \Psi)\mathbf{w}_i}{\mathbf{w}_i'\mathbf{H}\mathbf{w}_i}, \quad i = 1, \ldots, k.$$

Maximizing $\alpha_i, i = 1, \ldots, k$ with respect to $\mathbf{w}_i, i = 1, \ldots, k$ such that $f_i, i = 1, \ldots, k$ are uncorrelated is equivalent to maximizing $\lambda_i, i = 1, \ldots, k$ subject to the same restriction. The solution to this in turn is given by the roots of the determinantal equation

$$|\mathbf{H}^{-1/2}(\Sigma - \Psi)\mathbf{H}^{-1/2} - \lambda\mathbf{I}| = 0. \tag{4.17}$$

In practice, the estimation of Σ, $\mathbf{H}$, and Ψ will be used. Let these respectively be $\hat{\Sigma}$, $\hat{\mathbf{H}}$, and $\hat{\Psi}$. Then Equation 4.17 becomes

$$|\hat{\mathbf{H}}^{-1/2}(\hat{\Sigma} - \hat{\Psi})\hat{\mathbf{H}}^{-1/2} - \lambda\mathbf{I}| = 0. \tag{4.18}$$

Suppose $\hat{\lambda}_1 \geq \cdots \geq \hat{\lambda}_k$ are the roots of Equation 4.18, and Γ_1 is the matrix of the eigenvectors of $\hat{\mathbf{H}}^{-1/2}(\hat{\Sigma} - \hat{\Psi})\hat{\mathbf{H}}^{-1/2}$, corresponding to $\hat{\lambda}_1, \ldots, \hat{\lambda}_k$. Then the factor loading matrix $\mathbf{L}$ is estimated as

$$\hat{\mathbf{L}} = \mathbf{H}^{1/2}\Gamma_1\hat{\Lambda}_1^{1/2},$$

where $\Lambda_1^{1/2} = \text{diag}\,(\sqrt{\hat{\lambda}_1}, \ldots, \sqrt{\hat{\lambda}_k})$. The resulting solution is scale invariant and hence instead of $\hat{\Sigma}$, the correlation matrix $\mathbf{R}$ can also be used. This is what is done in practice except that instead of $\mathbf{R}$, the pseudo correlation matrix $\mathbf{R}_a$ is used.

Kaiser and Derflinger (1990) have presented some contrasts between maximum likelihood factor analysis and alpha factor analysis. Their conclusion is that the maximum likelihood factor analysis is appropriate if the factor analysis is considered as an inferential method, whereas the alpha factor analysis is appropriate if the factor analysis is treated as a exploratory method.

The alpha factor analysis can be performed using the SAS code

```
proc factor method=alpha;
```

The default prior estimates of the communalities used for getting $\mathbf{R}_a$ corresponds to the choice PRIORS=SMC.

EXAMPLE 7 ***Alpha Factor Analysis, Ecology Data (continued)*** We return to ecology data and perform the alpha factor analysis. Suppose we want to extract three factors (NFACT=3 option). The corresponding code is included in Program 4.2. Output 4.2 follows, which we briefly discuss next.

Output 4.2 *(continued)*

```
                    Output 4.2
          Factor Analysis of Ecology Data
         Alpha Method, nfact=3, HEY option

                The FACTOR Procedure
      Initial Factor Method: Alpha Factor Analysis

          Prior Communality Estimates: SMC

     grade        moist         dock         acar         chey

0.33503944   0.45499985   0.23850409   0.10786071   0.25234462
```

glyc	lars	cryp	psoc
0.31596140	0.41044427	0.11466990	0.17043474

Preliminary Eigenvalues: Total = 9 Average = 1

	Eigenvalue	Difference	Proportion	Cumulative
1	7.20891253	4.92036227	0.8010	0.8010
2	2.28855026	0.73353167	0.2543	1.0553
3	1.55501858	0.64965427	0.1728	1.2281
4	0.90536432	0.80426438	0.1006	1.3286
5	0.10109994	0.34144282	0.0112	1.3399
6	-.24034288	0.33730826	-0.0267	1.3132
7	-.57765114	0.22526769	-0.0642	1.2490
8	-.80291884	0.63511393	-0.0892	1.1598
9	-1.4380328		-0.1598	1.0000

3 factors will be retained by the NFACTOR criterion.

Iteration	Change	Communalities				
1	0.1478	0.48282	0.56326	0.28591	0.11868	0.31355
		0.41940	0.53151	0.13531	0.17546	
2	0.0839	0.56675	0.57757	0.29397	0.12229	0.33194
.						
.						
.						
20	0.0011	0.76822	0.57481	0.24210	0.13173	0.33990
		0.61865	0.57340	0.12198	0.15867	
21	0.0009	0.76912	0.57490	0.24178	0.13176	0.33995
		0.61932	0.57330	0.12183	0.15861	

Convergence criterion satisfied.

Eigenvalues of the Weighted Reduced Correlation
Matrix: Total = 9 Average = 1

	Eigenvalue	Difference	Proportion	Cumulative
1	5.68808990	3.80898432	0.6320	0.6320
2	1.87910558	0.44662378	0.2088	0.8408
3	1.43248180	0.51764877	0.1592	1.0000
4	0.91483303	0.57557206	0.1016	1.1016
5	0.33926097	0.20137943	0.0377	1.1393
6	0.13788155	0.26446961	0.0153	1.1546
7	-.12658807	0.15981390	-0.0141	1.1406
8	-.28640197	0.69226084	-0.0318	1.1087
9	-.97866280		-0.1087	1.0000

Coefficient Alpha for Each Factor

Factor1	Factor2	Factor3
0.92721832	0.52631092	0.33964971

Root Mean Square Off-Diagonal Residuals: Overall = 0.04671565

The iteration process converged (the maximum absolute change in the communalities is less than the default value of 0.001) in 21 iterations. The details of the iteration steps are suppressed to save space. Only three common factors are selected using the alpha factor analysis method because we have used the NFACT=3 option in the program. The first three eigenvalues (see Output 4.2) of the weighted reduced correlation matrix $\hat{\mathbf{H}}^{-1/2}(\mathbf{R}_a - \hat{\boldsymbol{\Psi}})\hat{\mathbf{H}}^{-1/2}$ are 5.6881, 1.8791, and 1.4325. The values of Cronbach's *alpha* $(\alpha_1, \alpha_2, \alpha_3)$ are printed in SAS output, under the heading "Coefficient Alpha for Each Factor." These values are 0.9272, 0.5263, and 0.3397. Total communality estimate is 3.5318. The overall root mean square, that is, RMS_overall, is 0.0467.

4.6 Heywood Cases

In factor analysis, a certain objective function representing distance between the sample variance-covariance (or correlation) matrix $\mathbf{S}$ and the corresponding population covariance matrix $\boldsymbol{\Sigma} = \mathbf{LL}' + \boldsymbol{\Psi}$ is minimized to obtain the estimates $\hat{\mathbf{L}}$ and $\hat{\boldsymbol{\Psi}}$ subject to the condition that $\boldsymbol{\Psi}$ is positive definite. The solutions obtained may sometimes lead to improper solutions. An example of such case was presented in Section 4.2. Such a case is often termed a Heywood case. Specifically, a Heywood case occurs if, during the iterative process, at least one of the estimates of specific variances $\hat{\psi}_i$ turns out to be nonpositive. Since $\hat{\psi}_i$, $i = 1, \ldots, k$ represent estimates of nonzero variances, such solutions are not feasible and hence are unacceptable.

A Heywood case may occur in practice because a matrix $\boldsymbol{\Sigma} = \mathbf{LL}' + \boldsymbol{\Psi}$ may still be positive definite even though one or more elements of $\boldsymbol{\Psi}$ may be zero or negative. This can be seen from the following construction of a positive definite matrix due to Heywood (1931).

We choose numbers $\ell_1, \ell_2, \ldots \ell_p$ satisfying the conditions $\ell_1 > 0, 0 < \ell_i < 1, i = 2, \ldots, p$ and

$$\ell_1^2 \leq 1 + \frac{1}{\frac{\ell_2^2}{1-\ell_2^2} + \cdots + \frac{\ell_p^2}{1-\ell_p^2}},$$

and let $\psi_1, \psi_2, \ldots \psi_p$ be such that $\ell_1^2 + \psi_1 = 1, \ell_2^2 + \psi_2 = 1, \ldots, \ell_p^2 + \psi_p = 1$. Then the matrix

$$\begin{bmatrix} \ell_1^2 + \psi_1 & \ell_1\ell_2 & \ldots & \ell_1\ell_p \\ \ell_1\ell_2 & \ell_2^2 + \psi_2 & \ldots & \ell_2\ell_p \\ \vdots & & & \\ \ell_1\ell_p & \ell_2\ell_p & \ldots & \ell_p^2 + \psi_p \end{bmatrix}$$

will be positive definite. Note that ψ_1 here can be ≤ 0. This construction of a positive definite matrix differs from the factor analytic solution of the one factor model in the sense that whenever $\psi_1 < 0$, we must have $\ell_1^2 \geq 1$, whereas that in factor analysis ℓ_1^2 must be less than one. Thus, a Heywood case may give an improper solution to a factor analysis problem. However, it is not true that all improper solutions are due to Heywood cases. There have been studies by Van Driel (1978), Anderson and Gerbing (1984), Boomsma (1985), Bollen (1987) and others for determining circumstances under which improper solutions for a factor analysis problem are likely to occur. Using the empirical studies, some of the causes identified for improper solutions are

- small sample size,
- small number of variables,

- too many common factors extracted,
- too few common factors extracted,
- poor choice of prior communality estimates,
- inappropriate factor model for the data in hand,
- presence of outliers in the data.

The occurrence of improper solutions should not always be considered in a negative light. If possible, causes for the improper solutions should be identified and attempts made to rectify the problem. Some of the cures proposed in the literature are

- ignore the problem when the estimate is not significantly different from a proper solution,
- drop the problematic variables,
- fix the improper solution to a plausible value,
- obtain a larger sample,
- increase the number of variables,
- use an inequality restricted minimization to prevent an improper solution,
- identify and drop outliers before applying the factor analysis methods to a data set.

In the FACTOR procedure of SAS, if a communality h_i^2 is equal to one, and thus the specific variance $\psi_i = 1 - h_i^2$ is zero, the situation is referred to as a Heywood case. If a communality exceeds one, the specific variance will be negative. Such a situation is referred to as an Ultra-Heywood case. For any of the four iterative procedures discussed in the previous section, a Heywood or Ultra-Heywood case may occur. In the case of using the default choice, if a Heywood or an Ultra-Heywood case occurs during the iterative process the process will stop without producing a solution. However, when the HEYWOOD (or HEY) option in PROC FACTOR is specified, the iteration continues by setting any communality greater than one equal to one. By contrast, the ULTRAHEYWOOD (or ULTRA) option allows the iterative process to continue despite the fact that a communality exceeds one. This option, however, can cause nonconvergence problems. However, the final solution obtained with the HEYWOOD or ULTRAHEYWOOD options may still be improper.

4.7 Comparison of the Methods

Several simulation studies to compare various factor analysis methods are evaluated in the literature (see Browne, 1968, Linn, 1968, and Velicer, 1977). Different methods in general yield different factor solutions. However, when the sample size and the number of variables are large all the methods appear to yield similar solutions. Nevertheless, Acito and Anderson (1980) concluded through simulation studies that when the sample size and number of variables are small, the image and alpha factor analysis methods are relatively more accurate than the principal components method in recovering the factors.

To compare various methods and illustrate features of these methods on a data set, we will consider the ecology data and analyze using these different factor analysis methods. The methods in consideration are the principal component method (PRINCIPAL), the principal factor method with the three choices of estimates of prior communalities (PF-SMC, PF-MAX, PF-ASMC), image analysis (IMAGE), Harris' canonical factor analysis (HARRIS), maximum likelihood method (ML), unweighted least squares (ULS), iterated principal factor method (PRINIT), and the alpha factor analysis (ALPHA). For each method we allow the FACTOR procedure to pick the default options (such as the number of factors selected and the prior communality estimates). Table 4.4 presents the methods used, number of factors selected (NFACT), criterion used to select the number of factors (CRITERION), the total or sum of the estimated communalities (TOTCOM), and the square root of the mean squares of the off-diagonal elements of the residual matrix (RMS_overall). Also presented in the same table, within parentheses, are the values of TOTCOM and RMS_overall when a method was forced to select exactly three factors using the NFACT=3 option. The default choice of prior estimates of the communalities for the ML, ULS, and ALPHA procedures is SMC. For all the other methods this option does not apply. The number of maximum iterations in the entire comparison was limited to 30. The SAS statements required to produce the quantities in Table 4.4 are

```
* Non-iterative methods;
proc factor data=ecology method=principal res;
proc factor data=ecology method=principal nfact=3 res;
proc factor data=ecology method=principal priors=smc res;
proc factor data=ecology method=principal priors=smc nfact=3 res;
proc factor data=ecology method=principal priors=max res;
proc factor data=ecology method=principal priors=max nfact=3 res;
proc factor data=ecology method=principal priors=asmc res;
proc factor data=ecology method=principal priors=asmc nfact=3 res;
proc factor data=ecology method=image res;
proc factor data=ecology method=image nfact=3 res;
proc factor data=ecology method=harris res;
proc factor data=ecology method=harris nfact=3 res;

* Iterative methods;
proc factor data=ecology method=ml hey res;
proc factor data=ecology method=ml hey nfact=3 res;
proc factor data=ecology method=uls hey res;
proc factor data=ecology method=uls hey nfact=3 res;
proc factor data=ecology method=prinit hey res;
proc factor data=ecology method=prinit hey nfact=3 res;
proc factor data=ecology method=alpha hey res;
proc factor data=ecology method=alpha hey nfact=3 res;
```

Table 4.4 is presented to point out the vast differences that may occur when different methods of extracting the factors are applied to the same data. The estimates of total communality (TOTCOM) and RMS_overall, jointly are used for selecting the best method.

TABLE 4.4 Factor Analysis of Ecology Data

Method	NFACT	CRITERION	TOTCOM (for nfact=3**)	RMS (for nfact=3**)
PRINCIPAL	3	MINEIGEN	5.1769 (5.1769)	.1101 (.1101)
PF-SMC	2	PROPORTION	2.6868 (3.3860)	.0618 (.0455)
PF-MAX	4	PROPORTION	3.6740 (3.3860)	.0618 (.0455)
PF-ASMC	4	PROPORTION	3.6618 (3.4186)	.0261 (.0385)
IMAGE	2	MINEIGEN	4.0107 (4.4486)	.0994 (.1170)
HARRIS	1	MINEIGEN	2.0834 (3.0200)	.0945 (.0461)
ML	2	PROPORTION	2.9444 (3.8622)	.0591 (.0357)
ULS	2	PROPORTION	2.9568 (3.8509)	.0579 (.0336)
PRINIT	3	MINEIGEN	3.3729*(3.7329)	.0340 (.0340)
ALPHA	2	PROPORTION	3.0718 (3.5317)	.0759 (.0467)

*The number of iterations exceeded 30.

**The values in parentheses are for the indicated option, NFACT=3.

Normally, we expect that a good method will have a large value of TOTCOM and a small value of RMS_overall.

From Table 4.4, it appears that the ULS and ML method with three factors fare best for analyzing ecology data. The TOTCOM and RMS values for the ML method respectively are 3.8622 and 0.0357 and these values for ULS method are 3.8509 and 0.0336. Another choice may be the principal factor method with ASMC as the prior communality estimates for which these values are 3.4186 and 0.0385, respectively.

4.8 Factor Rotation

As was observed in Section 4.2, the factor loadings given by $\mathbf{L}$ are determined only up to an orthogonal transformation matrix $\mathbf{\Gamma}$ in the sense that if $\mathbf{L}$ is a set of factor loadings then so is $\mathbf{L}^* = \mathbf{L\Gamma}$, where $\mathbf{\Gamma}$ is an orthogonal matrix. This apparent nonuniqueness of the solutions can be a blessing in disguise for the practitioners of factor analysis. If the factor loadings obtained are not readily interpretable, it is customary to transform these loadings by post multiplication using an orthogonal matrix so that a meaningful interpretation of the factor with new loadings is possible. This process is often referred to as *factor rotation* since instead of original factors $\mathbf{f}$, now the transformed or rotated factors $\mathbf{f}^* = \mathbf{\Gamma}'\mathbf{f}$ are used.

Generally, the factors are rotated so that only a few variables get very large absolute loadings while rest of the variables get small or zero loadings. This kind of pattern will enable the practitioner to interpret various factors. For example, suppose, in a practical problem, the factor loadings corresponding to only the first three original variables are large (in their absolute values) for the first factor. Then the first common factor can be interpreted as a linear combination of only the first three original variables.

In the literature certain nonorthogonal linear transformations or oblique rotations have also been suggested.

4.8.1 Orthogonal Rotations

Several methods for determining the appropriate orthogonal matrix introducing the rotation have been proposed in the literature. These methods optimize certain objective functions to determine an orthogonal transformation.

Quartimax rotation. One popular approach is to design an orthogonal transformation so that the variance computed from the squared loadings of the transformed factors is maximized. That is, if $\mathbf{L}$ is the p by k matrix of the factor loadings then we want to find a desirable orthogonal matrix $\mathbf{\Gamma}$ such that with $\mathbf{L}^* = \mathbf{L\Gamma}$,

$$\begin{aligned}&\frac{1}{pk}\sum_i\sum_j \ell_{ij}^{*4} - (\frac{1}{pk}\sum_i\sum_j \ell_{ij}^{*2})^2,\\ &= \frac{1}{pk}\sum_i\sum_j \ell_{ij}^{*4} - (\frac{1}{pk}\sum_i h_i^{*2})^2 \qquad (4.19)\end{aligned}$$

is maximized. Since the communalities $h_i^{*2} = \sum_{j=1}^k \ell_{ij}^{*2} = \sum_{j=1}^k \ell_{ij}^2 = h_i^2,\ i = 1, \ldots, p$ are all constants and independent of the transformation $\mathbf{\Gamma}$, maximizing the the expression in Equation 4.19 is the same as maximizing only the first term $\sum_i \sum_j \ell_{ij}^{*4}$. The orthogonal transformation obtained by maximizing $\sum_i \sum_j \ell_{ij}^{*4}$ is referred to as the *quartimax* rotation (Neuhaus and Wrigley (1954)).

Raw varimax rotation. Kaiser (1958) suggested that the variances of squared loading elements corresponding to each column be computed and the sum of these variances, that is,

$$\sum_{j=1}^{k}\left\{\frac{1}{p}\sum_{i=1}^{p}(\ell_{ij}^{*2})^2-\left[\frac{1}{p}\sum_{i=1}^{p}\ell_{ij}^{*2}\right]^2\right\} \tag{4.20}$$

be maximized to get the orthogonal transformation $\mathbf{\Gamma}$. The solution obtained is the *raw varimax* rotation (Kaiser, 1958).

Varimax rotation. The *varimax* procedure that is most frequently used in applications corresponds to the orthogonal transformation obtained by maximizing

$$\sum_{j=1}^{k}\left\{\frac{1}{p}\sum_{i=1}^{p}\left(\frac{\ell_{ij}^{*2}}{h_i}\right)^2-\left[\frac{1}{p}\sum_{i=1}^{p}\left(\frac{\ell_{ij}^{*2}}{h_i}\right)\right]^2\right\}, \tag{4.21}$$

which is essentially a normalized version of the objective function given in Expression 4.20.

Other rotations. Harman (1976) considered maximizing a linear combination of the objective functions corresponding to the quartimax and raw varimax rotations. He suggested that

$$\frac{1}{p}\sum_{j=1}^{k}\left\{\sum_{i=1}^{p}\ell_{ij}^{*4}-\frac{\gamma}{p}\left(\sum_{i=1}^{p}\ell_{ij}^{*2}\right)^2\right\}, \tag{4.22}$$

be maximized, where γ is a known constant. Also see Mulaik (1972) for additional discussion on combining several criteria. The criterion given in Expression 4.22 is quite general since by choosing different values for γ, different orthogonal transformations can be obtained. The choice $\gamma = 0$ corresponds to quartimax solution, $\gamma = 1$ results in raw varimax solution, and $\gamma = \frac{k}{2}$ leads to the *equimax* solution of Saunders (1962). Finally, $\gamma = p(k-1)/(p+k-2)$ gives the *parsimax* solution of Crawford (1967). We will not discuss the equimax and parsimax solutions in detail here.

In SAS, instead of Expression 4.22, its normalized form is adopted for determining orthogonal rotations using the various methods of rotation described above. Specifically, in Expression 4.22, instead of ℓ_{ij}^{*2} the term similar to that in Expression 4.21, that is, $\frac{\ell_{ij}^{*2}}{h_i}$ is taken. In other words, the quantity being maximized is

$$\sum_{j=1}^{k}\left\{\frac{1}{p}\sum_{i=1}^{p}\left(\frac{\ell_{ij}^{*2}}{h_i}\right)^2-\left[\frac{\gamma}{p}\sum_{i=1}^{p}\left(\frac{\ell_{ij}^{*2}}{h_i}\right)\right]^2\right\}. \tag{4.23}$$

In SAS any one of these aforementioned orthogonal rotations can be computed by specifying a value for γ in the GAMMA = option in the PROC FACTOR statement. Any specific cases such as QUARTIMAX, EQUIMAX, PARSIMAX, and VARIMAX can also be specified in the ROTATE = option, in which case the option GAMMA = is not required. Further, the ROTATE=ORTHOMAX option without the GAMMA = option (which defaults to $\gamma = 1$), results in varimax rotation.

Whenever the FACTOR procedure is adopted to get a factor solution using a certain specified estimation method, the matrix of factor loadings $\mathbf{L}$, which is also the matrix of the covariances (matrix of correlations if the starting point matrix is a correlation matrix) between the original variables and the common factors, is printed in the corresponding

SAS output under the heading "Factor Pattern." If a certain factor rotation is used then, along with **L**, the orthogonal transformation matrix $\mathbf{\Gamma}$ and $\mathbf{L}^*$ under the heading "Rotated Factor Pattern" is also printed in the output. This distinction is important in order to be able to interpret results correctly.

Generally, different rotation methods will lead to different rotated factor patterns. If most of these methods yield similar patterns with similar interpretations then we can be fairly certain that the factors have been identified correctly. However, in practice it will rarely happen that most of the methods will yield similar results. In such cases, the appropriate choice should be based on the particular context and on subject matter knowledge.

EXAMPLE 8 ***Factor Rotation, Ecology Data (continued)*** To illustrate different factor rotations we return to the ecology data of Example 5. We apply the principal factor method with prior communalities equal to ASMC and want to extract three common factors. In Program 4.3 various rotations have been implemented. Partial output follows the program.

```
/* Program 4.3 */

options ls=64 ps=45 nodate nonumber;
title1 'Output 4.3';
title2 'Factor Analysis of the Ecology Data';

data ecology (type=corr);
infile cards missover;
_type_='CORR';
if _n_=1 then _type_='N';
input _name_ $ grade moist dock acar chey glyc lars cryp psoc;
lines;
n 165
grade 1.0
moist .441 1.0
dock .441 .342 1.0
acar .107 .250 .040 1.0
chey .194 .323 .060 .180 1.0
glyc .105 .400 .082 .123 .220 1.0
lars .204 .491 .071 .226 .480 .399 1.0
cryp .197 .158 .051 .019 .138 -.114 .154 1.0
psoc -.236 -.220 -.073 -.199 -.084 -.304 -.134 -.096 1.00
;

proc factor data=ecology method=prin priors=asmc
nfact=3 rotate=varimax;
title3 'Principal Factor Method, Varimax Rotation';
run;
proc factor data=ecology method=prin priors=asmc
nfact=3 rotate=quartimax;
title3 'Principal Factor Method, Quartimax Rotation';
run;
proc factor data=ecology method=prin priors=asmc
nfact=3 rotate=equamax;
title3 'Principal Factor Method, Equamax Rotation';
run;
proc factor data=ecology method=prin priors=asmc
nfact=3 rotate=parsimax;
title3 'Principal Factor Method, Parsimax Rotation';
run;
```

```
proc factor data=ecology method=ml hey nfact=3
rotate=varimax;
title3 'ML Method, Varimax Rotation';
run;
proc factor data=ecology method=ml hey nfact=3
rotate=quartimax;
title3 'ML Method, Quartimax Rotation';
run;
proc factor data=ecology method=ml hey nfact=3
rotate=equamax;
title3 'ML Method, Equamax Rotation';
run;
proc factor data=ecology method=ml hey nfact=3
rotate=parsimax;
title3 'ML Method, Parsimax Rotation';
run;

proc factor data=ecology method=prin mineigen=1
rotate=varimax;
title3 'Little Jiffy: PC method with Varimax Rotation';
run;

proc factor data=ecology method=prin priors=asmc
nfact=3 rotate=hk;
title3 'Principal Factor Method, Oblique Rotation (HK)';
run;
proc factor data=ecology method=prin priors=asmc
nfact=3 rotate=promax;
title3 'Principal Factor Method, Oblique Rotation (PROMAX)';
run;

proc factor data=ecology method=ml hey nfact=3 rotate=hk;
title3 'ML Method, Oblique Rotation (HK)';
run;
proc factor data=ecology method=ml hey nfact=3 rotate=promax;
title3 'ML Method, Oblique Rotation (PROMAX)';
run;
```

Output 4.3

```
                          Output 4.3
               Factor Analysis of the Ecology Data
             Principal Factor Method, Varimax Rotation

                     The FACTOR Procedure
              Initial Factor Method: Principal Factors

                Prior Communality Estimates: ASMC
```

grade	moist	dock	acar	chey
0.48784854	0.66252202	0.34728410	0.15705521	0.36743719

glyc	lars	cryp	psoc
0.46006913	0.59764497	0.16697002	0.24816881

Eigenvalues of the Reduced Correlation Matrix:
Total = 3.495 Average = 0.38833333

	Eigenvalue	Difference	Proportion	Cumulative
1	2.23843032	1.49722682	0.6405	0.6405
2	0.74120350	0.30226042	0.2121	0.8525
3	0.43894308	0.19574465	0.1256	0.9781
4	0.24319843	0.15452839	0.0696	1.0477
5	0.08867004	0.02284109	0.0254	1.0731
6	0.06582895	0.10637630	0.0188	1.0919
7	-.04054735	0.04932374	-0.0116	1.0803
8	-.08987109	0.10098482	-0.0257	1.0546
9	-.19085591		-0.0546	1.0000

3 factors will be retained by the NFACTOR criterion.

Factor Pattern

	Factor1	Factor2	Factor3
grade	0.53823	0.47650	0.02508
moist	0.78421	0.09193	-0.04324
dock	0.35539	0.46445	-0.08199
acar	0.30943	-0.09874	-0.02195
chey	0.49448	-0.21455	0.27817
glyc	0.50520	-0.30743	-0.38293
lars	0.67365	-0.34444	0.21146
cryp	0.19390	0.14494	0.30642
psoc	-0.34722	-0.00652	0.25810

Rotation Method: Varimax

Orthogonal Transformation Matrix

	1	2	3
1	0.65756	0.51905	0.54608
2	-0.39878	0.85474	-0.33225
3	0.63921	-0.00071	-0.76903

Rotated Factor Pattern

	Factor1	Factor2	Factor3
grade	0.17993	0.68663	0.11631
moist	0.45137	0.48566	0.43095
dock	-0.00394	0.58151	0.10281
acar	0.22881	0.07623	0.21865
chey	0.58851	0.07308	0.12738
glyc	0.21002	-0.00028	0.67251
lars	0.71549	0.05510	0.31969
cryp	0.26557	0.22431	-0.17792
psoc	-0.06074	-0.18598	-0.38594

```
Principal Factor Method, Quartimax Rotation

              The FACTOR Procedure
          Rotation Method: Quartimax

      Orthogonal Transformation Matrix

                   1              2              3

  1          0.88608        0.46135        0.04496
  2         -0.46123        0.86785        0.18465
  3          0.04617       -0.18436        0.98177


Principal Factor Method, Equamax Rotation

              The FACTOR Procedure
          Rotation Method: Equamax

      Orthogonal Transformation Matrix

                   1              2              3

  1          0.64590        0.52369        0.55549
  2         -0.40059        0.85190       -0.33734
  3          0.64988        0.00464       -0.76002


Principal Factor Method, Parsimax Rotation

              The FACTOR Procedure
          Rotation Method: Parsimax

      Orthogonal Transformation Matrix

                   1              2              3

  1          0.64190        0.52649        0.55746
  2         -0.40320        0.85015       -0.33865
  3          0.65222        0.00739       -0.75799
```

From Output 4.3, we see that the estimate $\hat{\mathbf{L}}$ of the factor loadings matrix $\mathbf{L}$ printed under the heading "Factor Pattern" is given by

$$\hat{\mathbf{L}} = \begin{bmatrix} .5382 & .4765 & .0251 \\ .7842 & .0919 & -.0432 \\ .3554 & .4645 & -.0820 \\ .3094 & -.0987 & -.0220 \\ .4945 & -.2146 & .2782 \\ .5052 & -.3074 & -.3829 \\ .6737 & -.3444 & .2115 \\ .1939 & .1449 & .3064 \\ -.3472 & -.0065 & .2581 \end{bmatrix}$$

Note that the loadings assigned by all the variables to the first factor are quite large, except perhaps by the variable CRYP, which has a loading of 0.1939. This kind of pattern of loadings makes it difficult to represent the common factors in terms of only a few original variables. Hoping to improve this situation, we use various orthogonal transformations introduced in this section. The appropriate SAS code is provided in Program 4.3.

In Table 4.5, extracted from Output 4.3, we have summarized the rotated factor loadings produced by the four different orthogonal transformations, namely the varimax, quartimax, equamax, and parsimax. For each rotation method, SAS also prints the orthogonal matrix that was used to rotate the original factor loadings **L**. For example, for the varimax rotation, the orthogonal transformation matrix (printed under title "Rotation Method: Varimax") is

$$\Gamma = \begin{bmatrix} .6576 & .5191 & .5461 \\ -.3988 & .8547 & -.3323 \\ .6392 & -.0007 & -.7690 \end{bmatrix}.$$

See Output 4.3. In Table 4.5, the three common factors are identified as F1, F2, and F3.

TABLE 4.5 Rotated Factor Loadings for Ecology Data

Rotation →	Varimax			Quartimax			Equamax			Parsimax		
Variables	F1	F2	F3	F1	F2	F3	F1	F2	F3	F1	F2	F3
GRADE	.12	.69	.12	.26	.66	.14	.17	.69	.12	.17	.69	.12
MOIST	.45	.49	.43	.65	.45	.01	.44	.49	.44	.44	.49	.44
DOCK	-.00	.58	.10	.10	.58	.02	-.01	.58	.10	-.01	.58	.10
ACAR	.23	.08	.22	.32	.06	-.03	.23	.08	.22	.22	.08	.22
CHEY	.59	.07	.13	.55	-.01	.26	.59	.08	.14	.59	.08	.14
GLYC	.21	-.00	.67	.57	.04	-.41	.20	.00	.68	.20	.00	.68
LARS	.72	.06	.32	.77	-.03	.17	.71	.06	.33	.71	.06	.33
CRYP	.27	.22	-.18	.12	.16	.34	.27	.23	-.17	.27	.23	-.17
PSOC	-.06	-.19	-.39	-.29	-.21	.24	-.05	-.19	-.39	-.05	-.19	-.39

For discussion, let us first consider the factor loadings corresponding to the varimax rotation. The first factor (F1) receives high loadings, namely 0.59 and 0.72, from the variables CHEY and LARS, respectively. Hence, the first factor may be interpreted as the variable that represents the growth of (the arthropod counts of) cheyletus (CHEY) and larsonemus (LARS). The second factor (F2) receives high loadings, namely 0.69, 0.49, and 0.58, respectively from the variables GRADE, MOIST, and DOCK. Thus, the second factor may be interpreted as the grain environment factor. The third factor (F3) receives high loading of 0.67 from only one variable, that is, GLYC. Hence, the third factor may be interpreted as the factor that represents the growth of glycychagus (GLYC). Thus, for the present set of growth environment variables, the other arthropods except the three mentioned, namely CHEY, LARS and GLYC, are not prominent in their growth.

The equamax and parsimax methods also provide similar interpretations for the three factors. The quartimax rotation method, however, produces the factors with some other interesting interpretations. The first factor seems to indicate that the three types of arthropods, CHEY (with loading 0.55), GLYC (loading 0.57), and LARS (loading 0.77) grow well in a moist (MOIST with loading 0.65) environment. The second factor seems to represent grain quality (GRADE with loading 0.66), and dockage (DOCK with loading 0.58). This combination perhaps makes sense, since low-grade grain (represented by higher number) tends to contain more dockage. The third factor cannot be interpreted.

In Program 4.3 we also have included the code to perform a similar analysis using the ML method including various rotations.

Output 4.3
(continued)

```
                         Output 4.3
             Factor Analysis of the Ecology Data
                ML Method, Varimax Rotation

                    The FACTOR Procedure
          Initial Factor Method: Maximum Likelihood

             Prior Communality Estimates: SMC

     grade        moist         dock         acar         chey

0.33503944   0.45499985   0.23850409   0.10786071   0.25234462

       glyc            lars            cryp            psoc

 0.31596140      0.41044427      0.11466990      0.17043474


             Preliminary Eigenvalues: Total =
             3.60340214   Average = 0.40037802

       Eigenvalue    Difference    Proportion    Cumulative

  1    3.22852154    2.36126459        0.8960        0.8960
  2    0.86725695    0.43397155        0.2407        1.1366
  3    0.43328540    0.23823513        0.1202        1.2569
  4    0.19505027    0.17827358        0.0541        1.3110
  5    0.01677670    0.12972367        0.0047        1.3157
  6    -.11294697    0.17563764       -0.0313        1.2843
  7    -.28858461    0.03966041       -0.0801        1.2042
  8    -.32824502    0.07946711       -0.0911        1.1131
  9    -.40771213                     -0.1131        1.0000

  3 factors will be retained by the NFACTOR criterion.


                      Factor Pattern

                Factor1         Factor2         Factor3

  grade         0.10500         0.59141        -0.46411
  moist         0.40000         0.61633        -0.10863
  dock          0.08200         0.37190        -0.44973
  acar          0.12300         0.26041         0.07534
  chey          0.22000         0.46093         0.26014
  glyc          1.00000        -0.00000        -0.00000
  lars          0.39900         0.60695         0.42403
  cryp         -0.11400         0.32700         0.00769
  psoc         -0.30400        -0.15998         0.15682
```

Rotation Method: Varimax

Orthogonal Transformation Matrix

	1	2	3
1	0.25305	0.03464	0.96683
2	0.76440	0.60540	-0.22176
3	0.59301	-0.79516	-0.12671

Rotated Factor Pattern

	Factor1	Factor2	Factor3
grade	0.20342	0.73072	0.02918
moist	0.50792	0.47337	0.26382
dock	0.03834	0.58560	0.05380
acar	0.27486	0.10201	0.05162
chey	0.56227	0.07982	0.07752
glyc	0.25305	0.03464	0.96683
lars	0.81637	0.04410	0.19744
cryp	0.22567	0.18790	-0.18371
psoc	-0.10622	-0.23208	-0.27831

ML Method, Quartimax Rotation

The FACTOR Procedure
Rotation Method: Quartimax

Orthogonal Transformation Matrix

	1	2	3
1	0.34429	0.07067	0.93620
2	0.78147	0.53110	-0.32747
3	0.52035	-0.84436	-0.12763

ML Method, Equamax Rotation

The FACTOR Procedure
Rotation Method: Equamax

Orthogonal Transformation Matrix

	1	2	3
1	0.20785	0.97816	-0.00026
2	0.76341	-0.16205	0.62525
3	0.61156	-0.13016	-0.78042

```
                    ML Method, Parsimax Rotation

                        The FACTOR Procedure
                    Rotation Method: Parsimax

                Orthogonal Transformation Matrix

                        1               2               3

        1         0.18389         0.98275        -0.01959
        2         0.76294        -0.13014         0.63323
        3         0.61976        -0.13139        -0.77372
```

From Output 4.3, the matrix of estimated factor loadings **L**, when the maximum likelihood method is used for estimation, is

$$\hat{\mathbf{L}}_{ML} = \begin{bmatrix} .1050 & .5914 & -.4641 \\ .4000 & .6163 & -.1086 \\ .0820 & .3719 & -.4497 \\ .1230 & .2604 & .0753 \\ .2200 & .4609 & .2601 \\ 1.000 & -.0000 & -.0000 \\ .3990 & .6070 & .4240 \\ -.1140 & .3270 & .0077 \\ -.3040 & -.1600 & .1568 \end{bmatrix}.$$

As noted earlier in Example 5, a Heywood case has occurred while determining the ML solution. However, as the HEYWOOD option was used here, the iterative process continued, resulting in the above factor pattern. It is difficult to interpret the factors with these loadings. Hence, this factor pattern was rotated using different rotation methods discussed earlier. The SAS code is included in Program 4.3, and a summary is provided in Table 4.6.

TABLE 4.6 Rotated Factor Loadings (ML Method)

Rotation →	Varimax			Quartimax			Equamax			Parsimax		
Variables	F1	F2	F3	F1	F2	F3	F1	F2	F3	F1	F2	F3
GRADE	.20	.73	.03	.26	.71	-.04	.19	.07	.73	.18	.09	.73
MOIST	.51	.47	.26	.56	.45	.19	.49	.31	.47	.48	.33	.47
DOCK	.04	.59	.05	.08	.58	.01	.03	.08	.58	.02	.09	.58
ACAR	.27	.10	.05	.29	.08	.02	.27	.07	.10	.27	.08	.10
CHEY	.56	.08	.08	.57	.04	.02	.56	.11	.09	.55	.12	.09
GLYC	.25	.03	.97	.34	.07	.94	.21	.98	-.00	.18	.98	-.02
LARS	.82	.04	.20	.83	-.01	.12	.81	.24	.05	.80	.26	.05
CRYP	.23	.19	-.18	.22	.16	-.21	.23	-.17	.20	.23	-.16	.20
PSOC	-.11	-.23	-.28	-.15	-.24	-.25	-.09	-.29	-.22	-.08	-.30	-.22

Under all the rotation alternatives used, the variables provide very similar loadings to the three factors and thus the factors have similar interpretations under these rotations. For example, for varimax rotation, the first factor (F1) represents the growth of arthropods, cheyletus (CHEY with loading 0.56) and larsoneums (LARS with loading 0.82), in moisture (MOIST with loading 0.51). The second factor (F2) represents a combination of quality (GRADE with loading 0.73) and dockage (DOCK with loading 0.59) and the last factor (F3) represents the growth of glycychagus (GLYC with loading 0.97) in the environment considered by the experimenters. The interpretation of the factors produced by the ML method closely agrees with the quartimax rotated factors of the principal factor method with ASMC priors (see Table 4.5).

Little jiffy method. Kaiser (1970) suggested using the principal component method for factor extraction, selecting all the factors corresponding to the eigenvalues ≥ 1 and rotating the resulting factors by varimax rotation. This method is applicable only to a factor analysis using the correlation matrix. It is based on the idea that the selected factors must have more information (that is, larger variance) than that for any of the original variables. If a correlation matrix is used, then the variance of any original variable is one and hence the justification of using the criterion of "eigenvalues ≥ 1" for selecting the factors. This method is the *little jiffy* method. The SAS code for this analysis is also included in Program 4.3. We summarize the results produced by this code (output not shown) in Table 4.7.

TABLE 4.7 Results from the Little Jiffy Method

	Factor Pattern			Rotated Factor Pattern		
Variables	FACT1	FACT2	FACT3	FACT1	FACT2	FACT3
GRADE	.6007	.5731	-.0696	.1800	.8133	-.0154
MOIST	.8029	.0802	-.0460	.5480	.5370	.2540
DOCK	.4184	.6526	-.2735	-.0946	.8143	.0606
ACAR	.4042	-.2428	-.0462	.3811	.0523	.2766
CHEY	.5768	-.2740	.4303	.7679	.0252	-.0514
GLYC	.5617	-.4311	-.4075	.3926	.0725	.7127
LARS	.7072	-.3480	.2697	.8145	.0767	.1571
CRYP	.2521	.3370	.6750	.3961	.2662	-.6364
PSOC	-.4390	.0507	.3837	-.1552	-.2961	-.4804

This method gives an interpretation similar to the ML method for the first two factors. The last factor appears to represent a contrast in growth between GLYC and CRYP and PSOC instead of just the growth of GLYC.

4.8.2 Oblique Rotations

Sometimes even after an orthogonal transformation on a factor loading matrix is made, the resulting factors may still not be interpretable. In this case, certain oblique rotations leading to correlated factors are suggested. The idea is to find a transformation of the original factor loading axes so that the new transformed axes pass through the clusters of the factor loadings more closely than those corresponding to an orthogonal transformation. Specifically, for a given factor loadings matrix $\mathbf{L}$ we want to obtain a k by k matrix $\mathbf{T}$, not necessarily an orthogonal matrix such that the loadings $\mathbf{L}^* = \mathbf{LT}$ have a more meaningful interpretation than $\mathbf{L}$.

HK rotation. Harris and Kaiser (1964) proposed that the matrix $\mathbf{T}$ we are seeking may be taken as a function of some simple orthonormal and diagonal matrices. For example, the transformation matrix $\mathbf{T}$ may be

$$\mathbf{T} = \mathbf{\Gamma}_2 \mathbf{D}_2 \mathbf{\Gamma}_1 \mathbf{D}_1,$$

where Γ_2 and Γ_1 are the orthonormal matrices, and $\mathbf{D}_2$ and $\mathbf{D}_1$ are diagonal. One useful class of Harris-Kaiser oblique rotation corresponds to the case when $\Gamma_2 = \mathbf{I}_k$, in which case,

$$\mathbf{T} = \mathbf{D}_2\Gamma_1\mathbf{D}_1.$$

Here $\mathbf{D}_2$ and Γ_1 are selected first and then $\mathbf{D}_1$ is chosen to provide an appropriate scaling so that the matrix of factor correlations, $\Phi = \mathbf{D}_1^{-1}\Gamma_1'\mathbf{D}_2^{-2}\Gamma_1\mathbf{D}_1^{-1}$ is a correlation matrix. Here Φ is such that the matrix we have selected for factoring is expressed as $(\mathbf{LT})\Phi(\mathbf{LT})'$. Harris and Kaiser (1964) suggested first using the principal factor method to obtain $\mathbf{L}$. Suppose Λ is the diagonal matrix of the eigenvalues of the pseudo correlation matrix used for the principal factor method. Then suggested selections of the matrices $\mathbf{D}_2$ and Γ_1 are $\mathbf{D}_2 = \Lambda^{-1/2}$ and Γ_1, the quartimax rotation (orthogonal transformation) matrix.

For determining this $\mathbf{T}$, the corresponding SAS option in PROC FACTOR is ROTATE=HK. While the orthogonal transformation $\mathbf{L}^* = \mathbf{L}\Gamma$ leaves the factors uncorrelated, the oblique transformation $\mathbf{L}^* = \mathbf{LT}$ leads to a new set of factors that are correlated. The correlation matrix of the new factors is $(\mathbf{T}'\mathbf{T})^{-1}$. The FACTOR procedure with the ROTATE=HK option prints these correlations between the factors under the heading "Inter-factor correlations"; the oblique rotated factors $\mathbf{LT}$, under the heading "Rotated Factor Pattern (Standardized Regression Coefficients)," and finally the correlation matrix between the original variables and the factors under the heading "Factor Structure (Correlations)." When the correlation matrix is used as a starting point for factor analysis and an orthogonal rotation is used, the correlation matrix between the original variables and the factors is $\mathbf{L}$ itself. However, when an oblique rotation $\mathbf{L}^* = \mathbf{LT}$ is used, the correlation matrix is given by $\mathbf{L}(\mathbf{T}')^{-1}$. These correlations can be useful for an in-depth study of the selected factors. However, the rotated factor pattern matrix should be used to find the correct interpretation of the selected factors.

PROMAX rotation. Another oblique transformation, called the *promax rotation*, was proposed by Hendrickson and White (1964). They assume that the simple structure for the loadings produced by an orthogonal transformation is close to the simple structure produced by an oblique rotation. Hence, they start with an orthogonally transformed factor loadings matrix. The orthogonal transformation usually corresponds to the varimax rotation. Thus, for the following discussion the factor rotation is assumed to be varimax.

Let $\mathbf{L}^* = \mathbf{L}\Gamma$ be the varimax rotated factor loadings. We construct a p by k matrix $\mathbf{Q} = (q_{ij})$ commonly called as the target matrix, which is presumably simply structured, such that

$$q_{ij} = |\ell_{ij}^{*^{m-1}}|\ell_{ij}^*,$$

where m is an integer greater than one. Thus, q_{ij} and ℓ_{ij}^* have the same sign and $|q_{ij}| = |\ell_{ij}^{*^m}|$.

The rationale of taking q_{ij} in terms of the mth power of ℓ_{ij}^* is to force the elements with smaller magnitudes to approach zero rapidly. Hendrickson and White (1964) recommend the target matrix $\mathbf{Q}$ is to be approximated by an oblique transformation of the matrix $\mathbf{L}^*$. The promax rotation is given by the matrix

$$\mathbf{T} = \mathbf{U}\{\text{diag}(\mathbf{U}'\mathbf{U})^{-1/2}\},$$

where the k by k matrix $\mathbf{U} = (\mathbf{u}_1 : \mathbf{u}_2 : \cdots : \mathbf{u}_k)$, $\mathbf{u}_j$ being the jth column of $\mathbf{U}$ is obtained by minimizing the quadratic forms

$$(\mathbf{q}_j - \mathbf{L}^*\mathbf{u}_j)'(\mathbf{q}_j - \mathbf{L}^*\mathbf{u}_j)$$

with respect to columns $\mathbf{u}_j$, $j = 1, \ldots, k$ of $\mathbf{U}$. The vectors $\mathbf{q}_j$, $j = 1, \ldots, k$ are the columns of the target matrix $\mathbf{Q}$. The solutions to the k minimization problems can be jointly expressed as

$$\mathbf{U} = (\mathbf{L}^{*'}\mathbf{L}^{*})^{-1}\mathbf{L}^{*'}\mathbf{Q}.$$

The promax rotation is adopted by specifying ROTATE=PROMAX (or simply ROTATE=P) as an option in FACTOR procedure. Instead of the default choice of varimax rotation for matrix $\mathbf{L}^*$, other orthogonal rotations can be chosen using the PREROTATE=option in conjunction with the ROTATE=PROMAX option. Different values of m can be suggested using the POWER=option. The default value of m is 3. The target matrix $\mathbf{Q}$ is printed under the title "Target Matrix for Procrustean Transformation." Also printed are the oblique transformation matrix $\mathbf{T}$, the interfactor correlation matrix $(\mathbf{T}'\mathbf{T})^{-1}$, the estimated rotated factor pattern matrix $\hat{\mathbf{L}}\mathbf{T}$, and the estimated factor structure (correlations) matrix $\hat{\mathbf{L}}(\mathbf{T}')^{-1}$.

EXAMPLE 9 ***Factor Rotation, Ecology Data (continued)*** For illustration, let us again consider the ecology data of Example 5. In the last section and in Example 8, we considered this data set and performed the factor analysis using the principal factor method and ML method with several orthogonal transformations. Here, we will use the same methods for estimation, but we will use the oblique rotations instead of orthogonal rotations. Specifically, we use the two oblique rotations discussed, namely HK and PROMAX. The SAS code for performing this analysis is provided in Program 4.3 and parts of the output are provided in the continuation of Output 4.3.

Output 4.3 *(continued)*

```
                         Output 4.3
              Factor Analysis of the Ecology Data
         Principal Factor Method, Oblique Rotation (HK)

                    The FACTOR Procedure
            Initial Factor Method: Principal Factors

              Prior Communality Estimates: ASMC

      grade         moist          dock          acar          chey

 0.48784854    0.66252202    0.34728410    0.15705521    0.36743719

        glyc            lars            cryp            psoc

  0.46006913      0.59764497      0.16697002      0.24816881


         Eigenvalues of the Reduced Correlation Matrix:
               Total = 3.495  Average = 0.38833333

         Eigenvalue    Difference    Proportion    Cumulative

    1    2.23843032    1.49722682        0.6405        0.6405
    2    0.74120350    0.30226042        0.2121        0.8525
    3    0.43894308    0.19574465        0.1256        0.9781
    4    0.24319843    0.15452839        0.0696        1.0477
    5    0.08867004    0.02284109        0.0254        1.0731
    6    0.06582895    0.10637630        0.0188        1.0919
    7    -.04054735    0.04932374       -0.0116        1.0803
    8    -.08987109    0.10098482       -0.0257        1.0546
    9    -.19085591                     -0.0546        1.0000
```

3 factors will be retained by the NFACTOR criterion.

Factor Pattern

	Factor1	Factor2	Factor3
grade	0.53823	0.47650	0.02508
moist	0.78421	0.09193	-0.04324
dock	0.35539	0.46445	-0.08199
acar	0.30943	-0.09874	-0.02195
chey	0.49448	-0.21455	0.27817
glyc	0.50520	-0.30743	-0.38293
lars	0.67365	-0.34444	0.21146
cryp	0.19390	0.14494	0.30642
psoc	-0.34722	-0.00652	0.25810

Rotation Method: Harris-Kaiser

Oblique Transformation Matrix

	1	2	3
1	0.26388384	0.4646937	0.50384783
2	-0.415886	1.04554511	-0.6205889
3	-1.1107383	-0.0980572	1.00279999

Inter-Factor Correlations

	Factor1	Factor2	Factor3
Factor1	1.00000	0.37005	0.55494
Factor2	0.37005	1.00000	0.47460
Factor3	0.55494	0.47460	1.00000

Rotated Factor Pattern (Standardized Regression Coefficients)

	Factor1	Factor2	Factor3
grade	-0.08399	0.74585	0.00063
moist	0.21673	0.46478	0.29471
dock	-0.00830	0.65880	-0.19139
acar	0.14709	0.04270	0.19517
chey	-0.08926	-0.02181	0.66123
glyc	0.68651	-0.04912	0.06133
lars	0.08614	-0.06782	0.76523
cryp	-0.34947	0.21160	0.31503
psoc	-0.37560	-0.19347	0.08792

Factor Structure (Correlations)

	Factor1	Factor2	Factor3
grade	0.19236	0.71507	0.30800
moist	0.55227	0.68485	0.63557
dock	0.12927	0.56489	0.11666
acar	0.27120	0.18976	0.29707
chey	0.26961	0.25897	0.60135
glyc	0.70236	0.23403	0.41898
lars	0.48570	0.32723	0.78084
cryp	-0.09634	0.23179	0.22152
psoc	-0.39841	-0.29074	-0.21234

Principal Factor Method, Oblique Rotation (PROMAX)

Prerotation Method: Varimax

Orthogonal Transformation Matrix

	1	2	3
1	0.65756	0.51905	0.54608
2	-0.39878	0.85474	-0.33225
3	0.63921	-0.00071	-0.76903

Rotated Factor Pattern

	Factor1	Factor2	Factor3
grade	0.17993	0.68663	0.11631
moist	0.45137	0.48566	0.43095
dock	-0.00394	0.58151	0.10281
acar	0.22881	0.07623	0.21865
chey	0.58851	0.07308	0.12738
glyc	0.21002	-0.00028	0.67251
lars	0.71549	0.05510	0.31969
cryp	0.26557	0.22431	-0.17792
psoc	-0.06074	-0.18598	-0.38594

Rotation Method: Promax

Target Matrix for Procrustean Transformation

	Factor1	Factor2	Factor3
grade	0.01714	0.91108	0.00486
moist	0.20361	0.24262	0.18610
dock	-0.00000	1.00000	0.00607
acar	0.38018	0.01345	0.34841
chey	1.00000	0.00183	0.01065
glyc	0.02900	-0.00000	1.00000
lars	0.82711	0.00036	0.07748
cryp	0.34435	0.19849	-0.10874
psoc	-0.00303	-0.08316	-0.81588

Procrustean Transformation Matrix

	1	2	3
1	1.44149206	-0.228206	-0.396635
2	-0.212361	1.35075523	-0.1617789
3	-0.3716827	-0.1020911	1.51769275

Normalized Oblique Transformation Matrix

	1	2	3
1	0.51705561	0.39454094	0.36408986
2	-0.515583	1.01918398	-0.3643604
3	0.98364832	-0.0544161	-1.0686343

Inter-Factor Correlations

	Factor1	Factor2	Factor3
Factor1	1.00000	0.38773	0.52639
Factor2	0.38773	1.00000	0.30646
Factor3	0.52639	0.30646	1.00000

Rotated Factor Pattern (Standardized Regression Coefficients)

	Factor1	Factor2	Factor3
grade	0.05729	0.69663	-0.00445
moist	0.31555	0.40545	0.29823
dock	-0.13636	0.61804	0.04779
acar	0.18931	0.02264	0.17209
chey	0.63991	-0.03871	-0.03906
glyc	0.04305	-0.09317	0.70517
lars	0.73391	-0.09677	0.14480
cryp	0.32694	0.20755	-0.30967
psoc	0.07770	-0.15768	-0.39986

Factor Structure (Correlations)

	Factor1	Factor2	Factor3
grade	0.32505	0.71748	0.23920
moist	0.62974	0.61920	0.58859
dock	0.12843	0.57982	0.16542
acar	0.28868	0.14878	0.27868
chey	0.60434	0.19743	0.28592
glyc	0.37812	0.13963	0.69928
lars	0.77261	0.23216	0.50146
cryp	0.24441	0.23941	-0.07396
psoc	-0.19392	-0.25009	-0.40728

ML Method, Oblique Rotation (HK)

The FACTOR Procedure
Rotation Method: Harris-Kaiser

Oblique Transformation Matrix

	1	2	3
1	0.06086	0.17602	0.97879
2	0.53535	0.63691	-0.18388
3	-0.94641	0.86417	-0.10850

Inter-Factor Correlations

	Factor1	Factor2	Factor3
Factor1	1.00000	0.39234	-0.04851
Factor2	0.39234	1.00000	0.01350
Factor3	-0.04851	0.01350	1.00000

Rotated Factor Pattern (Standardized Regression Coefficients)

	Factor1	Factor2	Factor3
grade	0.76224	-0.00591	0.04438
moist	0.45711	0.36907	0.28997
dock	0.62971	-0.13734	0.06067
acar	0.07560	0.25262	0.06433
chey	0.01396	0.55709	0.10235
glyc	0.06086	0.17602	0.97879
lars	-0.05209	0.82323	0.23292
cryp	0.16084	0.19485	-0.17255
psoc	-0.25257	-0.01989	-0.28515

Factor Structure (Correlations)

	Factor1	Factor2	Factor3
grade	0.75777	0.29374	0.00733
moist	0.58784	0.55233	0.27278
dock	0.57288	0.11053	0.02827
acar	0.17159	0.28314	0.06408
chey	0.22756	0.56395	0.10920
glyc	0.08245	0.21312	0.97822
lars	0.25960	0.80594	0.24657
cryp	0.24565	0.25562	-0.17772
psoc	-0.24654	-0.12283	-0.27317

ML Method, Oblique Rotation (PROMAX)

The FACTOR Procedure
Prerotation Method: Varimax

Orthogonal Transformation Matrix

	1	2	3
1	0.25305	0.03464	0.96683
2	0.76440	0.60540	-0.22176
3	0.59301	-0.79516	-0.12671

Rotated Factor Pattern

	Factor1	Factor2	Factor3
grade	0.20342	0.73072	0.02918
moist	0.50792	0.47337	0.26382
dock	0.03834	0.58560	0.05380
acar	0.27486	0.10201	0.05162
chey	0.56227	0.07982	0.07752
glyc	0.25305	0.03464	0.96683
lars	0.81637	0.04410	0.19744
cryp	0.22567	0.18790	-0.18371
psoc	-0.10622	-0.23208	-0.27831

Rotation Method: Promax

Target Matrix for Procrustean Transformation

	Factor1	Factor2	Factor3
grade	0.02039	0.90918	0.00006
moist	0.33877	0.26383	0.04959
dock	0.00029	1.00000	0.00084
acar	0.83382	0.04101	0.00577
chey	1.00000	0.00275	0.00274
glyc	0.01716	0.00004	1.00000
lars	0.96873	0.00015	0.01431
cryp	0.29294	0.16268	-0.16507
psoc	-0.02358	-0.23658	-0.44294

Procrustean Transformation Matrix

	1	2	3
1	1.50223663	-0.2367917	-0.2268435
2	-0.2908758	1.33067586	-0.0154005
3	-0.3810734	-0.0162936	1.0930852

Normalized Oblique Transformation Matrix

	1	2	3
1	0.00123368	-0.0236279	1.00301052
2	0.80055882	0.50191654	-0.4268772
3	0.88669465	-0.9559345	-0.2618585

Inter-Factor Correlations

	Factor1	Factor2	Factor3
Factor1	1.00000	0.37748	0.45195
Factor2	0.37748	1.00000	0.16217
Factor3	0.45195	0.16217	1.00000

Rotated Factor Pattern (Standardized Regression Coefficients)

	Factor1	Factor2	Factor3
grade	0.0620615	0.73801644	-0.025611
moist	0.3975724	0.40374031	0.16655589
dock	-0.1009458	0.61463352	0.04125814
acar	0.27543169	0.05578006	-0.0075228
chey	0.59993226	-0.0225227	-0.0442162
glyc	0.00123368	-0.0236279	1.00301052
lars	0.86237172	-0.1101315	0.030074
cryp	0.2684633	0.15946328	-0.255946
psoc	0.01060166	-0.2230261	-0.2776871

Factor Structure (Correlations)

	Factor1	Factor2	Factor3
grade	0.32907	0.75729	0.12212
moist	0.62525	0.58083	0.41171
dock	0.14971	0.58322	0.09531
acar	0.29309	0.15853	0.12600
chey	0.57145	0.19677	0.22327
glyc	0.44563	0.13949	0.99974
lars	0.83439	0.22027	0.40197
cryp	0.21298	0.21930	-0.10875
psoc	-0.19909	-0.26406	-0.30906

The rotated factor loadings obtained using oblique transformations, Harris-Kaiser (results of ROTATE=HK option), and promax (results of ROTATE=PROMAX option) for both the principal factor method and the ML method are summarized in Table 4.8.

TABLE 4.8 Oblique Rotated Factor Loadings

	Principal Factor Method (ASMC)						ML Method					
	HK			Promax			HK			Promax		
Variables	F1	F2	F3	F1	F2	F3	F1	F2	F3	F1	F2	F3
GRADE	-.08	.75	.00	.06	.70	-.00	.76	- .01	.04	.06	.74	-.03
MOIST	.22	.46	.29	.32	.41	.30	.46	.37	.29	.39	.40	.17
DOCK	-.01	.66	-.19	-.14	.62	.05	.63	- .14	.06	-.10	.61	.04
ACAR	.15	.04	.20	.19	.02	.17	.08	.25	.06	.28	.06	-.01
CHEY	-.09	-.02	.66	.64	-.04	-.04	.01	.56	.10	.60	-.02	-.04
GLYC	.69	-.04	.06	.04	-.09	.71	.06	.18	.98	.00	-.02	1.00
LARS	.09	-.07	.77	.73	-.10	.14	-.05	.82	.23	.86	-.11	.03
CRYP	-.35	.21	.32	.33	.21	-.31	.16	.19	-.17	.27	.16	-.26
PSOC	-.38	-.19	.09	.08	-.16	-.40	-.25	- .02	-.29	.01	-.22	-.28

Factor loadings from all these four analyses are very similar in magnitude as well as in their signs. It is comforting to note that all the four analyses essentially result in similar interpretations of the three factors. One factor represents the grain and the other two factors, the arthropods. For example, for the principal factor method with HK rotation, the first factor (Table 4.8, Column 2) essentially represents GLYC (with a loading of 0.69), whereas when the PROMAX rotation is used, the third factor (Column 7) represents GLYC (with a loading of 0.71). The second factor represents grain environment. For HK the loadings for GRADE, MOIST, and DOCK, respectively, are 0.75, 0.46, and 0.66 (Column 3), whereas for PROMAX these loadings for the second factor are 0.70, 0.41, and 0.62 (Column 6). The third factor can be represented as the growth of the arthropods CHEY and LARS. The corresponding loadings for the HK rotation are 0.66 and 0.77 (Column 4). For the PROMAX rotation, the first factor can be represented as the growth of the arthropods CHEY and LARS, with the corresponding loadings 0.64 and 0.73 (Column 5). A similar interpretation is noted from the results of the ML solution. Among the six arthropods, only GLYC (one of the factors), CHEY and LARS seem to be prominent in their growth in the present environment.

4.9 Estimation of Factor Scores

Predicted (estimated) values of common factors corresponding to each of the original (raw) observational units or individuals are called *factor scores* for the particular observation. These values can be used as reduced data for any further statistical analyses. For example, they are often used for the diagnostic purposes. The factor scores are calculated after an appropriate factor model has been determined and the estimate of the parameters has been computed. Generally, the final estimates of the factor loadings obtained after an appropriate factor rotation are used to determine the factor scores.

We here present two methods of predicting the factor scores. Both procedures assume that raw data are available so that these factor scores can be calculated.

4.9.1 Weighted Least Squares Method

This method (Bartlett, 1937) determines the factor score corresponding to observation $\mathbf{x}_i$ using the weighted least squares formula

$$\hat{\mathbf{f}}_i = (\hat{\mathbf{L}}'\hat{\mathbf{\Psi}}^{-1}\hat{\mathbf{L}})^{-1}\hat{\mathbf{L}}'\hat{\mathbf{\Psi}}^{-1}(\mathbf{x}_i - \bar{\mathbf{x}}) = \hat{\mathbf{\Delta}}^{-1}\hat{\mathbf{L}}'\hat{\mathbf{\Psi}}^{-1}(\mathbf{x}_i - \bar{\mathbf{x}}),\ i = 1, \dots, n,$$

where $\bar{\mathbf{x}}$ is the sample mean vector of the data. The estimators $\hat{\mathbf{L}}$ and $\hat{\mathbf{\Psi}}$ are obtained using the method of maximum likelihood. If the sample correlation matrix is used (that is, variables have been standardized) for the factor analysis, then $\mathbf{z}_i = diag(1/\sqrt{s_{11}}, \ldots, 1/\sqrt{s_{pp}})$ $(\mathbf{x}_i - \bar{\mathbf{x}})$, instead of $(\mathbf{x}_i - \bar{\mathbf{x}})$ will be used in the above formula.

4.9.2 Regression Method

This method estimates the factor score corresponding to $\mathbf{x}_i$ using the formula corresponding to the linear regression coefficient formula

$$\hat{\mathbf{f}}_i = \hat{\mathbf{L}}'\mathbf{S}^{-1}(\mathbf{x}_i - \bar{\mathbf{x}}),\ i = 1, \ldots, n.$$

If the sample correlation matrix (or its variation as in the principal factor method with a certain PRIORS= option) is used, then $\mathbf{z}_i$ defined in the previous subsection will be used instead of $(\mathbf{x}_i - \bar{\mathbf{x}})$ and $\mathbf{R}$, the sample correlation matrix (or its variation) will be used, in place of $\mathbf{S}$, in the above formula.

The FACTOR procedure in SAS calculates the factor scores using only the regression method. These are computed by specifying the NFACTORS= and OUT= options in the PROC FACTOR statement. The scores are then stored in the SAS data set specified in the OUT = option. Whenever these options are used, SAS also prints out the scoring coefficients estimated by regression $\hat{\mathbf{L}}'\mathbf{S}^{-1}$ or the standardized scoring coefficients estimated by regression $\hat{\mathbf{L}}'\mathbf{R}^{-1}$ (or its variation). The estimated factor scores stored in the data set can be printed by using the PROC PRINT statement.

EXAMPLE 10 *Factor Scores, The World Bank Data* Data on eight economic indices related to the state of a country's development for 12 countries are taken from The World Bank World Development Report (1988). The variables observed are population in millions (POP), area in thousands of square kilometers (AREA), gross national product in US dollars (GNP), life expectancy (LIFE), number of radios per one thousand persons (RADIO), number of tourists in thousands (TOURIST), average available calories per person per day (FOOD), and percentage enrollment in schools of children between the ages 6 to 17 (SCHOOL). The raw data for these variables for all twelve countries are provided in Program 4.4 as a data set named ECONOMY. Using the SAS code included in Program 4.4, three factors are extracted by the principal factor method with the ASMC option.

```
/* Program 4.4 */

options ls=64 ps=45 nodate nonumber;
title1 'Output 4.4';
title2 'Factor Scores: World Bank Data';

data economy;
input country $ pop area gnp life radio tourist
food school;
lines;
Canada 25.6 9976 14120 76 758 12854 3404 98
USA 241.6 9363 17480 75 2133 20441 3632 99
Haiti 6.1 28 330 54 21 167 1906 48
Brazil 138.4 8512 1810 65 355 1420 2575 78
Austria 7.6 84 9990 74 475 14482 3479 80
Iceland .24 103 13410 77 593 78 3122 100
Spain 38.7 505 4860 76 274 25583 3325 97
UK 56.7 245 8870 75 986 12499 3210 96
Gambia 0.77 11 230 43 120 37 2217 34
India 781.4 3288 290 57 56 1305 2031 54
```

```
Malaysia 16.1 330 1830 69 415 1050 2569 77
Austrlia 16.0 7687 11920 78 1159 944 3044 89
;
/* Source: The World Bank World Development Report (1988).
Reprinted by permission of the World Bank Office of Publisher.*/

proc factor data=economy method=prin prior=asmc
nfact=3 rotate=varimax out=fscore;
var pop area gnp life radio tourist food school;
run;

proc print data=fscore;
run;

data labels;
set fscore;
retain xsys '2' ysys '2';
length text $8 function $8;
text=country;
y=factor2;
x=factor1;
size=1.2;
position='8';
function = 'LABEL';
run;
filename gsasfile "prog44a.graph";
goptions reset=all gaccess=gsasfile autofeed dev=psl;
goptions horigin=1in vorigin=2in;
goptions hsize=6in vsize=8in;
title1 h=1.2 'World Bank Data: Plot of Scores on the
First and Second Factors';
title2 j=l 'Output 4.4';
proc gplot data=fscore;
plot factor2*factor1/annotate=labels vaxis=axis2 haxis=axis1;
axis1 label=(h=1.2 'First Factor');
axis2 label=(h=1.2 a=90 r=0  'Second Factor');
run;

data labels;
set fscore;
retain xsys '2' ysys '2';
length text $8 function $8;
text=country;
y=factor3;
x=factor1;
size=1.2;
position='8';
function = 'LABEL';
run;
filename gsasfile "prog44b.graph";
goptions reset=all gaccess=gsasfile autofeed dev=psl;
goptions horigin=1in vorigin=2in;
goptions hsize=6in vsize=8in;
title1 h=1.2 'World Bank Data: Plot of Scores on the
First and Third Factors';
title2 j=l 'Output 4.4';
proc gplot data=fscore;
plot factor3*factor1/annotate=labels vaxis=axis2 haxis=axis1;
axis1 label=(h=1.2 'First Factor');
```

```
axis2 label=(h=1.2 a=90 r=0  'Third Factor');
run;
```

Output 4.4

Output 4.4
Factor Scores: World Bank Data

The FACTOR Procedure
Initial Factor Method: Principal Factors

Prior Communality Estimates: ASMC

pop	area	gnp	life
0.36188524	0.42386635	0.90957244	0.95944332

radio	tourist	food	school
0.75336911	0.79687081	0.96544468	0.96287293

Eigenvalues of the Reduced Correlation Matrix:
Total = 6.13332489 Average = 0.76666561

	Eigenvalue	Difference	Proportion	Cumulative
1	4.71337668	3.90915549	0.7685	0.7685
2	0.80422119	0.32204964	0.1311	0.8996
3	0.48217156	0.17298051	0.0786	0.9782
4	0.30919105	0.30700180	0.0504	1.0286
5	0.00218925	0.02910527	0.0004	1.0290
6	-.02691602	0.00820757	-0.0044	1.0246
7	-.03512360	0.08066163	-0.0057	1.0189
8	-.11578522		-0.0189	1.0000

3 factors will be retained by the NFACTOR criterion.

Factor Pattern

	Factor1	Factor2	Factor3
pop	-0.22262	0.42535	0.29090
area	0.42271	0.54169	0.02207
gnp	0.90876	0.18667	-0.15131
life	0.90752	-0.16462	-0.18121
radio	0.77892	0.39295	0.01625
tourist	0.65403	-0.24920	0.56443
food	0.96266	-0.19201	0.09064
school	0.93516	-0.12060	-0.11942

Rotation Method: Varimax

Orthogonal Transformation Matrix

	1	2	3
1	0.69828	0.55231	0.45537
2	0.69037	-0.35148	-0.63234
3	-0.18919	0.75593	-0.62672

Rotated Factor Pattern

	Factor1	Factor2	Factor3
pop	0.08316	-0.05255	-0.55266
area	0.66496	0.05975	-0.16387
gnp	0.79207	0.32192	0.39061
life	0.55434	0.42210	0.63093
radio	0.81211	0.30437	0.09603
tourist	0.17787	0.87548	0.10166
food	0.52250	0.66769	0.50297
school	0.59233	0.46861	0.57695

Scoring Coefficients Estimated by Regression

Squared Multiple Correlations of the Variables with Each Factor

Factor1	Factor2	Factor3
0.86452622	0.85079096	0.75904482

Standardized Scoring Coefficients

	Factor1	Factor2	Factor3
pop	0.09423369	0.16414323	-0.2189832
area	0.18514269	-0.0166734	-0.2269241
gnp	0.80074987	-0.6751939	-0.1702235
life	-0.0138119	-0.344438	0.45157563
radio	0.33881108	0.01128343	-0.2881192
tourist	-0.0251394	0.4307033	-0.568259
food	-0.6855882	1.22071803	0.75332179
school	0.32840538	0.05888674	0.15417957

Factor Scores: World Bank Data

Obs	country	pop	area	gnp	life	radio	tourist
1	Canada	25.60	9976	14120	76	758	12854
2	USA	241.60	9363	17480	75	2133	20441
3	Haiti	6.10	28	330	54	21	167
4	Brazil	138.40	8512	1810	65	355	1420
5	Austria	7.60	84	9990	74	475	14482
6	Iceland	0.24	103	13410	77	593	78
7	Spain	38.70	505	4860	76	274	25583
8	UK	56.70	245	8870	75	986	12499
9	Gambia	0.77	11	230	43	120	37
10	India	781.40	3288	290	57	56	1305
11	Malaysia	16.10	330	1830	69	415	1050
12	Austrlia	16.00	7687	11920	78	1159	944

Obs	food	school	Factor1	Factor2	Factor3
1	3404	98	0.87909	0.30122	0.24656
2	3632	99	1.88624	0.98487	-0.90367
3	1906	48	-0.69034	-1.33194	-0.80453
4	2575	78	-0.23207	-0.24674	-0.17523
5	3479	80	-0.60338	1.01132	0.83521
6	3122	100	0.63887	-0.81053	1.39147
7	3325	97	-0.94808	1.77139	0.33123
8	3210	96	0.12025	0.54310	0.49437
9	2217	34	-1.19960	-0.39108	-0.97892
10	2031	54	-0.26788	-0.53497	-1.50789
11	2569	77	-0.61781	-0.46019	0.52440
12	3044	89	1.03470	-0.83645	0.54699

Output 4.4 (*continued*) World Bank Data: Plot of Scores on the First and Second Factors

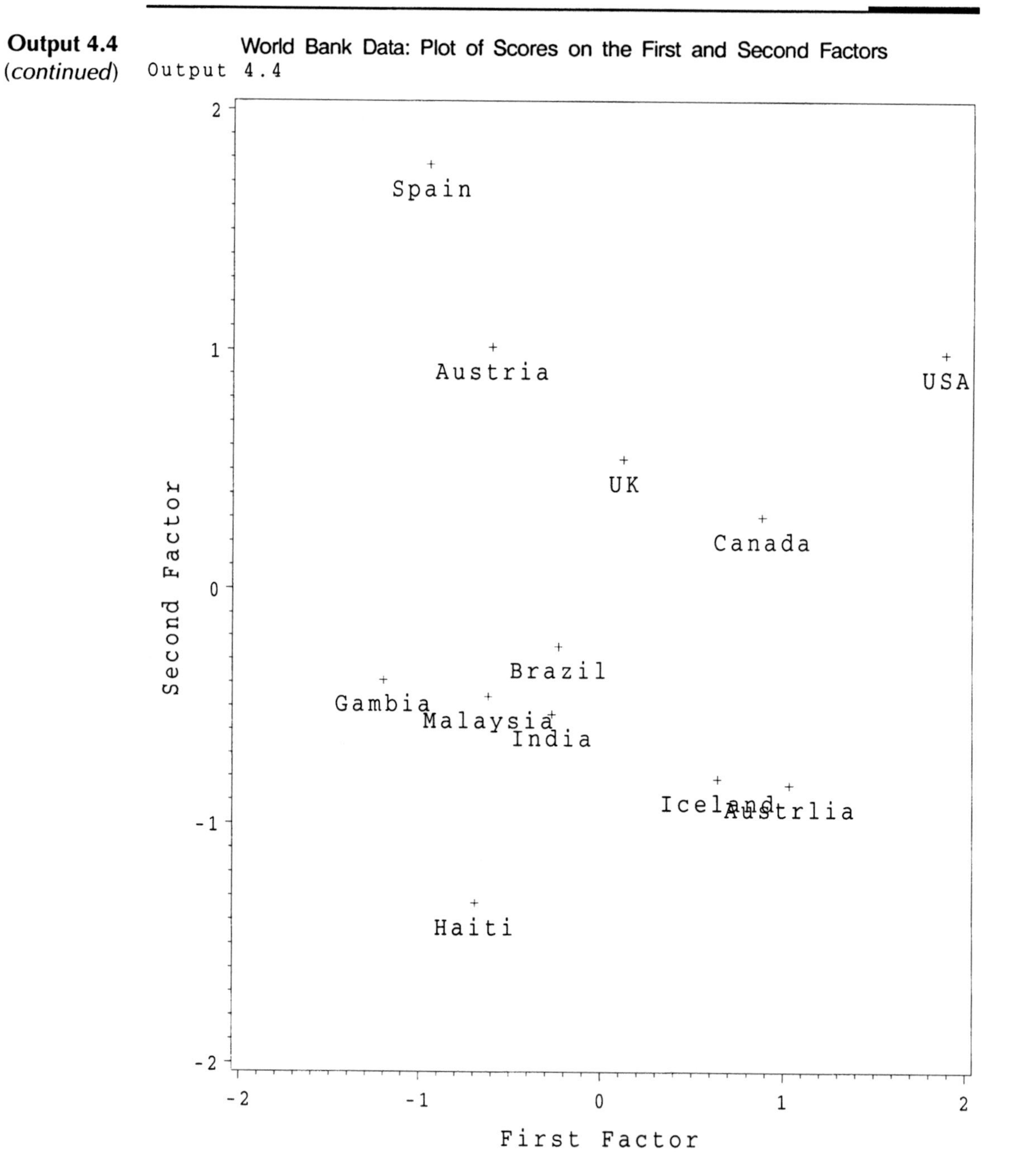

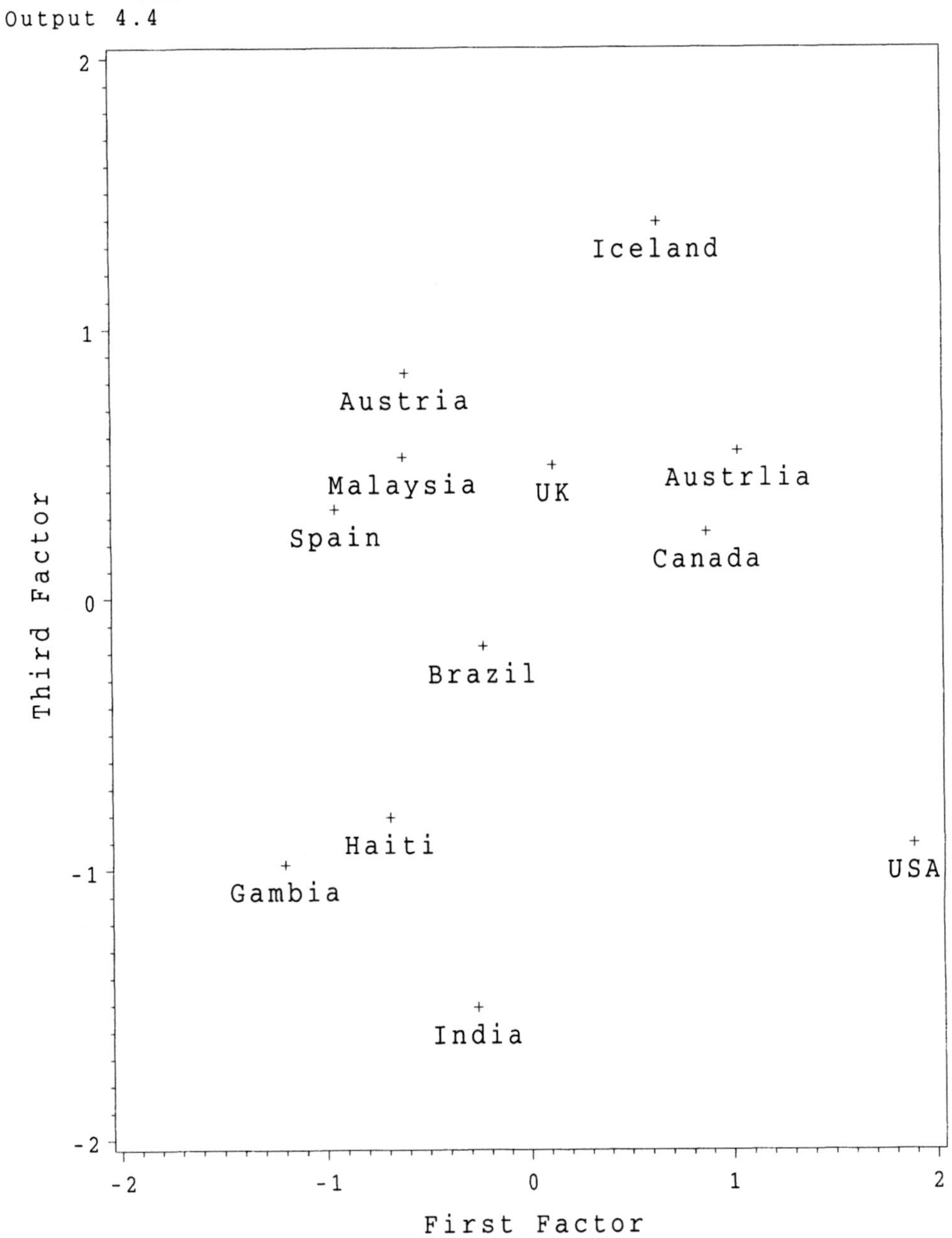
World Bank Data: Plot of Scores on the First and Third Factors
Output 4.4
Iceland
Austria
Malaysia
UK
Austrlia
Spain
Canada
Brazil
Haiti
Gambia
USA
India
Third Factor
First Factor
2
1
0
-1
-2
-2
-1
0
1
2

The output is stored in Output 4.4. First we will interpret the three common factors based on the factor loadings after they have been rotated by the VARIMAX option. The variables AREA, GNP, RADIO, and SCHOOL, respectively, assign the loadings of 0.6650, 0.7921, 0.8121, and 0.5923 to the first factor (see the heading "Rotated Factor Pattern" in Output 4.4). Hence, the first factor can be interpreted as a measure of overall development of a country. The second factor receives very high loadings from the variables TOURIST (0.8755) and FOOD (0.6677) and therefore can be taken as a measure of tourist attraction and availability of food in a country. These two factors explain about 90% (see Output 4.4 under the column: Cumulative) of the variability in the data. The variable POP loads the third factor with a value of -0.5527, and the variable LIFE loads it with 0.6309. Hence, the third factor can be interpreted as a measure of a contrast between the life expectancy (LIFE) and the overcrowding of the population (POP) in a country.

Because we have used the OUT=FSCORE option in the PROC FACTOR statement, the factor scores for the three factors are stored in a SAS data set named FSCORE as the variables FACTOR1, FACTOR2, and FACTOR3, respectively, and the standardized scoring coefficients (by regression) are printed as a standard output. Also given in the output are the squared multiple correlation coefficients of the variables with each factor. For example these values for the three factors, respectively, are 0.8645, 0.8508, and 0.7590 (see Output 4.4). The SAS data set FSCORE contains all the original variables and the factor scores (see Output 4.4). If we are interested in storing and using various statistics, these can be stored using the option, OUTSTAT= in the FACTOR procedure. This is often convenient. For example, if this information is stored in a SAS data set named FACT (by using OUTSTAT=FACT), then running the SCORE procedure as follows on our raw data set ECONOMY will produce the factor scores, stored in a SAS data set named SCORES:

```
proc score data=economy score=fact out=scores;
```

Suppose we plot the factor scores corresponding to the first two factors for all the 12 countries (see Output 4.4). We can examine this plot for identifying outliers and also to find groups or clusters among the observations, if any. By examining the plot we see that the United States (USA) stands out in terms of having the highest value for FACTOR1, indicating its most developed status. Spain, with a very high value for FACTOR2, stands out as the most toured of these countries. This fact is confirmed by noting that the largest number of visitors (25,583) went to Spain. At the same time it also has one of the higher values corresponding to the variable FOOD, which is a prominent variable in FACTOR2. If we examine a similar plot of FACTOR3 versus FACTOR1, India stands out as having lowest value for FACTOR3 indicating a lower life expectancy, perhaps due to overcrowding of the population. On the other hand, Iceland has the highest value for FACTOR3 indicating a higher life expectancy due to less crowding.

4.10 Factor Analysis Using Residuals

Sometimes it may be meaningful to perform factor analysis on the residuals rather than on the original variables. Such a need may arise when the data have time or some other type of trend or must be adjusted for certain covariates. For example, in the crime data considered by Ahamad (1967), the number of offenses committed in a category over a period of 14 years may have a trend. This trend over the years may be represented by a polynomial of appropriate degree or by some other function of time. In these problems, we can use a regression model (perhaps nonlinear) to first obtain the residuals. Factor analysis then can be performed on the residuals.

We may view this process of performing factor analysis on the residuals as a process where the effect of time or year has been eliminated or partialed out, before the analysis. This amounts to performing the factor analysis by taking the matrix of partial correlation coefficients (after eliminating trends or the effects of covariates) as the starting point instead of the original correlation matrix of the raw data. The PARTIAL statement can be used in the FACTOR procedure to perform this analysis.

EXAMPLE 11 ***Elimination of Trend, Residual Analysis, Crime Data*** The numbers of offenses committed in each of the following 18 categories, Homicide (X1), Woundings (X2), Homosexual offenses (X3), Heterosexual offenses (X4), Breaking and entering (X5), Robbery (X6), Larceny (X7), Fraud and false pretense (X8), Receiving (X9), Malicious injury to property (X10), Forgery (X11), Blackmail (X12), Assault (X13), Malicious damage (X14), Revenue laws (X15), Intoxication laws (X16), Indecent exposure (X17), and Taking motor vehicle without consent (X18), are available for England and Wales for the period 1950 to 1963. The raw data taken from Ahamad (1967) are provided in Program 4.5.

The data are unusual in many respects. First, the number of variables is more than the number of observations. This makes the covariance or the correlation matrix of the data singular. The singularity of the variance-covariance matrix prevents us from using many of the iterative procedures that require the existence of the inverse of this matrix. Also for this reason squared multiple correlation (SMC) and adjusted squared multiple correlation (ASMC) options cannot be used to perform factor analysis by the principal factor method. Second, the data are counts. Also, the offense categories are very different in terms of the seriousness of the crime. Finally, the data are observed over a period of time leading to the possibility that there may be a time trend relating the observations to the time component.

To overcome some of these problems, we first transform the counts to log-counts in an effort to normalize the data by taking the natural logarithm. Next, we use the variance-covariance matrix (instead of the correlation matrix) of the transformed data for factor analysis to preserve the heterogeneity of the variances of the variables among different categories. For illustration purposes, we will use the principal factor method with maximum correlation for the factor analysis by using the PRIOR = MAX option in the PROC FACTOR statement. To eliminate the time trend before performing the analysis, we partial out a second degree polynomial time effect using the PARTIAL statement. Specifically, we create variables YEAR, which takes the values 1 to 14 for the years 1950 to 1963, and S_YEAR, which is the square of the variable YEAR and use the statement

```
partial year s_year;
```

to first partial out the effect of the time component. Thus, the factor analysis is performed on residuals after eliminating the second-order time trend from the data. The detailed SAS code is provided in Program 4.5 and the corresponding output is provided in Output 4.5.

```
/* Program 4.5 */

options ls=64 ps=45 nodate nonumber;
title1 'Output 4.5';
title2 'Factor Analysis of Residuals';
data crime;
input x1 x2 x3 comma5. x4-x12;
input x13-x18;
```

```
array x{18} x1-x18;
array u{18} u1-u18;
do i=1 to 18;
u{i}=log(x{i});
end;
year =_n_;
year_s=year*year;
cards;
529 5258 4,416 8178 92839 1021 301078 25333 7586 4518 3790 118
20844 9477 24616 49007 2786 3126
455 5619 4,876 9223 95946 800 355407 27216 9716 4993 3378 74
19963 10359 21122 55229 2739 4595
555 5980 5,443 9026 97941 1002 341512 27051 9188 5003 4173 120
19056 9108 23339 55635 2598 4145
456 6187 5,680 10107 88607 980 308578 27763 7786 5309 4649 108
17772 9278 19919 55688 2639 4551
487 6586 6,357 9279 75888 812 285199 26267 6468 5251 4903 104
17379 9176 20585 57011 2587 4343
448 7076 6,644 9953 74907 823 295035 22966 7016 2184 4086 92
17329 9460 19197 57118 2607 4836
477 8433 6,196 10505 85768 965 323561 23029 7215 2559 4040 119
16677 10997 19064 63289 2311 5932
491 9774 6,327 11900 105042 1194 360985 26235 8619 2965 4689 121
17539 12817 19432 71014 2310 7148
453 10945 5,471 11823 131132 1692 409388 29415 10002 3607 5376 164
17344 14289 24543 69864 2371 9772
434 12707 5,732 13864 133962 1900 445888 34061 10254 4083 5598 160
18047 14118 26853 69751 2544 11211
492 14391 5,240 14304 151378 2014 489258 36049 11696 4802 6590 241
18801 15866 31266 74336 2719 12519
459 16197 5,605 14376 164806 2349 531430 39651 13777 5606 6924 205
18525 16399 29922 81753 2820 13050
504 16430 4,866 14788 192302 2517 588566 44138 15783 6256 7816 250
16449 16852 34915 89709 2614 14141
510 18655 5,435 14722 219138 2483 635627 45923 17777 6935 8634 257
15918 17003 40434 89149 2777 22896
;
/* Source: Ahamad (1967). Reprinted by permission of
the Royal Statistical Society.*/

proc factor data=crime cov prior=max rotate=quartimax nfact=3;
var u1-u18;
partial year year_s;
run;
```

Output 4.5

```
                    Output 4.5
            Factor Analysis of Residuals

                The FACTOR Procedure
         Initial Factor Method: Principal Factors

          Prior Communality Estimates: MAX

        u1            u2            u3            u4            u5

0.53746472    0.88696134    0.78941886    0.70242540    0.87618142
```

u6	u7	u8	u9	u10
0.81154407	0.93441570	0.86978446	0.93441570	0.86978446

u11	u12	u13	u14	u15
0.77771761	0.74098285	0.88696134	0.85227068	0.74098285

u16	u17	u18
0.58921738	0.51580729	0.69472557

Eigenvalues of the Reduced Covariance Matrix:
Total = 0.17440942 Average = 0.00968941

	Eigenvalue	Difference	Proportion	Cumulative
1	0.09232727	0.04575695	0.5294	0.5294
2	0.04657032	0.01573132	0.2670	0.7964
3	0.03083900	0.02408279	0.1768	0.9732
4	0.00675621	0.00209009	0.0387	1.0119
5	0.00466613	0.00163452	0.0268	1.0387
6	0.00303161	0.00132376	0.0174	1.0561
7	0.00170785	0.00098696	0.0098	1.0659
8	0.00072089	0.00061749	0.0041	1.0700
9	0.00010340	0.00038613	0.0006	1.0706
10	-.00028272	0.00021870	-0.0016	1.0690
11	-.00050143	0.00018777	-0.0029	1.0661
12	-.00068920	0.00021719	-0.0040	1.0622
13	-.00090639	0.00009272	-0.0052	1.0570
14	-.00099911	0.00032980	-0.0057	1.0512
15	-.00132890	0.00057541	-0.0076	1.0436
16	-.00190431	0.00030358	-0.0109	1.0327
17	-.00220790	0.00128542	-0.0127	1.0200
18	-.00349332		-0.0200	1.0000

3 factors will be retained by the NFACTOR criterion.

Rotation Method: Quartimax

Orthogonal Transformation Matrix

	1	2	3
1	0.75762	0.64876	0.07160
2	0.65269	-0.75249	-0.08803
3	-0.00323	0.11343	-0.99354

Rotated Factor Pattern

	Factor1	Factor2	Factor3
u1	-0.04074	0.08042	0.56112
u2	0.87159	-0.12030	0.00844
u3	-0.83115	0.06165	-0.20674
u4	0.67944	0.13469	-0.30204
u5	0.94563	0.16071	-0.14170

u6	0.85356	0.13669	0.29805
u7	0.83289	0.13294	-0.46831
u8	0.47072	0.83261	-0.17869
u9	0.66569	0.17187	-0.56771
u10	0.18842	0.91221	-0.05046
u11	-0.01636	0.86147	0.42718
u12	0.51136	0.23295	0.68176
u13	0.72791	-0.07784	0.10053
u14	0.82941	-0.24127	-0.18582
u15	0.57677	0.36067	0.43302
u16	0.21533	0.07383	-0.45995
u17	-0.06203	0.36879	-0.00294
u18	0.55615	0.04771	-0.53770

From Output 4.5, we see that the first three factors explain more than 97% of the total variation. An examination of the rotated factors reveals some interesting interpretations of the three factors. The third factor receives high positive loadings, respectively, 0.5611 and 0.6818 from the two serious crime categories, homicide and blackmail. At the same time, it also has the two highly negative loadings, corresponding to two relatively minor crimes—namely, receiving and taking a motor vehicle without consent. Hence, this factor may be viewed as a contrast between the two *very serious* crimes and two relatively minor crimes. The second factor receives high loadings, respectively, 0.8326, 0.9121, and 0.8615 from the variables representing fraud and false pretense, injury to property, and forgery. All these are less serious crimes in the sense that they do not inflict injuries to human beings. Hence, this factor may be classified as a *less serious* crimes factor. Finally, the variables representing the serious crimes assign high loadings to the first factor. For example, woundings, breaking and entering, robbery, malicious damage, all these assign high loadings to this factor. Hence, the first factor may be classified as the *serious* crime factor. The negative high loading of -0.8311 assigned by the variables representing homosexual offenses could not be interpreted.

4.11 Some Applications

In this section we will illustrate several applications of factor analysis in a variety of fields. Our objective is to provide a good understanding of the methods used and familiarity with the different steps involved in the analysis. All data sets are real and are taken from various fields. For each example we present the results for only one of the estimation methods and only one particular factor rotation. Other methods of estimation and different rotations may also be tried and results compared with those presented here.

EXAMPLE 12 ***Sleep in Mammals*** Allison and Cicchetti (1976) make certain interesting observations about sleep in mammals using the factor analysis. The discussion here is largely based on their work.

Sleep requirements vary substantially from species to species. The reasons for such differences in sleep requirements are not very obvious and may depend on the particular characteristics of the species. By studying the interrelationships between sleep, certain constitutional characteristics, and certain ecological influences, it may be possible to understand the significance of sleep in the life of mammals. Allison and Cicchetti (1976) collected data on 39 species for various characteristics. Two sleep variables were considered. The first is slow-wave sleep (SWS), characterized by high-amplitude slow waves in the electroencephalogram and by behavioral and autonomic nervous system quiescence.

The second is paradoxical sleep (PS), which is characterized by a low-voltage electroencephalogram, brief movements of the extremities and facial muscles, autonomic irregularity, and dreaming (in man). The four constitutional variables considered were life-span in years (L), body weight in kilograms (WB), brain weight in grams (WBR), and gestation time in days (TG). The three ecological variables considered were predation index (P), measured in a five-point scale on the extent to which the mammals are preyed upon (a score of 5 indicated maximum predation and a score of 1 indicated minimum predation), sleep exposure index (S), measured in a five-point scale based on the extent of exposure while sleeping (species that sleep in maximally exposed places were assigned a value of 5 and species that usually sleep in a barrow, den, or other well-protected place were assigned a value of 1), and overall danger (D), again measured in a five-point scale providing a general estimate of predatory danger. In fact, all the three ecological variables P, S, and D measure the extent of danger to species.

The nine by nine correlation matrix of these variables is given in Program 4.6. Factor analysis of the data using the little jiffy method is performed. The code is included in Program 4.6.

```
/* Program 4.6 */

options ls=64 ps=45 nodate nonumber;
title1 'Output 4.6';
title2 'Factor Analysis of Sleep in Mammals';

data sleep (type=corr);
infile cards missover;
_type_='CORR';
if _n_=1 then _type_='N';
input _name_ $ sws ps l wb wbr tg p s d;
lines;
n 39
sws 1.000
ps .582 1.000
l -.377 -.342 1.000
wb -.712 -.370 .685 1.000
wbr -.679 -.435 .777 .945 1.000
tg -.589 -.651 .682 .692 .781 1.000
p -.369 -.536 .018 .253 .192 .158 1.000
s -.580 -.591 .518 .662 .624 .588 .680 1.000
d -.542 -.686 .226 .432 .377 .363 .930 .819 1.00
;
/* Source: Allison and Cicchetti (1976). Reprinted by
permission of American Association for the Advancement of
Science.*/

proc factor method=principal mineigen=1 rotate=varimax;
var sws ps l wb wbr tg p s d;
run;
```

Output 4.6

```
                         Output 4.6
              Factor Analysis of Sleep in Mammals

                     The FACTOR Procedure
           Initial Factor Method: Principal Components

               Prior Community Estimates: ONE
   Eigenvalues of the Correlation Matrix: Total = 9  Average = 1

          Eigenvalue    Difference    Proportion    Cumulative

     1    5.38756498    3.53874552        0.5986        0.5986
     2    1.84881946    1.20237047        0.2054        0.8040
     3    0.64644899    0.08378986        0.0718        0.8759
     4    0.56265913    0.34937178        0.0625        0.9384
     5    0.21328735    0.03918051        0.0237        0.9621
     6    0.17410684    0.07025330        0.0193        0.9814
     7    0.10385354    0.06503005        0.0115        0.9930
     8    0.03882350    0.01438728        0.0043        0.9973
     9    0.02443621                      0.0027        1.0000

     2 factors will be retained by the MINEIGEN criterion.

                    Rotated Factor Pattern

                        Factor1         Factor2

          sws          -0.63031        -0.47855
          ps           -0.39207        -0.68891
          l             0.86152         0.01678
          wb            0.88207         0.25781
          wbr           0.94179         0.19831
          tg            0.84154         0.24437
          p            -0.04405         0.95734
          s             0.52310         0.73747
          d             0.18969         0.96365
```

From Output 4.6 we see that only two factors corresponding to the eigenvalues greater than one are selected since the option MINEIGEN=1 was used. We examine the varimax rotated factor loadings (see Output 4.6). The first factor has received large loadings from the constitutional variables related to body size and hence can be called the *size* factor. This factor also received a large negative loading from SWS. This is not surprising since it is known that the large-sized species spend less time in slow wave sleep (SWS). The second factor received large loadings from ecological variables related to the danger, and hence it can be called the *danger* factor. This factor has large negative loadings from PS, indicating that greater predatory danger is associated with less paradoxical sleep (PS).

In conclusion, this study indicates that slow-wave sleep is negatively correlated with a factor related to body size, suggesting that large amounts of this type of sleep are disadvantageous in large species. On the other hand, paradoxical sleep is associated with a factor related to predatory danger, which suggests that large amounts of this type of sleep are disadvantageous to small species, who are preyed upon.

EXAMPLE 13 ***Factor Analysis of Photographic Data*** Jackson and Bradley (1966) have presented an interesting example involving photographic products in the context of statistical quality control. Quality of photographic products and processes are usually monitored by means of a special piece of film that contains a series of graduated exposures designed to represent the entire range of exposures used in actual practice. For the data under consideration, optical densities are obtained at 14 steps in even increments of log exposure, after the strip is exposed and processed. The resulting curve is a function of exposure and displays the photographic quality of the product. A typical curve starts at a high position (high density) for low log exposure and dies down as log-exposure becomes large. The high density portion of the curve represents shadow areas in the picture, the middle density represents the average areas in the picture, and the low density represents light areas. A sample of 232 film strips was used, and densities corresponding to 14 log-exposure values on each strip were collected. The resulting variance-covariance matrix of these 14 log-exposures is given in Program 4.7.

The objective is to obtain the factors that may represent different levels (low, normal, and high) of exposure. Since all 14 variables are measured in the same unit, it is more meaningful to use the covariance matrix rather than the correlation matrix to conduct any statistical analysis for cases when the factor analysis method is not invariant of scaling. We will illustrate this analysis using the principal component method that is not invariant of this choice. Program 4.7 is used for this purpose. The NFACT=3 option was specified with the hope that the first three factors selected may satisfy our basic objective to categorize low, normal, and high levels of exposure. The COV option in the PROC FACTOR statement has been used to specify that the covariance matrix rather than correlation matrix should be used in the analysis. The OUTSTAT=SAVEALL option in the PROC FACTOR statement instructs that all output produced be stored in the SAS data set SAVEALL.

```
/* Program 4.7 */

options ls=64 ps=45 nodate nonumber;
title1 'Output 4.7';
title2 'Factor Analysis of Photographic Data';

data density (type=cov);
infile cards missover;
_type_='cov';
if _n_=1 then _type_='N';
input _name_ $ x1-x14;
cards;
n 232
x1 6.720
x2 6.669 6.856
x3 6.504 6.862 7.260
x4 6.329 6.876 7.586 8.387
x5 5.769 6.490 7.477 8.591 9.238
x6 4.866 5.648 6.767 8.046 8.919 8.929
x7 3.526 4.239 5.299 6.534 7.451 7.649 6.875
x8 2.276 2.807 3.634 4.635 5.431 5.738 5.348 4.464
x9 1.117 1.442 1.998 2.678 3.249 3.546 3.409 2.956 2.108
x10 .498 .662 .967 1.350 1.695 1.892 1.876 1.682 1.239 .811
x11 .167 .215 .323 .486 .652 .781 .825 .822 .654 .450 .321
x12 -.015 -.022 .003 .073 .155 .254 .330 .408 .369 .281 .223 .208
x13 -.061 -.091 -.111 -.094 -.062 .007 .091 .224 .251 .216 .198 .179 .211
x14 -.152 -.200 -.256 -.278 -.277 -.204 -.095 .083 .158 .157 .175 .176 .192 .219
;
/* Source: Jackson and Bradley (1966, p. 515). Reprinted by permission of
the Academic Press.*/
```

```
proc factor data=density (type=cov) cov method=prin mineigen=1
rotate=varimax outstat=saveall;
var x1-x14;
run;
proc print data=saveall;
run;
data temp1(type=factor);
set saveall;
if _type_='PATTERN' then output;
run;
proc transpose data=temp1 out=temp2;
run;
data fin;
set temp2;
obser=_n_;
run;
filename gsasfile "prog47.graph";
goptions reset=all gaccess=gsasfile autofeed dev=psl;
goptions horigin=1in vorigin=2in;
goptions hsize=6in vsize=8in;
title1 h=1.2 'Plots of Factor Loadings for Photographic Data';
title2 j=l 'Output 4.7';
proc gplot data=fin;
plot factor1*obser factor2*obser factor3*obser/overlay
vaxis=axis1 haxis=axis2 legend=legend1;
axis1 label=(a=90 h=1.2 'The Factor Loadings');
axis2 offset=(2) label=(h=1.2 'Photo Sequence');
symbol1 i=join v=star c=red;
symbol2 i=join v=+ c=blue;
symbol3 i=join v=x c=green;
legend1 across=3;
run;
```

Output 4.7

```
                    Output 4.7
         Factor Analysis of Photographic Data

                The FACTOR Procedure
       Initial Factor Method: Principal Components

         Prior Communality Estimates: ONE

         Eigenvalues of the Covariance Matrix:
         Total = 62.607   Average = 4.47192857
```

	Eigenvalue	Difference	Proportion	Cumulative
1	52.0358294	43.6023048	0.8312	0.8312
2	8.4335246	7.0086325	0.1347	0.9659
3	1.4248921	1.1701643	0.0228	0.9886
4	0.2547277	0.1387806	0.0041	0.9927
5	0.1159471	0.0410643	0.0019	0.9945
6	0.0748829	0.0148470	0.0012	0.9957
7	0.0600359	0.0099497	0.0010	0.9967
8	0.0500862	0.0092230	0.0008	0.9975
9	0.0408632	0.0094845	0.0007	0.9981
10	0.0313787	0.0064667	0.0005	0.9986
11	0.0249120	0.0014847	0.0004	0.9990

```
12      0.0234273      0.0037528        0.0004         0.9994
13      0.0196745      0.0028559        0.0003         0.9997
14      0.0168186                       0.0003         1.0000

 3 factors will be retained by the MINEIGEN criterion.

                    Rotated Factor Pattern

                  Factor1         Factor2         Factor3

  x1              0.11511         0.98941         0.01310
  x2              0.23988         0.96739        -0.03863
  x3              0.42601         0.89700        -0.08618
  x4              0.60281         0.78669        -0.10050
  x5              0.74539         0.65597        -0.09318
  x6              0.83936         0.53665        -0.03880
  x7              0.90271         0.41518         0.06866
  x8              0.90370         0.30815         0.26665
  x9              0.87413         0.19232         0.42225
  x10             0.78809         0.12010         0.55662
  x11             0.54385         0.05048         0.77210
  x12             0.27366        -0.04478         0.84455
  x13             0.06280        -0.05036         0.85382
  x14            -0.08331        -0.10792         0.84577
```

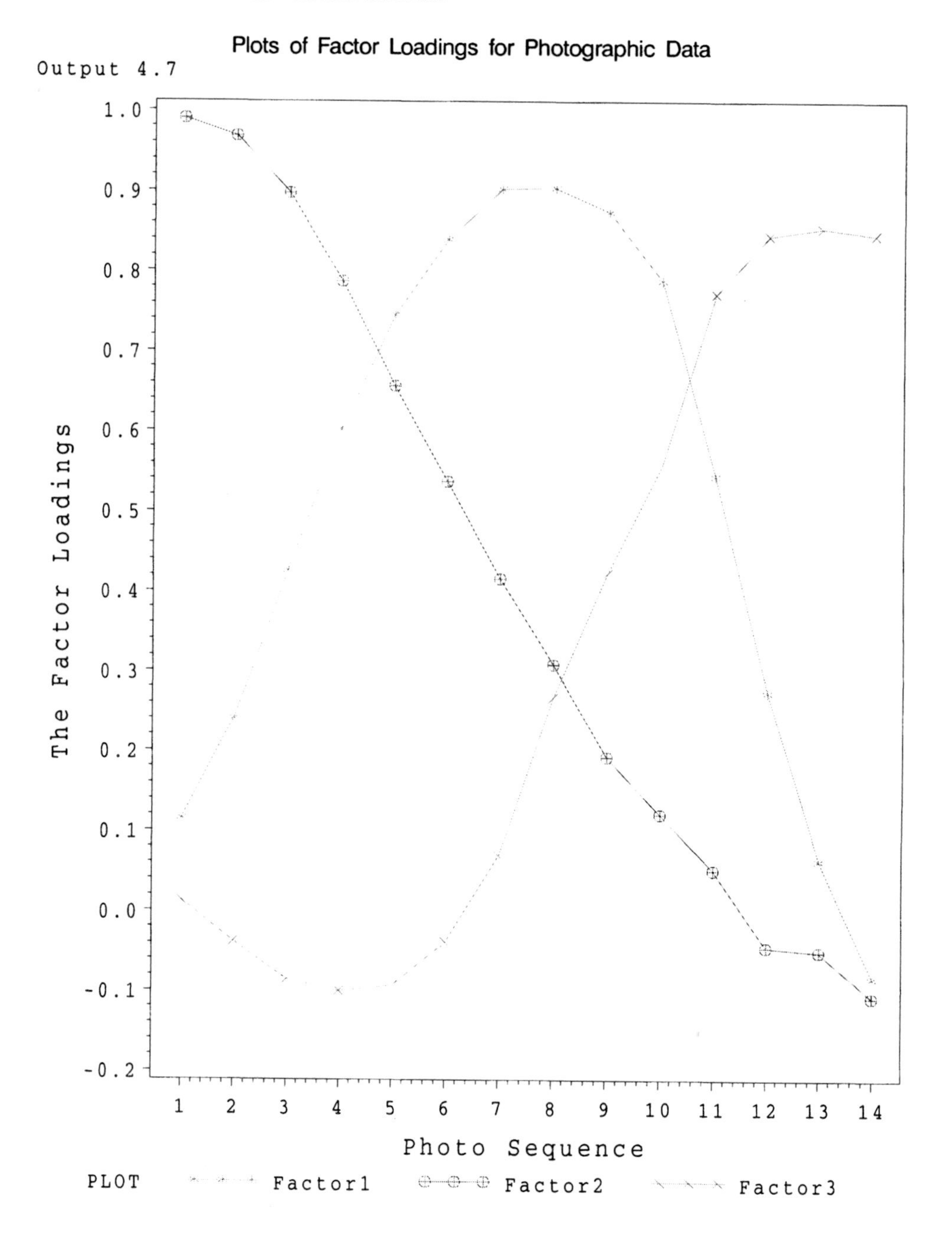
Plots of Factor Loadings for Photographic Data
Output 4.7
1.0
0.9
0.8
0.7
0.6
0.5
0.4
0.3
0.2
0.1
0.0
-0.1
-0.2
The Factor Loadings
1 2 3 4 5 6 7 8 9 10 11 12 13 14
Photo Sequence
PLOT
Factor1
Factor2
Factor3

From the cumulative eigenvalues presented in Output 4.7 we see that the first three factors account for almost 99% of the variation. Next, an examination of the varimax rotated factor loadings reveals the following:

- the first factor has very high loadings for the variables corresponding to medium exposure range, namely X5, X6, X7, X8, X9, and X10, and hence it represents the normal density.
- the second factor with very high loadings for the variables corresponding to low exposure range, namely X1, X2, X3, and X4 represents high density.
- the third factor with very high loadings for the variables X11, X12, X13, and X14, all of which correspond to high exposure range, represents the low density.

A plot of the factor loadings corresponding to these three factors is also provided in Output 4.7. The SAS data set SAVEALL stores the rotated factor loadings under _TYPE_=PATTERN in rows. In Program 4.7 we extract these loadings and then transpose them using PROC TRANSPOSE before plotting. The plot clearly indicates how these three factors represent different densities.

EXAMPLE 14 ***Application in Marketing Research, Diabetes Data*** Data on diabetic patients' attitude toward the disease, provided by Dr. G. R. Patwardhan of Strategic Marketing Corporation, are given in the Appendix (see DIABETIC.DAT). The opinions of 122 patients on 25 questions were recorded in a five-point scale. The scales used are 1 (Strongly disagree), 2 (Somewhat disagree), 3 (Neither disagree nor agree) 4 (Somewhat agree), and 5 (Strongly agree). This is a typical Likert scale, often used in surveys, psychological research, and other applications. Responses to these 25 questions are denoted by variables X1 to X25, respectively. The questions were

1. I am concerned about the long-term effects of diabetes.
2. I don't want anyone to know that I have diabetes.
3. I would test my blood sugar more frequently if I didn't have to prick my finger.
4. Testing is not really that important; I could get by without it.
5. Pricking my finger to test is no big deal.
6. I feel reassured if I test my blood sugar more often.
7. Hourly blood sugar readings would make me nervous.
8. I would like to have more daily readings of my blood sugar level to control my diabetes.
9. Testing my blood sugar interferes with my lifestyle.
10. My doctor would like me to test more often than I do now.
11. There are times I would like to test more, but it's too inconvenient.
12. There are some days that I don't test my blood sugar at all.
13. I would like to have tighter control of my blood sugar, but I'm afraid of hypoglycemia.
14. I feel that testing my blood sugar allows me to have a better quality of life.
15. Hypoglycemia could be avoided if I rested more often.
16. I would test more if the test strips were cheaper.
17. When I am sick or don't feel well, I test my blood sugar more often.
18. Carrying around my blood sugar testing materials is a nuisance.
19. The Diabetes Control and Complication Trial has made a difference in how often I test my blood sugar.
20. Diabetes is financial burden.
21. More blood sugar readings would not make any difference to me.
22. Hourly glucose readings over a 24-hour period would be very useful.

23. More blood sugar readings than I have now are unnecessary.
24. Blood sugar testing is critical for managing diabetes.
25. Knowing what my blood sugar levels are while I sleep would give me peace of mind.

The objective of the study is to form groups among these questions such that each group variable measures a certain aspect of the disease. We perform a factor analysis on these data using the principal factor method with the PRIORS = ASMC option and the VARIMAX option. The code is given in Program 4.8, and the output is given in Output 4.8.

```
/* Program 4.8 */

options ls=64 ps=45 nodate nonumber;
title1 'Output 4.8';
title2 'Factor Analysis of Marketing Data';

data diab;
infile 'diabetic.dat' firstobs=3 obs=125;
input id $ x1-x25;
run;
proc factor data=diab n=5 method=prin priors=asmc
rotate=varimax;
var x1-x25;
run;
```

Output 4.8

```
                    Output 4.8
          Factor Analysis of Marketing Data

                The FACTOR Procedure
        Initial Factor Method: Principal Factors

     Eigenvalues of the Reduced Correlation Matrix:
        Total = 10.1455605  Average = 0.40582242

        Eigenvalue    Difference    Proportion    Cumulative

   1    4.29532757    1.90882584        0.4234        0.4234
   2    2.38650172    1.38317944        0.2352        0.6586
   3    1.00332228    0.09494445        0.0989        0.7575
   4    0.90837783    0.08728159        0.0895        0.8470
   5    0.82109625    0.13909771        0.0809        0.9280
   6    0.68199854    0.07690699        0.0672        0.9952
   7    0.60509155    0.03269249        0.0596        1.0548
   8    0.57239906    0.15536289        0.0564        1.1112
   9    0.41703617    0.14235293        0.0411        1.1523
  10    0.27468324    0.05300032        0.0271        1.1794
  11    0.22168292    0.06786819        0.0219        1.2013

                Rotated Factor Pattern

        Factor1     Factor2     Factor3     Factor4     Factor5

x1     -0.05162     0.41142     0.01548    -0.00284     0.16195
x2      0.11088    -0.25240     0.28936     0.13014    -0.17716
x3      0.10806     0.15632     0.65088     0.17195    -0.05959
x4     -0.04644    -0.05858     0.24899     0.09325    -0.34480
x5     -0.08362    -0.01149    -0.28504    -0.15245     0.38359
x6      0.64244     0.11227     0.08021    -0.08318     0.35726
```

x7	-0.27245	-0.29289	0.40535	0.06698	0.06880
x8	0.26636	0.62155	0.08864	0.18607	0.15275
x9	-0.08671	0.07705	0.05412	0.62594	-0.11129
x10	0.15900	0.25849	0.49343	0.22532	-0.18215
x11	0.14870	0.31935	0.51129	0.41603	0.03921
x12	-0.06560	-0.03037	0.60034	-0.02458	-0.06365
x13	0.09073	-0.01981	0.25592	0.39421	0.09079
x14	0.41451	0.01611	-0.21725	-0.05372	0.35809
x15	0.19512	0.35260	0.09821	0.14198	0.06918
x16	0.45273	0.08614	0.02363	0.27375	-0.10261
x17	0.51878	0.18301	0.07721	0.00501	-0.14434
x18	0.17527	0.15874	0.28509	0.58993	0.02703
x19	0.44136	0.09006	0.18650	0.01381	0.19027
x20	0.42919	-0.10718	-0.17314	0.40071	-0.08043
x21	-0.30186	-0.31422	0.08490	0.25885	0.06740
x22	0.56840	0.45875	0.10401	0.17905	0.03511
x23	-0.22418	-0.65952	-0.01363	0.02510	0.12773
x24	0.08082	0.15181	0.09583	0.18119	0.55211
x25	0.55089	0.48493	-0.04804	0.18679	0.03681

Because of some missing values, only 109 observations out of 122 were used in the analysis. We select five factors that account for about 93% of the total variability (see Output 4.8). Examining the output under the heading "Rotated Factor Pattern" we see that the first common factor with high loadings for the variables X6, X17, X22, and X25 represents a component that represents the peace of mind derived from the blood sugar level testing. We may call this variable "Usefulness of testing blood sugar level." This factor accounts for about 42% of the total variability. The second factor that explains about 23% of the variability has high positive loading for variable X8 and high negative loading for X23. These opposite signs are consistent with the mutually opposite directions (with respect to the phrase used) of questions 8 and 23. Thus the second factor represents a component representing willingness to have more daily readings of the blood sugar level. The other three factors explain about 10%, 9%, and 8% variability, respectively. The third common factor has high loadings for the variables X3, X10, X11, and X12. This factor may be described as "apathy for testing." The fourth factor has high loadings for X9 and X18, representing a variable that affects the lifestyle of the patient. Finally, the fifth common factor has high loading for only one variable, X24, which indicates the importance of blood sugar testing for managing diabetes.

Usually we want each common factor to have high loadings for several (around four) variables. Also, we want the common factors to be nearly uncorrelated. In the present example, several factors have only a small number of variables with high loadings, and the extracted factors appear to be mildly correlated. This indicates the need for a better questionnaire that includes a number of questions that are likely to be associated with the five broad categories of variables identified by the five common factors.

It may be pointed out that the problem can also be viewed as that of clustering variables. Techniques for this approach using the CLUSTER and VARCLUS procedures are discussed in Chapter 6.

EXAMPLE 15 *Factor Analysis of Longitudinal Data, Mice Data* Data on the weights of 33 male mice (given in Appendix as MICE.DAT) measured in three groups over the 21 days from birth to weaning are analyzed by Izenman and Williams (1989) using the factor analysis techniques. We present part of their analyses here to illustrate an exploratory analysis of growth curves using a factor analytic approach. Data consist of three groups of mice. Group 1 contains the data on weights of 11 mice at ages 0, 3, 6, 9, 12, 15, and 18 days following birth; data in Group 2 contain the weights of 10 mice at ages 1, 4, 7, 10, 13, 16, and 19 days following birth; and Group 3 contains the weights of 12 mice at ages 2, 5, 8, 11, 14, 17, and 20 days following birth.

The objective of the study is to use these growth data to identify the factors representing the prenatal environment, the postnatal environment, and the genetic aspect of the early part of the growth. In Program 4.9, factor analysis of each of these three data sets is separately conducted, and three factors in each case (using NFACT=3) are selected. The three factor-loading matrices thus obtained are then stacked to form one big matrix of factor loadings such that the loadings corresponding to the ages 0 through 20 are in a sequence. Program 4.9 contains the SAS code and Output 4.9 contains the results.

```
/* Program 4.9 */

options ls=64 ps=45 nodate nonumber;
title1 'Output 4.9';

data mice1;
infile 'mice.dat' firstobs=3 obs=13;
input x1-x7;
run;
proc factor data=mice1 cov method=prin nfact=3 rotate=varimax
outstat=save1;
title2 'Factor Loadings of Group 1 Data';
run;
data temp1(type=factor);
set save1;
if _type_='PATTERN' then output;
run;
proc transpose data=temp1 out=out1;
run;

data mice2;
infile 'mice.dat' firstobs=16 obs=25;
input x1-x7;
run;
proc factor data=mice2 cov method=prin nfact=3 rotate=varimax
outstat=save2;
title2 'Factor Loadings of Group 2 Data';
run;
data temp2(type=factor);
set save2;
if _type_='PATTERN' then output;
run;
proc transpose data=temp2 out=out2;
run;
data mice3;
infile 'mice.dat' firstobs=28 obs=39;
input x1-x7;
run;
proc factor data=mice3 cov method=prin nfact=3 rotate=varimax
outstat=save3;
title2 'Factor Loadings of Group 3 Data';
run;
data temp3(type=factor);
set save3;
if _type_='PATTERN' then output;
run;
proc transpose data=temp3 out=out3;
run;
```

```
data plot;
set out1 out2 out3;
pos=mod(_n_-1,7);
proc sort;
by pos;
run;
data plot;
set plot;
obser=_n_-1;
pos_env=factor1**2;
pre_env=factor2**2;
genetic=factor3**2;
run;
proc print data=plot;
var obser pos_env pre_env genetic;
title2 'Squared Factor Loadings from the Three Groups';
run;

filename gsasfile "prog49.graph";
goptions reset=all gaccess=gsasfile autofeed dev=psl;
goptions horigin=1in vorigin=2in;
goptions hsize=6in vsize=8in;
title1 h=1.2 'Plots of Squares of Factor Loadings for Mice Data';
title2 j=l 'Output 4.9';
proc gplot data=plot;
plot pos_env*obser pre_env*obser genetic*obser/overlay
vaxis=axis1 haxis=axis2 legend=legend1;
axis1 label=(a=90 h=1.2 'Proportion of Variation of Weight');
axis2 offset=(2) label=(h=1.2 'Number of Days After Birth');
symbol1 v=none i=sm65 v=star c=red;
symbol2 v=none i=sm65 v=+ c=blue;
symbol3 v=none i=sm65 v=x c=green;
legend1 across=3;
run;
```

Output 4.9

```
                    Output 4.9
           Factor Loadings of Group 1 Data

                The FACTOR Procedure
      Initial Factor Method: Principal Components

          Prior Communality Estimates: ONE

      Eigenvalues of the Covariance Matrix: Total
          = 0.03990425  Average = 0.00570061
```

	Eigenvalue	Difference	Proportion	Cumulative
1	0.03156078	0.02624922	0.7909	0.7909
2	0.00531155	0.00351587	0.1331	0.9240
3	0.00179569	0.00086767	0.0450	0.9690
4	0.00092801	0.00073088	0.0233	0.9923
5	0.00019714	0.00011847	0.0049	0.9972
6	0.00007866	0.00004624	0.0020	0.9992
7	0.00003242		0.0008	1.0000

3 factors will be retained by the NFACTOR criterion.

Rotated Factor Pattern

	Factor1	Factor2	Factor3
x1	-0.21166	0.67412	-0.04303
x2	0.29437	0.86928	-0.04013
x3	0.49693	0.80547	0.31258
x4	0.87743	0.34009	0.24682
x5	0.96915	0.06693	0.12545
x6	0.98164	-0.15644	-0.02434
x7	0.86602	0.39302	-0.29913

Factor Loadings of Group 2 Data

The FACTOR Procedure
Initial Factor Method: Principal Components

Prior Communality Estimates: ONE

Eigenvalues of the Covariance Matrix: Total
= 0.01560856 Average = 0.00222979

	Eigenvalue	Difference	Proportion	Cumulative
1	0.01038378	0.00730842	0.6653	0.6653
2	0.00307535	0.00187868	0.1970	0.8623
3	0.00119667	0.00042403	0.0767	0.9390
4	0.00077264	0.00064436	0.0495	0.9885
5	0.00012828	0.00008753	0.0082	0.9967
6	0.00004075	0.00002967	0.0026	0.9993
7	0.00001108		0.0007	1.0000

3 factors will be retained by the NFACTOR criterion.

Rotated Factor Pattern

	Factor1	Factor2	Factor3
x1	0.10933	0.08811	0.44064
x2	0.05892	0.82255	0.10785
x3	0.31143	0.94091	0.00585
x4	0.72510	0.43394	-0.53195
x5	0.91042	0.37751	0.02252
x6	0.81958	0.35296	0.21604
x7	0.85804	-0.24450	0.39371

Factor Loadings of Group 3 Data

The FACTOR Procedure
Initial Factor Method: Principal Components

Prior Communality Estimates: ONE

Eigenvalues of the Covariance Matrix: Total
= 0.05845062 Average = 0.00835009

	Eigenvalue	Difference	Proportion	Cumulative
1	0.05160666	0.04816779	0.8829	0.8829
2	0.00343888	0.00158896	0.0588	0.9417
3	0.00184992	0.00091745	0.0316	0.9734
4	0.00093247	0.00056168	0.0160	0.9893
5	0.00037079	0.00015106	0.0063	0.9957
6	0.00021973	0.00018754	0.0038	0.9994
7	0.00003218		0.0006	1.0000

3 factors will be retained by the NFACTOR criterion.

Rotated Factor Pattern

	Factor1	Factor2	Factor3
x1	-0.04228	0.89126	-0.00566
x2	0.21391	0.92210	0.02338
x3	0.66963	0.69734	0.04028
x4	0.91526	0.30684	-0.14631
x5	0.97687	0.11978	-0.06017
x6	0.99057	-0.03514	0.07479
x7	0.90880	0.19249	0.36643

Squared Factor Loadings from the Three Groups

Obs	obser	pos_env	pre_env	genetic
1	0	0.04480	0.45443	0.00185
2	1	0.01195	0.00776	0.19417
3	2	0.00179	0.79434	0.00003
4	3	0.08665	0.75564	0.00161
5	4	0.00347	0.67659	0.01163
6	5	0.04576	0.85028	0.00055
7	6	0.24694	0.64879	0.09771
8	7	0.09699	0.88532	0.00003
9	8	0.44840	0.48628	0.00162
10	9	0.76988	0.11566	0.06092
11	10	0.52577	0.18830	0.28298
12	11	0.83770	0.09415	0.02141
13	12	0.93925	0.00448	0.01574
14	13	0.82887	0.14251	0.00051
15	14	0.95427	0.01435	0.00362
16	15	0.96361	0.02447	0.00059
17	16	0.67172	0.12458	0.04667
18	17	0.98124	0.00123	0.00559
19	18	0.74999	0.15447	0.08948
20	19	0.73624	0.05978	0.15501
21	20	0.82591	0.03705	0.13427

Output 4.9
(continued)

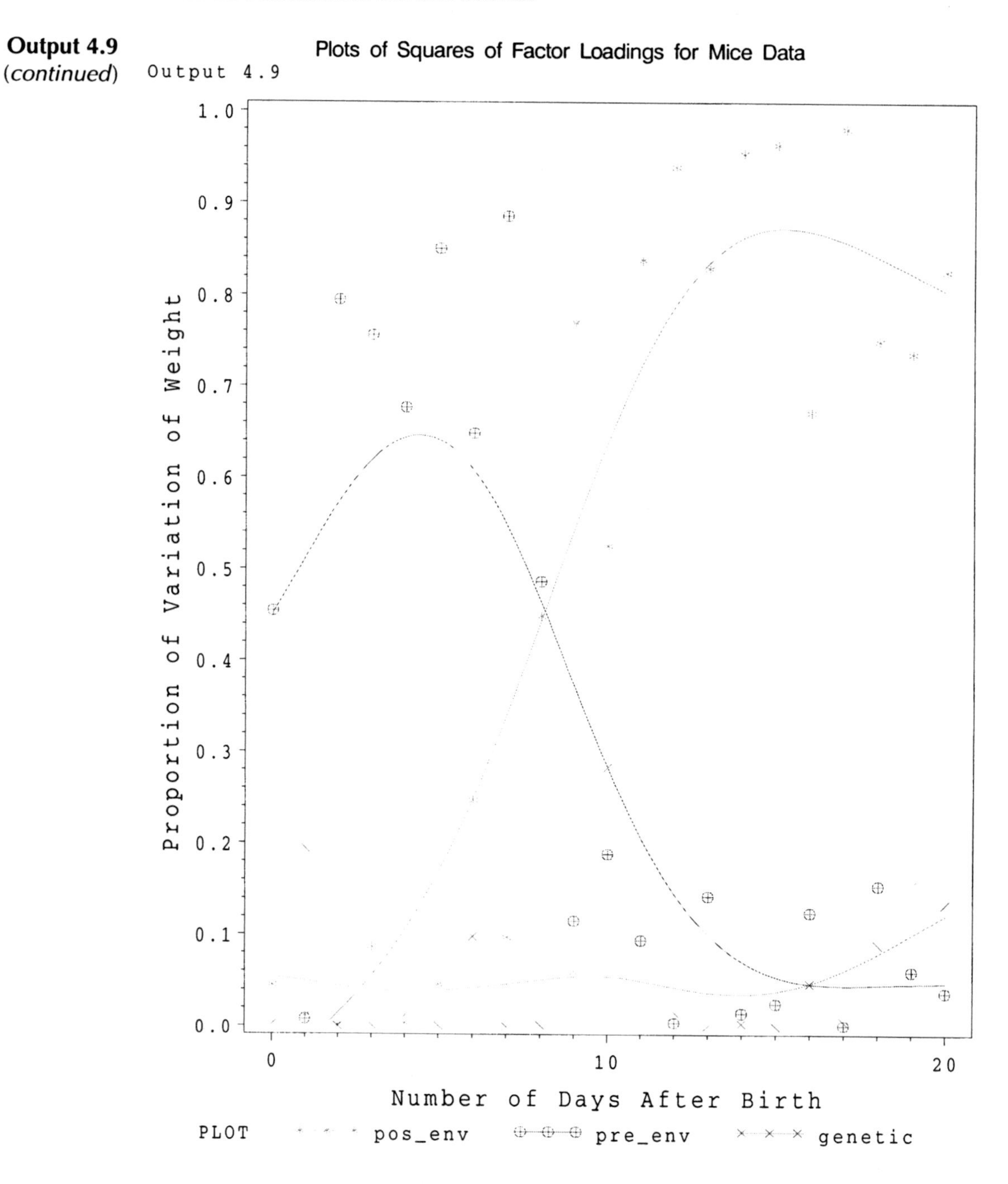

Corresponding to any variable, the square of the factor loading value for any factor is the contribution of that factor to the communality (common variance) of that variable. For each of the 21 variables we calculate the portion of the variances contributed by each of the 3 factors.

Using the information available in Output 4.9, the contributions by each of the three factors are plotted on the same graph against the identity of the variables (listed from 0 to 20). Smooth curves over the plotted points are drawn using the option I=SM65 in the SYMBOL statement in PROC GPLOT. This plot given in Output 4.9 depicts an interesting biological phenomenon observed by El Oksh, Sutherland, and Williams (1967) and illustrated by Izenman and Williams (1989). The contribution of the variance by the first factor corresponds to the effects of post-natal environment, the contribution of the variance by the second factor corresponds to the effects of pre-natal environment, and that by the third component corresponds to the effect of the genetic factor.

EXAMPLE 16 *Geological Data* Sediments on the beaches exhibit wide variations in the grain size. These differences over short stretches of the coastal area reflect on the differential energy prevailing in the near shore zones. Perhaps this contributes in a significant way to the variations in the grain sizes over time along with the source and supply of sediments. The beach receives material from offshore or mainland sources or as a linear longshore drift primarily due to wave-induced flows (Nayak, 1993). In order to understand these variations, we use factor analysis on a part of data provided by Dr. G. N. Nayak of Goa University, India. Our objective is to demonstrate how factor analysis can be used for understanding some of these phenomena in sedimentology and other geological data.

The data were collected at Arabian Sea beaches of Karwar in Karnataka, India, for a period of 13 months from December 1982 to December 1983. The smaller data set that we consider here contains seven sizes of sediment grains (X1 to X7), collected at ten stations (STATION with values 1 to 10) along the beaches, at high tide (TIDE=HIGH), middle tide (TIDE=MID) and low tide (TIDE=LOW) for the three months, April 1983 (SEASON=APR_83), July 1983 (SEASON=JUL_83), and September 1983 (SEASON=SEP_83). These three months were selected because April represents the premonsoon season, July the monsoon season and September the retreating phase of monsoon season.

Hoping to sort out the variation in the grain sizes for different seasons and the tide levels, we perform factor analysis using the principal component method with varimax rotation for each of the three months and for each of the three values of the variable TIDE. Thus we have factor analysis results for nine cases. For each of these cases, more than 94% of the total variation is explained by the first three factors (output not shown). Hence, these three factors may suffice for understanding the variations in the sediment grain sizes. See Nayak and Chavadi (1988) for a similar analysis.

The data are provided in the Appendix in a data file SEDIMNT.DAT and the SAS code is provided in Program 4.10. While the numerical output has been suppressed, certain graphs of the factor loadings for the first factor are provided in Output 4.10.

```
/* Program 4.10 */

options ls=64 ps=45 nodate nonumber;
title1 'Output 4.10';
title2 'Factor Analysis of Sediments Data';
data phi;
infile 'sedimnt.dat' firstobs=3 obs=93;
input id season$ tide$ station$ x1-x7;
run;
data phi;
set phi;
proc sort;
by season tide;
```

```
run;
proc factor data=phi cov method=prin n=3 rotate=varimax
outstat=save1;
var x1-x7;
by season tide;
run;
data temp1(type=factor);
set save1;
if _type_='PATTERN' and tide='high' then output;
run;
data temp1;
set temp1;
proc sort;
by season;
run;
proc transpose data=temp1 out=out1;
by season;
run;
data plot;
set out1;
obser=mod(_n_-1,7);
if season='Apr_83' then April=factor1;
if season='Jul_83' then July=factor1;
if season='Sep_83' then Sept=factor1;
run;
filename gsasfile "prog410a.graph";
goptions reset=all gaccess=gsasfile autofeed dev=psl;
goptions horigin=1in vorigin=2in;
goptions hsize=6in vsize=8in;
title1 h=1.2 'Plot of Loadings for Factor 1: High Tide';
title2 j=l 'Output 4.10';
proc gplot data=plot;
plot April*obser July*obser Sept*obser/overlay
vaxis=axis1 haxis=axis2 legend=legend1;
axis1 label=(a=90 h=1.2 'Factor 1: Loadings');
axis2 offset=(2) label=(h=1.2 'Grain Size: High Tide');
symbol1 i=join v=star c=red;
symbol2 i=join v=dot c=blue;
symbol3 i=join v=x c=green;
legend1 across=3;
footnote j=l 'Sedimentology Data: Courtesy: Dr. G. N. Nayak';
run;
data temp2(type=factor);
set save1;
if _type_='PATTERN' and tide='mid' then output;
run;
data temp2;
set temp2;
proc sort;
by season;
run;
proc transpose data=temp2 out=out2;
by season;
run;
data plot;
set out2;
obser=mod(_n_-1,7);
if season='Apr_83' then April=factor1;
if season='Jul_83' then July=factor1;
```

```
if season='Sep_83' then Sept=factor1;
run;
filename gsasfile "prog410b.graph";
goptions reset=all gaccess=gsasfile autofeed dev=psl;
goptions horigin=1in vorigin=2in;
goptions hsize=6in vsize=8in;
title1 h=1.2 'Plot of Loadings for Factor 1: Middle Tide';
title2 j=l 'Output 4.10';
proc gplot data=plot;
plot April*obser July*obser Sept*obser/overlay
vaxis=axis1 haxis=axis2 legend=legend1;
axis1 label=(a=90 h=1.2 'Factor 1: Loadings');
axis2 offset=(2) label=(h=1.2 'Grain Size: Middle Tide');
symbol1 i=join v=star c=red;
symbol2 i=join v=dot c=blue;
symbol3 i=join v=x c=green;
legend1 across=3;
footnote j=l 'Sedimentology Data: Courtesy: Dr. G. N. Nayak';
run;
data temp3(type=factor);
set save1;
if _type_='PATTERN' and tide='low' then output;
run;
data temp3;
set temp3;
proc sort;
by season;
run;
proc transpose data=temp3 out=out3;
by season;
run;
data plot;
set out3;
obser=mod(_n_-1,7);
if season='Apr_83' then April=factor1;
if season='Jul_83' then July=factor1;
if season='Sep_83' then Sept=factor1;
run;
filename gsasfile "prog410c.graph";
goptions reset=all gaccess=gsasfile autofeed dev=pslmono;
goptions horigin=1in vorigin=2in;
goptions hsize=6in vsize=8in;
title1 h=1.2 'Plot of Loadings for Factor 1: Low Tide';
title2 j=l 'Output 4.10';
proc gplot data=plot;
plot April*obser July*obser Sept*obser/overlay
vaxis=axis1 haxis=axis2 legend=legend1;
axis1 label=(a=90 h=1.2 'Factor 1: Loadings');
axis2 offset=(2) label=(h=1.2 'Grain Size: Low Tide');
symbol1 i=join v=star c=red;
symbol2 i=join v=dot c=blue;
symbol3 i=join v=x c=green;
legend1 across=3;
footnote j=l 'Sedimentology Data: Courtesy: Dr. G. N. Nayak';
run;
```

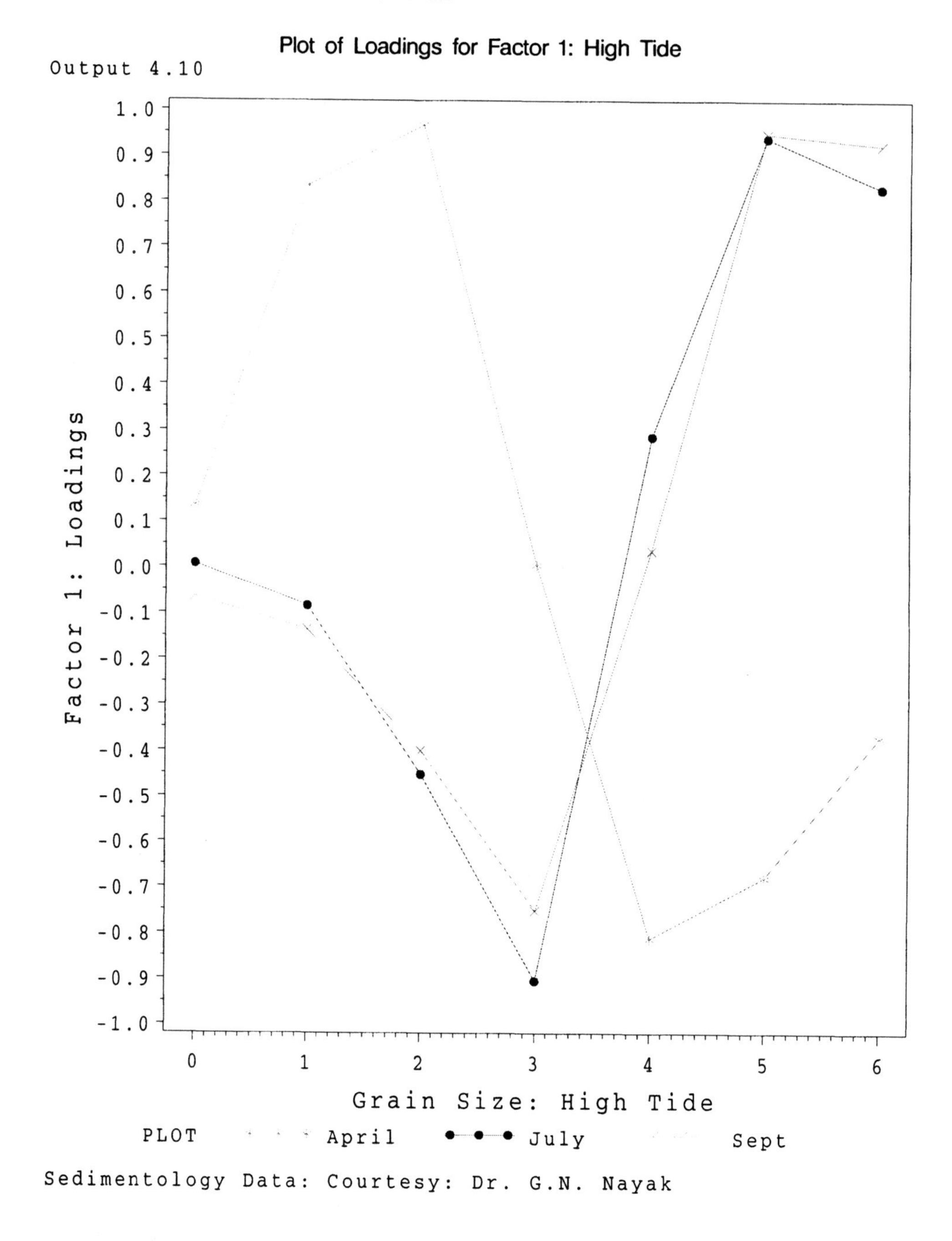
Plot of Loadings for Factor 1: High Tide
Output 4.10
1.0
0.9
0.8
0.7
0.6
0.5
0.4
0.3
0.2
0.1
0.0
-0.1
-0.2
-0.3
-0.4
-0.5
-0.6
-0.7
-0.8
-0.9
-1.0
Factor 1: Loadings
0
1
2
3
4
5
6
Grain Size: High Tide
PLOT
April
July
Sept
Sedimentology Data: Courtesy: Dr. G.N. Nayak

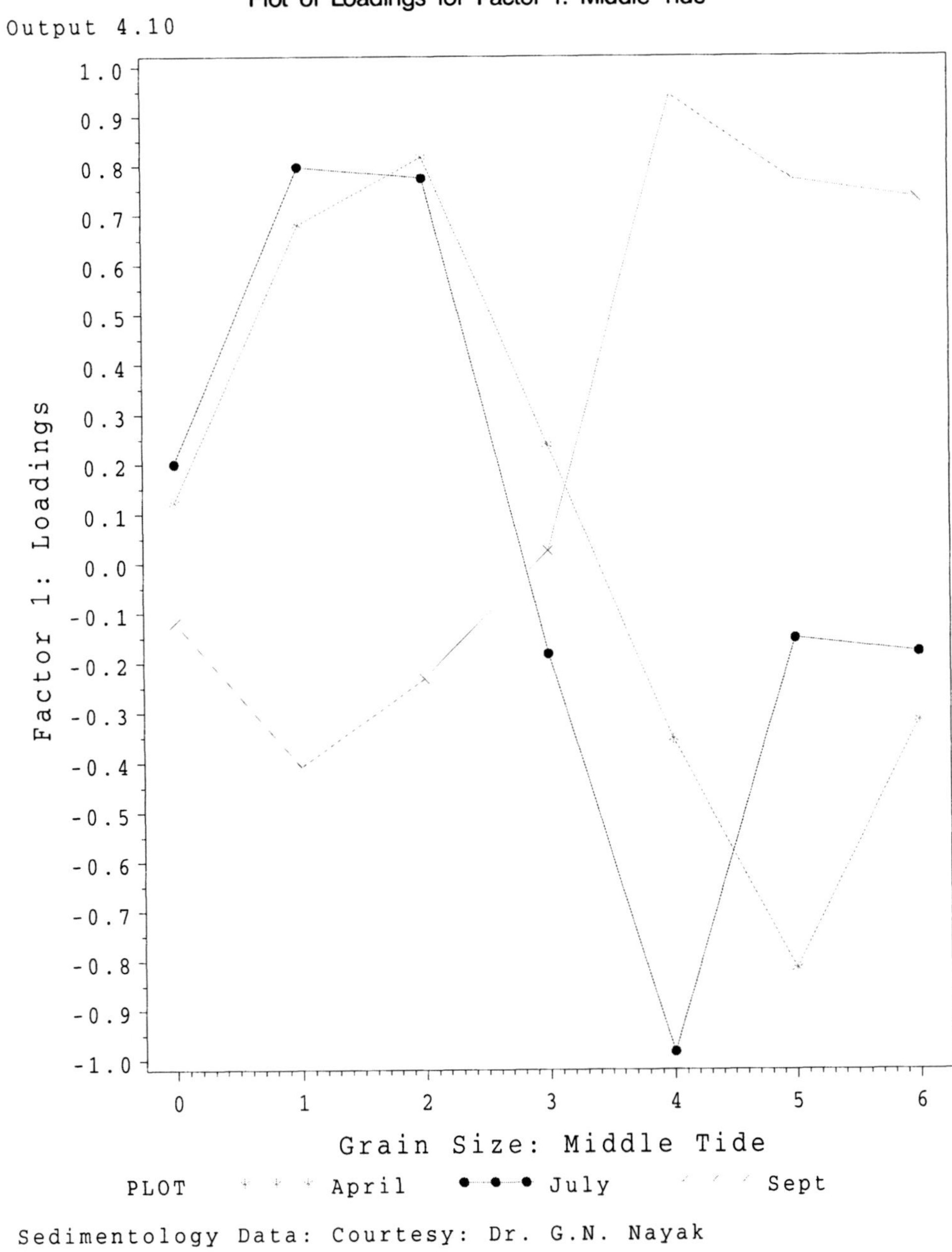

Sedimentology Data: Courtesy: Dr. G.N. Nayak

Output 4.10

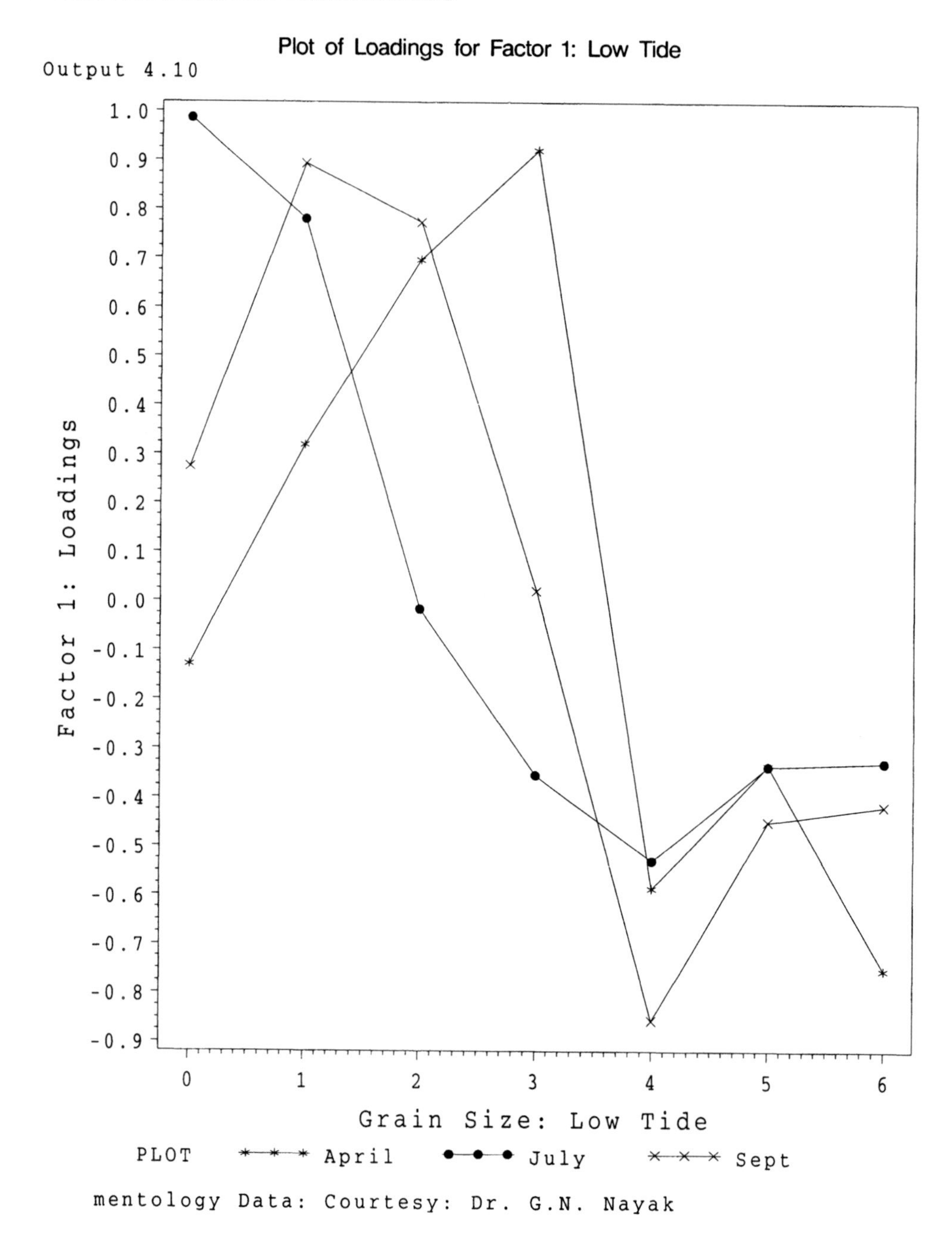
Plot of Loadings for Factor 1: Low Tide
Factor 1: Loadings
1.0
0.9
0.8
0.7
0.6
0.5
0.4
0.3
0.2
0.1
0.0
-0.1
-0.2
-0.3
-0.4
-0.5
-0.6
-0.7
-0.8
-0.9
0
1
2
3
4
5
6
Grain Size: Low Tide
PLOT
April
July
Sept
mentology Data: Courtesy: Dr. G.N. Nayak

The factor loadings for the first common factor are plotted for each of the three seasons on the same plot for high tide and similarly on the same plots for the middle and low tides. As it is usually done, we take the mode area/bandwidth as the guidelines for making the following conclusions from these plots. See Nayak and Chavadi (1988).

The grain size variability is very high at high tide during July and September, perhaps as a result of relatively high wave energy prevailing. The variation is relatively less in April, indicating slightly smaller range sediments during the fair-weather season. At middle-tide level, more variability is observed during September than that corresponding to July and April. For low-tide level the variability in grain size is more during July as compared to the other two months.

For the high tide, the plots for July and September behave the same way but these are very different from the corresponding plot for April. This indicates that the variations in grain sizes are different for the premonsoon period but similar during the middle of the monsoon season and up to the end of the season. For medium tides, the behavior is quite different from that observed for high tide, in that the variations for April and July are similar, whereas the variation is different for September. However, for the low tide, all these seasons seem to depict similar variations in the grain sizes.

The plots for the second common factor (not shown) also show a varied degree of variation in grain sizes for different seasons and tide levels. However, the third common factor shows similar variation in the grain sizes for all tide levels as well as seasons.

4.12 Concluding Remarks

Generally, factor analysis is performed as an exploratory data analysis. The aim is to see whether the original variables can be expressed in terms of a few common factors, thus reducing the dimensionality of the original data. In this chapter we have discussed various methods for performing factor analysis and illustrated these methods with examples that use SAS.

As we have observed in the examples discussed, by adopting certain optimal orthogonal or oblique transformation (factor rotation) of the factor loadings, that it is generally possible to determine the common factors that have high loadings for disjoint sets of variables. We have discussed many available methods for both orthogonal and oblique factor rotations and illustrated them in many real-life data sets. However, it may not always be possible to achieve the objective of factor rotation. That is, even after using different rotations we may not be able to determine common factors that have high loadings for disjoint sets of variables. When this happens, we may look for alternative methods for analyzing the data.

If the factor analysis as described is successful, then sometimes as an additional analysis another type of factor analysis called *confirmatory factor analysis* is performed on the data. This is a form of factor analysis in which the factor model is hypothesized, and large sample significance tests using maximum likelihood method are performed for factor loadings, the specific variances, and covariances between the common factors. The significance of these parameters will confirm the hypothesized factor model as the correct model to adopt. Most of these tasks of performing the significance tests can be done by using the SAS CALIS procedure. We have not illustrated PROC CALIS here, but a detailed account of confirmatory factor analysis and other applications of the CALIS procedure can be found in Hatcher (1994).

Chapter 5

Discriminant Analysis

5.1 Introduction

Discriminant analysis deals with a special type of multivariate analysis problem. It focuses on separating distinct sets of units into two or more populations and then allocating the units not readily known to belong to a particular population into one of these populations. This allocation is of interest in many scientific studies, and numerous applications can be found in a variety of scientific disciplines. For example, a medical researcher may be interested in classifying diseases by symptoms or a geologist may be interested in identifying the type of rock by its chemical composition. Numerous applications of discriminant analysis in chemometrics, chemistry, anthropology, biology, genetics, education, sociology, and other sciences can easily be found.

The allocation of units, or classification into one of several populations, is a multivariate problem in that for each unit a number of possibly correlated responses have been collected. A single response out of these, by itself, may not describe the population sufficiently and adequately, and what may be needed is to produce a single index or a well-defined criterion that can be used as the classification rule to classify an object into one of several populations. Of course, since none of such decision rules are expected to be perfect, there will always be certain probabilities of misclassifications. These are either to be controlled or to be minimized.

To fix the ideas, suppose we have k distinct populations $\Pi_1, \Pi_2, \cdots, \Pi_k$, and the probability density function of a p by 1 random vector $\mathbf{x}$ from the tth population is given by $f_t(\mathbf{x})$. Suppose $R_1, R_2, \cdots, R_k$ are the k disjoint regions such that their union constitutes the entire p dimensional space. The random observation $\mathbf{x}$ is classified into the tth population Π_t if $\mathbf{x}$ belongs to R_t. Under this rule, each observation can be classified into one of the k populations. However, it is possible that an object that may actually belong to the sth population Π_s may have been classified into the tth population $\Pi_t, t \neq s$. In view of this, a misclassification may occur in $k(k-1)$ different ways. The probability that an object coming from the sth population Π_s is classified into the tth population Π_t is equal to the conditional probability $P(\mathbf{x} \in R_t | \mathbf{x} \in \Pi_s)$, $s \neq t$. This probability is often denoted in short by $P(t|s)$. A good classification rule should have minimal misclassification probabilities that would happen if the regions $R_1, R_2, \cdots, R_k$ differ from each other markedly. This, in essence, leads to a search for the good decision rules, or decision indexes, that can define the regions $R_t, t = 1, \cdots k$ as different from each other as possible. It may also be noted that the effectiveness of the partition of the p-dimensional space into the regions R_t also depends on how different the populations $\Pi_t, t = 1, \cdots, k$ are from each other.

Often the populations are assumed to have some specific probability distributions. The multivariate normal distribution is a common assumption for the populations; however some parametric discriminant analysis can be done under certain other distributions as well. Another approach is that of nonparametrics when no distributional assumptions are made and the densities of the populations are estimated nonparametrically. Yet another (third) approach to discriminant analysis is using the logistic regression, where the response variable is defined in terms of k populations and is then regressed on the p measurements through a logistic function. Further, there is a surge of interest in certain other discrimination techniques such as classification trees and neural networks. Many of these approaches enable us to also deal with the situations where some or all the variables are categorical.

5.2 Multivariate Normality

If we want to use discrimination procedures that assume the multivariate normality of the individual populations, we must first examine the validity of this assumption. Tests for multivariate normality based on Mardia's multivariate kurtosis are given in Chapter 1 of Khattree and Naik (1999). Graphically, multivariate normality can be examined using $Q-Q$ plots (Khattree and Naik, 1999). These should be used to examine the validity of this assumption. It may be remembered that in many nonnormal situations, the use of multivariate normality-based discriminant analysis could result in very misleading results. If some of the variables are continuous, yet nonnormal, the Box-Cox type of transformations can be made on these variables to attain (at least) univariate normality for the marginal distributions of these transformed variables. A macro for finding the optimal Box-Cox transformation is available in *SAS/QC Software: Reference, Version 6, First Edition*. It is also available in the SAS online documentation in the *SAS/QC User's Guide*.

EXAMPLE 1 ***Testing Multivariate Normality, Flea Beetles Data*** Lubischew (1962) considers a problem of discrimination between three species of flea beetles, namely, *Chaetocnema concinna* (Π_1), *Chaetocnema heikertingeri* (Π_2), and *Chaetocnema heptapotamica* (Π_3), based on various physical measurements. Data on the maximal width of the aedeagus in the forepart ($X1$) in microns and the front angle of the aedeagus ($X2$) in units of 7.5 degrees were subsequently analyzed by Seber (1984). We consider the data on $X1$ and $X2$ only and will examine these for marginal normality, as well as bivariate normality for each of the three species.

To examine univariate normality marginally for $X1$ and $X2$ in each of the three populations of species, we resort to graphical methods. Using the CHART procedure, we can obtain the individual charts, as well as an overlay of the three populations in a single chart. These plots for variable $X2$ resulting from the first part of Program 5.1 are given in Output 5.1. By and large they seem to convey a reasonable impression of symmetry. Although they are not shown here, an overlaid plot for three species suggests (at least as a precursor) that the beetles from species Π_2 can probably be well discriminated from Π_1 based on $X1$ only. Of course, a formal discriminant analysis needs to be performed before anything more definite can be asserted.

To assess bivariate normality, we resort to $Q - Q$ plots based on the Mahalanobis distance as explained in Chapter 2 of Khattree and Naik (1999). The $Q - Q$ plots presented in Output 5.1, although not perfect, seem to advocate the assumption of bivariate normality. The formal tests (Output 5.1) for multivariate normality, based on Mardia's multivariate skewness and kurtosis, also accepts this hypothesis (p values for skewness are 0.8922, 0.9879 and 0.1413 and for kurtosis, these are 0.3213, 0.3450 and 0.7179, respectively for these species). Therefore, it seems reasonable to assume that the three populations of flea beetles are individually distributed as the bivariate normal.

```
/* Program 5.1 */

filename gsasfile "prog51.graph";
goptions reset=all gaccess=gsasfile gsfmode = append
autofeed dev = pslmono;
goptions horigin =1in vorigin = 2in;
goptions hsize=6in vsize = 8in;
options ls = 64 ps=45 nodate nonumber;

title1 ' ';
title2 j = 1  'Output 5.1';
data beetles;
infile 'beetle.dat' firstobs=7;
input species  x1 x2;

proc format;
value specname
1 = 's1'
2 = 's2'
3 = 's3' ;
value specchar
1 = '1'
2 = '2'
3 = '3' ;
run;

proc gchart data = beetles;
by species ;
vbar x1 /subgroup = species midpoints = 115 to 160 ;
format species specchar. ;
title1 h = 1.2 'Histogram for X1';
title2 j = 1 'Output 5.1';
run;

proc gchart data = beetles;
vbar x1 /subgroup = species midpoints = 115 to 160 ;
format species specchar. ;
title1 h = 1.2 'Histogram for X1';
title2 j = 1 'Output 5.1';
run;

proc gchart data = beetles;
by species ;
vbar x2 /subgroup = species midpoints = 7 to 21 ;
format species specchar. ;
title1 h = 1.2 'Histogram for X2';
title2 j = 1 'Output 5.1';
run;
```

```
proc gchart data = beetles;
vbar x2 /subgroup = species midpoints = 7 to 21 ;
format species specchar. ;
title1 h = 1.2 'Histogram for X2';
title2 j = 1 'Output 5.1';
run;


/* Graphical Test for multivariate Normality */

data s1;set beetles;if species = 1 ;
data s2;set beetles;if species = 2 ;
data s3;set beetles;if species = 3 ;

proc princomp data=s1 n= 2 cov std out=b1 noprint;
var x1 x2;
data chiq1;
set b1;
dsq=uss(of prin1-prin2);
proc sort;
by dsq;
proc means noprint;
var dsq;
output out=chiq1n n=totn;
data chiqq1;
if(_n_=1) then set chiq1n;
set chiq1;

novar=2; /* novar=number of variables. */
chisq=cinv(((_n_-.5)/ totn),novar);
if mod(_n_,2)=0 then chiline=chisq;

proc gplot;
plot dsq*chisq chiline*chisq/overlay vaxis = axis1 haxis = axis2 ;
     axis1 label = (a = 90 h= 1.2 'Mahalanobis D Square');
axis2 offset = (2) label = (h= 1.2 'Chi-Square Quantile');
symbol1 v=star;
symbol2 i=join v=+;
title1 h=1.2 'Q-Q Plot for Assessing Normality';
title2 j = 1 'Output 5.1';
title3 'SPECIES = 1';
run;

proc princomp data=s2 n= 2 cov std out=b2 noprint;
var x1 x2;
data chiq2;
set b2;

dsq=uss(of prin1-prin2);
proc sort;
by dsq;
proc means noprint;
var dsq;
output out=chiq2n n=totn;
data chiqq2;
if(_n_=1) then set chiq2n;
set chiq2;
```

```
novar=2; /* novar=number of variables. */
chisq=cinv(((_n_-.5)/ totn),novar);
if mod(_n_,2)=0 then chiline=chisq;


proc gplot;
plot dsq*chisq chiline*chisq/overlay vaxis = axis1 haxis = axis2 ;
     axis1 label = (a = 90 h= 1.2 'Mahalanobis D Square');
axis2 offset = (2) label = (h= 1.2 'Chi-Square Quantile');
symbol1 v=star;
symbol2 i=join v=+;
title1 h=1.2 'Q-Q Plot for Assessing Normality';
title2 j = l 'Output 5.1';
title3 'SPECIES = 2';
run;

proc princomp data=s3 n= 2 cov std out=b3 noprint;
var x1 x2;
data chiq3;
set b3;

dsq=uss(of prin1-prin2);
proc sort;
by dsq;
proc means noprint;
var dsq;
output out=chiq3n n=totn;
data chiqq3;
if(_n_=1) then set chiq3n;
set chiq3;

novar=2; /* novar=number of variables. */
chisq=cinv(((_n_-.5)/ totn),novar);
if mod(_n_,2)=0 then chiline=chisq;


proc gplot;
plot dsq*chisq chiline*chisq/overlay vaxis = axis1 haxis = axis2 ;
     axis1 label = (a = 90 h= 1.2 'Mahalanobis D Square');
axis2 offset = (2) label = (h= 1.2 'Chi-Square Quantile');
symbol1 v=star;
symbol2 i=join v=+;
title1 h=1.2 'Q-Q Plot for Assessing Normality';
title2 j = l 'Output 5.1';
title3 'SPECIES = 3';
run;

/* In this program we are testing for the multivariate
normality of Beetles Data using the Mardia's skewness
and kurtosis measures */

title1 'Output 5.1';
title2 h = 1.5 "Mardia's Tests for Skewness and Kurtosis";
title3 ' ';
proc iml ;
y1 ={
150 15,
147 13, 144 14, 144 16, 153 13,
140 15, 151 14, 143 14, 144 14,
```

```
142 15, 141 13, 150 15, 148 13,
154 15, 147 14, 137 14, 134 15,
157 14, 149 13, 147 13, 148 14
} ;

y2 ={
120 14,
123 16, 130 14, 131 16, 116 16,
122 15, 127 15, 132 16, 125 14,
119 13, 122 13, 120 15, 119 14,
123 15, 125 15, 125 14, 129 14,
130 13, 129 13, 122 12, 129 15,
124 15, 120 13, 119 16, 119 14,
133 13, 121 15, 128 14, 129 14
} ;

y3 ={
124 13, 129 14, 145 8, 140 11,
140 11, 131 10, 139 11, 139 10,
136 12, 129 11, 140 10, 137 9,
141 11, 138 9, 143 9, 142 11,
144 10, 138 10, 140 10, 130 9,
137 11, 137 10, 136 9, 140 10
};
do times = 1 to 3;
if times = 1 then do ;
n = nrow(y1) ;
p = ncol(y1) ;
dfchi = p*(p+1)*(p+2)/6 ;
q = i(n) - (1/n)*j(n,n,1) ;
s = (1/(n))*y1`*q*y1 ;
s_inv = inv(s) ;
g_matrix = q*y1*s_inv*y1`*q ;
end ;
else if times = 2 then do ;
n = nrow(y2) ;
p = ncol(y2) ;
dfchi = p*(p+1)*(p+2)/6 ;
q = i(n) - (1/n)*j(n,n,1) ;
s = (1/(n))*y2`*q*y2 ;
s_inv = inv(s) ;
g_matrix = q*y2*s_inv*y2`*q ;
end ;
else if times = 3 then do ;
n = nrow(y3) ;
p = ncol(y3) ;
dfchi = p*(p+1)*(p+2)/6 ;
q = i(n) - (1/n)*j(n,n,1) ;
s = (1/(n))*y3`*q*y3 ;
s_inv = inv(s) ;
g_matrix = q*y3*s_inv*y3`*q ;
end ;
beta1hat = (  sum(g_matrix#g_matrix#g_matrix)  )/(n*n)  ;
beta2hat =trace(  g_matrix#g_matrix  )/n ;
kappa1 = n*beta1hat/6 ;
kappa2 = (beta2hat - p*(p+2) ) /sqrt(8*p*(p+2)/n) ;
pvalskew = 1 - probchi(kappa1,dfchi) ;
pvalkurt = 2*( 1 - probnorm(abs(kappa2))  ) ;
```

```
print 'Species = ' times ;
print s ;
print s_inv ;
print beta1hat kappa1  pvalskew ;
print beta2hat  kappa2  pvalkurt ;
end ;
```

Output 5.1

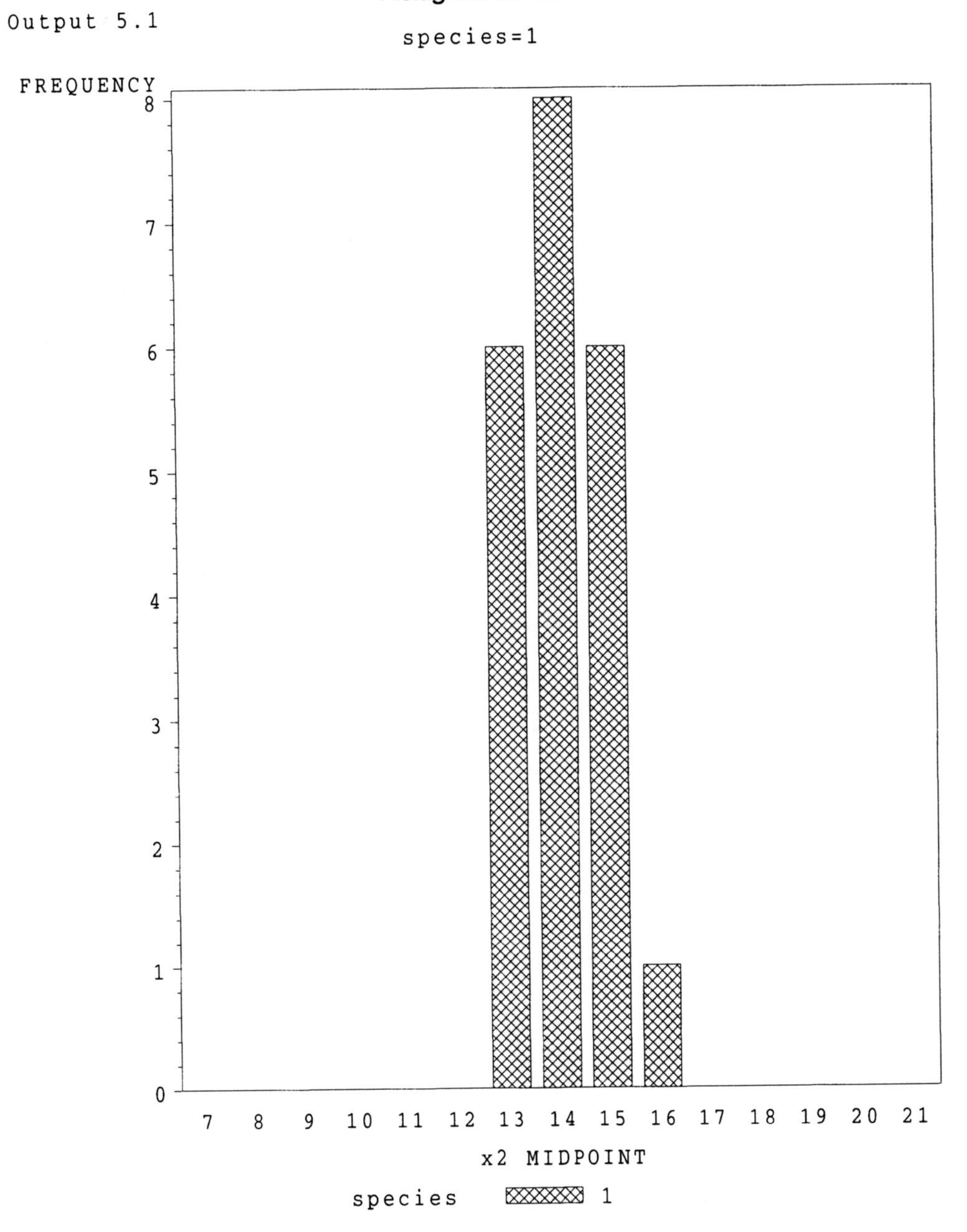

Output 5.1

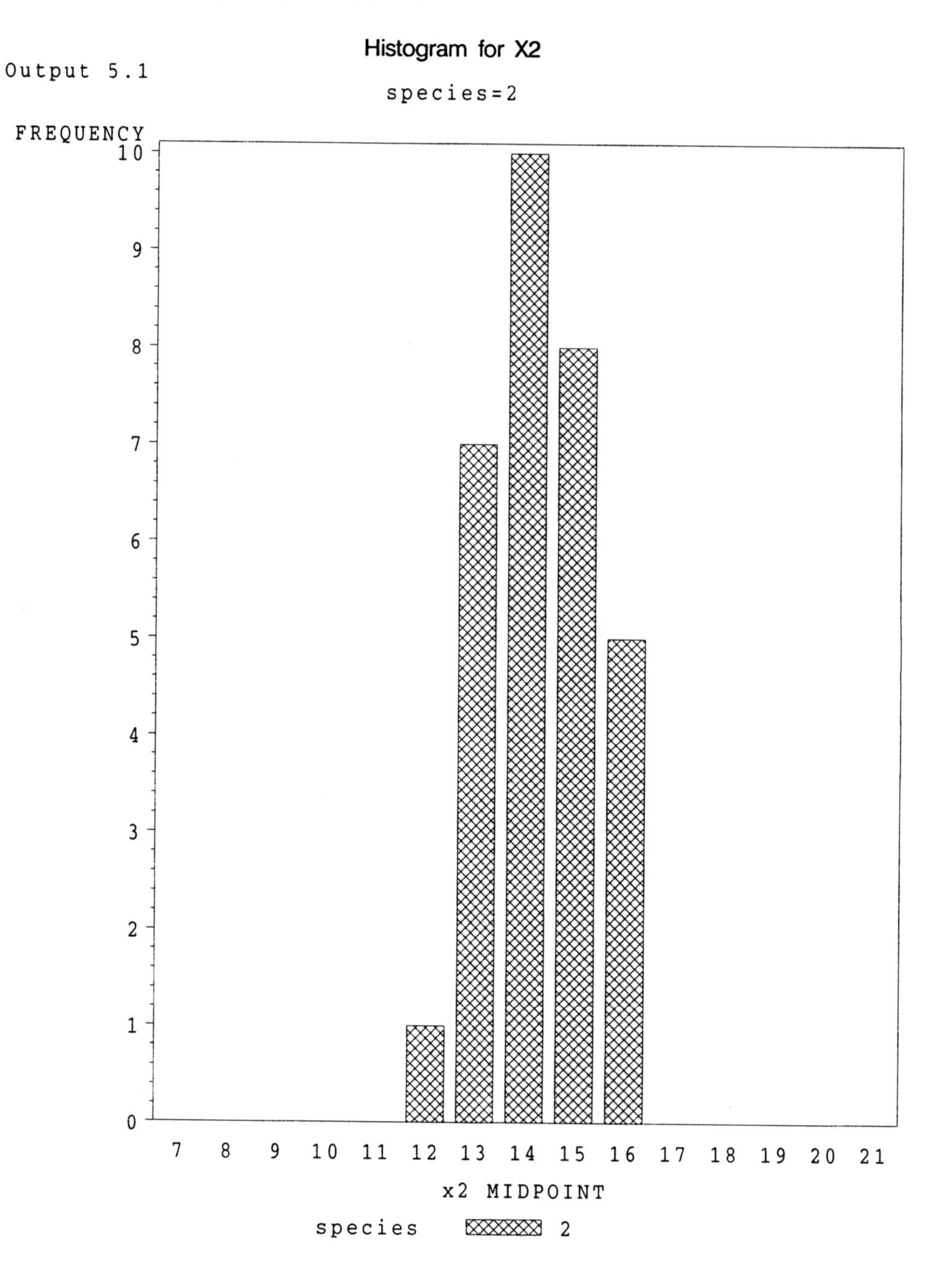
Histogram for X2
species=2
FREQUENCY
10
9
8
7
6
5
4
3
2
1
0
7 8 9 10 11 12 13 14 15 16 17 18 19 20 21
x2 MIDPOINT
species 2

Output 5.1

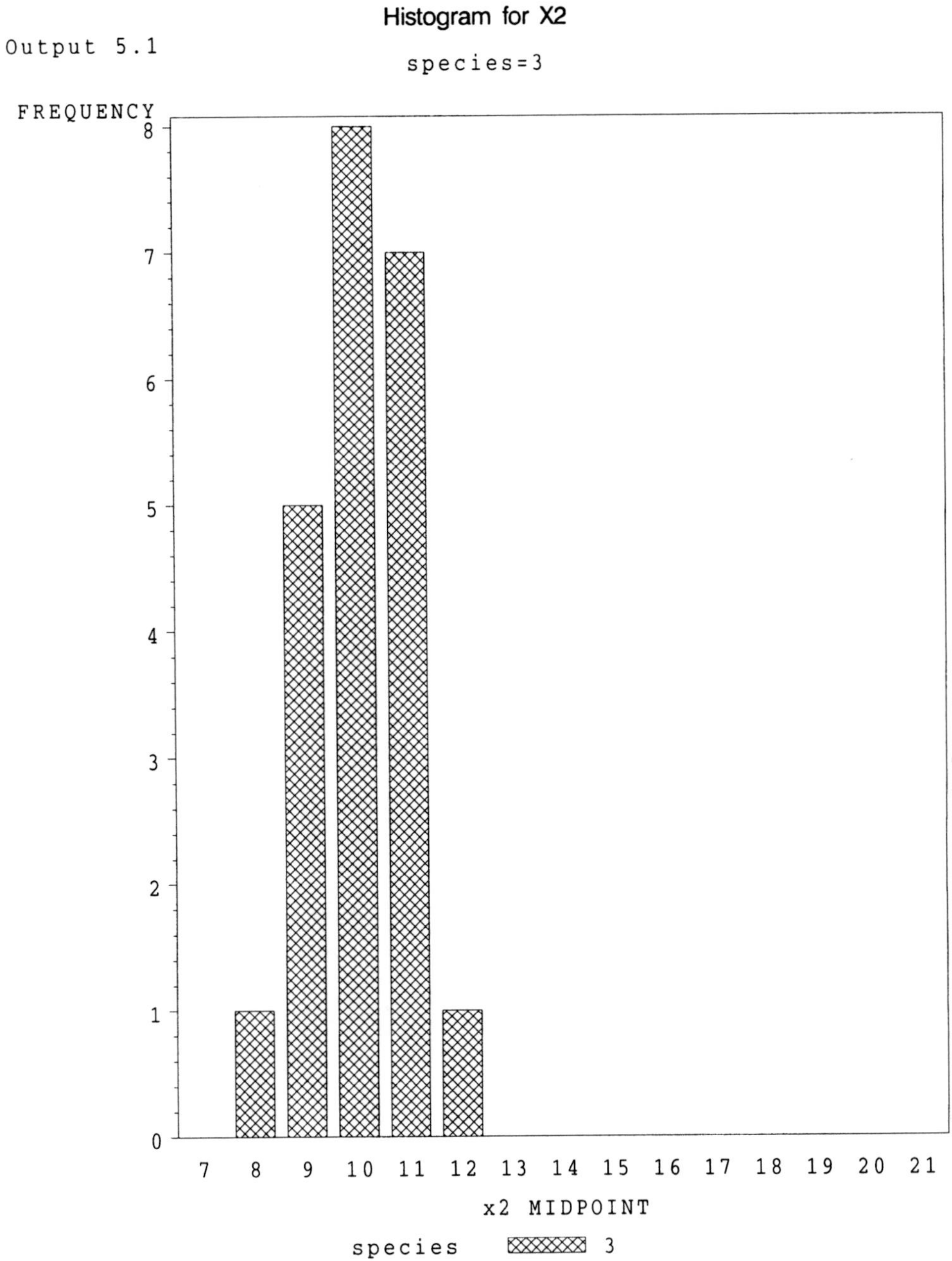
Histogram for X2
species=3
FREQUENCY
8
7
6
5
4
3
2
1
0
7 8 9 10 11 12 13 14 15 16 17 18 19 20 21
x2 MIDPOINT
species 3

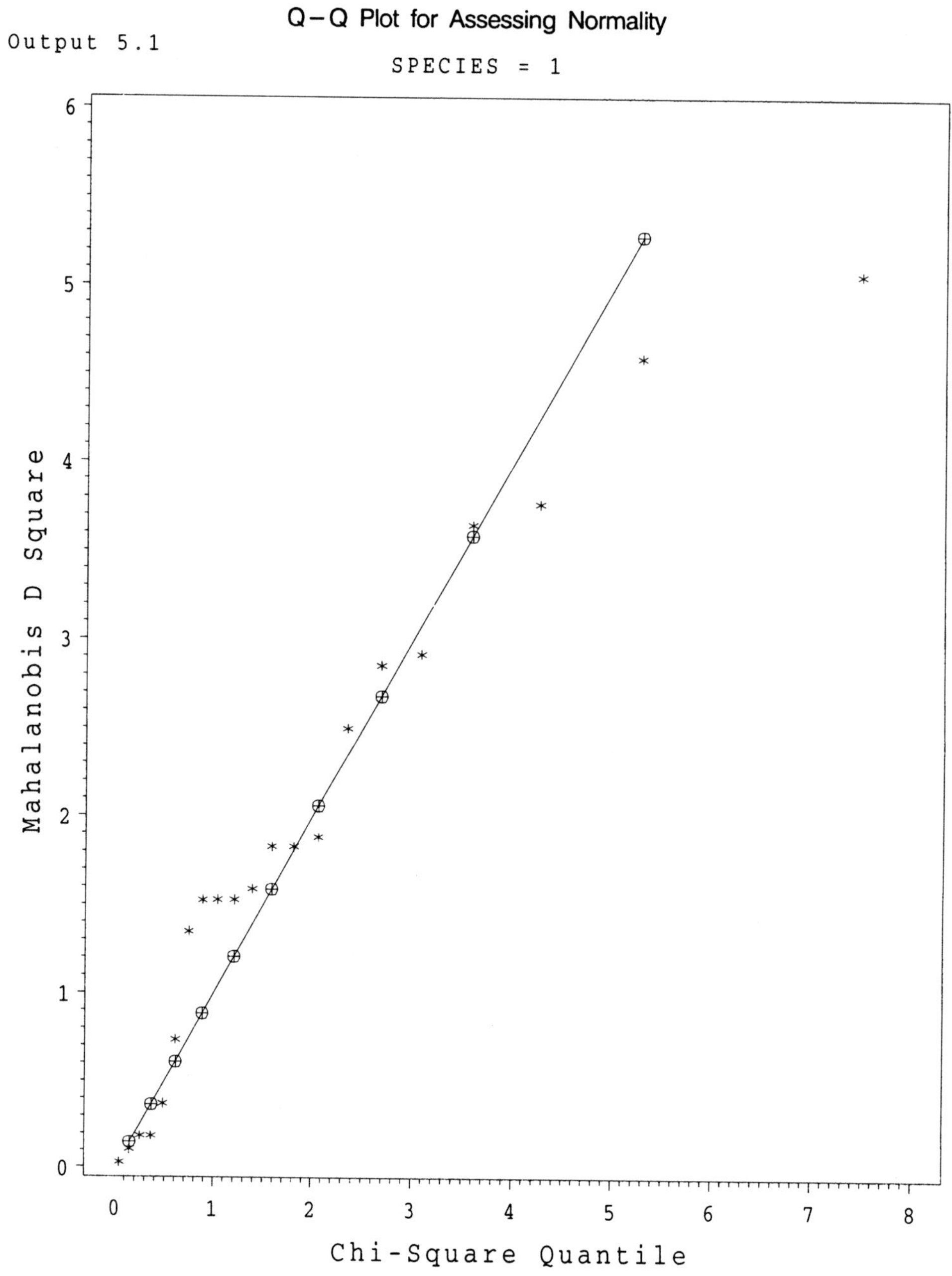
Output 5.1
Q-Q Plot for Assessing Normality
SPECIES = 1
Mahalanobis D Square
Chi-Square Quantile
0
1
2
3
4
5
6
7
8

Output 5.1

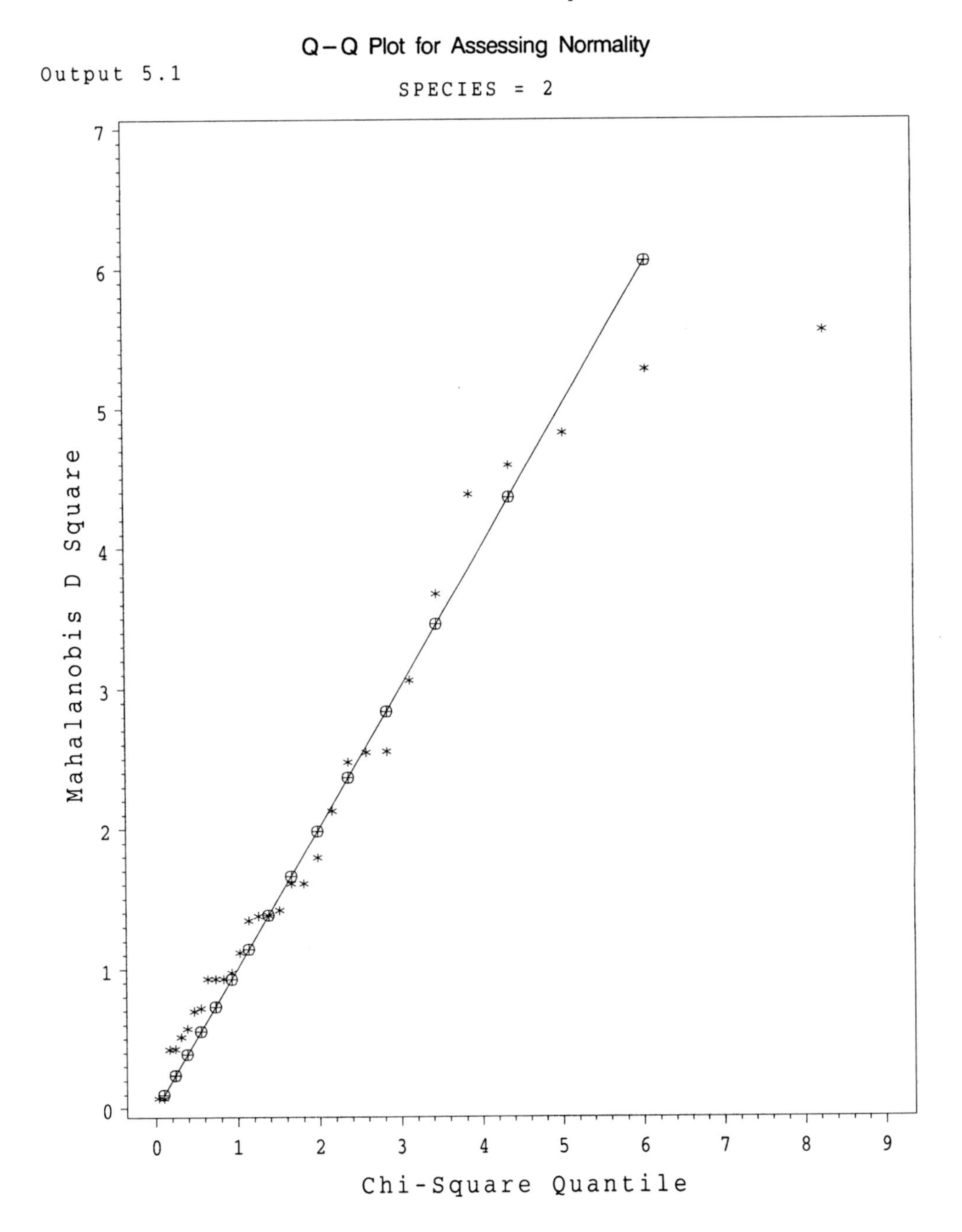
Q-Q Plot for Assessing Normality
SPECIES = 2
Mahalanobis D Square
Chi-Square Quantile
0
1
2
3
4
5
6
7
8
9

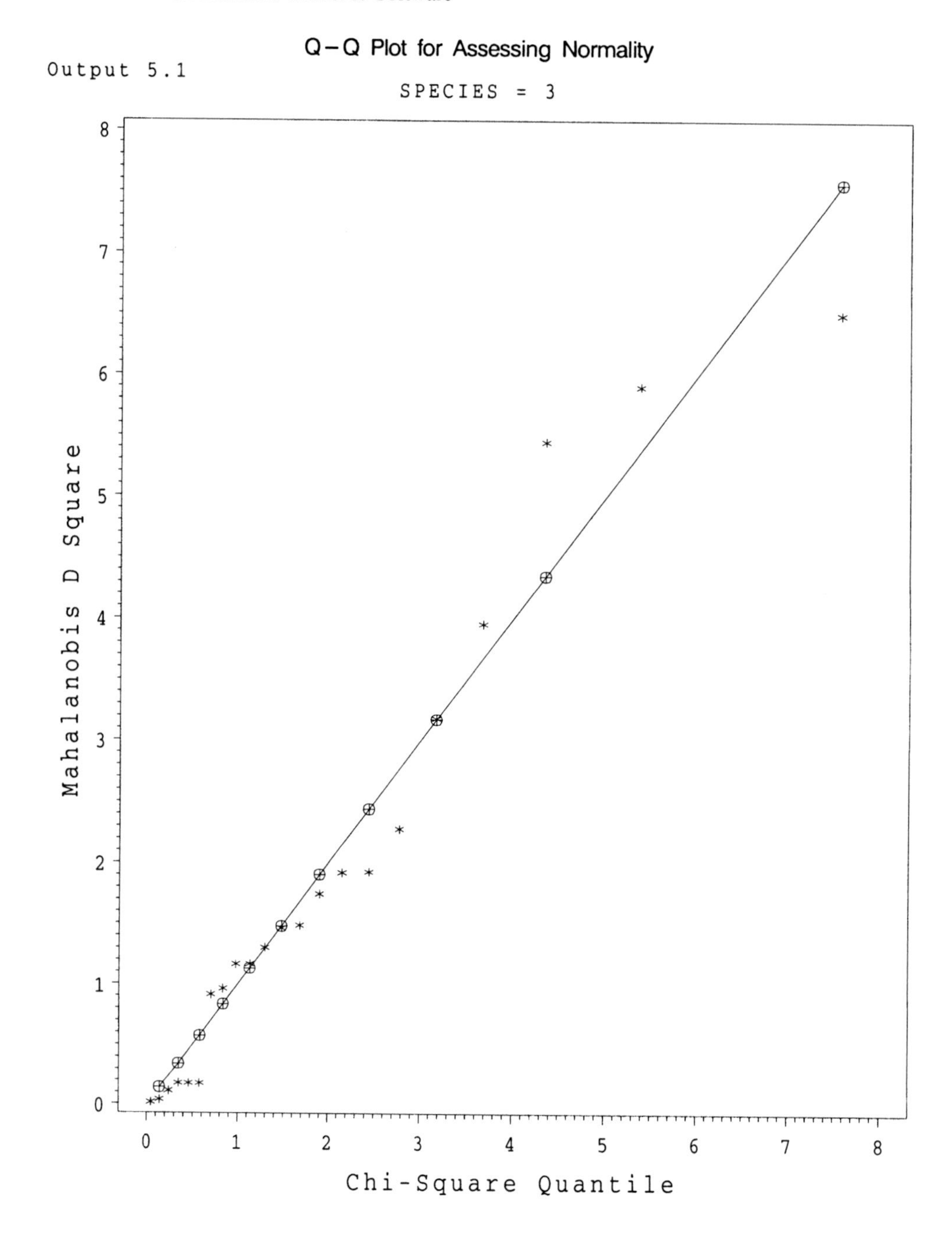
Output 5.1
Q-Q Plot for Assessing Normality
SPECIES = 3
Mahalanobis D Square
Chi-Square Quantile
0
1
2
3
4
5
6
7
8

Output 5.2
(continued)

```
                    Output 5.1
     Mardia's Tests for Skewness and Kurtosis

                        TIMES

              Species =         1

                    S

              30.154195 -0.922902
              -0.922902 0.7528345

                  S_INV

              0.0344557 0.0422393
              0.0422393 1.3800946

          BETA1HAT     KAPPA1   PVALSKEW

          0.3179654 1.1128789 0.8922235

          BETA2HAT     KAPPA2   PVALKURT

          6.2687387 -0.991704 0.3213417

                        TIMES

              Species =         2

                    S

              21.422117  -0.31629
               -0.31629 1.1914388


     Mardia's Tests for Skewness and Kurtosis

                  S_INV

              0.0468644 0.0124411
              0.0124411 0.8426241

          BETA1HAT     KAPPA1   PVALSKEW

          0.0679703 0.3285231 0.9878993
```

```
BETA2HAT     KAPPA2  PVALKURT

6.5970771 -0.944371 0.3449799


                    TIMES

Species =              3


            S

26.123264 -3.401042
-3.401042  1.734375


          S_INV

0.0514034 0.1008001
0.1008001 0.7742417


BETA1HAT     KAPPA1  PVALSKEW

1.7249788 6.8999153 0.1412727



Mardia's Tests for Skewness and Kurtosis


BETA2HAT     KAPPA2  PVALKURT

8.5898672 0.3612184 0.7179362
```

EXAMPLE 2 ***Crystals in Urine Data*** This data set collected by J. S. Elliot and available in Andrews and Herzberg (1985) as data set 44, provides measurements on the physical characteristics of the urine of the two groups of patients defined by the presence and absence of the crystals in the urine. We will examine this data set for normality. The univariate frequency charts obtained by using Program 5.2 for three characteristics, namely, amount of calcium, pH value, and amount of urea for each of the two groups provide overwhelming evidence of the nonnormality in each of the univariate populations (Output 5.2). A further investigation using $Q - Q$ plots or skewness and kurtosis tests is really not required at this stage.

What can we do then to perform discriminant analysis on this data set? Two options are available. We may either attempt to make transformations to attain the approximate normality or, better yet, opt for the nonparametric approach to discriminant analysis. As the third option, we may also attempt the logistic regression methods.

```
/* Program 5.2 */

filename gsasfile "prog52.graph";
goptions reset=all gaccess=gsasfile gsfmode = append
autofeed dev = pslmono;
goptions horigin =1in vorigin = 2in;
goptions hsize=6in vsize = 8in;
```

```
options ls = 64 ps=45 nodate nonumber;
title1 ' ';
title2 j = 1  'Output 5.2';

data urine;
infile 'urine.dat' firstobs=7;
input crystals $ sg ph mosm mmho urea calcium ;
patient = _n_ ;
sg100 = 100*sg;
ph10=100*ph;
mmho10 =10*mmho ;
cal100=100*calcium ;

proc format;
value specchar
1 = 'yes'
0 = 'no'
run;

proc gchart data = urine ;by crystals ;
vbar calcium /subgroup = crystals midpoints = 0 to 15 ;
title1 h= 1.2 ' Urine Data: Univariate Frequency Chart for Calcium' ;
title2 j = 1 'Output 5.2';
run;

proc gchart data = urine ;by crystals ;
vbar ph /subgroup = crystals midpoints = 4 to 8 by .5;
title1 h= 1.2 ' Urine Data: Univariate Frequency Chart for pH' ;
title2 j = 1 'Output 5.2';
run;

proc gchart data = urine ;by crystals ;
vbar urea /subgroup = crystals midpoints = 100 to 500 by 25;
title1 h = 1.2 ' Urine Data: Univariate Frequency Chart for Urea' ;
title2 j = 1 'Output 5.2';
run;
```

Output 5.3

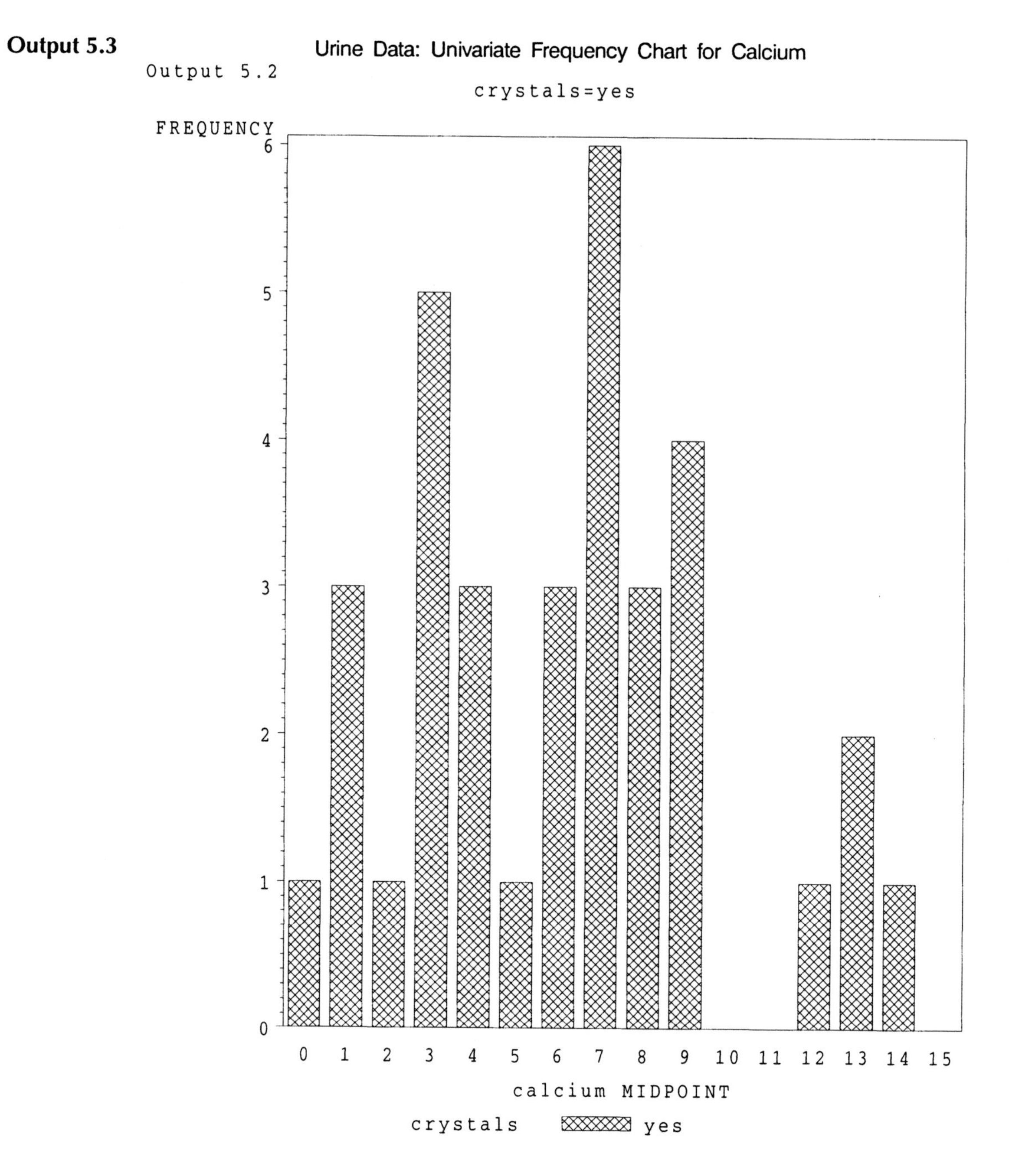
Urine Data: Univariate Frequency Chart for Calcium
Output 5.2
crystals=yes
FREQUENCY
6
5
4
3
2
1
0
0 1 2 3 4 5 6 7 8 9 10 11 12 13 14 15
calcium MIDPOINT
crystals
yes

Output 5.2

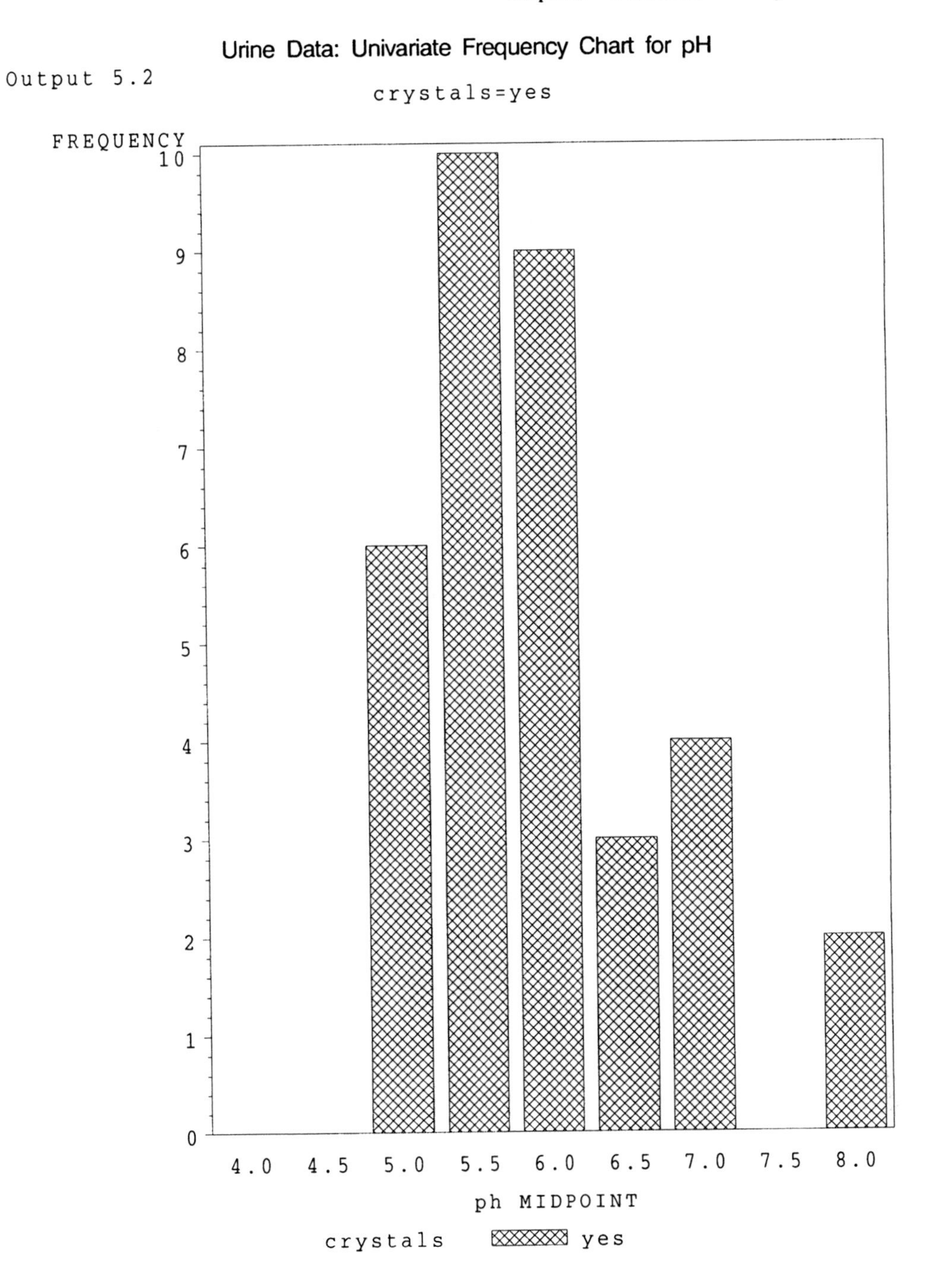
Urine Data: Univariate Frequency Chart for pH
crystals=yes
FREQUENCY
10
9
8
7
6
5
4
3
2
1
0
4.0
4.5
5.0
5.5
6.0
6.5
7.0
7.5
8.0
ph MIDPOINT
crystals
yes

Output 5.2

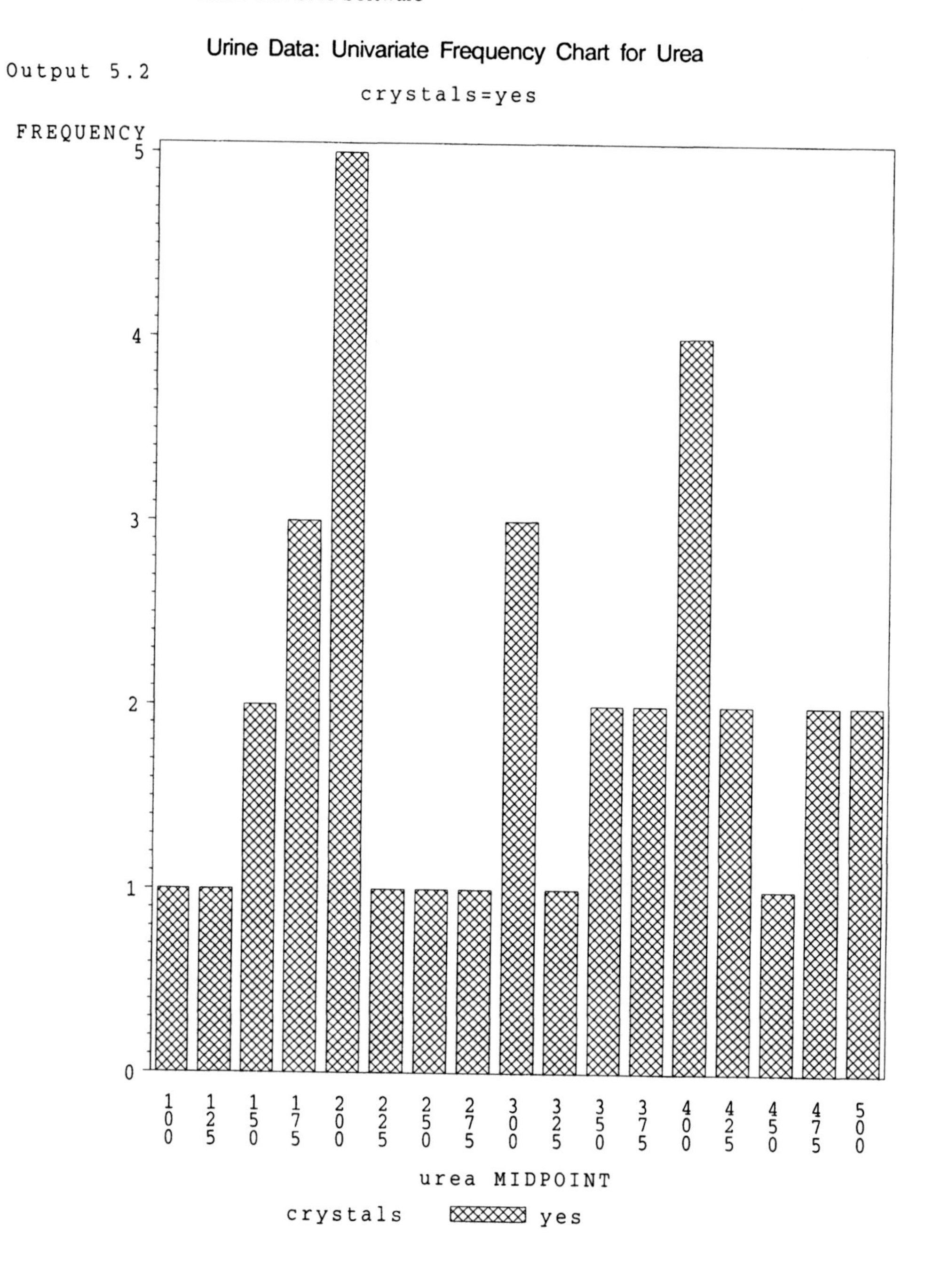

Urine Data: Univariate Frequency Chart for Urea
crystals=yes
FREQUENCY
5
4
3
2
1
0
100 125 150 175 200 225 250 275 300 325 350 375 400 425 450 475 500
urea MIDPOINT
crystals
yes

Output 5.2

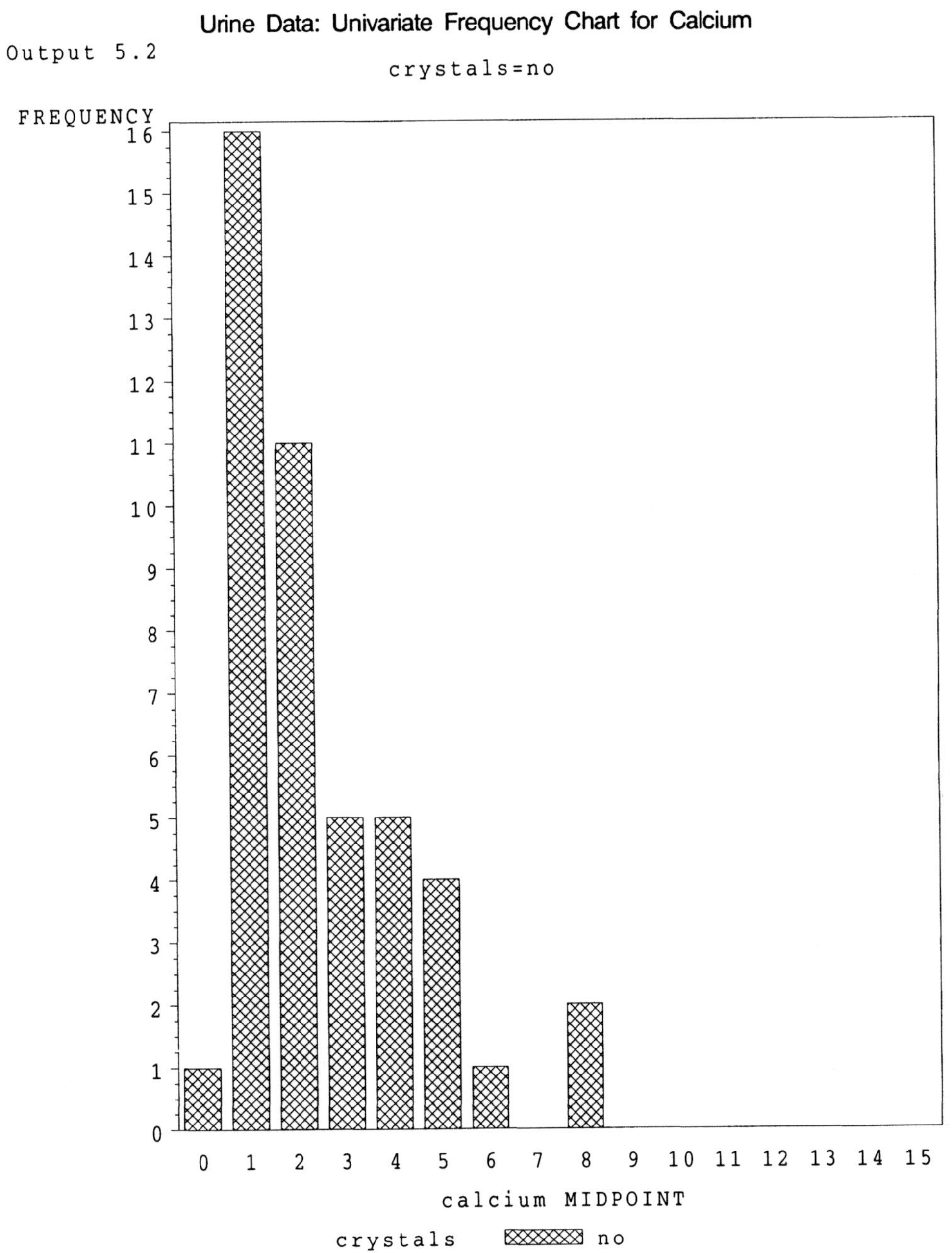
Urine Data: Univariate Frequency Chart for Calcium
crystals=no
FREQUENCY
16
15
14
13
12
11
10
9
8
7
6
5
4
3
2
1
0
0 1 2 3 4 5 6 7 8 9 10 11 12 13 14 15
calcium MIDPOINT
crystals no

Output 5.2

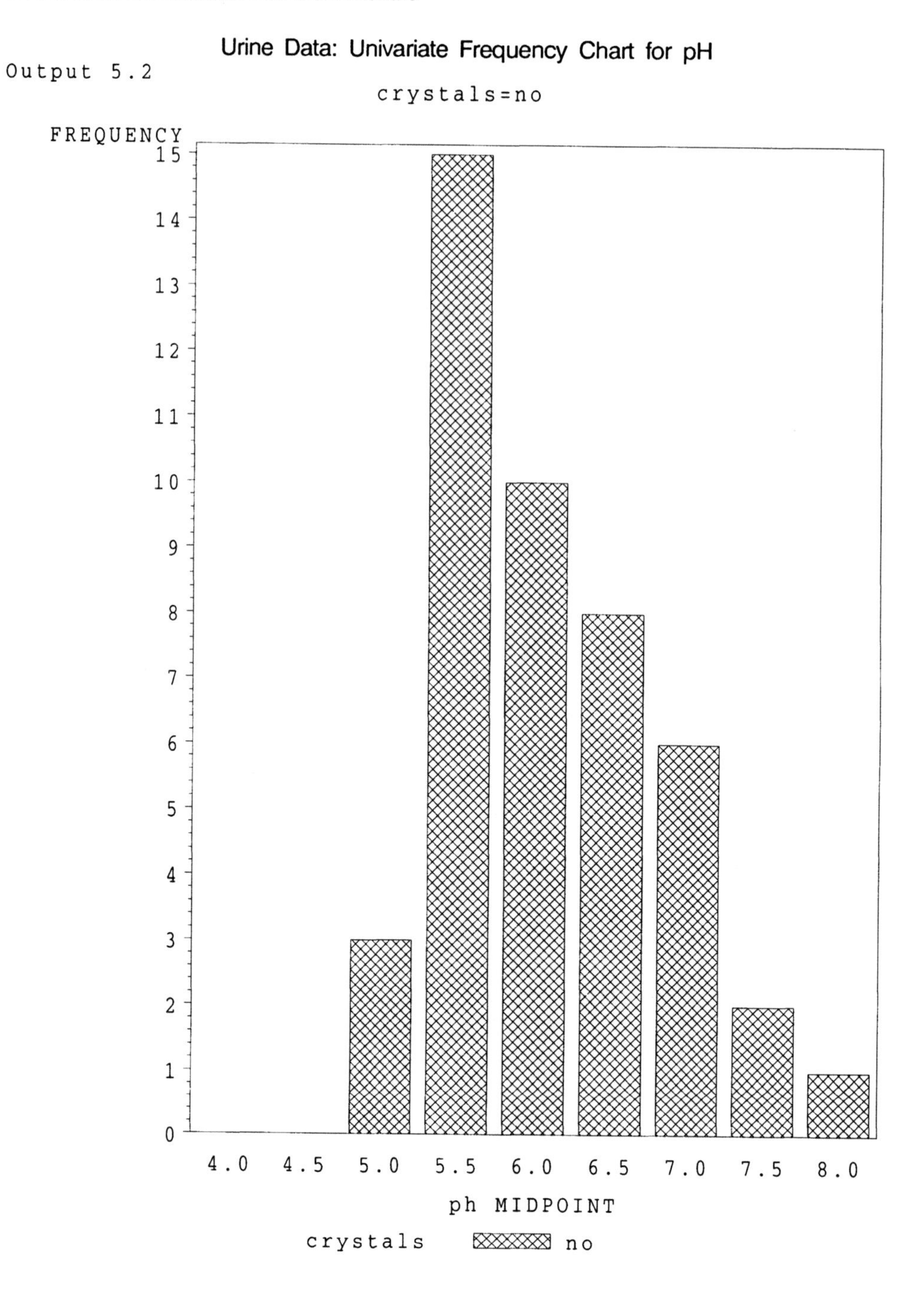

Urine Data: Univariate Frequency Chart for pH
crystals=no
FREQUENCY
15
14
13
12
11
10
9
8
7
6
5
4
3
2
1
0
4.0
4.5
5.0
5.5
6.0
6.5
7.0
7.5
8.0
ph MIDPOINT
crystals
no

It may be remarked that if some of the populations are multivariate normal while others are not, the discriminant analysis based on the multivariate normality assumption cannot be used. The recommendation in such cases is either to use logistic discrimination or one of the several nonparametric discrimination alternatives. These have been discussed in details in the later sections.

5.3 Statistical Tests for Relevance

When the k populations are assumed to have a multivariate normal distribution with t^{th} population being $N_p(\boldsymbol{\mu}_t, \boldsymbol{\Sigma}_t)$, $t = 1, \cdots, k$, we may first want to examine if the populations are indeed different enough for the discriminant analysis to be meaningful. The populations may differ from each other either due to different mean vectors or different variance-covariance matrices or both. Thus, the following hypotheses may be tested before performing any discriminant analysis:

$$
\begin{aligned}
H_0^{(1)} &: \ \boldsymbol{\mu}_1 = \cdots = \boldsymbol{\mu}_k \ \text{ and } \ \boldsymbol{\Sigma}_1 = \cdots = \boldsymbol{\Sigma}_k, \\
H_0^{(2)} &: \ \boldsymbol{\Sigma}_1 = \cdots = \boldsymbol{\Sigma}_k, \\
H_0^{(3)} &: \ \boldsymbol{\mu}_1 = \cdots = \boldsymbol{\mu}_k \ \text{ given } \ \boldsymbol{\Sigma}_1 = \cdots = \boldsymbol{\Sigma}_k = \boldsymbol{\Sigma} \ \text{(say), and} \\
H_0^{(4)} &: \ \boldsymbol{\mu}_1 = \cdots = \boldsymbol{\mu}_k \ \text{ given not all } \ \boldsymbol{\Sigma}_i \ \text{ are equal.}
\end{aligned}
$$

If $H_0^{(1)}$ is true, we may not need to do any discriminant analysis as the k populations themselves are indistinguishable. If $H_0^{(2)}$ is true, then to see if the populations are indeed different, $H_0^{(3)}$ may be tested as the next step. Acceptance of $H_0^{(3)}$ would then exclude any hopes of discriminant analysis being meaningful. However, if $H_0^{(2)}$ is false, then $H_0^{(4)}$ may be tested. In case of the acceptance of $H_0^{(4)}$ although the k populations are still different, the discrimination in most cases may not be meaningful because the populations differ from each other only in variances and covariances but are the same with respect to population averages. Examples, however, can be found to the contrary. Okamoto (1961) has developed some theory for this situation and applied it to an interesting example involving the zygosity of the pairs of twins. Bartlett and Please (1963) and Kshirsagar (1972) also provide some theories for this situation. We will not discuss this aspect of discriminant analysis here. If $H_0^{(3)}$ or $H_0^{(4)}$ is rejected, the discriminant analysis may be meaningful. However, the respective discriminant functions appropriate in the two situations are different. Specifically, in the case of the variance-covariance matrices being unequal, the quadratic discriminant function may be needed, while, if these matrices are assumed to be equal, the corresponding discriminant function turns out to be linear. In other words (at least in the case of multivariate normal populations), whether one should apply the discriminant analysis or not and if applied, what is the choice of the discriminant function, depends on the assumptions implied by $H_0^{(1)}$ through $H_0^{(4)}$. Hence, testing these hypotheses should be the first step in discriminant analysis. These are discussed below.

Let $\mathbf{x}_{t1} \cdots, \mathbf{x}_{tn_t}$ be the random sample from the tth population, $N_p(\boldsymbol{\mu}_t, \boldsymbol{\Sigma}_t)$, $t = 1, \cdots, k$. Let $\bar{\mathbf{x}}_t = n_t^{-1} \sum_{j=1}^{n_t} \mathbf{x}_{tj}$

$$\mathbf{S}_t = (n_{t-1})^{-1} \sum_{j=1}^{n_t} (\mathbf{x}_{tj} - \bar{\mathbf{x}}_t)(\mathbf{x}_{tj} - \bar{\mathbf{x}}_t)'$$

$$\bar{\mathbf{x}} = \frac{n_1\bar{\mathbf{x}}_1 + \cdots + n_k\bar{\mathbf{x}}_k}{\sum_{t=1}^{k} n_t}$$

$$\mathbf{S} = \frac{(n_1 - 1)\mathbf{S}_1 + \cdots + (n_k - 1)\mathbf{S}_k}{\sum_{t=1}^{k} n_t - k}$$

$$\mathbf{B} = \sum_{t=1}^{k} n_t(\bar{\mathbf{x}}_t - \bar{\mathbf{x}})(\bar{\mathbf{x}}_t - \bar{\mathbf{x}})',$$

then the likelihood ratio test to test $H_0^{(1)}$ is approximately equivalent to

$$\text{Reject } H_0 \ \text{ if } \ \eta_1 = -2ln\,\lambda_1 > \chi^2_\alpha(\nu_1),$$

where $\chi^2_\alpha(\nu_1)$ is the cutoff point from a chi-square distribution with ν_1 df at the $100\alpha\%$ level of significance, and where

$$\nu_1 = p(k-1) + \frac{p(p+1)(k-1)}{2}$$

and

$$\lambda_1 = \frac{\prod_{t=1}^{k} \left| \frac{(n_t-1)}{n_t} \mathbf{S}_t \right|^{n_t/2}}{\left| \frac{1}{\sum n_t} \{(\sum n_t - k)\mathbf{S} + \mathbf{B}\} \right|^{\frac{\sum n_t}{2}}}. \tag{5.1}$$

To test $H_0^{(2)}$, the likelihood ratio test suggests that we

$$\text{reject } H_0^{(2)} \text{ if } \eta = -2ln\lambda_2 > \chi^2_\alpha(\nu_2)$$

at $100\alpha\%$ level of significance, where

$$\nu_2 = \frac{p(p+1)(k-1)}{2}$$

and

$$\lambda_2 = \frac{\prod_{t=1}^{k} \left| \frac{(n_t-1)}{n_t} \mathbf{S}_t \right|^{n_t/2}}{\left| \frac{1}{\sum n_t} \{(\sum n_t - k)\mathbf{S}\} \right|^{\sum n_t/2}}. \tag{5.2}$$

For $H_0^{(3)}$, the likelihood ratio test leads to

$$\text{reject } H_0^{(3)} \text{ if } \eta = -2\,ln\,\lambda_3 > \chi^2_\alpha(\nu_3)$$

at $100\alpha\%$ level of significance, where $\nu_3 = \nu_1 - \nu_2$ and

$$\lambda_3 = \lambda_1/\lambda_2. \tag{5.3}$$

Muirhead (1982) suggests Bartlett's correction for all three hypotheses.

The hypotheses $H_0^{(2)}$ can be tested as part of the computations in the DISCRIM procedure by specifying POOL=TEST. Bartlett's correction is applied in this case to make the test unbiased and improve the approximation. Upon acceptance, $H_0^{(3)}$ can be tested using the MANOVA statement in the GLM procedure or in fact, by just using the MANOVA option in the DISCRIM procedure. Since the test statistics λ_2 and λ_3 can be calculated rather directly (from SAS output and then from the back calculations), λ_1 can be computed easily by using $\lambda_1 = \lambda_2\lambda_3$ and corresponding df ν_1 can be obtained as $\nu_1 = \nu_2 + \nu_3$.

EXAMPLE 3 ***Preliminary Statistical Tests, Flea Beetles Data*** We have already examined the bivariate normality assumption (and accepted it) for the flea beetles data. It may then be worth pursuing further analysis and asking if for the three species, the variance-covariance matrices are the same. That is, we want to test $H_0^{(2)}$. Here $k = 3$ and $p = 2$. Using Equation 5.2, the value of the likelihood ratio test statistic can be computed, and then we can use the chi-square approximation to obtain the corresponding p value. Fortunately, a modified version of this calculation (by making Bartlett's correction to make the test unbiased) can be performed by specifying the POOL=TEST option in PROC DISCRIM. The corresponding SAS statements are

```
proc discrim data = beetles method = normal pool = test
wcov bcov pcov;
class species;
var x1 x2;
```

Using the METHOD=NORMAL option, the assumption of multivariate normality is invoked, and the test for the equality of the covariance matrices for the three species is requested using the POOL=TEST option. If such a test is not needed (e.g., if the past data suggest that the covariance matrices can or cannot be assumed to be equal), we can indicate accordingly whether to pool (POOL=YES) or not to pool (POOL=NO). Sample variance-covariance matrices for each population can be obtained by specifying the WCOV option. Similarly, the PCOV option provides the pooled sample variance-covariance matrix that would estimate the true variance-covariance matrix if a common variance-covariance matrix for all populations is assumed. The BCOV option provides the *between sample variance-covariance matrix*, which is nothing but the *between sum of squares and cross-product (SS&CP)* matrix divided by the corresponding degrees of freedom, namely, $(k-1)$. There are similar options to obtain the within, pooled, and between-SS&CP or correlation matrices.

In order to specify a categorical variable whose values would refer to the different populations, a CLASS statement is needed. In the present context, this variable is SPECIES, which takes values 1, 2, and 3 corresponding to *Chaetocnema concinna, Chaetocnema heikertingeri* and *Chaetocnema heptapotamica,* respectively. We could well have used a more descriptive but also more cumbersome set of nonnumeric values instead of 1, 2, and 3. Finally, the VAR statement is used to list the variables whose variance-covariance matrices are to be compared (and that will later be used to construct the appropriate discriminant function). The SAS statements given above are included in Program 5.3. The resulting output is included as Output 5.3.

```
/* Program 5.3 */

options ls = 64 ps=45 nodate nonumber;
title1  'Output 5.3';

data beetles;
infile 'beetle.dat' firstobs=7;
input species  x1 x2 ;

proc discrim data = beetles method = normal pool =test
                                  wcov pcov bcov manova ;
class species ;
var x1 x2 ;
title2 'Testing the Equality of Normal Population Parameters';
run;
```

Output 5.3

```
                      Output 5.3
     Testing the Equality of Normal Population Parameters

                  The DISCRIM Procedure
             Within-Class Covariance Matrices

                species = 1,      DF = 20

        Variable                  x1                x2

        x1               31.66190476       -0.96904762
        x2               -0.96904762        0.79047619
```

```
---------------------------------------------------------------

                species = 2,      DF = 30

          Variable              x1              x2

          x1             21.36989247     -0.32688172
          x2             -0.32688172      1.21290323

---------------------------------------------------------------

                species = 3,      DF = 21

          Variable              x1              x2

          x1             17.16017316     -0.50216450
          x2             -0.50216450      0.94372294

---------------------------------------------------------------

       Pooled Within-Class Covariance Matrix,      DF = 71

          Variable              x1              x2

          x1             23.02392262     -0.55961773
          x2             -0.55961773      1.01429299

          Between-Class Covariance Matrix,     DF = 2

          Variable              x1              x2

          x1             125.4051252      -7.4281689
          x2              -7.4281689       5.3305071


       Test of Homogeneity of Within Covariance Matrices

Notation: K     = Number of Groups

          P     = Number of Variables

          N     = Total Number of Observations - Number of Groups

          N(i)  = Number of Observations in the i'th Group - 1

                    __                    N(i)/2
                    ||   |Within SS Matrix(i)|
          V     = -----------------------------------
                                       N/2
                       |Pooled SS Matrix|

                       _                  _     2
                       |      1       1   |   2P + 3P - 1
          RHO   = 1.0 - | SUM -----  -  --- | -------------
                       |_     N(i)     N  _|  6(P+1)(K-1)
```

```
                DF   = .5(K-1)P(P+1)

                                          _                     _
                                          |     PN/2            |
                                          |   N          V      |
Under the null hypothesis:     -2 RHO ln  | ------------------- |
                                          |   __         PN(i)/2|
                                          |_  ||  N(i)         _|

is distributed approximately as Chi-Square(DF).


              Chi-Square        DF    Pr > ChiSq

                3.288462         6        0.7719

Since the Chi-Square value is not significant at the 0.1 level,
a pooled covariance matrix will be used in the discriminant func
tion.
Reference: Morrison, D.F. (1976) Multivariate Statistical Method
s p252.

         Multivariate Statistics and F Approximations

                    S=2     M=-0.5     N=34

Statistic                        Value   F Value   Num DF   Den DF

Wilks' Lambda               0.04730797    125.92        4      140
Pillai's Trace              1.56067142    126.11        4      142
Hotelling-Lawley Trace      7.28656510    126.93        4   82.971
Roy's Greatest Root         4.29288815    152.40        2       71

         Multivariate Statistics and F Approximations

                    S=2     M=-0.5     N=34

                Statistic                Pr > F

                Wilks' Lambda            <.0001
                Pillai's Trace           <.0001
                Hotelling-Lawley Trace   <.0001
                Roy's Greatest Root      <.0001

  NOTE: F Statistic for Roy's Greatest Root is an upper bound.
         NOTE: F Statistic for Wilks' Lambda is exact.
```

From Output 5.3, it is easily seen that

$$S_1 = \begin{bmatrix} 31.6619 & -0.9690 \\ -0.9690 & 0.7904 \end{bmatrix}, S_2 = \begin{bmatrix} 21.3699 & -0.3269 \\ -0.3269 & 1.3129 \end{bmatrix}$$

and

$$S_3 = \begin{bmatrix} 17.1602 & -0.5022 \\ -0.5022 & 0.9437 \end{bmatrix}.$$

Also, upon pooling, we obtain

$$\mathbf{S} = \begin{bmatrix} 23.0239 & -0.5596 \\ -0.5596 & 1.0143 \end{bmatrix}.$$

The corresponding likelihood ratio test statistic $\eta_2 = -2\rho_2(ln\lambda_2)$ approximately follows a chi-square distribution with $6\,df$. The coefficient ρ_2 has been defined in Output 5.3 and represents Bartlett's correction factor. With an observed value of 3.2885 for this data set, the corresponding p value is 0.7719. Thus, the test statistic η_2 is not found to be statistically significant. Hence, we may assume that the variance-covariance matrices for the three species are equal. As we will see later, the assertion of multivariate normality with equal variance-covariance matrices leads to a much simpler and relatively more interpretable discrimination scheme.

Having tested that the variance-covariance matrices for the three populations are equal, it may be worth examining to see if the three populations (of species) are indeed different. In other words, we must now test the null hypothesis,

$$H_0 : \boldsymbol{\mu}_1 = \boldsymbol{\mu}_2 = \boldsymbol{\mu}_3,$$

where $\boldsymbol{\mu}_1$, $\boldsymbol{\mu}_2$, and $\boldsymbol{\mu}_3$ are 2 by 1 vectors of the true means of $X1$ and $X2$ for the three species. In view of an equal variance-covariance assumption, the above null hypothesis is nothing but a testing problem for equality of means in the multivariate one-way classification set up. These types of hypotheses were considered in detail in Khattree and Naik (1999). Specifically, the following SAS commands could achieve our purpose:

```
proc glm data = beetles;
class species;
model x1 x2 = species/nouni;
manova h = species;
```

These statements have not been included in Program 5.3 because the MANOVA option in the DISCRIM procedure can fortunately provide the same analysis without running an additional procedure, namely GLM. The value of the exact F statistic, corresponding to Wilks' Λ for the hypothesis stated above, is 112.9166 with $df(4,140)$. In view of the very small p value (< 0.0001) for this, as well as for all the other multivariate tests, the hypothesis is rejected.

The rejection of $H_0 : \boldsymbol{\mu}_1 = \boldsymbol{\mu}_2 = \boldsymbol{\mu}_3$ indicates that at least two of the three species are probably different in their population means, and hence a discriminant analysis may be worth pursuing. Had the above hypothesis been accepted, the three species would be deemed identical in their probability distributions (hypotheses of multivariate normality, equal variance-covariance, and then finally equal mean are all accepted) and hence no meaningful discrimination between the three species would have been possible.

The procedure to test $H_0^{(4)}$ is more complicated and less satisfactory. $H_0^{(4)}$ is, in fact, the multivariate and k population generalization of what is popularly known as the two-sample Behrens-Fisher problem. Anderson (1963) gives a rather indirect way of testing $H_0^{(4)}$ by making a suitable linear transformation. We describe the procedure below.

Without loss of generality, let us assume that the smallest sample size corresponds to the kth population Π_k. If we transform the data $\mathbf{x}_{t1}, \mathbf{x}_{t2}, \cdots, \mathbf{x}_{tn_k}$ (the remaining data from each population are thus ignored) to $\mathbf{y}_{t1} \cdots \mathbf{y}_{tn_k}$ as

$$\mathbf{y}_{tj} = a_{tk}\mathbf{x}_{kj} + \sum_{g=1}^{k-1} a_{tg}\left(\frac{n_k}{n_g}\right)^{\frac{1}{2}}\left\{\mathbf{x}_{gj} - \bar{\mathbf{x}}_{g(*)} + \left(\frac{n_g}{n_k}\right)^{\frac{1}{2}}\bar{\mathbf{x}}_g\right\},$$

$$t = 1, \cdots, k-1,\ j = 1, \cdots, n_k,$$

where $\bar{\mathbf{x}}_{g(*)}$ is the mean of the data $\mathbf{x}_{g1} \cdots \mathbf{x}_{gn_k}$, which were retained for the purpose of transformation in the above equation, then $\bar{\mathbf{x}}_g$ is the sample mean of all observations from the gth population $g = 1, \cdots, k-1$.

The coefficients $a_{tg}, t = 1, \cdots, k-1, g = 1, \cdots, k$ are such that, for every fixed t, $a_{t1} + \cdots + a_{tk} = 0$. Any choice of a_{tg} satisfying this requirement can be taken. For convenience, we take for a fixed $t = 1, \cdots, k-1, a_{t1} = \cdots = a_{t,t-1} = a_{t,t+1} = \cdots = a_{t,k-1} = 0$ and $a_{tt} = -1, a_{tk} = +1$. This choice is implemented in Program 5.4. The transformation of data from the kth population is not needed since it would transform all of them to a value of zero. However, if $H_0^{(4)}$ is true, then all $\mathbf{y}_{tj}, j = 1, \cdots, n_k, t = 1, \cdots, k-1$ would have zero population mean, and $\mathbf{v}_j = (\mathbf{y}'_{1j} : \mathbf{y}'_{2j} : \cdots : \mathbf{y}'_{k-1,j})', j = 1, \cdots, n_k$ will all be independently distributed as $(k-1)p$-variate multivariate normal with common variance-covariance matrix.

If the estimate of the variance-covariance matrix of $\mathbf{v}_j$ is $\mathbf{S}_v = (n_k - 1)^{-1} \sum_{j=1}^{n_k} (\mathbf{v}_j - \bar{\mathbf{v}})(\mathbf{v}_j - \bar{\mathbf{v}})'$ where $\bar{\mathbf{v}} = n_k^{-1} \sum_{j=1}^{n_k} \mathbf{v}_j$, then the quantity

$$T^2 = n_k \bar{\mathbf{v}}' \mathbf{S}_v^{-1} \bar{\mathbf{v}} \tag{5.4}$$

follows the Hotelling's T^2, and the quantity $\frac{[n_k - p(k-1)]}{[(n_k-1)p(k-1)]} T^2$ has the F distribution with $p(k-1)$ and $n_k - p(k-1)$ degrees of freedom. It is, of course, assumed that $n_k - p(k-1)$ is nonnegative.

EXAMPLE 4 ***Preliminary Statistical Tests, Admission Data*** Johnson and Wichern (1998) present a data set on the undergraduate GPA and GMAT scores of students who applied for admission to a business school graduate program. Three groups of students are available: those who were admitted, those who were not admitted, and those on the borderline. The problem of interest is to examine if the three groups are indeed different with respect to GPA and GMAT scores. In other words, we want to test the equality of the means for these three strata of applicants, which we assume to be normally distributed. Of course, a natural question to ask before such a test is performed is if the variance-covariance matrices for the three groups can be assumed to be equal.

An application of PROC DISCRIM similar to that in Example 3 with the POOL=TEST option indicates that the three variance-covariance matrices may not be assumed to be equal (the p value for the corresponding likelihood ratio test is 0.0134). In view of this, the validity of multivariate analysis of variance tests used in testing the equality of mean vectors is questionable, and it may be worth to resort to the test suggested by Anderson (1963).

Note that sample sizes for the groups (the corresponding class variable is termed as STATUS) of applicants are $n_1 = 31$ (admitted), $n_2 = 28$ (not admitted), and $n_3 = 26$ (borderline). Since $\min(n_1, n_2, n_3) = n_3 = 26$, we will need to define, using Equation 5.4, 26 $p(k-1) = 2(3-1) = 4$ dimensional vectors, which under the null hypothesis (of equality of group means) would form a sample from a multivariate population with a zero mean vector.

In Program 5.4, we have used the first 26 observations from the groups YES (i.e., those admitted) and NO (not admitted) to compute the respective partial averages $\bar{\mathbf{x}}_{t(*)}$. Undoubtedly, there is some arbitrariness here in that any set of 26 observations from each group could have been taken, and hence the results are dependent on this somewhat arbitrary selection. The sample means of three *entire samples* are computed as usual using PROC MEANS along with the BY STATUS option on the original data set (DATA = ADMSION). The partial means for the observations in the three groups (called (YGPA YGMAT), (NGPA NGMAT), and (BGPA, BGMAT)) are computed by first creating a data set ALL as given in Program 5.4, and then by using

```
proc means data = all;
var ygpa ygmat ngpa ngmat bgpa bgmat;
```

The values of full sample means (Y_TAVE1, Y_TAVE2, etc.) and of partial sample means (Y_PAVE1, Y_PAVE2, etc.) are used to define $\min(n_1, n_2, n_3) = 26$ observations of $p(k-1) = 2(3-1) = 4$ dimensional vector (YES1, YES2, NO1, NO2)$'$ using Equation 5.4.

Testing the null hypothesis of equality of three group means is equivalent to testing the hypothesis that the population mean of the above four-dimension vector is zero. To test this null hypothesis, we use the following SAS code:

```
proc glm data = all;
class status;
model yes1 yes2 no1 no2 = /nouni;
manova h = intercept;
```

Thus, the multivariate null hypothesis is on the intercept (i.e., the four-dimension mean vector) and tests if this mean vector is zero. All four multivariate tests presented in Output 5.4 show that this hypothesis can be summarily rejected (F(4,22) statistic=793.77, p value < 0.0001), and thus we conclude that the three groups of applicants do not have equal mean GPA and GMAT scores. As a result, if we want to construct a discriminant function to classify the future applicants into one of the three groups, it can be appropriately done. However, such a construction should take into account that the variance-covariance matrices for the three groups are possibly unequal.

```
/* Program 5.4 */

options ls = 64 ps=45 nodate nonumber;
title1  'Output 5.4';

/* Following program tests the equality of  means
assuming unequal Covariance Matrices */

data admsion;
infile 'admission.dat' firstobs=7;
input status $ gpa gmat ;

*Here we test the Equality of Covariance matrices
assuming multivariate normality for each of the
three populations ;

proc discrim data = admsion method = normal pool =test wcov pcov bcov ;
class status ;
var gpa gmat ;
prior equal ;
title2 'Test the Equality of Covariance Matrices' ;
run;

* Here we test the Equality of mean vectors assuming
the Unequal Covariance Matrices, using the
Anderson's (1963) test under multivariate normality of
the populations;

title2 'Test the Equality of Mean vectors' ;
proc sort data = admsion ;
by status;
proc means data = admsion;
by status;
var gpa gmat;
```

```
data yes;
set admsion ;
if status = 'yes';
ygpa = gpa;
ygmat = gmat;
sr=_n_;
drop gpa gmat;
run;

data no;
set admsion ;
if status = 'no';
ngpa = gpa;
ngmat = gmat;
sr=_n_;
drop gpa gmat;
run;

data border;
set admsion ;
if status = 'border';
bgpa = gpa;
bgmat = gmat;
sr=_n_;
drop gpa gmat;
run;

proc sort data = yes ;
by sr;

proc sort data = no ;
by sr;

proc sort data = border ;
by sr;

data all;
set yes;
set no;
set border;
run;

data all;
set all;
n1 = 31;
n2=28;
n3 = 26;
smallest = min(n1,n2,n3) ;

y_tave1 = 340.3871 ;
y_tave2 = 561.2258 ;

y_pave1 = 337.5000 ;
y_pave2 = 561.3846 ;

yes1 =
bgpa -(sqrt(smallest/n1))*( ygpa - y_tave1
+ (sqrt(n1/smallest))*y_tave2  ) ;
```

```
yes2 =
bgmat-(sqrt(smallest/n1))*( ygmat - y_pave1
+ (sqrt(n1/smallest))*y_pave2 ) ;

n_tave1 = 248.2500 ;
n_tave2 = 447.0714 ;

n_pave1 = 245.2308 ;
n_pave2 = 452.0385 ;

no1 =
bgpa -(sqrt(smallest/n2))*( ngpa - n_tave1
+ (sqrt(n2/smallest))*n_tave2 ) ;
no2 =
bgmat-(sqrt(smallest/n2))*( ngmat - n_pave1
+ (sqrt(n2/smallest))*n_pave2 ) ;
run;

proc glm data = all; class status;
model yes1 yes2 no1 no2 = /nouni;
manova h = intercept/ printe printh;
title2 'Test the Equality of Mean Vectors: Unequal Covariance' ;
run;
```

An important general remark about testing the equality of mean vectors (whether variance-covariance matrices are equal or unequal) is in order. A rejection of null hypothesis does not suggest that all the k population are distinguishable from each other with respect to means. It says only that at least two mean vectors, say $\boldsymbol{\mu}_i$ and $\boldsymbol{\mu}_j$, are different from each other. Despite the rejection of the null hypothesis, some of the populations may be indistinguishable from each other. In that case, we would expect a large number of misclassifications among these populations when a discriminant analysis is performed. One way to circumvent this problem is to perform certain pairwise multivariate tests (using, say, Hotelling's T^2), especially when the number of populations k is small.

Output 5.4

```
                    Output 5.4
      Test the Equality of Covariance Matrices

               The DISCRIM Procedure

              Class Level Information

         Variable                                              Prior
status   Name        Frequency      Weight  Proportion   Probability

border   border             26     26.0000    0.305882      0.333333
no       no                 28     28.0000    0.329412      0.333333
yes      yes                31     31.0000    0.364706      0.333333


      Test the Equality of Covariance Matrices


         Chi-Square         DF     Pr > ChiSq

          16.074476          6         0.0134
```

```
Since the Chi-Square value is significant at the 0.1 level, the
within covariance matrices will be used in the discriminant func
tion.
Reference: Morrison, D.F. (1976) Multivariate Statistical Method
s p252.

                    The GLM Procedure
             Multivariate Analysis of Variance

    Test the Equality of Mean Vectors: Unequal Covariance

       MANOVA Test Criteria and Exact F Statistics for
        the Hypothesis of No Overall Intercept Effect
           H = Type III SSCP Matrix for Intercept
                   E = Error SSCP Matrix

                   S=1      M=1      N=10

Statistic                      Value  F Value  Num DF  Den DF

Wilks' Lambda             0.00688127   793.77       4      22
Pillai's Trace            0.99311873   793.77       4      22
Hotelling-Lawley Trace  144.32191483   793.77       4      22
Roy's Greatest Root     144.32191483   793.77       4      22

       MANOVA Test Criteria and Exact F Statistics for
        the Hypothesis of No Overall Intercept Effect
           H = Type III SSCP Matrix for Intercept
                   E = Error SSCP Matrix

                   S=1      M=1      N=10

                Statistic                Pr > F

                Wilks' Lambda            <.0001
                Pillai's Trace           <.0001
                Hotelling-Lawley Trace   <.0001
                Roy's Greatest Root      <.0001
```

5.4 Discriminant Analysis: Fisher's Approach

R. A. Fisher (1936, 1940) considered the problem of obtaining a discriminant function when there are only two populations ($k = 2$) by finding the linear decision rule with respect to which the two populations would be most different. It was assumed that the two populations have the same variance-covariance matrices. Let Π_t be the $N_p(\boldsymbol{\mu}_t, \boldsymbol{\Sigma})$, $t = 1$, 2, and suppose $\mathbf{x}$ is a p-dimensional observation to be classified into one of the two populations. If $\mathbf{a}'\mathbf{x}$ is a linear decision rule in $\mathbf{x}$ (i.e., a linear combination of the components of $\mathbf{x}$), then we are looking for that $\mathbf{a}$ for which the distance between $E(\mathbf{a}'\mathbf{x})$ in Π_1 and that in Π_2 is maximum. Since distance can be quantitatively made arbitrarily large by choosing $c\mathbf{a}$ instead of $\mathbf{a}$, for some $c \neq 0$, and since $\mathbf{a}'\mathbf{x}$ and $c \cdot \mathbf{a}'\mathbf{x}$ would essentially classify the object in the same way, it is appropriate to remove this ambiguity by setting an additional condition on $\mathbf{a}$. For this, we may require that $\mathbf{a}$ is standardized and that $\mathbf{a}'\mathbf{x}$ has unit variance. That is,

$$var\,(\mathbf{a}'\mathbf{x}) = \mathbf{a}'\boldsymbol{\Sigma}\mathbf{a} = 1.$$

Thus, Fisher's approach is to choose an $\mathbf{a}$ for which the distance between $E(\mathbf{a}'\mathbf{x}) = \mathbf{a}'\boldsymbol{\mu}_1$ in Π_1 and $E(\mathbf{a}'\mathbf{x}) = \mathbf{a}'\boldsymbol{\mu}_2$ in Π_2, that is, $|\mathbf{a}'\boldsymbol{\mu}_1 - \mathbf{a}'\boldsymbol{\mu}_2|$ is the maximum subject to $\mathbf{a}'\boldsymbol{\Sigma}\mathbf{a} = 1$.

From a matrix theory result (Rao, 1973, p. 60), we arrive at the optimum choice of $\mathbf{a}$ to be proportional to $\boldsymbol{\Sigma}^{-1}(\boldsymbol{\mu}_1 - \boldsymbol{\mu}_2)$. Since the objective is classification, ignoring the constant of proportionality, $\mathbf{a}$ can be taken as

$$\mathbf{a} = \boldsymbol{\Sigma}^{-1}(\boldsymbol{\mu}_1 - \boldsymbol{\mu}_2),$$

and we classify $\mathbf{x}$ into Π_1, if $\mathbf{a}'\mathbf{x} \geq h$, and in Π_2 otherwise where $h = \mathbf{a}'(\boldsymbol{\mu}_1 + \boldsymbol{\mu}_2)/2$. This choice of h essentially amounts to choosing the rule which classifies $\mathbf{x}$ into the closer of the two populations when the distance is taken to be Eucledean distance on $\mathbf{a}'\mathbf{x}$ scale.

5.4.1 Implementation of Fisher's Approach

Of course, in practice, $\boldsymbol{\mu}_1$, $\boldsymbol{\mu}_2$, and $\boldsymbol{\Sigma}$ are all unknown and hence we use the estimates of $\boldsymbol{\mu}_1$, $\boldsymbol{\mu}_2$ and $\boldsymbol{\Sigma}$, namely,

$$\hat{\boldsymbol{\mu}}_1 = \bar{\mathbf{x}}_1$$

$$\hat{\boldsymbol{\mu}}_2 = \bar{\mathbf{x}}_2$$

and

$$\hat{\boldsymbol{\Sigma}} = \mathbf{S} = \frac{1}{n_1 + n_2 - 2}\left[\sum_{j=1}^{n_1}(\mathbf{x}_{1j} - \bar{\mathbf{x}}_1)(\mathbf{x}_{1j} - \bar{\mathbf{x}}_1)' + \sum_{j=1}^{n_2}(\mathbf{x}_{2j} - \bar{\mathbf{x}}_2)(\mathbf{x}_{2j} - \bar{\mathbf{x}}_2)'\right]$$

in the above choice of $\mathbf{a}$, leading to the appropriate estimate of $\mathbf{a}$ as

$$\mathbf{b} = \mathbf{S}^{-1}(\bar{\mathbf{x}}_1 - \bar{\mathbf{x}}_2).$$

The quantity h is approximated by c, where

$$c = \frac{1}{2}\mathbf{b}'(\bar{\mathbf{x}}_1 + \bar{\mathbf{x}}_2) = \frac{1}{2}(\bar{\mathbf{x}}_1 - \bar{\mathbf{x}}_2)'\mathbf{S}^{-1}(\bar{\mathbf{x}}_1 + \bar{\mathbf{x}}_2).$$

As mentioned earlier, no classification procedure is likely to be perfect. If $\boldsymbol{\mu}_1$, $\boldsymbol{\mu}_2$, and $\boldsymbol{\Sigma}$ were indeed known and $\mathbf{a}'\mathbf{x}$ was used for classification, then assuming that the two populations were distributed as the multivariate normal, the probabilities of two misclassification errors are

$$P(2|1) = \Phi\left(\frac{h - \mathbf{a}'\boldsymbol{\mu}_1}{\Delta_p}\right) = \Phi\left(-\frac{1}{2}\Delta_p\right)$$

and

$$P(1|2) = \Phi\left(\frac{\mathbf{a}'\boldsymbol{\mu}_2 - h)}{\Delta_p}\right) = \Phi\left(-\frac{1}{2}\Delta_p\right),$$

and the overall chance of misclassifications is, assuming no specific preferences (that is, any observation is equally likely to come from either of two populations),

$$\frac{1}{2}P(2|1) + \frac{1}{2}P(1|2) = \Phi(-\frac{1}{2}\Delta_p),$$

where $\Phi(\cdot)$ indicates the cumulative probability distribution function of a (univariate) standard normal distribution, and $\Delta_p^2 = (\boldsymbol{\mu}_1 - \boldsymbol{\mu}_2)'\boldsymbol{\Sigma}^{-1}(\boldsymbol{\mu}_1 - \boldsymbol{\mu}_2)$. In case $\hat{\boldsymbol{\mu}}_1$, $\hat{\boldsymbol{\mu}}_2$, and $\hat{\boldsymbol{\Sigma}}$ and hence $\mathbf{b}'\mathbf{x}$ are used, the true overall probability of misclassification will be somewhat higher than that given above, which itself will be approximated by $\Phi(-\frac{1}{2}(\bar{\mathbf{x}}_1-\bar{\mathbf{x}}_2)'\mathbf{S}^{-1}(\bar{\mathbf{x}}_1-\bar{\mathbf{x}}_2))$. The use of estimated parameters, instead of true ones, results in the further approximation (due to sampling error) in the calculations of probabilities and consequently this substitution of estimates in place of true parameters will lead to the underestimation of the misclassification probabilities.

It may be pointed out that strictly speaking, Fisher's approach to obtain the discriminant function does not require the normality assumption. Normality is needed, however, to evaluate the two probabilities of misclassification.

Fisher's approach for classification may be adopted in the DISCRIM procedure by appropriately specifying that under normality (the METHOD = NORMAL option) a linear discriminant function will be used using the pooled (the POOL = YES option) sample variance-covariance matrix. To do so, the appropriate SAS statement is as follows:

```
proc discrim method = normal pool=yes;
```

Of course, we also need to indicate a class variable, which is used to define the two (in general k) populations. If the corresponding variable is named POPLN (say, taking values $PI1$ and $PI2$), then the next SAS statement will be

```
class popln;
```

To specify what variables would be used (in a linear way) to construct Fisher's discriminant function, we need to list the corresponding variables in the VAR statement. Specifically, for the two variables $X1$ and $X2$ the SAS statement is

```
var x1 x2;
```

In addition, there are a number of other features that can be used to obtain a variety of other information. These will be shown through an example.

EXAMPLE 5 ***Fisher's Discriminant Function, Flea Beetles Data Revisited*** To illustrate the methodology and its application, let us reconsider the flea beetles data and confine ourselves to only the first two species, namely, *Chaetocnema concinna* and *Chactocnema heikertingeri*. As the variance-covariance matrices can be assumed to be equal and population means are indeed statistically different, we may use the Fisher's linear discriminant function to discriminate between these two species.

The data on the first two species are saved in the SAS data set named FISHRLDF. With the METHOD = NORMAL and POOL = YES options in PROC DISCRIM, we analyze the data on $X1$ and $X2$. The specification of two populations is attained through the CLASS statement where the values of CLASS variable SPECIES indicate the corresponding species. Fisher's approach assumes that any object is equally likely to come from either of the two populations. This fact can be indicated by specifying another statement

```
priors equal;
```

but is optional (since it is a default in PROC DISCRIM) and can be skipped. The corresponding SAS code thus becomes

```
proc discrim method = normal pool = yes;
class species;
var x1 x2;
```

In the PROC DISCRIM statement, many other options to obtain or print additional output are available. A few important ones, especially relevant in the present context, are PCOV, which prints the pooled variance-covariance matrix; OUT=FOUT, which saves the output of classifications in a data set termed as FOUT; OUTSTAT=FSTAT, which saves the detailed summary statistics in the data set FSTAT; LIST, which essentially lists all the details of the classifications done on the data; and MANOVA, which performs the multivariate analysis of variance on the data to test if the population means are equal. Note that the multivariate analysis of variance assumes the equality of the variance-covariance matrices, and hence, in general, the option MANOVA should be used only when such a test has already been performed and the hypothesis accepted. This in turn implicitly means that the MANOVA option should be used only in conjunction with the POOL = YES option. In Program 5.5, we have used all of these options. Output 5.5 corresponds to Program 5.5. The 2 by 2 pooled sample variance-covariance matrix is

$$\mathbf{S} = \begin{bmatrix} 25.4867 & -0.5837 \\ -0.5837 & 1.0439 \end{bmatrix}.$$

This matrix is different from the $\mathbf{S}$ found in Output 5.3 since, unlike the former one, the latter was computed by pooling the sample variance-covariance matrices of the data from all three species. The pairwise squared distance between the centers of two populations, namely, $(\boldsymbol{\mu}_1 - \boldsymbol{\mu}_2)'\boldsymbol{\Sigma}^{-1}(\boldsymbol{\mu}_1 - \boldsymbol{\mu}_2)$ is estimated by $(\bar{\mathbf{x}}_1 - \bar{\mathbf{x}}_2)'\mathbf{S}^{-1}(\bar{\mathbf{x}}_1 - \bar{\mathbf{x}}_2)$ and, as shown in the Output 5.5, is equal to 18.2998. This quantity would be especially useful when there are more than two populations. It is used to indicate the relative dissimilarities between various populations. Greater squared distances between two populations can be interpreted as the corresponding populations being more dissimilar and hence, in a pairwise classification scheme, we would expect fewer misclassifications.

In Output 5.5, the values of $-0.5\bar{\mathbf{x}}_j'\mathbf{S}^{-1}\bar{\mathbf{x}}_j$ and $\mathbf{S}^{-1}\bar{\mathbf{x}}_j$ are listed for $j = 1, 2$. From these we can easily compute the coefficients of the linear discriminant function. Since an observation $\mathbf{x}$ is classified into Π_1, if $\mathbf{b}'\mathbf{x} \geq c$ and in Π_2, otherwise, where $\mathbf{b} = \mathbf{S}^{-1}(\bar{\mathbf{x}}_1 - \bar{\mathbf{x}}_2)$ and $c = \frac{1}{2}(\bar{\mathbf{x}}_1 - \bar{\mathbf{x}}_2)'\mathbf{S}^{-1}(\bar{\mathbf{x}}_1 + \bar{\mathbf{x}}_2)$, we have

$$\mathbf{b} = \mathbf{S}^{-1}\bar{\mathbf{x}}_1 - \mathbf{S}^{-1}\bar{\mathbf{x}}_2 = \begin{bmatrix} 6.12363 \\ 16.92628 \end{bmatrix} - \begin{bmatrix} 5.27165 \\ 16.63674 \end{bmatrix} = \begin{bmatrix} 0.85198 \\ 0.28954 \end{bmatrix}$$

and

$$c = \frac{1}{2}\bar{\mathbf{x}}_1'\mathbf{S}^{-1}\bar{\mathbf{x}}_1 - \frac{1}{2}\bar{\mathbf{x}}_2'\mathbf{S}^{-1}\bar{\mathbf{x}}_2 = 566.89836 - 447.41475 = 119.48361.$$

In fact, the quantities $\mathbf{S}^{-1}\bar{\mathbf{x}}_1, \mathbf{S}^{-1}\bar{\mathbf{x}}_2, -\frac{1}{2}\bar{\mathbf{x}}_1'\mathbf{S}^{-1}\bar{\mathbf{x}}_1$, and $-\frac{1}{2}\bar{\mathbf{x}}_2'\bar{\mathbf{S}}^{-1}\bar{\mathbf{x}}_2$ are also available as observations 50 through 53 in the data set FSTAT (and printed as the last four lines of Output 5.5) obtained by using the OUTSTAT= option.

We classify an observation $(\mathbf{x}_1, \mathbf{x}_2)'$ into Π_1 if

$$0.85198x_1 + 0.28954x_2 \geq 119.48361$$

and in Π_2, otherwise. As an illustration, we see that the seventeenth observation in the sample of species 1, namely $(134, 15)'$, yields the value of $\mathbf{b}'\mathbf{x} = 0.85218(134) + 0.28954(15) = 118.53$, which is less than $c = 119.48361$. Hence this observation is (mis)classified in Π_2.

```
/* Program 5.5 */

options ls = 64 ps=45 nodate nonumber;
title1 'Output 5.5';

data beetles;
infile 'beetle.dat' firstobs=7;
input species  x1 x2 ;
```

```
/* Here we perform Fisher's discriminant analysis for
two groups of species  namely species 1 and 2 */

proc sort data = beetles; by species;
data fishrldf;set beetles; if species ne 3 ;

proc discrim data = fishrldf out = fout outstat =fstat
method = normal list pool =yes pcov manova ;
class species ;
priors equal ;
var x1 x2 ;
title2 'Discriminant Analysis of Beetles Data: Two Groups';

proc print data = fout;
title2 'Classification Details';

proc print data = fstat;
title2 'Summary Statistics: Two Group Discriminant Analysis';
```

Output 5.5

```
                         Output 5.5
         Discriminant Analysis of Beetles Data: Two Groups

                    The DISCRIM Procedure

    Observations       52          DF Total                51
    Variables           2          DF Within Classes       50
    Classes             2          DF Between Classes       1


                    Class Level Information

          Variable                                               Prior
species   Name        Frequency      Weight   Proportion   Probability

      1   _1                 21     21.0000     0.403846      0.500000
      2   _2                 31     31.0000     0.596154      0.500000

        Pooled Within-Class Covariance Matrix,       DF = 50

           Variable                  x1                  x2

           x1               25.48669739         -0.58374808
           x2               -0.58374808          1.04393241


               Pooled Covariance Matrix Information

                                   Natural Log of the
                     Covariance    Determinant of the
                     Matrix Rank    Covariance Matrix

                               2              3.26826

       Pairwise Generalized Squared Distances Between Groups

                    2           _   _         -1  _   _
                   D (i|j) = (X - X )' COV    (X - X )
                               i   j            i   j
```

Generalized Squared Distance to species

From species	1	2
1	0	18.29984
2	18.29984	0

Multivariate Statistics and Exact F Statistics

S=1 M=0 N=23.5

Statistic	Value	F Value	Num DF	Den DF
Wilks' Lambda	0.17914729	112.26	2	49
Pillai's Trace	0.82085271	112.26	2	49
Hotelling-Lawley Trace	4.58199902	112.26	2	49
Roy's Greatest Root	4.58199902	112.26	2	49

Multivariate Statistics and Exact F Statistics

S=1 M=0 N=23.5

Statistic	Pr > F
Wilks' Lambda	<.0001
Pillai's Trace	<.0001
Hotelling-Lawley Trace	<.0001
Roy's Greatest Root	<.0001

Linear Discriminant Function

$$\text{Constant} = -.5\ \bar{X}_j'\ \text{COV}^{-1}\ \bar{X}_j \qquad \text{Coefficient Vector} = \text{COV}^{-1}\ \bar{X}_j$$

Linear Discriminant Function for species

Variable	1	2
Constant	-566.89836	-447.41475
x1	6.12363	5.27165
x2	16.92628	16.63674

Classification Results for Calibration Data: WORK.FISHRLDF
Resubstitution Results using Linear Discriminant Function

Generalized Squared Distance Function

$$D_j^2(X) = (X-\bar{X}_j)'\ \text{COV}^{-1}\ (X-\bar{X}_j)$$

Posterior Probability of Membership in Each species

$$\Pr(j|X) = \exp(-.5\ D_j^2(X))\ /\ \text{SUM}_k \exp(-.5\ D_k^2(X))$$

Posterior Probability of Membership in species

Obs	From species	Classified into species	1	2
1	1	1	1.0000	0.0000
2	1	1	0.9999	0.0001
3	1	1	0.9993	0.0007
.				
.				
.				
14	1	1	1.0000	0.0000
15	1	1	0.9999	0.0001
16	1	1	0.7845	0.2155
17	1	2 *	0.2740	0.7260
18	1	1	1.0000	0.0000
19	1	1	1.0000	0.0000
20	1	1	0.9999	0.0001
.				
.				
.				
50	2	2	0.0040	0.9960
51	2	2	0.0000	1.0000
52	2	2	0.0040	0.9960

* Misclassified observation

Classification Summary for Calibration Data: WORK.FISHRLDF
Resubstitution Summary using Linear Discriminant Function

Generalized Squared Distance Function

$$D_j^2(X) = (X-\bar{X}_j)' \, COV^{-1} \, (X-\bar{X}_j)$$

Posterior Probability of Membership in Each species

$$Pr(j|X) = \exp(-.5\, D_j^2(X)) \,/\, SUM_k \exp(-.5\, D_k^2(X))$$

Number of Observations and Percent Classified into species

From species	1	2	Total
1	20	1	21
	95.24	4.76	100.00
2	0	31	31
	0.00	100.00	100.00
Total	20	32	52
	38.46	61.54	100.00
Priors	0.5	0.5	

Error Count Estimates for species

	1	2	Total
Rate	0.0476	0.0000	0.0238
Priors	0.5000	0.5000	

Classification Details

Obs	species	x1	x2	_1	_2	_INTO_
1	1	150	15	1.00000	0.00000	1
2	1	147	13	0.99993	0.00007	1
3	1	144	14	0.99929	0.00071	1
	.					
	.					
	.					
14	1	154	15	1.00000	0.00000	1
15	1	147	14	0.99995	0.00005	1
16	1	137	14	0.78452	0.21548	1
17	1	134	15	0.27404	0.72596	2
18	1	157	14	1.00000	0.00000	1
19	1	149	13	0.99999	0.00001	1
20	1	147	13	0.99993	0.00007	1
	.					
	.					
	.					
50	2	129	14	0.003975	0.99602	2
51	2	124	13	0.000042	0.99996	2
52	2	129	14	0.003975	0.99602	2

Summary Statistics: Two Group Discriminant Analysis

Obs	species	_TYPE_	_NAME_	x1	x2
1	.	N		52.00	52.0000
2	1	N		21.00	21.0000
3	2	N		31.00	31.0000
4	.	MEAN		133.35	14.2115
5	1	MEAN		146.19	14.0952
6	2	MEAN		124.65	14.2903
7	1	PRIOR		0.50	0.5000
8	2	PRIOR		0.50	0.5000
9	1	CSSCP	x1	633.24	-19.3810
10	1	CSSCP	x2	-19.38	15.8095
11	2	CSSCP	x1	641.10	-9.8065
12	2	CSSCP	x2	-9.81	36.3871
13	.	PSSCP	x1	1274.33	-29.1874
14	.	PSSCP	x2	-29.19	52.1966
15	.	BSSCP	x1	5811.43	-52.6203
16	.	BSSCP	x2	-52.62	0.4765
17	.	CSSCP	x1	7085.77	-81.8077
18	.	CSSCP	x2	-81.81	52.6731
19	.	RSQUARED		0.82	0.0090
20	1	COV	x1	31.66	-0.9690
21	1	COV	x2	-0.97	0.7905
22	2	COV	x1	21.37	-0.3269
23	2	COV	x2	-0.33	1.2129
24	.	PCOV	x1	25.49	-0.5837

```
25      .   PCOV       x2           -0.58      1.0439
26      .   BCOV       x1          223.52     -2.0239
27      .   BCOV       x2           -2.02      0.0183
28      .   COV        x1          138.94     -1.6041
29      .   COV        x2           -1.60      1.0328
30      1   STD                      5.63      0.8891
31      2   STD                      4.62      1.1013
32      .   PSTD                     5.05      1.0217
33      .   BSTD                    14.95      0.1354
34      .   STD                     11.79      1.0163
35      1   CORR       x1            1.00     -0.1937
36      1   CORR       x2           -0.19      1.0000
37      2   CORR       x1            1.00     -0.0642
38      2   CORR       x2           -0.06      1.0000
39      .   PCORR      x1            1.00     -0.1132
40      .   PCORR      x2           -0.11      1.0000
41      .   BCORR      x1            1.000    -1.000
42      .   BCORR      x2           -1.000     1.000
43      .   CORR       x1            1.000    -0.134
44      .   CORR       x2           -0.134     1.000
45      1   STDMEAN                    1.090    -0.114
46      2   STDMEAN                   -0.738     0.078
47      1   PSTDMEAN                   2.544    -0.114
48      2   PSTDMEAN                  -1.724     0.077
49      .   LNDETERM                   3.268     3.268
50      1   LINEAR     _LINEAR_        6.124    16.926
51      1   LINEAR     _CONST_      -566.898  -566.898
52      2   LINEAR     _LINEAR_        5.272    16.637
53      2   LINEAR     _CONST_      -447.415  -447.415
```

We can equivalently classify an observation $\mathbf{x}$ to whichever of the two populations is closest, where the distance of $\mathbf{x}$ from population Π_j is measured as $d_j(\mathbf{x}) = \{(\mathbf{x} - \bar{\mathbf{x}}_j)'\mathbf{S}^{-1}(\mathbf{x} - \bar{\mathbf{x}}_j)\}^{\frac{1}{2}}$. It is further equivalent to classifying $\mathbf{x}$ into the population Π_j with the higher posterior probability, which is defined as

$$P(j|\mathbf{x}) = \frac{e^{-\frac{1}{2}d_j^2(\mathbf{x})}}{e^{-\frac{1}{2}d_1^2(\mathbf{x})} + e^{-\frac{1}{2}d_2^2(\mathbf{x}))}}.$$

As an illustration, for $\mathbf{x} = (134, 15)'$ used earlier,

$$\begin{aligned} d_1^2 &= \mathbf{x}'\mathbf{S}^{-1}\mathbf{x} - 2\mathbf{x}'\mathbf{S}^{-1}\bar{\mathbf{x}}_1 + \bar{\mathbf{x}}_1'\mathbf{S}^{-1}\bar{\mathbf{x}}_1 \\ &= \mathbf{x}'\mathbf{S}^{-1}\mathbf{x} - 2\left[(134, 15)\begin{pmatrix} 6.12363 \\ 16.92628 \end{pmatrix} - 566.89836\right] \\ &= \mathbf{x}'\mathbf{S}^{-1}\mathbf{x} - 1015.1245 \end{aligned}$$

and, similarly,

$$\begin{aligned} d_2^2 &= \mathbf{x}'\mathbf{S}^{-1}\mathbf{x} - 2\mathbf{x}'S^{-1}\bar{\mathbf{x}}_2 + \bar{\mathbf{x}}_2 S^{-1}\bar{\mathbf{x}}_2 \\ &= \mathbf{x}'\mathbf{S}^{-1}\mathbf{x} - 1017.0749. \end{aligned}$$

Naturally, d_2^2 is (barely) smaller than d_1^2 for $\mathbf{x} = (134, 15)'$, and hence this observation is classified into Π_2. Also, using these values of d_1^2 and d_2^2,

$$\begin{aligned} P(1|\mathbf{x}) &= \frac{e^{-\frac{1}{2}(\mathbf{x}'\mathbf{S}^{-1}\mathbf{x} - 1015.1245)}}{e^{-\frac{1}{2}(\mathbf{x}'\mathbf{S}^{-1}\mathbf{x} - 1015.1245)} + e^{-\frac{1}{2}(\mathbf{x}'\mathbf{S}^{-1}\mathbf{x} - 1017.0749)}} \\ &= 0.2740 \end{aligned}$$

and hence $P(2|\mathbf{x}) = 1 - P(1|\mathbf{x}) = 0.7260$. Since $P(2|\mathbf{x})$ is larger than $P(1|\mathbf{x})$, the observation (134, 15) is again classified as that from the population of the second species.

In Output 5.5, reclassifications for all the data using the approach described above are provided. Only one observation coming from the population of species 1 has been misclassified into the population of the second species. This information is presented in the next table of Output 5.5. Thus, for the first species (for which $n_1 = 21$), the error count estimate is $1/21 = 0.0476$ (reported as 4.76% in the output) and that for the second species is 0. Consequently, in view of the fact that an observation is a priori equally likely to come from either of the two populations, the total error count estimate is $\frac{1}{2}(0.0476) + \frac{1}{2}(0) = 0.0238$. This suggests (pretending that sampling errors are small and that the resubstitution has minimal effect) that about 2.38% of the data would be misclassified by this procedure. Outputs generated by the use of the OUT= and OUTSTAT= options are printed next. Items reported on lines 50 through 53 of the summary statistics were used to construct the linear discriminant function.

5.4.2 Fisher's Approach and Multiple Regression

Fisher's discriminant function for two populations can also be computed by means of multiple regression. This connection was first observed by Fisher (1936). Flury and Riedwyl (1985) provide a detailed discussion of this connection.

Following Flury and Riedwyl (1985), given the data $\mathbf{x}$ on an object, we define a binary dependent variable y as

$$y = \begin{cases} \frac{n_2}{n_1+n_2} & \text{if } \mathbf{x} \text{ belongs to population } \Pi_1 \\ \frac{-n_1}{n_1+n_2} & \text{if } \mathbf{x} \text{ belongs to population } \Pi_2 \end{cases}$$

and fit the linear regression model

$$y = \beta_0 + \boldsymbol{\beta}_1'\mathbf{x} + \epsilon$$

The resulting least square estimate $\hat{\boldsymbol{\beta}}_1$ of $\boldsymbol{\beta}_1$ is indeed proportional to $\mathbf{b} = \mathbf{S}^{-1}(\bar{\mathbf{x}}_1 - \bar{\mathbf{x}}_2)$ and is given by

$$\hat{\boldsymbol{\beta}}_1 = \left\{\left(\frac{1}{n_1} + \frac{1}{n_2}\right)(n_1 + n_2 - 2) + \mathbf{d}'\mathbf{S}^{-1}\mathbf{d}\right\}^{-1} \mathbf{b},$$

where $\mathbf{d} = \bar{\mathbf{x}}_1 - \bar{\mathbf{x}}_2$ and $\mathbf{S}$ is the pooled variance-covariance matrix of data on $\mathbf{x}$.

Since the two vectors $\hat{\boldsymbol{\beta}}_1$ and $\mathbf{b}$ are proportional to each other, we can alternatively use $\hat{\boldsymbol{\beta}}_1'\mathbf{x}$ as Fisher's discriminant function. The quantity c used as the classification boundary also changes by the same constant of proportionality. Strictly speaking, the binary dependent variable y does not have to be defined exactly as above. The fact that $\hat{\boldsymbol{\beta}}_1$ is proportional to $\mathbf{b}$ remains true even if we define the binary variable y to indicate the population membership (e.g., 1 if $\mathbf{x}$ belongs to Π_1 and 0 if $\mathbf{x}$ belongs to Π_2). Of course the constant of proportionality will be different in that case.

EXAMPLE 6 ***Multiple Regression Approach, Flea Beetles Data Revisited*** We reconsider the flea beetles data used in Example 5 and analyze it using the multiple regression of y defined as

$$y = \begin{cases} \frac{n_2}{n_1+n_2} = \frac{31}{52} & \text{if the species = Chaetocnema Concinna} \\ \frac{-n_1}{n_1+n_2} = -\frac{21}{52} & \text{if the species = Chactocnema heikertingeri} \end{cases}$$

on independent variables $X1$ and $X2$ (defined in Example 1). The following SAS statements would achieve our purpose:

```
proc reg;
model y = x1 x2;
```

where the values of dummy dependent variable y can be defined in the DATA step as

```
data beetles;
input species x1 x2;
if species = 1 then y = 31/52;
if species = 2 then y = -21/52;
lines;
```

The complete SAS program is presented as Program 5.6. The corresponding output is available in Output 5.6. From Output 5.6, we see that

$$\hat{\boldsymbol{\beta}}_1 = \begin{bmatrix} 0.0382 \\ 0.0130 \end{bmatrix},$$

which is proportional to $\mathbf{b} = \begin{bmatrix} 0.85198 \\ 0.28954 \end{bmatrix}$, as found in Example 5. The constant of proportionality is approximately equal to .0448. As in the case of $\mathbf{b}$, the quantity γ, which is the cutoff used for the classification rule, can be analogously computed as $\gamma = \frac{1}{2}\hat{\boldsymbol{\beta}}_1'(\bar{\mathbf{x}}_1 + \bar{\mathbf{x}}_2)$. The quantity γ estimates $h = \mathbf{a}'(\boldsymbol{\mu}_1 + \boldsymbol{\mu}_2)/2$ defined earlier, where $\hat{\boldsymbol{\beta}}_1$ estimates $\mathbf{a}$ (up to a constant of proportionality) and $(\boldsymbol{\mu}_1 + \boldsymbol{\mu}_2)/2$ is estimated by $(\bar{\mathbf{x}}_1 + \bar{\mathbf{x}}_2)/2$. In fact, instead of γ, another estimate of h can be taken as $-\hat{\beta}_0$. In this case, $\hat{y} = \hat{\beta}_0 + \hat{\boldsymbol{\beta}}_1'\mathbf{x}$ can be used as the classification function with the classification rule

$$\text{classify } \mathbf{x} \text{ into } \Pi_1 \text{ if } \hat{y} > 0 \text{ and into } \Pi_2 \text{ if } \hat{y} \le 0.$$

Unless the sample sizes n_1 and n_2 are equal, γ and $\hat{\beta}_0$ are not necessarily equal even though both estimate h. It is so, since in the case of γ as an estimate of h, $(\boldsymbol{\mu}_1 + \boldsymbol{\mu}_2)/2$ is estimated by $(\bar{\mathbf{x}}_1 + \bar{\mathbf{x}}_2)/2$. However, it follows from Searle (1971, p.85 Equation 37) that $\hat{\beta}_0 = \hat{\boldsymbol{\beta}}_1'\bar{\mathbf{x}}$ where $\bar{\mathbf{x}} = \frac{(n_1\bar{\mathbf{x}}_1 + n_2\bar{\mathbf{x}}_2)}{(n_1+n_2)}$.

In other words, in this case, $(\boldsymbol{\mu}_1 + \boldsymbol{\mu}_2)/2$ is estimated by the pooled sample mean $\bar{\mathbf{x}}$, which is not equal to $(\bar{\mathbf{x}}_1 + \bar{\mathbf{x}}_2)/2$ unless $n_1 = n_2$. This is precisely the reason that, as shown in Output 5.6, the regression approach classifies the seventeenth observation $(x_1, x_2)' = (134, 15)'$ into Π_1 (since $\hat{y} > 0$), even though, as shown in Output 5.5, this observation was earlier (mis)classified into Π_2. This, however, does not necessarily imply that the regression approach is more correct.

```
/* Program 5.6 */

options ls = 64 ps=45 nodate nonumber;
title1 'Output 5.6';

data beetles;
infile 'beetle.dat' firstobs=7;
input species  x1 x2 ;

/* y = n2/(n1+n2) for species 1 */
if species = 1 then y = 31/52;

/* y =-n1/(n1+n2) for species 2 */
if species = 2 then y = -21/52;
```

```
proc sort data = beetles; by species;
data fishrldf;set beetles; if species ne 3 ;

proc reg data =beetles;
model y = x1 x2;
output out = b p=yhat;
title2 'Regression Approach to Discrimination: Beetles Data';

proc print data = b;
title2 'Classification: Dummy Response Variable';
```

Output 5.6

```
                          Output 5.6
        Regression Approach to Discrimination: Beetles Data

                       The REG Procedure
                        Model: MODEL1
                     Dependent Variable: y

                      Analysis of Variance

                                   Sum of           Mean
Source                   DF       Squares         Square    F Value

Model                     2      10.27644        5.13822     112.26
Error                    49       2.24279        0.04577
Corrected Total          51      12.51923

                      Analysis of Variance

                   Source                Pr > F

                   Model                 <.0001
                   Error
                   Corrected Total


      Root MSE                0.21394    R-Square      0.8209
      Dependent Mean      1.13157E-16    Adj R-Sq      0.8135
      Coeff Var           1.890659E17


                      Parameter Estimates

                       Parameter      Standard
Variable      DF        Estimate         Error    t Value    Pr > |t|

Intercept      1        -5.28060       0.57902      -9.12      <.0001
x1             1         0.03822       0.00256      14.90      <.0001
x2             1         0.01299       0.02975       0.44      0.6643

              Classification: Dummy Response Variable

      Obs    species      x1     x2          y          yhat

        1        1       150     15    0.59615       0.64669
        2        1       147     13    0.59615       0.50607
        3        1       144     14    0.59615       0.40441
        4        1       144     16    0.59615       0.43038
```

5	1	153	13	0.59615	0.73537
6	1	140	15	0.59615	0.26453
7	1	151	14	0.59615	0.67192
8	1	143	14	0.59615	0.36619
9	1	144	14	0.59615	0.40441
10	1	142	15	0.59615	0.34096
11	1	141	13	0.59615	0.27677
12	1	150	15	0.59615	0.64669
13	1	148	13	0.59615	0.54428
14	1	154	15	0.59615	0.79956
15	1	147	14	0.59615	0.51905
16	1	137	14	0.59615	0.13689
17	1	134	15	0.59615	0.03523
18	1	157	14	0.59615	0.90122
19	1	149	13	0.59615	0.58250
20	1	147	13	0.59615	0.50607
21	1	148	14	0.59615	0.55727
22	2	120	14	-0.40385	-0.51279
23	2	123	16	-0.40385	-0.37217
24	2	130	14	-0.40385	-0.13063
25	2	131	16	-0.40385	-0.06643
26	2	116	16	-0.40385	-0.63968
27	2	122	15	-0.40385	-0.42337
28	2	127	15	-0.40385	-0.23229
29	2	132	16	-0.40385	-0.02822
30	2	125	14	-0.40385	-0.32171
31	2	119	13	-0.40385	-0.56399
32	2	122	13	-0.40385	-0.44935
33	2	120	15	-0.40385	-0.49980
34	2	119	14	-0.40385	-0.55101
35	2	123	15	-0.40385	-0.38515
36	2	125	15	-0.40385	-0.30872
37	2	125	14	-0.40385	-0.32171
38	2	129	14	-0.40385	-0.16884
39	2	130	13	-0.40385	-0.14361
40	2	129	13	-0.40385	-0.18183
41	2	122	12	-0.40385	-0.46233
42	2	129	15	-0.40385	-0.15585
43	2	124	15	-0.40385	-0.34694
44	2	120	13	-0.40385	-0.52578
45	2	119	16	-0.40385	-0.52503
46	2	119	14	-0.40385	-0.55101
47	2	133	13	-0.40385	-0.02896
48	2	121	15	-0.40385	-0.46159
49	2	128	14	-0.40385	-0.20706
50	2	129	14	-0.40385	-0.16884
51	2	124	13	-0.40385	-0.37291
52	2	129	14	-0.40385	-0.16884
53	3	145	8	.	0.36470
54	3	140	11	.	0.21258
55	3	140	11	.	0.21258
56	3	131	10	.	-0.14436
57	3	139	11	.	0.17436
58	3	139	10	.	0.16137
59	3	136	12	.	0.07270
60	3	129	11	.	-0.20780
61	3	140	10	.	0.19959
62	3	137	9	.	0.07195
63	3	141	11	.	0.25079

64	3	138	9	.	0.11017
65	3	143	9	.	0.30125
66	3	142	11	.	0.28901
67	3	144	10	.	0.35245
68	3	138	10	.	0.12316
69	3	140	10	.	0.19959
70	3	130	9	.	-0.19556
71	3	137	11	.	0.09793
72	3	137	10	.	0.08494
73	3	136	9	.	0.03374
74	3	140	10	.	0.19959

The regression approach used here was presented not as a substitute to the usual Fisher's discriminant analysis, but with an objective to illustrate the equivalence of the two approaches. As we shall see later in Section 5.7, this equivalence will be found to be very useful in assessing the importance of a set of discriminatory variables.

5.5 Discriminant Analysis for k Normal Populations

Suppose we have $k(\geq 2)$ normal populations and let t^{th} population be $N_p(\boldsymbol{\mu}_t, \boldsymbol{\Sigma}_t)$, $t = 1, \cdots, k$. We want to obtain a discrimination rule to classify any future observation into one of these k populations. Since there may be situations in which that observation is more likely to come from one population than others, we may also assume that there is a certain prespecified prior probability π_t for an observation coming from the tth population. Of course, we must have $\pi_1 + \pi_2 + \cdots + \pi_k = 1$. Further, certain misclassifications may be less desirable than others and hence there may also be a cost component that should be taken into account when constructing a discriminant function. Suppose $c(s|t)$ is the cost of classifying an item to Π_s when it actually belongs to $\Pi_t, t, s = 1, \cdots, k$. Of course, $c(t|t) = 0$ since when $t = s$, there is no misclassification and hence the cost should be zero. Also let $P(s|t)$ be the probability of classifying an object actually coming from Π_t into Π_s. In this case, we must divide the entire p-dimensional space into k disjoint regions, say $R_1, \cdots, R_k$, so that if the object, say $\mathbf{x}$, falls in R_t, it is classified in Π_t. A meaningful criterion to obtain a decision rule, in view of different π_t, and $c(s|t)$ values, $t, s = 1, \cdots, k$, is to minimize the *expected cost of misclassification*, namely,

$$\sum_{s=1}^{k} \pi_s \sum_{t=1}^{k} P(t|s)c(t|s).$$

If $f_t(\mathbf{x})$ is the multivariate pdf of tth population, then the above criterion yields the discrimination rule

$$\text{classify } \mathbf{x} \text{ into } \Pi_t \text{ if the quantity } \sum_{s=1}^{k} \pi_s f_s(x)c(t|s) \text{ is smallest.} \tag{5.5}$$

In case of a tie, $\mathbf{x}$ can be assigned to any of the tied populations.

5.5.1 Linear Discriminant Function

When the k populations have the same variance-covariance matrices, say $\boldsymbol{\Sigma}$, and the costs of misclassification are assumed to be equal, Rule 5.5 simplifies considerably and, in fact, can be written in terms of squared distances, defined as

$$D_t^2(\mathbf{x}) = (\mathbf{x} - \boldsymbol{\mu}_t)'\boldsymbol{\Sigma}^{-1}(\mathbf{x} - \boldsymbol{\mu}_t) - 2ln(\pi_t).$$

The object $\mathbf{x}$ is classified to the closest population in terms of the distance metric given above. That is, $\mathbf{x}$ is classified into the tth population if

$$D_t^2(\mathbf{x}) = \min_{j=1,\cdots,k}(D_j^2(\mathbf{x})). \tag{5.6}$$

When $\boldsymbol{\mu}_t, t = 1, \cdots, k$ and $\boldsymbol{\Sigma}$ are unknown, we can substitute their sample estimates $\bar{\mathbf{x}}_t, t = 1, \cdots, k$ and $\mathbf{S}$ to obtain the estimates of $D_t^2(\mathbf{x})$ say, $d_t^2(\mathbf{x})$ as

$$d_t^2(\mathbf{x}) = (\mathbf{x} - \bar{\mathbf{x}}_t)'\mathbf{S}^{-1}(\mathbf{x} - \bar{\mathbf{x}}_t) - 2ln(\pi_t) \tag{5.7}$$

and use the classification rule prescribed above for allocation. It must be remembered, however, that the use of estimates may increase the actual probabilities of misclassification. It may be remarked that the formula for $d_t^2(x)$ given above is the modified version of that used in Fisher's discriminant function in that it has now been modified by adding an extra term $-2ln(\pi_t)$ to account for different prior probabilities.

As in the previous section, the criterion for classifying an observation to the closest population is equivalent to classifying it to the population with maximum posterior probability given the observation. In the case of k populations, these probabilities are defined as

$$P(t|\mathbf{x}) = e^{-\frac{1}{2}D_t^2(\mathbf{x})} \Big/ \sum_{j=1}^{k} e^{-\frac{1}{2}D_j^2(\mathbf{x})}, t = 1, \cdots, k. \tag{5.8}$$

Obviously, for a given observation $\mathbf{x}$, sum of $P(t|\mathbf{x})$ over t is 1. The sample version of $P(t|\mathbf{x})$ is obtained by replacing the quantity $D_j^2(\mathbf{x})$ by $d_j^2(\mathbf{x})$, $j = 1, \cdots, k$ in Equation 5.8.

5.5.2 Error Rate Estimation

There are a number of error rate estimation techniques. These can be used for the evaluation of the discrimination scheme adopted. However, two specific error rate estimates based on the data provided for such estimation will be described here.

Error count estimates. Suppose a data set with known membership of its observations to various populations is available to evaluate the performance of a discriminant function. Such a data set is often called the test data. The number (and hence percentage) of observations from the tth population but misclassified into s^{th} population can provide some idea about the $P(s|t)$. From these we can also compute the proportion of misclassifications from the tth population as $\hat{ER}(t) = \sum_{s=1, s\neq t}^{k} \hat{P}(s|t)$. This later quantity provides an estimate for the error rate for t^{th} population.

We should make a few remarks about the error rate estimates. First of all, if the test data set used to estimate the error rate is different from that used to obtain the discriminant function, the error rate estimates are indeed unbiased. However, if the performance is being examined on the very same (training) data set, the estimates are biased (downward), and hence these estimates are overly optimistic. Whether unbiased or not, however, these estimates tend to have large variances, especially when the test data set is small.

The overall error rate for the given discrimination procedure is defined as

$$\sum_{t=1}^{k} \pi_t \sum_{\substack{s=1 \\ s \neq t}}^{k} P(s|t)$$

and is estimated by $\sum_{t=1}^{k} \pi_t \hat{ER}(t)$. This is also called the apparent error rate.

Estimation of error rate by cross-validation. One way to circumvent the problem of underestimation is to divide the data set into two parts: one part to construct the discriminant function and the other to evaluate it. As Lachenbruch (1975) and Hand (1981) point out, this suggestion is wasteful of data and needs large samples to arrive at the satisfactory estimates. Further, one would ultimately prefer to use the discriminant function constructed from the entire data, but the method evaluates a discriminant function, which is constructed only from the part of the data available, and the estimated coefficients in the two may not always be comparable.

Lachenbruch (1975) suggests an alternative method to estimate the error rate through cross-validation. What it amounts to is to leave one observation out and construct a discriminant rule from the rest of the data. This rule is used to classify the observation that was left out. This procedure is then repeated for every observation, and finally the numbers of misclassified observations for each population are counted and the individual error rates are computed as the respective proportions. The overall error rate can be computed as the weighted average of these proportions with prior probabilities $\pi_1, \cdots, \pi_k$ as the weights. This procedure leads to estimates that have a considerably smaller bias (and are almost unbiased). The concern that in each iteration a different discriminant function (and different from the one that we would obtain if the entire data were used) is being evaluated may not be a major issue since elimination of one observation is unlikely to affect the discriminant function significantly unless there are a few outliers in the data. Error rates, using the cross-validation described above, can be estimated using SAS.

Estimation of posterior probability error rate. These error rates for individual populations are computed using the corresponding posterior probabilities given the observation. For a given group, say t^{th}, we may estimate the error rate $e_t = 1 - \int_{R_t} f_t(\mathbf{x})\, dx$ using the following two estimators,

$$PPER1_t = 1 - \frac{1}{\pi_t \sum_{j=1}^{k} n_j} \sum_{R_t} \hat{p}(t|\mathbf{x})$$

and

$$PPER2_t = 1 - \frac{1}{\pi_t} \sum_{j=1}^{k} \frac{\pi_j}{n_j} \sum_{R_{jt}} \hat{p}(t|\mathbf{x}),$$

where $f_t(\mathbf{x})$ is the probability density function for t^{th} population, $p(t|\mathbf{x})$ is the posterior probability of $\mathbf{x}$ for population t, R_t is the set of observations such that the posterior probability of belonging to population t is the largest, and R_{jt} is the set of observations from population j such that the posterior probability of belonging to population t is the largest. The quantity PPER2_t is often referred to as the stratified posterior probability error rate estimate since it takes into account the relative sizes or likelihood of the k populations (through the presence of π_j, $j = 1, \cdots, k$ in the sum). In contrast, PPER1_t is termed the unstratified or simple posterior probability error rate estimation. SAS provides the estimation of these error rates as options in PROC DISCRIM.

EXAMPLE 7 ***Error Rate Estimation, Flea Beetles Data Revisited*** We reconsider the flea beetles data for the discriminant analysis, but this time, between three species, *Chaetocnema concinna* (Π_1), *Chaetocnema heikertingeri* (Π_2) and *Chaetocnema heptapotamica* (Π_3). The samples from the three populations are of respective sizes, $n_1 = 21$, $n_2 = 31$ and $n_3 = 22$. For illustration, we will assume that the prior probabilities are proportional to the sample sizes. Thus, $\pi_1 = 21/(21 + 31 + 22) = 21/74 = 0.2838$, $\pi_2 = 31/74 = 0.4189$, and $\pi_3 = 22/74 = 0.2973$. In SAS, the specification of proportional prior probabilities is achieved by specifying the following statement after the VAR statement,

```
priors prop;
```

Alternatively, we could specify the three prior probabilities calculated above by explicitly listing them in the statement

```
priors 'Pi1' =0.2838 'Pi2' =0.4189 'Pi3' =0.2973;
```

We assume that the costs of misspecification are all equal. In view of the multivariate normality assumption and the acceptance of the null hypothesis of equal variance-covariance matrices for the three species, we may use the linear discriminant function for classification. Hence, **S**, the pooled estimate of the variance-covariance matrix, will be used. This goal is achieved by using the METHOD=NORMAL and POOL = YES options in the PROC DISCRIM statement. The pooled variance-covariance matrix **S** can be printed by specifying PCOV in the same statement. Additionally, the hypothesis $H_0^{(3)}$, which was earlier tested by using PROC GLM can be tested by using the MANOVA option. This facility in turn eliminates the need for another run of PROC GLM. It may be remembered, however, that the use of this option is relevant only in conjunction with METHOD=NORMAL and POOL=YES, as the multivariate analysis of variance has these as the underlying assumptions.

PROC DISCRIM can print many calculations and save a variety of outputs in different data sets. In Program 5.7, which performs the linear discriminant analysis on flea beetles data, we have used the OUT=BEETOUT option to save in a data set named BEETOUT the output on the posterior probabilities $P(j|x)$, $j = 1, \cdots, k$ for the k populations and the classification results for the data being analyzed. These results are obtained by resubstituting the individual observations in the estimated discriminant function. The OUT= option thus eliminates the need for the LIST option, which prints the classifications resulting by resubstitution for each observation. However, it does not provide the estimates of various error rates.

Various statistics such as the sample means, standard deviations, correlation, covariances, etc. can be saved in a data set, say BSTAT, by specifying OUTSTAT=BSTAT. This data set may be useful as an input file for classifying new observations, as it contains all essential information needed for classification and can be read in the SAS DATA steps. For example, in case classification is to be performed on a new data set using the linear discriminant function, the data set BSTAT would need to be specified as TYPE=LINEAR. A discussion of similar applications will be provided in Section 5.5.4.

```
/* Program 5.7 */

options ls = 64 ps=45 nodate nonumber;
title1 'Output 5.7';

data beetles;
infile 'beetle.dat' firstobs=7;
input species x1 x2 ;

/* Here we use LDF for the discriminant analysis on
the three groups of Flea Beetles */
```

```
proc discrim data = beetles out = beetout outstat =bstat
method = normal list pool =yes pcov manova ;
class species ;
priors prop ;
var x1 x2 ;
title2 'Discriminant Analysis of Beetles Data: Three Groups';

data grid;
do x1 = 110 to  160 by .1;
do x2 = 6 to 20 by .05 ;
output;end;end;

title2 h = 1.5 'Linear Discriminant Functions: Beetles Data';
proc discrim data = beetles method = normal pool = yes
           testdata = grid testout = plotc short noclassify;
class species;
var x1 x2;
priors prop;
run;

proc plot data = plotc;
 plot x1*x2 = _into_ ;
title2 h = 1.5 'Linear Discriminant Functions: Beetles Data';
```

Output 5.7

```
                         Output 5.7
         Discriminant Analysis of Beetles Data: Three Groups

                      The DISCRIM Procedure

     Observations       74          DF Total                 73
     Variables           2          DF Within Classes        71
     Classes             3          DF Between Classes        2


                      Class Level Information

          Variable                                               Prior
 species  Name       Frequency     Weight  Proportion  Probability

       1  _1                21    21.0000    0.283784     0.283784
       2  _2                31    31.0000    0.418919     0.418919
       3  _3                22    22.0000    0.297297     0.297297

        Pooled Within-Class Covariance Matrix,      DF = 71

           Variable                x1                  x2

           x1             23.02392262         -0.55961773
           x2             -0.55961773          1.01429299


                 Pooled Covariance Matrix Information

                                 Natural Log of the
                  Covariance     Determinant of the
                  Matrix Rank     Covariance Matrix

                            2               3.13722
```

Pairwise Generalized Squared Distances Between Groups

$$D^2(i|j) = (\bar{X}_i - \bar{X}_j)' \, COV^{-1} \, (\bar{X}_i - \bar{X}_j) - 2 \ln PRIOR_j$$

Generalized Squared Distance to species

From species	1	2	3
1	2.51909	22.00972	22.74968
2	22.78865	1.74016	25.44453
3	22.84272	24.75865	2.42605

Discriminant Analysis of Beetles Data: Three Groups

The DISCRIM Procedure

Linear Discriminant Function

$$\text{Constant} = -.5 \, \bar{X}_j' \, COV^{-1} \, \bar{X}_j + \ln PRIOR_j \qquad \text{Coefficient Vector} = COV^{-1} \, \bar{X}_j$$

Linear Discriminant Function for species

Variable	1	2	3
Constant	-621.00583	-488.15389	-506.83153
x1	6.77817	5.83441	6.33234
x2	17.63635	17.30798	13.44247

Classification Results for Calibration Data: WORK.BEETLES
Resubstitution Results using Linear Discriminant Function

Generalized Squared Distance Function

$$D_j^2(X) = (X-\bar{X}_j)' \, COV^{-1} \, (X-\bar{X}_j) - 2 \ln PRIOR_j$$

Posterior Probability of Membership in Each species

$$Pr(j|X) = \exp(-.5 \, D_j^2(X)) \, / \, SUM_k \exp(-.5 \, D_k^2(X))$$

Posterior Probability of Membership in species

Obs	From species	Classified into species	1	2	3
1	1	1	1.0000	0.0000	0.0000
2	1	1	0.9972	0.0000	0.0028
3	1	1	0.9994	0.0005	0.0002

4	1	1	0.9998	0.0002	0.0000
5	1	1	0.9998	0.0000	0.0002
6	1	1	0.9852	0.0148	0.0000
7	1	1	1.0000	0.0000	0.0000
8	1	1	0.9985	0.0012	0.0002
9	1	1	0.9994	0.0005	0.0002
10	1	1	0.9977	0.0023	0.0000
11	1	1	0.9508	0.0107	0.0385
12	1	1	1.0000	0.0000	0.0000
13	1	1	0.9982	0.0000	0.0018
14	1	1	1.0000	0.0000	0.0000
15	1	1	0.9999	0.0000	0.0000
16	1	1	0.7370	0.2604	0.0027
17	1	2 *	0.1881	0.8119	0.0000
18	1	1	1.0000	0.0000	0.0000
19	1	1	0.9989	0.0000	0.0011
20	1	1	0.9972	0.0000	0.0028
21	1	1	1.0000	0.0000	0.0000
22	2	2	0.0000	1.0000	0.0000
23	2	2	0.0000	1.0000	0.0000
24	2	2	0.0038	0.9959	0.0003
25	2	2	0.0186	0.9814	0.0000
26	2	2	0.0000	1.0000	0.0000
27	2	2	0.0000	1.0000	0.0000
28	2	2	0.0003	0.9997	0.0000
29	2	2	0.0465	0.9535	0.0000
30	2	2	0.0000	0.9999	0.0000
31	2	2	0.0000	0.9999	0.0001
32	2	2	0.0000	0.9997	0.0003
33	2	2	0.0000	1.0000	0.0000
34	2	2	0.0000	1.0000	0.0000
35	2	2	0.0000	1.0000	0.0000
36	2	2	0.0000	1.0000	0.0000
37	2	2	0.0000	0.9999	0.0000
38	2	2	0.0015	0.9983	0.0002
39	2	2	0.0027	0.9825	0.0148
40	2	2	0.0011	0.9899	0.0090
41	2	2	0.0000	0.9868	0.0132
42	2	2	0.0021	0.9979	0.0000
43	2	2	0.0000	1.0000	0.0000
44	2	2	0.0000	0.9999	0.0001
45	2	2	0.0000	1.0000	0.0000
46	2	2	0.0000	1.0000	0.0000
47	2	2	0.0420	0.8979	0.0601
48	2	2	0.0000	1.0000	0.0000
49	2	2	0.0006	0.9993	0.0001
50	2	2	0.0015	0.9983	0.0002
51	2	2	0.0000	0.9992	0.0008
52	2	2	0.0015	0.9983	0.0002
53	3	3	0.0000	0.0000	1.0000
54	3	3	0.0036	0.0002	0.9962
55	3	3	0.0036	0.0002	0.9962
56	3	3	0.0000	0.0004	0.9996
57	3	3	0.0023	0.0003	0.9974
58	3	3	0.0000	0.0000	1.0000
59	3	3	0.0362	0.0633	0.9005
60	3	3	0.0000	0.0459	0.9541
61	3	3	0.0001	0.0000	0.9999
62	3	3	0.0000	0.0000	1.0000

63	3	3	0.0056	0.0001	0.9943
64	3	3	0.0000	0.0000	1.0000
65	3	3	0.0000	0.0000	1.0000
66	3	3	0.0087	0.0001	0.9912
67	3	3	0.0003	0.0000	0.9997
68	3	3	0.0000	0.0000	1.0000
69	3	3	0.0001	0.0000	0.9999
70	3	3	0.0000	0.0000	1.0000
71	3	3	0.0009	0.0009	0.9982
72	3	3	0.0000	0.0000	1.0000
73	3	3	0.0000	0.0000	1.0000
74	3	3	0.0001	0.0000	0.9999

* Misclassified observation

Classification Summary for Calibration Data: WORK.BEETLES
Resubstitution Summary using Linear Discriminant Function

Generalized Squared Distance Function

$$D_j^2(X) = (X-\bar{X}_j)' \, COV^{-1} \, (X-\bar{X}_j) - 2 \ln PRIOR_j$$

Posterior Probability of Membership in Each species

$$Pr(j|X) = \exp(-.5 \, D_j^2(X)) \, / \, SUM_k \exp(-.5 \, D_k^2(X))$$

Number of Observations and Percent Classified into species

From species	1	2	3	Total
1	20	1	0	21
	95.24	4.76	0.00	100.00
2	0	31	0	31
	0.00	100.00	0.00	100.00
3	0	0	22	22
	0.00	0.00	100.00	100.00
Total	20	32	22	74
	27.03	43.24	29.73	100.00
Priors	0.28378	0.41892	0.2973	

```
        Plot of x1*x2.  Symbol is value of _INTO_.

  x1 |
160.0 +333333333333333333311111111111111111111111111111111111111
      |333333333333333333331111111111111111111111111111111111111
156.8 +333333333333333333331111111111111111111111111111111111111
      |333333333333333333333111111111111111111111111111111111111
153.6 +333333333333333333333111111111111111111111111111111111111
      |333333333333333333333311111111111111111111111111111111111
150.4 +333333333333333333333311111111111111111111111111111111111
      |333333333333333333333331111111111111111111111111111111111
147.2 +333333333333333333333333111111111111111111111111111111111
      |333333333333333333333333111111111111111111111111111111111
144.0 +333333333333333333333333311111111111111111111111111111111
      |333333333333333333333333331111111111111111111111111111111
140.8 +333333333333333333333333331111111111111111111111111111111
      |333333333333333333333333333111111111111111111111111111111
137.6 +333333333333333333333333333311111111111111111111111111111
      |333333333333333333333333333222222222222221111111111111111
134.4 +333333333333333333333333332222222222222222222222222222222
      |333333333333333333333333332222222222222222222222222222222
131.2 +333333333333333333333333322222222222222222222222222222222
      |333333333333333333333333222222222222222222222222222222222
128.0 +333333333333333333333332222222222222222222222222222222222
      |333333333333333333333322222222222222222222222222222222222
124.8 +333333333333333333333222222222222222222222222222222222222
      |333333333333333333333222222222222222222222222222222222222
121.6 +333333333333333333332222222222222222222222222222222222222
      |333333333333333333322222222222222222222222222222222222222
118.4 +333333333333333333222222222222222222222222222222222222222
      |333333333333333332222222222222222222222222222222222222222
115.2 +333333333333333332222222222222222222222222222222222222222
      |333333333333333322222222222222222222222222222222222222222
112.0 +333333333333333222222222222222222222222222222222222222222
      |333333333333332222222222222222222222222222222222222222222
108.8 +
      -+-------+-------+-------+-------+-------+-------+-------+
       6       8       10      12      14      16      18      20

                                 x2

NOTE: 138957 obs hidden.
```

The selected parts of the output resulting from a run of Program 5.7 are presented in Output 5.7. The estimated squared distances $d^2(i|j)$ between the populations defined as

$$d^2(i|j) = (\bar{\mathbf{x}}_i - \bar{\mathbf{x}}_j)'\mathbf{S}^{-1}(\bar{\mathbf{x}}_i - \bar{\mathbf{x}}_j) - 2\,ln\,\pi_j$$

are computed first. In a strict sense, *these are not distances in that $d^2(i|i)$ may not be equal to zero.* Also, the distance from i to j may not be the same as that from j to i, and some so-called squared distances can well be negative. However, regardless of this defect in the definition, these quantities do provide a feel for the relative positions of the populations. For instance, as seen in Output 5.7, the species Π_2 and Π_3 seem to be more separated from each other than Π_1 and Π_2 or Π_1 and Π_3. As a result, we may hope for relatively fewer misclassifications between Π_2 and Π_3 than for the other two scenarios.

An observation is classified into the tth population according to the criterion given in Equations 5.6 and 5.7. As in Section 5.3, this criterion is equivalent to classifying an observation into the population with the highest posterior probability. As in the case of discrimination between two populations Π_1 and Π_2 only, this time also the seventeenth observation from Π_1 has been misclassified into Π_2. The reason for such misclassification is clearly evident as the posterior probability $P(\Pi_2|\mathbf{x} = (134, 15)') = 0.8119$ is higher than the other two posterior probabilities given $\mathbf{x} = (134, 15)'$. Apparently, this is the only misclassification that was encountered in the entire data set. Thus, the error count estimate for species Π_1 is

$$\frac{\textit{number of misclassifications out of } \Pi_1}{n_1} = \frac{1}{21} = 0.0476.$$

For the other two species there are no misclassifications and hence the corresponding estimated error count rates are zero. Consequently, the total error count estimate is

$$\sum_{t=1}^{3} \pi_t \hat{ER}(t) = (0.2838)(0.0476) + 0 + 0 = 0.0135\,.$$

What is the linear discriminant function in this case? In other words, for a given observation we may like to compute its score in each of the three populations and then classify it accordingly in the appropriate population. Assuming that the assignment is done in the population for which this score (commonly referred to as the discriminant score) is highest, appropriate linear functions can be obtained using the following simple calculations.

The observation $\mathbf{x}$ is classified into Π_t for which

$$d_t^2(\mathbf{x}) = \min_{j=1,\cdots,k} (d_j^2(\mathbf{x})),$$

that is, for which $d_t^2(\mathbf{x}) \le d_s^2(\mathbf{x})$ for all $s = 1, \cdots, k, s \ne t$. Using Equation 5.7, it simplifies to

classify $\mathbf{x}$ into Π_t if

$$\left(-\frac{1}{2}\bar{\mathbf{x}}_t'\mathbf{S}^{-1}\bar{\mathbf{x}}_t + ln\,\pi_t\right) + \mathbf{x}'\mathbf{S}^{-1}\bar{\mathbf{x}}_t \ge \left(-\frac{1}{2}\bar{\mathbf{x}}_s'\mathbf{S}^{-1}\bar{\mathbf{x}}_s + ln\pi_s\right) + \mathbf{x}'\mathbf{S}^{-1}\bar{\mathbf{x}}_s$$

$$\text{for all } s = 1, \cdots, k, s \ne t$$

or

$$(\mathbf{b}_c^{(t)'} : \mathbf{b}_v^{(t)'}) \begin{pmatrix} 1 \\ \mathbf{x} \end{pmatrix} \ge (\mathbf{b}_c^{(s)'} : \mathbf{b}_v^{(s)'}) \begin{pmatrix} 1 \\ \mathbf{x} \end{pmatrix}$$

or

$$\mathbf{b}^{(t)'}\mathbf{x}^* \ge \mathbf{b}^{(s)'}\mathbf{x}^*, s = 1, \cdots, k, s \ne t$$

where

$$\mathbf{x}^*_{(p+1)\times 1} = \begin{bmatrix} 1 \\ \mathbf{x} \end{bmatrix} \text{ and } \mathbf{b}^{(s)}_{(p+1)\times 1} = \begin{bmatrix} -\frac{1}{2}\bar{\mathbf{x}}_s'\mathbf{S}^{-1}\bar{\mathbf{x}}_s + ln\,\pi_s \\ \mathbf{S}^{-1}\bar{\mathbf{x}}_s \end{bmatrix}, s = 1, \cdots, k.$$

Thus, for each population Π_s, a *discriminant score* $L_s(\mathbf{x}^*) = \mathbf{b}^{(s)'}\mathbf{x}^*$ is computed and the object is assumed to belong to that population, say Π_t, for which this score is the highest.

In SAS, the vectors $\mathbf{b}^{(s)}$ are computed for each $s = 1, \cdots, k$. The quantities $\mathbf{b}_c^{(s)}$ are referred to as the CONSTANT while the vectors $\mathbf{b}_v^{(s)}$ are called the COEFFICIENT VECTORS in the SAS output. For the flea beetles data these are (see Output 5.7)

$$\mathbf{b}^{(1)} = \begin{bmatrix} -621.0058 \\ 6.7782 \\ 17.6364 \end{bmatrix}, \mathbf{b}^{(2)} = \begin{bmatrix} -488.1539 \\ 5.8344 \\ 17.3080 \end{bmatrix} \text{ and } \mathbf{b}^{(3)} = \begin{bmatrix} -506.8315 \\ 6.3323 \\ 13.4425 \end{bmatrix}.$$

An advantage of this information is that for the small data sets of additional observations, classification can be done using a hand-held calculator. This is especially helpful since the computations of $d_j^2(\mathbf{x})$ or $P(j|\mathbf{x})$ would have required an explicit evaluation of $\mathbf{S}^{-1}$. For illustration, let us assume that a new observation $\mathbf{x} = (134,\ 15)'$ (which is the same as the seventeenth observation in $S1$) is to be classified into either $S1$ or $S2$ or $S3$. With $\mathbf{x}^* = (1,\ 134,\ 15)'$, we have

$$\mathbf{b}^{(1)\prime}\mathbf{x}^* = (-621.0058 \quad 6.7782 \quad 17.6364)\begin{pmatrix}1\\134\\15\end{pmatrix} = 551.8190,$$

$$\mathbf{b}^{(2)\prime}\mathbf{x}^* = (-488.1539 \quad 5.8344 \quad 17.3080)\begin{pmatrix}1\\134\\15\end{pmatrix} = 553.2757,$$

$$\mathbf{b}^{(3)\prime}\mathbf{x}^* = (-506.8315 \quad 6.3323 \quad 13.4425)\begin{pmatrix}1\\134\\15\end{pmatrix} = 543.3342.$$

Since the discriminant score corresponding to $S2$, $\mathbf{b}^{(2)\prime}\mathbf{x}^* = 553.2757$ is the largest, this observation will be classified into Π_2.

A two-dimensional plot of three classification regions is also included in Output 5.7. The plot is obtained by first simulating a data set titled GRID and then using it as a test data using the TESTDATA= option in the DISCRIM procedure. The output of this test data set is stored in another data set called PLOTC using the TESTOUT= option. Finally, the data set PLOTC is plotted using PROC PLOT. Quite often, such a plot can be helpful in readily classifying a future observation into one of the several populations. However, such plots are not possible for higher dimensional data sets.

5.5.3 Quadratic Discriminant Function

When the variance-covariance matrices for the populations are not equal, the classification rule about the k populations is more complex even under the multivariate normality assumption. In this case, the discrimination rule depends on individual variance-covariance matrices $\boldsymbol{\Sigma}_1, \cdots, \boldsymbol{\Sigma}_k$, or if the sample data are used, on their estimates $\mathbf{S}_1, \cdots, \mathbf{S}_k$. The classification of an observation $\mathbf{x}$ is done in the population, say t^{th}, for which $D_t^2(\mathbf{x}) = \min_{j=1,\cdots,k}(D_j^2(\mathbf{x}))$, where the squared distances $D_j^2(\mathbf{x})$ are defined (differently from the case of linear discriminant function) as

$$D_j^2(\mathbf{x}) = (\mathbf{x}-\boldsymbol{\mu}_j)'\boldsymbol{\Sigma}_j^{-1}(\mathbf{x}-\boldsymbol{\mu}_j) + ln|\boldsymbol{\Sigma}_j| - 2\,ln\,\pi_j,\ j = 1,\cdots,k \tag{5.9}$$

along with their sample version defined by replacing $\boldsymbol{\mu}_j$ by $\bar{\mathbf{x}}_j$ and $\boldsymbol{\Sigma}_j$ by $\mathbf{S}_j$, as

$$d_j^2(\mathbf{x}) = (\mathbf{x}-\bar{\mathbf{x}}_j)'\mathbf{S}_j^{-1}(\mathbf{x}-\bar{\mathbf{x}}_j) + ln|\mathbf{S}_j| - 2\,ln\,\pi_j. \tag{5.10}$$

Apart from the addition of a constant (so that the distance of a point from itself is zero), $D_j^2(\mathbf{x})$ is indeed a distance function. With $D_j^2(\mathbf{x})$, $j = 1, \cdots, k$ defined in Equation 5.9 the posterior probability that, given $\mathbf{x}$, the object will be classified into j^{th} population is given in terms of $D_j^2(\mathbf{x})$ by Equation 5.8. The estimated posterior probabilities are obtained by replacing $D_j^2(\mathbf{x})$ by its estimate $d_j^2(\mathbf{x})$, $j = 1, \cdots, k$ as defined in Equation 5.10.

In a spirit similar to Equation 5.10, the estimated pairwise generalized squared distance of group i from group j is defined as

$$d^2(i|j) = (\mathbf{x}_i - \mathbf{x}_j)'|\mathbf{S}_j|^{-1}(\mathbf{x}_i - \mathbf{x}_j) + ln\ |\mathbf{S}_j| - 2\ ln\ \pi_j.$$

The quantities $d^2(i|j)$ can take negative values. Further, the estimated generalized square distance of group i from group j is not the same as that of group j from group i. Nonetheless, larger distances do indicate that the corresponding groups are well separated.

EXAMPLE 8 ***Discrimination for Cushing's Syndrome Data*** Cushing's Syndrome is a rare hypersensitive disorder with three common types of syndromes: adenoma (A), bilateralhyperplasia (B) and carcinoma (C). Aitchison and Dunsmore (1975) consider the problem of distinguishing the types on the basis of two measurements on the urinary excretion rates of metabolites tetrahydrocortisone (TETRA) and pregnanetriol (PREG). Patients in each group defined by the CLASS variable TYPE were observed with respective sample sizes as $n_1 = 6, n_2 = 10$ and $n_3 = 5$. To attain the multivariate normality, the logarithms of variables TETRA and PREG, respectively, denoted by LOGTETRA and LOGPREG, are considered for further analysis. The values for these variables are stored in the data set CUSHING. In addition, we have the measurements on a set of six undiagnosed patients stored in the data UNDGNOSD. These six patients are to be appropriately classified into the three populations defined by the types of Cushing's syndrome, according to an appropriate classification rule. We will assume that three populations are each bivariate normal in variables LOGTETRA and LOGPREG and attempt to find an appropriate classification procedure under normal theory using PROC DISCRIM. As in the previous example, we need to specify a CLASS variable that would define the three populations and then specify the list of predictor variables that will be used to construct the appropriate discrimination rule. The CLASS variable here is TYPE, and the variables for analysis are LOGTETRA and LOGPREG, respectively, specified in the VAR statement).

With an assumption of multivariate normality (for data sets so small formal tests of multivariate normality may be meaningless), using POOL=TEST in PROC DISCRIM shows that for this data set the three groups of patients differ in their variance-covariance matrices. From Output 5.8, which resulted from running Program 5.8, the chi-square test statistic with $6df$ is equal to 19.2405 and results in a very small p value of 0.0038. Thus, it is reasonable to use quadratic discriminant analysis, assuming that the populations are distinguishable from each other with respect to their population means (that this is a reasonable assumption can be intuitively determined by looking at the sample means $\begin{bmatrix} 1.0433 \\ -0.6034 \end{bmatrix}$, $\begin{bmatrix} 2.0073 \\ -0.2060 \end{bmatrix}$ and $\begin{bmatrix} 2.7097 \\ 1.5998 \end{bmatrix}$, respectively).

Since the p value ($= 0.0038$) is smaller than the level of significance (specified by default)$= 0.10$, SAS would choose the quadratic discriminant function for further analysis. Hence, the sample estimates of $\mathbf{\Sigma}_1$, $\mathbf{\Sigma}_2$, and $\mathbf{\Sigma}_3$, namely, $\mathbf{S}_1$, $\mathbf{S}_2$, and $\mathbf{S}_3$, would be used. We could use any other significance level by specifying it in the SLPOOL= option. For example, if we want the level of significance $\alpha = 0.001$, we can use the SLPOOL= 0.001 option in the PROC DISCRIM statement. Under this value of α, the decision would have been to pool the three sample variance-covariance matrices and use the linear discriminant function. Further, if we want to use the quadratic discriminant function regardless of the acceptance or rejection of the null hypothesis of equality of variance-covariance matrices, we can do it by specifying POOL=NO as an option (along with METHOD=NORMAL) in the PROC DISCRIM statement. Of course, under this choice no test of the equality of variance-covariance matrices will be performed.

```
/* Program 5.8 */

options ls = 64 ps=45 nodate nonumber;
title1  'Output 5.8';

data cushing;
infile 'cushing.dat' firstobs = 7;
input type $ tetra preg ;
logtetra = log(tetra);logpreg =log(preg);

data undgnosd;
input tetra preg ;
logtetra = log(tetra);logpreg =log(preg);
lines;
```

```
5.1 0.4
12.9 5.0
13.0 0.8
2.6 0.1
30.0 0.1
20.5 0.8
;
/*
Data from Aitchison and Dunsmore (1975),
reprinted with permission from Cambridge University Press
*/

proc discrim data = Cushing pool=test out = cushout
outstat = cushstat method = normal testdata = undgnosd
list testlist wcov;
class type ;
priors prop ;
var logtetra logpreg ;
title2 'Cushing Syndrome Data: Quadratic Discrimination';

data grid;
do logtetra = 0 to log(56) by .1;
do logpreg = log(.1) to log(18) by .05 ;
output;end;end;

title2 h = 1.5 'Quadratic Discriminant Functions: Cushing Data';
proc discrim data = cushing method = normal pool = no
testdata = grid testout = plotc short noclassify;
class type;
var logtetra logpreg;
priors prop;
run;

proc plot data = plotc;
plot logtetra * logpreg = _into_ ;
run;

title2 h = 1.5 'Linear Discriminant Functions: Cushing Data';
proc discrim data = cushing method = normal pool = yes
testdata = grid testout = plotclin short noclassify;
class type;
var logtetra logpreg;
priors prop;
run;

proc plot data = plotclin;
plot logtetra * logpreg = _into_ ;
```

Output 5.8

```
                         Output 5.8
           Cushing Syndrome Data: Quadratic Discrimination

                     The DISCRIM Procedure

                    Class Level Information

        Variable                                              Prior
 type   Name        Frequency     Weight   Proportion   Probability

 a      a                   6     6.0000     0.285714      0.285714
 b      b                  10    10.0000     0.476190      0.476190
 c      c                   5     5.0000     0.238095      0.238095

               Within-Class Covariance Matrices

                     type = a,      DF = 5

            Variable          logtetra             logpreg

            logtetra       0.110684524         0.123890550
            logpreg        0.123890550         4.089097740

-----------------------------------------------------------------


                     type = b,      DF = 9

            Variable          logtetra             logpreg

            logtetra      0.2118714934        0.3241321837
            logpreg       0.3241321837        0.7202960925

-----------------------------------------------------------------


                     type = c,      DF = 4

            Variable          logtetra             logpreg

            logtetra      0.5552159646        -.2422352228
            logpreg       -.2422352228        0.2885006407

-----------------------------------------------------------------


        Test of Homogeneity of Within Covariance Matrices


              Chi-Square         DF     Pr > ChiSq

               19.240469          6         0.0038

Since the Chi-Square value is significant at the 0.1 level, the
within covariance matrices will be used in the discriminant func
tion.
Reference: Morrison, D.F. (1976) Multivariate Statistical Method
s p252.
```

Classification Results for Calibration Data: WORK.CUSHING
Resubstitution Results using Quadratic Discriminant Function

Generalized Squared Distance Function

$$D_j^2(X) = (X-\bar{X}_j)' \, COV_j^{-1} \, (X-\bar{X}_j) + \ln |COV_j| - 2 \ln PRIOR_j$$

Posterior Probability of Membership in Each type

$$Pr(j|X) = \exp(-.5 \, D_j^2(X)) \; / \; SUM_k \exp(-.5 \, D_k^2(X))$$

Posterior Probability of Membership in type

Obs	From type	Classified into type		a	b	c
1	a	a		0.6521	0.0000	0.3479
2	a	a		0.9996	0.0004	0.0000
3	a	a		0.8646	0.1354	0.0000
4	a	a		0.9988	0.0012	0.0000
5	a	a		0.8675	0.1325	0.0000
6	a	a		0.9981	0.0019	0.0000
7	b	b		0.0011	0.9989	0.0000
8	b	b		0.2995	0.7005	0.0000
9	b	b		0.4857	0.5143	0.0000
10	b	b		0.0024	0.9975	0.0001
11	b	b		0.0009	0.9991	0.0000
12	b	c	*	0.0000	0.4433	0.5567
13	b	b		0.0052	0.9926	0.0022
14	b	b		0.0133	0.9867	0.0000
15	b	b		0.0267	0.9733	0.0000
16	b	b		0.0000	0.9792	0.0208
17	c	c		0.0004	0.0103	0.9894
18	c	c		0.0013	0.0008	0.9979
19	c	b	*	0.0012	0.5745	0.4244
20	c	c		0.0000	0.0000	1.0000
21	c	c		0.0000	0.0895	0.9105

* Misclassified observation

Classification Summary for Calibration Data: WORK.CUSHING
Resubstitution Summary using Quadratic Discriminant Function

Number of Observations and Percent Classified into type

From type	a	b	c	Total
a	6	0	0	6
	100.00	0.00	0.00	100.00

```
b                0          9          1         10
              0.00      90.00      10.00     100.00

c                0          1          4          5
              0.00      20.00      80.00     100.00

Total            6         10          5         21
             28.57      47.62      23.81     100.00

Priors     0.28571    0.47619     0.2381


              Error Count Estimates for type

                      a          b          c      Total

Rate             0.0000     0.1000     0.2000     0.0952
Priors           0.2857     0.4762     0.2381

        Classification Results for Test Data: WORK.UNDGNOSD

          Posterior Probability of Membership in type

             Classified
      Obs    into type         a          b          c

        1    b            0.0515     0.9485     0.0000
        2    c            0.0000     0.1511     0.8489
        3    b            0.0001     0.9997     0.0002
        4    a            0.7614     0.2386     0.0000
        5    a            0.9996     0.0000     0.0004
        6    b            0.0000     0.7413     0.2587
```

```
              Quadratic Discriminant Functions: Cushing Data

          Plot of logtetra*logpreg.  Symbol is value of _INTO_.

        logtetra |
                 |
               4 +  cccccccccccccccccccccccccccccccccbb
                 |  cccccccccccccccccccccccccccccccccbb
                 |  aaacccccccccccccccccccccccccccccbb
                 |  aaaaaccccccccccccccccccccccccccccbb
                 |  aaaaaabbbbbbccccccccccccccccccccccc
               3 +  aaaaaabbbbbbbbbbbbbbbbbcccccccccccc
                 |  aaaaabbbbbbbbbbbbbbbbbbcccccccccccc
                 |  aaaaabbbbbbbbbbbbbbbbbbcccccccccccc
                 |  aaaabbbbbbbbbbbbbbbbbbbcccccccccccc
                 |  aaaabbbbbbbbbbbbbbbbbbccccccccccccc
               2 +  aaabbbbbbbbbbbbbbbbbbbccccccccccccc
                 |  aaabbbbbbbbbbbbbbbbaaaacccccccccccc
                 |  aaabbbbbbbbbbbbbaaaaaaaaaaacccccccc
                 |  aaabbbbbbbbaaaaaaaaaaaaaaaaaaaaaaaa
                 |  aaaaaaaaaaaaaaaaaaaaaaaaaaaaaaaaaaa
               1 +  aaaaaaaaaaaaaaaaaaaaaaaaaaaaaaaaaaa
                 |  aaaaaaaaaaaaaaaaaaaaaaaaaaaaaaaaaaa
                 |  aaaaaaaaaaaaaaaaaaaaaaaaaaaaaaaaaaa
                 |  aaaaaaaaaaaaaaaaaaaaaaaaaaaaaaaaaac
                 |  aaaaaaaaaaaaaaaaaaaaaaaaaaaaaaaaccc
               0 +  aaaaaaaaaaaaaaaaaaaaaaaaaaaaaaacccc
                 |
                 ---+---------+---------+---------+---------+--
                 -2.303    -0.803     0.697     2.197     3.697

                                   logpreg

NOTE: 3529 obs hidden.
```

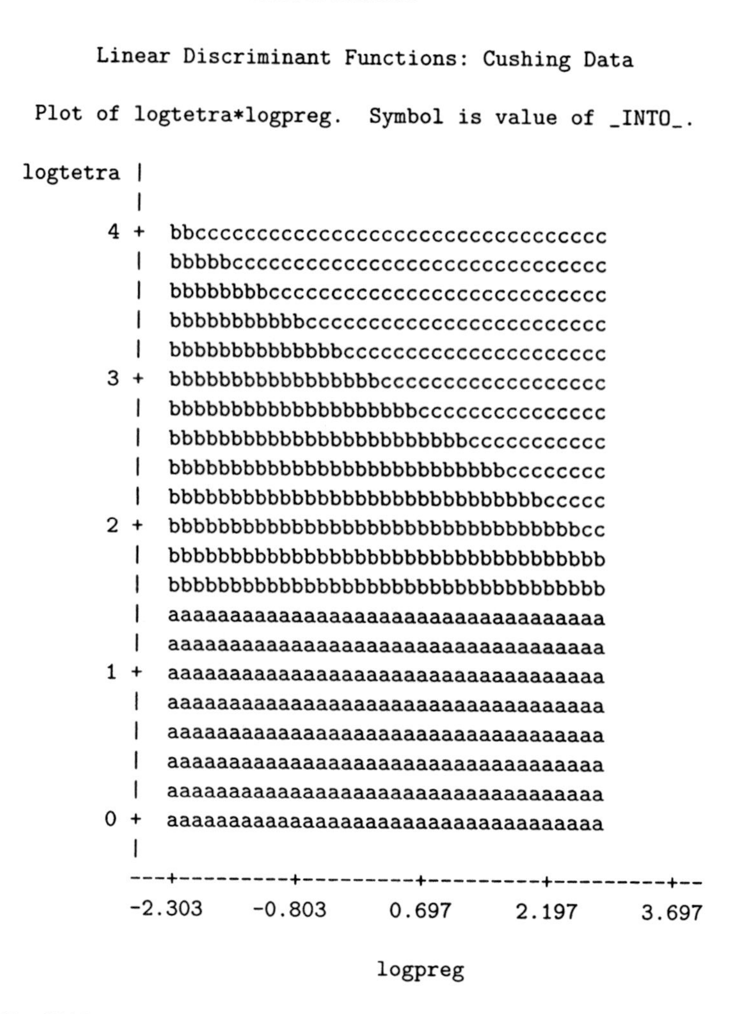

From Output 5.8, the three sample variance-covariance matrices are

$$\mathbf{S}_1 = \begin{bmatrix} 0.1107 & 0.1239 \\ 0.1239 & 4.0891 \end{bmatrix}, \mathbf{S}_2 = \begin{bmatrix} 0.2119 & 0.3241 \\ 0.3241 & 0.7203 \end{bmatrix},$$

and

$$\mathbf{S}_3 = \begin{bmatrix} 0.5552 & -.2422 \\ -.2422 & 0.2885 \end{bmatrix}.$$

Let us assume that the prior probabilities for the three groups of patients are proportional. This is a reasonable assumption for the data, which are observational in nature, as we hope that the number of cases in the respective samples are proportional to their actual relative frequencies in the populations. With the assumption that the more prevalent type of Cushing's Syndrome would have a higher number of cases observed in the sample, we choose the prior probabilities as

$$\pi_1 = \frac{n_1}{n_1 + n_2 + n_3} = \frac{6}{6 + 10 + 5} = 0.2857$$

and similarly

$$\pi_2 = 0.4762 \text{ and } \pi_3 = 0.2381.$$

In SAS, this specification can be done either explicitly as

```
priors 'a' =0.2857 'b' =0.4762 'c' =0.2381;
```

or implicitly by stating that the PRIORS are proportional as in

```
priors prop;
```

The matrix of estimated-pairwise-generalized squared distances between the groups indicates that the group of patients with TYPE A syndrome is well separated from those with TYPE B and TYPE C syndromes. The patients are assigned to the population with highest posterior probability given their observed values of LOGTETRA and LOGPREG. Apparently two patients out of a total of 21 have been misclassified. The sixth patient in the group of patients with TYPE B Cushing's Syndrome and the third patient in the group with TYPE C Syndrome are respectively misclassified as falling in group C, (P(C|TETRA= 15.4, PREG= 3.60) = 0.5567), and in group B, (P(B|TETRA= 9.6, PREG= 3.10) = 0.5745), respectively. Thus, the error count estimates for the three groups are 0, 0.10, and 0.20, respectively. This results in the total error count estimate as $0 \cdot (\pi_1) + 0.1(\pi_2) + 0.2(\pi_3) \cong 0 + 0.0476 + 0.0476 = 0.0952$. In other words, approximately an (under)estimated 10% of the patients will be misclassified. Of course, we may note that the discriminant functions as well as the above misclassification rate are obtained from the three samples that are very small, and only so much can be expected from small samples.

Aitchison and Dunsmore present measurements on six additional patients whose memberships in the three groups are yet to be determined. The discrimination criterion obtained above can be used to classify these patients into one of the three groups. In Program 5.8, measurements (and their natural logarithms) on these patients are stored in the data set UNDGNOSD. To classify these patients using the discrimination criterion obtained from the set of diagnosed patients (whose measurements are given in the data set CUSHING), we use the data set UNDGNOSD as the test data by using the option TESTDATA by specifying

```
proc discrim data=cushing testdata=undgnosd;
```

in the PROC DISCRIM statement. Of course, other relevant options and further statements as described earlier and presented in Program 5.8 are also to be included.

Upon classification, the patients numbered 4 and 5 are classified as having TYPE A Cushing Syndrome. Patients 1, 3 and 6 appear to have TYPE B, while patient 2 is classified as having TYPE C Syndrome. A look at the posterior probabilities indicates that these are considerably higher for the assigned type than those corresponding to the other types for the respective patients, and hence provide a considerable degree of confidence on the classification.

Proportion of total classifications into the respective groups viz. 0.3333, 0.5000, and 0.1667, are relatively close (given that the data set has only six patients) to their respective prior probabilities, 0.2857, 0.4762, and 0.2381. For a random sample taken from the appropriate mixture of these three populations, we would expect that the two sets of numbers are comparable to each other.

The regions defined by the discriminant procedure are also plotted in Output 5.8. The SAS code is essentially similar to that used for plotting the regions in the case of linear discriminant function. It is easy to observe that the boundaries are considerably nonlinear.

What are the practical consequences of using the linear discriminant function when we should really use the quadratic discriminant function (that is, when we pool the sample variance-covariance matrices even though such a pooling is clearly inappropriate)? The answer is obviously more misclassifications. We may verify that in this case 6 of the total 21 patients in the data set CUSHING are misclassified even though the very same data set was used to arrive at the discrimination criterion. The SAS statements for this analysis and the plot of the corresponding classification regions are included in Program 5.8. While the numerical output has been suppressed, for comparison's sake, we present a plot of the classification regions defined by linear discriminant functions. We can easily observe the obvious differences between the regions defined by this method and those defined by the quadratic discriminant function.

Cost considerations. In all the analyses so far, we have assumed that the costs of misclassification are all equal. However, various misclassifications may have different cost consequences. For example, classifying a person with a certain disease as disease-free has considerably more serious consequences of denying him or her treatment than classifying a disease-free person as carrying the disease. The classification rule given in Equation 5.6 can be used for classification provided that for all $t, s = 1, \cdots, k$, the cost of misclassifying an observation into population t when it actually comes from population s, $c(t|s)(\geq 0)$ is known. By definition, $c(t|t) = 0$ for all $t = 1, \cdots, k$.

PROC DISCRIM does not have any options to allow for different costs of misclassifications. However, at least for the two-population case ($k = 2$), it can be indirectly manipulated. The two costs of misclassifications, in this case, are $c(2|1)$ and $c(1|2)$, and the classification rule in Equation 5.6 becomes

$$\text{classify } \mathbf{x} \text{ into } \Pi_1 \text{ if } \pi_2 f_2(x) c(1|2) \leq \pi_1 f_1 c(2|1)$$

and into Π_2, otherwise. If we call $\psi_1 = \frac{\pi_1 c(2|1)}{\pi_1 c(2|1)+\pi_2 c(1|2)}$ and $\psi_2 = \frac{\pi_2 c(1|2)}{\pi_1 c(2|1)+\pi_2 c(1|2)}$, then $\psi_1 + \psi_2 = 1$ and the classification rule given above can be restated as

$$\text{classify } \mathbf{x} \text{ into } \Pi_1 \text{ if } \psi_2 f_2(\mathbf{x}) \leq \psi_1 f_1(\mathbf{x})$$

and into Π_2 otherwise. In other words, instead of π_1 and π_2, we would use ψ_1 and ψ_2 as the corresponding prior probabilities, which also account for costs, and specify these in the PRIORS statement of the DISCRIM procedure. The quantities ψ_1 and ψ_2 can be viewed as the cost-adjusted priors. Of course, we may then need to remember that the discriminant analysis has been (implicitly) performed under the minimum total cost criterion and hence various error rates (which depend on costs through ψ_1 and ψ_2) printed in the output, are really not the error rates. They may be interpreted as the measure of costs incurred in the two types of misclassifications.

EXAMPLE 9 ***Cost Considerations, Explosions vs. Earthquakes Data*** Shumway (1988) considers the problem of correctly classifying nuclear explosions and earthquakes using two seismological features, namely, body wave magnitude (BODY) and surface wave magnitude (SURFACE). Twenty observations for earthquakes (Π_1) and nine for nuclear explosions (Π_2) are available as the samples. We take the prior probabilities $\pi_1 = 0.8$ and $\pi_2 = 0.2$. Since the interest is in detecting an explosion, presumably identified as a violation of an underground nuclear test ban treaty, $P(2|1)$ can be identified as a false alarm probability, whereas $1 - P(1|2)$ is a measure of explosion detection probability. With an argument for being able to correctly detect an explosion more often than correctly being able to detect an earthquake, we may assign a higher cost for making an error in detecting an explosion. For illustration, let the cost of misclassifying a nuclear explosion as an earthquake $c(1|2)$ be three times that of $c(2|1)$ (we have no idea if this ratio is a reasonable one, and it is taken only to make a point). Thus, we may take $c(1|2) = 3$ and $c(2|1) = 1$ and hence

$$\begin{aligned}
\psi_1 &= \frac{c(2|1)\pi_1}{c(2|1)\pi_1 + c(1|2)\pi_2} = \frac{.8}{.8+.6} = 0.5714 \\
\psi_2 &= \frac{c(1|2)\pi_2}{c(2|1)\pi_1 + c(1|2)\pi_2} = \frac{.6}{.8+.6} = 0.4286 .
\end{aligned}$$

Assuming bivariate normality, an application of POOL=TEST in PROC DISCRIM suggests that the variance-covariance matrices of BODY and SURFACE in the two populations may not be equal and hence a quadratic discriminant analysis is appropriate. The corresponding SAS statements to perform the above test and then to obtain the classification rule are given by

```
proc discrim data=shumway method = normal pool = test list;
class popn;
var body surface;
priors 'equake' = 0.5714 'explosn' = 0.4286;
```

Note that in the PRIOR statement, we have specified the respective values of the cost-adjusted priors, ψ_1 and ψ_2. The complete code along with data is presented in Program 5.9. The corresponding selected output is included as Output 5.9.

```
/* Program 5.9 */

options ls = 64 ps=45 nodate nonumber;
title1  'Output 5.9';

data Shumway;
infile 'earthquake.dat' firstobs = 7;
input popn $  body surface;
if popn = 'equake' then populatn = 'q';
if popn = 'explosn' then populatn = 'x';

proc discrim data = shumway method = normal pool =test wcov list;
class popn ;
var body surface;
priors 'equake' = .5714 'explosn' = .4286 ;
title2 "Discriminant Analysis with Cost Considerations";
title3 "Earth Quakes vs. Nuclear Explosions";

proc sort data = shumway ; by populatn;

data grid;
do body = 5 to 10 by .1;
do surface = 3 to 6 by .1 ;
output;end;end;

title2 h = 1.5 'Quadratic Discriminant Functions';
title3 h = 1.5 'Earth Quakes vs. Nuclear Explosions';

proc discrim data = shumway method = normal pool = no
testdata = grid testout = plotc short noclassify;
class populatn;
var body surface;
priors 'q' = .5714 'x' = .4286 ;
run;

proc plot data = plotc;
plot surface * body = _into_ ;
title2 h = 1.5 'Plot of Quadratic Discriminant Regions';
title3 h = 1.5 'Earth Quakes vs. Nuclear Explosions';
```

Output 5.9

```
                         Output 5.9
          Discriminant Analysis with Cost Considerations
               Earth Quakes vs. Nuclear Explosions

        Test of Homogeneity of Within Covariance Matrices


               Chi-Square        DF    Pr > ChiSq

                13.692277         3        0.0034

Since the Chi-Square value is significant at the 0.1 level, the
within covariance matrices will be used in the discriminant func
tion.
Reference: Morrison, D.F. (1976) Multivariate Statistical Method
s p252.

     Classification Results for Calibration Data: WORK.SHUMWAY
   Resubstitution Results using Quadratic Discriminant Function


           Posterior Probability of Membership in popn

              From       Classified
        Obs   popn       into popn        equake      explosn

          1   equake     explosn   *      0.1655       0.8345
          2   equake     equake           0.9752       0.0248
          3   equake     equake           1.0000       0.0000
          4   equake     equake           0.9997       0.0003
          5   equake     equake           1.0000       0.0000
          6   equake     equake           0.9999       0.0001
          7   equake     equake           1.0000       0.0000
          8   equake     equake           1.0000       0.0000
          9   equake     equake           1.0000       0.0000
         10   equake     equake           1.0000       0.0000
         11   equake     equake           0.9998       0.0002
         12   equake     equake           1.0000       0.0000
         13   equake     equake           0.9993       0.0007
         14   equake     equake           0.9620       0.0380
         15   equake     equake           1.0000       0.0000
         16   equake     equake           1.0000       0.0000
         17   equake     equake           1.0000       0.0000
         18   equake     equake           1.0000       0.0000
         19   equake     equake           1.0000       0.0000
         20   equake     equake           1.0000       0.0000
         21   explosn    explosn          0.0004       0.9996
         22   explosn    explosn          0.0022       0.9978
         23   explosn    explosn          0.0102       0.9898
         24   explosn    explosn          0.0038       0.9962
         25   explosn    explosn          0.0021       0.9979
         26   explosn    explosn          0.0000       1.0000
         27   explosn    explosn          0.0133       0.9867
         28   explosn    explosn          0.0004       0.9996
         29   explosn    explosn          0.0002       0.9998

                  * Misclassified observation
```

```
                  Plot of Quadratic Discriminant Regions
                   Earth Quakes vs. Nuclear Explosions

          Plot of surface*body.  Symbol is value of _INTO_.

surface |
    6.0 +  qqqqqqqqqqqqqqqqqqqqqqqqqqqqqqqqqqqqqqqqqqqqqqqqqqq
    5.9 +  qqqqqqqqqqqqqqqqqqqqqqqqqqqqqqqqqqqqqqqqqqqqqqqqqqq
    5.8 +  qqqqqqqqqqqqqqqqqqqqqqqqqqqqqqqqqqqqqqqqqqqqqqqqqqq
    5.7 +  qqqqqqqqqqqqqqqqqqqqqqqqqqqqqqqqqqqqqqqqqqqqqqqqqqq
    5.6 +  qqqqqqqqqqqqqqqqqqqqqqqqqqqqqqqqqqqqqqqqqqqqqqqqqqq
    5.5 +  qqqqqqqqqqqqqqqqqqqqqqqqqqxxxxxxxxxxxxqqqqqqqqqqqqq
    5.4 +  qqqqqqqqqqqqqqqqqqqqqqqxxxxxxxxxxxxxxxxxqqqqqqqqqqq
    5.3 +  qqqqqqqqqqqqqqqqqqqqxxxxxxxxxxxxxxxxxxxxqqqqqqqqqqq
    5.2 +  qqqqqqqqqqqqqqqqqqxxxxxxxxxxxxxxxxxxxxxxqqqqqqqqqqq
    5.1 +  qqqqqqqqqqqqqqqqxxxxxxxxxxxxxxxxxxxxxxxxqqqqqqqqqqq
    5.0 +  qqqqqqqqqqqqqqxxxxxxxxxxxxxxxxxxxxxxxxxxqqqqqqqqqqq
    4.9 +  qqqqqqqqqqqqqxxxxxxxxxxxxxxxxxxxxxxxxxxqqqqqqqqqqqq
    4.8 +  qqqqqqqqqqqxxxxxxxxxxxxxxxxxxxxxxxxxxxxqqqqqqqqqqqq
    4.7 +  qqqqqqqqqqxxxxxxxxxxxxxxxxxxxxxxxxxxxxqqqqqqqqqqqqq
    4.6 +  qqqqqqqqqxxxxxxxxxxxxxxxxxxxxxxxxxxxxqqqqqqqqqqqqqq
    4.5 +  qqqqqqqqxxxxxxxxxxxxxxxxxxxxxxxxxxxxqqqqqqqqqqqqqqq
    4.4 +  qqqqqqqxxxxxxxxxxxxxxxxxxxxxxxxxxxxqqqqqqqqqqqqqqqq
    4.3 +  qqqqqqxxxxxxxxxxxxxxxxxxxxxxxxxxxqqqqqqqqqqqqqqqqqq
    4.2 +  qqqqqxxxxxxxxxxxxxxxxxxxxxxxxxxxqqqqqqqqqqqqqqqqqqq
    4.1 +  qqqqqxxxxxxxxxxxxxxxxxxxxxxxxxqqqqqqqqqqqqqqqqqqqqq
    4.0 +  qqqqxxxxxxxxxxxxxxxxxxxxxxxxxqqqqqqqqqqqqqqqqqqqqqq
    3.9 +  qqqqxxxxxxxxxxxxxxxxxxxxxxxqqqqqqqqqqqqqqqqqqqqqqqq
    3.8 +  qqqqxxxxxxxxxxxxxxxxxxxxxqqqqqqqqqqqqqqqqqqqqqqqqqq
    3.7 +  qqqqqxxxxxxxxxxxxxxxxxqqqqqqqqqqqqqqqqqqqqqqqqqqqqq
    3.6 +  qqqqqqxxxxxxxxxxxxxqqqqqqqqqqqqqqqqqqqqqqqqqqqqqqqq
    3.5 +  qqqqqqqqqxxxxxqqqqqqqqqqqqqqqqqqqqqqqqqqqqqqqqqqqqq
    3.4 +  qqqqqqqqqqqqqqqqqqqqqqqqqqqqqqqqqqqqqqqqqqqqqqqqqqq
    3.3 +  qqqqqqqqqqqqqqqqqqqqqqqqqqqqqqqqqqqqqqqqqqqqqqqqqqq
    3.2 +  qqqqqqqqqqqqqqqqqqqqqqqqqqqqqqqqqqqqqqqqqqqqqqqqqqq
    3.1 +  qqqqqqqqqqqqqqqqqqqqqqqqqqqqqqqqqqqqqqqqqqqqqqqqqqq
    3.0 +  qqqqqqqqqqqqqqqqqqqqqqqqqqqqqqqqqqqqqqqqqqqqqqqqqqq
        |
        ---+---------+---------+---------+---------+---------+--
           5         6         7         8         9        10

                                  body
```

As shown in Output 5.9, the hypothesis of the equality of two variance-covariance matrices (obtained by using the POOL=TEST option) can be rejected. The corresponding chi-square test statistic with 3 df results in a value of 13.6923 and yields a p value of 0.0034. In view of this, a quadratic discriminant analysis will be more appropriate. Recognizing this at the default level of significance of 0.10, SAS automatically makes the choice of quadratic discriminant analysis.

The classification results indicate only one misclassification. The first observation from the sample of earthquake data has been classified as a nuclear explosion. None of the nuclear explosions have been misidentified as earthquakes. Of course, for this example, the estimates of error rates as reported in the SAS program have no meaning since $\psi_1 = 0.5714$ and $\psi_2 = 0.4286$ are really not the prior probabilities.

A plot presenting the two classification regions (q = Earthquake and x = Explosion) is also shown as part of Output 5.9.

5.5.4 When Raw Data Are Not Available

There are situations when the raw data on the samples from various populations are not available. The appropriate discriminant rule using SAS can still be computed if the descriptive summary statistics such as sample sizes, sample means, and sample variance-covariance matrices are available for each of these populations. This discriminant rule can still be applied to any test data (the TESTDATA= option) for classification of objects into various populations. However, since the original raw data are unavailable, no estimates of error rates are possible. This approach can best be illustrated using an example.

EXAMPLE 10 *Using Descriptive Statistics, Analysis of Color Tolerances* Numerical color tolerances provide a quantitative basis for better color matches between the assembled components of a colored product. In an automotive color tolerance study, Vance (1983, 1989) attempts to develop a numerical color difference tolerance region to determine if a color sample is acceptable as compared to a given standard. For each sample, a number of visual judgments by a number of observers were obtained. The observers were asked to classify each sample as acceptable (GRP1) or unacceptable (GRP2). Thus, the two populations under consideration are GRP1 and GRP2, respectively.

To measure the color differences numerically, numerical readings on color differences in three color components termed L, A, and B were measured, using a spectrophotometer, for each of the samples. Specifically, component L represents the measurement on the light-dark axis, A represents that on the red-green direction, and B represents the measurement on the yellow-blue axis. Thus, the vector $\mathbf{x} = (L, A, B)'$ forms our vector of discriminatory variables. Multivariate normality of $\mathbf{x}$ in each of the two populations was assumed.

The number of the visually accepted (GRP1) units was $n_1 = 40$ and, based on these samples, the following sample mean vector and sample variance-covariance matrix were computed,

$$\bar{\mathbf{x}}_1 = \begin{bmatrix} -0.146 \\ -0.543 \\ -0.073 \end{bmatrix}, \quad \mathbf{S}_1 = \begin{bmatrix} 0.102 & 0.018 & -0.029 \\ 0.018 & 0.229 & 0.039 \\ -0.029 & 0.039 & 0.093 \end{bmatrix}.$$

The number of units that were visually unacceptable (GRP2) was $n_2 = 71$. These samples yielded the following descriptive statistics:

$$\bar{\mathbf{x}}_2 = \begin{bmatrix} -0.248 \\ -0.622 \\ -0.004 \end{bmatrix}, \quad \mathbf{S}_2 = \begin{bmatrix} 0.337 & 0.338 & 0.034 \\ 0.338 & 0.879 & 0.077 \\ 0.034 & 0.077 & 0.271 \end{bmatrix}.$$

Given this summary data, we want to obtain an appropriate classification rule to classify a sample either into GRP1 (that is, as acceptable) or into GRP2 (as unacceptable).

Since raw data are unavailable and various summary variables are of different data types, we need to present this information carefully in the DATA step by clearly specifying the type of the particular data entity.

For illustration, we will present the information for GRP1 here. For variables in vector $\mathbf{x}$, that is L, A, and B, the sample variance-covariance matrix is to be specified as TYPE='COV', and the respective sample means as TYPE='MEAN'. The common sample sizes for L, A, and B will be specified as TYPE = 'N'. The counting variable _N_ will be used to indicate the exact locations for these quantities. Specifically, for GRP1, the DATA steps given below will read the necessary quantities into a data set called V_ACCEP. This data set was defined as TYPE=COV. Such a specification will be needed later when we perform discriminant analysis using PROC DISCRIM.

```
data v_accep(type=cov);
infile cards missover;
input l a b;
length _name_ _type_ $ 8.;
_type_ = 'cov';
if _n_ = 1 then _name_ = 'l';
if _n_ = 2 then _name_ = 'a';
if _n_ = 3 then _name_ = 'b';
if _n_ = 4 then _type_='n';
 else if _n_ = 5 then _type_ ='mean';
 else _type_ ='cov';
lines;
```

The data set V_UNACCE is created in a similar way. For the analysis, we need to present all the information on GRP1 and GRP2 as a single data set. A single data set named CONCAT is therefore created by concatenating these two data sets one below the other, using the following SAS statements:

```
data concat; set v_accep(in = gp1) v_unacce(in = gp2);
if gp1 then group = 'grp1';
else group = 'grp2';
```

In order to combine the two data sets, V_ACCEP and V_UNACCE, in the format we want, we use the IN= option in the SET statement. For each of the two data sets it creates and names a variable that indicates whether the particular data set has contributed data to the current observation. Within the DATA step, the value of the respective variable is 1 if the data set contributed to the current observation, and 0, otherwise. Using these two variables, GP1 and GP2, respectively, we have created a new class variable called GROUP by taking the value GRP1 for the values of descriptive statistics coming from the data set V_ACCEP and GRP2 for those coming from data set V_UNACCE.

We now perform the discriminant analysis on DATA=CONCAT. However, since sample variance-covariance matrices (and not the raw data) are being supplied, we need to specify this by using the DATA=CONCAT(TYPE=COV) option. If the correlation matrices instead of the variance-covariance matrices were supplied, we would have correspondingly used the DATA = CONCAT (TYPE=CORR) option. Assuming equal prior probabilities and specifying so by the PRIORS EQUAL statement, we allow the program to choose the appropriate analysis by specifying the POOL=TEST option. The computed statistical quantities are stored in the data set C_STAT, obtained by using the OUTSTAT=C_STAT option. The corresponding SAS statements are given below.

```
proc discrim data=concat (type=cov) method=normal pool=no
outstat = c_stat;
class group ;
priors equal ;
var l a b ;
```

The entire SAS program comprising all the steps of data manipulations and statistical analysis is systematically presented as Program 5.10. The selected SAS output is reported as Output 5.10, which will be discussed now.

```
/* Program 5.10 */

options ls = 64 ps=45 nodate nonumber;
title1  'Output 5.10';

data v_accep(type=cov) ;
infile cards missover;
```

```
input l a b ;
length _name_ _type_ $  8.;
_type_ = 'cov' ;
if _n_ = 1 then _name_ = 'l' ;
if _n_ = 2 then _name_ = 'a' ;
if _n_ = 3 then _name_ = 'b' ;
if _n_=4 then _type_='n';
else if _n_=5 then _type_='mean'; else _type_='cov';
lines;
.102
.018 .229
-.029 .039 .093
40 40 40
-.146 -.543 -.073
;

data v_unacce(type=cov) ;
infile cards missover;
input l a b ;
length _name_ _type_ $  8.;
_type_ = 'cov' ;
if _n_ = 1 then _name_ = 'l' ;
if _n_ = 2 then _name_ = 'a' ;
if _n_ = 3 then _name_ = 'b' ;
if _n_=4 then _type_='n';
 else if _n_=5 then _type_='mean'; else _type_='cov';
lines;
.337
.338 .879
.034 .077 .271
71 71 71
-.248 -.622 -.004
;
/*
Data from Vance (1983),
printed with permission from Modern Paint and Coatings,
Intertec Publishing Corp., Atlanta.
*/

data concat; set v_accep(in = gp1) v_unacce(in = gp2);
if gp1 then group = 'grp1';
else group = 'grp2';

proc print data=concat;
title2 'Summary Statistics on both Groups';

proc discrim data=concat(type=cov) method=normal pool=test
outstat = c_stat;
class group ;
priors equal ;
var l a b  ;
title2 'Discriminant Analysis for Numerical Color Tolerances';

proc print data = c_stat;
title2 'Coefficients for the Quadratic Discriminant function';
```

Output 5.10

Output 5.10
Discriminant Analysis for Numerical Color Tolerances

Test of Homogeneity of Within Covariance Matrices

Chi-Square	DF	Pr > ChiSq
46.047876	6	<.0001

Since the Chi-Square value is significant at the 0.1 level, the within covariance matrices will be used in the discriminant func tion.
Reference: Morrison, D.F. (1976) Multivariate Statistical Method s p252.

Coefficients for the Quadratic Discriminant function

Obs	group	_TYPE_	_NAME_	1	a	b
41	grp2	STD		0.58052	0.93755	0.52058
42		PSTD		0.50291	0.80401	0.45532
43		BSTD		0.06926	0.05364	0.04685
44		STD		0.50303	0.80125	0.45446
45	grp1	CORR	1	1.00000	0.11778	-0.29775
46	grp1	CORR	a	0.11778	1.00000	0.26724
47	grp1	CORR	b	-0.29775	0.26724	1.00000
48	grp2	CORR	1	1.00000	0.62102	0.11251
49	grp2	CORR	a	0.62102	1.00000	0.15777
50	grp2	CORR	b	0.11251	0.15777	1.00000
51		PCORR	1	1.00000	0.55276	0.05004
52		PCORR	a	0.55276	1.00000	0.17320
53		PCORR	b	0.05004	0.17320	1.00000
54		BCORR	1	1.00000	1.00000	-1.00000
55		BCORR	a	1.00000	1.00000	-1.00000
56		BCORR	b	-1.00000	-1.00000	1.00000
57		CORR	1	1.00000	0.55414	0.04251
58		CORR	a	0.55414	1.00000	0.16906
59		CORR	b	0.04251	0.16906	1.00000
60	grp1	STDMEAN		-0.14600	-0.54300	-0.07300
61	grp2	STDMEAN		-0.24800	-0.62200	-0.00400
62	grp1	PSTDMEAN		-0.29031	-0.67537	-0.16033
63	grp2	PSTDMEAN		-0.49313	-0.77362	-0.00879
64		LNDETERM		-3.78270	-3.78270	-3.78270
65	grp1	LNDETERM		-6.34602	-6.34602	-6.34602
66	grp2	LNDETERM		-3.03506	-3.03506	-3.03506
67	grp1	QUAD	1	-5.63832	0.79973	-2.09356
68	grp1	QUAD	a	0.79973	-2.46477	1.28299
69	grp1	QUAD	b	-2.09356	1.28299	-6.56720
70	grp1	QUAD	_LINEAR_	-1.08354	-2.25590	-0.17680
71	grp1	QUAD	_CONST_	2.47498	2.47498	2.47498
72	grp2	QUAD	1	-2.41597	0.92549	0.04015
73	grp2	QUAD	a	0.92549	-0.93788	0.15037
74	grp2	QUAD	b	0.04015	0.15037	-1.89278
75	grp2	QUAD	_LINEAR_	-0.04669	-0.70647	0.19183
76	grp2	QUAD	_CONST_	1.29241	1.29241	1.29241

The null hypothesis of equality of two variance-covariance matrices is rejected (chi-square test statistics on 6 $df = 46.0479$ with a p value < 0.0001). Thus, a quadratic discriminant analysis is more appropriate. Since no raw data were provided, the analysis would not provide any reclassification results. However, information about the classification rule is stored in the output data set C_STAT. Using

```
proc print data = c_stat;
```

this data set can be printed. The coefficients of the quadratic discriminant functions are presented as observations 67 through 76 within this data set. From these, we obtain the quadratic functions for GRP1 and GRP2, respectively, as

$$Q_1 = 2.4750 + [-1.0835 \quad -2.2559 \quad -0.1768] \begin{bmatrix} L \\ A \\ B \end{bmatrix} + [L\ A\ B] \begin{bmatrix} -5.6383 & 0.7997 & -2.0936 \\ 0.7997 & -2.4648 & 1.2830 \\ -2.0936 & 1.2830 & -6.5672 \end{bmatrix} \begin{bmatrix} L \\ A \\ B \end{bmatrix},$$

and

$$Q_2 = 1.2924 + [-0.0467 \quad -0.7065 \quad -0.1918] \begin{bmatrix} L \\ A \\ B \end{bmatrix} + [L\ A\ B] \begin{bmatrix} -2.4160 & 0.9255 & 0.0402 \\ 0.9255 & -0.9379 & 0.1504 \\ 0.0402 & 0.1504 & -1.8928 \end{bmatrix} \begin{bmatrix} L \\ A \\ B \end{bmatrix},$$

and we will classify any future unit $[L\ A\ B]'$ as acceptable if $Q_1 > Q_2$ and as unacceptable if $Q_1 \leq Q_2$. In case a data set were available for classification purposes, it could be specified in the TESTDATA= option. In fact, Vance (1989) points out that this decision rule is such that very few samples that it rejects would be acceptable to human observers. But there is a very high chance (about 40%) that a sample accepted by the statistical method will be judged unacceptable to human observers. Thus, this analysis can be used at a first stage screening when the ultimate judgment is to be done by humans. Such a practice will significantly reduce manually performed screening thereby resulting in substantial economy.

The analysis illustrated in the above example can also be used for the case when the true means $\boldsymbol{\mu}_i$ and the variance-covariance matrices $\boldsymbol{\Sigma}_i, i = 1, \ldots, k$ are known, and the interest is in obtaining the population discriminant functions. In this case, the only change would be to artificially define the sample sizes as some very large numbers (e.g., $n_i = 10^5$). Such computations may be of interest in specifying different population covariance structure scenarios in simulation studies involving discriminant analysis.

5.6 Canonical Discriminant Analysis

Canonical discriminant analysis is a dimensionality reduction technique similar to principal component analysis and canonical correlation analysis. This technique is, however, specialized to the context of discriminant analysis and is used to provide a representation of various populations in a subspace of smaller dimensions. Based on a large number of possibly correlated characteristics on which measurements are taken, the analysis attempts to obtain only a few new variables that can help describe the differences between various populations. These new variables obtained as certain linear combinations of original measurements are termed *canonical variables*. In general, the extraction of various canonical variables does not require any distributional assumption such as multivariate normality. This assumption is needed, however, when certain statistical tests, such as the determination of the number of important canonical variables, are to be performed. However, with or without any distributional assumption, the canonical discriminant analysis is applicable only when all populations have a common variance-covariance matrix.

Let $\Pi_1, \ldots, \Pi_g$ be g independent populations with respective population mean vectors $\boldsymbol{\mu}_1, \ldots, \boldsymbol{\mu}_g$ and a common variance-covariance matrix $\boldsymbol{\Sigma}$. To describe any differences between these g populations, it may be appropriate to use a measure in terms of differences of individual population mean vectors $\boldsymbol{\mu}_1, \ldots, \boldsymbol{\mu}_g$ from their overall center of gravity $\bar{\boldsymbol{\mu}} = \frac{1}{g}\sum_{i=1}^{g} \boldsymbol{\mu}_i$. We thus define a matrix $\boldsymbol{\Delta}$, termed the true between-population sums of squares and cross-product, as a function of $\boldsymbol{\mu}_1 - \bar{\boldsymbol{\mu}}, \ldots, \boldsymbol{\mu}_g - \bar{\boldsymbol{\mu}}$ as

$$\boldsymbol{\Delta} = \sum_{i=1}^{g}(\boldsymbol{\mu}_i - \bar{\boldsymbol{\mu}})(\boldsymbol{\mu}_i - \bar{\boldsymbol{\mu}})'$$

and compare it with the common variance-covariance matrix $\boldsymbol{\Sigma}$ in some meaningful way. To do so, let $\mathbf{x}$ be a p by 1 observation from one, say i^{th}, of these populations. If $\mathbf{a}$ is a nonzero, p by 1 vector of coefficients, then corresponding to this multivariate population, we can define a univariate population so that $u_1 = \mathbf{a}'\mathbf{x}$ is viewed as an observation from this univariate population. This univariate population will have the population mean $\mathbf{a}'\boldsymbol{\mu}_i$ and the population variance $\mathbf{a}'\boldsymbol{\Sigma}\mathbf{a}$. Thus, we can obtain from g multivariate populations $\Pi_1, \ldots, \Pi_g$, another set of g univariate populations with respective population means $\mathbf{a}'\boldsymbol{\mu}_1, \ldots, \mathbf{a}'\boldsymbol{\mu}_g$ and the common population variance $\mathbf{a}'\boldsymbol{\Sigma}\mathbf{a}$. We would like to choose the vector $\mathbf{a}$ in such a way that when plotted on a real line, say on x axis, these univariate populations are as far apart as possible from each other. This amounts to choosing the vector $\mathbf{a}$ such that $\mathbf{a}'\boldsymbol{\Delta}\mathbf{a}/\mathbf{a}'\boldsymbol{\Sigma}\mathbf{a}$ is as large as possible. The choice of vector $\mathbf{a}$, say, $\mathbf{u}_1$, resulting from the solution of this optimization problem yields the *first canonical variable* $u_1 = \mathbf{u}_1'\mathbf{x}$, which may be interpreted as the single best linear discriminator of these g populations. The *second canonical variable*, say u_2, is chosen in the same way but subject to the additional restriction that it is uncorrelated with the first canonical variable. It may be interpreted as the next best linear discriminator uncorrelated to the first canonical variable. The corresponding vector is denoted by $\mathbf{u}_2$. Proceeding in a similar way, we can obtain a list of $r = \min(p, g-1)$ canonical variables, uncorrelated with each other. Similar interpretation can also be assigned to these variables. Usually only a few canonical variables can adequately describe the differences in the g populations. For example, a two-dimensional plot u_1 and u_2 will often illustrate these differences very clearly and effectively.

In practice, independent samples $\{\mathbf{x}_{11}, \ldots, \mathbf{x}_{1n_1}\}, \ldots, \{\mathbf{x}_{g1}, \ldots, \mathbf{x}_{gn_g}\}$ of sizes $n_1, \ldots, n_g$ are available from the respective populations and $\boldsymbol{\mu}_1, \ldots, \boldsymbol{\mu}_g$ and $\boldsymbol{\Sigma}$ are all unknown. In this case, we estimate $\boldsymbol{\mu}_i$ by $\bar{\mathbf{x}}_i = \frac{1}{n_i}\sum_{j=1}^{n_i} \mathbf{x}_{ij}$ $i = 1, \ldots, g$, $\bar{\boldsymbol{\mu}}$ by $\bar{\mathbf{x}} = \frac{1}{\sum_{i=1}^{g} n_i}\sum_{i=1}^{g}\sum_{j=1}^{n_i} \mathbf{x}_{ij} = \frac{1}{\sum_{i=1}^{g} n_i}\sum_{i=1}^{g} n_i\bar{\mathbf{x}}_i$ and the common variance-covariance matrix $\boldsymbol{\Sigma}$ by the pooled sample variance-covariance matrix $\mathbf{S} = \mathbf{E}/\left(\sum_{i=1}^{g} n_i - g\right)$, where $\mathbf{E}$ is the pooled within-class sums of squares and the cross-product (SS&CP) matrix, given by

$$\mathbf{E} = \sum_{i=1}^{g}\sum_{j=1}^{n_i}(\mathbf{x}_{ij} - \bar{\mathbf{x}}_i)(\mathbf{x}_{ij} - \bar{\mathbf{x}}_i)'.$$

Also, we estimate $(\sum n_i)\boldsymbol{\Delta}$ by $\mathbf{D}$, which is given by

$$\mathbf{D} = \sum_{i=1}^{g}\sum_{j=1}^{n_i}(\bar{\mathbf{x}}_i - \bar{\mathbf{x}})(\bar{\mathbf{x}}_i - \bar{\mathbf{x}})' = \sum_{i=1}^{g} n_i(\bar{\mathbf{x}}_i - \bar{\mathbf{x}})(\bar{\mathbf{x}}_i - \bar{\mathbf{x}})',$$

and correspondingly find vectors $\mathbf{v}_1, \mathbf{v}_2, \ldots$ (which are the sample counterparts of vectors $\mathbf{u}_1, \mathbf{u}_2 \ldots$) that respectively solve the optimization problems,

$$\max_{\mathbf{b}\neq\mathbf{0}} \frac{\mathbf{b}'\mathbf{D}\mathbf{b}}{\mathbf{b}'\mathbf{S}\mathbf{b}}.$$

This is subject to the restrictions that $\mathbf{v}_1, \mathbf{v}_2, \ldots$ are all such that $\mathbf{v}_i'\mathbf{S}\mathbf{v}_j = 0$ for $i \neq j$ (these restrictions are the *sample versions* of the restrictions requiring zero correlations among the sample versions of $u_1, u_2, \ldots$). The vectors $\mathbf{v}_1, \mathbf{v}_2, \ldots$ so obtained are the coefficients of (sample versions of) the respective canonical variables, which we now denote by $v_1, v_2, \ldots$. It is customary to scale the vectors $\mathbf{v}_1, \mathbf{v}_2 \ldots$ so that $\mathbf{v}_i'\mathbf{S}\mathbf{v}_i = 1$. The corresponding canonical variables $v_i = \mathbf{v}_i'\mathbf{x}$, $i = 1, \ldots, r$ will then have the unit (estimated) variances. It may be pointed out that the vectors $\mathbf{v}_1, \mathbf{v}_2, \ldots$ are nothing but the eigenvectors of the nonsymmetric matrix $\mathbf{S}^{-1}\mathbf{D}$. The corresponding eigenvalues, say $w_i = \frac{\mathbf{v}_i'\mathbf{D}\mathbf{v}_i}{\mathbf{v}_i'\mathbf{S}\mathbf{v}_i} = \mathbf{v}_i'\mathbf{D}\mathbf{v}_i$, then have the interpretation of the estimated variances of $\mathbf{v}_i$, $i = 1, \ldots, r$. The canonical correlations (see Chapter 3) r_i can be obtained as $r_i = \left(\frac{w_i}{1+w_i}\right)^{\frac{1}{2}}$. By definition, these are nonnegative.

As we mentioned earlier, the analysis described above results in $r = \min(p, g-1)$ canonical variables $v_1, v_2, \ldots, v_r$. Since the basic interest here is in the reduction of dimensionality of the data, we ask, what is the minimum number of canonical variables that can adequately describe the data and adequately discriminate among g groups? This question can be addressed by asking how many of these r canonical correlations are significantly different from zero. We have addressed this problem in Chapter 3 assuming multivariate normality of samples. Under the same assumption and by sequentially testing the hypotheses of the type $H_0^{(j)} : \lambda_j = \lambda_{j+1} = \ldots = \lambda_r = 0$, for $j = 1, 2, \ldots, r$, and stopping as soon as $H_0^{(j)}$ is accepted, we can determine the number of important canonical variables and hence the corresponding dimensionality of data. For example, suppose $H_0^{(1)}, \ldots, H_0^{(j-1)}$ are all rejected, but $H_0^{(j)}$, $j = 1, \ldots, r+1$, is accepted. This suggests that only $(j-1)$ canonical variables $v_1, \ldots, v_{j-1}$ are sufficient to describe the data and hence the appropriate smallest dimension of the space in which the data can be presented is $(j-1)$.

In SAS, the canonical discriminant analysis described above can be performed using the CANDISC procedure. We will describe this analysis, as well as its interpretation in the context of an example.

EXAMPLE 11 ***Canonical Discrimination, Mapping of Lithology of Arid Areas*** Fox (1993) presents an extensive study to map the lithology of arid areas using Lansat Thematic Mapper satellite data. The objective of the study was to develop a set of coefficients for certain geological measurements, which could then be used to discriminate between the rocks and minerals. Such an analysis is of great utility in the development of geological maps in mineral exploration or in land-use analyses.

The study was carried out for the Frenchman Mountain area. Six groupings (populations) of rock and mineral types were defined. These are denoted as LIMESTON, GYPSUM, SANDSTON, CARBONAT, M_BASALT, and I_DACITE. In the SAS program, these six populations of various sample types are the values of the CLASS variable SAMLTYPE. A total of 51 samples were available for analysis. On each sample, six different geological measurements, termed $TM1, \ldots, TM5$ and $TM7$, were taken. Definitions of these measurements are given in Fox (1993, p.31). The data are read as the data set FOX in Program 5.11 and are given in the Appendix as file SATELLITE.DAT. Selected output follows the program.

```
/* Program 5.11 */


filename gsasfile "prog511.graph";
goptions reset=all gaccess=gsasfile gsfmode = append
autofeed dev = pslmono;
goptions horigin =1in vorigin = 2in;
goptions hsize=6in vsize = 8in;
options ls = 64 ps=45 nodate nonumber;
title1  'Output 5.11';

data fox;
infile 'satellite.dat' firstobs = 7;
input smpltype $ tm1 tm2 tm3 tm4 tm5 tm7 ;

if smpltype = 'limeston' then mark = 'L';
if smpltype = 'gypsum' then mark = 'G';
if smpltype = 'sandston' then mark = 'S';
if smpltype = 'carbonat' then mark = 'C';
if smpltype = 'm_basalt' then mark = 'M';
if smpltype = 'i_dacite' then mark = 'I';

proc candisc data = fox out = outcan bsscp pcov pcorr ;
class smpltype;
var tm1 tm2 tm3 tm4 tm5 tm7 ;
title2 'Canonical Discriminant Analysis: Landsat Thematic Mapper Data';

proc sort data = outcan;
by mark;
proc print data = outcan ;
var smpltype mark can1 can2 can3 can4 can5;
title2 'Scores on Canonical Variables: Landsat Thematic Mapper Data';
run;


symbol1 value = C ;
symbol2 value = G ;
symbol3 value = I ;
symbol4 value = L ;
symbol5 value = M ;
symbol6 value = S ;

proc gplot data = outcan;
where mark = "C" or mark = "G" or mark = "I" or
mark = "L" or mark = "M" or mark = "S";
plot  can2*can1 = mark;
title1 h = 1.2 'Plot of First Two Canonical Variables';
title2 j= l 'Output 5.11';
run;

proc gplot data = outcan;
where mark = "C" or mark = "G" or mark = "I" or
mark = "L" or mark = "M" or mark = "S";plot can3*can1 = mark ;
title1 h = 1.2 'Plot of First and Third Canonical Variables';
title2 j= l 'Output 5.11';
run;
```

```
proc gplot data = outcan;
where mark = "C" or mark = "G" or mark = "I" or
mark = "L" or mark = "M" or mark = "S";
plot can4*can1 = mark ;
title1 h = 1.2 'Plot of First and Fourth Canonical Variables';
title2 j= l 'Output 5.11';
run;

proc gplot data = outcan;
where mark = "C" or mark = "G" or mark = "I" or
mark = "L" or mark = "M" or mark = "S";
plot can3*can2 = mark ;
title1 h = 1.2 'Plot of Second and Third Canonical Variables';
title2 j= l 'Output 5.11';
run;

proc gplot data = outcan;
where mark = "C" or mark = "G" or mark = "I" or
mark = "L" or mark = "M" or mark = "S";
plot can4*can2 = mark ;
title1 h = 1.2 'Plot of Second and Fourth Canonical Variables';
title2 j= l 'Output 5.11';
run;

proc gplot data = outcan;
where mark = "C" or mark = "G" or mark = "I" or
mark = "L" or mark = "M" or mark = "S";
plot can4*can3 = mark ;
title1 h = 1.2 'Plot of Third and Fourth Canonical Variables';
title2 j= l 'Output 5.11';
run;
```

Output 5.11

Output 5.11
Canonical Discriminant Analysis: Landsat Thematic Mapper Data

The CANDISC Procedure

Between-Class SSCP Matrix

Variable	tm1	tm2	tm3
tm1	80620.8065	53021.0210	74687.7824
tm2	53021.0210	35210.3894	50097.2717
tm3	74687.7824	50097.2717	72472.1027
tm4	66321.7597	44656.8221	65040.7211
tm5	91828.7833	62831.3109	93540.9461
tm7	43972.0446	31900.5448	48690.0682

Between-Class SSCP Matrix

Variable	tm4	tm5	tm7
tm1	66321.7597	91828.7833	43972.0446
tm2	44656.8221	62831.3109	31900.5448
tm3	65040.7211	93540.9461	48690.0682
tm4	58597.0169	85997.8473	45086.5454
tm5	85997.8473	146274.8316	80403.4261
tm7	45086.5454	80403.4261	56875.3779

Pooled Within-Class Covariance Matrix, DF = 45

Variable	tm1	tm2	tm3
tm1	316.0692240	194.5084656	220.7961199
tm2	194.5084656	126.2719577	154.1164021
tm3	220.7961199	154.1164021	224.1319224
tm4	160.8946649	117.0544974	184.7417108
tm5	90.0797619	75.6867725	154.9981481
tm7	-23.7656966	-4.8513228	34.8416226

Pooled Within-Class Covariance Matrix, DF = 45

Variable	tm4	tm5	tm7
tm1	160.8946649	90.0797619	-23.7656966
tm2	117.0544974	75.6867725	-4.8513228
tm3	184.7417108	154.9981481	34.8416226
tm4	163.9939594	183.2957672	67.1957231
tm5	183.2957672	607.6477513	376.0284392
tm7	67.1957231	376.0284392	284.0713404

Pooled Within-Class Correlation Coefficients / Pr > |r|

Variable	tm1	tm2	tm3	tm4	tm5	tm7
tm1	1.00000	0.97363	0.82956	0.70670	0.20555	-0.07931
		<.0001	<.0001	<.0001	0.1706	0.6003
tm2	0.97363	1.00000	0.91610	0.81343	0.27324	-0.02561
	<.0001		<.0001	<.0001	0.0662	0.8658
tm3	0.82956	0.91610	1.00000	0.96361	0.42000	0.13808
	<.0001	<.0001		<.0001	0.0037	0.3601
tm4	0.70670	0.81343	0.96361	1.00000	0.58065	0.31132
	<.0001	<.0001	<.0001		<.0001	0.0352
tm5	0.20555	0.27324	0.42000	0.58065	1.00000	0.90507
	0.1706	0.0662	0.0037	<.0001		<.0001
tm7	-0.07931	-0.02561	0.13808	0.31132	0.90507	1.00000
	0.6003	0.8658	0.3601	0.0352	<.0001	

	Canonical Correlation	Adjusted Canonical Correlation	Approximate Standard Error	Squared Canonical Correlation
1	0.964392	0.956190	0.009892	0.930051
2	0.942031	0.937500	0.015921	0.887422
3	0.767955	0.741949	0.058017	0.589755
4	0.583393	0.567609	0.093289	0.340347
5	0.136775	0.074359	0.138776	0.018707

Eigenvalues of Inv(E)*H
= CanRsq/(1-CanRsq)

	Eigenvalue	Difference	Proportion	Cumulative
1	13.2962	5.4134	0.5743	0.5743
2	7.8827	6.4452	0.3405	0.9148
3	1.4376	0.9216	0.0621	0.9769
4	0.5159	0.4969	0.0223	0.9992
5	0.0191		0.0008	1.0000

Test of H0: The canonical correlations in the
current row and all that follow are zero

	Likelihood Ratio	Approximate F Value	Num DF	Den DF	Pr > F
1	0.00209118	19.85	30	162	<.0001
2	0.02989585	12.88	20	136.93	<.0001
3	0.26555684	6.04	12	111.41	<.0001
4	0.64731258	3.48	6	86	0.0040
5	0.98129273	0.42	2	44	0.6600

Raw Canonical Coefficients

Variable	Can1	Can2	Can3
tm1	-.0478989618	0.1193468214	-.1113981061
tm2	0.2432480812	-.2504234501	0.0986763096
tm3	-.1843660101	0.0580768528	0.2006196140
tm4	0.1866564864	0.0084166586	-.2617287009
tm5	-.1225598948	0.0756323496	0.0711086819
tm7	0.1695657061	-.0884889416	-.0194820236

Raw Canonical Coefficients

Variable	Can4	Can5
tm1	-.0216267538	-.0606111212
tm2	-.1589697755	0.3118127492
tm3	0.1031845240	-.5537013571
tm4	0.0657402654	0.4848380974
tm5	-.0170063413	-.0258289849
tm7	-.0105628927	-.0018006904

Output 5.11
(continued)

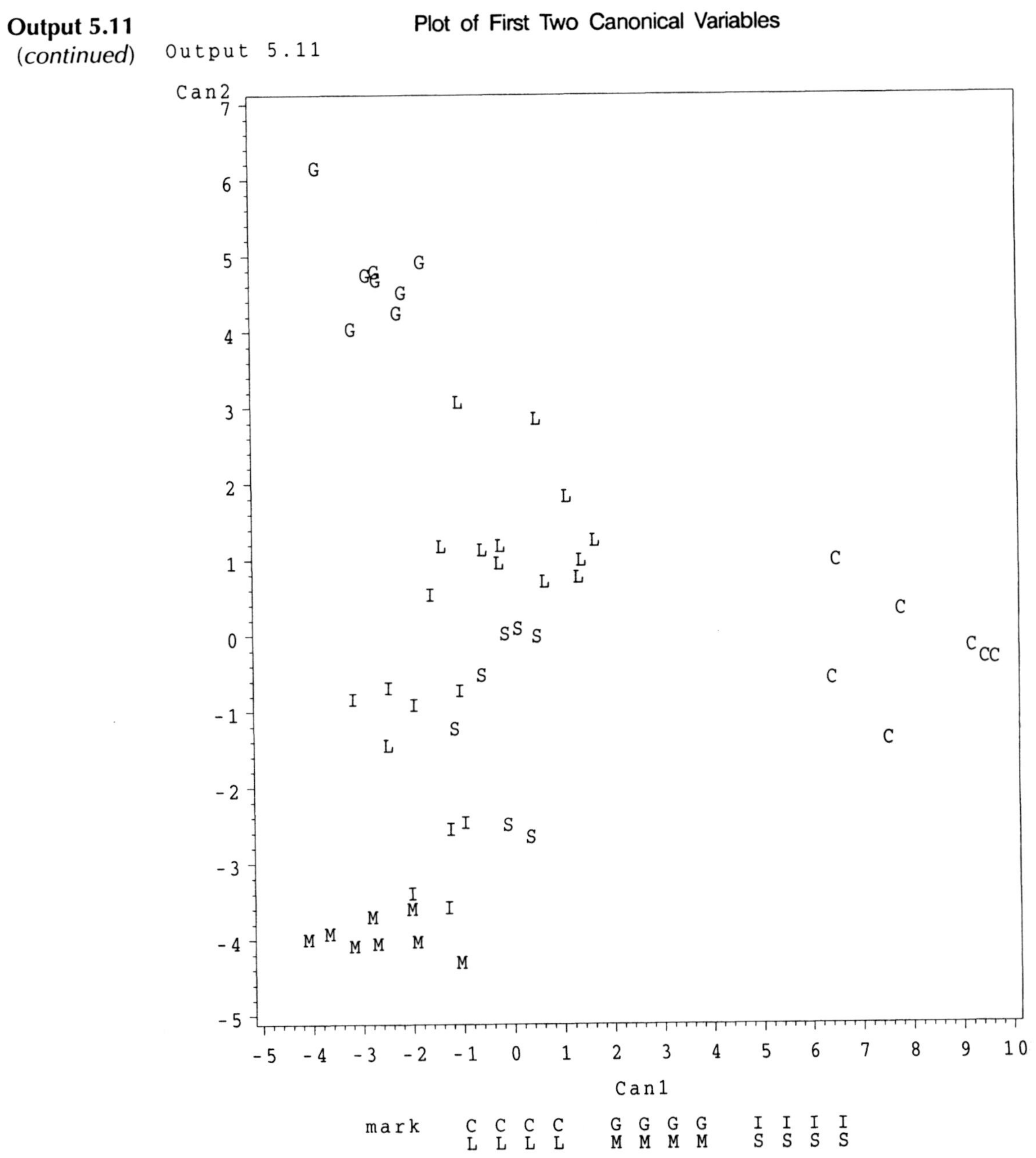

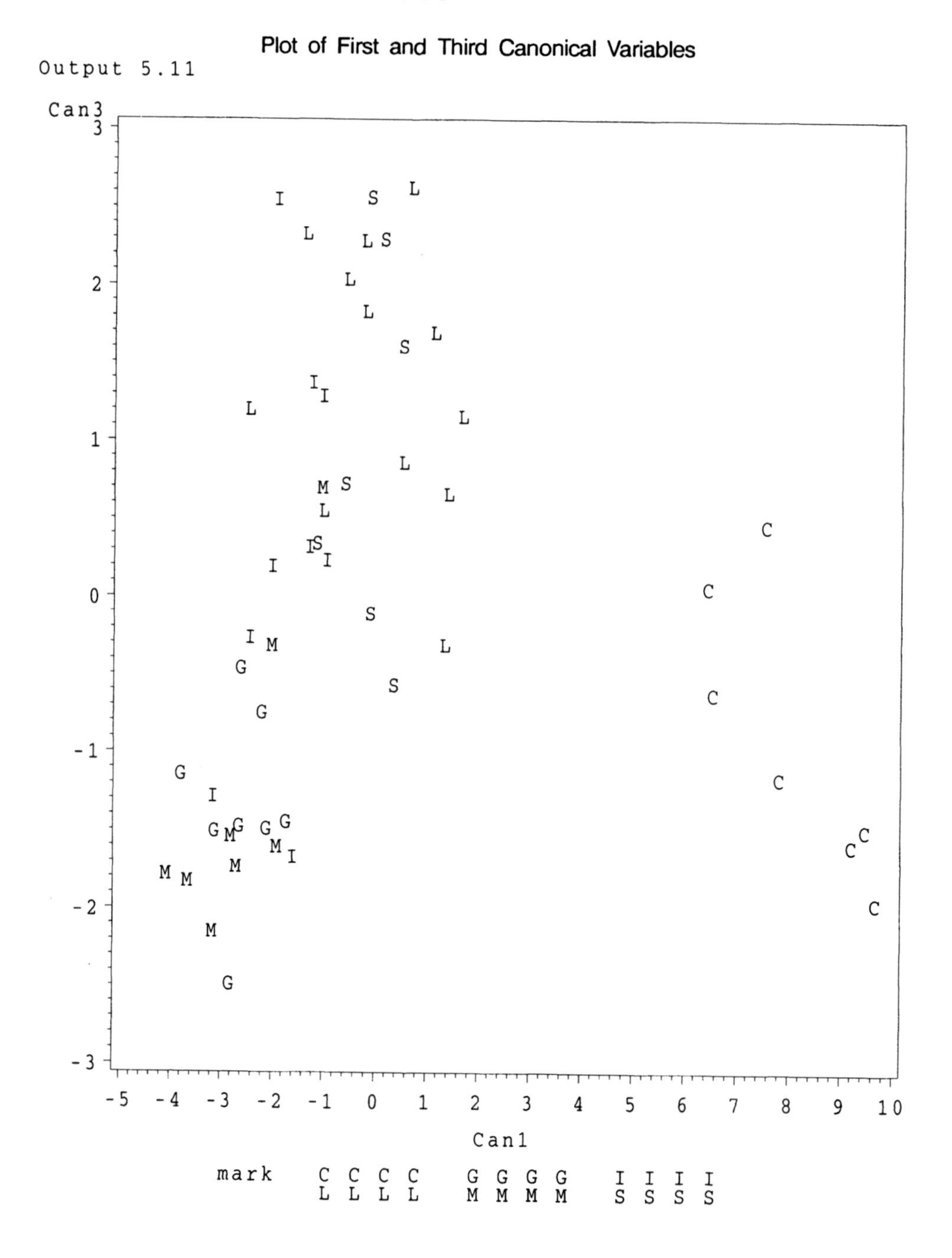

Plot of First and Third Canonical Variables
Output 5.11
Can3
3
2
1
0
-1
-2
-3
-5 -4 -3 -2 -1 0 1 2 3 4 5 6 7 8 9 10
Can1
mark C C C C G G G G I I I I
L L L L M M M M S S S S

Output 5.11

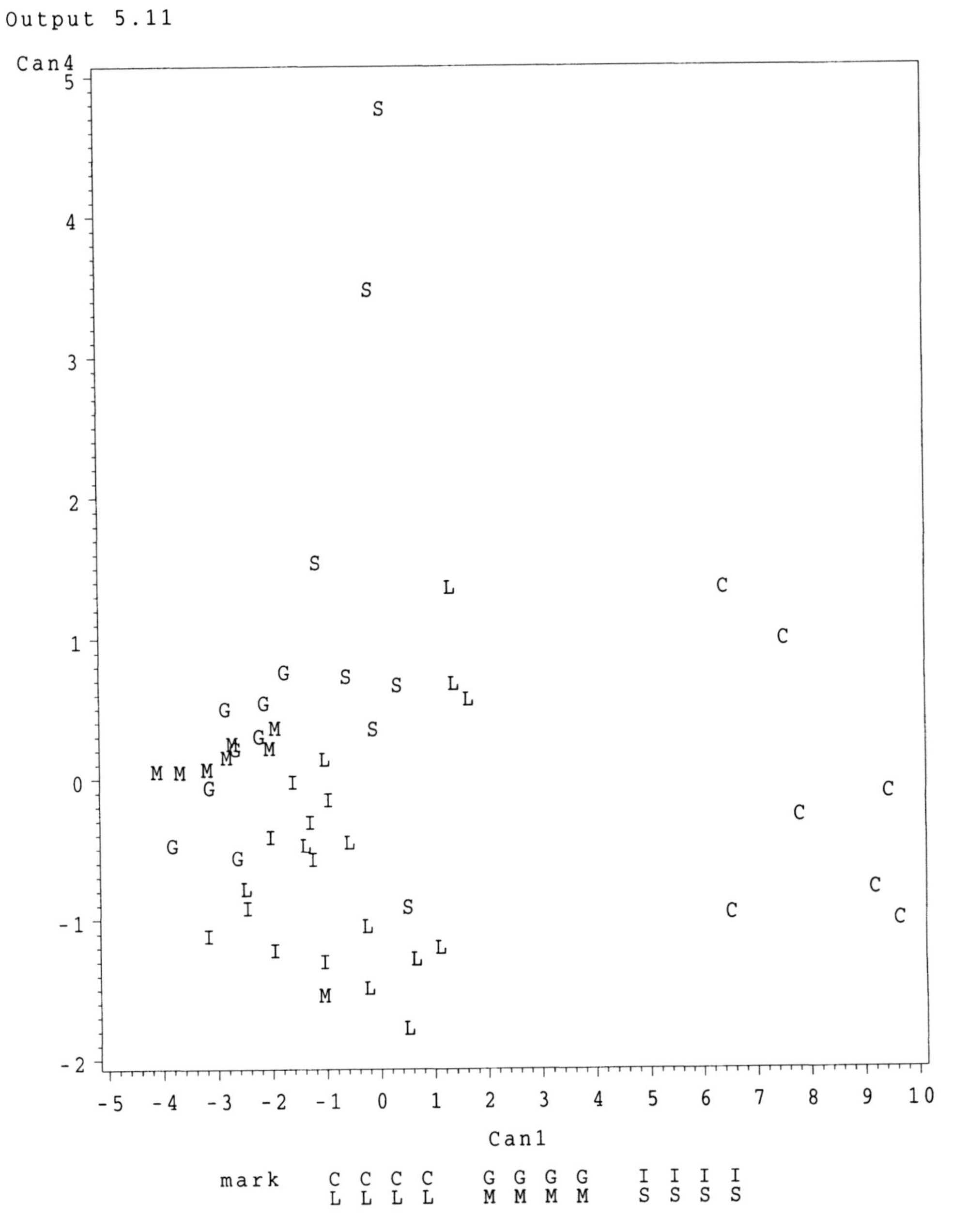
Plot of First and Fourth Canonical Variables
Can4
5
4
3
2
1
0
-1
-2
-5 -4 -3 -2 -1 0 1 2 3 4 5 6 7 8 9 10
Can1
mark C C C C G G G G I I I I
L L L L M M M M S S S S

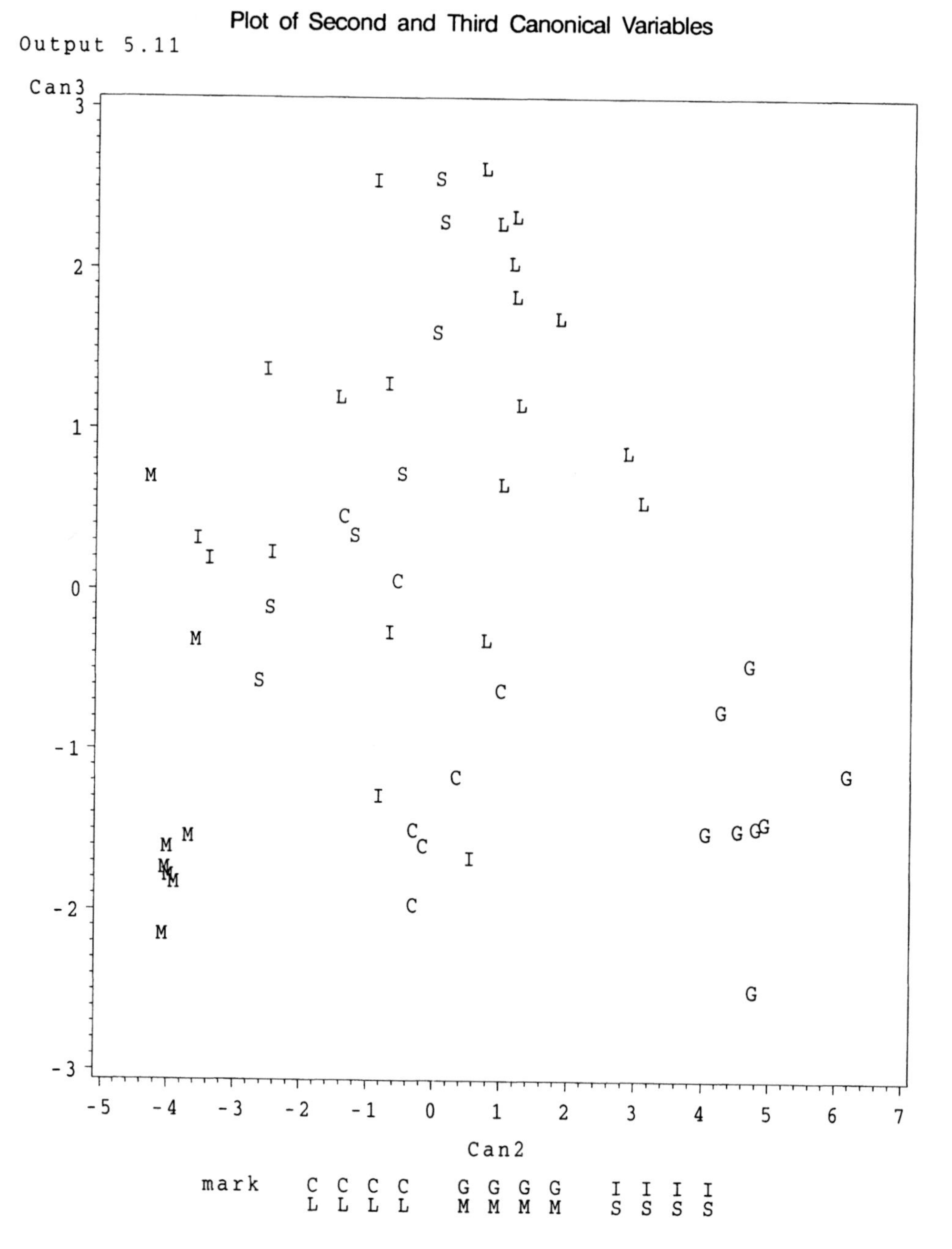
Plot of Second and Third Canonical Variables
Output 5.11
Can3
Can2
mark
C C C C
L L L L
G G G G
M M M M
I I I I
S S S S

Plot of Second and Fourth Canonical Variables

Output 5.11

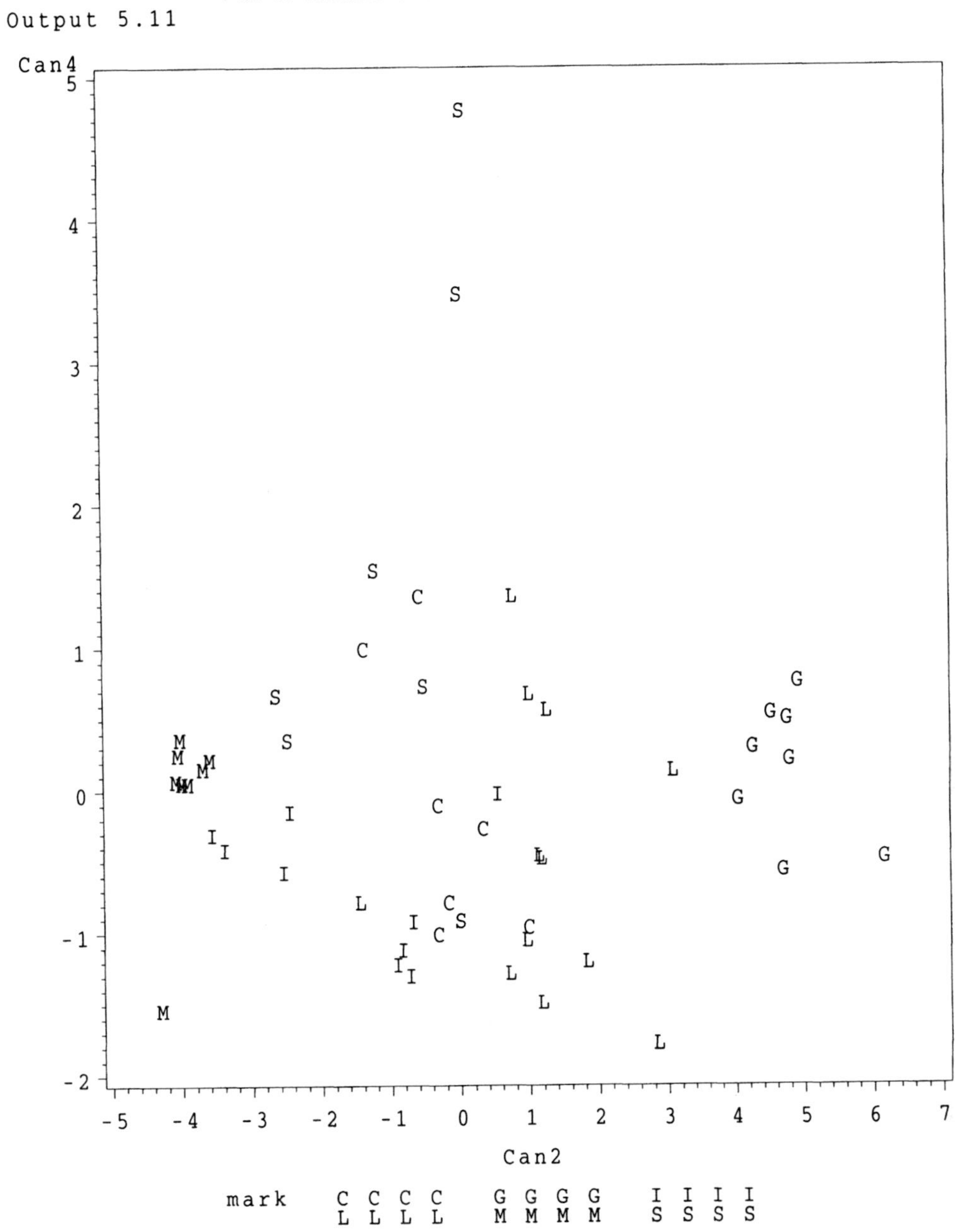

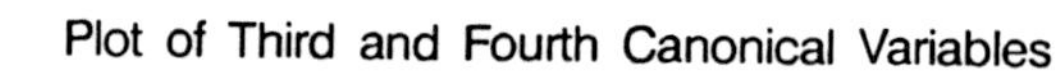
Plot of Third and Fourth Canonical Variables

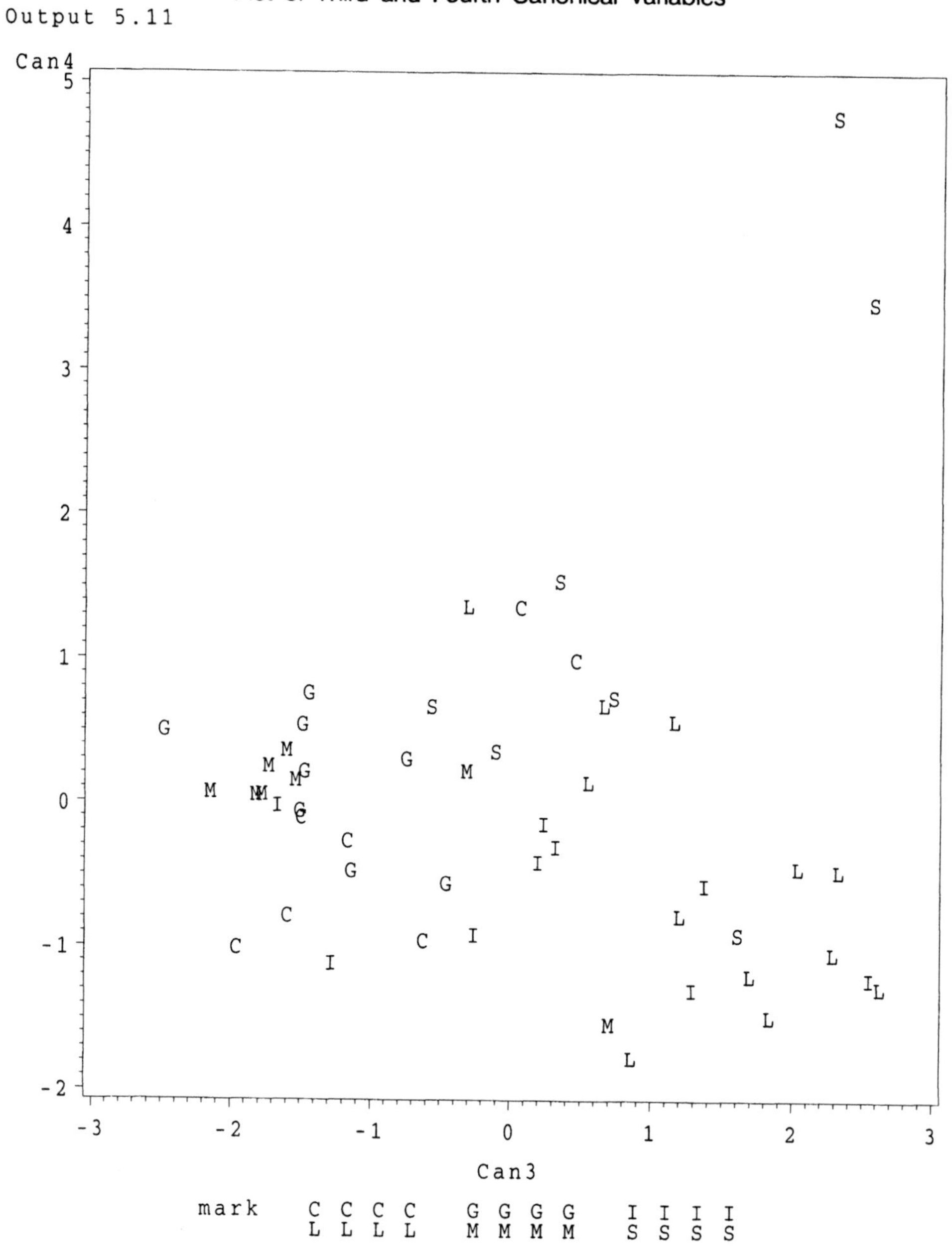
Output 5.11
Can4
5
4
3
2
1
0
-1
-2
-3
-2
-1
0
1
2
3
Can3
mark
CL CL CL CL
GM GM GM GM
IS IS IS IS

As indicated earlier, the need here is to develop a set of coefficients that can help to discriminate between these six rock or mineral groups. The canonical discriminant analysis appears to be an ideal way to deal with this problem. The following SAS code will perform the statistical analysis:

```
proc candisc data = fox out = outcan bsscp pcov pcorr;
class smpltype;
var tm1 tm2 tm3 tm4 tm5 tm7;
```

The BSSCP, PCOV, and PCORR options result respectively in the "between-SS&CP" matrix **D**, the pooled sample variance-covariance matrix **S**, and the pooled sample correlation matrix for the six variables. The class variable SMPLTYPE indicating the membership of a particular sample into one of the six populations is specified in the CLASS statement, and the six geological measurements are listed in the VAR statement. Also, since we may need the output for further analysis (such as for plotting the canonical variables), the option OUT= in the PROC CANDISC statement is used to store the output in a data set titled OUTCAN. All these statements are included as part of Program 5.11.

The resulting output is presented as Output 5.11. First, the **B** and **S** matrices are presented and then the pooled sample correlation matrix is also provided. These matrices can provide some meaningful information in this particular context. For instance, it is easy to observe from the output that the first four variables are highly correlated with each other. Hence there is a possibility that one or more variables can be dropped using one of the methods to be described in Section 5.7; in fact, such an analysis under multivariate normality assumptions recommends dropping the variable TM3. This analysis is not included here.

Next, in the output are the values of $r = \min(g - 1, p) = 5$ canonical correlations, r_i, their squares, and their approximate standard errors. The eigenvalues w_i, which are the estimated variances of respective canonical variables, are listed next. It is observed (by looking at the Cumulative column) that the first two variables can explain more than 91% of the total variance. While this does suggest that the first two canonical variables would do a satisfactory job of describing the data, statistical tests, under the assumption of multivariate normality, will probably provide more information. These tests are performed and their output shown in the next table within Output 5.11. Following a similar argument as provided in Chapter 3, we conclude that we may actually need four canonical variables and hence the reduced dimension of this data set is four.

The coefficients of canonical variables $\mathbf{v}_1, \ldots, \mathbf{v}_5$ discussed above are listed as Raw Canonical Coefficients. These are scaled to have $\mathbf{v}_i'\mathbf{S}\mathbf{v}_i = 1$. There are several other choices of canonical variables available in the output and described in Chapter 16 of *SAS/STAT User's Guide, Version 6, Fourth Edition, Volume 2*, or in the SAS online documentation (the CANDISC procedure). We will, however, confine our discussion to the raw canonical variables only. Specifically, the first two (raw) canonical variables are

$$\begin{aligned} v_1 = &-0.0479 * TM1 + 0.2432 * TM2 - 0.1844 * TM3 + 0.1867 * TM4 \\ &- 0.1226 * TM5 + 0.1696 * TM7, \end{aligned}$$

and

$$\begin{aligned} v_2 = &\ 0.1193 * TM1 - 0.2504 * TM2 + 0.0581 * TM3 + 0.0084 * TM4 \\ &+ 0.0756 * TM5 - 0.0885 * TM7. \end{aligned}$$

The first canonical variable with alternating signs appears to represent a contrast between $\{TM1, TM3, TM5\}$ and $\{TM2, TM4, TM7\}$. The second canonical variable appears to be largely influenced by $TM2$ and $TM1$.

The scores on all five canonical variables for each of the 51 samples are stored as the data set OUTCAN. Using the PROC PRINT statement as in Program 5.11 they can be printed if such a need arises.

To see how good is the separation of the six populations using the first two canonical variables, we can obtain a plot of the score on these two canonical variables. We do so by using the SAS statements

```
proc gplot data = outcan;
plot can2*can1=mark;
```

where the variable MARK taking values L, G, S, C, M, and I for six respective populations has been defined early in the program in the DATA step. The resulting plot presented in Output 5.11 suggests that certain populations can be discriminated more easily than the rest using only one or both of these canonical variables. For example, while the populations CARBONAT, GYPSUM, and LIMESTON (except for one point) are quite different from each other and from the rest, the first canonical variable will probably suffice to discriminate between CARBONAT and the rest. Also, the population GYPSUM distinctly differs from the rest of the populations in the scores of the the second canonical variable. As far as the rest of the populations are concerned, it is interesting to note that while the separation is quite good between the populations (except perhaps between SANDSTON and I_DACITE), all these populations (except CARBONAT) differ in the scores of the second canonical variable. The scores of the first canonical variable for the five populations except CARBONAT are comparable.

As we observe in this plot, the separation between SANDSTON and I_DACITE, using the first canonical variable, is not very clear-cut. However, the plots of the scores on the first and third and the first and fourth canonical variables make this separation clearer. These plots are also included as part of the output. This observation further reinforces the point that even though the first two canonical variables explained more than 91% of the total variation, we may still need the next two canonical variables to understand the nature of our data.

To sum up, for this data set, four canonical variables can describe the differences between these six populations. While CARBONAT substantially differs from the rest in its scores of the first canonical variable and GYPSUM differs from the rest in its scores of the second canonical variable, the next two canonical variables may additionally be needed to discriminate between the other populations.

In closing this section, we make two important remarks. First, the same canonical analysis can alternatively be performed by using the DISCRIM procedure and by specifying the CANONICAL option in the corresponding SAS statement. However, a discrimination criterion (and hence additional output) is also derived in PROC DISCRIM. Therefore, if the interest is only in the dimensionality reduction and not in actual classification of objects, then PROC CANDISC is the more efficient choice of the two. Second, the canonical discriminant analysis described here and the classification rule described in Section 5.5.1 are closely related, and the discriminant scores can be expressed in terms of canonical scores. This relationship, in mathematical form, is expressed in Johnson and Wichern (1998). As a result, given the canonical scores for an observation resulting from PROC CANDISC, we can easily compute the corresponding discriminant scores to classify that observation into one of the several populations.

5.7 Variable Selection in Discriminant Analysis

An important query in discriminant analysis is whether all the variables on which measurements are obtained contain useful information and if only some of them may suffice for discrimination. Since the measurements collected on the same unit are likely to be correlated, it can be intuitively argued that they may contain overlapping pieces of information about their membership in the particular population. Hence, if we have decided to use a carefully chosen subset of some of these variables using some meaningful criterion, others may not contain any substantial useful additional information and may be deemed redundant. This necessitates a need to devise a procedure that can be used to decide which particular subset of variables should be used for discrimination and which of the variables can be viewed as being redundant in the presence of this subset of variables. A case for such a selection can be made further by pointing out that by increasing the number of variables we do not necessarily ensure an increase in the discriminatory power. In fact, the discriminatory power may substantially decrease if too many variables are used for the discriminant analysis. Another observation in support of the case of variable selection is that for the traditional discriminant analysis, the greater the number of variables p, the bigger the sample needed to achieve the same level of precision. Hence for smaller samples, the use of a large number of variables is not recommended.

Using the Rao's test for additional information as given in Rao (1970), we can devise a number of methods to determine the redundancy of variables. However, what may be more preferable is to have an algorithm that can sequentially select or eliminate the variables one at a time. Based on this idea, we can devise three approaches.

- Start by selecting the most important variable and continue selecting the important variables one at a time, using a well-defined criterion. One such criterion involves the attainment of a desired prespecified level of significance. Another criterion involves the attainment of the prespecified level of the squared partial correlation for predicting the variable under consideration from the categorical variable defining various populations while controlling the effect of the variables that have been already selected at previous stages. The process stops only when none of the remaining variables meets the prespecified criterion used. Such a procedure is called a *forward selection* procedure.
- Start with the biggest possible model. At each step, discard the most unimportant variable, one at a time, using one of the criteria similar to that specified above for forward selection. Continue until no variables can be discarded. Since the intent here is to trim the biggest possible model as much as possible by eliminating the redundant variables, such a procedure is termed a *backward selection* procedure.
- Combine the two approaches. Another procedure, called *stepwise selection*, can be devised as a combination of the above two: at any step, a variable can enter or exit from the list of important variables, according to a certain prespecified criterion for these choices.

All three procedures are similar in spirit to the variable selection procedures used in the multiple regression analysis. A good theoretical discussion of these in the context of regression analysis is given in Draper and Smith (1998). For some discussion with special reference to discriminant analysis, see Seber (1984, p.337), Jobson (1992, p.274) and Rencher (1995, p.319).

In SAS, the three methods of variable selection can be implemented using the STEPDISC procedure. The underlying assumption for such a selection is the joint multivariate normality of all variables with a common covariance matrix. The two criteria for including or discarding a variable from the list of important variables are those briefly indicated above. Specifically, these are

- the significance level of an F-test. This F-test corresponds to that in an analysis of covariance model where the variables already chosen (or not yet discarded) are treated as covariates, and a categorical (CLASS) variable specifying the membership in the particular population is used in the model as the main variable of interest, and
- the squared partial correlation of the discriminatory variable under consideration with the CLASS variable described above, while discounting for the effects of the other discriminatory variables already chosen (if forward selection is used) or not yet discarded (if backward selection is used) in an analysis of covariance model.

As we mentioned earlier, in all three selection procedures we need to prespecify some well-defined criteria for including a new variable in the list or eliminating a variable from the list. For F-test criteria, these are specified by the SLENTRY= and SLSTAY= options (in the PROC STEPDISC statement), respectively. A variable (which is not already in the list of selected variables) with the smallest p value less than the specified value of SLENTRY will be included in the list. A variable (which is already in the list) with the largest p value bigger than SLSTAY will be deleted from the list of selected variables. As a result, for forward selection, we need only specify SLENTRY, and for the backward selection, we need to specify a value of SLSTAY. Both of these are needed for the stepwise method of variable selection. Costanza and Afifi (1979) recommend the use of a value between 0.10 and 0.25 for SLENTRY, as well as for SLSTAY. If neither of the two values is specified in the PROC STEPDISC statement, then a default choice of 0.15 is used for both of these.

Although we have described the application of an F-test in selecting the variables, we have not clearly indicated how the value of the F statistics in question is obtained. Specifically, let $\Lambda_F(j)$ be the value of Wilks' Λ for a one-way classification MANOVA based on a set of j variables (with these variables treated as dependent variables and a CLASS variable identifying the membership of the observation in one of the populations as the variable defining the *fixed effect*) already selected. $\Lambda_F(j+1)$ is that with these j variables plus the additional variable (whose inclusion is currently being examined). By definition, $\Lambda_F(0)$ is taken to be equal to 1. With g as the number of populations and n as the total number of (nonmissing valued) observations in the data set, the quantity

$$F = \frac{(n-g-j)}{(g-1)} \cdot \left(\frac{\Lambda_F(j)}{\Lambda_F(j+1)} - 1 \right) \tag{5.11}$$

follows an F distribution with df $(g-1, n-g-j)$. Thus, according to Equation 5.11, the variable in question will be deemed important if $F > F_{g-1,n-g-j}(\alpha)$ where α is the prespecified level of significance. As per our forward selection method, the variable (among those not yet in the model) that corresponds to the maximum of F and has this value larger than $F_{g-1,n-g-1}(\alpha)$, where α is the prespecified value of SLENTRY, will be entered at this stage.

For the backward procedure, we need to calculate F statistics for all the variables currently in the one-way classification MANOVA model described above and discard the one that is most insignificant and has the corresponding p value larger than a given α, which is specified by the SLSTAY option. Thus, if $\Lambda_B(j)$ is the Wilks' Λ based on j variables currently in the model, and $\Lambda_B(j-1)$ is that based on all but one particular variable (whose elimination is currently being examined), then the corresponding F statistic

$$F = \frac{(n-g-j+1)}{(g-1)} \cdot \left(\frac{\Lambda_B(j-1)}{\Lambda_B(j)} - 1 \right) \tag{5.12}$$

approximately follows an F distribution with $df(g-1, n-g-j+1)$ for large samples. Now since the problem here is to eliminate the most insignificant variable, the above F statistic is calculated for all these j variables currently in the model, and the one that corresponds to the minimum of these F statistics and has the corresponding p value bigger than SLSTAY=α will be discarded.

For the criterion based on squared partial correlation, instead of SLENTRY and SLSTAY, we need to specify the squared partial correlation values for entry and stay by using the PR2ENTRY= and PR2STAY= options, respectively. The case for using one of these or both is the same as that detailed previously. We recommend using a value in the neighborhood of .25 for both of these. There is no default value used for PROC STEPDISC in this case, since if none of these are specified (nor SLENTRY and SLSTAY specified) then the procedure defaults to F-test criterion with SLENTRY=.15 and SLSTAY=.15.

EXAMPLE 12 *Variable Selection, Characterization of Soil Types* Horton, Russell, and Moore (1968) consider a problem of statistical discrimination in a study of an area of gilgaied soil at Meandarra, Queensland. A feature of gilgaied soils is the pronounced microtopography that is classified into three categories, namely, top, slope, and depression. Each of these three microtopographic areas was divided into four blocks, respectively defined by four levels of soil layers (0-10 cm, 10-30 cm, 30-60 cm, 60-90 cm). This two-way categorization in essence defines a set of 12 populations (denoted as 1, ..., 12). The objective of the study was to identify important soil variables that could be used to construct a useful linear discriminant function to be eventually used to classify any new soil sample as coming from one of these twelve populations.

As for data collection, four soil samples were taken from each of these twelve populations and nine variables, namely, pH value (PH_VALUE), total nitrogen in % (NITRO), bulk density in gm/cm^3 (BULK_DCT), total phosphorus in parts per million (PHOSPHO), calcium in me/100 gms (CA), magnesium in me/100 gms (MG), potassium in me/100 gms (K), sodium in me/100 gms (NA), and conductivity in mmhos/cm at 25 degrees centigrade (CONDCTVT) were observed for all these samples. It is clear from this long list of discriminatory variables that there is a genuine need for variable selection as there are only four samples per population and in all there are a total of twelve populations of soil types. We will assume that all these populations have a common variance-covariance matrix (testing such an assumption is not possible for this data set since for each population, the corresponding variance-covariance matrix is of order p by p or 9 by 9 but there are only $n_i = 4$ observations from each population). We will illustrate the analysis using the backward selection method. For variables to be discarded, the significance level will be set at 0.15. Thus, at every stage of elimination, the most insignificant variable, i.e., the one with the largest p value, will be discarded so long as this largest p value exceeds 0.15. No variable will be discarded, and the process of elimination will stop, as soon as the p values of all variables currently under consideration are less than 0.15. Thus, in the corresponding SAS statement we will use the SLSTAY=.15 option. The following SAS statements will achieve the result we want:

```
proc stepdisc data=horton bcov pcov method=backward slstay=.15;
class soil_grp;
var ph_value nitro bulk_dct phospho ca mg k na condctvt;
```

The twelve populations are numbered as 1, ..., 12, which are the values taken by the variable SOIL_GRP specified in the CLASS statement. The BCOV and PCOV options are used to print between-population covariances and the pooled-within-variance-covariance matrix (which is the unbiased estimate of the common variance-covariance matrix of these twelve populations). The complete SAS program is presented as Program 5.12 and parts of its output are listed as Output 5.12.

```
/* Program 5.12 */

options ls = 64 ps=45 nodate nonumber;
title1 j=l 'Output 5.12';

data horton;
infile 'soil.dat' firstobs = 7;
input soil_grp $ ph_value nitro bulk_dct phospho
ca mg k na condctvt;

proc stepdisc data = horton bcov pcov pcorr method = backward
slstay =.15;
class soil_grp ;
var ph_value nitro phospho bulk_dct ca mg k na condctvt;
title2 "Backward Selection of Variables: Soil Data" ;
run;
```

Output 5.12

Output 5.12
Backward Selection of Variables: Soil Data

The STEPDISC Procedure
Backward Elimination: Step 0

All variables have been entered.

Multivariate Statistics

Statistic	Value	F Value	Num DF	Den DF
Wilks' Lambda	0.000561	3.99	99	208.9
Pillai's Trace	3.307899	1.90	99	324
Average Squared Canonical Correlation	0.300718			

Statistic	Pr > F
Wilks' Lambda	<.0001
Pillai's Trace	<.0001
Average Squared Canonical Correlation	

Backward Elimination: Step 1

Statistics for Removal, DF = 11, 28

Variable	Partial R-Square	F Value	Pr > F
ph_value	0.5707	3.38	0.0045
nitro	0.3917	1.64	0.1415
phospho	0.6139	4.05	0.0013
bulk_dct	0.3668	1.47	0.1963
ca	0.3821	1.57	0.1613
mg	0.6480	4.69	0.0005
k	0.4661	2.22	0.0432
na	0.4429	2.02	0.0648
condctvt	0.1970	0.62	0.7927

Variable condctvt will be removed.

Variable(s) that have been Removed

condctvt

Multivariate Statistics

Statistic	Value	F Value	Num DF	Den DF
Wilks' Lambda	0.000699	4.60	88	199.63
Pillai's Trace	3.160285	2.14	88	288
Average Squared Canonical Correlation	0.287299			

Statistic	Pr > F
Wilks' Lambda	<.0001
Pillai's Trace	<.0001
Average Squared Canonical Correlation	

Backward Elimination: Step 2

Statistics for Removal, DF = 11, 29

Variable	Partial R-Square	F Value	Pr > F
ph_value	0.6405	4.70	0.0004
nitro	0.3805	1.62	0.1454
phospho	0.6130	4.18	0.0010
bulk_dct	0.3512	1.43	0.2138
ca	0.3634	1.50	0.1832
mg	0.6703	5.36	0.0001
k	0.4262	1.96	0.0725
na	0.8140	11.54	<.0001

Variable bulk_dct will be removed.

Variable(s) that have been Removed

bulk_dct condctvt

Multivariate Statistics

Statistic	Value	F Value	Num DF	Den DF
Wilks' Lambda	0.001077	5.17	77	187.22
Pillai's Trace	2.953964	2.39	77	252
Average Squared Canonical Correlation	0.268542			

Statistic	Pr > F
Wilks' Lambda	<.0001
Pillai's Trace	<.0001
Average Squared Canonical Correlation	

Backward Elimination: Step 3

Statistics for Removal, DF = 11, 30

Variable	Partial R-Square	F Value	Pr > F
ph_value	0.6466	4.99	0.0002
nitro	0.3692	1.60	0.1506
phospho	0.6480	5.02	0.0002
ca	0.3522	1.48	0.1896
mg	0.6628	5.36	0.0001
k	0.4431	2.17	0.0455
na	0.8008	10.96	<.0001

Variable ca will be removed.

Variable(s) that have been Removed

bulk_dct ca condctvt

Multivariate Statistics

Statistic	Value	F Value	Num DF	Den DF
Wilks' Lambda	0.001662	5.99	66	171.33
Pillai's Trace	2.787288	2.84	66	216
Average Squared Canonical Correlation	0.253390			

Statistic	Pr > F
Wilks' Lambda	<.0001
Pillai's Trace	<.0001
Average Squared Canonical Correlation	

Backward Elimination: Step 4

Statistics for Removal, DF = 11, 31

Variable	Partial R-Square	F Value	Pr > F
ph_value	0.7089	6.86	<.0001
nitro	0.5361	3.26	0.0047
phospho	0.6660	5.62	<.0001
mg	0.6292	4.78	0.0003
k	0.4598	2.40	0.0274
na	0.8032	11.50	<.0001

No variables can be removed.

No further steps are possible.

Backward Elimination Summary

Step	Number In	Removed	Partial R-Square	F Value	Pr > F	Wilks' Lambda	Pr < Lambda
0	9		.	.	.	0.00056105	<.0001
1	8	condctvt	0.1970	0.62	0.7927	0.00069867	<.0001
2	7	bulk_dct	0.3512	1.43	0.2138	0.00107685	<.0001
3	6	ca	0.3522	1.48	0.1896	0.00166232	<.0001

Step	Number In	Removed	Average Squared Canonical Correlation	Pr > ASCC
0	9		0.30071806	<.0001
1	8	condctvt	0.28729863	<.0001
2	7	bulk_dct	0.26854218	<.0001
3	6	ca	0.25338980	<.0001

As seen in Output 5.12, we start with the biggest possible one-way classification MANOVA model that contains all nine variables as the dependent variables and the variable SOIL_GRP defining the twelve populations as the nominal values taken by the CLASS variable. The corresponding value of Wilks' Λ is $\Lambda_B(9) = 0.00056105$. Now models with eight dependent variables in each model are analyzed and the values of corresponding Wilks' Λ are calculated. For example, as presented in the output, the model with all variables except CONDCTVT results in a value of Wilks' Λ as $\Lambda_B(8) = 0.00069867$. Similarly, values of $\Lambda_B(8)$ for all nine such models are calculated. All these values are, however, not presented in the output. Using these $\Lambda_B(8)$ values and $\Lambda_B(9)$, we can calculate the F statistics using the formula given in Equation 5.12. With $n = 48$, $g = 12$, and $j = 9$, these reduce to

$$F = \frac{48 - 12 - 9 + 1}{12 - 1}\left(\frac{\Lambda_B(8)}{\Lambda_B(9)} - 1\right) = \frac{28}{11}\left(\frac{\Lambda_B(8)}{\Lambda_B(9)} - 1\right),$$

which follows an F distribution with $(11,28)df$. The values of F statistics for all nine 8-variable models are listed under STEP1 in the output. For example, when the 8-variable model without CONDCTVT is compared with the 9-variable model to test for the elimination of CONDCTVT, the corresponding F value is equal to

$$F = \frac{28}{11}\left(\frac{0.00056105}{0.00069867} - 1\right) = 0.624.$$

As the output shows, this particular F value also corresponds to the smallest of all F values, and the corresponding p value =0.7927 is also larger than the value of SLSTAY=.15. Thus, CONDCTVT will be removed from the model. Using a similar calculation, at STEP2 with

$$F = \frac{48 - 12 - 8 + 1}{12 - 1}\left(\frac{\Lambda_B(7)}{\Lambda_B(8)} - 1\right) = \frac{29}{11}\left(\frac{\Lambda_B(7)}{\Lambda_B(8)} - 1\right),$$

and then at STEP3, with

$$F = \frac{48 - 12 - 7 + 1}{12 - 1}\left(\frac{\Lambda_B(6)}{\Lambda_B(7)} - 1\right) = \frac{30}{11}\left(\frac{\Lambda_B(6)}{\Lambda_B(7)} - 1\right),$$

the variables BULK_DCT and CA are also removed. However, since at STEP4, all $F = \frac{31}{11}\left(\frac{\Lambda_B(5)}{\Lambda_B(6)} - 1\right)$ have p values smaller than LSSTAY=.15, no more elimination of variables is possible. The algorithm therefore stops at this stage. As a result, we conclude that the other six variables, namely, PH_VALUE, NITRO, PHOSPHO, MG, K, and NA should be used in the construction of the linear discriminant function.

It may be worth mentioning that, as observed previously, the procedure at every step implicitly relies on the comparison of two one-way classification (M)ANOVA models. For example, at Step2 in Output 5.12, the value of the F statistic =1.427 for the eliminated variable BULK_DCT is the same as that corresponding to the CLASS variable SOIL_GRP when BULK_GRP and all other seven variables are taken as covariates. This test and the corresponding F statistic can be alternatively obtained by fitting an analysis of the covariance model using the SAS statements

```
proc glm;
class soil_grp;
model bulk_dct= ph_value nitro phospho
ca mg k na soil_grp/ss3;
```

We have not included the corresponding output here.

It must be emphasized that the variable selection methods, although intuitively attractive and often useful, have weak theoretical foundation or justification. Also, it is neither necessary nor does it frequently occur in practice that the three variable selection methods would result in the same set of variables in the final model. It is, in general, difficult to make a recommendation about the preference of a particular method. Further, choices of variables arrived at by these methods are tentative and not necessarily optimal. Consequently, we recommend that the final selection should not always be taken on its face value and its validity should be subjected to further examination using personal knowledge of the particular subject areas. In fact, in the particular data set discussed here, the forward and backward methods not only result in different choices but also appear to present some contradictions. A run of the forward selection method (SAS code and output not presented here) would show that the variable CONDCTVT is deemed as the most significant and enters at the very first step. On the contrary, as Output 5.12 shows, it was deemed the least significant at STEP1 when the backward elimination procedure is adopted. There is, however, no contradiction since by itself CONDCTVT may be most significant (as the forward selection shows). However, because it is correlated with other variables in the presence of these other variables, it has very little additional information to provide (as the backward elimination suggests).

It may be pointed out that in their original work Horton, Russell, and Moore (1968) used the forward selection procedure. They used the chi-square approximation of Wilks' Λ instead of the F approximation adopted in PROC STEPDISC. Their list of finally selected variables is, however, identical to what would be obtained using the FORWARD option in the PROC STEPDISC statement. There are several other options available in the STEPDISC procedure that allow us to perform some more problem-specific analyses of the data. One especially appealing feature is the ability to force a specific set of variables to appear in the discriminant function. This can be very useful when a partial list of potentially important variables is a priori known. The desired task can be easily carried out by using the INCLUDE= option in the PROC STEPWISE statement. See the *SAS/STAT User's Guide, Version 6, Fourth Edition, Volume 2* for details or the SAS online documentation.

While concluding this section, we remark on the possibility of simulteneously discarding a set of variables. If we want to consider that only a smaller group of q variables may suffice and the remaining $(p - q)$ variables do not add any significant extra discriminatory power, then a statistical test based on the likelihood ratio and its chi-square or Rao's F approximation can be devised. See Seber (1984, p. 340) for details. However, the question of how to choose the potentially unimportant $p - q$ variables to be discarded, to form the statistical hypothesis, has no adequate answer. The algorithmic approach given above although not perfect, circumvents this problem.

5.8 When Dimensionality Exceeds Sample Size

There are frequent situations when it is expensive to sample observations from the populations. However, once a unit is sampled, a number of different characters or measurements on it can be obtained relatively inexpensively. One such situation arises when data are collected using satellites (for example, in remote-sensing applications). In such data, it is therefore quite possible that $n = \sum_{i=1}^{g} n_i$, the total sample size, is less than p, the number of variables that are measured. When attempting to do the linear discriminant analysis (or in general most multivariate analyses), this will result in difficulty as the pooled sample variance-covariance matrix will no longer be nonsingular and hence cannot be inverted as needed to construct the linear discriminant function. Although SAS procedures like DISCRIM and CANDISC attempt to circumvent this problem by replacing the inverse with a corresponding generalized inverse computed by using spectral decomposition (see the *SAS/STAT User's Guide, Version 6, Fourth Edition, Volume* 2, pp. 392, 691), we would like to avoid this problem altogether, if possible. Kshirsagar, Kocherlakota, and Kocherlakota (1990) have suggested a stepwise discriminant analysis approach to deal with such situations. The idea behind the stepwise discriminant analysis is to partition the set of variables into several subsets so that for each subset n is bigger than the number of variables in the subset. This will ensure that the pooled sample variance-covariance matrix for each of these subsets of variables is nonsingular and hence invertible.

At the first stage, separate canonical discriminant analyses are performed using the variables in each of these several subsets. Since the first canonical variable is the best linear combination of the variables in the particular subset, to describe the differences between the populations, it seems appropriate that at the second stage a single discriminant analysis is performed using the canonical scores on these first canonical variables. Of course, to ensure the nonsingularity of the pooled sample variance-covariance matrix at the second stage, we must again have the number of subsets no more than $n - 1$.

The linear discriminant function so arrived at can now be used for classification purposes. It is a linear function of all first canonical variables, each of which, in turn, is a linear function of some of the original variables. As a result, the linear discriminant function so obtained uses the information available on each of the original variables. We will illustrate the procedure using part of the data given in Horton, Russell, and Moore (1968).

EXAMPLE 13 ***Stepwise Discrimination, Characterization of Soil Layers*** We reconsider the data presented in Horton, Russell, and Moore (1968), but now consider only the first four soil layers (SOIL_GRP). These all correspond to the top microtopographic position of above the 60-cm contour, but for the four different soil layers described earlier. Thus, our present objective is to derive a linear discriminant function to discriminate the four soil layers. Let us suppose that only the first two observations are available on each soil layer, and all nine measurements described earlier are taken on each soil sample.

With $p = 9$ and $n_i = 2$ for all i, we only have $n = 8$ observations. As $n < p$, this would result in a singular pooled variance-covariance matrix. To alleviate this problem, we will follow the stepwise discrimination scheme described earlier. Nine variables are divided into three subsets, namely, {PH_VALUE, NITRO, BULK_DCT}, {PHOSPHO, CA, MG} and {K, NA, CONDCTVT} of three variables each. Using PROC CANDISC, we perform the three separate canonical discriminant analyses for the above three subsets of variables. The scores on the first canonical variable (CAN1) are first saved in three separate data sets using the OUT= option. These are then combined in a single data set titled STAGE2. The first canonical variables obtained from the three data sets are named D1, D2, and D3, respectively. Next the discriminant analysis is performed using variables D1, D2, and D3. Results of classification using D1, D2, and D3 are contained in the data set called RESULTS, and various statistical information is available in the data set STATALL. The complete SAS program is presented in Program 5.13, and selected parts of the output are presented in Output 5.13.

From Output 5.13 (titled Statistics from Subset 1, Subset 2 and Subset 3, respectively) we observe that the first canonical variables are

$$D1 = 10.1658 * \text{PH\_VALUE} + 1.4444 * \text{NITRO} - 3.1542 * \text{BULK\_DCT},$$

$$D2 = 0.0485 * \text{PHOSPHO} + 0.2092 * \text{CA} - 0.6054 * \text{MG},$$

and

$$D3 = 79.2828 * \text{K} - 1.1856 * \text{NA} - 5.4532 * \text{CONDCTVT}.$$

From $D1$, $D2$, and $D3$, the linear discriminant function can be constructed using the method illustrated in Section 5.5. The corresponding SAS code is given as part of Program 5.13. As in Section 5.5, using the quantities referred to in the output as CONSTANT and COEFFICIENT VECTORS, in the SAS output (last page, lines 95-102) we obtain the four discriminant scores

$$L_1 = -705.95 + 13.31 * D1 + 24.79 * D2 + 43.38 * D3,$$

$$L_2 = -362.65 + 7.87 * D1 + 15.72 * D2 + 31.73 * D3,$$

$$L_3 = -19.43 - 4.37 * D1 - 4.22 * D2 - 5.30 * D3,$$

and

$$L_4 = -1755.68 - 16.81 * D1 - 36.29 * D2 - 69.81 * D3,$$

and classify any new observation $(D1, D2, D3)'$ into the tth if

$$L_t \geq L_s \text{ for } s = 1, \ldots, 4, \ s \neq t.$$

```
/* Program 5.13 */

options ls = 64 ps=45 nodate nonumber;
title1  'Output 5.13';

data horton;
input soil_grp $ ph_value nitro bulk_dct phospho
ca mg k na condctvt;
lines;
1 5.40 .188 .92 215 16.35 7.65 .72 1.14 1.09
1 5.65 .165 1.04 208 12.25 5.15 .71 .94 1.35

2 5.14 .164 1.12 174 14.17 8.12 .70 2.17 1.85
2 5.10 .094 1.22 129 8.55 6.92 .81 2.67 3.18

3 4.37 .112 1.07 121 8.85 10.35 .74 5.74 5.73
3 4.39 .058 1.54 115 4.73 6.91 .77 5.85 6.45

4 3.88 .077 1.25 127 6.41 10.96 .56 9.67 10.64
4 4.07 .046 1.54 91 3.82 6.61 .50 7.67 10.07
;
/*
Data from Horton et. al (1968),
reprinted with permission from International Biometric Society
*/

proc candisc data = horton out =a1 ncan =1;
class soil_grp;
```

```
var ph_value nitro bulk_dct ;
title2 'Discrimination of Soil Layers';
title3 "Statistics from Subset 1";

proc candisc data = horton out =a2 ncan =1;
class soil_grp;
var phospho ca mg ;
title2 'Discrimination of Soil Layers';
title3 "Statistics from Subset 2";

proc candisc data = horton out =a3 ncan =1;
class soil_grp;
var k na condctvt ;
title2 'Discrimination of Soil Layers';
title3 "Statistics from Subset 3";

data a1;
set a1;
d1 =can1;
keep soil_grp  d1;

data a2;
set a2;
d2 = can1;
keep soil_grp  d2;

data a3;
set a3;
d3 = can1;
keep soil_grp  d3;

data stage2;
merge a1 a2 a3 ;
by soil_grp;

proc discrim data = stage2
method =normal pool=yes short outstat = statall out = results;
class soil_grp;
var d1 d2 d3 ;
title2
'Discrimination Using 1st Canonical Variables';
title3 " ";
proc print data = results;
proc print data = statall;
```

Output 5.13

```
                    Output 5.13
           Discrimination of Soil Layers


              Statistics from Subset 1

             Raw Canonical Coefficients

             Variable               Can1

             ph_value        10.16580254
             nitro            1.44441621
             bulk_dct        -3.15421217
```

```
                    Output 5.13
            Discrimination of Soil Layers

           Canonical Discriminant Analysis


              Statistics from Subset 1

            Raw Canonical Coefficients

                                  CAN1

            PH_VALUE       10.16580254
            NITRO           1.44441621
            BULK_DCT       -3.15421217


              Statistics from Subset 2

            Raw Canonical Coefficients

            Variable              Can1

            phospho       0.0484536607
            ca            0.2091640997
            mg            -.6053881986


              Statistics from Subset 3

            Raw Canonical Coefficients

            Variable              Can1

            k              79.28284360
            na             -1.18559347
            condctvt       -5.45319679


      Discrimination Using 1st Canonical Variables

Obs  soil_grp  _TYPE_    _NAME_        d1        d2        d3

  1            N                      8.00     8.000      8.00
  2     1      N                      2.00     2.000      2.00
  3     2      N                      2.00     2.000      2.00

  .
  .
  .

 95     1      LINEAR    _LINEAR_     13.31     24.79     43.38
 96     1      LINEAR    _CONST_    -705.95   -705.95   -705.95
 97     2      LINEAR    _LINEAR_      7.87     15.72     31.73
 98     2      LINEAR    _CONST_    -362.65   -362.65   -362.65
 99     3      LINEAR    _LINEAR_     -4.37     -4.22     -5.30
```

```
100     3     LINEAR    _CONST_       -19.43     -19.43     -19.43
101     4     LINEAR    _LINEAR_      -16.81     -36.29     -69.81
102     4     LINEAR    _CONST_     -1755.68   -1755.68   -1755.68


                 Canonical Discriminant Analysis

                   Raw Canonical Coefficients

                                        CAN1

                   PHOSPHO        0.0484536607
                   CA             0.2091640997
                   MG             -.6053881986


                    Statistics from Subset 3

                 Canonical Discriminant Analysis

                   Raw Canonical Coefficients

                                        CAN1

                   K                79.28284360
                   NA               -1.18559347
                   CONDCTVT         -5.45319679


            Discrimination Using 1st Canonical Variables

OBS  SOIL_GRP  _TYPE_     _NAME_          D1         D2         D3

  1            N                        8.00      8.000       8.00
  2     1      N                        2.00      2.000       2.00
  3     2      N                        2.00      2.000       2.00

.
.
.

 95     1     LINEAR    _LINEAR_       13.31      24.79      43.38
 96     1     LINEAR    _CONST_      -705.95    -705.95    -705.95
 97     2     LINEAR    _LINEAR_        7.87      15.72      31.73
 98     2     LINEAR    _CONST_      -362.65    -362.65    -362.65
 99     3     LINEAR    _LINEAR_       -4.37      -4.22      -5.30
100     3     LINEAR    _CONST_       -19.43     -19.43     -19.43
101     4     LINEAR    _LINEAR_      -16.81     -36.29     -69.81
102     4     LINEAR    _CONST_     -1755.68   -1755.68   -1755.68
```

Of course, the classification is done automatically when PROC DISCRIM is used, as illustrated in Section 5.5. Kshirsagar, Kocherlakota, and Kocherlakota (1990) also present an alternative approach to dealing with the problem of an excessive number of variables. This second approach, although similar in spirit, employs principal component analysis. Instead of forming subsets of variables and obtaining the first canonical variables, the first few most important principal components (say, $q < n$ of them) of the pooled sample variance-covariance matrix are identified. The discriminant analysis is then performed using these q

principal components, and a linear discriminant function is identified. In SAS, this analysis can be implemented using the PRINCOMP and DISCRIM procedures.

Using PROC PRINCOMP, we can find the principal components of the pooled sample variance-covariance matrix. However, since PROC PRINCOMP would routinely compute the total sample variance-covariance matrix, some data manipulation is needed. Note that

$$\mathbf{S} = \frac{1}{n-g}\sum_{i=1}^{g}\sum_{j=1}^{n_i}(\mathbf{x}_{ij}-\bar{\mathbf{x}}_i)(\mathbf{x}_{ij}-\bar{\mathbf{x}}_i)'.$$

Thus, we define $\mathbf{z}_{ij} = \mathbf{x}_{ij} - \bar{\mathbf{x}}_i$. The total sample variance-covariance matrix of transformed data, $\mathbf{z}_{ij}$ is (apart from a constant, which does not affect the final result in the present context) the pooled sample variance-covariance matrix of $\mathbf{x}_{ij}$. By finding the sample means for samples from different populations a priori and then using PROC STANDARD as in Program 5.14, we obtain the transformed data on $\mathbf{z}_{ij}$. The following SAS statements will compute the coefficients of all principal components and the corresponding principal component scores for various observations:

```
proc sort data= horton;
by soil_grp;
proc standard data= horton mean=0 out=pc;
by soil_grp;
var z1-z9;
proc princomp data=pc cov out=pcoutput;
var z1-z9;
```

The variables $Z1, \ldots, Z9$, which before the execution of PROC STANDARD were the same as the nine original variables, are standardized within SOIL_GRP to have the zero mean. These standardized variables with the same name $Z1, \ldots, Z9$ are stored in an output data set named PC. This data set is used as input for PROC PRINCOMP. We want to compute the principal components of the variance-covariance matrix and not the correlation matrix. This is achieved by using the COV option in the PROC PRINCOMP statement (see Section 2.3). The principal component scores are stored in the output file titled PCOUTPUT. This program is presented as the first part of Program 5.14.

The output of this program is presented in the early parts of Output 5.14. As expected from the fact that here, $n < p$, and as evident from certain zero eigenvalues, the matrix **S** is singular. Moreover, the first two principal components seem to be sufficient to explain more than 99% of the variability as measured by the total variance (see Section 2.2). Therefore, coefficients from the first two principal components will be used to create the new variables as the linear combinations of the nine original variables. Specifically, these are given as

$$\begin{aligned} PC1 \;&= -.002049 * \text{PH\_VALUE} + 0.001392 * \text{NITRO} - .005503 * \text{BULK\_DCT} \\ &+ 0.990179 * \text{PHOSPHO} + 0.117297 * \text{CA} + 0.073206 * \text{MG} - .000844 * \text{K} \\ &+ 0.014601 * \text{NA} - .013352 * \text{CONDCTVT} \end{aligned}$$

and

$$\begin{aligned} PC2 \;&= -.028268 * \text{PH\_VALUE} + 0.006893 * \text{NITRO} - .066610 * \text{BULK\_DCT} \\ &- .137925 * \text{PHOSPHO} + 0.760426 * \text{CA} + 0.624900 * \text{MG} + 0.000041 * \text{K} \\ &+ 0.005790 * \text{NA} - .083135 * \text{CONDCTVT}. \end{aligned}$$

Linear discriminant analysis will be performed on these two variables. The SAS statements to do so are given below.

```
proc discrim data = horton;
method =normal pool=yes short outstat = statall out = results;
class soil_grp;
var pc1 pc2;
```

Parts of the resulting output are presented in Output 5.14. Based on this output, the discriminant scores can be obtained as

$$L_1 = -75.2832 + 0.4873 * PC1 - 3.2493 * PC2,$$

$$L_2 = -33.6366 + 0.3496 * PC1 - 1.7732 * PC2,$$

$$L_3 = -20.6620 + 0.2724 * PC1 - 1.4235 * PC2,$$

$$L_4 = -18.6655 + 0.2514 * PC1 - 1.4831 * PC2.$$

```
/* Program 5.14 */

options ls = 64 ps=45 nodate nonumber;
title1  'Output 5.14';

data horton;
input soil_grp $ ph_value nitro bulk_dct phospho
ca mg k na condctvt;
z1 =ph_value;
z2= nitro;
z3 = bulk_dct;
z4=phospho;
z5 =ca;
z6 =mg;
z7 = k;
z8 =na;
z9 =condctvt;
lines;
1 5.40 .188 .92 215 16.35 7.65 .72 1.14 1.09
1 5.65 .165 1.04 208 12.25 5.15 .71 .94 1.35

2 5.14 .164 1.12 174 14.17 8.12 .70 2.17 1.85
2 5.10 .094 1.22 129 8.55 6.92 .81 2.67 3.18

3 4.37 .112 1.07 121 8.85 10.35 .74 5.74 5.73
3 4.39 .058 1.54 115 4.73 6.91 .77 5.85 6.45

4 3.88 .077 1.25 127 6.41 10.96 .56 9.67 10.64
4 4.07 .046 1.54 91 3.82 6.61 .50 7.67 10.07
;
/*
Data from Horton et. al (1968),
reprinted with permission from International Biometric Society.
*/

proc sort data = horton;
by soil_grp;
proc standard data = horton mean = 0 out =pc;
by soil_grp;
var z1-z9;
Title2 'Data on Nine Variables: Standardized for Group Means';
proc princomp data = pc cov out = pcoutput;
var z1-z9;
```

```
data horton;
set horton;
pc1 = -.002049*ph_value+  0.001392*nitro -.005503*bulk_dct
+ 0.990179*phospho+0.117297*ca+0.073206*mg-.000844*k
+0.014601*na -.013352*condctvt;

pc2=  -.028268*ph_value + 0.006893*nitro -.066610*bulk_dct
-.137925*phospho + 0.760426*ca + 0.624900*mg + 0.000041*k
+ 0.005790*na -.083135 *condctvt;

proc discrim data =horton
method =normal pool=yes short outstat = statall out = results;
class soil_grp;
var pc1 pc2 ;
title2
'Discrimination Using First Two Principal Components';
proc print data = results;
proc print data = statall;
```

Output 5.14

Output 5.14
Data on Nine Variables: Standardized for Group Means

The PRINCOMP Procedure

Eigenvalues of the Covariance Matrix

	Eigenvalue	Difference	Proportion	Cumulative
1	248.085908	245.515965	0.9856	0.9856
2	2.569942	1.544222	0.0102	0.9958
3	1.025720	0.995798	0.0041	0.9999
4	0.029923	0.029923	0.0001	1.0000
5	0.000000	0.000000	0.0000	1.0000
6	0.000000	0.000000	0.0000	1.0000
7	0.000000	0.000000	0.0000	1.0000
8	0.000000	0.000000	0.0000	1.0000
9	0.000000		0.0000	1.0000

Eigenvectors

	Prin1	Prin2	Prin3	Prin4	Prin5
z1	-.002049	-.028268	-.039799	-.287058	-.008969
z2	0.001392	0.006893	-.005690	-.031127	0.999475
z3	-.005503	-.066610	-.038740	0.320565	0.010230
z4	0.990179	-.137925	0.008294	-.015563	-.000865
z5	0.117297	0.760426	-.488415	0.379319	0.003625
z6	0.073206	0.624900	0.631339	-.375075	-.012499
z7	-.000844	0.000041	0.030107	0.053718	0.001845
z8	0.014601	0.005790	0.491258	0.475905	0.017558
z9	-.013352	-.083135	0.342687	0.547439	0.019592

Eigenvectors

	Prin6	Prin7	Prin8	Prin9
z1	0.016666	0.956479	0.000000	0.000000
z2	0.000000	0.000000	0.000000	0.000000
z3	-.016105	0.092992	0.939286	0.000000
z4	0.001432	-.006313	0.002332	0.012798
z5	-.005621	0.116375	-.106644	-.024521
z6	0.001202	-.067810	0.205661	0.133483
z7	0.998100	0.000000	0.000000	0.000000
z8	-.040452	0.164342	-.158817	-.691200
z9	-.039844	0.176949	-.197088	0.709690

Discrimination Using First Two Principal Components

Obs	soil_grp	_TYPE_	_NAME_	pc1	pc2
1		N		8.00	8.00
2	1	N		2.00	2.00
3	2	N		2.00	2.00
.					
.					
.					
74	1	LINEAR	_LINEAR_	0.49	-3.249
75	1	LINEAR	_CONST_	-75.28	-75.283
76	2	LINEAR	_LINEAR_	0.35	-1.773
77	2	LINEAR	_CONST_	-33.64	-33.637
78	3	LINEAR	_LINEAR_	0.27	-1.423
79	3	LINEAR	_CONST_	-20.66	-20.662
80	4	LINEAR	_LINEAR_	0.25	-1.483
81	4	LINEAR	_CONST_	-18.6655	-18.6655

It may, however, be observed from the error count estimates that the discriminatory power in this case is not very good. This is not very surprising, as Kshirsagar, Kocherlakota, and Kocherlakota (1990) point out, principal component ordering often has no bearing on the ability of the new variables to discriminate between the populations, and the criteria, other than the magnitude of % variability explained, should be used to select appropriate principal components for the purpose of discrimination. Of course, the added problem in the present case is that the extremely small samples are taken from each of the four populations.

Finally, in closing this section, we may point out another way of dealing with the problem of dimensionality exceeding the sample size: it may be possible to do a forward selection of variables and use only the selected variables for discrimination. This may result in a shorter list of variables not exceeding the sample size. It is difficult to say in all certainty which among the three suggested methods is superior to the other two, but we believe that the stepwise discriminant analysis may work better than the other two in most situations of this type.

5.9 Logistic Discrimination

In the previous discussion of linear or quadratic discriminant functions, one of the assumptions made was that all the explanatory variables are continuous. There may be situations when some or all of these variables are categorical. It has been observed in the literature that the linear or quadratic discriminant functions become relatively less effective when some of the explanatory variables are categorical or even in situations when the joint distribution of these variables significantly departs from multivariate normality. Efron (1975) has shown that although the discriminant function approach is quite efficient in the ideal situation of multivariate normality, under nonnormality, an alternative approach termed logistic regression may be preferred. See Press and Wilson (1978), Crawley (1979), Byth and McLachlan (1980) or O'Hara, Hosmer, Lemeshow, and Hartz (1982) for more details. Also see Stokes, Davis, and Koch (1995) for a SAS oriented approach. We will confine our discussion to the two-population case only.

5.9.1 Logistic Regression Model

The logistic discrimination approach consists of performing the *logistic regression* of the categorical variable, indicating the population membership, on various explanatory variables. Some of these explanatory variables may themselves be qualitative and hence may have been represented in the model by certain dummy variables. The particular function to be treated as a response variable in the model is written in terms of the *logit function*, defined as

$$logit\,(\pi) = \log_e\left(\frac{\pi}{1-\pi}\right),$$

where π is the probability of a unit belonging to population 1 and $1-\pi$ is that for population 2. We define the dummy variable y to indicate the membership of a unit in one of the two populations as

$$y = \begin{cases} 1 & \text{if the unit actually belongs to population 1} \\ 0 & \text{if the unit actually belongs to population 2} \end{cases}.$$

Thus, the distribution of y is binomial with $P(y = 1) = \pi$ and $P(y = 0) = 1 - \pi$. We assume the logit function $logit\,(P(y = 1)) = logit(\pi)$ to be a linear function, of k explanatory variables,

$$logit\,(\pi) = \beta_0 + \beta_1 x_1 + \ldots + \beta_k x_k, \tag{5.13}$$

which provides the basic underlying model for the estimation of unknown parameters $\beta_0, \ldots, \beta_k$ and subsequently, the decision rule for discrimination and classification. The estimation is carried out under the maximum likelihood criterion. The closed form expressions for the parameter estimates are not possible, and the maximization of likelihood function is carried out iteratively.

To assess the model fit, we could use *-2 times log of likelihood function* as a criterion, which under the null hypothesis that none of the explanatory variable is significant, approximately follows a chi-square distribution with k df. Statistical significance of individual explanatory variables is judged by the square of the ratio of the corresponding parameter estimate and its standard error, which approximately follows a chi-square distribution with 1 df. The corresponding test statistic is often referred to as the Wald's chi-square statistic.

In SAS, the logistic regression model can be fitted using the LOGISTIC, CATMOD, GENMOD, PROBIT or NLIN procedures. We will, however, confine our discussion to the first choice (as it can also provide certain relevant information about correct classifications or misclassifications). As indicated earlier, we define a response variable y taking values 1 and 0. In keeping with the common usage, the membership in population 1 (i.e., when $y = 1$) will be referred to as the occurrence of the EVENT. The membership in population 2 (i.e., when $y = 0$) is referred to as NONEVENT.

EXAMPLE 14 ***Logistic Discrimination, Crystals in Urine Data*** The data were collected by J. S. Elliot of Stanford University School of Medicine and were contributed in the book by Andrews and Herzberg (1985) by D. P. Byar of the National Cancer Institute. A total of 77 urine samples were collected and analyzed for six physical characteristics namely, specific gravity (SG), pH value (PH), osmolarity (MOSM), conductivity (MMHO), urea concentration (UREA) and calcium concentration (CALCIUM). Subjects were classified according to the presence of the calcium oxalate crystals. An objective of the study may be to determine the possible presence of crystals in urine using these physical characteristics. The two populations for possible classification of a subject are thus defined by the presence of the crystals (PRESENCE=YES or NO). In the present analysis, two subjects with missing values for at least one variable have been deleted.

The plots of data (not shown here) on individual characteristics in the individual population indicate that these measurements do not follow the normal distribution. In view of this lack of normality, we may want to apply logistic discrimination instead of the usual linear or quadratic discriminant analysis. This is done using the LOGISTIC procedure.

```
/* Program 5.15 */

options ls = 64 ps=45 nodate nonumber;
title1  'Output 5.15';

data urine;
infile 'urine.dat' firstobs=7;
input presence $ sg ph mosm mmho urea calcium ;
patient = _n_ ;
if presence = 'yes' then crystals  = 1;
if presence = 'no' then crystals  = 0;

proc logistic data = urine descending outest = est
covout nosimple ;
model crystals = sg ph mosm mmho urea calcium
/ risklimits link=logit covb ctable pprob=.5 ;
output out=predict predprobs = cross p = phat lower = lcl
upper = ucl  / alpha =.05;
title2 'Logistic Regression: Crystals Data';
run;

proc print data = predict ;
*var presence phat lcl ucl;
title2 'Predicted Probabilities: Logistic Regression';
title3 ' Crystals Data';
run;
```

Output 5.15

```
                    Output 5.15
        Logistic Regression: Crystals Data

              The LOGISTIC Procedure

                 Model Information

Data Set                          WORK.URINE
Response Variable                 crystals
Number of Response Levels         2
Number of Observations            77
Link Function                     Logit
Optimization Technique            Fisher's scoring


                 Response Profile

      Ordered                          Total
        Value       crystals       Frequency

            1              1              33
            2              0              44

NOTE: 2 observations were deleted due to missing values for the
      response or explanatory variables.


              Model Convergence Status

    Convergence criterion (GCONV=1E-8) satisfied.


               Model Fit Statistics

                                     Intercept
                      Intercept            and
   Criterion               Only     Covariates

   AIC                  107.168         71.560
   SC                   109.512         87.967
   -2 Log L             105.168         57.560


   Testing Global Null Hypothesis: BETA=0

Test                 Chi-Square       DF     Pr > ChiSq

Likelihood Ratio        47.6079        6         <.0001
Score                   32.4981        6         <.0001
Wald                    15.5886        6         0.0161
```

Analysis of Maximum Likelihood Estimates

Parameter	DF	Estimate	Standard Error	Chi-Square	Pr > ChiSq
Intercept	1	-355.3	222.8	2.5443	0.1107
sg	1	355.9	222.1	2.5681	0.1090
ph	1	-0.4957	0.5698	0.7569	0.3843
mosm	1	0.0168	0.0178	0.8904	0.3454
mmho	1	-0.4328	0.2512	2.9679	0.0849
urea	1	-0.0320	0.0161	3.9443	0.0470
calcium	1	0.7837	0.2422	10.4717	0.0012

Odds Ratio Estimates

Effect	Point Estimate	95% Wald Confidence Limits	
sg	>999.999	<0.001	>999.999
ph	0.609	0.199	1.861
mosm	1.017	0.982	1.053
mmho	0.649	0.396	1.061
urea	0.968	0.938	1.000
calcium	2.190	1.362	3.520

Association of Predicted Probabilities and Observed Responses

Percent Concordant	89.9	Somers' D	0.798
Percent Discordant	10.1	Gamma	0.799
Percent Tied	0.1	Tau-a	0.396
Pairs	1452	c	0.899

Wald Confidence Interval for Adjusted Odds Ratios

Effect	Unit	Estimate	95% Confidence Limits	
sg	1.0000	>999.999	<0.001	>999.999
ph	1.0000	0.609	0.199	1.861
mosm	1.0000	1.017	0.982	1.053
mmho	1.0000	0.649	0.396	1.061
urea	1.0000	0.968	0.938	1.000
calcium	1.0000	2.190	1.362	3.520

Estimated Covariance Matrix

Variable	Intercept	sg	ph	mosm
Intercept	49627.23	-49473.9	-3.14303	2.680354
sg	-49473.9	49334.96	1.078041	-2.66383
ph	-3.14303	1.078041	0.324636	-0.00176
mosm	2.680354	-2.66383	-0.00176	0.000317
mmho	-27.9131	27.64902	0.030243	-0.00428
urea	-1.08165	1.065143	0.002291	-0.00025
calcium	-7.54316	7.627181	-0.01733	0.00114

Estimated Covariance Matrix

Variable	mmho	urea	calcium
Intercept	-27.9131	-1.08165	-7.54316
sg	27.64902	1.065143	7.627181
ph	0.030243	0.002291	-0.01733
mosm	-0.00428	-0.00025	0.00114

Estimated Covariance Matrix

Variable	mmho	urea	calcium
mmho	0.06312	0.003708	-0.02545
urea	0.003708	0.00026	-0.00187
calcium	-0.02545	-0.00187	0.058651

Classification Table

	Correct		Incorrect		Percentages				
Prob Level	Event	Non-Event	Event	Non-Event	Correct	Sensi-tivity	Speci-ficity	False POS	False NEG
0.500	21	36	8	12	74.0	63.6	81.8	27.6	25.0

We consider all the six physical characteristics as our explanatory variables. Using the categorical variable PRESENCE, we define the response variable CRYSTALS, taking value 1 if PRESENCE=YES and value 0 if PRESENCE=NO.

The following SAS code achieves the task:

```
proc logistic data = urine descending nosimple;
model crystals = sg ph mosm mmho urea calcium/
risklimits link=logit covb ctable pprob=.5;
```

where the option NOSIMPLE in the PROC LOGISTIC statement is used to suppress the printing of simple descriptive statistics. The DESENDING option is chosen to ensure that the presence of crystals (CRYSTALS=1) is defined as the EVENT and the absence is termed NONEVENT in the SAS output. It is so, since the DESCENDING option reverses the sorting order of the response variable from higher numerical values to lower numerical values or reverse alphabetical order, whichever may be the case. Other choices of such assignments are also possible by using the ORDER=option. These are given in Chapter 27 of the *SAS/STAT User's Guide, Version 6, Fourth Edition, Volume 2* and are also discussed in Scheuchenpflug and Blettner (1996). In the MODEL statement, we have specified our logistic model, namely,

$$\text{LOGIT (CRYSTALS =1)} = \beta_0 + \beta_1 * \text{SG} + \beta_2 * \text{PH} + \beta_3 * \text{MOSM} + \beta_4 * \text{MMHO} + \beta_5 * \text{UREA} + \beta_6 * \text{CALCIUM},$$

by using the (default) option LINK=LOGIT. Several other options are also included in the MODEL statement. The option COVB is used to calculate and print the estimated variance-covariance matrix of $\hat{\beta}_0, \ldots, \hat{\beta}_6$. The CTABLE option is used to request the printing of the classification table in order to have an idea about the misclassifications. The PPROB= option is used to specify the probability cutoff value. Any observation with the estimated probability of its membership in the population corresponding to EVENT bigger than this cutoff will be classified as the EVENT. In our example (and for illustration purposes only), we have taken the value of PPROB=.5. However, as Seber (1984, p.312) points out, this cutoff point depends on the sampling design as well as the sample size(s). Thus, its value should be appropriately chosen. However, rather than determining the appropriate value for this cutoff, SAS provides the option of using any other choice of cutoff. The option RISKLIMITS prints the $100(1-\alpha)\%$ confidence intervals for the conditional odds ratios $\exp(\beta_i)$ for all independent variables. The value of α is specified by the ALPHA= option. The default value of α is .05, which is also used in our example. The LINK = LOGIT option, which is also the default choice, uses the log-odd function $logit(\pi) = log_e(\frac{\pi}{1-\pi})$. The complete SAS code is presented in Program 5.15.

Output 5.15 presents the output resulting from the execution of Program 5.15. Out of 77 observations, a total of $n_1 = 33$ correspond to population 1 (CRYSTALS=1), and the remaining $n_2 = 44$ correspond to population 2 (CRYSTALS=0). As indicated earlier, $-2\,log_e L$, the minus two times the log of maximized likelihood follows, under the null hypothesis, a chi-square distribution with k degrees of freedom. The value of the corresponding chi-square test statistic (= 47.6079 with 6 df) is highly significant (p value < 0.0001) leading us to reject the null hypothesis that none of the six physical characteristics are important. The score test also rejects the null hypothesis. The other two criteria, AIC and SC, are useful when comparing several models used for the same set of raw data. In general, the smaller the values of AIC and SC among these competing models, the better is the model.

From Output 5.15, the fitted model is seen to be

$$\begin{aligned}\text{LOGIT } (\hat{\text{P}} \text{ (CRYSTALS=1)}) = -355.3 &+ 355.9 * SG - 0.4957 * PH \\ &+ 0.0168 * MOSM - 0.4328 * MMHO \\ &- 0.0320 * UREA + 0.7837 * CALCIUM.\end{aligned}$$

The estimated standard errors (se) of various parameter estimates are also reported. Using these, we can obtain Wald's chi-square test statistic

$$W = \{\hat{\beta}_i/\text{se}(\hat{\beta}_i)\}^2,$$

which under the null hypothesis of $H_0 : \beta_i = 0$ approximately follows a chi-square distribution with $1df$. Wald's test suggests that perhaps variables SG, PH, MOSM, and MMHO (especially the middle two variables) are not significant (their respective p values are quite large) and hence may not be very informative in identifying the individuals with crystal formation in their urine. However, based on such observations, it may not always be appropriate to discard variables from the model. Instead, a variable selection scheme, to be discussed in Section 5.9.3, is recommended.

In the present context, our main interest is in the performance of the logistic model with respect to classification. The classification table given at the end of Output 5.15 and generated by the specification of the CTABLE and PPROB= options provides this information. The estimated probability of classifying an observation as EVENT, say $\widehat{\pi}$, is estimated using the logistic model described earlier in Equation 5.13. The observation is classified as an EVENT if $\widehat{\pi}$ is greater than the probability value specified in the PPROB= option. In our example, this value is specified as 0.5. Of the 33 EVENT cases, 21 are correctly classified; however, the 12 remaining cases have been misclassified as NONEVENT. Similarly, out of a total 44 NONEVENT cases, only 36 are classified correctly, whereas 8 cases have been incorrectly classified as the EVENT cases. Thus, only $100(21+36)/77 = 74\%$ cases have been correctly classified. It may be mentioned in passing that the situation would have been somewhat better if as suggested by Seber (1984), a value of $\frac{n_1}{n_1+n_2} = 0.5714$) instead of .5 was used for PPROB.

In view of the number of correct and incorrect classifications given in the Classification Table in Output 5.15, the *sensitivity* of this logistic discriminant, defined as the percentage of correctly predicted EVENT responses, is estimated as $100 \times (21/33) = 63.6\%$.

Likewise, *specificity* is defined as that, but for NONEVENT responses. As there are 36 correctly predicted NONEVENT responses out of a total of 44, the specificity of this logistic discriminant is estimated as $100 \times (36/44) = 81.8\%$.

False-positive and false-negative rates are defined a little differently. More specifically, the *false-positive rate* is defined as the percentage of misclassifications among those classified as EVENT. The Classification Table in Output 5.15 shows that a total of 21+8=29 cases are classified as EVENT, and 8 of these are incorrectly classified. Thus, the estimated false-positive rate is equal to $100 \times (8/29) = 27.6\%$. The *false-negative rate* is defined essentially in the same spirit, but for the NONEVENT cases. Its estimated value for our data is equal to $100 \times [12/(12 + 36)] = 25\%$.

More succinctly, all these quantities can be expressed in terms of simple algebraic formulas in terms of the following table, usually referred to as the *confusion matrix*. The numbers used in the table are adopted from the Classification Table of Output 5.15.

	Classification	
Actual Response	EVENT	NONEVENT
EVENT	$a_{11}(= 21)$	$a_{10}(= 12)$
NONEVENT	$a_{01}(= 8)$	$a_{00}(= 36)$

$$
\begin{array}{rrcl}
(i) & \%\text{ Correct classification} & = & 100(a_{11} + a_{00})/(a_{11} + a_{10} + a_{01} + a_{00}), \\
(ii) & \text{Sensitivity} & = & 100a_{11}/(a_{11} + a_{10}), \\
(iii) & \text{Specificity} & = & 100a_{00}/(a_{01} + a_{00}), \\
(iv) & \text{False-positive rate} & = & 100a_{01}/(a_{01} + a_{11}). \\
(v) & \text{False-negative rate} & = & 100a_{10}/(a_{10} + a_{00}).
\end{array}
$$

The values of predicted probability $\hat{\pi}$ for various observations and corresponding confidence limits can be obtained by using the OUTPUT statement,

```
output out=predict p = phat lower = lcl upper = ucl/alpha=0.05;
```

The above OUTPUT statement, which should appear after the MODEL statement in the SAS code, saves the output on $\hat{\pi}$ (denoted in output by PHAT) and on the corresponding lower and upper 95% (specified by the ALPHA=0.05 option after the /) confidence intervals in a data set named PREDICT. By using the PROC PRINT statement, we can print this information. These values are included as part of Output 5.15.

The association between the predicted probabilities and the observed response can be measured by various rank order correlation statistics. SAS reports four such statistics, namely, Somers' D, Gamma, Tau-a, and c. These are computed by first identifying the concordant, discordant, and tied pairs of observations and then computing the corresponding percentages. These percentages are also reported in the SAS output. If n_1 and n_2 are the respective sample sizes from the two populations, then the total number of pairs are $n_1 n_2$ which, in our example, is equal to $33 \times 44 = 1452$. Thus, any pair will have one observation from Π_1 (EVENT) and the other from Π_2 (NONEVENT). A pair is concordant if $\hat{\pi}$ corresponding to the EVENT-observation is greater than that corresponding to the NONEVENT-observation. A pair is discordant if the inequality is in the opposite direction. The cases of equality lead to the tied pairs. For our examples, the percentages of concordant, discordant, and tied pairs are 89.9, 10.1, and 0.1%, respectively. In theory, these percentages should obviously add to 100, but in our example this sum exceeds 100 due to some round-off error.

5.9.2 Identification of Misclassified Cases

Often interest may be not only in error rates but also in the cases that have been misclassified. In particular, we are interested in identifying the individual false positive and false negative cases. When the critical probability value PPROB has been specified, we classify an observation as EVENT if the corresponding probability $\widehat{\pi} >$ PPROB and as NON-EVENT if $\widehat{\pi} \leq$ PPROB. However, Pregibon (1981) has indicated that since the same data set has been used to estimate the logistic regression coefficients, these predicted probabilities are biased and hence using these for classification will result in the underestimation of error rates. In view of this, SAS does not use these probabilities to calculate the error count estimates. Instead, it applies a bias correction on $\widehat{\pi}$ to obtain the bias-corrected estimate $\widehat{\pi}^*$. Pregibon (1981) suggests that the error count estimates should be obtained by applying the *leave one out* method. In this method, the observation to be classified is removed from the data, parameters are estimated from the remaining data, and this model is then used to classify the particular observation.

This procedure is repeated for every observation in the data set. However, since such a procedure may be too resource-intensive, a one-step approximation is adopted to obtain the new parameter estimates. This approximation is also given by Pregibon (1981) and is also reported in the *SAS/STAT User's Guide, Version 6, Fourth Edition, Volume 2* (Chapter 27, p. 1092). These parameters are then used to obtain the bias-corrected estimate of π, denoted by $\widehat{\pi}^*$. In fact, the classification table printed by using the CTABLE and PPROB= options is obtained by using $\widehat{\pi}^*$ and not the original estimates $\widehat{\pi}$. This is precisely the reason that in Output 5.15, even though there are only 4 out of 44 NONEVENT cases with $\widehat{\pi} > PPROB = 0.5$ (when physically counted; the corresponding table of Output 5.15 has been suppressed as the same information is also available in the second column of Output 5.16), the classification table reports 8 and not 4 false positive cases. It is so since after bias correction there are four more cases for which $\widehat{\pi}^* >$ PPROB $= 0.5$. A similar argument can be made for false-negative cases. Unfortunately, in spite of the fact that $\widehat{\pi}^*$ values are actually computed and used for classification, they are not automatically reported anywhere as part of the output, thereby making it impossible to readily identify the cases that have been misclassified.

If there is a need, the values of $\widehat{\pi}^*$ can be computed with a little bit of back calculation. Specifically, using the quantities call DFBETAs (not defined here) used for diagnostic purposes and reported as output, we can compute the new parameter estimates corrected for bias as

$$\hat{\beta}_j^* = \hat{\beta}_j - se(\hat{\beta}_j) \cdot \text{DFBETA}_j, \; j = 0, \ldots, k,$$

where $se(\hat{\beta}_j)$ is the estimated standard error of $\hat{\beta}_j$. Using these bias-corrected estimates, the bias-corrected logit function is obtained as

$$\text{logit}\,(\hat{\pi}^*) = \hat{\beta}_0^* + \hat{\beta}_1^* x_1 + \cdots + \hat{\beta}_k^* x_k,$$

from which we can obtain $\hat{\pi}^*$ as

$$\hat{\pi}^* = e^{\text{logit}\,(\hat{\pi}^*)}/(1 + e^{\text{logit}\,(\hat{\pi}^*)}).$$

In the LOGISTIC procedure, $\hat{\pi}^*$ can be calculated by using the PREDPROBS = CROSS option in the OUTPUT statement. This option makes the cross-validated (leave one out method) version of predicted probabilities, viz. $\hat{\pi}^*$ available. This is supposedly a superior estimate of π than $\hat{\pi}$, and thus its use is generally preferred for classification purposes and subsequently for the estimation of error rates.

For illustration, we will identify the misclassified cases for the data set discussed in Example 14 and dealing with the presence of crystals in urine.

EXAMPLE 14 ***(continued)*** We return to Example 14. In order to obtain $\hat{\pi}^*$ (and 1 - $\hat{\pi}^*$), we use the PREDPROBS = CROSS option in the OUTPUT statement (see Program 5.16). This results in the computation of $\hat{\pi}^*$ and 1 - $\hat{\pi}^*$, which are reported as XP_1 and XP_0, respectively, in Output 5.16. Similarly, values of $\hat{\pi}$ for various observations are obtained using the option P = PHAT (and these are reported in the output as PHAT).

The classification rule is this: Classify an observation as EVENT if ($\hat{\pi}^*$ =) XP_1 >P_PROB and as NONEVENT, otherwise. Output 5.16 provides the complete classification details including all false-positive and false-negative cases. Consistent with the Classification Table in Output 5.15, there are 8 false-positive and 12 false-negative cases (after ignoring the missing value case, namely the 55th observation in the list). Selected output from Program 5.16 follows the code.

```
/* Program 5.16 */

options ls = 64 ps=45 nodate nonumber;
title1  'Output 5.16';

data urine;
infile 'urine.dat' firstobs=7;
input presence $ sg ph mosm mmho urea calcium ;
patient = _n_ ;

sg100 = 100*sg;
ph10=100*ph;
mmho10 =10*mmho;
cal100=100*calcium ;

if presence = 'no' then crystals  = 0;
if presence = 'yes' then crystals  = 1;

proc logistic data = urine descending nosimple ;
model crystals = sg ph mosm mmho urea calcium
/ risklimits link=logit ctable pprob=.5 ;
output out=predict predprobs = cross p = phat
dfbetas = dfbeta0  dfbeta1  dfbeta2  dfbeta3
dfbeta4  dfbeta5  dfbeta6 ;

title2 'Logistic Regression: Crystals Data';

data predict;
set predict;
p_prob =.5;
if (xp_1 > p_prob and crystals = 1)
 then decision = 'true_pos';
if (xp_1 <= p_prob and crystals = 1)
 then decision = 'fals_neg';
if (xp_1 <= p_prob and crystals = 0)
 then decision = 'true_neg';
if (xp_1 > p_prob and crystals = 0)
 then decision = 'fals_pos';

proc print data = predict ;
var phat xp_1 xp_0 presence decision  ;
title2 'Predicted Probabilities: Logistic Regression';
title3 ' Crystals Data';
run;
```

Output 5.16

Output 5.16
Predicted Probabilities: Logistic Regression
Crystals Data

Obs	phat	XP_1	XP_0	presence	decision
1	.	.	.	no	true_neg
2	0.24944	0.28315	0.71685	no	true_neg
3	0.07230	0.07668	0.92332	no	true_neg
4	0.09192	0.09703	0.90297	no	true_neg
5	0.04961	0.05161	0.94839	no	true_neg
6	0.48561	0.54410	0.45590	no	fals_pos
7	0.05547	0.05716	0.94284	no	true_neg
8	0.55680	0.71833	0.28167	no	fals_pos
9	0.16522	0.18008	0.81992	no	true_neg
10	0.02982	0.03087	0.96913	no	true_neg
11	0.25390	0.26755	0.73245	no	true_neg
12	0.02590	0.02707	0.97293	no	true_neg
13	0.05459	0.05726	0.94274	no	true_neg
14	0.06689	0.07090	0.92910	no	true_neg
15	0.09988	0.10732	0.89268	no	true_neg
16	0.42159	0.48361	0.51639	no	true_neg
17	0.04715	0.04890	0.95110	no	true_neg
18	0.04249	0.04543	0.95457	no	true_neg
19	0.16309	0.29802	0.70198	no	true_neg
20	0.02014	0.02051	0.97949	no	true_neg
21	0.41494	0.45352	0.54648	no	true_neg
22	0.05064	0.05470	0.94530	no	true_neg
23	0.04119	0.04428	0.95572	no	true_neg
24	0.14630	0.17448	0.82552	no	true_neg
25	0.39107	0.49766	0.50234	no	true_neg
26	0.29037	0.34685	0.65315	no	true_neg
27	0.46936	0.79189	0.20811	no	fals_pos
28	0.14908	0.16841	0.83159	no	true_neg
29	0.11389	0.12286	0.87714	no	true_neg
30	0.03977	0.04139	0.95861	no	true_neg
31	0.00680	0.00692	0.99308	no	true_neg
32	0.73140	0.78185	0.21815	no	fals_pos
33	0.50809	0.59691	0.40309	no	fals_pos
34	0.47795	0.56332	0.43668	no	fals_pos
35	0.09296	0.10194	0.89806	no	true_neg
36	0.28111	0.29606	0.70394	no	true_neg
37	0.02910	0.02974	0.97026	no	true_neg
38	0.54896	0.61753	0.38247	no	fals_pos
39	0.20793	0.23274	0.76726	no	true_neg
40	0.49898	0.54775	0.45225	no	fals_pos
41	0.08677	0.09346	0.90654	no	true_neg
42	0.07370	0.07747	0.92253	no	true_neg
43	0.21601	0.23067	0.76933	no	true_neg
44	0.25910	0.27480	0.72520	no	true_neg
45	0.09974	0.10926	0.89074	no	true_neg
46	0.75453	0.73343	0.26657	yes	true_pos
47	0.99978	0.99978	0.00022	yes	true_pos
48	0.72713	0.70151	0.29849	yes	true_pos
49	0.51188	0.43702	0.56298	yes	fals_neg
50	0.95083	0.94792	0.05208	yes	true_pos
51	0.99999	0.99999	0.00001	yes	true_pos
52	0.35460	0.30090	0.69910	yes	fals_neg

```
53   0.16626   0.11521   0.88479       yes   fals_neg
54   0.02314   0.00794   0.99206       yes   fals_neg
55   .         0.00794   0.99206       yes   fals_neg
56   0.55570   0.48522   0.51478       yes   fals_neg
57   0.85840   0.81911   0.18089       yes   true_pos
58   0.96593   0.96249   0.03751       yes   true_pos
59   0.99997   0.99997   0.00003       yes   true_pos
60   0.69575   0.64109   0.35891       yes   true_pos
61   0.93381   0.93037   0.06963       yes   true_pos
62   0.55461   0.45415   0.54585       yes   fals_neg
63   0.99990   0.99989   0.00011       yes   true_pos
64   0.99949   0.99949   0.00051       yes   true_pos
65   0.71099   0.66903   0.33097       yes   true_pos
66   0.23778   0.19110   0.80890       yes   fals_neg
67   0.94554   0.94132   0.05868       yes   true_pos
68   0.38680   0.26307   0.73693       yes   fals_neg
69   0.12206   0.04782   0.95218       yes   fals_neg
70   0.49882   0.39180   0.60820       yes   fals_neg
71   0.99966   0.99966   0.00034       yes   true_pos
72   0.98786   0.98761   0.01239       yes   true_pos
73   0.55479   0.46968   0.53032       yes   fals_neg
74   0.95349   0.95023   0.04977       yes   true_pos
75   0.99292   0.99272   0.00728       yes   true_pos
76   0.40635   0.11843   0.88157       yes   fals_neg
77   0.98290   0.98236   0.01764       yes   true_pos
78   0.99361   0.99344   0.00656       yes   true_pos
79   0.99769   0.99767   .002326884    yes   true_pos
```

5.9.3 Variable Selection in Logistic Discrimination

As in the case of linear discriminant analysis, the problem of selection of the important variables arises in logistic discrimination as well. Further, as earlier, the inclusion of too many variables may actually sometimes decrease the discriminatory power, and hence due care must be taken to select an appropriate model before it is used for classification purposes.

The methods of variable selection and the underlying philosophy for logistic regression are the same in spirit as the variable selection methods described earlier in this chapter. The only difference is that all intermediate models are being fitted for the logit (π) as the response variables. Dummy variables can be included in the list of potential variables.

EXAMPLE 15 ***Logistic Variable Selection, Crystals in Urine Data*** We return to the data set used in Example 14 and apply backward selection on the six explanatory variables, namely, SG, PH, MOSM, MMHO, UREA, and CALCIUM. The response variable used is (the logit function of) CRYSTALS. It takes, respectively, values 1 and 0 corresponding to the presence and absence of the formation of crystals in the urine. The following SAS code will achieve the goal of variable selection using the BACKWARD option:

```
proc logistic data = urine descending outest = betacoef
covout nosimple;
model crystals = sg ph mosm mmho urea calcium /
selection =backward ctable pprob =.5 link=logit
slstay =.15 details;
```

Estimates of the parameters in the finally selected model are stored in the data set BETACOEF, using the OUTEST= option. The COVOUT option is used to also output the estimated variance-covariance matrix of these estimated coefficients in the same data set. The value of SLSTAY is chosen as 0.15, and hence any variable with a p value less than 0.15, at a particular selection stage will stay in the model at that stage. Finally, a classification table corresponding to the finally selected model has been requested using the CTABLE option in the MODEL statement. The observation with estimated probability greater than PPROB= 0.5 will be classified as EVENT by the logistic model finally selected by the backward elimination procedure. The complete program is presented as Program 5.17 and its output is given as Output 5.17.

```
/* Program 5.17 */

options ls = 64 ps=45 nodate nonumber;
title1  'Output 5.17';

data urine;
infile 'urine.dat' firstobs=7;
input presence $ sg ph mosm mmho urea calcium ;
patient = _n_ ;

sg100 = 100*sg; ph10=100*ph;  mmho10 =10*mmho ;cal100=100*calcium ;

if presence = 'no' then crystals  = 0;
if presence = 'yes' then crystals  = 1;

proc logistic data = urine descending outest = betacoef covout nosimple ;
model  crystals = sg ph mosm mmho urea calcium
/ selection =backward ctable pprob =.5 link=logit slstay =.15 details ;
title2 'Backward Variable Selection; Logistic Regression';
run;

proc logistic data = urine ;
model  crystals = sg ph mosm mmho urea calcium
/ selection =score best =2 link=logit ;
title2 'Selection of Variables in Logistic Regression';
run;

proc logistic data = urine descending nosimple;
model crystals = sg mosm calcium ;
title2 'The Logistic Model Selected: Score Statistics';
run;
```

Output 5.17

```
                        Output 5.17
          Backward Variable Selection; Logistic Regression

                   The LOGISTIC Procedure

                     Model Information

          Data Set                        WORK.URINE
          Response Variable               crystals
          Number of Response Levels       2
          Number of Observations          77
          Link Function                   Logit
          Optimization Technique          Fisher's scoring
```

Response Profile

Ordered Value	crystals	Total Frequency
1	1	33
2	0	44

NOTE: 2 observations were deleted due to missing values for the response or explanatory variables.

Backward Elimination Procedure

Step 0. The following effects were entered:

Intercept sg ph mosm mmho urea calcium

Model Convergence Status

Convergence criterion (GCONV=1E-8) satisfied.

Model Fit Statistics

Criterion	Intercept Only	Intercept and Covariates
AIC	107.168	71.560
SC	109.512	87.967
-2 Log L	105.168	57.560

Testing Global Null Hypothesis: BETA=0

Test	Chi-Square	DF	Pr > ChiSq
Likelihood Ratio	47.6079	6	<.0001
Score	32.4981	6	<.0001
Wald	15.5886	6	0.0161

Analysis of Maximum Likelihood Estimates

Parameter	DF	Estimate	Standard Error	Chi-Square	Pr > ChiSq
Intercept	1	-355.3	222.8	2.5443	0.1107
sg	1	355.9	222.1	2.5681	0.1090
ph	1	-0.4957	0.5698	0.7569	0.3843
mosm	1	0.0168	0.0178	0.8904	0.3454
mmho	1	-0.4328	0.2512	2.9679	0.0849
urea	1	-0.0320	0.0161	3.9443	0.0470
calcium	1	0.7837	0.2422	10.4717	0.0012

Odds Ratio Estimates

Effect	Point Estimate	95% Wald Confidence Limits	
sg	>999.999	<0.001	>999.999
ph	0.609	0.199	1.861
mosm	1.017	0.982	1.053
mmho	0.649	0.396	1.061
urea	0.968	0.938	1.000
calcium	2.190	1.362	3.520

Association of Predicted Probabilities and Observed Responses

Percent Concordant	89.9	Somers' D	0.798
Percent Discordant	10.1	Gamma	0.799
Percent Tied	0.1	Tau-a	0.396
Pairs	1452	c	0.899

Analysis of Effects in Model

Effect	DF	Wald Chi-Square	Pr > ChiSq
sg	1	2.5681	0.1090
ph	1	0.7569	0.3843
mosm	1	0.8904	0.3454
mmho	1	2.9679	0.0849
urea	1	3.9443	0.0470
calcium	1	10.4717	0.0012

Step 1. Effect ph is removed:

Model Convergence Status

Convergence criterion (GCONV=1E-8) satisfied.

Model Fit Statistics

Criterion	Intercept Only	Intercept and Covariates
AIC	107.168	70.331
SC	109.512	84.394
-2 Log L	105.168	58.331

Testing Global Null Hypothesis: BETA=0

Test	Chi-Square	DF	Pr > ChiSq
Likelihood Ratio	46.8365	5	<.0001
Score	32.3529	5	<.0001
Wald	15.7526	5	0.0076

Analysis of Maximum Likelihood Estimates

Parameter	DF	Estimate	Standard Error	Chi-Square	Pr > ChiSq
Intercept	1	-364.6	226.7	2.5859	0.1078
sg	1	362.0	226.1	2.5632	0.1094
mosm	1	0.0145	0.0173	0.7059	0.4008
mmho	1	-0.3962	0.2398	2.7297	0.0985
urea	1	-0.0291	0.0154	3.5770	0.0586
calcium	1	0.7710	0.2362	10.6511	0.0011

Odds Ratio Estimates

Effect	Point Estimate	95% Wald Confidence Limits	
sg	>999.999	<0.001	>999.999
mosm	1.015	0.981	1.050
mmho	0.673	0.420	1.077
urea	0.971	0.942	1.001
calcium	2.162	1.361	3.435

Association of Predicted Probabilities and Observed Responses

Percent Concordant	89.3	Somers' D	0.787
Percent Discordant	10.6	Gamma	0.788
Percent Tied	0.1	Tau-a	0.391
Pairs	1452	c	0.894

Residual Chi-Square Test

Chi-Square	DF	Pr > ChiSq
0.7685	1	0.3807

Analysis of Effects in Model

Effect	DF	Wald Chi-Square	Pr > ChiSq
sg	1	2.5632	0.1094
mosm	1	0.7059	0.4008
mmho	1	2.7297	0.0985
urea	1	3.5770	0.0586
calcium	1	10.6511	0.0011

Step 2. Effect mosm is removed:

Model Convergence Status

Convergence criterion (GCONV=1E-8) satisfied.

Model Fit Statistics

Criterion	Intercept Only	Intercept and Covariates
AIC	107.168	69.071
SC	109.512	80.790
-2 Log L	105.168	59.071

Testing Global Null Hypothesis: BETA=0

Test	Chi-Square	DF	Pr > ChiSq
Likelihood Ratio	46.0968	4	<.0001
Score	32.2168	4	<.0001
Wald	15.7729	4	0.0033

Analysis of Maximum Likelihood Estimates

Parameter	DF	Estimate	Standard Error	Chi-Square	Pr > ChiSq
Intercept	1	-500.0	161.9	9.5413	0.0020
sg	1	497.1	161.3	9.4947	0.0021
mmho	1	-0.2055	0.0710	8.3637	0.0038
urea	1	-0.0178	0.00723	6.0798	0.0137
calcium	1	0.7223	0.2200	10.7820	0.0010

Odds Ratio Estimates

Effect	Point Estimate	95% Wald Confidence Limits	
sg	>999.999	>999.999	>999.999
mmho	0.814	0.708	0.936
urea	0.982	0.969	0.996
calcium	2.059	1.338	3.169

Association of Predicted Probabilities and Observed Responses

Percent Concordant	89.3	Somers' D	0.787
Percent Discordant	10.6	Gamma	0.788
Percent Tied	0.1	Tau-a	0.391
Pairs	1452	c	0.894

Residual Chi-Square Test

Chi-Square	DF	Pr > ChiSq
1.4594	2	0.4821

Analysis of Effects in Model

Effect	DF	Wald Chi-Square	Pr > ChiSq
sg	1	9.4947	0.0021
mmho	1	8.3637	0.0038
urea	1	6.0798	0.0137
calcium	1	10.7820	0.0010

NOTE: No (additional) effects met the 0.15 significance level for removal from the model.

Summary of Backward Elimination

Step	Effect Removed	DF	Number In	Wald Chi-Square	Pr > ChiSq
1	ph	1	5	0.7569	0.3843
2	mosm	1	4	0.7059	0.4008

Classification Table

	Correct		Incorrect		Percentages				
Prob Level	Event	Non-Event	Event	Non-Event	Correct	Sensi-tivity	Speci-ficity	False POS	False NEG
0.500	23	39	5	10	80.5	69.7	88.6	17.9	20.4

Selection of Variables in Logistic Regression

Model Information

Data Set	WORK.URINE
Response Variable	crystals
Number of Response Levels	2
Number of Observations	77
Link Function	Logit
Optimization Technique	Fisher's scoring

Response Profile

Ordered Value	crystals	Total Frequency
1	0	44
2	1	33

NOTE: 2 observations were deleted due to missing values for the response or explanatory variables.

Regression Models Selected by Score Criterion

Number of Variables	Score Chi-Square	Variables Included in Model
1	22.4462	calcium
1	13.7879	sg
2	24.4454	sg calcium
2	24.0674	mmho calcium
3	31.4084	sg mosm calcium
3	29.4344	sg mmho calcium
4	32.2168	sg mmho urea calcium
4	31.5336	sg ph mosm calcium
5	32.3619	sg ph mmho urea calcium
5	32.3529	sg mosm mmho urea calcium
6	32.4981	sg ph mosm mmho urea calcium

The Logistic Model Selected: Score Statistics

The LOGISTIC Procedure

Model Information

Data Set	WORK.URINE
Response Variable	crystals
Number of Response Levels	2
Number of Observations	78
Link Function	Logit
Optimization Technique	Fisher's scoring

Response Profile

Ordered Value	crystals	Total Frequency
1	1	33
2	0	45

NOTE: 1 observation was deleted due to missing values for the response or explanatory variables.

Model Convergence Status

Convergence criterion (GCONV=1E-8) satisfied.

Model Fit Statistics

Criterion	Intercept Only	Intercept and Covariates
AIC	108.277	70.907
SC	110.634	80.333
-2 Log L	106.277	62.907

Testing Global Null Hypothesis: BETA=0

Test	Chi-Square	DF	Pr > ChiSq
Likelihood Ratio	43.3708	3	<.0001
Score	31.9016	3	<.0001
Wald	17.1123	3	0.0007

Analysis of Maximum Likelihood Estimates

Parameter	DF	Estimate	Standard Error	Chi-Square	Pr > ChiSq
Intercept	1	-504.3	165.9	9.2401	0.0024
sg	1	501.3	165.5	9.1774	0.0025
mosm	1	-0.0142	0.00482	8.6676	0.0032
calcium	1	0.6117	0.1803	11.5122	0.0007

Odds Ratio Estimates

Effect	Point Estimate	95% Wald Confidence Limits	
sg	>999.999	>999.999	>999.999
mosm	0.986	0.977	0.995
calcium	1.844	1.295	2.625

Association of Predicted Probabilities and Observed Responses

Percent Concordant	88.4	Somers' D	0.769
Percent Discordant	11.5	Gamma	0.770
Percent Tied	0.1	Tau-a	0.380
Pairs	1485	c	0.885

The output indicates that only two variables, namely, PH and MOSM from the list of six, can be discarded. The remaining four, namely, SG, MMHO, UREA, and CALCIUM, all have their p values (corresponding to Wald's chi-square test) smaller than SLSTAY=0.15. The estimated model finally arrived at is

$$\begin{aligned} \text{logit}\,(\hat{\pi}) &= -500.0 + 497.1 * \text{SG} - 0.2055 * MMHO \\ &\quad - 0.0178 * UREA + .0.7223 * CALCIUM. \end{aligned}$$

A comparison of the Classification Table in Output 5.17 with that in Output 5.15 indicates that perhaps this model also has slightly better discriminatory power. The numbers of both types of misclassifications have slightly decreased. This observation also illustrates another important point that the models with more explanatory variables are not necessarily better and sometimes the discriminatory performance may actually decrease with the inclusion of redundant variables.

In addition to the usual forward, backward, and stepwise variable selection procedures, another choice for the variable selection is that based on the score statistic. This can be adopted in SAS by using the SELECTION=SCORE option. Under this method, we can obtain a specified number of best models (by using the option BEST=) for all possible model sizes. The relative superiority of any model is determined by the magnitude of the corresponding score statistic.

In the later part of Program 5.17, we have used the score statistic criterion to arrive at an appropriate model, using the SAS statements

```
proc logistic data = urine;
model crystals = sg ph mosm mmho urea calcium
    / selection =score best=2 link=logit;
```

This results in various models with one, two,..., and up to all six independent variables. The BEST=2 option is used to print only the two best models, determined by the highest value of score statistics for any given model size. These are presented as part of Output 5.17.

The value of score statistics always increases if more variables are included in the model. As a result, the highest value of the score statistics would always correspond to the biggest models. In our example, the biggest model with all six independent variables yields the value of the score statistic equal to 32.4981. However, it is observed that the model containing only three of these six variables, namely, SG, MOSM, and CALCIUM, produces a value of the score statistic equal to 31.4084, which is quite close to the full model value of 32.4981. This suggests that the model

$$\text{logit}(\hat{\pi}) = \hat{\beta}_0 + \hat{\beta}_1 * \text{SG} + \hat{\beta}_2 * \text{MOSM} + \hat{\beta}_3 * \text{CALCIUM}$$

may be adequate for discrimination purposes. A separate SAS program fitting the above model results in $\hat{\beta}_0 = -504.3$, $\hat{\beta}_1 = 501.3$, $\hat{\beta}_2 = -0.0142$, and $\hat{\beta}_3 = 0.6117$.

It may be pointed out that the model selected by using the score statistic criterion is quite different from the earlier one that was selected by using the backward selection procedure. This is not surprising, however, since not only are the criteria different and unrelated but also both of these criteria are based on intuitive heuristics.

5.10 Nonparametric Discrimination

In the discriminant analysis, the decision rule to classify an object in one of the several populations is

Classify $\mathbf{x}$ into population Π_t if the quantity

$$\sum_{s=1}^{k} \pi_s f_s(\mathbf{x}) c(t \mid s)$$

is smallest, where π_s and $f_s(x), s = 1, \ldots, k$ are respectively the prior probabilities and the density functions for the corresponding populations and $c(t \mid s)$ is the cost of misclassifications when the object is classified into t^{th} population Π_t when it actually came from Π_s.

So far, whenever needed, the density functions $f_s(\mathbf{x})$ were assumed to be multivariate normal. However, often the functional form for these densities is unknown (or suggested to be nonnormal by the data at hand). Many times some suitable transformations of the variables may lead to multivariate normality to enable us to use one of the linear or quadratic discriminant analyses. However, sometimes such attempts may fail or transformation of the data may not be advisable. This necessitates a search for alternative approaches where the density functions themselves are to be estimated from the data available as training sets. Since in this case no parametric forms for densities are assumed, this approach is called the nonparametric approach.

Another approach that is also nonparametric is that referred to as the *nearest-neighbor method.* This approach is completely based on certain criteria involving distances from immediate neighbors and hence bypasses the need for a density altogether. However, we will first describe the approach based on nonparametric estimation of densities.

5.10.1 Nonparametric Discrimination: Kernel Method

Suppose $\mathbf{x}_1, \cdots, \mathbf{x}_{n_t}$ is a random sample from population Π_t, and $\mathbf{x}$ is an additional observation from population Π_t which has an (unknown) probability density function $f_t(\mathbf{x})$. The unknown density $f_t(\mathbf{x})$ is estimated by

$$\hat{f}_t(\mathbf{x}) = \frac{1}{n_t} \sum_{i=1}^{n_t} K_t(\mathbf{x} - \mathbf{x}_i),$$

where the function $K_t(\mathbf{z})$ is a function defined for the p-dimensional vector $\mathbf{z}$ satisfying

$$\int_{\mathbf{R}^p} K_t(\mathbf{z}) d\mathbf{z} = 1.$$

The quantity $K_t(\mathbf{z})$ is called the *kernel* function. We often assume that $K_t(\mathbf{z})$ is also nonnegative. Thus, any multivariate density can be a prospective choice for the kernel function. In SAS, the following choices for the kernel function are available:

Uniform kernel

$$\begin{aligned} K_t(\mathbf{z}) &= \tfrac{1}{v_r(t)} \quad \text{if } \mathbf{z}'\mathbf{V}_t^{-1}\mathbf{z} \le r^2 \\ &= 0 \quad \text{elsewhere.} \end{aligned}$$

where $v_r(t)$ is the volume of the ellipsoid $\{\mathbf{z} : \mathbf{z}'\mathbf{V}_t^{-1}\mathbf{z} \le r^2\}$.

Normal kernel (with mean $\mathbf{0}$, variance $r^2\mathbf{V}_t$)

$$K_t(\mathbf{z}) = \frac{1}{c_0(t)} \exp(-0.5\mathbf{z}'\mathbf{V}_t^{-1}\mathbf{z}/r^2),$$

where $c_0(t) = (2\pi)^{p/2} r^p \mid \mathbf{V}_t \mid^{1/2}$.

Epanechnikov kernel

$$\begin{aligned} K_t(\mathbf{z}) &= c_1(t)(1 - \mathbf{z}'\mathbf{V}_t^{-1}\mathbf{z}/r^2) \quad \text{if } \mathbf{z}'\mathbf{V}_t^{-1}\mathbf{z} \le r^2 \\ &= 0 \quad \text{elsewhere,} \end{aligned}$$

where $c_1(t) = (1 + p/2)v_r(t)$.

Biweight kernel

$$K_t(\mathbf{z}) = c_2(t)(1 - \mathbf{z}'\mathbf{V}_t^{-1}\mathbf{z}/r^2)^2 \quad \text{if } \mathbf{z}'\mathbf{V}_t^{-1}\mathbf{z} \le r^2$$
$$= 0 \quad \text{elsewhere}$$

where $c_2(t) = (1 + p/4)c_1(t)$.

Triweight kernel

$$K_t(\mathbf{z}) = c_3(t)(1 - \mathbf{z}'\mathbf{V}_t^{-1}\mathbf{z}/r^2)^3 \quad \text{if } \mathbf{z}'\mathbf{V}_t^{-1}\mathbf{z} \le r^2$$
$$= 0 \quad \text{elsewhere,}$$

where $c_3(t) = (1 + p/6)c_2(t)$.

In all the above expressions, the matrix $\mathbf{V}_t$ is used to assign an appropriate metric in the computation of distances and densities. In particular, the following five choices for $\mathbf{V}_t$ are suggested,

$$\mathbf{V}_t = \mathbf{S} \quad \text{the pooled variance-covariance matrix} \tag{5.14}$$

$$\mathbf{V}_t = \text{diag}\,(\mathbf{S}) \quad \text{the diagonal matrix of the pooled variance-covariance matrix} \tag{5.15}$$

$$\mathbf{V}_t = \mathbf{S}_t \quad \text{the variance-covariance matrix within group } t \tag{5.16}$$

$$\mathbf{V}_t = \text{diag}\,(\mathbf{S}_t) \quad \text{the diagonal matrix of the variance-covariance matrix within group } t \tag{5.17}$$

$$\mathbf{V}_t = \mathbf{I} \quad \text{the identity matrix.} \tag{5.18}$$

The nonparametric discriminant analysis using the Kernel method is performed by specifying the METHOD=NPAR option and by indicating the appropriate choices in the METRIC= and KERNEL= options in the PROC DISCRIM statement. Additionally, we also need to specify the radius r (using the $R =$ option) and the type of band widths (using the POOL= option), which are briefly described below. The radius r is often referred to as the smoothing parameter or window width.

We may subjectively choose the radius r, by trying out various choices of r on the data. Large values of r will usually produce smoother density estimates while the smaller values produce somewhat jagged density estimates. Rosenblatt (1956) presents a minimum mean integrated squared error criterion for choosing r. The resulting choice of r depends on the choice of the kernel, as well as on the density being estimated. Assuming that all populations have multivariate normal distributions, there are certain optimal choices of r given in the *SAS/STAT User's Guide, Version 6, Fourth Edition, Volume 2*, page 684. These are also available in the SAS online documentation. Some more related details on this issue are deferred to Chapter 6, Section 8. However, the merit of using the kernel method in discriminant analysis is valued when the densities are unknown and not necessarily multivariate normal. Further, based on Remme, Habbema, and Hermans (1980), we observe that in the case of multivariate normality, there is little extra to be gained by using the nonparametric density estimation approach in discriminant analysis. As a result, for the nonnormal densities, the only option left to us is to try various values of smoothing parameter r and choose the one that appears to work best. In fact, the choice of r appears to have more impact on the quality of the density estimate than the choice of the kernel itself. PROC DISCRIM uses only a single value of r for all populations. However, in different populations, different or equal band widths for density estimation can be used by using the POOL=NO or POOL=YES options, respectively. Often, for classification into one of the populations, we may want to specify a minimum posterior probability. An observation will be classified into one of the populations only if the largest of all posterior probabilities exceeds this threshold level. This can be done by using the THRESHOLD= option in PROC DISCRIM. The default choice is THRESHOLD=0, and this choice will be used in the example that follows. The observation is not classified, and it is reported as belonging to OTHER if the posterior probability is below the threshold or if there is a tie for the maximum posterior probability.

EXAMPLE 16 ***Nonparametric Kernel Discrimination, AIDS Data*** Everitt (1993, p.34) presents a data set consisting of the distribution of cell sizes in the brains of 11 patients who died of AIDS and 9 other patients who died from other causes. For illustration, we will use these cell sizes as the variables for the kernel discriminant analysis. The original data, as reported in its entirety in Program 5.18, have 13 cell sizes in increasing order that are denoted as $S1, S2, \ldots, S13$. We will, however, use only $S1$, $S2$, $S3$, and $S4$ here, mainly to avoid the possibility of resulting with a singular variance-covariance matrix (since $n_i < p = 13$). The logarithmic transformation $x = log(s)$ is applied on $S1$, $S2$, $S3$, and $S4$ since the original values represent the count data. The two patient groups are referred to as A and O for AIDS and Others, respectively. Since there are no cost considerations for this illustration, we have set $c(A \mid O) = c(O \mid A) = 1$. Also, equal prior probabilities ($\pi_O = \pi_A$) are assumed.

Using the GCHART procedure (or simply PROC CHART), we first attempt to assess the normality of the raw as well as of the transformed data. The corresponding SAS statements are presented as part of Program 5.18. These plots (included only for variable s1 here) show the clear indication of nonnormality, thereby eliminating the possibility of using the normal theory-based discriminant analysis. As a result (and barring the choice of logistic regression which, for this example, provides a poor fit, thereby implying that the logistic model may not hold), we resort to the nonparametric method using the kernel density estimation. The choice of kernel is specified as the Epanechnikov kernel (KERNEL=EPA), and the value of r is taken to be 0.3 ($r = .3$). The plots also show that there possibly is more variability in the A group compared to the O group. Thus, the POOL=NO option is used, which results in the use of the different sample variance-covariance matrices $\mathbf{S}_t$ as $\mathbf{V}_t$ in the expression of Epanechnikov kernel. That the full matrices, and not their diagonal parts only, are to be used is indicated by using the METRIC=FULL option. The corresponding SAS code, also included as the later part of Program 5.18, is

```
proc discrim data = aids method =npar kernel =epa
r = .3 pool =no metric=full testdata = aids
testout =aidsout testoutd=density;
class group;
var x1-x4;
```

Note that the original training data are also being used as the test data here, although, in practice, the test data should be different from the training or calibration data. The estimated densities corresponding to various observations and all other relevant information are stored in data set titled DENSITY using the TESTOUTD= option. It is observed that for this data set this approach results in perfect discrimination, with no misclassifications. Consequently, the resulting error rates are all zero. However, the normality-based linear and quadratic discriminant analyses (not shown) result in 28% and 9% overall error count estimates, respectively. We have already noted that for this data set the logistic model (not shown) did not fit satisfactorily and consequently resulted in more misclassifications than correct classifications. The fact that the nonparametric kernel-based discriminant analysis shows itself to be superior to linear and quadratic discriminant analysis in our example is consistent with the observations made by Remme, Habbema, and Hermans (1980) through certain detailed studies. Their findings are succinctly summarized by Silverman (1986, p.125) essentially with the recommendation to almost always use the kernel-based discriminant analysis as the overall safest bet, especially for the nonnormal data. Of course, we must still remember that the success of the kernel approach depends also on the value of r, and choosing the correct smoothing parameter can sometimes be a challenging problem in itself. Selected output follows Program 5.18.

```
/* Program 5.18 */

filename gsasfile "prog518.graph";
goptions reset=all gaccess=gsasfile gsfmode = append
autofeed dev = pslmono;
goptions horigin =1in vorigin = 2in;
goptions hsize=6in vsize = 8in;
options ls = 64 ps=45 nodate nonumber;
title1 ' ';
title2 j = 1  'Output 5.18' ;

data aids; input group $ s1-s13;
x1 =log(s1) ;
x2 = log(s2);
x3= log(s3);
x4 = log(s4);
lines;
a 8 8 20 10 9 5 3 1 4 1 1 3 0
a 2 11 10 12 5 3 2 2 2 3 0 0 0
a 18 12 10 18 14 10 8 5 1 3 2 2 0
a 19 12 7 11 9 6 3 1 3 1 2 1 0
a 14 17 8 8 2 4 3 3 1 3 1 2 0
a 10 16 9 4 6 4 0 1 2 0 0 0 0
a 3 1 4 3 3 2 2 0 2 0 2 1 1
a 27 22 16 7 6 5 2 2 4 2 0 0 0
a 12 16 16 17 9 7 4 5 1 2 0 0 1
a 10 14 10 10 7 3 1 0 0 2 1 1 0
a 2 3 12 7 4 9 6 2 3 3 2 3 1
o 21 19 9 4 3 2 1 0 0 0 0 0 0
o 20 17 11 7 2 6 1 0 1 0 0 0 0
o 16 27 22 12 9 2 3 3 2 4 2 0 0
o 11 15 13 14 9 9 5 2 2 0 0 1 0
o 10 13 5 12 5 9 5 1 1 2 3 3 0
o 23 13 14 8 4 6 2 3 1 1 2 1 2
o 15 14 15 10 7 12 5 4 0 1 1 2 0
o 14 15 13 11 7 4 1 2 1 0 0 0 0
o 10 19 16 19 8 3 4 1 1 1 1 1 0
;

/*
Data from Everitt (1993, p 34),
Reprinted with permission from Edward Arnold, London.
*/

proc sort data = aids ;
by group;

proc gchart data = aids ;
by group ;
vbar s1 /subgroup = group midpoints = 0 to 30 by 3;
title1 h = 1.2 'Histogram for s1' ;
title2 j = 1 'Output 5.18';
run;
```

```
proc discrim data = aids method = npar kernel =epa r = .3 pool =no
metric =full testdata = aids testout =aidsout testoutd =density;
class group ;
var x1-x4;
title1 'Output 5.18';
title2 'AIDS Data: Nonparametric Discrimination using EPA Kernel';
run;

proc print data = density;
var group s1 s2 s3 s4 a o ;
title1 'Output 5.18';
title2 'Density Estimates for (x1,x2,x3,x4) with x=log(s) ';
title3 'using EPA Kernel';
run;
```

Output 5.18

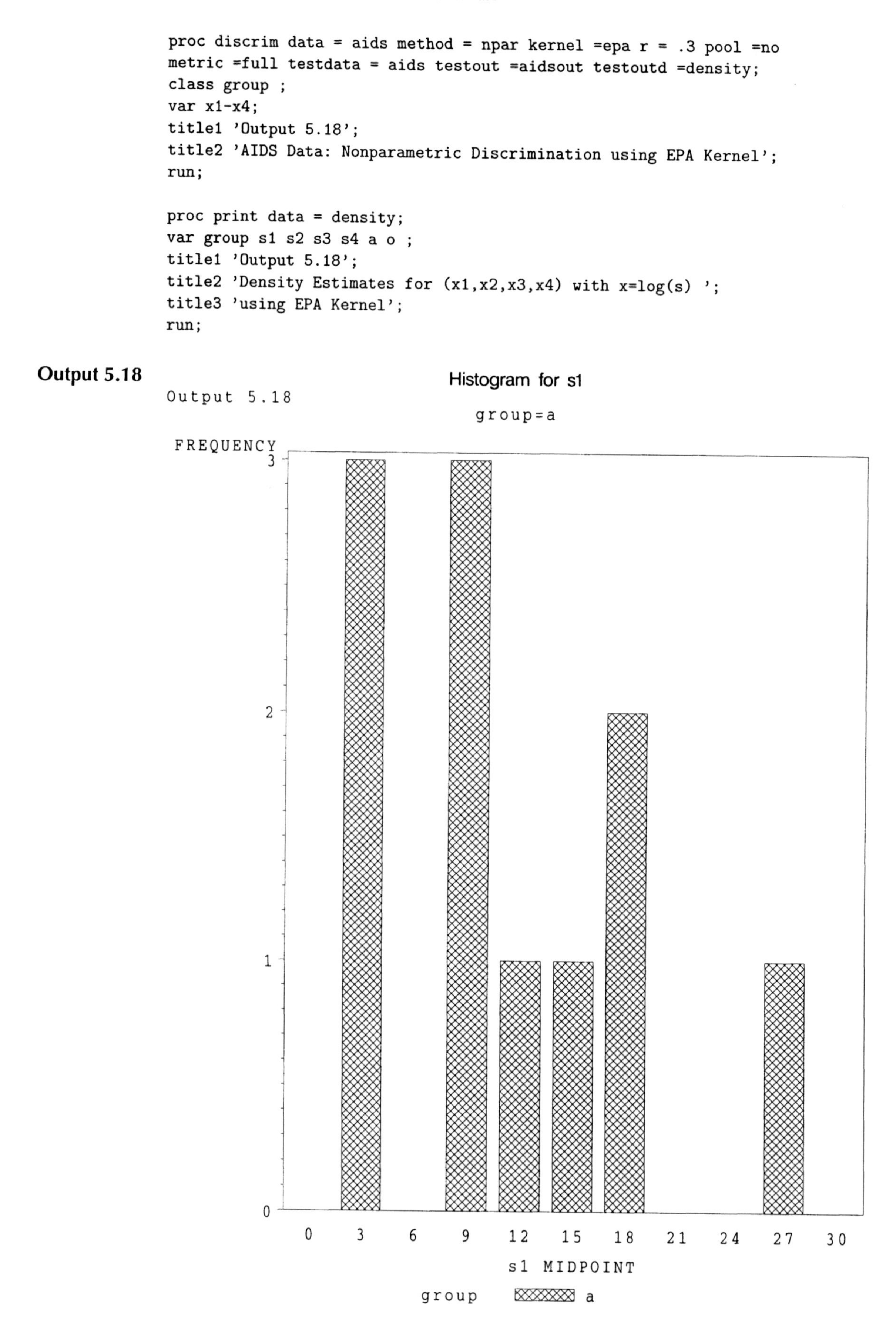

Output 5.18

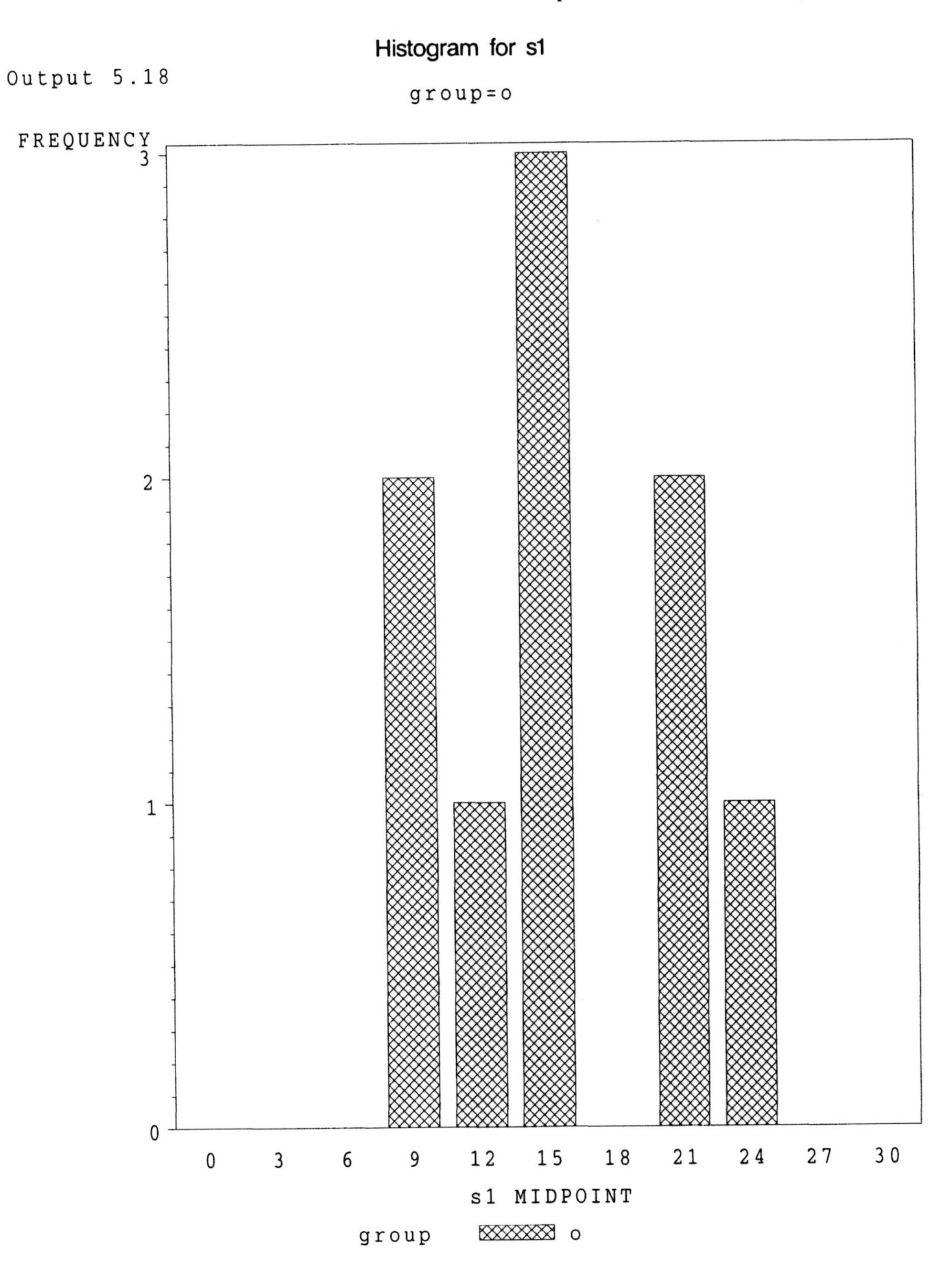
Histogram for s1
group=o
FREQUENCY
3
2
1
0
0 3 6 9 12 15 18 21 24 27 30
s1 MIDPOINT
group
o

Output 5.18 *(continued)*

```
                         Output 5.18
      Density Estimates for (x1,x2,x3,x4) with x=log(s)
                       using EPA Kernel
```

Obs	group	s1	s2	s3	s4	a	o
1	a	8	8	20	10	70.8131	0.00
2	a	2	11	10	12	70.8131	0.00
3	a	18	12	10	18	70.8131	0.00
4	a	19	12	7	11	70.8131	0.00
5	a	14	17	8	8	70.8131	0.00
6	a	10	16	9	4	70.8131	0.00
7	a	3	1	4	3	70.8131	0.00
8	a	27	22	16	7	70.8131	0.00
9	a	12	16	16	17	70.8131	0.00
10	a	10	14	10	10	70.8131	0.00
11	a	2	3	12	7	70.8131	0.00
12	o	21	19	9	4	0.0000	2217.57
13	o	20	17	11	7	0.0000	2217.57
14	o	16	27	22	12	0.0000	2217.57
15	o	11	15	13	14	0.0000	2217.57
16	o	10	13	5	12	0.0000	2217.57
17	o	23	13	14	8	0.0000	2217.57
18	o	15	14	15	10	0.0000	2217.57
19	o	14	15	13	11	0.0000	2217.57
20	o	10	19	16	19	0.0000	2217.57

5.10.2 *k*-Nearest-Neighbor Method

Another nonparametric method for discriminant analysis is based on the simple idea of assigning a new object to the same class as its nearest neighbor(s) defined in some meaningful sense. The nearness of an object to another can be measured using an appropriate distance function. Typically, the squared distance between two objects in group t with respective observation vectors $\mathbf{x}_1$ and $\mathbf{x}_2$ is given by

$$d_t^2(\mathbf{x}_1, \mathbf{x}_2) = (\mathbf{x}_1 - \mathbf{x}_2)'\mathbf{V}_t^{-1}(\mathbf{x}_1 - \mathbf{x}_2),$$

where the matrix $\mathbf{V}_t$ is suitably chosen. Various choices of $\mathbf{V}_t$ are given in Equations 5.14 through 5.18.

Suppose the object with observation $\mathbf{x}$ is to be assigned into one of the populations. Suppose k is a prespecified positive integer. Based on the squared distance defined above, we first find the k observations that are closest to $\mathbf{x}$. Suppose that, out of these k observations, k_t observations come from the tth population for which the prior probability is π_t. Then the posterior probability that the object actually belongs to t^{th} populations is estimated by

$$\frac{\pi_t \hat{f}_t(\mathbf{x})}{\sum_{s=1}^{g} \pi_s \hat{f}_s(\mathbf{x})} = \frac{\pi_t (k_t/n_t)}{\sum_{s=1}^{g} \pi_s (k_s/n_s)}.$$

The observation $\mathbf{x}$ will be classified into t^{th} population if the above posterior probability is highest for the tth population. Alternatively (since the denominator in the expression above is the same for all populations), $\mathbf{x}$ is classified into t^{th} population if $\pi_t k_t / n_t$ is bigger than $\pi_s k_s / n_s$ for every $s = 1, \ldots, g$. In case of a tie, randomization can be used to decide on the assignment.

It may be mentioned that the integer k is to be prespecified and there is no clear-cut way of choosing the best k. However, it has been found in various studies that in most cases, especially for large samples, the choice of k is immaterial.

In SAS, the k-nearest neighbor method can be conveniently implemented by using the METHOD=NPAR option along with the $K =$ option, where the choice of k is specified. All other relevant options, as discussed earlier, can also be specified. It may however be remembered that the K= option is applicable only with POOL =YES option. Also, the K= and R= options cannot be used simultaneously as they correspond to two different nonparametric methods. Further, it is sometimes possible that there is a tie, and two or more populations may have the same largest value of $\pi_t k_t / n_t$. In that case, the particular observation is classified as OTHER.

EXAMPLE 17 ***k-Nearest Neighbor Method, Cushing's Syndrome Data Revisited*** In Example 8 we considered the analysis of the Cushings' Syndrome data (Aitchison and Dunsmore (1975)). The problem there was to obtain a quadratic discriminant rule for classification into one of the three patient populations defined by three types of syndromes using the training data and to apply it on another test data of six undiagnosed patients. The three populations (TYPE) were referred to as A, B, and C. Logarithms of the original measurements were analyzed. We will now perform the discriminant analysis using the k-nearest neighbor method under the same basic setup.

As mentioned earlier, usually the choice of k in the $K =$ option is relatively unimportant. We will illustrate the procedure for the three-nearest neighbor method. To compute distances, full variance-covariance matrices will be used (METRIC=FULL). The following SAS code will attain the objective of the three-nearest neighbor discriminant analysis:

```
proc discrim data = cushing method =npar k=3
pool=yes metric=full testdata = undgnosd testout = cushout;
class type;
var logtetra logpreg;
run;
```

Data sets specified in the TESTDATA= and TESTOUT= options can be printed using the PROC PRINT statement.

The complete program is reported in Program 5.19, and the selected parts of the resulting output are produced in Output 5.19. The interpretation of most of the output is the same as earlier. It is seen that for training data there are a few misclassifications resulting in a total error count rate of 13.33%. For training data, as well as for the data set UNDGNOSD, we observed that some observations are not classified into any of the three populations but are classified as OTHER. This is due to the occurrences of ties for the respective largest values of $\pi_t k_t / n_t$. For instance, consider the fifth observation of the data set UNDGNOSD. In this example $n_1 = 6, n_2 = 10, n_3 = 5$ and $k = 3$. For the fifth observation of the data set UNDGNOSD, among the $k = 3$ nearest neighbors (from the data set CUSHING), none came from population A, 2 came from population B, and only 1 came from population C. Thus, $k_1 = 0$, $k_2 = 2$ and $k_3 = 1$. This results in the numerators of the three estimated posterior probabilities $\pi_1 k_1 / n_1 = 0$, $\pi_2 k_2 / n_2 = \frac{(1/3)(2)}{10} = 1/15$ and $\pi_3 k_3 / n_3 = \frac{(1/3)(1)}{5} = 1/15$. The estimated posterior probabilities are therefore 0, 0.5 and 0.5, respectively, as reported in the last part of Output 5.19. In view of this tie between the populations B and C, the particular observation is classified as OTHER. A similar situation arises for the three observations from the training data set CUSHING.

```
/* Program 5.19 */

options ls = 64 ps=45 nodate nonumber;
title1  'Output 5.19';

data cushing;
infile 'cushing.dat' firstobs = 7;
input type $ tetra preg ;
logtetra = log(tetra);
logpreg =log(preg);

data undgnosd;
input tetra preg ;
logtetra = log(tetra);
logpreg =log(preg);
lines;
5.1 0.4
12.9 5.0
13.0 0.8
2.6 0.1
30.0 0.1
20.5 0.8
;
/*
Data from Aitchison and Dunsmore (1975),
reprinted with permission from Cambridge University Press.
*/

proc discrim data = cushing method = npar k= 3 pool =yes
metric =full testdata = undgnosd testout =cushout;
class type ;
var logtetra logpreg;
title2 'Nonparametric Discrimination using k-NN Method';
title3 'with k=3';
run;

proc print data = cushout;
title2 'k-NN Classifications for the Undiagnosed Data';
title3 ' with k =3' ;
```

Output 5.19

```
                         Output 5.19
          Nonparametric Discrimination using k-NN Method
                          with k=3

                    The DISCRIM Procedure



   Classification Summary for Calibration Data: WORK.CUSHING
        Resubstitution Summary using 3 Nearest Neighbors

                  Squared Distance Function

                   2                   -1
                  D (X,Y) = (X-Y)' COV   (X-Y)

        Posterior Probability of Membership in Each type

        m (X) = Proportion of obs in group k in 3
         k      nearest neighbors of X
```

```
Pr(j|X) = m (X) PRIOR  / SUM ( m (X) PRIOR  )
           j         j    k     k         k
```

Number of Observations and Percent Classified into type

From type	a	b	c	Other	Total
a	6	0	0	0	6
	100.00	0.00	0.00	0.00	100.00
b	0	8	0	2	10
	0.00	80.00	0.00	20.00	100.00
c	0	0	4	1	5
	0.00	0.00	80.00	20.00	100.00
Total	6	8	4	3	21
	28.57	38.10	19.05	14.29	100.00
Priors	0.33333	0.33333	0.33333		

Error Count Estimates for type

	a	b	c	Total
Rate	0.0000	0.2000	0.2000	0.1333
Priors	0.3333	0.3333	0.3333	

Classification Summary for Test Data: WORK.UNDGNOSD
Classification Summary using 3 Nearest Neighbors

Number of Observations and Percent Classified into type

	a	b	c	Other	Total
Total	1	3	1	1	6
	16.67	50.00	16.67	16.67	100.00
Priors	0.33333	0.33333	0.33333		

k-NN Classifications for the Undiagnosed Data
with k =3

Obs	tetra	preg	logtetra	logpreg	a	b	c	_INTO_
1	5.1	0.4	1.62924	-0.91629	0.00000	1.00000	0.0	b
2	12.9	5.0	2.55723	1.60944	0.00000	0.20000	0.8	c
3	13.0	0.8	2.56495	-0.22314	0.00000	1.00000	0.0	b
4	2.6	0.1	0.95551	-2.30259	0.76923	0.23077	0.0	a
5	30.0	0.1	3.40120	-2.30259	0.00000	0.50000	0.5	
6	20.5	0.8	3.02042	-0.22314	0.00000	1.00000	0.0	b

5.11 Concluding Remarks

In this chapter, we have presented the discussion of various alternative methods of discriminant analysis. The natural question to ask is, which of these methods is best? While there is no decisive answer to this, we may conclude our discussion with the following recommendations.

- Use the normality-based discriminant analysis if we are comfortably sure of the multivariate normality assumptions. The choice of linear or quadratic discriminant function could depend on whether the population variance-covariance matrices can be assumed to be equal or unequal. In view of this, a test of homogeneity should always be performed. Further, the discussion here corresponds to the situation where populations are different at least with respect to their respective mean vectors. Hence, an appropriate test for the equality of means should also be performed. The discriminant analysis should be performed only if the means of the populations are found to be statistically significantly different.
- If the data exhibits nonnormality, then we have two choices. We may either use the logistic regression approach or an approach based on the nonparametrics. Often the logistic regression approach would result in a satisfactory model and reasonable error rates. Being simpler and more easily understood and involving less subjectivity than kernel or nearest neighbor approaches (e.g., in selecting the appropriate values of r or k, the choice of kernel, decision to pool or not to pool the variance-covariance matrices), we recommend the use of this approach. For instance, in Example 5.17, a logistic model may not even fit the data. In such cases the nonparametric approaches may be used.
- It is true that for data sets of reasonably large sizes, between the kernel method and nearest neighbor methods, the kernel methods would perform better provided that the value of r and the kernel being used are chosen carefully. There is, however, no satisfactory way as yet to arrive at these choices—at least in an automated form. On the other hand, the nearest neighbor method, which itself essentially is a density estimation-based approach using a uniform kernel, is relatively robust to the choice of k. In view of this, we recommend the nearest neighbor approach over the other kernel-based approaches. Also, for small training data sets, the nearest neighbor approach will usually perform better than other kernel-based approaches.
- Estimated error rates, especially when computed not just from the training sample used to obtain the discrimination rule, but also from test sample, can provide some insight into which of the several competing methods may be preferred. However, we should bear in mind that the error rates cannot be very reliable guidelines for such decisions, especially for small data sets. For moderately sized to large data sets they do provide valuable information.
- We have not explicitly discussed the situations where some or all of the variables are categorical or are discrete. The multivariate normality-based methods are obviously unsatisfactory in these situations (unless in case of discrete variables, transformations are made to attain the multivariate normality). While the kernel density-based approaches may be theoretically implemented, these are not available in most commercial software packages in general and in SAS in particular. Nearest neighbor-based methods are also not applicable for categorical data, as in such cases distances may not be appropriately defined. However, logistic regression methods can handle both of these situations without much difficulty (even though we have not considered such situations in this chapter and in our examples) and hence as the only readily available choice, their use is recommended.

In closing this chapter, we may add that the discriminant analysis is still a very active research area. We have not discussed many of the new emerging and often less statistical techniques such as the classification and regression trees, neural networks and other computationally intensive yet intuitive techniques used in data mining problems. The usefulness and relative merits of these tools will only be seen in the years to come.

Chapter

6

Cluster Analysis

6.1 Introduction

In the last chapter we considered various statistical approaches of classifying a multivariate observation into one of several known populations or groups. The groups were assumed to be well defined and, at least in philosophy, the idea was to classify the observation into the population where its membership seemed most plausible either in terms of posterior probability or in terms of some kind of distance measure.

Cluster analysis attempts to address a very different problem from a different point of view. For example, say that we are given a set of observations and, if possible, we want to form smaller subgroups of similar observations, so that in each subgroup, objects are similar to each other, but the subgroups themselves are very dissimilar to each other. The dissimilarities can be measured in terms of some meaningful index such as the Euclidean or some other distance, some kind of probabilistic index, or in terms of within- and between-cluster variabilities. Obviously, we would expect that not all clustering criteria will be mutually agreeable and what may be an appropriate criterion depends on the nature of observations and what kind of clustering we want as well as the objective of clustering.

The objective of any clustering is of crucial importance. The natural question to ask before clustering is what kind of similarity or dissimilarity are we looking for. For example, for the group of students enrolled in an undergraduate program at a university, some students may be very similar under one criterion while they may be quite dissimilar under another. Students can be clustered with respect to race, sex, age group, geographical region, high school GPA or by whether they are enrolled in any mathematics classes. Of course, different variables, quantitative or qualitative, seem to measure different things, and clustering with respect to one variable (or a set of variables) may represent only similarities or dissimilarities with respect to only that variable.

The above example clearly suggests that the choice of variables used is quite important and should be determined by the objective and from the very purpose of clustering. Another aspect of clustering that is crucial is how the dissimilarities between the objects are quantified. There are many ways to do so. Any reasonable index of dissimilarity between two objects or observations, say a or b, is an index $I(a, b)$ such that

(i) $I(a, b) \geq 0$

(ii) $I(a, a) = 0$

(iii) $I(a, b) = I(b, a)$

(iv) $I(a, b)$ increases as objects a and b become more and more dissimilar.

For example, the Euclidean (or any other weighted) distance functions or their squares can serve as a measure of dissimilarity. Further, we also need to measure the dissimilarities between the clusters (being) formed. In that situation, often but not always, the dissimilarity between two clusters will be measured as that between two or more reference observations or reference points coming from each of the two clusters.

It may be pointed out that several authors define a dissimilarity index in a more relaxed manner by requiring only (i) and/or (iv) of the above. For most but not all, clustering applications, (ii) and (iii) are also required. However, exceptions in point are certain indexes for binary data (such as Jaccard's dissimilarity coefficient) where they fail to obey (ii) and (iii).

To begin with, we will assume that all measurements are quantitative. Examples of dissimilarity indexes for qualitative (in fact, only binary) variables will be discussed in Section 6.3. However, first we will present some graphical methods of clustering using the data on quantitative variables.

6.2 Graphical Methods for Clustering

The simplest graphical approach for clustering is via profile plots of various observations. As discussed in Khattree and Naik (1999), profile plots are the polygonal display of p-dimensional observation vectors made on p quantitative variables. Many times it may be helpful to standardize these variables. Of course, profile plots are useful only for moderate sized data. We will illustrate the use of profile plots using an example.

EXAMPLE 1 ***Graphical Clustering of Rocks and Minerals, Satellite Data.*** Fox (1993) presents a set of six geological measurements on six groups of geological samples. These data were used in Chapter 5 for canonical discriminant analysis. Suppose, based on these measurements, we are interested in knowing which of the samples are similar and which are quite different from each other. Instead of working with a raw data set, which is rather large, let us just consider the sample mean vectors for the six groups. A profile plot of these sample mean vectors is displayed in Output 6.1, which is produced using the SAS code given in Program 6.1 (adapted with minor changes from Program 2.5 of Khattree and Naik, 1999).

```
/* Program 6.1 */

options ls=64 ps=45 nodate nonumber;
filename gsasfile "prog61.graph";
goptions reset=all gaccess=gsasfile autofeed dev=pslmono;
goptions horigin=1in vorigin=2in;
goptions hsize=6in vsize=8in;
```

```
symbol1 i=join v=c;
symbol2 i=join v=g;
symbol3 i=join v=i;
symbol4 i=join v=l;
symbol5 i=join v=m;
symbol6 i=join v=s;
legend1 across=3;

title1 h = 1.2  'Mean Profiles: Satellite Data';
title2 j = l  'Output 6.1';
data meanfox;
infile 'meanfox.dat' firstobs = 7;
input group $ tm1 tm2 tm3 tm4 tm5 tm7;

proc transpose data=meanfox
out=meanfox2 name=measure;
by group;
run;

proc gplot data=meanfox2(rename=(col1=values));
plot values*measure=group/
vaxis=axis1 haxis=axis2 legend=legend1;
axis1 label=(a=90 h=1.2 'Mean Values');
axis2 offset=(2) label=(h=1.2 'Rock/Mineral Measurements');
run;
```

Output 6.1

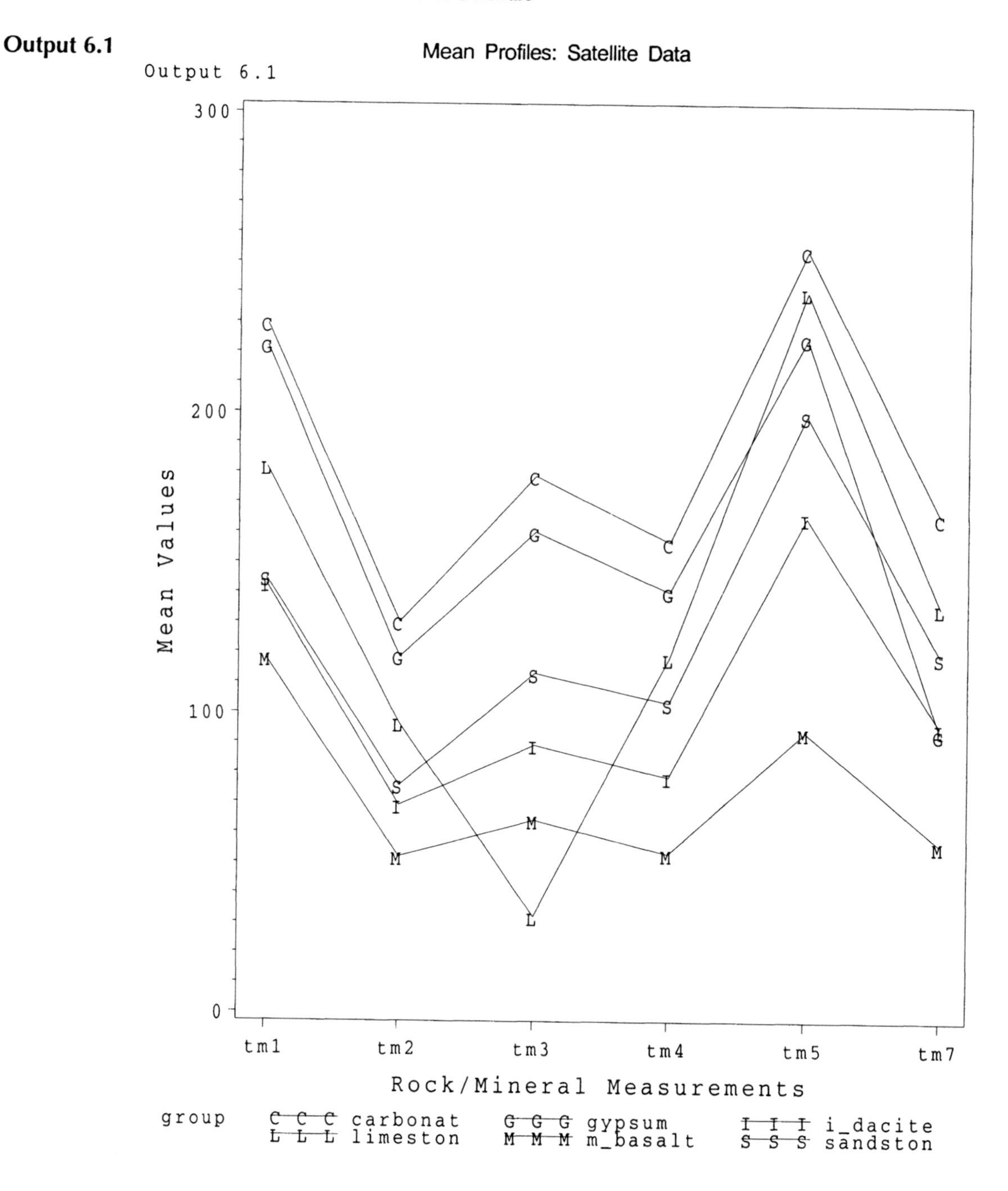

From the plot it is clear that the groups CARBONAT and GYPSUM are very similar and can be put in the same cluster. The same can be said about I_DACITE and SANDSTON. The group M_BASALT is slightly different from I_DACITE and SANDSTON in that the measurements are consistently smaller although the three profiles appear to be largely parallel. However, the profile of LIMESTON is very different from the rest of the profiles and it forms a cluster of its own.

Andrews plots (Andrews, 1972) and modified Andrews plots (Khattree and Naik, 1998) are other alternatives to profile plots, and frequently they can be very effective in identifying the clusters of observations. For a vector of measurements $\mathbf{x} = (x_1, \ldots, x_p)'$ the Andrews function and its modified versions given by Khattree and Naik (1998) are defined as

$$f_{\mathbf{x}}(t) = x_1/\sqrt{2} + x_2 \sin(t) + x_3 \cos(t) + x_4 \sin(2t) + x_5 \cos(2t) + \ldots, \; -\pi \leq t \leq \pi$$

and

$$g_{\mathbf{x}}(t) = \frac{1}{\sqrt{2}}\{x_1 + x_2[\sin(t) + \cos(t)] + x_3[\sin(t) - \cos(t)] + x_4[\sin(2t) + \cos(2t)] + x_5[\sin(2t) - \cos(2t)] + \cdots\}, \quad -\pi \le t \le \pi,$$

respectively. Both of these have some very useful and meaningful properties. See Khattree and Naik (1998, 1999) for details on these properties. One of the important implications is that similar observations will result in similar Andrews curves that may be lying in each other's vicinity, and hence these curves will cluster together in the plot. Thus the clusters can be readily identified. Like the profile plots, these plots also are the effective graphical tools when the number of observations is not too large. For large data sets it becomes increasingly difficult to observe any striking features in the clutter of the large number of curves.

EXAMPLE 1 *(Continued)* Using Program 2.6 of Khattree and Naik (1999) reported here with suitable changes as Program 6.2, we obtain the Andrews plots and modified Andrews plots for the group averages of Fox's data. The program will plot both curves. Plots given as Output 6.2 present both the Andrews plot as well as its modified version as suggested in Khattree and Naik (1999). For the reasons stated in Khattree and Naik (1998, 1999), the modified Andrews plots are more informative and this fact is evident from the two plots shown in Output 6.2. Modified Andrews plots clearly show the similarity within certain clusters. Specifically, the groups I_DACITE and SANDSTON form a cluster, and so do the groups CARBONAT and GYPSUM. However, LIMESTON and M_BASALT are individually different from the rest. These observations are consistent with what was observed in the profile plots of Output 6.1. The human eye tends to notice more the visual differences around the center of the plot, where t is close to zero, and hence for this reason any differences or similarities observed in Andrews plots are largely due to odd-numbered terms of the right-hand side of $f(t)$. The modified Andrews plots to some extent, reduce this drawback.

```
/* Program 6.2 */

options ls = 64 ps = 45 nodate nonumber ;

data meanfox;
infile 'meanfox.dat' firstobs= 7;
input group$ tm1 tm2 tm3 tm4 tm5 tm7;

filename gsasfile "prog62.graph";
goptions reset=all gaccess=gsasfile autofeed dev=pslmono
gsfmode=append;
goptions horigin=1in vorigin=2in;
goptions hsize=6in vsize=8in;

data andrews;
set meanfox;
pi=3.14159265;
s=1/sqrt(2);
inc=2*pi/100;
do t=-pi to pi by inc;

z=s*tm1+sin(t)*tm2+cos(t)*tm3+sin(2*t)*tm4+cos(2*t)*tm5+sin(3*t)*tm7;
```

```
mz=s*(tm1+(sin(t)+cos(t))*tm2+(sin(t)-cos(t))*tm3+(sin(2*t)+cos(2*t))*tm4
+(sin(2*t)-cos(2*t))*tm5+(sin(3*t)+cos(3*t))*tm7);
output;
end;

symbol1 i=join v=none l=1;
symbol2 i=join v=none l=2;
symbol3 i=join v=none l=3;
symbol4 i=join v=none l=4;
symbol5 i=join v=none l=5;
symbol6 i=join v=none l=6;

proc gplot data=andrews;
plot z*t=group/vaxis=axis1 haxis=axis2 ;
axis1 label=(a=90 h=1.5 f=duplex 'Andrews Function');
axis2 label=(h=1.5 f=duplex 't') offset=(2);
title1 h = 1.2 'Andrews Plots: Satellite Data';
title2 j = l 'Output 6.2';
run;

proc gplot data=andrews;
plot mz*t=group/vaxis=axis1 haxis=axis2;* anno=labels;
axis1 label=(a=90 h=1.5 f=duplex 'Modified Andrews Function');
axis2 label=(h=1.5 f=duplex 't') offset=(2);
title1 h = 1.2 'Modified Andrews Plots: Satellite Data';
title2 j = l 'Output 6.2';
run;
```

Output 6.2

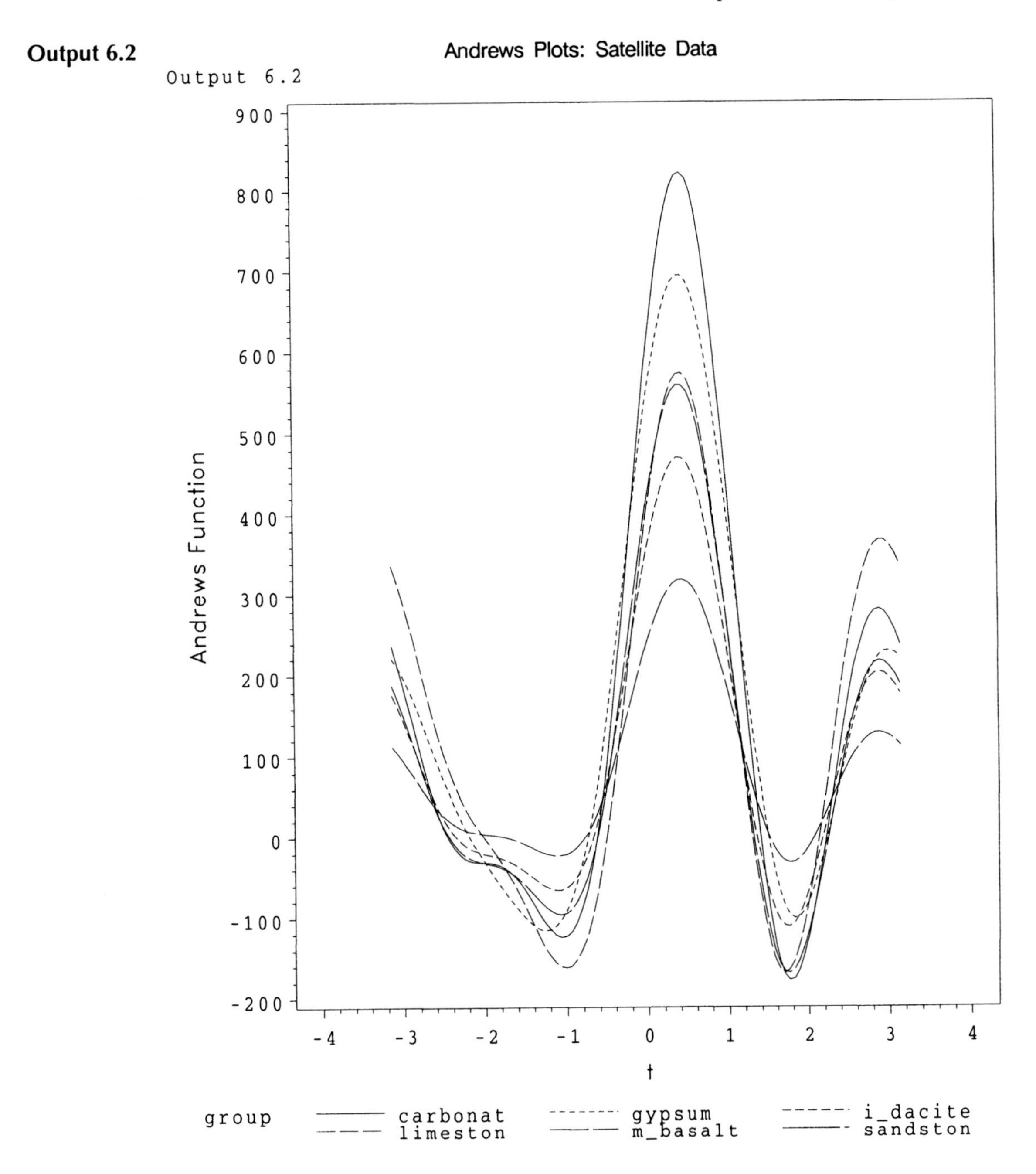

Andrews Plots: Satellite Data
Output 6.2
Andrews Function
900
800
700
600
500
400
300
200
100
0
-100
-200
-4
-3
-2
-1
0
1
2
3
4
t
group
carbonat
gypsum
i_dacite
limeston
m_basalt
sandston

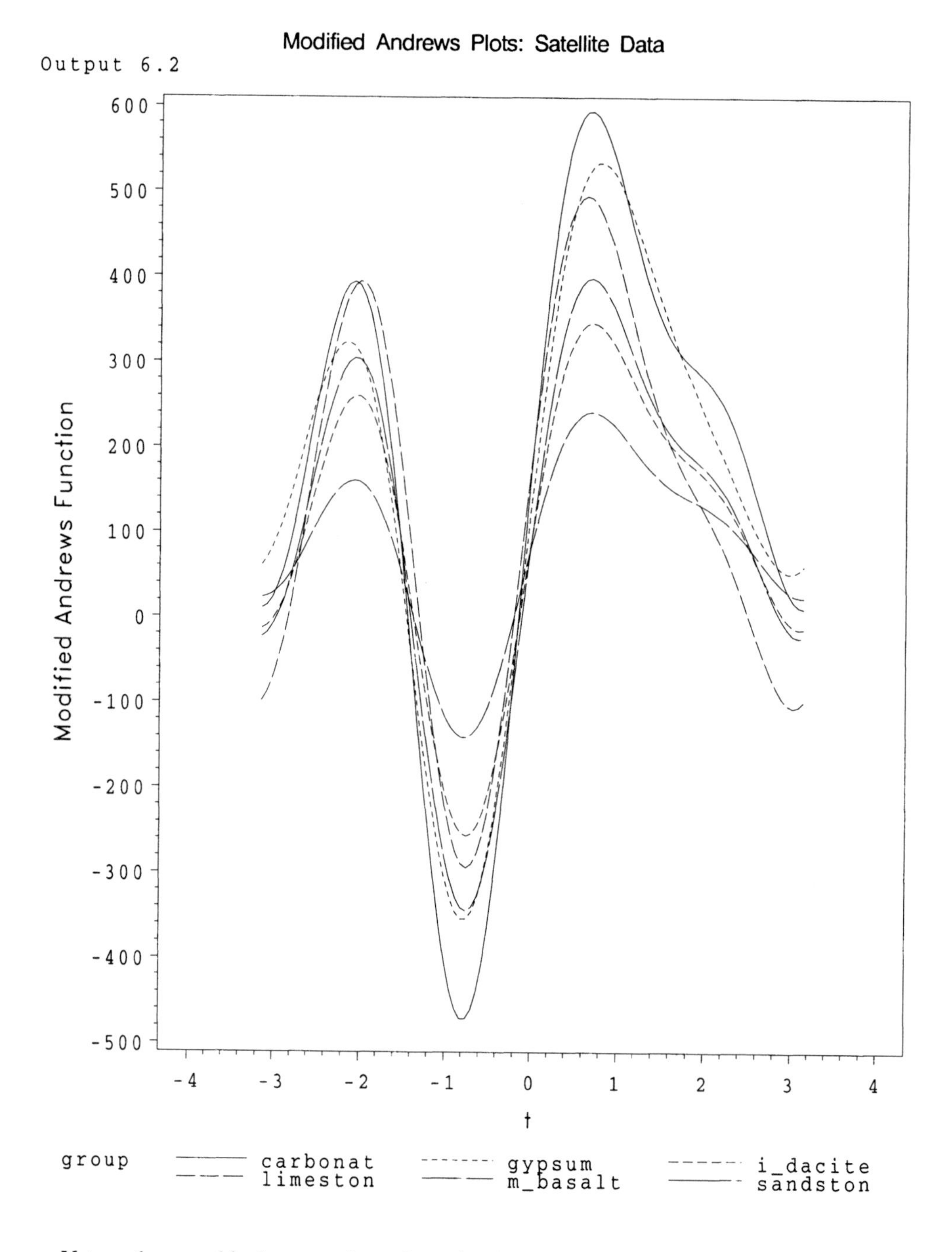

Yet another graphical approach to clustering is the use of biplots. As already described in Chapter 2, biplot methodology is a data dimensionality reduction technique that enables us to present each observation as a point in two- (or three-) dimensional space. Hence, these points can be plotted, and any possible clustering of observations can be determined by observing the corresponding clusters of points. The brief details of biplots along with a SAS program can be found in Khattree and Naik (1999). An extensive discussion of biplots is available in Gower and Hand (1996). It must, however, be remembered that there is some loss of information because of data/dimensionality reduction. Nothing can be said in advance about the magnitude of this loss. However, for most large data sets with a large number of variables if about 85% or more of the variability is still explained by the data, reduced to the smaller dimensional space, (usually two-dimensional) then biplots can be a very effective way of graphically summarizing the information and observing clusters.

EXAMPLE 2 ***Graphical Clustering of National Track Record Data*** Biplots for the national track record data were presented in Chapter 2 as Output 2.11. The correct choice used to obtain the most appropriate coordinates of the biplot in that context was (c), which corresponded to obtaining the principal component scores and the principal component weights on the same graph. For the purpose of clustering observations, selecting the pair of the first two principal component scores may be an appropriate option. Hence, the corresponding biplot in Output 2.11(c) will be used here. The biplot clearly shows various clusters largely formed along the x-axis. Specifically, the following clusters of countries (with positive first principal components) can be visually noticed.

{USA, USSR},
{GDR, CZECH, POLAND}.
{FINLAND, CANADA, GB, AUSTRALI, AUSTRIA, FRG, FRANCE},
{NORWAY}
{TAIPEI},

and

{PORTUGAL}.

In addition, there are two or three other large clusters of countries mainly along the x-axis, toward the right, middle, and left sides. It is obvious that the formation of clusters in this biplot is very subjective and thus somewhat arbitrary. It may be interesting to perform formal clustering using some algorithm on the original data or on the first two principal components.

Another natural question to ask here is, "among the seven track events considered, which of the events are similar to each other in nature?" In other words, we may also be interested in identifying any possible clustering of variables. The biplots can also provide this kind of information. Specifically, variables that are highly correlated result in points that make an angle close to zero (if the variables are positively correlated) or an angle close to π radians, that is, 180 degrees (if the variables are highly negatively correlated) at the origin. Variables depicting very weak correlation will make an angle close to $\frac{\pi}{2}$ radians (90 degrees). Thus, similarity or dissimilarity of variables could also be graphically determined by the biplots by considering the angles made at the origin by the points corresponding to these variables. In Section 6.5 we will consider certain nongraphical approaches to clustering the variables. This is an important problem, and any such clustering can provide us a way for dimensionality reduction.

EXAMPLE 2 ***(Continued)*** Although certain small angles made by various track events at the origin are clearly seen in the biplot corresponding to choice (c), an appropriate biplot emphasizing the clustering of variables corresponds to choice (b). As mentioned in Khattree and Naik (1999), in the corresponding biplot these angles can be interpreted as the cosines of the corresponding correlation coefficients. The biplot in Output 2.11(b) corresponds to this choice. The clusterings of small, medium, and large distance track events is obvious from this biplot. Specifically, the following clusters are subjectively identified,

{M 100, M 200, M400},
{M 800}

and

{M1500, M3000 MARATHON}.

6.3 Similarity and Dissimilarity Measures

Before we consider any nongraphical approach to clustering, a natural question to ask is how to decide if the two objects are similar. This, in effect, requires that one formally define a measure of similarity or dissimilarity between two objects. Having such a quantification allows us to speak about two items being relatively more similar than another two items, and hence can provide us with a nonambiguous and systematic approach to forming clusters.

It may be mentioned that it suffices to have only one index, either of similarity or of dissimilarity and by some suitable one-to-one transformation, an index of similarity can be converted to dissimilarity or vice versa. Hence, for the most part, in this chapter we will work with dissimilarity measures. It may also be noted that if raw data are not available, then most of the clustering procedures in SAS, except the VARCLUS procedure, work in the framework of dissimilarity measures only.

We have already specified in Section 6.1, the basic requirement imposed on a dissimilarity measure $I(a, b)$, measuring the dissimilarities between two objects a and b. An obvious example of a dissimilarity function is a distance function $d(a, b)$, measuring some appropriately defined distance between the objects a and b. In addition to satisfying the properties of a dissimilarity measure, namely,

(i) $d(a, b) \geq 0$

(ii) $d(a, a) = 0$

(iii) $d(a, b) = d(b, a)$

(iv) $d(a, b)$ increases as objects a and b become more and more dissimilar,

the distance functions also satisfy the triangular inequality, which states

(v) $d(a, c) \leq d(a, b) + d(b, c)$.

The most commonly used distance is the Euclidean distance, which is what is actually observed by human eyes. The Euclidean distance between two observations $\mathbf{x}$ and $\mathbf{y}$ is measured as

$$d(\mathbf{x}, \mathbf{y}) = \sqrt{(\mathbf{x} - \mathbf{y})'(\mathbf{x} - \mathbf{y})}$$

$$= \sqrt{\sum_{i=1}^{p} (x_i - y_i)^2}$$

A more general distance measure is the Minkowski distance defined as

$$d(\mathbf{x}, \mathbf{y}) = \left[\sum_{i=1}^{p} |x_i - y_i|^k \right]^{\frac{1}{k}}.$$

Of course the choice $k = 2$ corresponds to the Euclidean distance. It is also known as the L_2 distance. The choice $k = 1$ is often referred to as the L_1 distance and can be interpreted (when $p = 2$) as the actual distance traveled on the north-south and east-west streets to reach the point $\mathbf{y}$ from point $\mathbf{x}$. Thus it is also called the city-block or the Manhattan distance. It may be mentioned that the choice of $k > 1$, in defining the Minkowski distance, results in putting more emphasis on the large differences between the coordinates.

Often the distance between two points is defined as the weighted distance with weight matrix $\mathbf{A}$. Specifically,

$$d(\mathbf{x}, \mathbf{y}) = \sqrt{(\mathbf{x} - \mathbf{y})'\mathbf{A}(\mathbf{x} - \mathbf{y})},$$

where $\mathbf{A}$ is symmetric positive-definite matrix. The above distance is often useful in removing or reducing the effect of different scales for the components of $\mathbf{x}$ and $\mathbf{y}$. When $\mathbf{A} = \text{diag}(a_{11}, \ldots, a_{pp})$ is a diagonal matrix then the above reduces to

$$d(\mathbf{x}, \mathbf{y}) = \sqrt{\Sigma a_{ii}(x_i - y_i)^2}.$$

In many statistical problems, $\mathbf{x}$ and $\mathbf{y}$ are observations and often it is appropriate to measure the distance after accounting for the different variances and covariances. One popular distance measure that attempts to do so is

$$d(\mathbf{x}, \mathbf{y}) = \sqrt{(\mathbf{x} - \mathbf{y})'\mathbf{\Sigma}^{-1}(\mathbf{x} - \mathbf{y})},$$

where $\mathbf{\Sigma}$ is the common variance-covariance matrix of $\mathbf{x}$ and $\mathbf{y}$. Often an estimate of $\mathbf{\Sigma}$, say $\mathbf{S}$, is used in the above, resulting in what is commonly known as the Mahalanobis distance, defined as

$$d(\mathbf{x}, \mathbf{y}) = \sqrt{(\mathbf{x} - \mathbf{y})\mathbf{S}^{-1}(\mathbf{x} - \mathbf{y})}.$$

In much of the clustering work, distances are used as the measures of dissimilarity. Further, for ease of computation, L_2 distances are often computed as the squares of actual distances. This poses no problem because most clustering procedures require only comparisons of distances, Hence ignoring the square root presents no difficulty.

For a set of objects $\{\mathbf{x}_1, \mathbf{x}_2, \ldots, \mathbf{x}_n\}$, we can define a matrix $\mathbf{D} = (d_{ij})$ of distances, where $d_{ij} = d(\mathbf{x}_i, \mathbf{x}_j)$ for some appropriately defined distance measure. Clearly all the diagonal elements of $\mathbf{D}$ are zero as $d(\mathbf{x}_i, \mathbf{x}_i) = 0$ for all $\mathbf{x}_i$. The matrix $\mathbf{D}$ is used as the dissimilarity matrix for many distance-based cluster analysis applications, and it is easy to compute it using the IML procedure. For example, for the Euclidean distance

$$\begin{aligned} d_{ij} = d(\mathbf{x}_i, \mathbf{x}_j) &= \sqrt{(\mathbf{x}_i - \mathbf{x}_j)'(\mathbf{x}_i - \mathbf{x}_j)} \\ &= \sqrt{\mathbf{x}_i'\mathbf{x}_i - 2\mathbf{x}_i'\mathbf{x}_j + \mathbf{x}_j'\mathbf{x}_j}. \end{aligned}$$

If $\mathbf{X} = \begin{bmatrix} \mathbf{x}_1' \\ \mathbf{x}_2' \\ \vdots \\ \mathbf{x}_n' \end{bmatrix}$ is the n by p matrix of data with $\mathbf{x}_1, \ldots, \mathbf{x}_n$ arranged as rows, then d_{ij}^2 can be obtained as a linear combination of $(i, i)^{th}$, $(j, j)^{th}$, and $(i, j)^{th}$ elements of $\mathbf{XX}'$. Sample SAS code to compute d_{ij}^2 is given in Program 6.3. For a data set from Everitt (1989) on six patients with four measurements on each patient, the six by six distance matrix $\mathbf{D} = (d_{ij})$ is presented as Output 6.3.

```
/* Program 6.3 */


options ls = 64 ps = 45 nodate nonumber;
title1 'Output 6.3';
title2 'Distance Matrix: Acidosis Data';
```

```
/* Computation of Distance Matrix */

proc iml;
x = {
39.8 38.0 22.2 23.2,
53.7 37.2 18.7 18.5,
47.3 39.8 23.3 22.1,
41.7 37.6 22.8 22.3,
44.7 38.5 24.8 24.4,
47.9 39.8 22.0 23.3
};
/*
Data on Acidosis patients. Source: Everitt (1989),
Reprinted with permission from Edward Arnold, 1989.
*/

nrow = nrow(x);
xpx = x*t(x);
vdiag = vecdiag(xpx);
xi = j(1,nrow,1)@vdiag;
dist =sqrt(t(xi) - 2*xpx + xi);
print dist;
```

Output 6.3

```
                         Output 6.3
               Distance Matrix: Acidosis Data

                            DIST

        0 15.105959 7.8682908 2.2226111 5.6973678 8.3006024
15.105959         0 9.0465463 13.244244 12.438247 8.6214848
7.8682908 9.0465463         0 6.0406953 3.9987498 1.8681542
2.2226111 13.244244 6.0406953         0 4.2684892 6.7022384
5.6973678 12.438247 3.9987498 4.2684892         0  4.580393
8.3006024 8.6214848 1.8681542 6.7022384  4.580393         0
```

In order to compute any weighted distance with weights given by the positive definite matrix **A**,

$$d_{ij} = d(\mathbf{x}_i, \mathbf{x}_j) = \sqrt{(\mathbf{x}_i - \mathbf{x}_j)'\mathbf{A}(\mathbf{x}_i - \mathbf{x}_j)},$$

Program 6.3 can be easily modified. Specifically, since **A** is positive definite, it can be written in terms of its Cholesky decomposition (see Section 1.5.23),

$$\mathbf{A} = \mathbf{G}\mathbf{G}'$$

for some specific (lower triangular) matrix **G**. With $\mathbf{y}_i = \mathbf{G}'\mathbf{x}_i$, the distance

$$\begin{aligned} d(\mathbf{x}_i, \mathbf{x}_j) &= \sqrt{(\mathbf{x}_i - \mathbf{x}_j)'\mathbf{A}(\mathbf{x}_i - \mathbf{x}_j)} \\ &= \sqrt{(\mathbf{x}_i - \mathbf{x}_j)'\mathbf{G}\mathbf{G}'(\mathbf{x}_i - \mathbf{x}_j)} \\ &= \sqrt{(\mathbf{y}_i - \mathbf{y}_j)'(\mathbf{y}_i - \mathbf{y}_j)}. \end{aligned}$$

which is nothing but the Euclidean distance between $\mathbf{y}_i$ and $\mathbf{y}_j$. Thus Program 6.3 should be used on the data matrix

$$\mathbf{Y} = \begin{bmatrix} \mathbf{y}'_1 \\ \vdots \\ \mathbf{y}'_n \end{bmatrix} = \begin{bmatrix} \mathbf{x}'_1\mathbf{G} \\ \vdots \\ \mathbf{x}'_n\mathbf{G} \end{bmatrix} = \mathbf{XG}.$$

Once $\mathbf{A}$ is specified, the matrix $\mathbf{G}$ can be obtained using the ROOT function and once that is known, $\mathbf{Y}$ is obtained from the original data matrix $\mathbf{X}$ by $\mathbf{Y} = \mathbf{XG}$.

6.4 Hierarchical Clustering Methods

Methods discussed here are all based on the hierarchical approach of merging items and previously formed clusters one by one into new clusters by deciding their proximity to other clusters. These are often referred to as agglomerative methods. There are variations on how the proximity between two clusters is determined. However, the common theme is that we start by assuming n clusters, each containing one element each. The two closest clusters are then merged and then again the proximity between these new sets of $(n - 1)$ clusters is evaluated. Again the closest two clusters are merged and the procedure continues. It can go till we are left with a single cluster consisting of all items. In practice, if we want only k clusters then we must stop when k clusters have been obtained. However, often the optimal number of clusters k is not predetermined. In that case other statistical criteria, based on raw data, may be used to evaluate the potential solutions.

There are also a number of hierarchical divisive clustering methods, as opposed to the agglomerative methods, in which the above algorithm proceeds in reverse order. That is, we start with just one cluster containing all items and continue to divide clusters into two or more clusters using some well-defined rules. However, we do not present this approach here.

How do we measure the proximity between the clusters? Often we use some kind of dissimilarity measure and a relatively low magnitude of dissimilarity between clusters suggests that the clusters are relatively similar. A number of distance measures have been used by various authors to represent the dissimilarity. We briefly present some of them here.

6.4.1 Single Linkage or Nearest Neighbor Approach

As the name suggests, the clusters with the two nearest neighbors, each belonging to different clusters, should be merged. More specifically, for two clusters B_r and B_s, the distance between B_r and B_s, say $h(r, s)$, is defined as

$$h(B_r, B_s) = \min\{d(\mathbf{x}_i, \mathbf{x}_j);\ \mathbf{x}_i \text{ in } B_r \text{ and } \mathbf{x}_j \text{ in } B_s\}.$$

Clusters B_r^* and B_s^* are merged if $h(B_{r^*}, B_{s^*})$ is the smallest. Thus at any stage, the total number of clusters reduces by one. We will illustrate the method through an example.

EXAMPLE 3 *Single Linkage Clustering, Everitt's Data* Consider the distance matrix for six patients based on four measurements as reported in Output 6.3. To analyze these data we use Program 6.4. The following SAS statements produce the desired clustering.

```
proc cluster data=acidosis method=single;
id patient;
var patient1 patient2 patient3 patient4 patient5 patient6;
```

The METHOD=SINGLE option specifies that single linkage clustering should be performed. The data set ACIDOSIS used here contains distances and hence has been specified at the DATA step of Program 6.3 by clearly indicating within parenthesis TYPE=DISTANCE. The output is presented in the first part of Output 6.4. It may be appropriate to illustrate how the numbers in the output are obtained by illustrating the calculations for this small data set.

Given the distance matrix

$$\begin{matrix} 1 \\ 2 \\ 3 \\ 4 \\ 5 \\ 6 \end{matrix} \begin{bmatrix} 0 & 15.1060 & 7.8683 & 2.2226 & 5.6974 & 8.3006 \\ 15.1060 & 0 & 9.0465 & 13.2442 & 12.4382 & 8.6214 \\ 7.8683 & 9.0465 & 0 & 6.04070 & 3.9987 & 1.8682 \\ 2.2226 & 13.2442 & 6.04070 & 0 & 4.2685 & 6.7022 \\ 5.6974 & 12.4382 & 3.9987 & 4.2685 & 0 & 4.5804 \\ 8.3006 & 8.6214 & 1.8682 & 6.7022 & 4.5804 & 0 \end{bmatrix},$$

we begin with six clusters, with cluster C_i containing the ith patient. Thus, $C_i = \{i\}, i = 1, \ldots, 6$. The smallest distance (=1.8682) is observed as that between $C_3 = \{3\}$ and $C_6 = \{6\}$. Thus these two clusters are merged, and now we have only five clusters. This is indicated in the output in the first column. For future reference, in the output, this newly formed cluster $\{3, 6\}$, which leaves us with only five clusters, will be denoted by CL5. Thus $CL5 = \{3, 6\}$. Now, we compute the pairwise distances $h(C_i, C_j)$ between

$$C_1, \; C_2, \; C_4, \; C_5 \text{ and } CL5.$$

The new distance matrix is

$$\begin{matrix} 1 \\ 2 \\ 4 \\ 5 \\ 3,6 \end{matrix} \begin{bmatrix} 0 & 15.1060 & 2.2226 & 5.6974 & 7.8683 \\ 15.1060 & 0 & 13.2442 & 12.4383 & 8.6215 \\ 2.2226 & 13.2442 & 0 & 4.2685 & 6.0407 \\ 5.6974 & 12.4383 & 4.2685 & 0 & 3.9987 \\ 7.8683 & 8.6215 & 6.0407 & 3.9987 & 0 \end{bmatrix}$$

and the smallest distance (=2.2226) now corresponds to that between $C_1 = \{1\}$ and $C_4 = \{4\}$. Merging these will leave us with four clusters, so with appropriate notation, the new cluster will be $CL4 = \{1, 4\}$ and now we compute the distance matrix for

$$C_2, \; C_5, \; CL5, \; CL4,$$

which is given by

$$\begin{matrix} 2 \\ 5 \\ 3,6 \\ 1,4 \end{matrix} \begin{bmatrix} 0 & 12.4383 & 7.8683 & 13.2442 \\ 12.4383 & 0 & 3.9988 & 4.2685 \\ 7.8683 & 3.9988 & 0 & 6.0407 \\ 13.2442 & 4.2685 & 6.0407 & 0 \end{bmatrix}.$$

With the smallest distance (= 3.9988) between $CL5$ and C_5, we get $CL3 = \{3, 6, 5\}$ and now we have three clusters, namely,

$$C_2, \; CL3, \; CL4.$$

The distance matrix can be calculated similarly. At the next stage we can merge $CL4$ and $CL3$ and thus we get $CL2 = \{1, 3, 4, 5, 6\}$, and now there are only two clusters, namely,

$$C_2, \; CL2.$$

Finally, these two are also merged resulting in $CL1 = \{1, 2, 3, 4, 5, 6\}$. Thus we have a k-cluster solution for all $k = 1, \ldots, 6$.

Suppose we are interested in forming only three clusters. Then the corresponding choice is

$$C_2 = \{2\}, \; CL3 = \{3, 5, 6\} \text{ and } CL4 = \{1, 4\}.$$

The process of successively merging items to form the cluster can be graphically depicted in a tree diagram using the TREE procedure. Such a diagram is often called as a dendrogram. The corresponding statements of the TREE procedure to produce a dendrogram are given as part of Program 6.4. Specifically, the relevant data should be first stored in an output file named here TREE1, which is used to create the hierarchical tree. The resulting tree is presented in Output 6.4. The output of the TREE procedure is further stored in an output file called ACID_OUT using the option OUT=ACID_OUT. This file is also printed later using the PRINT procedure. (See the last part of Output 6.4.)

```
/* Program 6.4 */

options ls = 64 ps = 45 nodate nonumber;


title1 'Output 6.4';
data acidosis(type = distance);
input patient1 patient2 patient3 patient4 patient5 patient6 patient $  ;
lines;
          0 15.105959 7.8682908 2.2226111 5.6973678 8.3006024 patient1
  15.105959         0 9.0465463 13.244244 12.438247 8.6214848 patient2
  7.8682908 9.0465463         0 6.0406953 3.9987498 1.8681542 patient3
  2.2226111 13.244244 6.0406953         0 4.2684892 6.7022384 patient4
  5.6973678 12.438247 3.9987498 4.2684892         0  4.580393 patient5
  8.3006024 8.6214848 1.8681542 6.7022384  4.580393         0 patient6
;

/*
Distance matrix Computed from data on Acidosis patients
(Everitt, Statistical methods for medical Investigations,
London, Edward Arnold, 1989.
*/

proc cluster data = acidosis noeigen method = single
nonorm out =tree1;
id patient;
var patient1 patient2 patient3 patient4 patient5 patient6 ;
title2 'Acidosis Patient Data: Single Linkage Clustering';
run;

filename gsasfile "prog64tree.graph";
goptions reset=all gaccess=gsasfile autofeed dev=pslmono;
goptions horigin=1in vorigin=2in;
goptions hsize=6in vsize=8in;
proc tree data = tree1 out = acid_out nclusters = 3;
id patient;
title1 h = 1.2 'Dendrogram: Single Linkage Method';
title2 j = 1 'Output 6.4';
run;

proc sort data = acid_out;
by cluster ;
run;
```

```
proc print data = acid_out;
by cluster ;
title1 "Output 6.4";
title2 "3-Clusters Solution: Single Linkage Clustering" ;
title3 "Unstandardized Data: Using Euclidean Distance Matrix";
run;
```

Output 6.4

```
                         Output 6.4
        Acidosis Patient Data: Single Linkage Clustering

                   The CLUSTER Procedure
              Single Linkage Cluster Analysis

                      Cluster History
                                                          T
                                                   Min    i
  NCL    --Clusters Joined---      FREQ           Dist    e

    5    patient3    patient6         2         1.8682
    4    patient1    patient4         2         2.2226
    3    CL5         patient5         3         3.9987
    2    CL4         CL3              5         4.2685
    1    CL2         patient2         6         8.6215


        3-Clusters Solution: Single Linkage Clustering
      Unstandardized Data: Using Euclidean Distance Matrix

------------------------- CLUSTER=1 ---------------------------

                Obs    patient     CLUSNAME

                 1     patient3      CL3
                 2     patient6      CL3
                 3     patient5      CL3

------------------------- CLUSTER=2 ---------------------------

                Obs    patient     CLUSNAME

                 4     patient1      CL4
                 5     patient4      CL4

------------------------- CLUSTER=3 ---------------------------

                Obs    patient     CLUSNAME

                 6     patient2    patient2
```

Output 6.4
(*continued*)

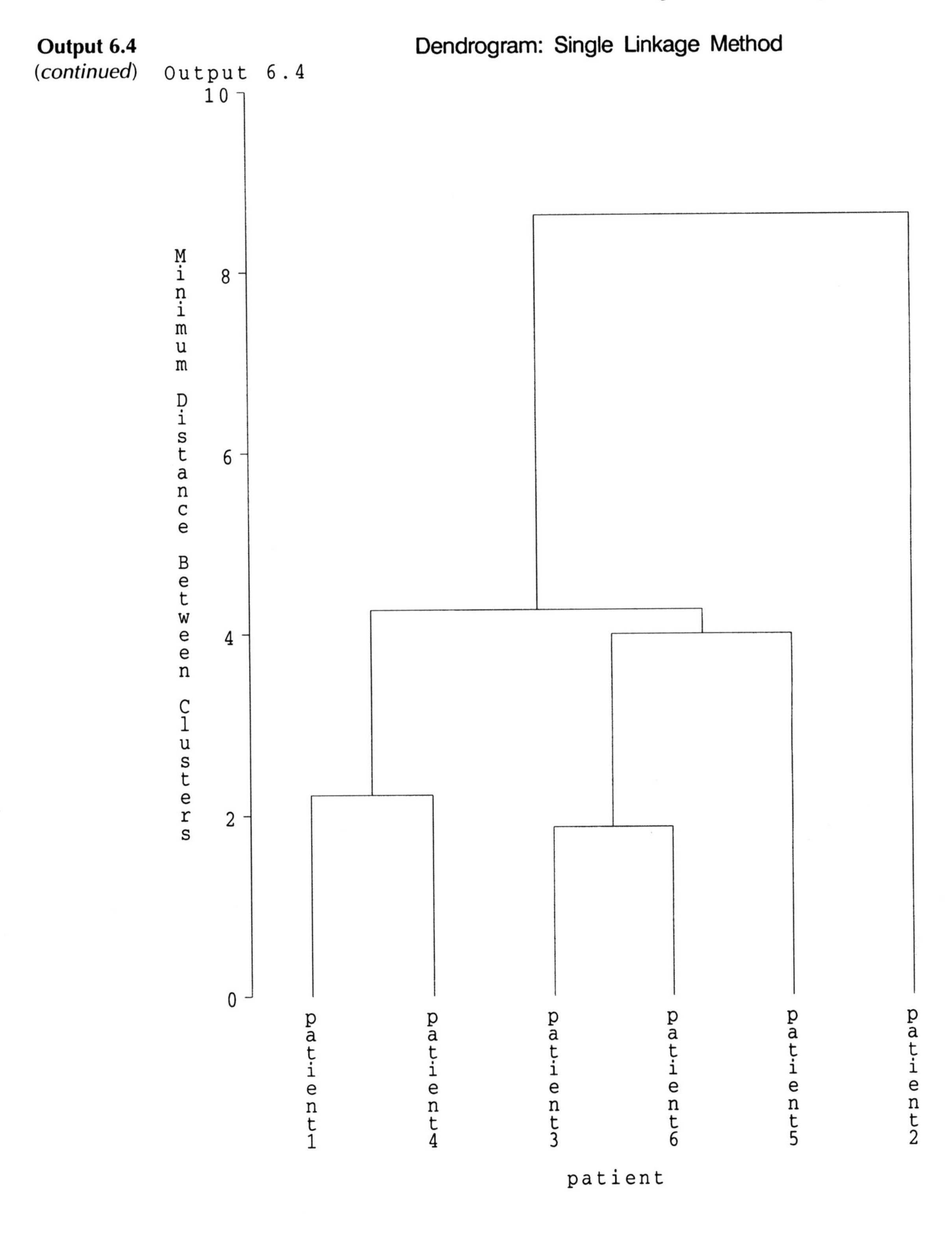

Instead of the single linkage approach, we can use certain other approaches. The basic difference is in the use of different distance measures defined to assess the distances between the clusters. In the CLUSTER procedure we can use them with the METHOD= option. We will briefly discuss some of these below. In all these approaches the rule for merging the clusters is the same, namely, at any stage of algorithm, the two clusters with the smallest cluster distance between them are merged.

6.4.2 Complete Linkage or Farthest Neighbor Approach

The complete linkage method is essentially identical in spirit except that the distance between two clusters B_r and B_s is defined as the largest possible distance between elements in B_r and B_s. More formally, this distance is defined as

$$h(B_r, B_s) = \max\{d(\mathbf{x}_i, \mathbf{x}_j) : \mathbf{x}_i \text{ in } B_r \text{ and } \mathbf{x}_j \text{ in } B_s\}.$$

At any stage of algorithm, two clusters B_{r^*} and B_{s^*} are merged if the distance $h(B_{r^*}, B_{s*})$ is smallest. Thus the rule of merging the clusters is the same while the distance $h(B_r, B_s)$ is defined differently. In the CLUSTER procedure, the corresponding option is METHOD= COMPLETE.

6.4.3 Centroid Approach

Here the distance between the clusters is defined as the Euclidean distance between the respective cluster means. Thus if $\overline{\mathbf{x}}_r$ and $\overline{\mathbf{x}}_s$ are the means of the observations in clusters B_r and B_s, then the cluster distance used in this approach is

$$h(B_r, B_s) = d(\overline{\mathbf{x}}_r, \overline{\mathbf{x}}_s).$$

The centroid of the newly formed cluster will be $\frac{n_r\overline{\mathbf{x}}_r + n_s\overline{\mathbf{x}}_s}{n_r+n_s}$, where n_r and n_s are the cluster sizes for B_r and B_s, respectively.

In the CLUSTER procedure, the corresponding option is METHOD= CENTROID. It may be remarked that distances reported are squared distances unless the NOSQUARE option is used. However, it does not affect the clustering results.

6.4.4 Median Approach

Suppose the cluster B_s was considerably smaller in size than the cluster B_r, that is $n_s << n_r$. It is possible that when B_r and B_s are merged, the centroid of the new cluster is not much different from $\overline{\mathbf{x}}_r$. Thus the cluster B_s is largely not represented in computing distances between the clusters for future merging operations. To avoid this, Gower (1967) suggests using the midpoint (i.e., median) between the recently merged clusters as the point for computing the future cluster distances. Thus initially when there are n clusters each containing a single observation, these observations themselves are the medians. At any stage when B_r and B_s are merged, the new median, say $\mathbf{m}_{\text{new}}$, is computed as the midpoint of the line between old medians, say, $\mathbf{m}_r$ and $\mathbf{m}_s$. That is

$$\mathbf{m}_{\text{new}} = \frac{\mathbf{m}_r + \mathbf{m}_s}{2}.$$

The cluster distances are defined as the distances between corresponding medians, and clusters with the smallest cluster distance are merged. In SAS, the appropriate option for the median approach within the CLUSTER procedure is METHOD=MEDIAN. Unless the NOSQUARE option is used, the distances reported in the output are squared distances. However, as earlier, the clustering outcome is unaffected.

6.4.5 Average Distance Approach

Here the distance between two clusters B_r and B_s is defined as the average of $n_r n_s$ distances computed between $\mathbf{x}_i$ and $\mathbf{x}_j$, where $\mathbf{x}_i$ belongs to B_r, and $\mathbf{x}_j$ belongs to B_s. That is

$$h(B_r, B_s) = \frac{1}{n_r n_s} \sum_{\mathbf{x}_i \text{ in } B_r} \sum_{\mathbf{x}_j \text{ in } B_s} d(\mathbf{x}_i, \mathbf{x}_j).$$

The rule for merging clusters is still the same as earlier approaches. The corresponding option to be used in the CLUSTER procedure is METHOD= AVERAGE. As earlier, the distances reported are actually squared distances. To obtain the actual distances the NOSQUARE option may be used. However, the clustering results are unaffected whether this option is used or not.

6.4.6 Flexible-Beta Approach

Lance and Williams (1967) considered a general distance formula for cluster distances. Since the new clusters are being formed by combining the two most similar clusters or by the union (denoted by $\cup$) of these, they define a distance measure between B_r and $B_s \cup B_t = B_q$ as

$$\begin{aligned} h(B_r, B_q) = \alpha_1 h(B_r, B_s) + \alpha_2 h(B_r, B_t) + \beta h(B_s, B_t) \\ + \gamma |h(B_r, B_s) - h(B_r, B_t)|, \end{aligned}$$

where initially when all clusters have just one element, the distance between clusters $B_r = \{r\}$ and $B_s = \{s\}$ is defined by viewing $\{s\}$ as $\{s\} \cup \{s\}$. Thus using the above definition of cluster distance

$$\begin{aligned} h(\{r\}, \{s\}) &= \alpha_1 h(\{r\}, \{s\}) + \alpha_2 h(\{r\}, \{s\}) + \beta h(\{s\}, \{s\}) \\ &\quad + \gamma |h(\{r\}, \{s\}) - h(\{r\}, \{s\})| \\ &= (\alpha_1 + \alpha_2) h(\{r\}, \{s\}). \end{aligned}$$

Thus, it seems natural to assume that $\alpha_1 + \alpha_2 = 1$. In fact, most of the approaches described earlier and Ward's approach (to be described later) are all special cases of the distance function defined here for certain specific choices of $(\alpha_1, \alpha_2, \beta, \gamma)$ values. These are shown in Table 6.1. Also, all of them, except Ward's approach, and the general Flexible-Beta approach satisfy the natural requirement $\alpha_1 + \alpha_2 = 1$. In centroid, average, and Ward's approaches, the values of the parameters α_1, α_2, β, and γ depend on cluster sizes. Additionally, as Seber (1984) points out, in Table 6.1 and in the case of centroid, median,

TABLE 6.1 $(\alpha_1, \alpha_2, \beta, \gamma)$ Values for Hierarchical Methods

Approach	α_1	α_2	β	γ
Single Linkage	$\frac{1}{2}$	$\frac{1}{2}$	0	$-\frac{1}{2}$
Complete Linkage	$\frac{1}{2}$	$\frac{1}{2}$	0	$\frac{1}{2}$
Centroid	$n_s/(n_s + n_t)$	$n_t/(n_s + n_t)$	$-n_s n_t/(n_s + n_t)^2$	0
Median	$\frac{1}{2}$	$\frac{1}{2}$	$-\frac{1}{4}$	0
Average	$n_s/(n_s + n_t)$	$n_t/(n_s + n_t)$	0	0
Ward's	$(n_r + n_s)/(n_r + n_q)$	$(n_r + n_t)/(n_r + n_q)$	$-n_r/(n_r + n_q)$	0
Flexible-Beta	$(1 - \beta)/2$	$(1 - \beta)/2$	$\beta(< 1)$	0
McQuitty	$\frac{1}{2}$	$\frac{1}{2}$	0	0

and Ward's approaches, the distances between the objects are taken as not the real distances or dissimilarities, but as the squares of these.

We will now describe the general flexible-beta approach.

The flexible-beta approach chooses the constants $\alpha_1, \alpha_2, \beta$ and γ such that

$$\alpha_1 + \alpha_2 + \beta = 1$$
$$\alpha_1 = \alpha_2$$
$$\gamma = 0$$
$$\beta < 1.$$

Thus, it suffices to choose only $\beta(< 1)$ and, accordingly, α_1 and α_2 are chosen as $\alpha_1 = \alpha_2 = (1 - \beta)/2$. Obviously, by definition, γ is zero here. Therefore, in this case

$$h(B_r, B_q) = \frac{1-\beta}{2}\{h(B_r, B_s) + h(B_r, B_t)\} + \beta h(B_s, B_t).$$

When the parameter β is chosen to be negative for the process of merging B_s and B_t, it amounts to discounting the distance of any cluster from the newly formed cluster. In fact, Lance and Williams (1967) recommend taking β to be negative yet still close to zero. In SAS, the flexible-beta approach can be implemented by using the METHOD=FLEXIBLE option, and the choice of β can be specified by the BETA= option. The default value of β is -0.25. If the data are likely to have several outliers (that is, clusters with a single element in them), then a smaller value of β, such as -0.5, may be more appropriate. The choice of β can be specified by using the options METHOD=FLEXIBLE and the BETA= in the PROC CLUSTER statement, where β is chosen as some appropriate number less than 1.

6.4.7 McQuitty's Approach

In the flexible-beta method, when β is chosen to be zero, the formula for the distance of cluster B_r from the newly formed cluster B_q at any stage becomes

$$h(B_r, B_q) = \frac{1}{2}\{h(B_r, B_s) + h(B_r, B_t)\}$$

which corresponds to McQuitty's similarity approach. It can be specified by the METHOD= MCQUITTY option. Thus, the statement

```
proc cluster method = flexible beta=0;
```

is equivalent to the statement

```
proc cluster method = mcquitty;
```

For all these clustering approaches, it suffices to provide the distance or dissimilarity matrix as the input information. This is what was done in Program 6.4. However, if the raw data are available and the dissimilarities used are the usual Euclidean distances, then it also suffices to just provide the raw data from which SAS can automatically compute the dissimilarities as Euclidean distances. The data are input in the usual format, as shown in the early part of Program 6.5. Note that we need not and should not use TYPE = DISTANCE to specify data type in this case. In fact, when raw data are available, then for certain approaches described above, we are also able to assess the quality of clustering using certain statistical measures which are computed from the data themselves. We will discuss these after we have introduced the Ward's minimum variance approach.

6.4.8 Ward's Minimum Variance Approach

Ward's minimum variance approach is more statistical in nature than the approaches previously described and requires raw data rather than distances. The basic idea is based on the fact that if two clusters are very similar and hence close to each other, then the *between cluster sum of squares* should be small. More formally, let $\bar{\mathbf{x}}_r$ and $\bar{\mathbf{x}}_s$ be the cluster mean vectors for B_r and B_s, and suppose $\bar{\mathbf{x}}_q = (n_r\bar{\mathbf{x}}_r + n_s\bar{\mathbf{x}}_s)/(n_r + n_s)$ is the cluster mean for that obtained by combining the two clusters, that is, of the cluster $B_q = B_r \cup B_s$. Then, the total squared distances of the observations from the corresponding cluster means can be written as

$$E_q = \sum_{i \text{ in } B_q} (\mathbf{x}_i - \bar{\mathbf{x}}_q)'(\mathbf{x}_i - \bar{\mathbf{x}}_q),$$

$$E_r = \sum_{i \text{ in } B_r} (\mathbf{x}_i - \bar{\mathbf{x}}_r)'(\mathbf{x}_i - \bar{\mathbf{x}}_r),$$

and

$$E_s = \sum_{i \text{ in } B_s} (\mathbf{x}_i - \bar{\mathbf{x}}_s)'(\mathbf{x}_i - \bar{\mathbf{x}}_s).$$

These are obviously the respective sums of squares corrected for the corresponding means, and it can be shown that

$$E_q = E_r + E_s + n_r(\bar{\mathbf{x}}_r - \bar{\mathbf{x}}_q)'(\bar{\mathbf{x}}_r - \bar{\mathbf{x}}_q) + n_s(\bar{\mathbf{x}}_s - \bar{\mathbf{x}}_q)'(\bar{\mathbf{x}}_s - \bar{\mathbf{x}}_q).$$

The last two terms in the above represent the between cluster sums of squared distances. Obviously, they will be small when $\bar{\mathbf{x}}_r$ and $\bar{\mathbf{x}}_s$ are close to each other, a situation that occurs when B_r and B_s are similar. Thus, the last two terms of the above expression, that is the difference $E_q-(E_r+E_s)$, can be taken as a measure of cluster dissimilarity. The algorithm for joining the clusters is based on this measure. Upon simplification, this measure reduces to $h(B_r, B_s) = \frac{n_r n_s}{n_r+n_s}(\bar{\mathbf{x}}_r - \bar{\mathbf{x}}_s)'(\bar{\mathbf{x}}_r - \bar{\mathbf{x}}_s)$. For clusters with a single element in each of them, it reduces to half to the squared Euclidean distance between them.

The terms E_r, E_s, and E_q can be viewed as the measures of respective *within cluster homogeneity*. By merging the two clusters, there is a loss in cluster homogeneity, and Ward's approach can be viewed as the process of merging those two clusters B_r and B_s for which this loss represented by $E_q - (E_r + E_s)$, with $B_q = B_s \cup B_s$, is the smallest. This is why the procedure is also known as the incremental sum of squares approach.

To start with, we have n clusters, each containing a single element, and the closest elements with the smallest $h(\{r\}, \{s\})$ are put together to form $(n - 1)$ clusters. Again, the two closest clusters, in the sense of the smallest $h(B_r, B_s)$, are fused together, and the process continues until the desired (or appropriately chosen by some meaningful criterion) number of clusters have been obtained.

EXAMPLE 3 ***(Continued) Ward's Approach*** We consider the raw data on six patients given by Everitt (1989) and discussed earlier, and use Ward's approach for clustering. A SAS program for this is reported as Program 6.5. Since raw data and not just the distances are available, certain options to assess the quality of clustering or to decide the number of clusters are also available. Specifically, these are SIMPLE, RMSSTD, RSQUARE, and NONORM. We will discuss their interpretations later in this section. The VAR statement identifies the variables on which measurements are made, and the ID statement identifies the name for the basic units, of which clusters are to be formed. These are patients in our example.

Ward's approach is implemented by using the METHOD= WARD option. When measurements are taken in different measurement units, then it is a common practice to standardize the data (usually to zero mean and standard deviation equal to 1). This can be done using the STANDARD procedure prior to applying the CLUSTER procedure. Alternatively, we can directly use the STANDARD option in the CLUSTER procedure. This standardization can affect the clustering results and different clustering may be obtained if the data were not standardized. However, for Ward's approach, we here present the output corresponding to the unstandardized data. This is shown as Output 6.5.

The SIMPLE option results in simple descriptive statistics for each variable. Viewing data as the sample from a certain population mixture, the estimated means, standard deviations and coefficients of skewness, kurtosis, and bimodality are obtained. The coefficient of bimodality is defined in terms of the coefficient of skewness and kurtosis $\hat{\beta}_1$ and $\hat{\beta}_2$, respectively, as

$$\hat{b} = \frac{\hat{\beta}_1^2 + 1}{\hat{\beta}_2 + \frac{3(n-1)^2}{(n-2)(n-3)}}.$$

A value of $\hat{b}$ larger than 0.555 suggests a possible case of more than one peak for the probability density.

For our data set, which is quite small, it is not appropriate to assess the skewness, kurtosis, and bimodality, as these estimates will have large sampling variability. However, generally, a value close to 0 for $\hat{\beta}_1$ suggests symmetry of distribution, and a value close to 0 for $\hat{\beta}_2$ suggests closeness to normality. SAS also reports the root mean square total standard deviation, $\sqrt{\frac{1}{(n-1)p}\sum_{j=1}^{p}\sum_{i=1}^{n}(x_{ij} - \overline{x}_{.j})^2}$, where $\mathbf{x}_i = (x_{i1}, \ldots, x_{ip})'$ is the ith p-dimensional data point, and $\overline{x}_{.j} = n^{-1}\sum_{i=1}^{n} x_{ij}$ is the mean over all data points for the jth variable. For our unstandardized data, its value is 2.9195.

Initially, the corrected within-cluster sum of squares for all clusters is zero since each of them contains only one element. As shown in Output 6.5, first the clusters {3} and {6} are joined, since the smallest between sum of squares, namely, 1.7450, corresponds to them. This is listed under the column BSS. Now there are five clusters, namely, {1}, {2}, {4}, {5} and {3, 6}. The corrected within sums of squares are all zero, except for the last cluster {3, 6}. The between corrected sum of squares is computed again for all pairs of clusters, and the smallest value is observed between {1} and {4}. Joining them results in four clusters at this stage. The process continues in a similar manner till we are left with a single cluster consisting of all six units.

```
/* Program 6.5 */

options ls = 64 ps = 45 nodate nonumber;

title1 'Output 6.5';
data acidosis;
input patient x1-x4;
lines;
1 39.8 38.0 22.2 23.2
2 53.7 37.2 18.7 18.5
3 47.3 39.8 23.3 22.1
4 41.7 37.6 22.8 22.3
5 44.7 38.5 24.8 24.4
6 47.9 39.8 22.0 23.3
;
/*
Data on Acidosis patients. Source: Everitt (1989),
Reprinted with permission from Edward Arnold, 1989.
*/
```

```
proc cluster data = acidosis
simple noeigen method = ward
rmsstd rsquare nonorm out =tree1;
id patient;
var x1 x2 x3 x4 ;
title2 "Acidosis Data: Ward's Clustering";
title3 "Unstandardized Data";
run;

filename gsasfile "prog65tree1.graph";
goptions reset=all gaccess=gsasfile autofeed dev=pslmono;
goptions horigin=1in vorigin=2in;
goptions hsize=6in vsize=8in;
proc tree data = tree1 out = un_acid2 nclusters = 3;
id patient;
copy x1 x2 x3 x4;
title1 h = 1.2 "Dendrogram: Ward's Method";
title2 j = 1 "Output 6.5";
run;

proc sort data = un_acid2;
by cluster ;
run;

proc print data = un_acid2;
by cluster ;
var patient x1 x2 x3 x4 ;
title1 'Output 6.5';
title2 "3-Clusters Solution: Ward's Approach" ;
title3 "Unstandardized Data";
run;

proc cluster data = acidosis
simple noeigen method = flexible beta = -.3 standard
rmsstd rsquare nonorm out =tree2;
id patient;
var x1 x2 x3 x4 ;
title1 'Output 6.5';
title2 'Acidosis Data: Flexible-Beta Clustering';
title3 "Standardized data";
run;

filename gsasfile "prog65tree2.graph";
goptions reset=all gaccess=gsasfile autofeed dev=pslmono;
goptions horigin=1in vorigin=2in;
goptions hsize=6in vsize=8in;
proc tree data = tree2 out = si_acid2 nclusters = 3;
id patient;
copy x1 x2 x3 x4;
title1 h = 1.2 "Dendrogram: Flexible-Beta Method";
title2 j = 1 "Output 6.5";
run;

proc sort data = si_acid2;
by cluster ;
run;
```

```
proc print data = si_acid2;
by cluster ;
var patient x1 x2 x3 x4 ;
title2 "3-Clusters Solution: Flexible-Beta Approach" ;
title2 "Standardized Data";
title3 ' ';
run;

filename gsasfile "prog65.graph";
goptions reset=all gaccess=gsasfile autofeed dev=pslmono;
goptions gsfmode=append;
goptions horigin=1in vorigin=2in;
goptions hsize=6in vsize=8in;

proc sort data = tree1;
by _ncl_;

data graph1;
set tree1;
tot_obs = 6;
if _ncl_ < tot_obs;
proc print data = graph1;
*var  _ncl_  _rsq_ ;

symbol1 i=join v=star;
legend1 across=3;

proc gplot data = graph1 ;
plot _sprsq_*_ncl_ /
vaxis=axis1 haxis=axis2 legend=legend1;
axis1 label=(a=90 h=1.2 'Semi-Partial R**2');
axis2 offset=(2) label=(h=1.2 'No. of Clusters');
title1 h = 1.2 'Semi-Partial R**2 vs. No. of Clusters';
title2 j = 1  'Output 6.5';
run;

proc gplot data = graph1 ;
plot _rsq_*_ncl_ /
vaxis=axis1 haxis=axis2 legend=legend1;
axis1 label=(a=90 h=1.2 ' R**2');
axis2 offset=(2) label=(h=1.2 'No. of Clusters');
title1 h = 1.2 ' R**2 vs. No. of Clusters';
title2 j = 1  'Output 6.5';
run;

proc gplot data = graph1 ;
plot _height_*_ncl_ /
vaxis=axis1 haxis=axis2 legend=legend1;
axis1 label=(a=90 h=1.2 'BSS');
axis2 offset=(2) label=(h=1.2 'No. of Clusters');
title1 h = 1.2 'BSS vs. No. of Clusters';
title2 j = 1  'Output 6.5';
run;
```

```
proc gplot data = graph1 ;
plot _rmsstd_*_ncl_ /
vaxis=axis1 haxis=axis2 legend=legend1;
axis1 label=(a=90 h=1.2 'RMSSTD');
axis2 offset=(2) label=(h=1.2 'No. of Clusters');
title1 h = 1.2 'RMSSTD vs. No. of Clusters';
title2 j = 1  'Output 6.5';
run;
```

Output 6.5

```
                         Output 6.5
              Acidosis Data: Ward's Clustering
                    Unstandardized Data

                   The CLUSTER Procedure
          Ward's Minimum Variance Cluster Analysis

Variable       Mean    Std Dev   Skewness  Kurtosis  Bimodality

x1          45.8500     4.9614     0.4937    0.0551      0.1973
x2          38.4833     1.1071     0.3647   -1.8762      0.2590
x3          22.3000     2.0278    -1.0770    2.3650      0.2507
x4          22.3000     2.0347    -1.5631    3.1279      0.3672


Root-Mean-Square Total-Sample Standard Deviation = 2.919446


                      Cluster History
                                                    RMS
   NCL  -----Clusters Joined------    FREQ          STD   SPRSQ
                                                     T
                                                     i
                                      RSQ     BSS    e

     5           3          6           2     0.6605  0.0102
                                    .990    1.745
     4           1          4           2     0.7858  0.0145
                                    .975     2.47
     3  CL5                 5           3     1.2984  0.0689
                                    .906   11.742
     2  CL4          CL3                5     1.9631  0.2681
                                    .638   45.703
     1  CL2                 2           6     2.9194  0.6383
                                    .000    108.8


            3-Clusters Solution: Ward's Approach
                    Unstandardized Data

------------------------- CLUSTER=1 --------------------------

        Obs    patient      x1       x2       x3       x4

         1        3       47.3     39.8     23.3     22.1
         2        6       47.9     39.8     22.0     23.3
         3        5       44.7     38.5     24.8     24.4
```

Output 6.5
(continued)

```
-------------------------- CLUSTER=2 ---------------------------

            Obs    patient     x1      x2      x3      x4

             4        1       39.8    38.0    22.2    23.2
             5        4       41.7    37.6    22.8    22.3


-------------------------- CLUSTER=3 ---------------------------

            Obs    patient     x1      x2      x3      x4

             6        2       53.7    37.2    18.7    18.5
```

Output 6.5
(*continued*)

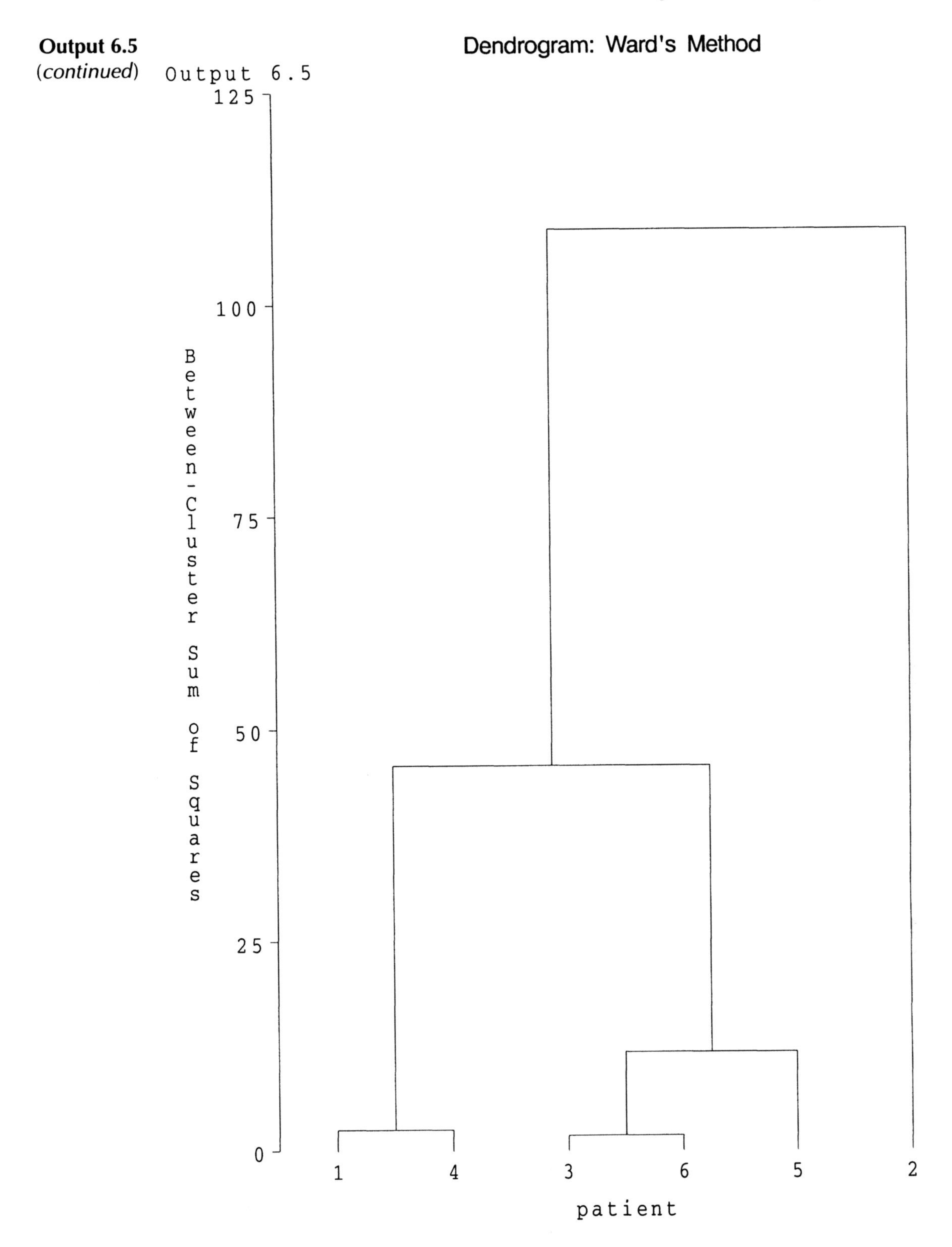

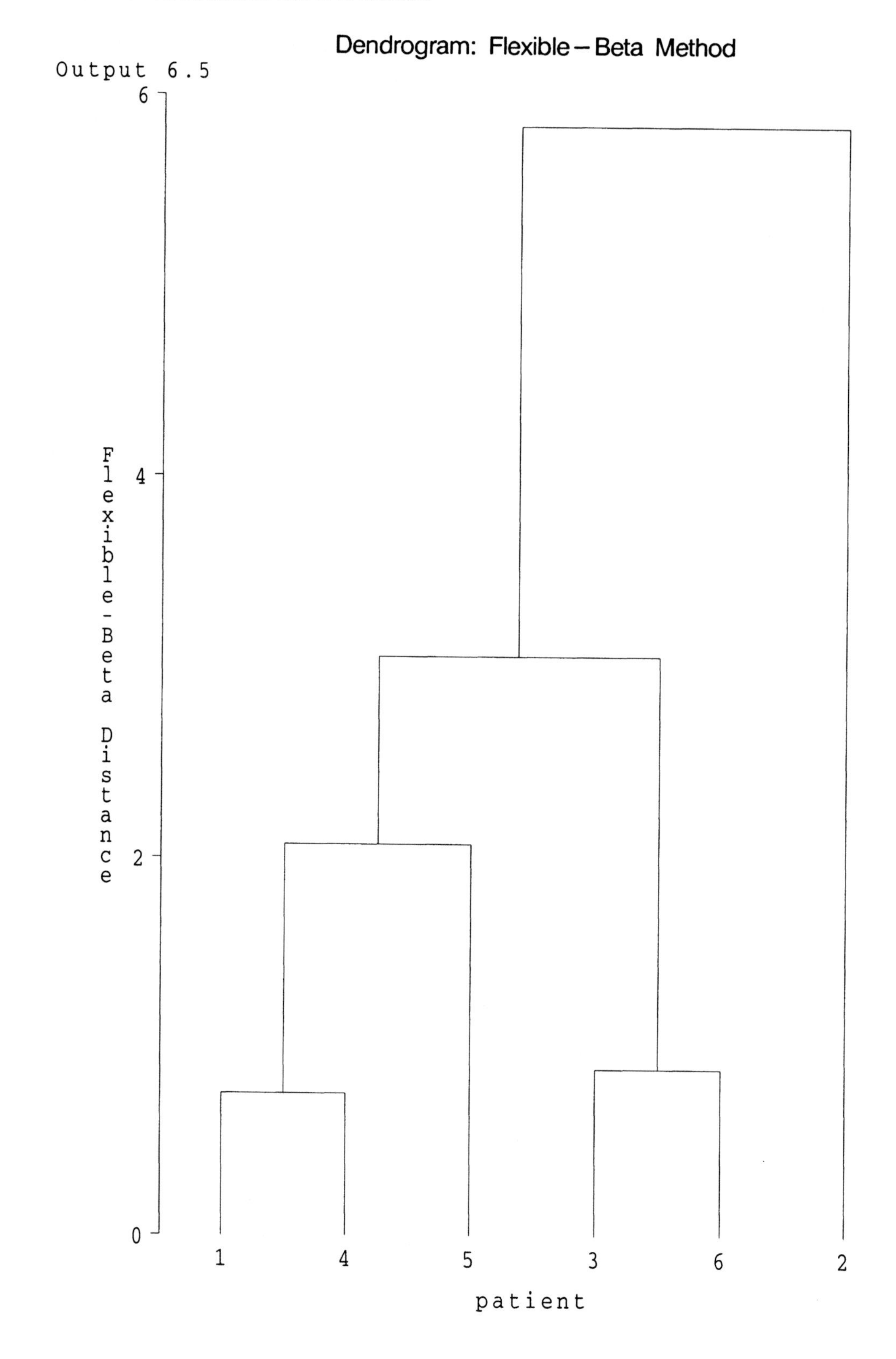
Dendrogram: Flexible – Beta Method
Output 6.5
6
4
2
0
Flexible-Beta Distance
1
4
5
3
6
2
patient

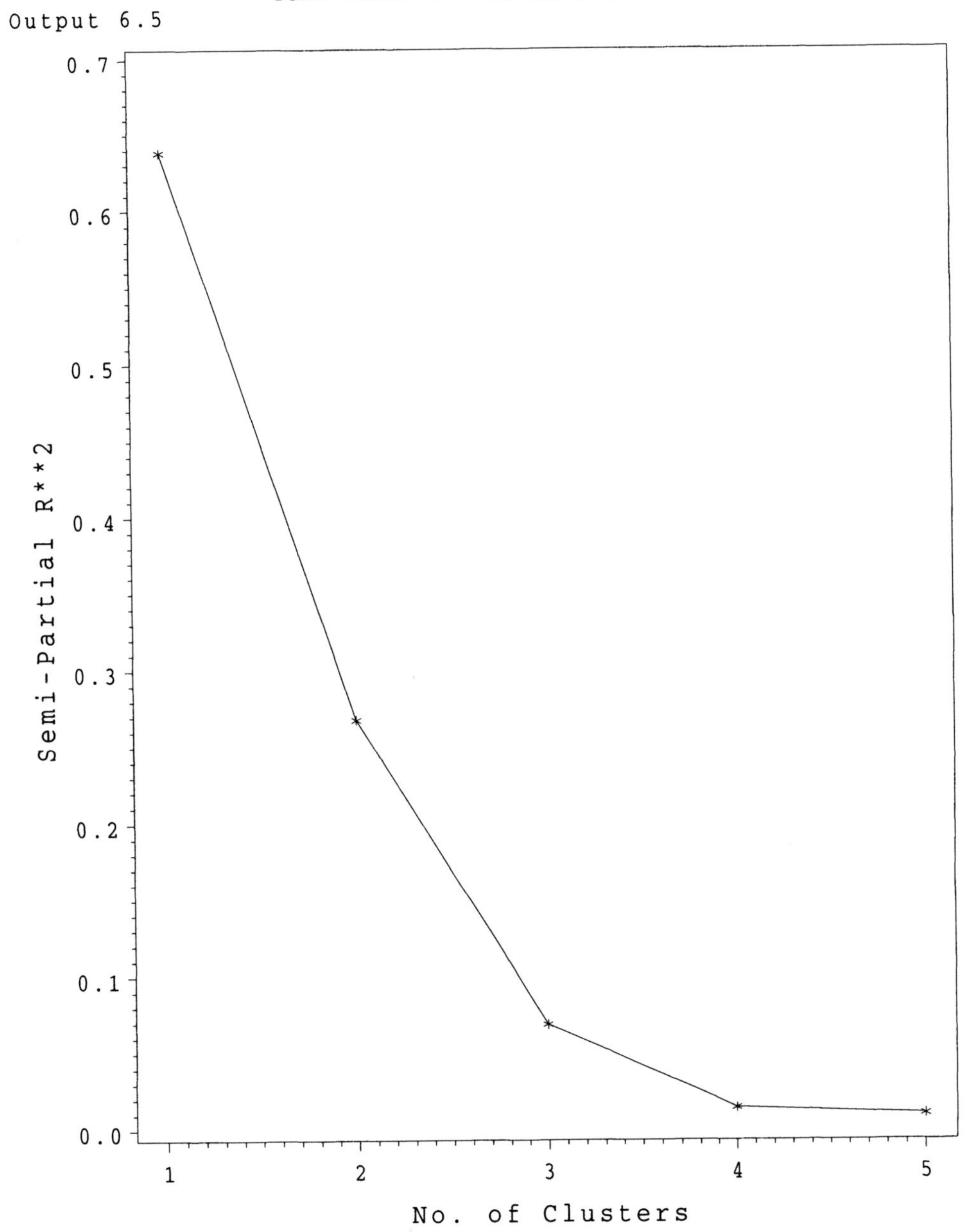
Semi-Partial R**2 vs. No. of Clusters
Output 6.5
0.7
0.6
0.5
0.4
0.3
0.2
0.1
0.0
Semi-Partial R**2
1
2
3
4
5
No. of Clusters

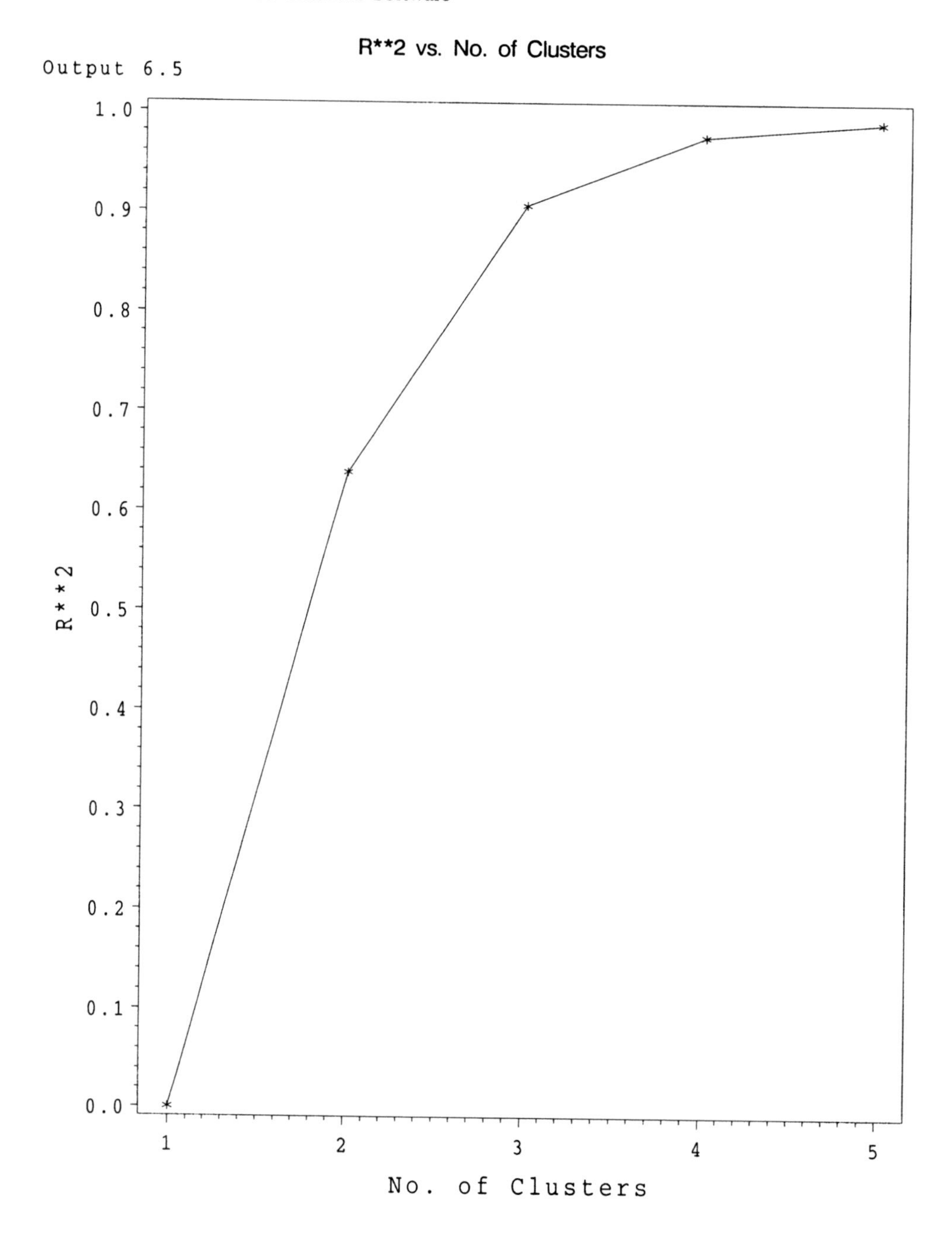
R**2 vs. No. of Clusters
Output 6.5
1.0
0.9
0.8
0.7
0.6
0.5
0.4
0.3
0.2
0.1
0.0
R**2
1
2
3
4
5
No. of Clusters

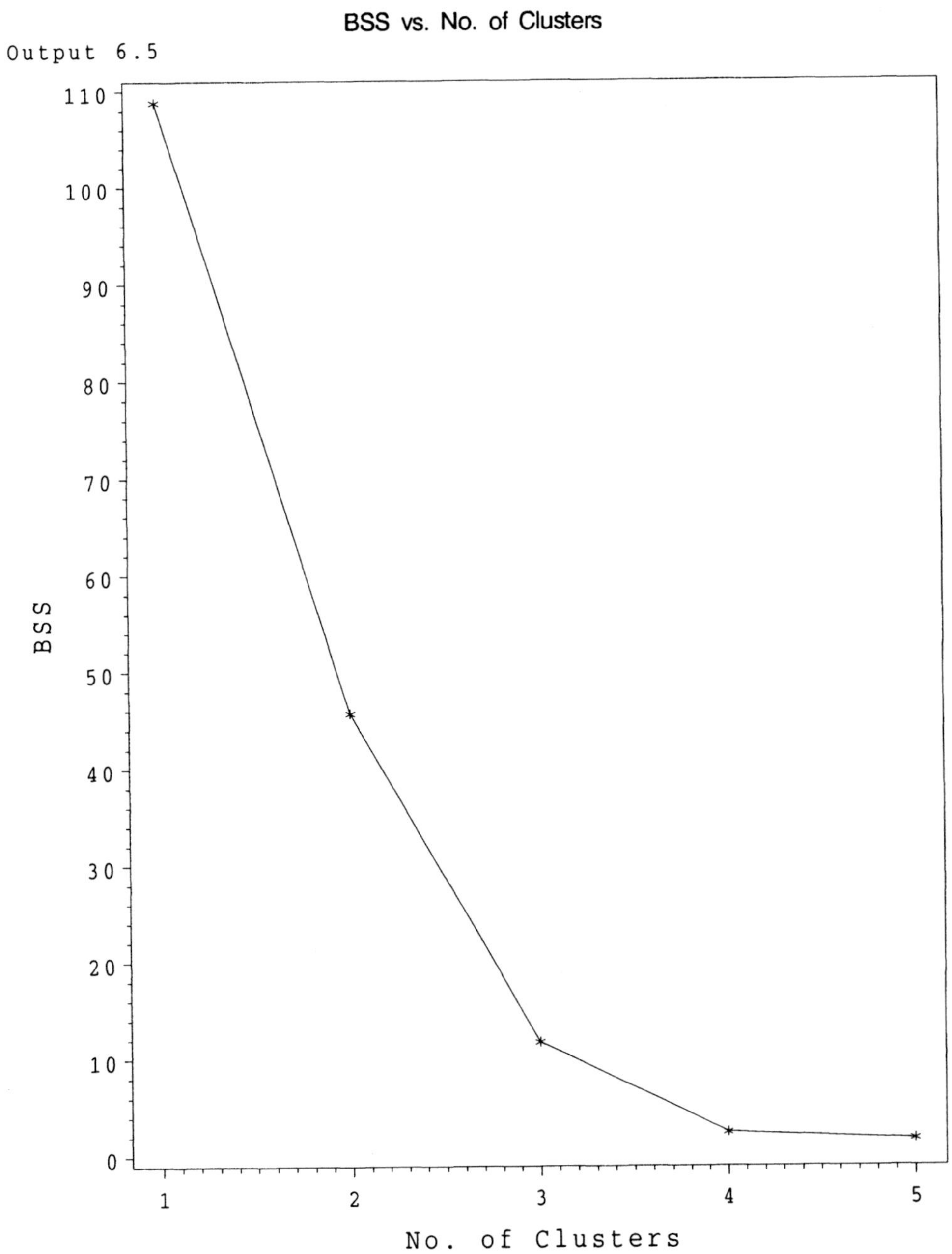
BSS vs. No. of Clusters
Output 6.5
110
100
90
80
70
60
50
40
30
20
10
0
BSS
1
2
3
4
5
No. of Clusters

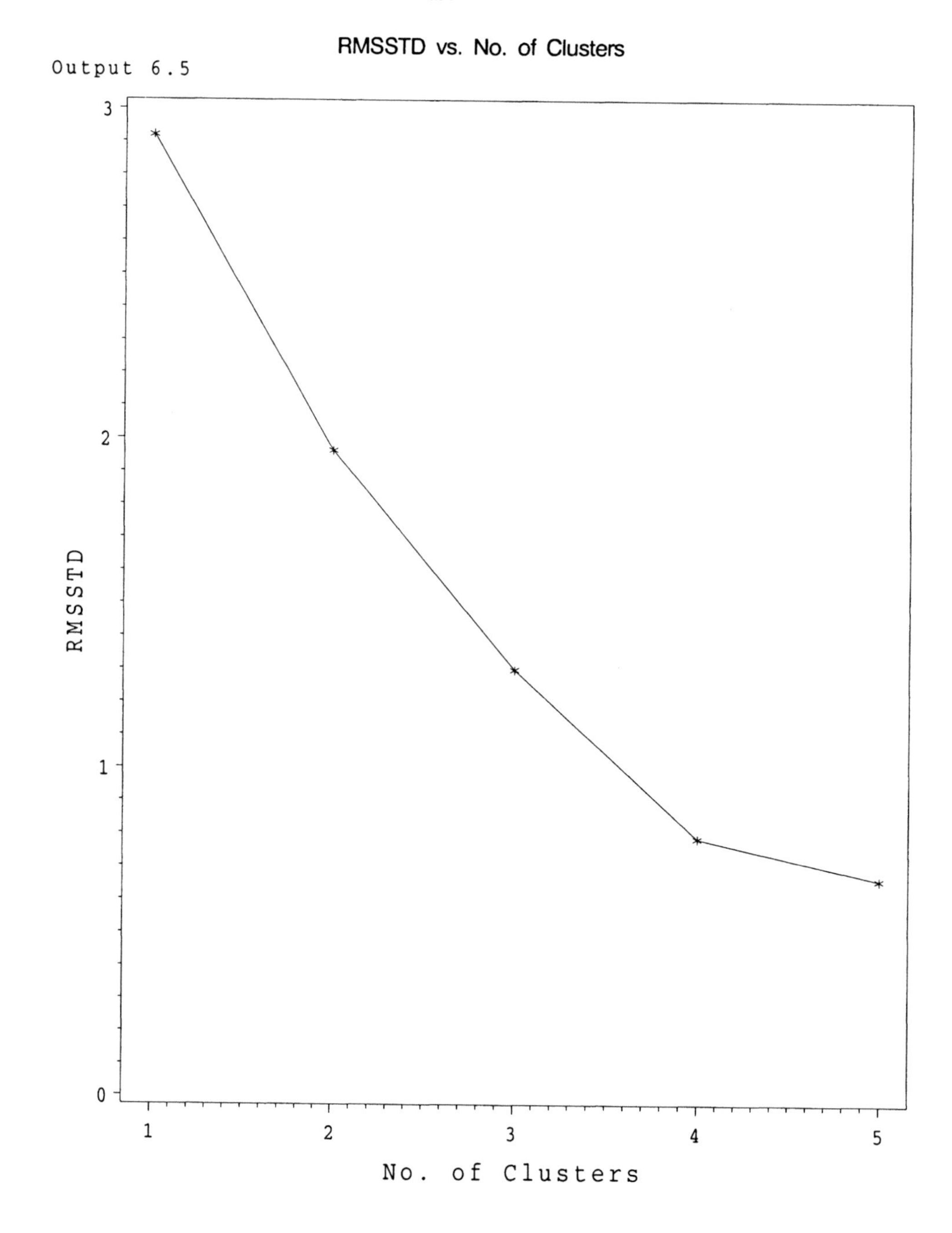

It should be remembered that in Ward's approach, BSS is used to measure the magnitude of dissimilarity between the clusters. However, different clustering methods use different measures. For example, in the case of the centroid approach it would be the squared Euclidean distance between the centroids of the respective clusters. For other approaches, the dissimilarity is measured accordingly. The NOSQUARE option may be used to prevent the distances from being squared.

All hierarchical procedures continue till all units have been put in one cluster. How do we measure the quality of clustering, and how do we decide when to stop clustering further? There are a few statistical measures that can be used to help us on these issues. These are the root mean square standard deviation (RMSSTD), R-square (RSQ), and semi-partial R-square (SPRSQ). For any of the approaches, their computations can be specified as options in the PROC CLUSTER statement, although R-square and semi-partial R-square are computed by default for the METHOD= WARD option.

The root mean squared standard deviation is the square root of the variance of the new cluster formed by merging two clusters and summed over all variables. Specifically, if the newly formed cluster B_q has respective variances $s_{q1}^2, s_{q2}^2, \ldots, s_{qp}^2$ for the p variables, then

$$RMSSTD_q = \sqrt{\frac{1}{p}\sum_{j=1}^{p} s_{qj}^2}.$$

Clearly, the more similar the two clusters that were merged to form B_q, the more homogeneous the cluster B_q, and consequently the smaller the value of $RMSSTD_q$.

R-square for cluster B_q is defined as

$$R_q^2 = 1 - \frac{SSW_q}{SST},$$

where SST represents the corrected total sum of squares of all units summed over all variables. The quantity SSW_q is the pooled corrected within-cluster sum of squares summed over all variables, at the stage when the cluster B_q was formed. Naturally, SST is constant for a given data set and SSW_q will change with q. As clusters are merged, the new cluster will be more heterogeneous than the two original ones, and hence SSW_q will continue to decrease. Since small values of SSW_q are desirable, we look for a high value of R_q^2, which lies between 0 and 1. Semi-partial R-square measures the loss of homogeneity by merging two clusters. It is defined as the reduction in the corresponding R-square value, when the two clusters are merged. When the newly formed cluster is still relatively homogeneous (as we want it to be), this loss should be small. Consequently, we look for the smaller values of semi-partial R-square.

All these quantities can be used to assess the quality of clustering. Thus, all of these together can be used to decide about how many clusters may be appropriate. As mentioned earlier, at any stage, a large value of R-square is desirable. However, RMSSTD, SPRSQ, and BSS should all be small. These criteria for evaluating the quality of clustering are applicable for other methods as well, except that BSS is relevant only when Ward's approach is used. For example, when the METHOD= CENTROID option is used, the appropriate quantity used and printed in the output will be centroid distances rather than BSS. Often it is helpful to plot these quantities against the number of clusters. The appropriate number of clusters can be determined by looking at these plots and identifying the point where the changes in these quantities are small.

EXAMPLE 3 ***(Continued) Determining the Number of Clusters*** We return to the data from Everitt (1989), and we use Output 6.5 to evaluate the quality of clustering at various stages. The relevant information has been summarized in Table 6.2. Note that more places after the decimal are represented here than in Output 6.5.

TABLE 6.2 Summary of Statistical Measures for Clustering

Stage	Clusters	No. of Clusters	BSS	RMSSTD	RSQ	SPRSQ
1	{1}, {2}, {4}, {5}, {3, 6}	5	1.7450	0.6605	0.9898	0.0102
2	{2}, {5}, {3, 6}, {1, 4}	4	2.4700	0.7858	0.9753	0.0145
3	{2}, {1, 4}, {3, 6, 5}	3	11.7417	1.2984	0.9064	0.0689
4	{2}, {1, 4, 3, 6, 5}	2	45.7033	1.9631	0.6383	0.2681
5	{1, 4, 3, 6, 5, 2}	1	108.8033	2.9195	0.0000	0.6382

The values in the last column, for SPRSQ, are derived from the previous column by taking the differences of successive RSQ values (the first value is $1 - 0.9898 = 0.0102$). Plots of these statistics against the number of clusters are obtained using the later part of Program 6.5, and these graphs are also included as part of Output 6.5.

It is seen from the graphs that BSS (in SAS output, the corresponding generic variable for plotting purposes is _HEIGHT_) and SPRSQ increase sharply when the number of clusters is reduced to two. A substantial drop in RSQ is also evident at that point. The change in RMSSTD is a bit more debatable in the sense that an increase in RMSSTD is seen slightly earlier, that is, when the number of clusters is reduced from four to three. Thus, we conclude that either three clusters or four clusters may be appropriate. The specific details about these clusters can be printed using the TREE procedure with the NCLUSTERS= option. For the choice NCLUSTERS=3, these are printed as part of Output 6.5.

In Program 6.5, we have also included the code for clustering using the flexible-beta method with BETA= -0.3. It is done for the standardized data using the STANDARD option. For comparison's sake, only the corresponding dendrogram has been included here and output has been eliminated. It is clear from the dendrogram that the clustering results in this case are different from those using the Ward's method. Another illustration of the flexible-beta method appears in Program 6.11.

Hierarchical clustering methods tend to be quite computation-intensive, especially for large data sets with a large number of variables. As alternatives there are various non-hierarchical approaches available. We will describe one of them, often known as k-means method, in Section 6.6.

6.5 Clustering of Variables

Sometimes we may be interested in clustering the variables rather than objects. The clustering of variables through graphical means has already been done earlier in the chapter. It is an important problem in that many times several variables may be measuring the same thing and thus are very similar. For any further modeling this may lead to multicollinearity problems, and hence it may be desirable to identify the clusters of similar variables. Also, in many situations the number of observations may be fewer than the number of variables. To reduce the number of variables, in order to perform the statistical analysis, we may choose to use fewer linear combinations (these may be just a few of these variables or just the first few principal components) of the similar variables. The similarity between variables is often measured by the correlation coefficients. Thus the correlation matrix may be used.

However, most of the methods and hence the related software (including PROC CLUSTER) are based on dissimilarity or distance. Thus, the correlation matrix must be converted to a dissimilarity matrix. This is often done by replacing any r_{ij} by $c - r_{ij}$ where c is any constant. A usual choice of c, which makes the dissimilarity of a variable with itself as zero, is $c = 1$. However, this may not always be appropriate, and we do not recommend this unless all r_{ij} values are nonnegative. Highly negative correlations also imply that variables are measuring the similar behavior or patterns, although the scale of measurement has been reversed, and hence highly negatively correlated variables may also be deemed to be very similar. Yet, the above choice will make them appear very dissimilar. Thus, in this case, r_{ij} are converted to $1 - |r_{ij}|$ as a measure of dissimilarity.

Since the correlation also has an interpretation as the cosine of the angle between the data vectors corresponding to the respective variables, it may be appropriate to express dissimilarities as

$$s_{ij} = \cos^{-1} r_{ij}, \qquad 0 \le r_{ij} \le 1.$$

Thus each s_{ij} takes values between 0 and $\frac{\pi}{2}$ radians and can be interpreted as angles. As r_{ij} increases, s_{ij} decreases. In fact, the same idea was implicit, while graphically assessing the clustering of variables through biplots, in Section 6.2. It may be remarked that when there are negative correlations, s_{ij} may be defined as $s_{ij} = \cos^{-1} |r_{ij}|$.

EXAMPLE 4 ***Clustering of Variables, National Track Record Data*** The national track record data for women have been previously used in this chapter for the purpose of graphically clustering seven track events. We will now do the same through analytic means.

The 7 by 7 correlations matrix for the data is given in the early part of Program 6.6 as input, and, for convenience, was separately computed using the CORR procedure. Alternatively, we could use the IML procedure to directly compute it from the data as follows:

```
data track;
infile 'womentrack.dat';
input x1-x7 country $ ;

proc iml;
use track;
read all var {x1 x2 x3 x4 x5 x6 x7} into w_matrix;

n= nrow(w_matrix);
p =ncol(w_matrix);
acosine = j(p,p,0);
sum = w_matrix[+, ];
sscp = t(w_matrix)*w_matrix - t(sum)*sum/n;
s = diag(1/sqrt(vecdiag(sscp)));
corr = s*sscp*s;
```

The computation of the correlation matrix is not given in the program. However, all correlations in this matrix are nonnegative. Later, this correlation matrix is read as TYPE=CORR data. The data set is named TRACK. The dissimilarities $s_{ij} = \cos^{-1}(r_{ij})$ are computed as TYPE=DISTANCE data, which is named as ARCOSINE. Using s_{ij} as distances, any of the methods listed in the previous section can be used to perform the cluster analysis. We have chosen to use the centroid approach for clustering the variables. The hierarchical tree is constructed for all possible number of clusters. However, if we are interested in just three clusters, then the NCLUSTERS=3 option can be used in the PROC TREE statement.

```
/* Program 6.6 */

options ls = 64 ps = 45 nodate nonumber;

title1 'Output 6.6';
data track(type = corr);
input
m100 m200 m400 m800 m1500 m3000 marathon  race $;
_type_ = 'corr';
lines;
1.0000  0.9528  0.8347  0.7277  0.7284  0.7417   0.6863 m100
0.9528  1.0000  0.8570  0.7241  0.6984  0.7099   0.6856 m200
0.8347  0.8570  1.0000  0.8984  0.7878  0.7776   0.7054 m400
0.7277  0.7241  0.8984  1.0000  0.9016  0.8636   0.7793 m800
0.7284  0.6984  0.7878  0.9016  1.0000  0.9692   0.8779 m1500
0.7417  0.7099  0.7776  0.8636  0.9692  1.0000   0.8998 m3000
0.6863  0.6856  0.7054  0.7793  0.8779  0.8998   1.000 marathon
;
```

```
data arcosine(type = distance);
set track;
m100 = arcos(m100);
m200 = arcos(m200);
m400 = arcos(m400);
m800 = arcos(m800);
m1500 = arcos(m1500);
m3000 = arcos(m3000);
marathon = arcos(marathon);
_type_ = 'distance';
race = race;

proc print data = arcosine;
title1 'Output 6.6';
title2 'Dissimilarity matrix: Arc Cos(Corr)';
run;

proc cluster data = arcosine
noeigen method = centroid rmsstd rsquare nonorm out =tree;
id race;
var  m100 m200 m400 m800 m1500 m3000 marathon;
title2 'Variable Clustering : Arc Cosine Transformed Correlations';
title3 'National Track Record Data';
run;


filename gsasfile "prog66tree.graph";
goptions reset=all gaccess=gsasfile autofeed dev=pslmono;
goptions horigin=1in vorigin=2in;
goptions hsize=6in vsize=8in;
proc tree data = tree out = trackout nclusters = 3;
id race;
title1 h = 1.2 'Dendrogram: Variable Clustering';
title2 j = l 'Output 6.6';
run;

proc sort;  by cluster ;

proc print data = trackout; by cluster ;
title1 "Output 6.6";
title2 "3-Clusters Solution: Centroid Method" ;
title3 'National Track Record Data';

proc factor data=track method=prin nfact=3 rotate=hk;
title1 'Output 6.6';
title2 'Clustering: Principal Factor, H-K Rotation';
title3 'National Track Record Data';
run;

proc varclus data = track maxc= 3 trace;
title1 'Output 6.6';
title2 'Cluster Component Analysis';
title3 'National Track Record Data';
run;
```

Output 6.6

Output 6.6
Dissimilarity matrix: Arc Cos(Corr)

Obs	m100	m200	m400	m800	m1500
1	0.00000	0.30847	0.58321	0.75583	0.75481
2	0.30847	0.00000	0.54138	0.76107	0.79764
3	0.58321	0.54138	0.00000	0.45468	0.66357
4	0.75583	0.76107	0.45468	0.00000	0.44734
5	0.75481	0.79764	0.66357	0.44734	0.00000
6	0.73519	0.78144	0.67996	0.52843	0.24883
7	0.81441	0.81537	0.78781	0.67725	0.49934

Obs	m3000	marathon	race	_type_
1	0.73519	0.81441	m100	dist
2	0.78144	0.81537	m200	dist
3	0.67996	0.78781	m400	dist
4	0.52843	0.67725	m800	dist
5	0.24883	0.49934	m1500	dist
6	0.00000	0.45149	m3000	dist
7	0.45149	0.00000	marathon	dist

Variable Clustering : Arc Cosine Transformed Correlations
National Track Record Data

The CLUSTER Procedure
Centroid Hierarchical Cluster Analysis

Cluster History

NCL	--Clusters Joined---		FREQ	RSQ	Cent Dist	RMS STD	Tie	SPRSQ
6	m1500	m3000	2	.975	0.2488	0.1760		0.0248
5	m100	m200	2	.937	0.3085	0.2181		0.0381
4	m400	m800	2	.854	0.4547	0.3215		0.0827
3	CL6	marathon	3	.742	0.4595	0.2930		0.1126
2	CL4	CL3	5	.452	0.5489	0.3989		0.2893
1	CL5	CL2	7	.000	0.6291	0.4564		0.4525

Output 6.6
(continued)

Dendrogram: Variable Clustering
3-Clusters Solution: Centroid Method
National Track Record Data

------------------------- CLUSTER=1 --------------------------

Obs	race	CLUSNAME
1	m1500	CL3
2	m3000	CL3
3	marathon	CL3

------------------------- CLUSTER=2 --------------------------

Obs	race	CLUSNAME
4	m100	CL5
5	m200	CL5

------------------------- CLUSTER=3 --------------------------

Obs	race	CLUSNAME
6	m400	CL4
7	m800	CL4

Clustering: Principal Factor, H-K Rotation
National Track Record Data

The FACTOR Procedure
Rotation Method: Harris-Kaiser

Oblique Transformation Matrix

	1	2	3
1	0.30037784	0.36655981	0.41153078
2	1.35006244	-0.0305665	-1.245104
3	0.92946721	-2.1968578	1.36727361

Inter-Factor Correlations

	Factor1	Factor2	Factor3
Factor1	1.00000	0.79805	0.70571
Factor2	0.79805	1.00000	0.84798
Factor3	0.70571	0.84798	1.00000

Output 6.6 *(continued)*

Rotated Factor Pattern (Standardized Regression Coefficients)

	Factor1	Factor2	Factor3
m100	0.9470793	-0.0309056	0.08614833
m200	0.96705133	0.03289899	-0.0056545
m400	0.28308846	0.95093245	-0.255747
m800	-0.1887722	1.04666995	0.10010023
m1500	-0.1183096	0.36805149	0.73899019
m3000	-0.020253	0.16845177	0.84974069
marathon	0.10909882	-0.2493621	1.09821066

Cluster Component Analysis
National Track Record Data

Oblique Principal Component Cluster Analysis

Observations	10000	PROPORTION	1
Variables	7	MAXEIGEN	0

Clustering algorithm converged.

Cluster summary for 1 cluster

Cluster	Members	Cluster Variation	Variation Explained	Proportion Explained	Second Eigenvalue
1	7	7	5.805706	0.8294	0.6536

Total variation explained = 5.805706 Proportion = 0.8294

Cluster 1 will be split.

Phase	Iteration	Variance Accounted For	m100	m200	m400	m800	m1500	m3000	marathon
Split	0	5.805706	1	1	1	1	1	1	1
NCS	1	6.412273	2	2	2	1	1	1	1

Cluster summary for 2 clusters

Cluster	Members	Cluster Variation	Variation Explained	Proportion Explained	Second Eigenvalue
1	4	4	3.648257	0.9121	0.2255
2	3	3	2.764016	0.9213	0.1900

Total variation explained = 6.412273 Proportion = 0.9160

Output 6.6
(continued)

Cluster	Variable	R-squared with Own Cluster	R-squared with Next Closest	1-R**2 Ratio
Cluster 1	m800	0.8600	0.6631	0.4157
	m1500	0.9649	0.5901	0.0856
	m3000	0.9569	0.5982	0.1074
	marathon	0.8665	0.5199	0.2780
Cluster 2	m100	0.9392	0.5701	0.1414
	m200	0.9541	0.5437	0.1006
	m400	0.8707	0.6872	0.4133

Standardized Scoring Coefficients

Cluster	1	2
m100	0.000000	0.350622
m200	0.000000	0.353393
m400	0.000000	0.337594
m800	0.254187	0.000000
m1500	0.269251	0.000000
m3000	0.268125	0.000000
marathon	0.255157	0.000000

Cluster Structure

Cluster	1	2
m100	0.755077	0.969125
m200	0.737380	0.976784
m400	0.828960	0.933116
m800	0.927341	0.814334
m1500	0.982295	0.768160
m3000	0.978190	0.773443
marathon	0.930880	0.721057

Inter-Cluster Correlations

Cluster	1	2
1	1.00000	0.80518
2	0.80518	1.00000

Cluster 1 will be split.

Cluster summary for 3 clusters

Cluster	Members	Cluster Variation	Variation Explained	Proportion Explained	Second Eigenvalue
1	2	2	1.8998	0.9499	0.1002
2	3	3	2.764016	0.9213	0.1900
3	2	2	1.9016	0.9508	0.0984

Output 6.6
(continued)

Total variation explained = 6.565416 Proportion = 0.9379

Cluster	Variable	R-squared with Own Cluster	R-squared with Next Closest	1-R**2 Ratio
Cluster 1	m3000	0.9499	0.8832	0.4291
	marathon	0.9499	0.7221	0.1803
Cluster 2	m100	0.9392	0.5575	0.1374
	m200	0.9541	0.5321	0.0981
	m400	0.8707	0.7476	0.5123
Cluster 3	m800	0.9508	0.7104	0.1699
	m1500	0.9508	0.8979	0.4820

Standardized Scoring Coefficients

Cluster	1	2	3
m100	0.000000	0.350622	0.000000
m200	0.000000	0.353393	0.000000
m400	0.000000	0.337594	0.000000
m800	0.000000	0.000000	0.512773
m1500	0.000000	0.000000	0.512773
m3000	0.513016	0.000000	0.000000
marathon	0.513016	0.000000	0.000000

Cluster Structure

Cluster	1	2	3
m100	0.732587	0.969125	0.746649
m200	0.715914	0.976784	0.729420
m400	0.760803	0.933116	0.864638
m800	0.842834	0.814334	0.975090
m1500	0.947592	0.768160	0.975090
m3000	0.974628	0.773443	0.939811
marathon	0.974628	0.721057	0.849768

Inter-Cluster Correlations

Cluster	1	2	3
1	1.00000	0.76670	0.91808
2	0.76670	1.00000	0.81146
3	0.91808	0.81146	1.00000

Output 6.6
(continued)

Number of Clusters	Total Variation Explained by Clusters	Proportion of Variation Explained by Clusters	Minimum Proportion Explained by a Cluster	Maximum Second Eigenvalue in a Cluster
1	5.805706	0.8294	0.8294	0.653632
2	6.412273	0.9160	0.9121	0.225519
3	6.565416	0.9379	0.9213	0.190005

Number of Clusters	Minimum R-squared for a Variable	Maximum 1-R**2 Ratio for a Variable
1	0.7751	
2	0.8600	0.4157
3	0.8707	0.5123

Output 6.6
(*continued*)

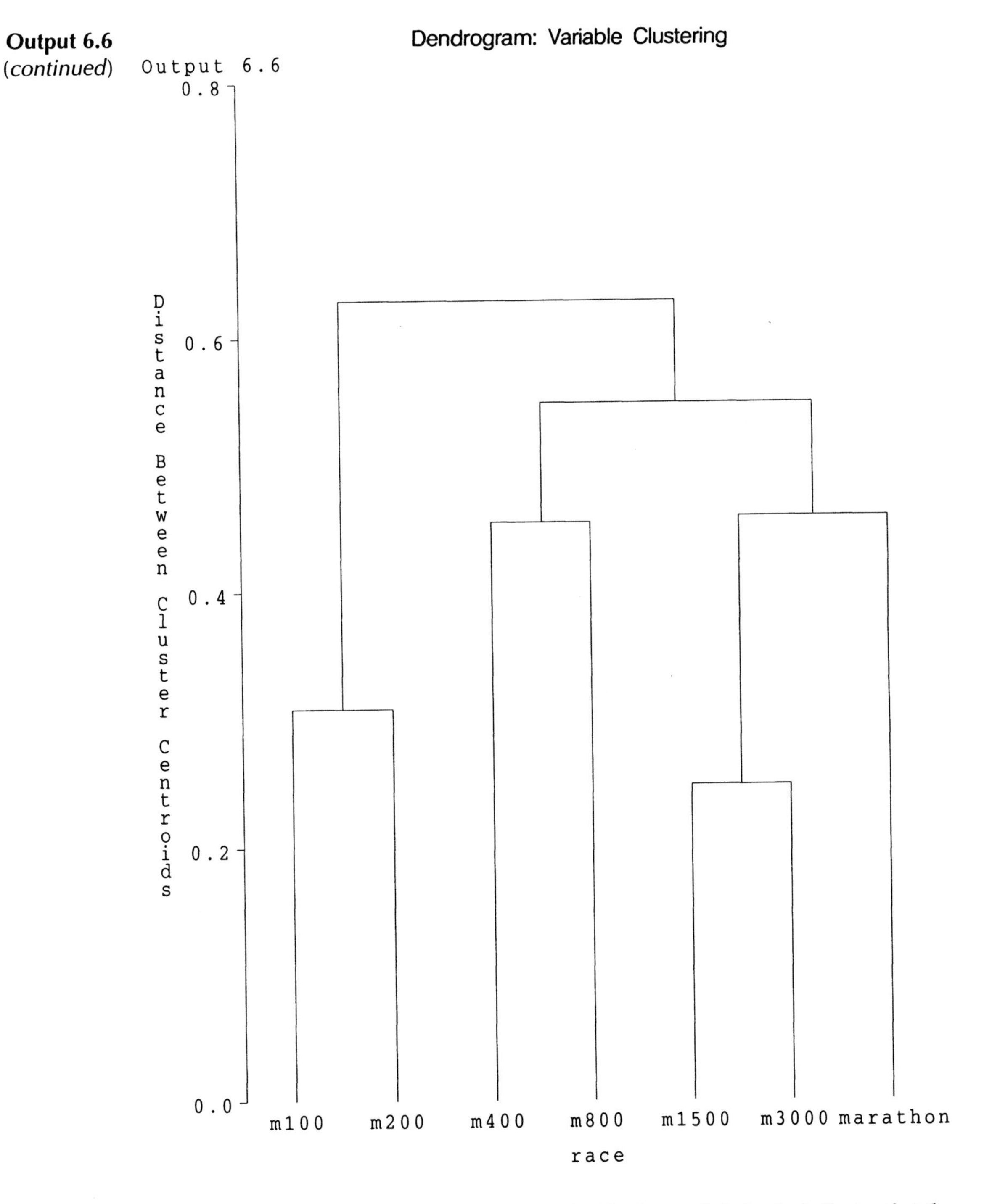

The part corresponding to the centroid method in Output 6.6 clearly indicates that the three clusters formed are {M1500, M3000, MARATHON}, {M100, M200}, and {M400, M800}, which respectively can be interpreted as clusters of long-distance, short-distance, and medium-distance races. It is natural to expect that countries will tend to fare similarly in similar races (long, short or medium) because usually, but not always, it is the same athlete taking part in all or most of the races in the same cluster. This clustering is different from what was previously observed in the biplot.

An alternative approach to clustering the variables may be to take a factor analytic route. Specifically motivated by Harris and Kaiser (1964), Harman (1976) suggests first identifying the important factors and then applying an ortho-oblique transformation to assign the higher loadings to a few of the variables. Variables with higher loadings in a given factor may be viewed as forming a cluster in that they all contribute to the particular factor in a significant way and hence can be argued to attempt to measure the same thing. With this interpretation, a factor analysis with an appropriate ortho-oblique transformation may be performed, and variables that load higher on a particular factor are clustered together. We have discussed the details of factor analysis in Chapter 4.

EXAMPLE 4 *(Continued)* For the national track record data on women, we have seven variables representing speed in various events. The correlation matrix of these will be used as data. We will analyze these data using the FACTOR procedure along with options METHOD=PRIN and ROTATE=HK (see Chapter 4). This allows us to first obtain the orthogonal factors that are rotated using the Harris-Kaiser rotation discussed in Chapter 4. Suppose we want to obtain three clusters and hence we use the NFACT=3 option. The corresponding SAS code has been included as part of Program 6.6. The output is presented as part of Output 6.6. The interpretation of similar outputs has already been discussed in Chapter 4. We view each factor as defining a cluster, and a variable is assigned to that cluster if it has the largest absolute loading on the corresponding rotated factor. For example, for M100, the largest of the absolute values of its loadings on the three factors (that is, the largest of 0.9471, 0.0309, and 0.0862) is 0.9471 (corresponding to Factor 1). Hence M100 is assigned to the corresponding cluster. Although we assign the variable to a cluster for which it has the highest absolute loading for practical applications, in reality the FACTOR procedure may often result in overlapping clusters because some of these loadings may be substantially high even in the case of other factors. In fact, that is the case for our example.

By looking at the rotated factor patterns we see that M100 and M200 load highest on Factor 1 (with coefficients of 0.9471 and 0.9671, respectively). The highest coefficients for M400 (= 0.9509) and M800 (= 1.0467) are observed for Factor 2. Finally, with coefficients 0.7390, 0.8497, and 1.0982 for M1500, M3000, and MARATHON, respectively, Factor 3 identifies long-distance races. Thus, the three clusters identified here are the same as what were found using the dissimilarities $s_{ij} = \cos^{-1} r_{ij}$.

The VARCLUS procedure is based on the factor analytic approach discussed above and presents a slight algorithmic variation of the same idea. The approach is summarized here. We will consider the approach based on the first principal component only, although it also provides, as an option, a centroid component approach.

The basic object for the analysis is the correlation or the variance-covariance matrix, and the objective is to maximize the sum across clusters of the variance of the original variables explained by the cluster components, which is either the first principal component or the centroid component. Of course, if a correlation matrix is used, each of the original (standardized) variables has variance 1. Since most of the time variables may be measuring different things, possibly on different units of measurements, using the correlation matrix is often more appropriate for clustering purposes.

The procedure relies on first obtaining the principal components. At that stage, we need some meaningful criterion for splitting a cluster. We can either specify

- that the cluster component (that is, for our discussion the first principal component obtained for the variables in the cluster) accounts for at least a specified percent of total cluster variance (in SAS, the option for this is PERCENT=, with a default value of zero for the principal component method)

or

- that the second eigenvalue, which is the estimated variance of the second principal component obtained for the variables in the cluster, does not exceed the average variance for all the variables (which is obviously 1 if the correlation matrix is used).

Both options cannot be used simultaneously.

The clustering procedure starts with a single cluster consisting of all variables. The following steps are used:

1. According to one of the above two criteria (whichever is chosen) a cluster is split into two clusters by finding the first and second principal components.
2. A raw quartimax ortho-oblique rotation (see Chapter 4) is then applied, and the squared correlation of every variable with each rotated component is calculated.
3. For a given variable, the rotated component with which it has the highest squared correlation is identified.
4. Each rotated component is viewed as an identifier for a cluster, and the particular variable of Step 3 is classified into the corresponding cluster.
5. Since the objective is to maximize the variance accounted for by cluster components, we may need to reassign the variables iteratively. To do so, after an assignment has been made, as in Step 4, using a certain search algorithm, each variable is reassigned to a different cluster to see if this assignment can increase the percent cluster variance explained. If this does happen, then the principal components for the two clusters affected by this reassignment are computed again as in Step 1.

The algorithm continues till no more improvement in the total explained variance can be made.

EXAMPLE 4 *(Continued)* We again consider the track record data, and this time we analyze it using PROC VARCLUS. The last part of Program 6.6 includes the appropriate statement. We are using the default option of PERCENT=0, and we want at most three clusters (the NCLUSTERS=3 option). The output is presented in Output 6.6.

Before we extensively interpret various statistical measures that are computed during the process, let us observe that initially we have all variables in a single cluster that is split into two clusters according to the criterion discussed earlier. These clusters are

Cluster 1: {M800, M1500, M3000, MARATHON}

and

Cluster 2: {M100, M200, M400}.

These can be viewed as the clusters of not-so-short distance races and the short distance races. Cluster 1 is split again into two clusters of {M3000, MARATHON} and {M800, M1500}, which can be interpreted as the clusters of long- and medium-distance races. Thus, we finally have

Cluster 1: {M3000, MARATHON}
Cluster 2: {M100, M200, M400}

and

Cluster 3: {M800, M1500}.

Much of the statistical information presented in the output of the VARCLUS procedure is in fact a standard factor analysis type summary. However, it is presented in a rather unusual format, and hence we must explain it here. Initially we started with one cluster with all seven variables in it. Since we are working with the correlation matrix, all seven variables are assumed to be standardized and hence all have variance 1. Thus the total cluster variance is equal to 7, and the first principal component (or as cluster component) has the estimated variance equal to 5.8057. It thus explains 82.94% of total cluster variation. The second eigenvalue is 0.6536, which is the estimated variance of the second principal component.

Since the default option of PERCENT=0 is used, this cluster will split anyway. The split that maximizes the explained total variation has already been indicated. Since the TRACE option is used, the output clearly traces which of the variables belong to which of the clusters. The first cluster consists of four variables, and the first principal component computed from these four variables explains $\frac{3.64826}{4} \times 100 = 91.21\%$ of the cluster variation. Similarly, the first principal component for the second cluster containing the other three variables explains $\frac{2.76402}{3} \times 100 = 92.13\%$ of cluster variation. Thus, across these two clusters, the corresponding first principal components together explain a total of

$$\frac{(3.64826 + 2.76402)}{(4+3)} \times 100 = \frac{6.412273}{7} \times 100 = 91.60\%$$

of total variation within two clusters.

If $\mathbf{x}$ is the vector of all seven cluster variables, then the two first principal components for the respective clusters can be written as $y_1 = \mathbf{a}_1'\mathbf{x}$ and $y_2 = \mathbf{a}_2'\mathbf{x}$, with some of the coefficients in vectors $\mathbf{a}_1$ and $\mathbf{a}_2$ zeros. Nonetheless, correlations (and squared correlations) of any variable x_i with y_1 and y_2 can be estimated by using the facts that

$$\text{cov}\,(y_1, x_i) = \text{cov}\,(\mathbf{a}_1'\mathbf{x}, x_i) = \mathbf{a}_1'\text{cov}\,(\mathbf{x}, x_i)$$

and

$$\text{cov}\,(y_2, x_i) = \text{cov}\,(\mathbf{a}_2'\mathbf{x}, x_i) = \mathbf{a}_2'\text{cov}\,(\mathbf{x}, x_i).$$

These correlations are calculated. Their squares are listed next in the output. A variable is expected to have higher squared correlation with its own cluster component than the cluster component corresponding to the other cluster. Thus, for the well-separated clusters, high values of R^2 in the own cluster column and the low values of R^2 in the next closest column serve as good indicators. The term *next closest* refers to the cluster with the next highest R^2 value. Since at the moment there are only two clusters, the choice is trivial; it is more relevant when we have three or more clusters. The last column gives the ratio of $1 - R^2$ values where the two R^2 values have been given in the previous two columns. Obviously, a low value of this ratio is desirable.

For our data set we see that R^2 values in the own cluster column are high. However, the next closest column values are not very small. Consequently, we may not expect a very good separation of clusters. The values of $1 - R^2$ ratio also support this observation. The (unsquared) correlations of variables with the two cluster components are listed under the title cluster structure. The fact that M100, M200, and M400, even though not present in the first cluster component, still have correlations with this cluster component which are higher than 0.70 confirms that the clusters are not well separated. The same can be said when we look at the correlations of M800, M1500, M3000, and MARATHON with the second cluster component.

The standardized scoring coefficients are the regression coefficients when the particular cluster component is regressed on the variables in the corresponding clusters.

Finally the correlation between the cluster components $\mathbf{a}_1'\mathbf{x}$ and $\mathbf{a}_2'\mathbf{x}$ (that is, $\mathbf{a}_1'\mathbf{R}\mathbf{a}_2/\sqrt{\mathbf{a}_1'\mathbf{R}\mathbf{a}_1 \cdot \mathbf{a}_2'\mathbf{R}\mathbf{a}_2}$) is estimated. This estimate is presented as the nondiagonal entry (= 0.8052) of the estimated inter-cluster correlation. Among the two clusters, Cluster 1 has the smallest percentage of total variation (=91.21%) explained. Consequently, for the next splitting, it is the chosen candidate, resulting in what are now termed as Cluster 1 (={M3000, MARATHON}) and Cluster 3 (={M800, M1500}). The statistical information that follows can be interpreted similarly. At this stage, the procedure will stop as the desired number of clusters (=3) have already been obtained. Thus, the seven variables have been split into three clusters as

{M100, M200, M400},
{M800, M1500}

and

{M3000, MARATHON}.

It may be noted that the clustering obtained here is different from that obtained by using the dissimilarities and by using the factor analytic approach described earlier. A reason for this discrepancy is the fact that, as already noted, the variables were not very well separated. This was also evident when the data were analyzed using the FACTOR procedure, where it was realized that, due to high loadings for many variables across all clusters, these clusters may be overlapping.

6.6 Nonhierarchical Clustering: *k*-Means Approach

The clustering procedures discussed so far were all hierarchical in nature in that clusters were being fused together in stages, in each case, by using a particular, well-defined criterion. From the very nature, such procedures have the disadvantages of

(a) being time-consuming since for a data set with n objects, the procedure offers n possible solutions, namely, one for every possible number of clusters.

(b) not being able to amend a previous assignment, in that the objects in the clusters to be merged will once and for all be assigned to the newly formed cluster and cannot be reassigned to any other cluster.

In case any one or both of the above drawbacks are undesirable, an alternative approach that can be used is the k-means clustering approach. The approach is relatively simple and is also reasonably fast. For large data sets with an objective of quick tentative clustering of objects, the method can be a very effective choice, although in general no claims for the optimal clustering can be made.

Suppose k nonoverlapping clusters are to be constructed using n observations. Also assume that the measure of dissimilarity used is the Euclidean distance (although in principle, any other measure can also be used). The approach can be summarized in the following steps, and in SAS, it is implemented as a refinement of the Nearest Centroid Sorting method as reported in Anderberg (1973). The corresponding SAS procedure is the FASTCLUS procedure.

1. Select seeds to act as the initial centroids for the k clusters using some appropriate method. One of the following methods for selecting seeds can be used:

   (a) Seeds are chosen independently (perhaps using some other clustering algorithm or from some previous prior knowledge). In this case they can be specified using the option SEED=MYDATA, where MYDATA is the name of the data set storing these seeds. No observation is taken as a seed unless its minimum distance to the previous seed is greater than a specified value determined by the RADIUS= option.

   (b) Seeds can be read from the data at hand as the first k complete observations, which are at least at a distance specified by the option RADIUS=.

   (c) Seeds can be taken as a random sample of k observations from the data at hand, so long as all k observations are at least at a distance specified by the RADIUS= option.

   The default value of RADIUS is zero. Subsequently, as clusters are formed, seeds are replaced by the cluster centroids, that is, the averages of all observations in the individual clusters.

2. An observation is assigned to the cluster with the nearest seed. Once the assignment is made, the cluster seeds are to be updated. It can be done using one of the following two options.
   (a) After all observations have been assigned to the respective clusters, all cluster seeds are updated by replacing old seeds with the new cluster centroid. In other words, we wait for this update of cluster seeds, till all observations have been examined using the old set of cluster seeds. This is the default choice in SAS. This step can be repeated until the changes in cluster seeds become very small or zero. The MAX-ITER= option can be used to specify the number of the maximum number of iterations for recomputing the seeds. The default value for this is 1.
   (b) A particular cluster seed is updated (as the new cluster centroid) as soon as a new observation joins the particular cluster, and for the subsequent assignment of the next observation, this new seed is used. The process of updating continues simultaneously along with the assignment of units and not just at the end of the assignment of all units. This is achieved using the DRIFT= option.
3. Finally, when all observations have been assigned to the clusters with seeds nearest to the corresponding observations, the final clusters result.

Hartigan (1975) points out that sometimes, the k-means clustering algorithm (in its simplest forms), may fail even for the "perfect data." By perfect data, we mean a set of observations in k clusters such that all distances between the observations within a cluster are smaller than all distances between two observations taken from two different clusters. Fortunately, the initialization method of PROC FASTCLUS ensures that the algorithm correctly identifies all k clusters, at least for the perfect data scenario.

The final clustering results in the k-means method can be very sensitive to the choice of initial seeds as they act as the cluster centroid in the beginning. This makes the seed selection a crucial step. Thus, if the units are arranged in some specific way that may affect the clustering, then the first k observations as seeds may result in some poorly organized clusters.

The k-means method of clustering can be a good quick approach to clustering large data, and thus for many commercial data, it is an effective tool. Certainly, the optimal clustering will not be often realized in the strict sense, but for practical purposes, the approach is very efficient. It is recommended that for such data sets seeds be chosen randomly rather than the first few observations. It may be noted that many recent data mining software products including SAS Enterprise Miner rely on the k-means approach for quick and efficient clustering.

On the other hand, k-means clustering can also be used as an extra step to improve upon hierarchical clustering methods. As we pointed out, one of the drawbacks of hierarchical methods is that any wrong or poor fusing of clusters taking place at a given stage is irreversible at any later stage. Thus, the suggestion is to first obtain a solution using a hierarchical method and then to improve upon this solution by taking the centroids of these clusters as initial seeds for the k-means clustering. Fortunately, in SAS, we can save these centroids as a data file, and we can subsequently use these in PROC FASTCLUS as initial seeds. In fact, this is what we do in the example that follows.

EXAMPLE 5 ***Nonhierarchical Clustering of Mutual Funds*** The data on the performance of various domestic (U.S.) mutual funds are taken from the newsletter *Fabian Premium Investment Resource* (1999, Volume 22, No. 9, p. 8–9). These mutual funds are listed there in three categories by the funds objectives: Aggressive Growth, Growth, and Growth and Income. For our purpose, we do not distinguish between these categories. We want to cluster these $n = 100$ mutual funds based on their 4 weeks (WK4), 8 weeks (WK8), 12 weeks (WK12) and year-to-date (YTD) performance (all as of August 27, 1999), which are expressed as a percent change in price over the appropriate periods. There is no valid justification to standardize the data (which in fact may mask the true performance features that can be useful while interpreting the clusters). Consequently, for us since there is considerably more variability in YTD, this variable may get more importance while forming the clusters. This extra importance imparted to YTD is also not unrealistic from the practical point of view.

Let us assume that we want to form a total of $k = 8$ clusters. We do this with the hope that this will help us pick some good funds out of the list and separate them from the bad ones and with an expectation that we may be able to assign some meaningful interpretations to some, if not all, of these clusters. Since in the data the aggressive growth funds (presumably these are similar in their growth objectives) are listed first, it would be inappropriate to take the initial seeds as the first eight funds that meet the distance-between-the-clusters criterion stated earlier. We thus first obtain an eight-cluster solution using Ward's approach, and the cluster means of these eight clusters will be taken as the initial seeds. In Program 6.7, these seeds are stored in the data file called MYSEEDS. These cluster means were obtained by using PROC MEANS on the data set NEWDATA to which the Ward's clustering results were output.

In PROC FASTCLUS, SEED=MYSEEDS specifies the seeds to be used. We choose to use option (a) of Step 2 of the algorithm specified earlier. The maximum number of iterations is specified as 20 (MAXITER=20) and the minimum required distance between seeds at any time is 0.001 (RADIUS= .001). Additionally, we can use a further criterion for replacing an old seed with a new one using the REPLACE= option. Details of various choices are given in *SAS/STAT User's Guide, Version 6, Fourth Edition, Volume 1*, p. 825, and also in the online documentation, and we skip these details. However, our recommendation is to use the choice REPLACE=FULL if initial seeds are being provided as a separate data set or as the first few observations from the raw data. The DISTANCE option requests the distances between the centroid of a given cluster and that of its closest neighbor to be printed. The output has been stored in a data set named CLUS_OUT. The LIST option can also be used to just list most of the output without storing it as a data set. Selected output is shown following Program 6.7.

```
/* Program 6.7 */

options ls = 64 ps = 45 nodate nonumber;
filename gsasfile "prog67.graph";
goptions reset=all gaccess=gsasfile autofeed dev=pslmono;
goptions gsfmode=append;
goptions horigin=1in vorigin=2in;
goptions hsize=6in vsize=8in;

symbol1 value =1 ;
symbol2 value =2 ;
symbol3 value =3 ;
symbol4 value =4 ;
symbol5 value =5 ;
symbol6 value =6;
symbol7 value =7;
symbol8 value =8;

data mfunds;
infile 'm_funds.dat' firstobs = 7;
input fund $ 1-33
price war39 wd39 wk4 wk8 wk12 ytd drawdown;
```

```
proc cluster data = mfunds method = ward noprint
noeigen nonorm std outtree = tree ;
var  wk4 wk8 wk12 ytd;
id fund;
title1  'Output 6.7';
title2 "Cluster Analysis Using Ward's Approach";
run;

proc tree data = tree out = newdata nclusters = 8 noprint;
id fund;
copy  wk4 wk8 wk12 ytd;
title1 'Output 6.7';
title2 "Tree Diagram of Clustering: Ward's Method";
run;

proc sort data = newdata;
by cluster;
run;

proc means data = newdata noprint;
by cluster;
output out = myseeds mean = wk4 wk8 wk12 ytd;
var wk4 wk8 wk12 ytd;
run;

proc fastclus data = mfunds maxiter = 20 maxclusters = 8 distance
radius = .001 replace = full seed = myseeds out = clus_out;
var wk4 wk8 wk12 ytd ;
id fund;
title1 'Output 6.7';
title2 'Seed Point Clustering of Mutual Funds: Seeds Supplied';
run;

proc sort data = clus_out;
by cluster distance;
run;

proc gplot data = clus_out;
where
cluster = 1 or cluster = 2 or cluster = 3 or
cluster = 4 or cluster = 5 or cluster = 6 or
cluster = 7 or cluster = 8;
plot wk8*wk4 = cluster ;
title1 h = 1.2 'Clusters in WK8 vs.WK4 Plots: Seeds Supplied';
title2 j = 1 'Output 6.7';
run;

proc gplot data = clus_out;
where
cluster = 1 or cluster = 2 or cluster = 3 or
cluster = 4 or cluster = 5 or cluster = 6 or
cluster = 7 or cluster = 8;
plot ytd*wk12 = cluster ;
title1 h = 1.2 'Clusters in YTD vs.WK12 Plots: Seeds Supplied';
title2 j = 1 'Output 6.7';
run;
```

```
proc print data = clus_out;
by cluster;
var fund wk4 wk8 wk12 ytd ;
title1 'Output 6.7';
title2 'Clustering Details: Seeds Supplied';
run;

proc fastclus data = mfunds maxiter = 20 maxclusters = 8 distance
radius = .001 replace = full out = c_out;
var wk4 wk8 wk12 ytd ;
id fund;
title1 'Output 6.7';
title2 'Clustering of Mutual Funds: Seeds from Data';
run;

proc sort data = c_out;
by cluster;
run;

proc gplot data = c_out;
where
cluster = 1 or cluster = 2 or cluster = 3 or
cluster = 4 or cluster = 5 or cluster = 6 or
cluster = 7 or cluster = 8;
plot wk8*wk4 = cluster ;
title1 h = 1.2 'Clusters in WK8 vs.WK4 Plots: Seeds as First k Obs';
title2 j = 1 'Output 6.7';
run;

proc gplot data = c_out;
where
cluster = 1 or cluster = 2 or cluster = 3 or
cluster = 4 or cluster = 5 or cluster = 6 or
cluster = 7 or cluster = 8;
plot ytd*wk12 = cluster ;
title1 h = 1.2 'Clusters in YTD vs.WK12 Plots: Seeds as First k Obs.';
title2 j = 1 'Output 6.7';
run;

proc print data = c_out;
by cluster;
var fund wk4 wk8 wk12 ytd ;
title1 'Output 6.7';
title2 'Seeds Taken from Data';
run;

proc princomp data = mfunds cov out = pc_out;
var wk4 wk8 wk12 ytd ;
run;

proc cluster data = pc_out method = ward noprint
noeigen nonorm std outtree = pctree ;
var prin1 prin2 ;
id fund;
title1 'Output 6.7';
title2 "Cluster Analysis (PC) Using Ward's Approach";
run;
```

```
proc sort data = pctree;
by _ncl_;

data pc_ccc;
set pctree;
if _ncl_ < = 20;

proc print data = pc_ccc;
var _ncl_ _freq_ _ccc_ _rmsstd_ _sprsq_;

proc fastclus data = pc_out maxiter = 20 maxclusters = 8 distance
radius = .001 replace = random random = 1229447 out = pc_clus;
var prin1 prin2 ;
id fund;
title1 'Output 6.7';
title2 'Seed point Clustering of Mutual Funds: First 2 PC Used';
run;

proc sort data = pc_clus;
by cluster;
run;

proc gplot data = pc_clus;
where
cluster = 1 or cluster = 2 or cluster = 3 or
cluster = 4 or cluster = 5 or cluster = 6 or
cluster = 7 or cluster = 8;
plot prin2*prin1 = cluster ;
title1 h = 1.2 'Clustering Using First Two PCs';
title2 j = 1 'Output 6.7';
run;

proc gplot data = pc_clus;
where
cluster = 1 or cluster = 2 or cluster = 3 or
cluster = 4 or cluster = 5 or cluster = 6 or
cluster = 7 or cluster = 8;
plot wk8*wk4 = cluster ;
title1 h = 1.2 'Clusters (Using PC) in WK8 vs.WK4 Plots';
title2 j = 1 'Output 6.7';
run;

proc gplot data = pc_clus;
where
cluster = 1 or cluster = 2 or cluster = 3 or
cluster = 4 or cluster = 5 or cluster = 6 or
cluster = 7 or cluster = 8;
plot ytd*wk12 = cluster ;
title1 h = 1.2 'Clusters (Using PC) in YTD vs.WK12 Plots ';
title2 j = 1 'Output 6.7';
run;
```

Output 6.7

Seed Point Clustering of Mutual Funds: Seeds Supplied

The FASTCLUS Procedure

Replace=FULL Radius=0.001 Maxclusters=8 Maxiter=20

Converge=0.02

Initial Seeds

Cluster	wk4	wk8	wk12	ytd
1	1.56900000	-2.58100000	2.97833333	9.60933333
2	-0.58735294	-4.58088235	-0.55705882	9.16235294
3	-2.00125000	-4.05625000	1.54000000	1.47000000
4	-2.89500000	-7.69375000	-5.02125000	2.83375000
5	1.87571429	-2.31428571	4.07142857	26.00142857
6	0.84285714	0.88714286	10.76142857	18.68285714
7	5.96200000	3.09000000	13.16800000	38.30800000
8	10.90000000	2.02000000	25.21000000	48.78000000

Minimum Distance Between Initial Seeds = 4.620409

Iteration History

		Relative Change in Cluster Seeds			
Iteration	Criterion	1	2	3	4
1	2.5493	0.2933	0.1266	0.3663	0.1494
2	2.3441	0.0833	0.0366	0.0675	0
3	2.3245	0.1230	0.1095	0.0727	0
4	2.3086	0.0720	0.1153	0	0
5	2.3022	0.0160	0.0565	0.0601	0

Iteration History

	Relative Change in Cluster Seeds			
Iteration	5	6	7	8
1	0.3643	0.4514	1.1101	1.4957
2	0.4013	0.1571	0	0
3	0	0	0	0
4	0	0	0	0
5	0	0	0	0

The FASTCLUS Procedure

Replace=FULL Radius=0.001 Maxclusters=8 Maxiter=20

Converge=0.02

Iteration History

		Relative Change in Cluster Seeds			
Iteration	Criterion	1	2	3	4
6	2.3007	0	0	0	0

Output 6.7
(continued)

Iteration History

	Relative Change in Cluster Seeds			
Iteration	5	6	7	8
6	0	0	0	0

Convergence criterion is satisfied.

Clustering Details: Seeds Supplied

-------------------------- Cluster=1 ---------------------------

Obs	fund	wk4	wk8	wk12	ytd
1	Vanguard Growth and Income	1.74	-2.76	3.08	12.59
2	Value Line Leveraged Growth	1.61	-2.73	3.64	11.81
3	Fidelity Magellan (clsd)	0.73	-3.84	1.50	11.07
4	Vanguard Index Trust 500	1.60	-2.90	1.82	10.56
5	Fidelity Spartan Market Index	1.58	-2.92	1.79	10.44
6	Schwab S&P 500	1.55	-2.92	1.75	10.34
7	Dreyfus Third Century	1.34	-2.79	3.82	10.76
8	Amer Century Ultra	1.60	-3.68	0.90	10.48
9	Warburg Pincus Capital App	0.87	-1.53	4.04	11.38
10	Lexington Corporate Leaders	1.91	-1.79	0.57	12.74
11	Columbia Common Stock	0.19	-4.06	1.09	10.45
12	Founders Growth	2.06	-3.16	2.34	9.46
13	Rydex Nova (sold 8/5/99)	2.13	-5.12	1.54	11.07
14	Selected American Shares	0.29	-5.26	0.99	12.01
15	Janus Growth & Income	0.91	-3.30	2.74	14.95
16	Safeco Equity	1.28	-1.78	0.83	9.39
17	Amer Cent Benhm Incm & Gwth	1.29	-2.33	2.98	8.56
18	Rob Stephens Value + Gwth	0.77	-2.57	5.74	10.96
19	Fremont Growth	1.53	-4.04	2.40	8.34
20	Dreyfus New Leaders	0.02	-2.81	0.49	15.19
21	Fidelity Capital Appreciation	0.47	-2.85	2.16	15.83
22	Fidelity Stock Selector	0.96	-0.46	5.07	13.34
23	Janus Equity Income	-1.02	-3.82	0.51	14.80
24	Fidelity Blue Chip Growth	2.03	-2.02	2.92	7.83
25	SIT Large Cap Growth	2.39	-2.93	3.24	7.44
26	Amer Cent Bnhm Eq Gwth	1.25	-1.42	4.13	7.77
27	Janus Fund	0.43	-3.53	3.20	16.79
28	Invesco Blue Chip Gw	4.74	-1.26	4.28	9.44
29	Janus Twenty	3.00	-2.87	2.44	16.74
30	Dreyfus Appreciation	2.96	-1.06	2.53	6.77
31	Strong Opportunity	-2.44	-3.35	0.53	16.53
32	Rob Stephs Mid Cap Opprtnties	-1.01	-4.95	1.39	19.00
33	Baron Growth	-2.03	-4.57	-1.44	18.38

Output 6.7 *(continued)*

-------------------------- Cluster=2 --------------------------

Obs	fund	wk4	wk8	wk12	ytd
34	Fidelity Equity-Income	-0.58	-4.62	-0.95	9.44
35	Price Growth & Income	-1.27	-4.78	-1.48	8.22
36	Scudder Large Co Value	-0.23	-5.40	-1.19	9.06
37	Price Equity-Income	-1.12	-4.19	-1.81	8.91
38	Fidelity Dividend Growth	0.00	-4.33	0.32	8.39
39	Fidelity Contrafund (clsd)	-1.18	-3.97	0.28	8.85
40	Fidelity Fund	0.13	-4.20	-0.07	7.33
41	American Century Value	-1.35	-4.07	-2.08	9.79
42	Columbia Growth	-0.47	-5.02	0.75	10.04
43	Tweedy Browne American Value	-0.90	-5.09	0.00	6.51
44	Invesco Industrial Income	-0.43	-3.62	1.37	8.67
45	Transamerica Equity Growth	-0.71	-6.30	-0.79	6.86
46	Schwab 1000	1.05	-3.49	1.00	8.92
47	Price Midcap Growth Stock	-3.10	-4.21	0.99	8.10
48	Amer Cent Eq Income	-0.90	-2.51	0.91	6.44
49	Babson Value	-0.62	-5.31	-4.13	8.37
50	Price Growth Stock	0.65	-3.66	1.04	5.96
51	Vanguard Equity Income	-0.77	-4.11	-1.53	4.97
52	Montgomery Growth	-2.09	-6.28	-1.75	11.49
53	Fidelity Equity-Income II	0.10	-4.04	-0.47	4.92
54	Dreyfus Core Value	-0.03	-5.98	-2.88	12.25
55	Fidelity Value	-1.98	-6.65	-3.63	15.15

-------------------------- Cluster=3 --------------------------

Obs	fund	wk4	wk8	wk12	ytd
56	Founders Grwth & Incm	1.23	-3.40	0.96	0.82
57	Warburg Pincus Emerging Grth	-1.08	-3.86	3.94	2.88
58	Price Small Cap Stock	-1.24	-2.63	3.62	3.22
59	Lexington Gwth & Incm	-0.58	-5.09	-0.84	2.15
60	Fidelity Growth & Income (clsd)	0.57	-3.59	0.41	4.18
61	Neuberger & Berman Manhattan	1.41	-3.47	4.79	2.43
62	Bramwell Growth	0.56	-4.32	1.24	4.68
63	Weiss Peck & Greer Gwth & Incm	1.96	-1.47	1.77	-1.13
64	Babson Enterprise II	-2.02	-4.00	2.54	4.71
65	Fidlty Low-Prcd Stck	-3.17	-4.55	-0.51	2.84
66	Schwab Small Cap Index	-1.78	-3.44	2.15	5.42
67	Vanguard US Growth	2.41	-2.80	2.94	5.49
68	Kaufmann Fund	-2.17	-4.76	2.27	-4.93
69	Emerging Growth	1.05	-1.92	6.52	-3.84
70	Safeco Growth	-1.45	-5.00	-2.68	-10.48

Output 6.7 (*continued*)

------------------------- Cluster=4 -------------------------

Obs	fund	wk4	wk8	wk12	ytd
71	Oakmark Fund	-3.21	-7.00	-5.61	3.43
72	Vanguard Windsor II	-1.09	-6.78	-4.55	4.69
73	Scudder Development	-2.23	-7.75	-3.17	4.65
74	Neuberger & Berman Partners	-2.32	-7.83	-4.97	5.65
75	Scudder Growth & Income	-1.64	-5.78	-3.13	5.95
76	Baron Asset	-3.08	-8.57	-6.98	7.84
77	SteinRoe Capital Opportunities	-1.71	-7.06	-2.35	-2.18
78	Oakmark Small Cap	-5.90	-7.30	-4.14	-2.78
79	Safeco Income	-2.86	-6.79	-4.75	-4.60
80	Longleaf Partners	-1.85	-9.25	-8.20	10.66

------------------------- Cluster=5 -------------------------

Obs	fund	wk4	wk8	wk12	ytd
81	Janus Olympus	2.25	-1.74	4.35	27.08
82	Janus Mercury	4.19	-1.75	6.55	32.93
83	Invesco Dynamics	0.77	-2.06	4.83	23.67
84	Fidelity Fifty	1.36	-2.90	0.59	24.65
85	Fidelity New Millenium (clsd)	0.22	-3.67	4.20	37.25

------------------------- Cluster=6 -------------------------

Obs	fund	wk4	wk8	wk12	ytd
86	Invesco Small Company	-0.66	0.30	9.74	16.75
87	US Global Bonnel Growth	4.28	0.97	10.20	17.08
88	Fidelity OTC	1.34	-1.21	5.54	19.69
89	Vanguard/Primecap (clsd)	1.85	0.95	8.13	21.50
90	Berger Growth and Income	0.19	-2.43	4.82	17.37
91	Managers Special Equity	0.38	0.13	8.48	12.17
92	Berger Small Compan Growth	3.10	1.31	12.86	23.02
93	Wasatch Micro-cap	-1.35	0.92	12.02	9.77

------------------------- Cluster=7 -------------------------

Obs	fund	wk4	wk8	wk12	ytd
94	Rydex OTC	6.01	2.36	14.05	30.95
95	Janus Enterprise	5.46	3.00	9.44	35.34
96	Fidelity Growth Company	6.00	5.47	13.82	28.13
97	Founders Discovery	-1.70	1.63	13.90	30.49
98	Fidelity Aggrssv Gw	6.92	3.43	14.28	42.00

------------------------- Cluster=8 -------------------------

Obs	fund	wk4	wk8	wk12	ytd
99	Rob Stephens Em Growth	5.42	1.19	14.25	55.12
100	Profund Ultra OTC (sold 2/8/99)	10.90	2.02	25.21	48.78

Output 6.7
(continued)

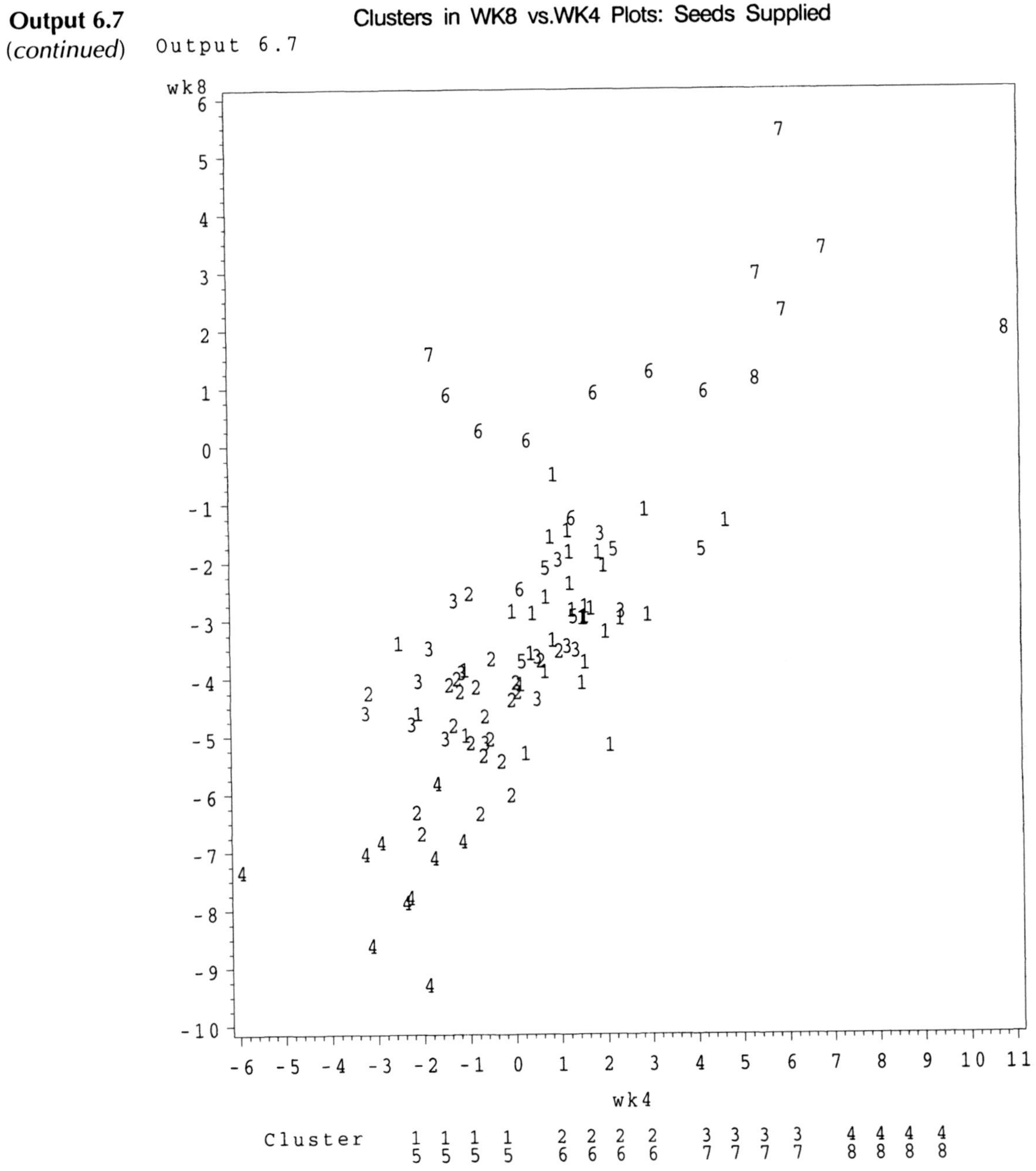

Output 6.7

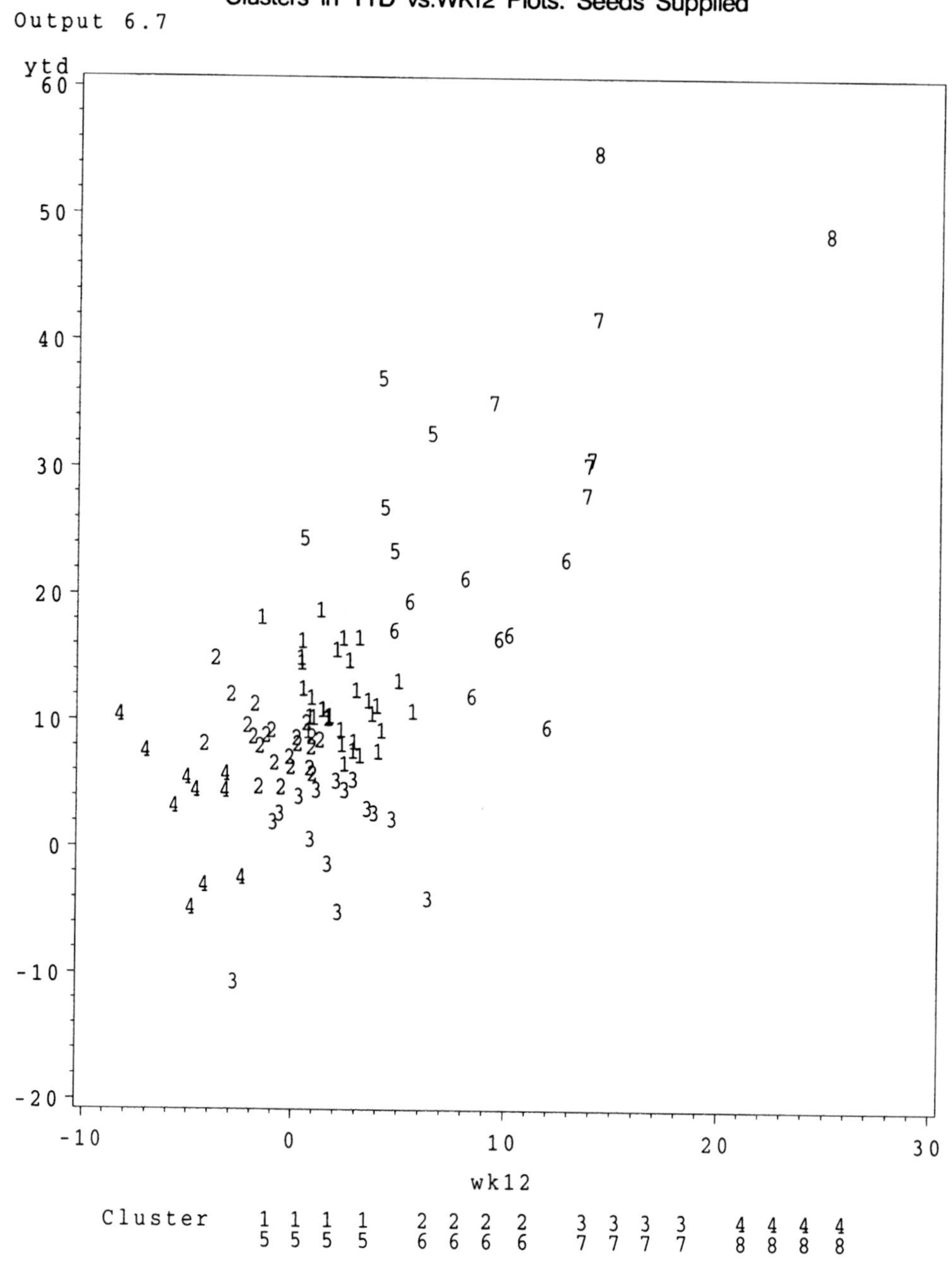
Clusters in YTD vs.WK12 Plots: Seeds Supplied
ytd
60
50
40
30
20
10
0
-10
-20
-10
0
10
20
30
wk12
Cluster 1 1 1 1 2 2 2 2 3 3 3 3 4 4 4 4
5 5 5 5 6 6 6 6 7 7 7 7 8 8 8 8

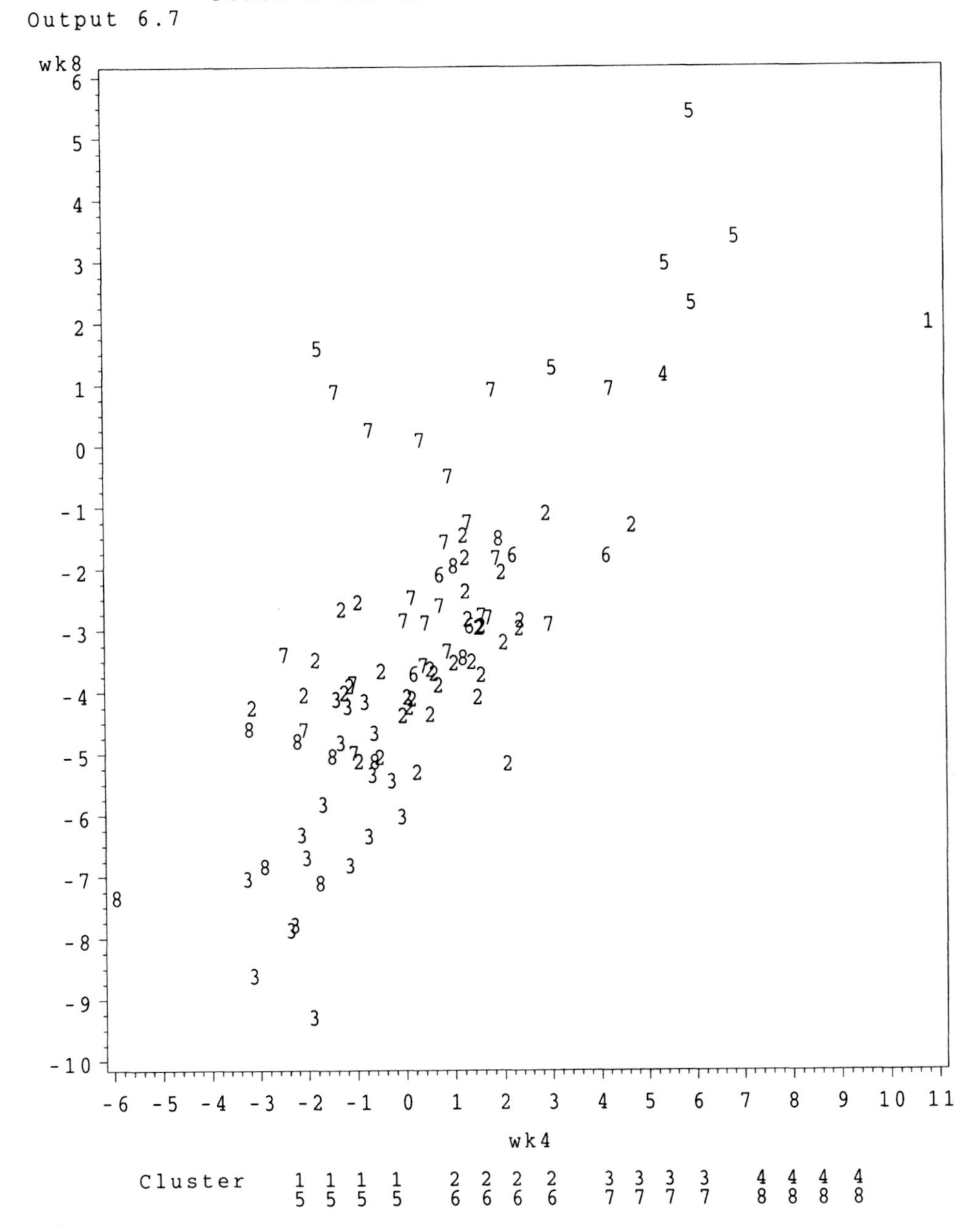
Clusters in WK8 vs.WK4 Plots: Seeds as First k Obs
Output 6.7
wk8
6
5
4
3
2
1
0
-1
-2
-3
-4
-5
-6
-7
-8
-9
-10
-6 -5 -4 -3 -2 -1 0 1 2 3 4 5 6 7 8 9 10 11
wk4
Cluster 1 1 1 1 2 2 2 2 3 3 3 3 4 4 4 4
5 5 5 5 6 6 6 6 7 7 7 7 8 8 8 8

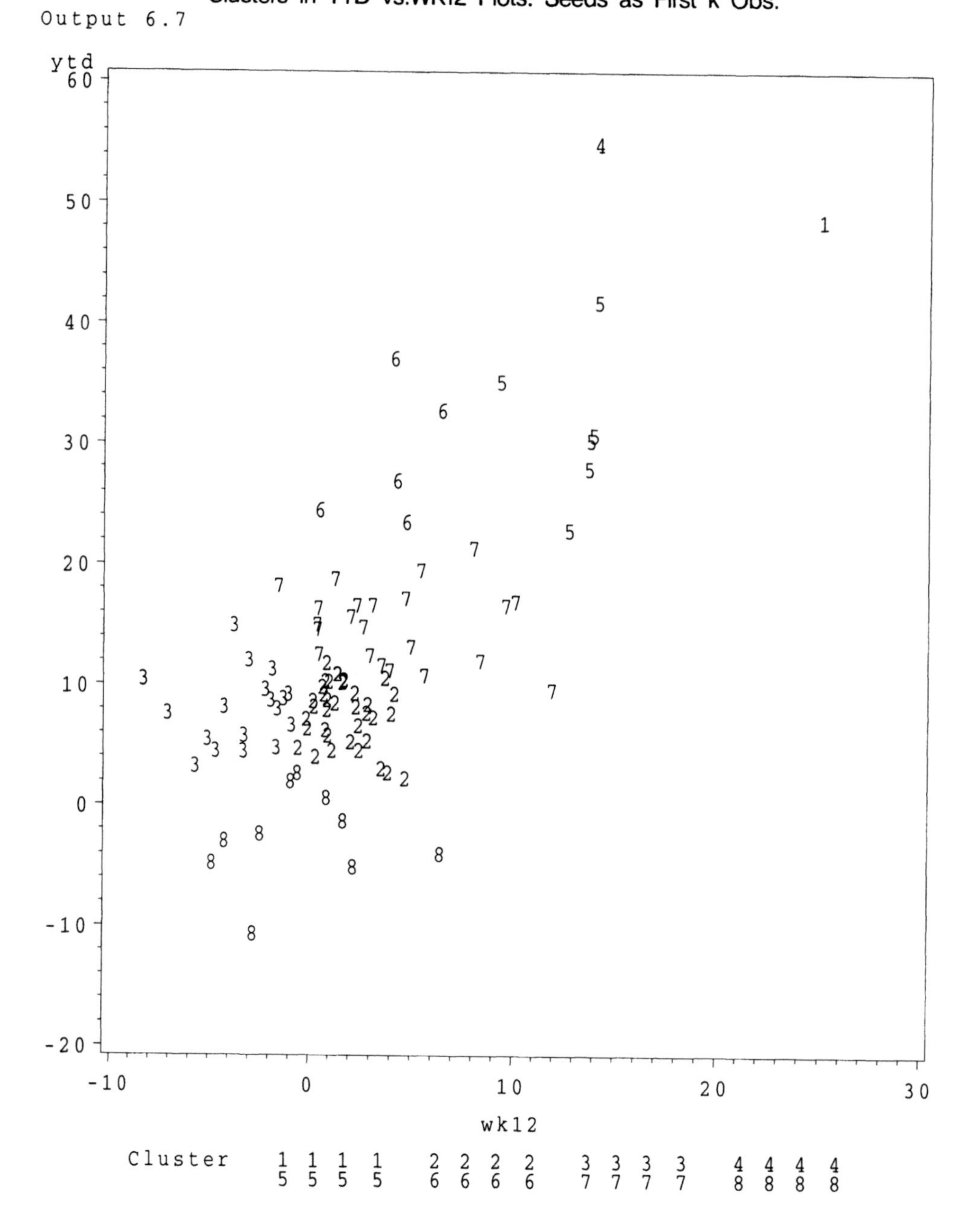
Clusters in YTD vs.WK12 Plots: Seeds as First k Obs.
Output 6.7
ytd
60
50
40
30
20
10
0
-10
-20
-10
0
10
20
30
wk12
Cluster
1 1 1 1 2 2 2 2 3 3 3 3 4 4 4 4
5 5 5 5 6 6 6 6 7 7 7 7 8 8 8 8

Output 6.7

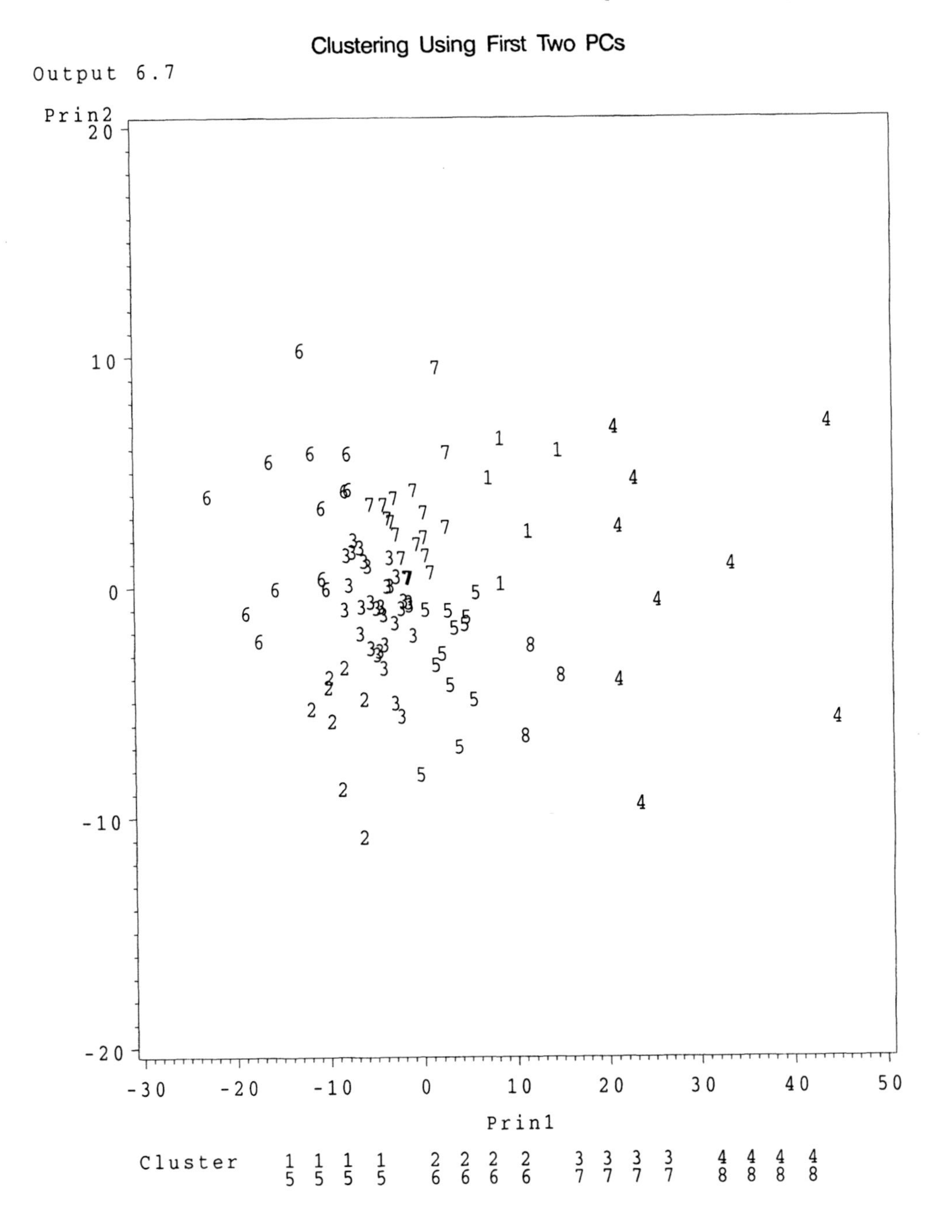
Clustering Using First Two PCs
Prin2
20
10
0
-10
-20
-30
-20
-10
0
10
20
30
40
50
Prin1
Cluster 1 5 1 5 1 5 1 5 2 6 2 6 2 6 2 6 3 7 3 7 3 7 3 7 4 8 4 8 4 8 4 8

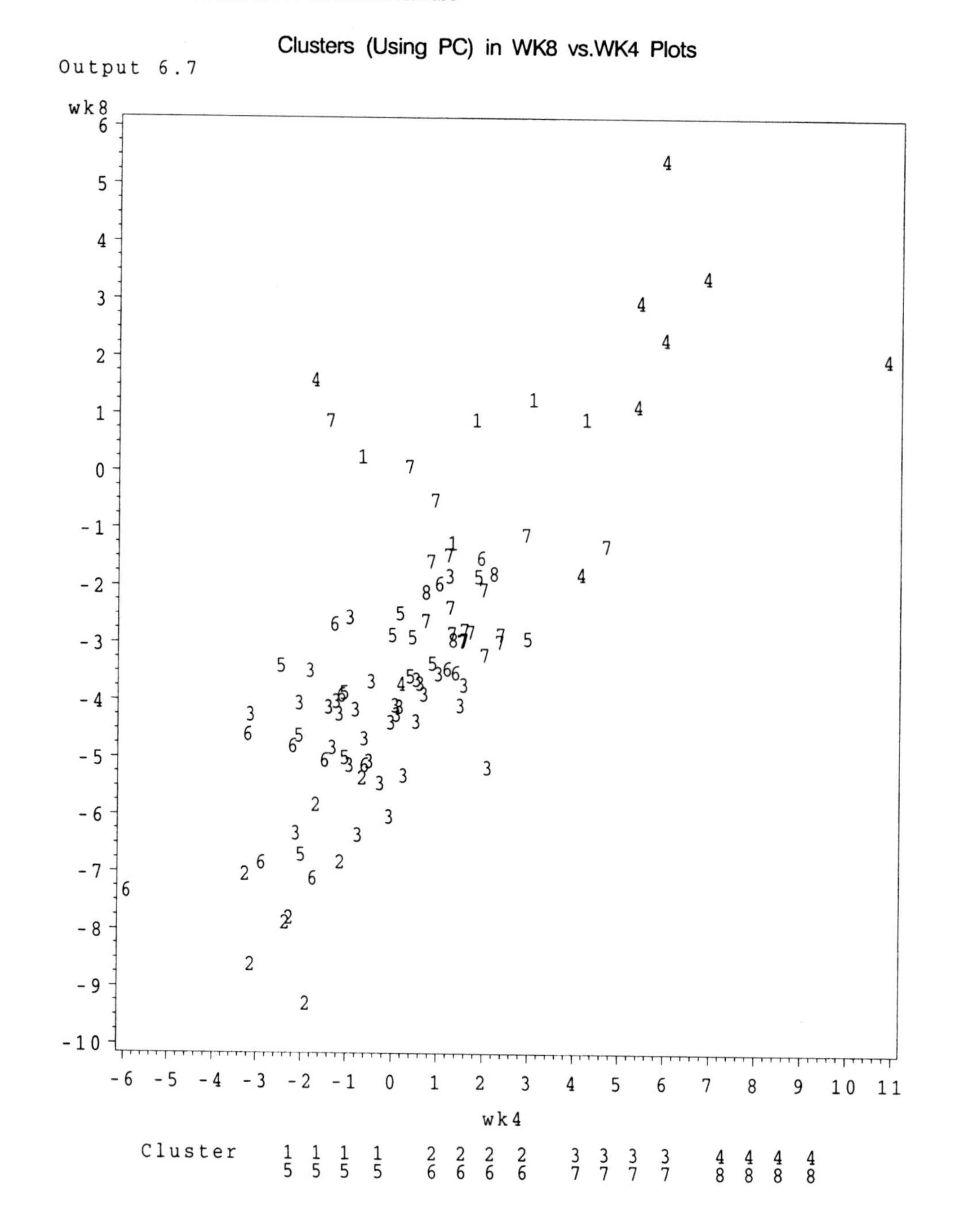
Clusters (Using PC) in WK8 vs.WK4 Plots
Output 6.7
wk8
6
5
4
3
2
1
0
-1
-2
-3
-4
-5
-6
-7
-8
-9
-10
-6 -5 -4 -3 -2 -1 0 1 2 3 4 5 6 7 8 9 10 11
wk4
Cluster 1 1 1 1 2 2 2 2 3 3 3 3 4 4 4 4
5 5 5 5 6 6 6 6 7 7 7 7 8 8 8 8

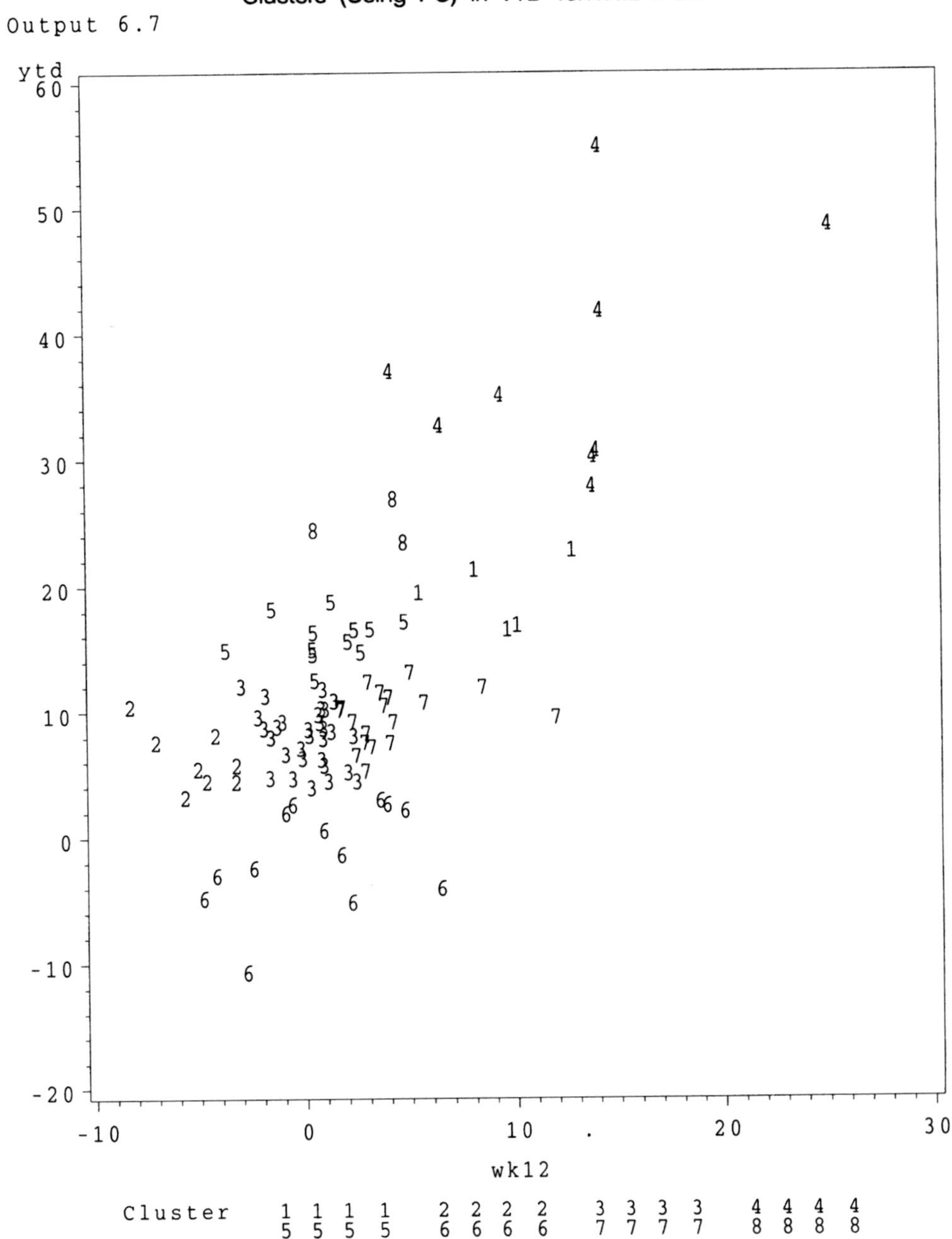

The corresponding output is included as the early part of Output 6.7. A glance at the initial seeds suggests that they are well separated from each other (except perhaps for the first two clusters). Convergence was obtained in six iterations.

A look at the summary statistics for all four variables suggests that there is considerable within-cluster variability for variables WK12 and YTD (the within-cluster standard deviations are, respectively, 2.0411 and 3.8766). This is not surprising, since for longer periods changes in prices were also considerably more. This indicates that, perhaps with respect to clustering, these two variables played a considerably larger role than the other two (this would perhaps not happen if the variables were standardized). The plot between WK12 and YTD clearly brings out this aspect of clustering, where all eight clusters are found to be reasonably well separated. Although separation of certain clusters is also evident in the plot between WK4 and WK8, it is not as distinctive.

The complete list of all eight clusters is also included as part of Output 6.7, and these clusters can be assigned appropriate interpretations. For example, Clusters 1 and 2 are different from each other largely with respect to their performances at 4, 8, and 12 weeks, although differences are also evident in the year-to-date performance as well. Cluster 3 is different from the previous two clusters in that it lags substantially in its year-to-date performance. Cluster 4 mainly contains those funds that have consistently been losing for the last 12 weeks and have shown a pretty minimal year-to-date performance. Clusters 5, 6, 7, and 8 represent funds with respectable 12-week and year-to-date performances, and thus can be clearly distinguished from the previous four clusters. Cluster 5 is different from Cluster 6 in that it consists of funds with a better year-to-date performance than that observed for funds in Cluster 6, yet an overall relatively worse short-term performance than those in Cluster 6. Finally, Clusters 7 and 8 stand out as the sets of overall best performers, with funds in Cluster 8 having a distinctive year-to-date record even when compared to those in Cluster 7.

It may be pointed out that if the seeds are chosen as the first k observations from data that meet the selection criterion, as opposed to being carefully specified as we did earlier, a new clustering of funds is obtained. However, this time the separation between the clusters is not as good. The SAS statements for this are included as part of Program 6.7 and, while the other output has been suppressed, two plots have been included for comparison's sake. It may be remembered that clusters have been renumbered, and thus the two resulting CLUSTER1 clusters are neither necessarily the same in memberships nor in interpretations.

The above discussion clearly points out the importance of seed selection. For our data, all eight initial seeds were the observations on the aggressive growth funds that are quite different from the funds in the other two categories (Growth funds, Growth and Income funds) in their very objectives.

It may be argued that the distances between observations may be calculated after transforming the data into uncorrelated variables. Some authors make an argument for using the principal components. In the process, the set of only those principal components that are able to explain a substantial amount of total (across all original variables) variance are retained. Using all principal components corresponding to the variance-covariance matrix is not all that useful anyway, because this is equivalent to using all of the original variables. It is so since the distances here are unaffected by any orthogonal transformation. If we want, we can achieve this by using PROC PRINCOMP as explained in Chapter 2. Unfortunately, the approach is somewhat faulty in that the estimated pooled within-cluster variance-covariance matrix must be used to obtain the coefficients of various principal components. However, this calculation itself requires knowledge of observations in various clusters. Using the total variance-covariance matrix instead of within-cluster variance-covariance is not only artificial, but also often misleading. We may verify that using the (first two) principal components of this matrix for our data results in a clustering that is difficult to interpret.

Art, Gnanadesikan, and Kettenring (1982) present an approach to indirectly and heuristically estimate the within-cluster covariance under the assumption that all clusters have the same unknown covariance matrix. In SAS, their approach is implemented as the ACECLUS procedure, with some minor refinements. The basic assumption is that the clusters, whatever and wherever they are, are random samples from independent multivariate normal populations, all with a common variance-covariance matrix. The algorithm attempts to iteratively predict the various cluster memberships and from these predicted clusters, estimates the unknown common variance-covariance matrix of these populations. Algorithmic details are given in the original reference and *SAS/STAT User's Guide, Version 6, Fourth Edition, Volume 1*, p. 189–207, and we do not discuss them here. We will illustrate its application using the mutual funds data.

EXAMPLE 5 *(Continued)* We again consider the mutual funds data and this time, first attempt to estimate the common variance-covariance matrix of the hypothetical populations of which the respective clusters are random samples. The appropriate SAS statements are included in Program 6.8. The common variance-covariance matrix is estimated as

$$\mathbf{S} = \begin{bmatrix} 0.8223 & 0.5850 & 1.0970 & 0.4522 \\ 0.5850 & 0.5976 & 0.9200 & 0.6500 \\ 1.0970 & 0.9200 & 1.9901 & 1.7266 \\ 0.4522 & 0.6500 & 1.7266 & 7.5303 \end{bmatrix},$$

and in Output 6.8 it is reported as ACE. Similarly, the total sample variance-covariance matrix **T** (reported as COV in Output 6.8) is

$$\mathbf{T} = \begin{bmatrix} 6.0315 & 4.7833 & 9.4373 & 16.6067 \\ 4.7833 & 6.8845 & 12.2360 & 17.1968 \\ 9.4373 & 12.2360 & 26.7883 & 37.7747 \\ 16.6067 & 17.1968 & 37.7747 & 114.0149 \end{bmatrix}$$

Thus, the between-cluster matrix is

$$\mathbf{D} = \mathbf{T} - \mathbf{S}.$$

PROC ACECLUS also provides the scores on canonical variables. The approach is the same as that described in Chapter 5 for canonical discrimination (PROC CANDISC). Specifically, canonical variables are defined as the linear combinations of original variables with coefficients of any linear combinations defined as the elements of the particular eigenvector of $\mathbf{S}^{-1}\mathbf{D}$. We can use these canonical variables for any clustering. This is what is done in the later part of Program 6.8. As earlier, we use PROC FASTCLUS. The interpretation of the resulting output is similar. The coefficients defining the canonical variables are presented in Output 6.8. However, it is difficult to assign any simple and intuitive meaning to these canonical variables. A plot of the first two canonical variables does, however, show good separation. The clusters formed here are not identical to those formed earlier using Program 6.7, and perhaps are a bit harder to interpret in terms of original variables. However, the plots of WK8 vs. WK4 and YTD vs. WK12 provide some impressions about the relative locations of the clusters.

```
/* Program 6.8 */

title1 'Output 6.8';

options ls = 64 ps = 45 nodate nonumber;

filename gsasfile "prog68.graph";
goptions reset=all gaccess=gsasfile autofeed dev=pslmono;
goptions gsfmode=append;
goptions horigin=1in vorigin=2in;
goptions hsize=6in vsize=8in;

data mfunds;
infile 'm_funds.dat' firstobs = 7;
input fund $ 1-33
price war39 wd39 wk4 wk8 wk12 ytd drawdown;
```

```
proc aceclus data = mfunds maxiter = 30 out = ace
p = .02;
var wk4 wk8 wk12 ytd ;
title1 'Output 6.8';
title2 'Estimation of Common Within Cluster Covariance Matrix';
run;

proc fastclus data = ace maxiter = 20 maxclusters = 8 distance
radius = .001 replace = full out = clus_ace ;
var can1 can2 can3 can4;
id fund;
title1 'Output 6.8';
title2 'Clustering Using Canonical Variables';
run;

proc sort data = clus_ace;
by cluster distance;
run;

symbol1 value = 1;
symbol2 value = 2;
symbol3 value = 3;
symbol4 value = 4;
symbol5 value = 5;
symbol6 value = 6;
symbol7 value = 7;
symbol8 value = 8;


proc gplot data = clus_ace;
where
cluster = 1 or cluster = 2 or cluster = 3 or
cluster = 4 or cluster = 5 or cluster = 6 or
cluster = 7 or cluster = 8 ;
plot can2*can1 = cluster ;
title1 h = 1.2 'Clustering Plot in First Two Canonical Variables';
title2  j = 1 'Output 6.8';
run;

proc gplot data = clus_ace;
where
cluster = 1 or cluster = 2 or cluster = 3 or
cluster = 4 or cluster = 5 or cluster = 6 or
cluster = 7 or cluster = 8 ;
plot wk8*wk4 = cluster ;
title1 h = 1.2 'Clusters in WK8 vs.WK4 Plots: Canonical Variables';
title2 j = 1 'Output 6.8';
run;

proc gplot data = clus_ace;
where
cluster = 1 or cluster = 2 or cluster = 3 or
cluster = 4 or cluster = 5 or cluster = 6 or
cluster = 7 or cluster = 8 ;
plot ytd*wk12 = cluster ;
title1 h = 1.2 'Clusters in YTD vs.WK12 Plots: Canonical Variables';
title2 j = 1  'Output 6.8';
run;
```

```
title1 'Output 6.8';
title2 ' ';

proc iml;
a = {
0.822298      0.584999      1.097001      0.452214,
0.584999      0.597609       0.91999      0.650016,
1.097001       0.91999      1.990096      1.726587,
0.452214      0.650016      1.726587      7.530282};

cov = {
6.031519      4.783282      9.437347      16.60669,
4.783282      6.884532        12.236      17.19683,
9.437347        12.236      26.78831      37.77468,
16.60669      17.19683      37.77468      114.0149};

call eigen(lam, p, a);
lhalfinv =inv( root(diag(lam)));
ahalfinv =  p*lhalfinv*t(p);

use mfunds;
read all into mmtx;
read all var {wk4 wk8 wk12 ytd} into mmtx;

mmtx2 = mmtx*p;
create newdata from  mmtx2;
append from mmtx2;
close newdata;

data newdata2;
merge newdata mfunds;
keep fund col1 col2 col3 col4 wk4 wk8 wk12 ytd;

proc standard data = newdata2 out = newdata2 m = 0 ;
var col1 col2 col3 col4;

proc fastclus data = newdata2 maxiter = 20 maxclusters = 8 distance
radius = .001 replace = full out = newout ;
var col1 col2 col3 col4;
id fund;
title1 'Output 6.8';
title2 'Clustering Using (Within Covariance) PC Scores';
run;

proc sort data = newout;
by cluster distance;
run;

proc gplot data = newout;
where
cluster = 1 or cluster = 2 or cluster = 3 or
cluster = 4 or cluster = 5 or cluster = 6 or
cluster = 7 or cluster = 8 ;
plot wk8*wk4 = cluster ;
title1 h = 1.2  'Clusters in WK8 vs.WK4 Plots: Within Covariance PC Scores';
title2 j = 1  'Output 6.8';
run;
```

```
proc gplot data = newout;
where
cluster = 1 or cluster = 2 or cluster = 3 or
cluster = 4 or cluster = 5 or cluster = 6 or
cluster = 7 or cluster = 8 ;
plot ytd*wk12 = cluster ;
title1 h= 1.2 'Clusters in YTD vs.WK12 Plots: Within Covariance PC Scores';
title2 j = 1  'Output 6.8';
run;
```

Output 6.8

Output 6.8
Estimation of Common Within Cluster Covariance Matrix

The ACECLUS Procedure

Approximate Covariance Estimation for Cluster Analysis

Observations	100	Proportion	0.0200
Variables	4	Converge	0.00100

Means and Standard Deviations

Variable	Mean	Standard Deviation
wk4	0.4767	2.4559
wk8	-3.1970	2.6238
wk12	2.3744	5.1757
ytd	11.8734	10.6778

COV: Total Sample Covariances

	wk4	wk8	wk12	ytd
wk4	6.0315193	4.7832817	9.4373470	16.6066901
wk8	4.7832817	6.8845323	12.2359988	17.1968271
wk12	9.4373470	12.2359988	26.7883077	37.7746839
ytd	16.6066901	17.1968271	37.7746839	114.0148954

ACE: Approximate Covariance Estimate Within Clusters

	wk4	wk8	wk12	ytd
wk4	0.822298365	0.584998846	1.097001346	0.452213846
wk8	0.584998846	0.597608750	0.919989712	0.650016442
wk12	1.097001346	0.919989712	1.990095962	1.726586635
ytd	0.452213846	0.650016442	1.726586635	7.530282212

Output 6.8
(continued)

Eigenvalues of Inv(ACE)*(COV-ACE)

	Eigenvalue	Difference	Proportion	Cumulative
1	29.8540	14.1859	0.5396	0.5396
2	15.6681	8.9707	0.2832	0.8228
3	6.6974	3.5914	0.1211	0.9439
4	3.1060		0.0561	1.0000

Eigenvectors (Raw Canonical Coefficients)

	Can1	Can2	Can3	Can4
wk4	-2.35214	0.276271	-0.27828	0.973995
wk8	0.82204	0.283791	2.48492	-.034865
wk12	1.38915	-.074303	-1.15193	0.083488
ytd	-0.25456	0.316745	0.02779	-.197624

Standardized Canonical Coefficients

	Can1	Can2	Can3	Can4
wk4	-5.77666	0.67850	-0.68342	2.39205
wk8	2.15691	0.74462	6.52002	-0.09148
wk12	7.18989	-0.38458	-5.96209	0.43211
ytd	-2.71817	3.38213	0.29669	-2.11019

Output 6.8
(continued)

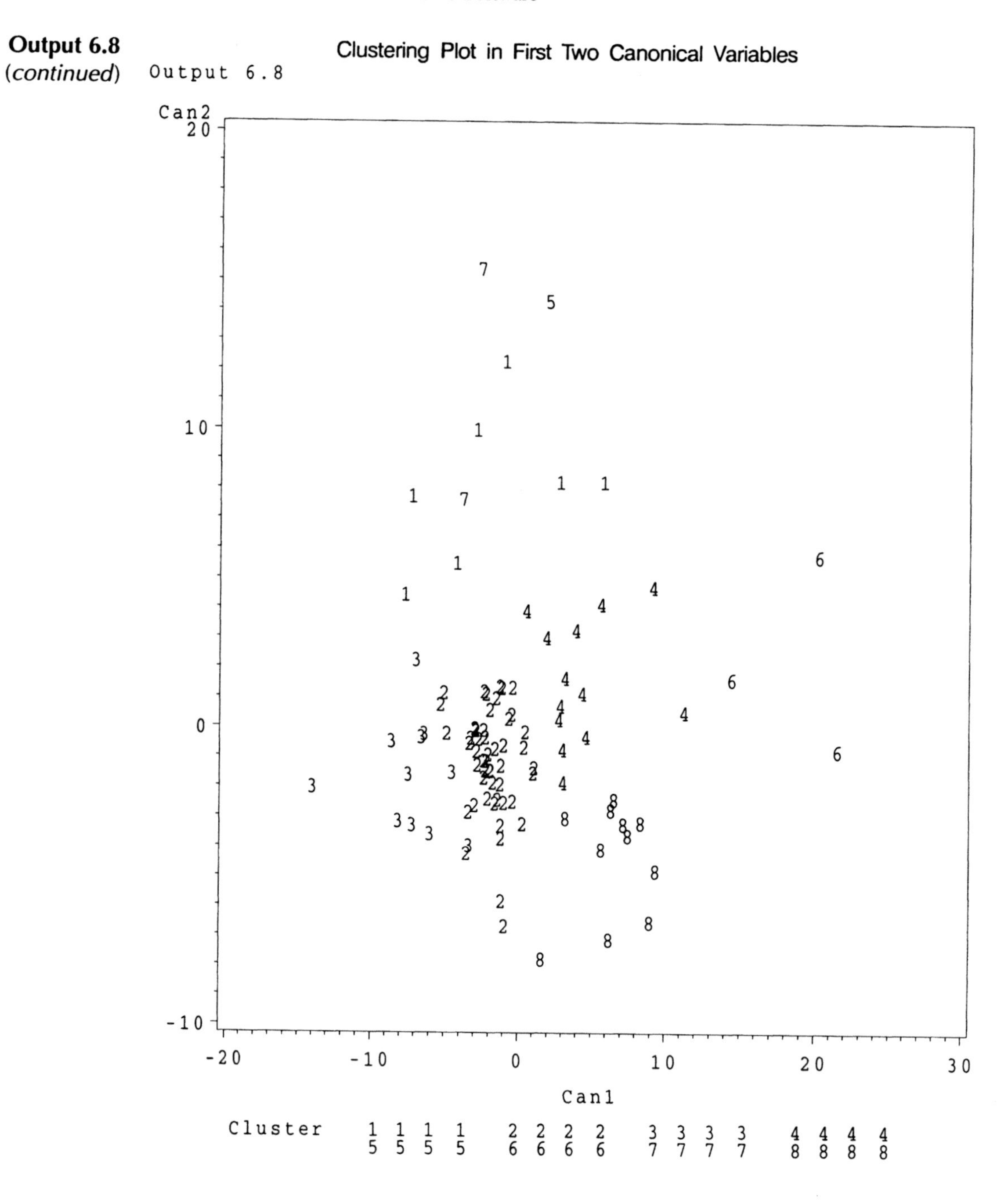

Clusters in WK8 vs.WK4 Plots: Canonical Variables

Output 6.8

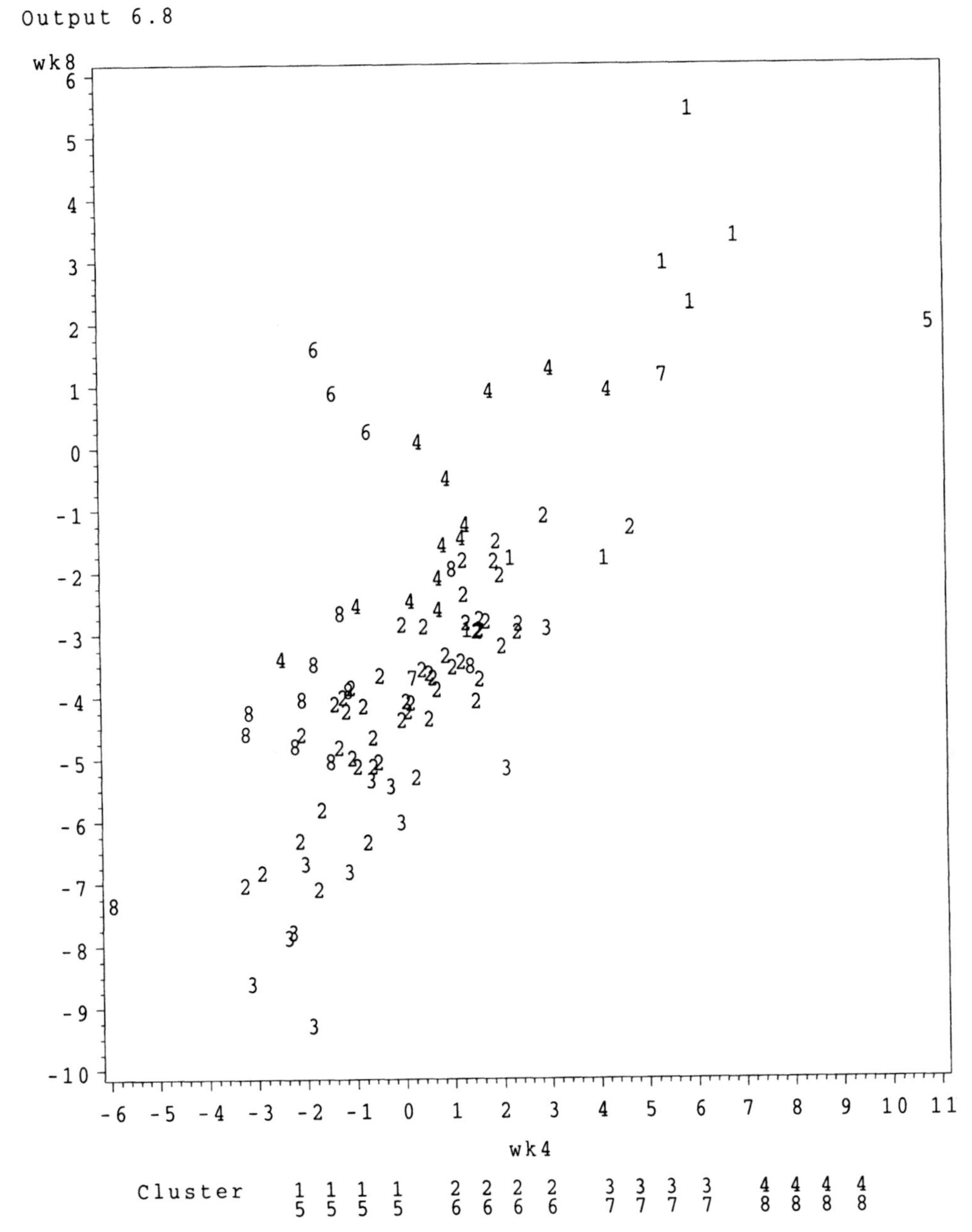

Output 6.8

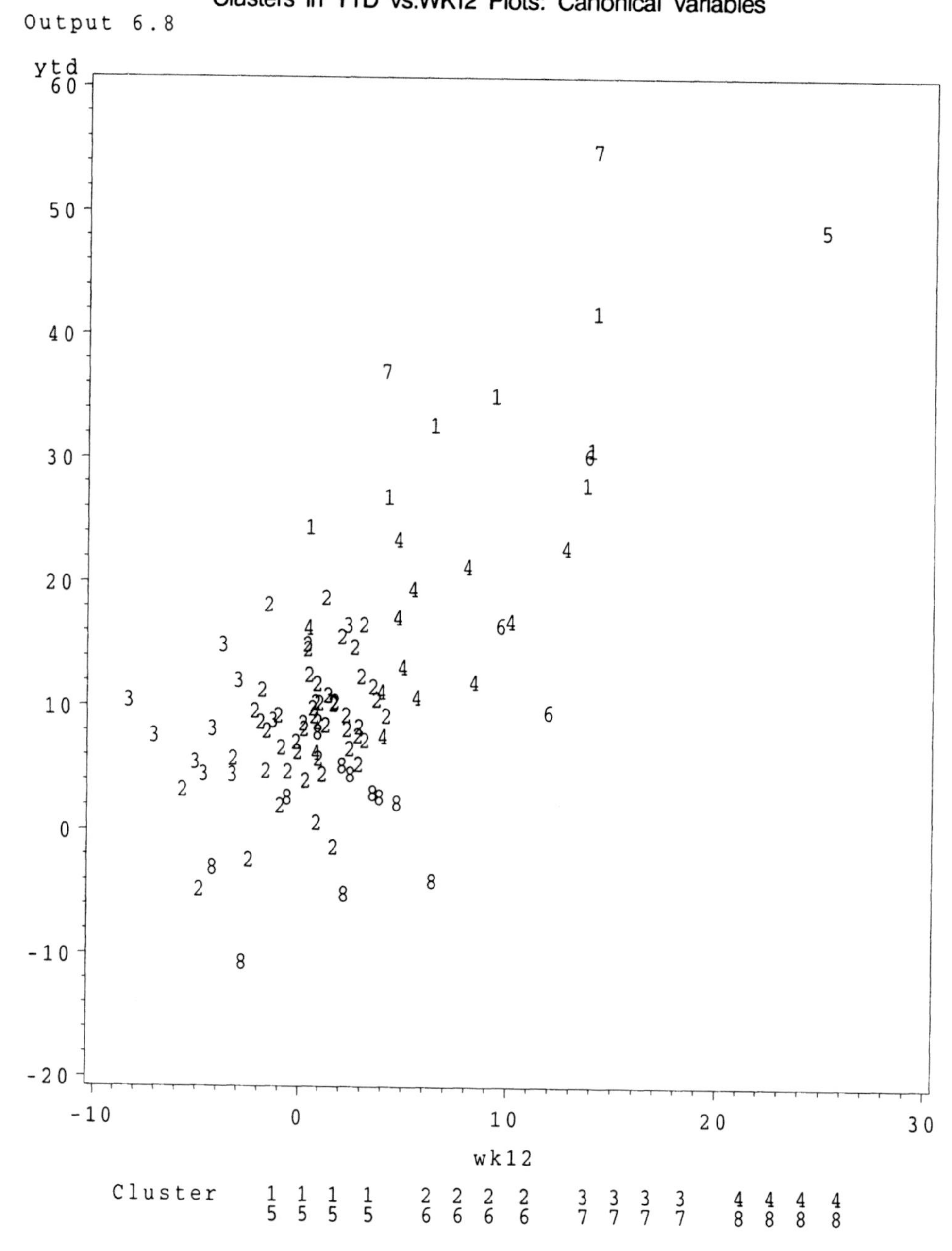
Clusters in YTD vs.WK12 Plots: Canonical Variables
ytd
60
50
40
30
20
10
0
-10
-20
-10
0
10
20
30
wk12
Cluster 1 5 2 6 3 7 4 8

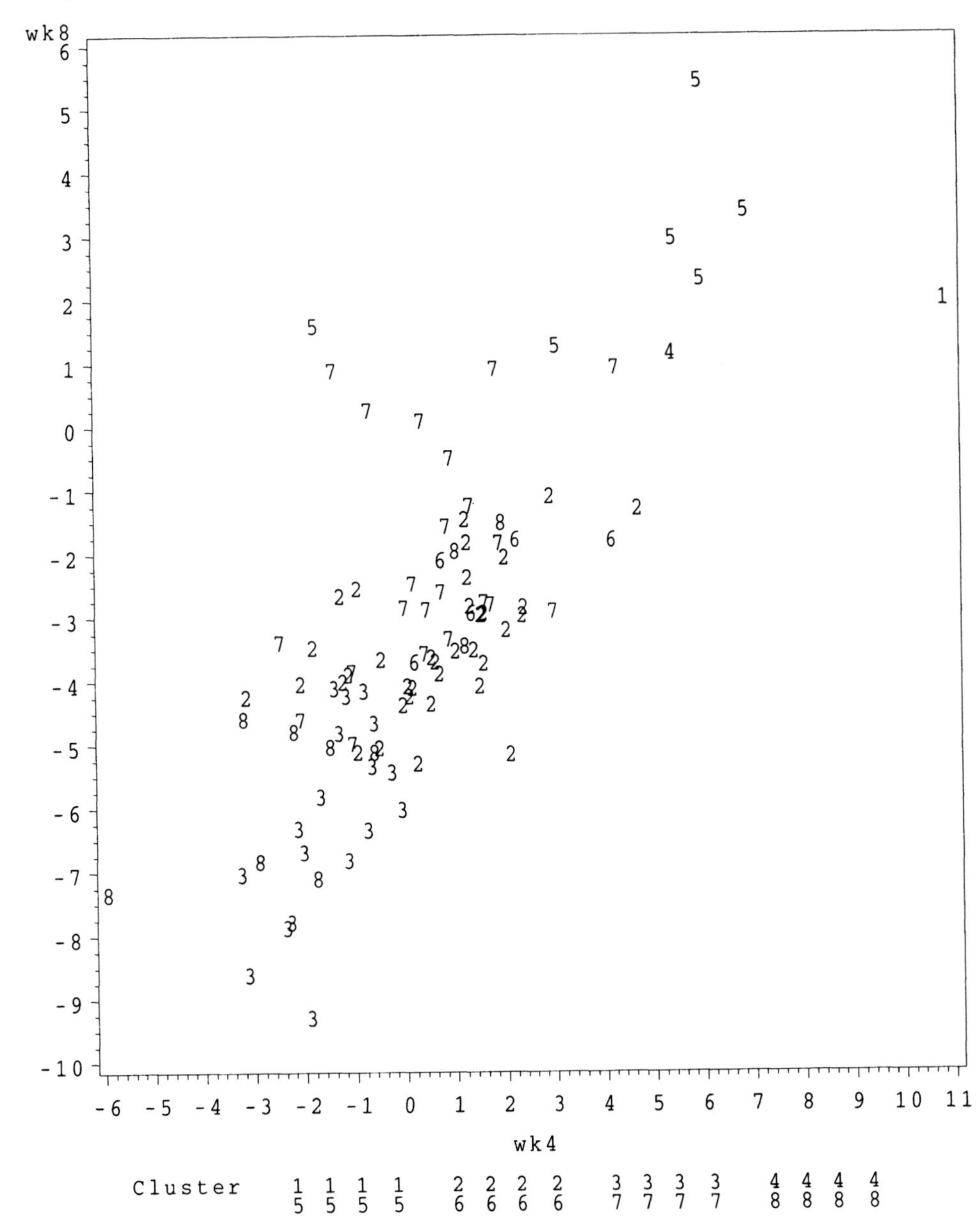
Clusters in WK8 vs.WK4 Plots: Within Covariance PC Scores
Output 6.8
wk8
6
5
4
3
2
1
0
-1
-2
-3
-4
-5
-6
-7
-8
-9
-10
-6 -5 -4 -3 -2 -1 0 1 2 3 4 5 6 7 8 9 10 11
wk4
Cluster 1 1 1 1 2 2 2 2 3 3 3 3 4 4 4 4
5 5 5 5 6 6 6 6 7 7 7 7 8 8 8 8

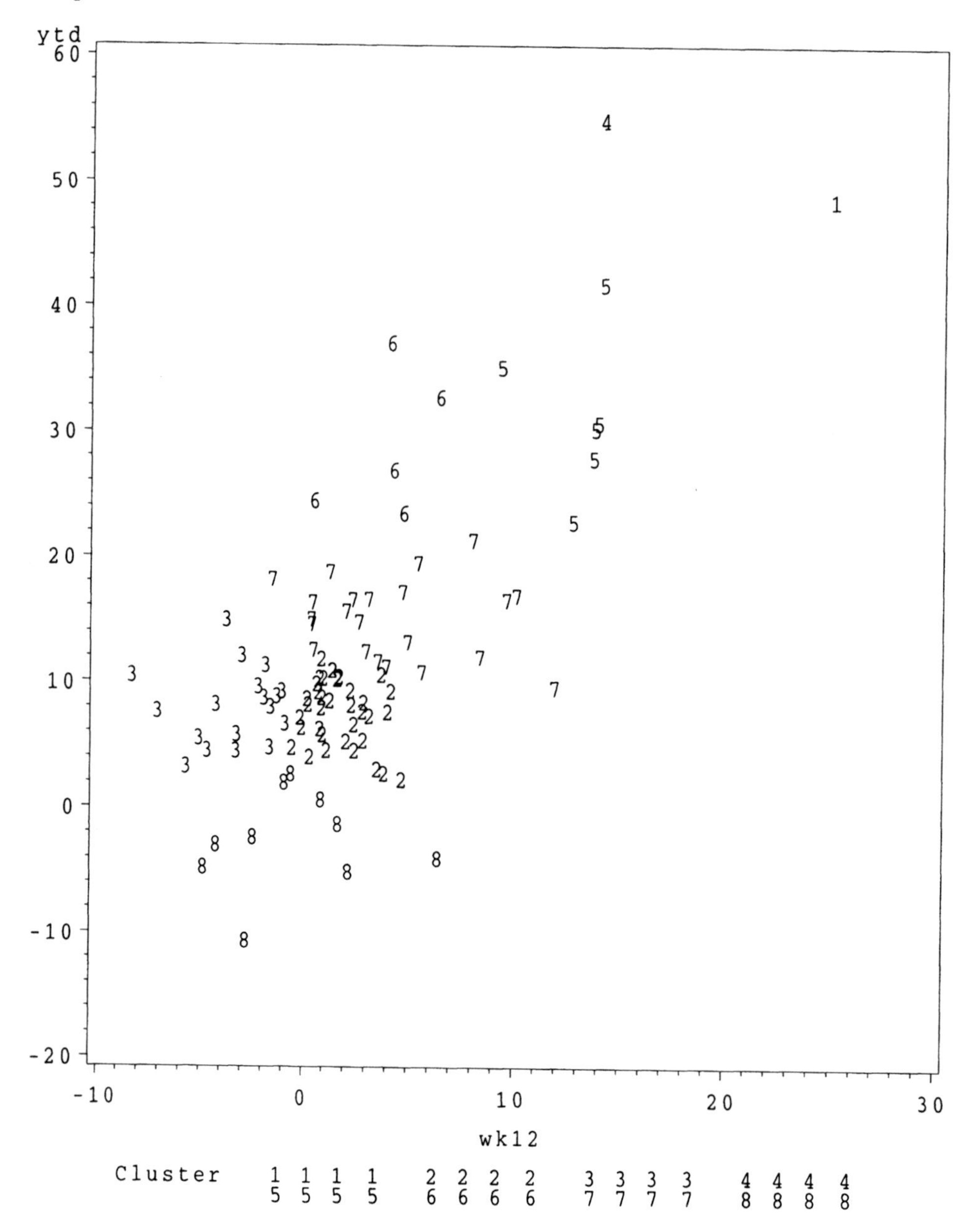

As an alternative, instead of canonical variables, we may want to use the principal components for clustering. For the present case, it is more appropriate to use the estimate of the common variance-covariance matrix rather than that of the common correlation matrix. Thus, the coefficients of the principal components are obtained as the eigenvectors of **S** listed above. We use the scores on all four principal components as data for clustering, as shown in Program 6.8. While suppressing the routine output, we include in Output 6.8 the two plots that show various clusterings in terms of original variables. By and large, the nature of clustering is similar but not identical to that produced by using the canonical variables.

It may be remarked that since the estimate of common variance-covariance matrix is available, instead of the Euclidean distance, we may want to use the (squared) Mahalanobis distance between two points **x** and **y**,

$$d^2(\mathbf{x}, \mathbf{y}) = (\mathbf{x} - \mathbf{y})'\mathbf{S}^{-1}(\mathbf{x} - \mathbf{y}).$$

However, using this for clustering is equivalent to using the principal component scores obtained by using the sample correlation matrix, and thus the same approach as described above can be used with obvious changes. In Program 6.8, it amounts to running PROC PRINCOMP without the COV option. Alternatively, the squared Mahalanobis distance between $\mathbf{x}$ and $\mathbf{y}$ can be calculated as the Euclidean distance between $\mathbf{x}_* = \mathbf{A}^{-1/2}\mathbf{x}$ and $\mathbf{y}_* = \mathbf{A}^{-1/2}\mathbf{y}$. However, we can mathematically observe that the coordinates computed this way and the canonical scores are related to each other by an orthogonal linear transformation, which amounts to the rotation of axes. Since the Euclidean distances are unaffected by any rotation of axes, the clustering obtained by using this transformation is also identical to that obtained by using the canonical scores. Thus, the three approaches, namely, using (all) canonical scores, using (all) principal components with correlation matrix, and using (all) the scores as $\mathbf{x}_* = \mathbf{A}^{-1/2}\mathbf{x}$ are equivalent. For this reason we have not shown the output corresponding to the last two analyses.

In closing this section, we must mention that we have not discussed here any of the approaches for the situations when the clusters may be overlapping. A good discussion of some of the methods for overlapping clusters may be found in Jardine and Sibson (1968).

6.7 How Many Clusters: Cubic Clustering Criterion

One of the important problems in any clustering investigation is to determine the possible number of clusters. This is not an easy problem and often the number of clusters is determined from various heuristic and tentative criteria. There have been several studies based on simulations that critically examine several procedures for determining the number of clusters. See Milligan and Cooper (1985) for one such extensive study. We have already discussed some of the procedures for determining the number of clusters based on R-square, semi-partial R-square, and root mean square standard deviations earlier.

One criterion which has been found useful in many, but not all, situations is the cubic clustering criterion or CCC introduced by Sarle (1983). The procedure basically assumes that clusters are obtained from a uniform distribution on a hyperbox or hypercubes of the same size. The problem is viewed as testing the above assumption as a null hypothesis with CCC as an approximate test statistic. This statistic can be viewed as the product of

$$ln\left[\frac{1 - E(R^2)}{1 - R^2}\right]$$

and

$$\frac{\sqrt{(np/2)}}{(0.001 + E(R^2))^{1.2}},$$

where R^2 represents the proportion of variance explained by clusters. $E(R^2)$ is its expected value under the null hypothesis, and p is an estimate of the dimensionality of the between-cluster variation. Higher values of CCC indicate good clustering.

Like any other criterion, CCC is not a perfect criterion and its use should also be accompanied by the use of other meaningful indices. As described in Sarle (1983), the following guidelines should be used in interpreting CCC.

1. A good clustering is indicated by $CCC > 3$. In general, higher values indicate better clustering.
2. For hierarchical data, several local high values of CCC may be observed. Nonhierarchical data will usually show a very distinctive global maximum for CCC, when plotted against the number of clusters.

3. For skewed distributions, CCC may be negative and will decrease further with the number of clusters.
4. If CCC continues to increase with the number of clusters, then it may indicate the possibility of excessive round-off in data, and the observations within a cluster may be forming the pockets of several subclusters.
5. For data with irregularly shaped or highly elongated clusters, CCC may not be an appropriate criterion (in fact, the same is also true for R^2).

EXAMPLE 6 ***Determining the Number of Clusters*** We consider the satellite data of Fox (1993) used in Section 6.2 and earlier in Chapter 5. Using Ward's method, we generate the clustering solutions for various numbers of clusters starting from one cluster to ten clusters. Various criteria such as R-square, semi-partial R-square, root mean square standard deviation, and the cubic clustering criterion (CCC) will all be taken into consideration. Program 6.9 computes these quantities, which are given as the first part of Output 6.9. A plot for cubic clustering criterion is also shown. Other plots can be obtained similarly. The criterion CCC attains its maximum (=43.3384) for the nine-clusters solutions. Although RMSSTD and SPRSQ continue to decrease, they are also reasonably small for this choice. The corresponding value of RSQ (=0.9591) is also large and thus the nine-cluster solution seems to be the best choice.

```
/* Program 6.9 */

options ls = 64 ps = 45 nodate nonumber;
data fox;
infile 'satellite.dat' firstobs = 7;
input group $ tm1 tm2 tm3 tm4 tm5 tm7;

proc cluster data = fox method = ward noprint
noeigen nonorm std outtree = tree ;
var  tm1 tm2 tm3 tm4 tm5 tm7;
id group;
title1 'Output 6.9';
title2   'Cluster Analysis: Satellite Data';
run;

proc sort data = tree;
by _ncl_;

data cccgraph;
set tree;
if _ncl_ < = 10;

proc print data = cccgraph;
var _ncl_  _rsq_ _sprsq_ _rmsstd_ _ccc_ ;
title1 'Output 6.9';
title2 'Criteria to Decide # Clusters: Satellite Data';
run;

filename gsasfile "prog69.graph";
goptions reset=all gaccess=gsasfile autofeed dev=pslmono;
goptions gsfmode=append;
goptions horigin=1in vorigin=2in;
goptions hsize=6in vsize=8in;

symbol1 i=join v=star;
legend1 across=3;

proc gplot data = cccgraph ;
plot _ccc_*_ncl_  /
```

```
vaxis=axis1 haxis=axis2 legend=legend1;
axis1 label=(a=90 h=1.2 'CCC Values');
axis2 offset=(2) label=(h=1.2 'No. of Clusters');
title1 h = 1.2 'CCC Values vs. No. of Clusters:Satellite Data';
title2 j = 1  'Output 6.9';
run;

data mfunds;
infile 'm_funds.dat' firstobs = 7;
input fund $ 1-33
price war39 wd39 wk4 wk8 wk12 ytd drawdown;

proc cluster data = mfunds method = ward noprint
noeigen nonorm std outtree = treem;
var  wk4 wk8 wk12 ytd;
id fund;
title1 'Output 6.9';
title2 "Cluster Analysis: Mutual Funds Data";
run;

proc sort data = treem;
by _ncl_;

data ccgraph1;
set treem;
if _ncl_ < = 20;

proc print data = ccgraph1;
var _ncl_  _rsq_ _sprsq_ _rmsstd_ _ccc_ ;
title1 'Output 6.9';
title2 'Criteria to Decide # Clusters: Mutual Funds Data';

proc gplot data = ccgraph1 ;
plot _ccc_*_ncl_  /
vaxis=axis1 haxis=axis2 legend=legend1;
axis1 label=(a=90 h=1.2 'CCC Values');
axis2 offset=(2) label=(h=1.2 'No. of Clusters');
title1 h= 1.2 'CCC Values vs. No. of Clusters: Mutual Funds Data';
title2 j = 1  'Output 6.9';
run;
```

Output 6.9

Output 6.9
Criteria to Decide # Clusters: Satellite Data

Obs	_NCL_	_RSQ_	_SPRSQ_	_RMSSTD_	_CCC_
1	1	0.00000	0.66080	1.00000	0.0000
2	2	0.66080	0.08935	0.59900	40.3424
3	3	0.75015	0.08280	0.57378	34.4018
4	4	0.83294	0.05353	0.49850	36.5738
5	5	0.88647	0.02815	0.41610	39.6237
6	6	0.91462	0.01862	0.40615	40.9470
7	7	0.93324	0.01532	0.32645	41.9951
8	8	0.94856	0.01057	0.31369	42.8241
9	9	0.95913	0.00579	0.30444	43.3384
10	10	0.96492	0.00507	0.25014	42.6018

Output 6.9
(continued)

Criteria to Decide # Clusters: Mutual Funds Data

Obs	_NCL_	_RSQ_	_SPRSQ_	_RMSSTD_	_CCC_
1	1	0.00000	0.40457	1.00000	0.0000
2	2	0.40457	0.23448	0.77002	13.6962
3	3	0.63905	0.06439	0.63606	19.5988
4	4	0.70344	0.05564	0.54459	16.2304
5	5	0.75907	0.03364	0.74155	14.4688
6	6	0.79271	0.02227	0.87231	14.1800
7	7	0.81499	0.01725	0.43125	13.6312
8	8	0.83223	0.01419	0.37719	13.1592
9	9	0.84642	0.01150	0.61807	12.7865
10	10	0.85793	0.00990	0.36549	12.4162
11	11	0.86783	0.00977	0.63239	12.1031
12	12	0.87760	0.00850	0.39405	12.0075
13	13	0.88611	0.00698	0.54202	11.9168
14	14	0.89308	0.00647	0.28860	11.7417
15	15	0.89955	0.00644	0.41795	11.6207
16	16	0.90600	0.00586	0.50436	11.6239
17	17	0.91186	0.00574	0.44710	11.6367
18	18	0.91760	0.00563	0.47287	11.7393
19	19	0.92323	0.00444	0.66317	11.9301
20	20	0.92767	0.00444	0.39376	11.9542

Output 6.9
(*continued*)

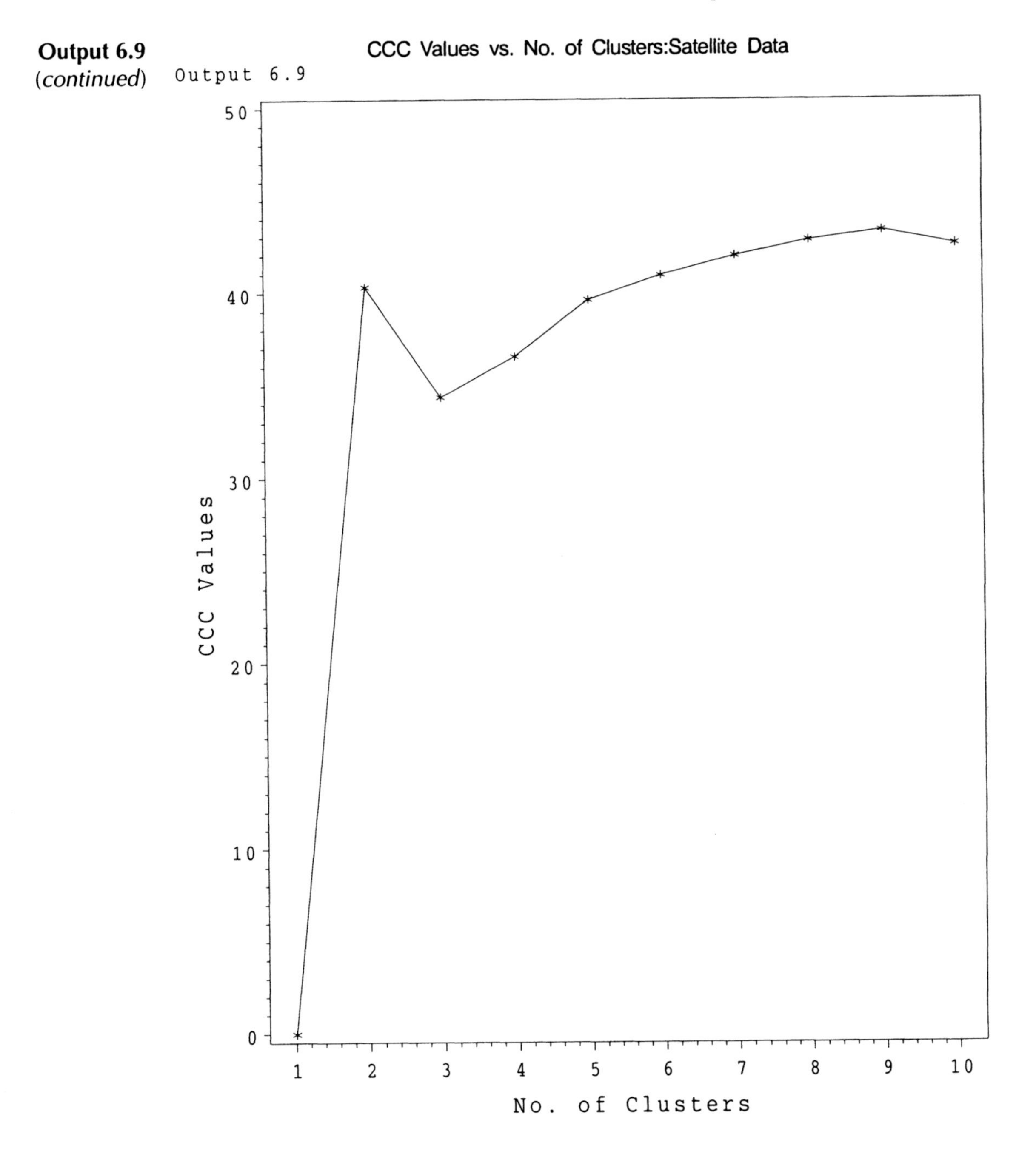

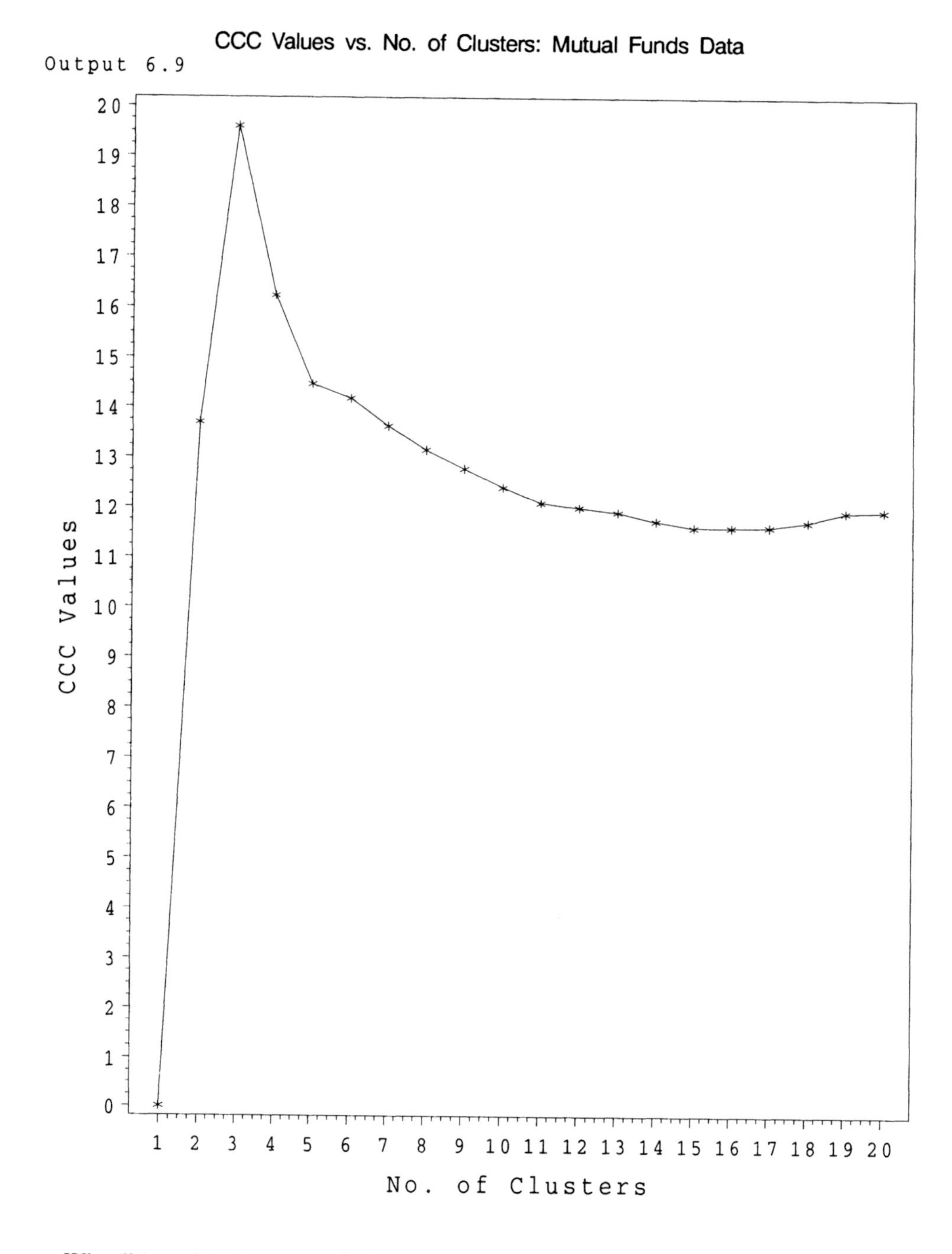

Why did we find the above choice to be best when in reality there are only six groups? One reason for this is that there are possibly certain subclusters within some of these groups. For the data at hand, this has been earlier pointed out by Khattree and Naik (1998) using Andrews plots. There is no way of identifying this nesting of subclusters within groups and hence since the algorithm finds them relatively dissimilar from the rest, it defines a new cluster for them separately. Thus, more clusters than the actual number of groups are identified.

The problem may occur in just the opposite way, in that the criterion may sometimes indicate the presence of very few clusters. In Output 6.9, we present a plot of CCC vs. number of clusters for the mutual funds data. In this case, CCC indicates that the optimal number of clusters is three. Other criteria seem to suggest the actual number of clusters somewhere between 8 to 14. This makes the point that for obtaining the optimal number of clusters, we must try various choices and see if some possible interpretation can be assigned to the nature of observations in the individual clusters. For this data set, the choice of eight clusters used in Program 6.7 was arrived at with such considerations.

6.8 Clustering Using Density Estimation

Most of the clustering procedures discussed earlier were based on distance or dissimilarity. An alternative approach suitable for large data sets is based on nonparametric density estimation. The approach is useful in many simple as well as complex clustering situations and can identify clusters of irregular shape, of unequal size and of different dispersion. Further, the density estimation methods tend to be less sensitive to scaling and linear transformation of data.

For density estimation-based clustering, we view a cluster as a region around as local maximum of the probability density function. In other words, the clusters may be viewed as the high-density regions in the space separated by low-density regions between them. Thus, the clustering problem can be interpreted as the one of identifying these high-density regions around the local maximum. This interpretation forms the basis for the density estimation approach to clustering. We present a brief working detail of clustering through this approach. An extensive discussion of density estimation can be found in Silverman (1986) and Scott(1992).

The densities at points are estimated using the kernel method, which was briefly discussed in Chapter 5. Specifically, a popular kernel is the uniform kernel defined by the kernel density,

$$K(\mathbf{z}) = \begin{cases} \frac{1}{v_r} & \text{if } \mathbf{z}'\mathbf{V}^{-1}\mathbf{z} \leq r^2 \\ 0 & \text{otherwise} \end{cases},$$

with fixed or variable radius r, which is often referred to as a smoothing parameter. In the above, $\mathbf{V}$ is some suitably chosen positive-definite symmetric matrix, and v_r is the volume of a p-dimensional ellipsoid given by $\{\mathbf{z} : \mathbf{z}'\mathbf{V}^{-1}\mathbf{z} \leq r^2\}$. The value of volume v_r is

$$v_r = \frac{\pi^{p/2} r^p |\mathbf{V}|^{1/2}}{\Gamma(\frac{p}{2} + 1)}.$$

The density estimate at a point is computed as the ratio of the number of observations within the ellipsoid and the product of the sample size and volume, that is, nv_r. In SAS, the MODECLUS procedure performs the clustering using the density estimation approach, and in this procedure the matrix $\mathbf{V}$ is taken as the p by p identity matrix. Thus, for the subsequent discussion, we will take the matrix $\mathbf{V}$ to be an identity matrix.

The choice of radius r, also known as the smoothing parameter, is important. Another useful smoothing parameter is k, which represents the number of neighbors to use if we want the k-nearest neighbor density estimation. The choice of k as a smoothing parameter is often desirable because it allows us to use the variable-radius kernels rather than the fixed ones. When k is specified, the radius is chosen as the smallest radius so that the corresponding sphere contains at least k observations (including the one at which the density is being estimated). Let this radius be r_k. Sometimes, we may want to use the maximum of r and r_k, because it allows more freedom to absorb certain distant observations into one of the clusters, thereby enabling us to avoid the clusters with one single outlying observation. One nice feature of density estimation-based clustering is that we do not need to specify the number of clusters we want. The appropriate number of clusters is determined by the method itself.

Now we make a few remarks about the selection of appropriate smoothing parameters. The selection of r is indeed a difficult problem. Often various values of r are attempted by trial and error. Alternatively, if p is the number of variables then, under the assumption of sampling from multivariate normality, the value of r can be approximated by

$$r = [\frac{2^{p+2}(p+2)\Gamma(\frac{1}{2p}+1)}{np^2}]^{\frac{1}{p+4}}.[\sum_{i=1}^{p} s_i^2]^{\frac{1}{2}},$$

where n is the total number of observations, and $s_1^2, \ldots, s_p^2$ are the sample variances computed from the data for p variables. The choice of r using above formula with all $s_i^2 = 1$ and for a few selected choices of n and p are given in the *SAS/STAT User's Guide, Version 6, Fourth Edition, Volume 1* (see the section on the CLUSTER procedure). The following SAS code (illustrated for $p = 3, n = 80, s_1^2 = s_2^2 = s_3^2 = 2$) readily computes the value of r.

```
data r;
input p n s1sq s2sq s3sq;
p2 = p+2;
halfp2 = p2/2;
a1 = (2**p2)*p2*gamma(halfp2);
a2 = n*p*p;
a3 = sqrt( sum(s1sq, s2sq, s3sq));
r = ( (a1/a2)**(1/ (p+4)) )*a3;
lines;
3 80 2 2 2
;
proc print data = r;
var p n s1sq s2sq s3sq r;
```

We must recall that the underlying assumption for the above formula was the sampling from the multivariate normal population. Thus, the calculated value of r is only a rough guess. To choose the value of k, the first guess is a k such that,

$$\frac{1}{10}n^{\frac{4}{p+4}} \leq k \leq n^{\frac{4}{p+4}}.$$

When p is large, a value of k toward the smaller side may be more suitable.

Finally, how are the clusters formed? The exact details of various methods, and there are several of them, are quite involved and technical. Intuitively, we compare the estimated densities of the observations with the estimated densities of the other observations in the neighborhood (defined by radius r and/or by k) or with the estimated densities of the cluster to which the particular neighbor belongs. If the densities are comparable, then the observation joins the cluster to which the responsible neighbor belonged. This procedure is continued for all observations.

The MODECLUS procedure of SAS performs clustering using the density estimation approach. Of the seven methods available there, six choices (that is, except METHOD = 0) are based on the approach briefly described above. The methods, however, differ from each other in algorithmic details and computational refinements. Usually, choosing METHOD = 1 or METHOD = 2 should perform satisfactorily for most problems with large data sets. We will illustrate clustering with the density estimation approach by using the mutual funds data.

EXAMPLE 7 ***Clustering of Mutual Funds, Density Estimation Approach*** The data on 100 mutual funds is large enough to use this approach. This time we will use the standardized variables for data analysis. This can be achieved by using the STANDARD (or simply STD) option in the MODECLUS procedure. The values of $k = 5$ and $r = 0.6$ were chosen by trial and error with the intent of avoiding too many or too few clusters. Both SAS options, namely,

K = 5 and R = 0.6, were used because this choice potentially reduces the possibility of getting clusters with a single observation in them. We have used the METHOD = 2 option, as illustrated in Mizoguchi and Shimura (1980). In fact, the method uses only the first stage of their algorithm. Program 6.10 presents the corresponding SAS code.

```
/* Output 6.10 */

options ls = 64 ps = 45 nodate nonumber;


title1 'Output 6.10';

data mfunds;infile 'm_funds.dat' firstobs = 7;
input fund $ 1-33
price war39 wd39 wk4 wk8 wk12 ytd drawdown;

proc modeclus data = mfunds method = 2 k =  5 r = .6
std  test out = output;
var wk4 wk8 wk12 ytd;
id fund;
title1 'Output 6.10';
title2 'Nonparametric Clustering of Mutual Funds Data';
run;

proc sort data = output;
by _k_ _r_;

filename gsasfile "prog610.graph";
goptions reset=all gaccess=gsasfile autofeed dev=pslmono;
goptions gsfmode=append;
goptions horigin=1in vorigin=2in;
goptions hsize=6in vsize=8in;

symbol1 value = 1;
symbol2 value = 2;
symbol3 value = 3;
symbol4 value = 4;

proc gplot data = output;
where
cluster = 1 or cluster = 2 or
cluster = 3 or cluster = 4;
plot wk8*wk4 = cluster ;
by _k_ _r_;
title1 h = 1.2 'Mutual Funds Data: Clustering in WK8 vs. WK4 Plot';
title2 j = 1 'Output 6.10';
run;

proc gplot data = output;
where
cluster = 1 or cluster = 2 or
cluster = 3 or cluster = 4;
plot ytd*wk12 = cluster ;
by _k_ _r_;
title1 h = 1.2 'Mutual Funds Data: Clustering in YTD vs. WK12 Plot';
title2 j = 1 'Output 6.10';
run;
```

```
proc sort data = output;
by cluster;
run;

proc print data = output ;
by  cluster;
var fund wk4 wk8 wk12 ytd;
title1 'Output 6.10';
title2 'Clusters of Mutual Funds';
run;
```

Output 6.10

Output 6.10
Nonparametric Clustering of Mutual Funds Data

The MODECLUS Procedure
K=5 R=0.6 METHOD=2

Cluster Statistics

Cluster	Frequency	Maximum Estimated Density	Boundary Frequency	Estimated Saddle Density
1	44	0.26581175	11	0.23453978
2	43	0.26581175	12	0.21890379
3	7	0.05592567	3	0.05592567
4	6	0.00215462	4	0.00215462

-------------Saddle Test: Version 92.7------------

Cluster	Mode Count	Saddle Count	Overlap Count	Z	Approx P-value
1	16	14	7	0.250	1
2	16	13	6	0.485	1
3	2	2	2	0.000	1
4	0	0	.	0.000	1

Cluster Summary

K	R	Number of Clusters	Frequency of Unclassified Objects
5	0.6	4	0

Clusters of Mutual Funds

------------------------- CLUSTER=1 ---------------------------

Obs	fund	wk4	wk8	wk12	ytd
1	Amer Century Ultra	1.60	-3.68	0.90	10.48
2	Berger Small Compan Growth	3.10	1.31	12.86	23.02
3	Dreyfus New Leaders	0.02	-2.81	0.49	15.19
4	Fidelity Capital Appreciation	0.47	-2.85	2.16	15.83

Output 6.10 *(continued)*

5	Fidelity Fifty	1.36	-2.90	0.59	24.65
6	Founders Discovery	-1.70	1.63	13.90	30.49
7	Invesco Dynamics	0.77	-2.06	4.83	23.67
8	Invesco Small Company	-0.66	0.30	9.74	16.75
9	Janus Olympus	2.25	-1.74	4.35	27.08
10	Managers Special Equity	0.38	0.13	8.48	12.17
11	Emerging Growth	1.05	-1.92	6.52	-3.84
12	Rob Stephens Value + Gwth	0.77	-2.57	5.74	10.96
13	US Global Bonnel Growth	4.28	0.97	10.20	17.08
14	Value Line Leveraged Growth	1.61	-2.73	3.64	11.81
15	Wasatch Micro-cap	-1.35	0.92	12.02	9.77
16	Amer Cent Bnhm Eq Gwth	1.25	-1.42	4.13	7.77
17	Dreyfus Appreciation	2.96	-1.06	2.53	6.77
18	Dreyfus Third Century	1.34	-2.79	3.82	10.76
19	Fidelity Blue Chip Growth	2.03	-2.02	2.92	7.83
20	Fidelity New Millenium (clsd)	0.22	-3.67	4.20	37.25
21	Fidelity OTC	1.34	-1.21	5.54	19.69
22	Fidelity Spartan Market Index	1.58	-2.92	1.79	10.44
23	Fidelity Stock Selector	0.96	-0.46	5.07	13.34
24	Founders Growth	2.06	-3.16	2.34	9.46
25	Fremont Growth	1.53	-4.04	2.40	8.34
26	Invesco Blue Chip Gw	4.74	-1.26	4.28	9.44
27	Janus Fund	0.43	-3.53	3.20	16.79
28	Janus Mercury	4.19	-1.75	6.55	32.93
29	Janus Twenty	3.00	-2.87	2.44	16.74
30	Neuberger & Berman Manhattan	1.41	-3.47	4.79	2.43
31	Schwab 1000	1.05	-3.49	1.00	8.92
32	Schwab S&P 500	1.55	-2.92	1.75	10.34
33	SIT Large Cap Growth	2.39	-2.93	3.24	7.44
34	Vanguard Index Trust 500	1.60	-2.90	1.82	10.56
35	Vanguard/Primecap (clsd)	1.85	0.95	8.13	21.50
36	Vanguard US Growth	2.41	-2.80	2.94	5.49
37	Warburg Pincus Capital App	0.87	-1.53	4.04	11.38
38	Amer Cent Benhm Incm & Gwth	1.29	-2.33	2.98	8.56
39	Berger Growth and Income	0.19	-2.43	4.82	17.37
40	Janus Growth & Income	0.91	-3.30	2.74	14.95
41	Lexington Corporate Leaders	1.91	-1.79	0.57	12.74
42	Safeco Equity	1.28	-1.78	0.83	9.39
43	Vanguard Growth and Income	1.74	-2.76	3.08	12.59
44	Weiss Peck & Greer Gwth & Incm	1.96	-1.47	1.77	-1.13

------------------------- CLUSTER=2 --------------------------

Obs	fund	wk4	wk8	wk12	ytd
45	Babson Enterprise II	-2.02	-4.00	2.54	4.71
46	Fidlty Low-Prcd Stck	-3.17	-4.55	-0.51	2.84
47	Kaufmann Fund	-2.17	-4.76	2.27	-4.93
48	Price Small Cap Stock	-1.24	-2.63	3.62	3.22
49	Safeco Growth	-1.45	-5.00	-2.68	-10.48
50	Schwab Small Cap Index	-1.78	-3.44	2.15	5.42
51	SteinRoe Capital Opportunities	-1.71	-7.06	-2.35	-2.18
52	Warburg Pincus Emerging Grth	-1.08	-3.86	3.94	2.88
53	Baron Growth	-2.03	-4.57	-1.44	18.38
54	Bramwell Growth	0.56	-4.32	1.24	4.68
55	Columbia Growth	-0.47	-5.02	0.75	10.04

Output 6.10 *(continued)*

Obs	fund	wk4	wk8	wk12	ytd
56	Dreyfus Core Value	-0.03	-5.98	-2.88	12.25
57	Fidelity Contrafund (clsd)	-1.18	-3.97	0.28	8.85
58	Fidelity Dividend Growth	0.00	-4.33	0.32	8.39
59	Fidelity Magellan (clsd)	0.73	-3.84	1.50	11.07
60	Fidelity Value	-1.98	-6.65	-3.63	15.15
61	Montgomery Growth	-2.09	-6.28	-1.75	11.49
62	Price Growth Stock	0.65	-3.66	1.04	5.96
63	Price Midcap Growth Stock	-3.10	-4.21	0.99	8.10
64	Rydex Nova (sold 8/5/99)	2.13	-5.12	1.54	11.07
65	Scudder Large Co Value	-0.23	-5.40	-1.19	9.06
66	Strong Opportunity	-2.44	-3.35	0.53	16.53
67	Transamerica Equity Growth	-0.71	-6.30	-0.79	6.86
68	Tweedy Browne American Value	-0.90	-5.09	0.00	6.51
69	Amer Cent Eq Income	-0.90	-2.51	0.91	6.44
70	American Century Value	-1.35	-4.07	-2.08	9.79
71	Babson Value	-0.62	-5.31	-4.13	8.37
72	Columbia Common Stock	0.19	-4.06	1.09	10.45
73	Fidelity Equity-Income	-0.58	-4.62	-0.95	9.44
74	Fidelity Equity-Income II	0.10	-4.04	-0.47	4.92
75	Fidelity Fund	0.13	-4.20	-0.07	7.33
76	Fidelity Growth & Income (clsd)	0.57	-3.59	0.41	4.18
77	Founders Grwth & Incm	1.23	-3.40	0.96	0.82
78	Invesco Industrial Income	-0.43	-3.62	1.37	8.67
79	Janus Equity Income	-1.02	-3.82	0.51	14.80
80	Lexington Gwth & Incm	-0.58	-5.09	-0.84	2.15
81	Price Equity-Income	-1.12	-4.19	-1.81	8.91
82	Price Growth & Income	-1.27	-4.78	-1.48	8.22
83	Rob Stephs Mid Cap Opprtnties	-1.01	-4.95	1.39	19.00
84	Scudder Growth & Income	-1.64	-5.78	-3.13	5.95
85	Selected American Shares	0.29	-5.26	0.99	12.01
86	Vanguard Equity Income	-0.77	-4.11	-1.53	4.97
87	Vanguard Windsor II	-1.09	-6.78	-4.55	4.69

------------------------- CLUSTER=3 --------------------------

Obs	fund	wk4	wk8	wk12	ytd
88	Baron Asset	-3.08	-8.57	-6.98	7.84
89	Oakmark Small Cap	-5.90	-7.30	-4.14	-2.78
90	Scudder Development	-2.23	-7.75	-3.17	4.65
91	Longleaf Partners	-1.85	-9.25	-8.20	10.66
92	Neuberger & Berman Partners	-2.32	-7.83	-4.97	5.65
93	Oakmark Fund	-3.21	-7.00	-5.61	3.43
94	Safeco Income	-2.86	-6.79	-4.75	-4.60

------------------------- CLUSTER=4 --------------------------

Obs	fund	wk4	wk8	wk12	ytd
95	Profund Ultra OTC (sold 2/8/99)	10.90	2.02	25.21	48.78
96	Rob Stephens Em Growth	5.42	1.19	14.25	55.12
97	Rydex OTC	6.01	2.36	14.05	30.95
98	Fidelity Aggrssv Gw	6.92	3.43	14.28	42.00
99	Fidelity Growth Company	6.00	5.47	13.82	28.13
100	Janus Enterprise	5.46	3.00	9.44	35.34

Output 6.10
(continued)

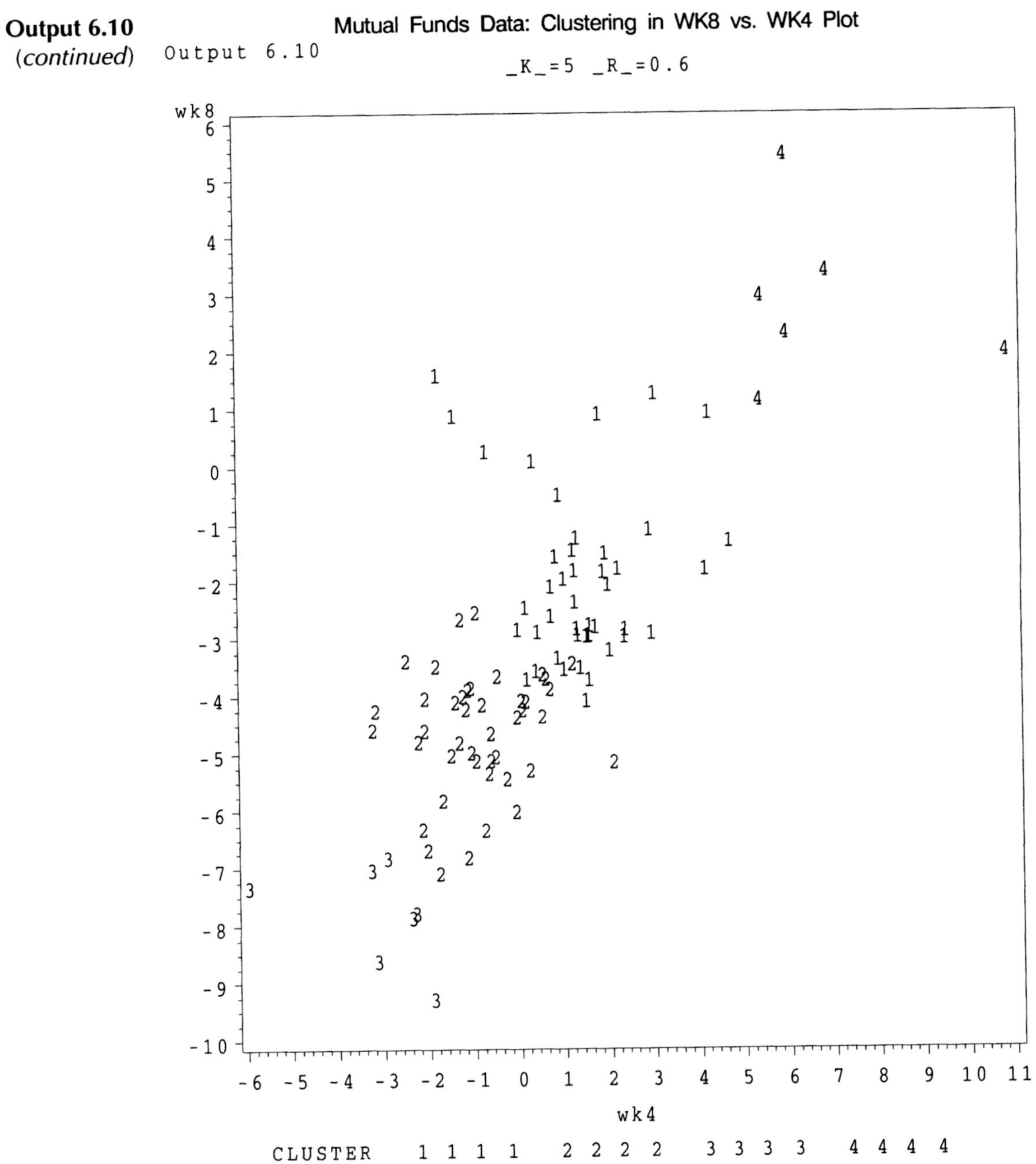

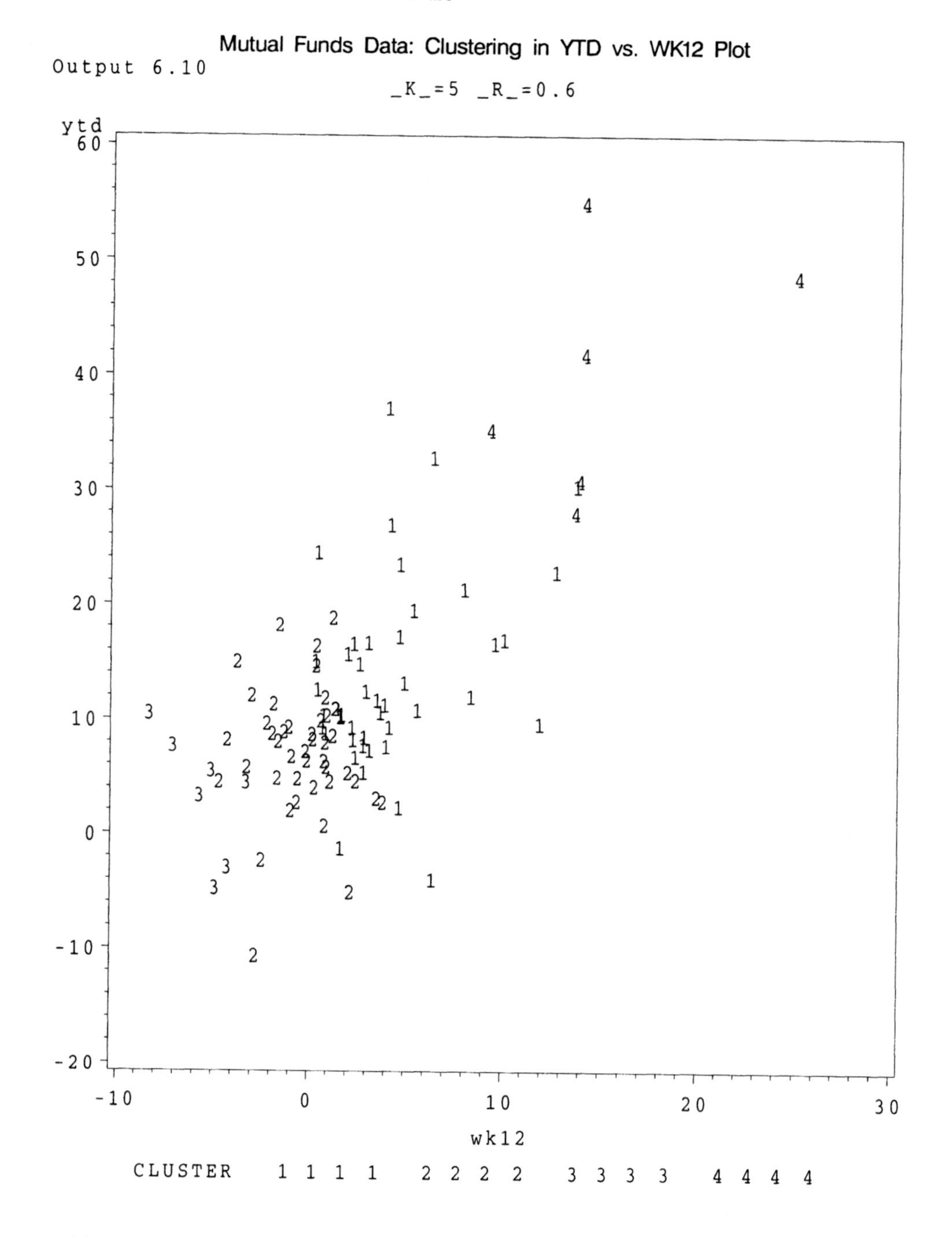

The output is shown in Output 6.10. Four clusters are obtained. Two of them are relatively large, and other two are quite small. Also, the first two clusters have the relatively larger values of the maximum estimated density. The boundary frequencies are not zero for any of the clusters. Their numbers indicate the number of observations in the cluster with a neighbor in some other cluster. Thus, for example, there is a total of 11 funds in the first cluster, each with at least one neighbor in one of the other three clusters. The large boundary frequencies suggest that the clusters are not *isolated.* For this reason, the estimated saddle densities are all positive (for isolated clusters, they would appear as the missing values).

The four clusters, as suggested by the method, are listed in Output 6.10 in their entirety. These were obtained by using the OUT = OUTPUT option. It is interesting to observe some special features of these four clusters and distinguish them from each other.

We start with the last cluster. Cluster 4 consists of six funds all with impressive performance throughout the period under consideration. Not only do the WK12 and YTD returns exhibit the distinguished records, but their four- and eight-week returns are also positive, in spite of the fact that during July and August of 1999, the domestic market as indicated by various leading indexes, was generally on the decline.

Cluster 3 represents funds with negative returns during the past 4, 8, and 12 weeks. These declines were generally higher than those for Clusters 1 and 2, which we discuss next.

In Cluster 2, the declines during the past eight weeks were moderate. The four-week changes are also negative or near zero. Cluster 1 is clearly quite different from Clusters 3 and 4 and can be distinguished from Cluster 2 in that although the eight-week returns were negative, they are not as bad as those found in Cluster 2. At the same time, for most of the funds in Cluster 1, a positive change is seen for other periods.

From the above description and from the WK8 vs. WK4 and YTD vs. WK12 scatter plots, it can be seen that the four clusters can be generally ordered in their descriptions, with Cluster 4 being the set of best performing funds, Cluster 1 as the next best, followed by Cluster 2. The worst performers are grouped as Cluster 3.

6.9 Clustering with Binary Data

So far we have assumed that all the measurements on the objects are quantitative. Hence distances could be calculated, and they were used as the measures of dissimilarities. When the variables are qualitative, distances as defined earlier have no meaning. Thus, we are faced with the problem of defining a measure of dissimilarity between the objects. Although the need for this definition exists for any kind of qualitative variable, here we will consider the case of binary variables only. Discussion for the more general case may be found in Sneath and Sokal (1973) and Gordon (1981). The situation in which some measurements are quantitative and other qualitative is considered in Gower (1971).

For binary data, with p binary variables, each taking value 0 and 1, we can compute, for objects r and s, the number of variables that take value 1 for both objects, the number of variables that take value 0 for both objects, and the respective numbers when one variable takes 0 values and the other 1. This information is presented in Table 6.3.

TABLE 6.3 Matches for the rth and s^{th} Objects

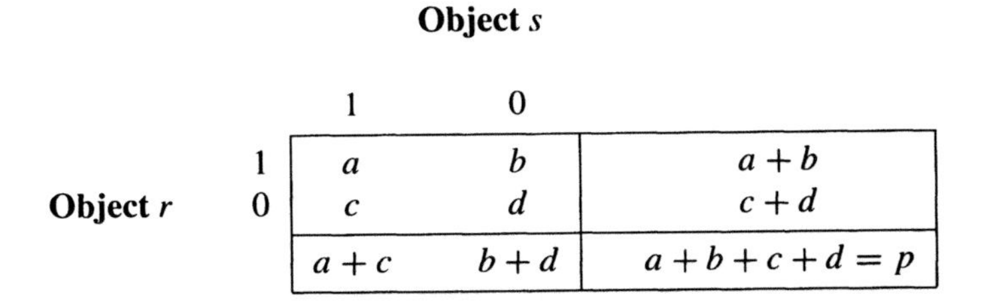

		Object s 1	Object s 0	
Object r	1	a	b	$a+b$
	0	c	d	$c+d$
		$a+c$	$b+d$	$a+b+c+d=p$

Based on this table, and with numbers a, d representing agreements and b, c representing disagreements, a number of similarity coefficients s_{rs} have been defined. Some of these are given in Table 6.4. Interpretations for most of these are briefly summarized in Anderberg (1973, p. 88). The dissimilarities can be defined as $d_{rs} = 1 - s_{rs}$. It may be pointed out that some of these may fail to satisfy some of the natural requirements a dissimilarity index is expected to satisfy (as mentioned in Section 6.1).

One particular index, $s_{rs} = \frac{a}{a+b+c}$ known as the Jaccard coefficient, is quite popular and represents the ratio of the number of variables that have value 1 for both r^{th} and s^{th} objects to the number of variables for which 1 is observed for at least one of these two objects. Clearly, in case of nonzero a and complete agreement, the Jaccard coefficient is

TABLE 6.4 Similarity Indexes for Binary Data

Simple matching coefficient	$s_{rs} = \frac{a+d}{a+b+c+d}$
Double matching coefficients	$s_{rs} = \frac{2(a+d)}{2(a+d)+(b+c)}$
Rogers-Tanimoto coefficient	$s_{rs} = \frac{a+d}{a+2b+2c+d}$
Jaccard coefficient	$s_{rs} = \frac{a}{a+b+c}$
Double unmatching Jaccard type	$s_{rs} = \frac{a}{a+2(b+c)}$
Jaccard type matches to mismatches	$s_{rs} = \frac{a}{b+c}$
Czekanowski-Sørensen-Dice coefficient	$s_{rs} = \frac{2a}{2a+b+c}$
Kulezynski coefficient	$s_{rs} = \frac{a}{a+b}$
Russell-Rao coefficient	$s_{rs} = \frac{a}{a+b+c+d}$
Hamman coefficient	$s_{rs} = \frac{a-(b+c)+d}{a+b+c+d}$
Ochiai coefficient	$s_{rs} = \frac{a}{[(a+b)(a+c)]^{\frac{1}{2}}}$
Yule coefficient	$s_{rs} = \frac{ad-bc}{ad+bc}$
Mountford coefficient	$s_{rs} = \frac{2a}{a(b+c)+2bc}$
Mozley-Margalef coefficient	$s_{rs} = \frac{a(a+b+c+d)}{(a+b)(a+c)}$
Phi coefficient	$s_{rs} = \frac{ad-bc}{[(a+b)(a+c)(b+d)(c+d)]^{\frac{1}{2}}}$

1, and it is 0 in case of absolutely no matches in (1, 1) category. When $a = b = c = 0$ but d is nonzero, the Jaccard coefficient is defined as zero, even though there is a complete match. It is so, because (0, 0) matches are considered totally irrelevant by this index. The dissimilarities, using the Jaccard coefficients, will be given by $d_{rs} = 1 - s_{rs} = \frac{b+c}{a+b+c}$. The dissimilarity matrix so constructed can be used for clustering the objects. It is worth noting that this dissimilarity index violates requirement (ii) as indicated in Section 6.1.

EXAMPLE 8 ***Binary Variables, Clustering the Cancer Hospitals*** The *U.S. News and World Report* lists the 50 best cancer hospitals in the nation. The rankings were obtained by using a particular index that was computed using various measurements. These were, reputational score (REPUT), which is the percentage of cancer specialists surveyed who named the hospital; mortality rate (MORTALIT), the ratio of actual to expected cancer deaths in this specialty; COTH member (COTH), indicating the membership in Council of Teaching Hospitals; technology score (TECHSCOR), indicating the availability of key technologies in this specialty; discharges (DISCHARG), which is the number of patients discharged in this specialty; and RNs to beds (RN_BED), which is the hospital-wide ratio of the number of registered nurses to the number of beds. We will use all of these variables except COTH (since its value is YES for all 50 hospitals) to divide these hospitals into several clusters.

One important feature of these data is that all variables measure very different things and are on different scales. Further, many of these variables may have larger values because of the sheer size of the corresponding hospitals (for example, RN_BED is a hospital-wide measure not necessarily limited to the specialty of cancer, and its values will be large for larger hospitals) or because of their geographic locations (for example, hospitals in bigger cities vs. smaller cities). Thus, the use of Euclidean distance either on original or even on standardized variables may not be appropriate. An alternative approach is to create a dichotomy for each variable and obtain the corresponding set of binary variables each taking values 0 and 1. While readily admitting to being somewhat arbitrary, we define the following five binary variables, in each case, the values 1 representing the more desirable scenario.

$$X1 = \begin{cases} 1 & \text{if REPUT} > 30 \\ 0 & \text{otherwise} \end{cases},$$

$$X2 = \begin{cases} 1 & \text{if MORTALIT} < 0.75 \\ 0 & \text{otherwise} \end{cases},$$

$$X3 = \begin{cases} 1 & \text{if TECHSCOR} > 6 \\ 0 & \text{otherwise} \end{cases},$$

$$X4 = \begin{cases} 1 & \text{if DISCHARG} > 100 \\ 0 & \text{otherwise} \end{cases}$$

and

$$X5 = \begin{cases} 1 & \text{if RN\_BED} > 1.5 \\ 0 & \text{otherwise} \end{cases}.$$

The Jaccard coefficients and the subsequent 50 by 50 dissimilarity matrix for all hospitals are computed using the code given in the first part of Program 6.11. Except for certain appropriate changes, the code is adapted from the section about the CLUSTER procedure in the *SAS/STAT User's Guide, Version 6, Fourth Edition, Volume 1*, pp. 537-538. Selected output is shown after Program 6.11.

```
/*Program 6.11 */

options ls=64 ps=45 nodate nonumber;

title1 'Output 6.11';

data hospital;
infile 'hospital.dat' firstobs = 7;
input Rank Hospital $ 6-31
index Reput Mortalit COTH $ techscor discharg  rn_bed;

data bin;
set hospital;
if reput < = 30 then x1 = 0;
if reput > 30 then x1 =1;
if mortalit <  .75 then x2 = 1;
if mortalit > = .75 then x2 = 0;
if  techscor < = 6 then x3 = 0;
if techscor > 6 then x3 = 1;
if discharg < = 1000 then x4 = 0;
if discharg > 1000 then x4 = 1;
if rn_bed < = 1.5 then x5 = 0;
if rn_bed > 1.5 then x5 = 1;
drop reput mortalit techscor discharg rn_bed index coth rank;
```

```
data jacc(type = distance);
array d(*) d1-d50;
retain d1-d50 .;

do row = 1 to 50;
set bin point = row;
array scores(*) x1--x5;
array save(*) save1-save5;

do g = 1 to 5;
save(g) =scores(g);
end;

do col = 1 to row;
set bin(drop = hospital) point = col;
num = 0;
den = 0;
do g = 1 to 5;
num = num+(scores(g) & save(g));
den = den+(scores(g) | save(g));
end;
if den then d(col) = 1 - num/den;
else d(col) =1;
end;
output;
end;
stop;
keep hospital d1-d50;
run;

proc print data = jacc;



proc cluster data = jacc
noeigen method = flexible beta = -0.5   nonorm out =tree1;
id hospital;
var d1-d50;
title2 'Clustering of Hospitals: Flexible-Beta (= -0.5) Approach';
proc tree data = tree1 out = output noprint nclusters = 4 ;
id hospital;

proc sort data = output;
by hospital;

proc sort data = bin;
by hospital;

proc sort data = hospital;
by hospital;

data all; merge hospital bin output;
by hospital;
run;

proc sort data = all;
by cluster;
run;
```

```
proc print data = all;
id hospital;
var x1-x5 reput mortalit techscor discharg rn_bed;
by cluster ;
title2 "4-Clusters Solution: Hospital Data" ;
run;
```

Output 6.11

Output 6.11
4-Clusters Solution: Hospital Data

------------------------- CLUSTER=1 -------------------------

Hospital	x1	x2	x3	x4	x5	Reput	Mortalit	techscor	discharg	rn_bed
Barnes-Jewish	0	0	1	1	1	3.3	0.77	6.5	1872	1.52
Baylor U.	0	0	0	1	1	1.9	0.82	6.0	1392	1.53
Cleveland Clinic	0	0	1	1	1	2.4	0.86	7.0	1686	1.69
Duke U. Medical	0	0	1	1	1	10.7	0.82	7.0	2523	1.63
M. D. Anderson	1	0	0	1	1	67.5	0.78	6.0	3683	1.87
Roswell Park Cancer	0	0	0	1	1	8.4	0.81	5.5	1481	2.34
Sloan-Ketterning	1	0	0	1	1	73.0	1.05	6.0	3544	1.85
U. Pennsylvania	0	0	0	1	1	5.5	0.96	6.0	1431	1.98

------------------------- CLUSTER=2 -------------------------

Hospital	x1	x2	x3	x4	x5	Reput	Mortalit	techscor	discharg	rn_bed
Arthur James Cancer	0	1	0	1	0	0.4	0.50	5.0	1611	1.10
Clarian Health	0	0	1	1	0	4.2	0.78	7.0	1139	1.48
Harper Hos. Detroit	0	1	0	1	0	0.4	0.68	5.5	1348	1.10
Jefferson U.	0	0	0	1	0	0.0	0.81	6.0	1310	1.47
Johns Hopkins	1	1	1	1	0	33.4	0.70	7.0	1278	1.33
Massachusetts Gen.	0	0	1	1	0	6.3	0.94	7.0	1547	1.35
Mayo Clinic	0	1	1	1	0	26.0	0.57	7.0	2589	1.10
NC Baptist Hospital	0	0	0	1	0	0.5	0.83	6.0	1364	1.34
NY Presbyterian	0	0	1	1	0	3.9	1.01	7.0	1923	1.13
U. Hos. Cleveland	0	0	1	1	0	0.0	0.81	7.0	1414	1.26
U. Iowa Hos.	0	0	1	1	0	0.9	0.75	7.0	1138	1.12
U. Michigan Medical	0	1	1	1	0	1.3	0.56	7.0	1221	1.46
U. Pittsburgh Medical	0	1	1	1	0	3.0	0.69	7.0	1730	1.34
U. Virginia Health	0	1	0	1	1	0.0	0.54	6.0	1026	1.91
University of Chicago	0	1	1	1	0	6.6	0.68	7.0	1116	1.26
Yale Hospital	0	0	1	1	0	1.0	0.77	6.5	1296	0.84

------------------------- CLUSTER=3 -------------------------

Hospital	x1	x2	x3	x4	x5	Reput	Mortalit	techscor	discharg	rn_bed
Allegheny U-Hos.	0	1	0	0	1	0.3	0.64	5.0	885	1.89
Baltimore Medical	0	1	0	0	0	0.0	0.54	4.0	903	1.47
Carolina Hospitals	0	1	0	0	0	0.0	0.69	6.0	751	1.49
Emory University	0	1	0	0	0	0.0	0.61	5.5	762	0.98
Evanston Hospital	0	1	0	0	0	0.0	0.61	5.0	914	0.81
Fox Chase Cancer	0	1	0	0	1	5.7	0.54	4.0	872	1.88
H. Lee Moffitt Cancer	0	1	0	0	0	0.8	0.61	3.0	980	1.43
Kettering, Ohio	0	1	0	0	0	0.0	0.57	6.0	512	0.94
Loma Linda	0	1	0	0	1	0.0	0.64	4.8	768	1.79
Loyola U.	0	1	0	0	0	0.5	0.63	6.0	683	1.39
Lutheran Gen.	0	1	0	0	0	0.4	0.60	5.0	935	0.92
Meritcare Health Fargo	0	1	0	0	0	0.4	0.64	6.0	627	0.41
SW Hos. Temple TX	0	1	0	0	0	0.0	0.60	3.5	716	0.92
U. Hos. Columbia MO	0	1	0	0	0	0.0	0.66	6.0	791	1.26
U. Hos. Portland OR	0	1	0	0	0	0.0	0.63	6.0	569	0.90
U. Kentucky	0	1	0	0	1	0.0	0.63	6.0	698	1.85
U. Wash. Medical	0	1	0	0	1	9.1	0.67	6.0	606	2.03

------------------------- CLUSTER=4 -------------------------

Hospital	x1	x2	x3	x4	x5	Reput	Mortalit	techscor	discharg	rn_bed
Brigham and Women's	0	0	0	0	0	2.9	0.80	6.0	939	1.30
Luke's Medical	0	1	1	0	0	0.0	0.67	7.0	820	1.01
Nebraska Health	0	1	1	0	0	0.0	0.74	7.0	648	1.28
Stanford U. Hos.	0	0	0	0	0	9.5	0.83	5.3	869	1.10
U. Florida	0	1	1	0	0	0.5	0.46	7.0	561	0.60
U. Wis. Madison	0	0	1	0	0	1.2	0.77	7.0	759	1.19
UC-Davis	0	0	1	0	1	0.0	0.84	7.0	543	1.89
UCLA Medical Center	0	0	0	0	0	8.9	0.80	6.0	825	0.91
Vanderbilt U.	0	1	1	0	1	1.4	0.71	7.0	779	1.88

The dissimilarity data are then analyzed using the flexible-beta method with a choice of $\beta = -0.5$. The choice of the method and the values of β was made in view of the fact that the data contain certain hospitals that clearly stand out and hence may result in clusters with a single hospital in each of them. As we have already mentioned earlier, the negative values of β help us avoid such clustering. The SAS code for this cluster analysis is given in the later part of Program 6.11. The output showing the four clusters is presented as Output 6.11.

There are a few striking features in the clustering that resulted. Cluster 1 represents those hospitals that have superior records with respect to discharges and ratio of number of nurses and the number of beds. Hospitals in Cluster 2 generally have the desirable feature in terms of number of discharges but they have a low RN_RATIO. This feature distinguishes them from Cluster 1. Hospitals in the third cluster have identical values for all binary variables except for X5. They all have low mortality rates. Cluster 4 consists of other hospitals with a low number of discharges.

6.10 Concluding Remarks

Among several clustering methods described here, which clustering methods are the best and under what circumstances? This is a question which has been extensively discussed in cluster analysts' circles, and there are a number of theoretical and simulation studies done by various authors. The preference should largely depend on the way observations are clustered in the sense of clusters' shapes, sizes, and presence of outliers. The choice of distance or dissimilarity also plays an important role. Sharma (1996) provides a detailed summary in a tabular form of various simulation studies done in this respect, and Seber (1984) provides a nice discussion on this topic. We limit ourselves to the following comments about the preferences and shortcomings of various methods. During our previous discussions, many of these have already been pointed out. We hope that these will provide some practical guidelines to select the appropriate approach for data analysis.

- As shown by Jardine and Sibson (1968), from the theoretical point of view, the single linkage approach is the best criterion in that it satisfies a number of desirable mathematical properties, which most of the other methods do not. However, the constraints imposed by these mathematical requirements can be viewed as too restrictive for practical applications. See Williams et. al. (1971), who support this later expression. Williams and Lance (1977) strongly advise against the use of single linkage method. Many other simulation studies also support this conviction.
- Most of the hierarchical approaches are prone to chaining effects. By this we mean that new observations tend to join the previous cluster in a chain-like fashion. Unless the original clusters themselves are elongated enough to look like a chain, this is clearly undesirable. The problem is especially acute for the single linkage method. At the same time, single linkage is perhaps the only method that can successfully identify the clusters with wavy shapes (such as the U- or S-shaped clusters).
- No one hierarchical approach is consistently better than others. However, any errors in distances affect the performance of the single linkage approach most severely. The effect on other hierarchical methods is not as substantial. At the same time, the single linkage method is robust to the presence of outliers in data.
- When there are outliers, the flexible-beta method with negative values for β can be very useful in avoiding the clusters with a single outlying observations. However, the value of β should be chosen judiciously. See Seber (1984) for further remarks.
- Ward's method tends to give clusters of nearly equal sizes. These clusters are relatively compact and spherical in shape. However, the method depends on ANOVA-type partitioning of the sum of squares, with implicit assumption of the equal cluster variances and covariances. Therefore, when the clusters are highly irregular in shape or highly

different from each other in elongation and/or compactness, the method may perform poorly. The same situation arises if the data had a few outliers. However, a number of comparative studies have shown that Ward's method, for most of the standard problems, outperforms other hierarchical methods. The Average method also performs favorably in most of these situations. Blashfield (1976) points out that when clusters are spherical, the single linkage method will also perform well.

- The choice of distance or dissimilarity wherever appropriate can alter the clustering results for the hierarchical as well as nonhierarchical clustering methods. This includes the case of scaling as well, because scaling essentially amounts to changing the distance function. The density-based methods are relatively less affected by such scalings.
- For approaches requiring initial seeds, the choice of seed is very important. For the k-means method, performance is far superior when the starting partitions are defined by initial seeds that were close to the true solution. In fact, in such a case, the k-means approach performs better than any of the hierarchical methods. Random seeds are not always very effective. It may be mentioned in passing that the SAS data mining software, Enterprise Miner, uses the k-means clustering as a tool during the data mining process.
- Density estimation-based methods are clearly inappropriate for small data sets and should not be used in those cases. For the most part, their use requires specifying smoothing parameters, which itself can be a tricky problem. At the same time, a definite advantage of these methods is that they do not require specifying the number of clusters to be formed.

Another approach to cluster analysis is to view a clustering problem as a discrete, nonlinear programming problem involving the assignments. This point of view is often promoted by computer scientists and others working in mathematical taxonomy. A nice brief introduction to this approach can be found in Arthanari and Dodge (1993).

We conclude this chapter by noting that as with discriminant analysis, there is renewed research interest in cluster analysis. Much of the recent work emphasizes the data mining applications, market segmentation, and medical and biological clusterings. New clustering approaches and refinements of old ones are likely to emerge in years to come.

Chapter

7

Correspondence Analysis

7.1 Introduction

In practice, we routinely come across data in the form of a contingency table in which the cells, formed by cross-classification of two or more categorical variables, contain the frequency counts of the individuals belonging to those cells. These data may be in the form of a two-way contingency table, which is formed by cross-classifying only two variables, or a multi-way contingency table, which is formed by cross-classifying more than two variables. The two-way table can also be viewed as a matrix in which the rows correspond to categories of one variable, columns correspond to the categories of the other variable, and the elements of the matrix are the cell frequencies. Data given in the form of a contingency table are usually analyzed using one of many categorical data analysis methods available in the literature. Pearson's chi-square test for association, a log-linear model, or a logit model are popular among these methods. These and many other statistical methods for analyzing categorical data, in general, are discussed in Agresti (1990, 1996). For a detailed treatment of this subject using SAS, see Stokes, Davis, and Koch (1995). If the cell frequencies in the contingency table are very small or zero (i.e., the contingency table is sparse), generally these statistical methods are not appropriate for the analysis. In that case, it would be convenient for a practitioner to have a method that will allow him to combine two or more categories of the variables, thus increasing the frequencies of the remaining cells. Correspondence analysis (CA) is such a method. However this is not the only objective of correspondence analysis.

In general, *correspondence analysis* is a graphical multivariate technique for performing an exploratory data analysis of a contingency table. It can be used on any two-way table, whether or not it is sparse. It projects the rows and columns of a data matrix as points into a graph in an Euclidean space. The graph is then used to gain an understanding of the data and to extract information from it. Correspondence analysis is routinely used to determine those categories of a variable that are similar, so that they may possibly be combined to form a single category. This in turn will reduce the number of categories in the variables, and at the same time it will increase the frequency counts in the cells formed by cross-classification of these new categories. Further, the graphs in correspondence analysis can be used to determine, to some extent, the possible association between the two sets of variables.

Correspondence analysis has been in use for many years in France, mainly because of the work of Benzecri (1992) and his team, and in Japan, because of the work of Hayashi (1950) and his associates. Historically, the ideas similar to the correspondence analysis are traced back to Richardson and Kuder (1933), Hirschfeld (later H. O. Hartley) (1935), Horst (1935), Fisher (1940), Guttman (1941), and Burt (1950). Some of the names that were used for these ideas as they developed include optimal scaling, reciprocal averaging, optimal scoring, appropriate scoring, quantification method, homogeneity analysis, dual scaling, and scalogram analysis. Some related topics are canonical coordinates (Rao, 1948; Rao, 1995), psychometric scaling (Nishisato, 1980), ecological ordination (Gauch, 1982), biplots (Gabriel, 1971; Gower, 1993), and canonical correspondence analysis (Ter Braak, 1986). Correspondence analysis became a popular tool among practitioners in the rest of the world after the release of Greenacre's (1984) book. The other useful books in this area are Benzecri (1992), Gifi (1990), Gower and Hand (1996), Greenacre (1993a), Lembart, Morineau, and Warwick (1984), and Nishisato (1980). Applications of correspondence analysis in numerous fields can be found in Greenacre (1984, 1993a). For its applications specifically in medical research see Greenacre (1992).

As noted above, the main problem of interest in correspondence analysis is that of graphically representing the rows and columns of a contingency table as points in a lower dimensional Euclidean space such that the affinities of the rows (or columns) in the higher dimensional space are preserved as much as possible in this lower dimensional Euclidean space. This problem can be viewed as a special case of a general problem of graphically representing n data vectors of p dimension, called *population profiles* (Rao, 1995), in a k dimensional ($k < p$) Euclidean space such that the relative positions in terms of the distance between the two data vectors in the original space are preserved, as much as possible, in the subspace. A solution to such a problem is given by a generalization of Eckart and Young's (1936) singular value decomposition of a matrix. See Section 1.5.24. Correspondence analysis can also be viewed as a form of principal component analysis (Chapter 2) and canonical correlation analysis (Chapter 3).

7.2 Correspondence Analysis

The correspondence analysis of a two-way contingency table is sometimes also called *simple correspondence analysis,* as opposed to a *multiple correspondence analysis*, which is correspondence analysis applied in the context of a multi-way contingency table. Suppose X and Y are two categorical variables with a and b categories respectively and suppose the observations made on these variables result in an a by b contingency table (or a matrix of order a by b) with a frequency $n_{ij} \geq 0$ in the $(i, j)^{th}$ cell. The matrix of relative frequencies, that is,

$$\mathbf{P}_{a \times b} = (p_{ij}) = (\frac{n_{ij}}{n}),$$

where $n = \sum_i \sum_j n_{ij}$, is called the *correspondence matrix*. The correspondence matrix is fundamental entity for all our subsequent analysis and discussions.

The correspondence matrix along with the row and column marginal totals can be displayed as

$$\begin{array}{cccc|c} p_{11} & p_{12} & \cdots & p_{1b} & p_{1\cdot} \\ p_{21} & p_{22} & \cdots & p_{2b} & p_{2\cdot} \\ \vdots & \vdots & \cdots & \vdots & \vdots \\ p_{a1} & p_{a2} & \cdots & p_{ab} & p_{a\cdot} \\ \hline p_{\cdot 1} & p_{\cdot 2} & \cdots & p_{\cdot b} & 1. \end{array}$$

For example, the following 4 by 3 matrix of relative frequencies whose elements all add up to one is a correspondence matrix. In the display, the column after the vertical line is the column of row sums and the row after the horizontal line is the row of column sums.

$$\begin{array}{ccc|c} .05 & .05 & .05 & .15 \\ .10 & .30 & .10 & .50 \\ .12 & .00 & .12 & .24 \\ .10 & .01 & .00 & .11 \\ \hline .37 & .36 & .27 & 1. \end{array} \tag{7.1}$$

Let the vector of row sums of $\mathbf{P}$ be

$$\mathbf{r} = \mathbf{P1} = (p_{1\cdot}, \ldots, p_{a\cdot})' = (n_{1\cdot}/n, \ldots, n_{a\cdot}/n)'$$

and the vector of column sums of $\mathbf{P}$ be

$$\mathbf{c} = \mathbf{P}'\mathbf{1} = (p_{\cdot 1}, \ldots, p_{\cdot b})' = (n_{\cdot 1}/n, \ldots, n_{\cdot b}/n)',$$

where $\mathbf{1} = (1, \ldots, 1)'$ is an array of an appropriate size (a by 1 for $\mathbf{r}$ and b by 1 for $\mathbf{c}$) with all entries as 1 and $n_{i\cdot} = \sum_{j=1}^{b} n_{ij},\ i = 1, \ldots, a,\ n_{\cdot j} = \sum_{i=1}^{a} n_{ij},\ j = 1, \ldots, b.$

For our numerical matrix in Display 7.1, the vector of row sums

$$\mathbf{r} = (0.15, 0.50, 0.24, 0.11)'$$

and that of column sums is

$$\mathbf{c} = (0.37, 0.36, 0.27)'.$$

Let

$$\mathbf{D}_r = diag(\mathbf{r}) = \begin{bmatrix} p_{1\cdot} & 0 & \cdots & 0 \\ 0 & p_{2\cdot} & \cdots & 0 \\ \vdots & \vdots & \ddots & \vdots \\ 0 & 0 & \cdots & p_{a\cdot} \end{bmatrix}$$

and

$$\mathbf{D}_c = diag(\mathbf{c}) = \begin{bmatrix} p_{\cdot 1} & 0 & \cdots & 0 \\ 0 & p_{\cdot 2} & \cdots & 0 \\ \vdots & \vdots & \ddots & \vdots \\ 0 & 0 & \cdots & p_{\cdot b} \end{bmatrix}$$

be the diagonal matrices of order a by a and b by b, respectively. Then the a rows of a by b matrix $\mathbf{R}$ defined as

$$\mathbf{R} = \mathbf{D}_r^{-1}\mathbf{P} = \begin{bmatrix} \frac{p_{11}}{p_{1\cdot}} & \frac{p_{12}}{p_{1\cdot}} & \cdots & \frac{p_{1b}}{p_{1\cdot}} \\ \frac{p_{21}}{p_{2\cdot}} & \frac{p_{22}}{p_{2\cdot}} & \cdots & \frac{p_{2b}}{p_{2\cdot}} \\ \vdots & \vdots & \cdots & \vdots \\ \frac{p_{a1}}{p_{a\cdot}} & \frac{p_{a2}}{p_{a\cdot}} & \cdots & \frac{p_{ab}}{p_{a\cdot}} \end{bmatrix} \tag{7.2}$$

are called *row profiles* in the b-dimensional space. Note that the sum of the elements of any row profile is one. We will denote the ith row profile by $\mathbf{r}_i'$, that is,

$$\mathbf{r}_i = (\frac{p_{i1}}{p_{i\cdot}}, \frac{p_{i2}}{p_{i\cdot}}, \ldots, \frac{p_{ib}}{p_{i\cdot}})'.$$

For example, the second row profile for our numerical example is calculated as $\mathbf{r}_2 = (0.10/0.50, 0.30/0.50, 0.10/0.50)' = (0.20, 0.60, 0.20)'$. The matrix $\mathbf{R}$ consisting of four row profiles for the matrix in Display 7.1 is

$$\mathbf{R} = \begin{bmatrix} .33 & .33 & .33 \\ .20 & .60 & .20 \\ .50 & .00 & .50 \\ .91 & .09 & .00 \end{bmatrix}. \tag{7.3}$$

Note that (except for the round-off error) the sum of the elements of each row profile is, as expected, one.

Similarly the b rows of b by a matrix $\mathbf{C}$ defined as

$$\mathbf{C} = \mathbf{D}_c^{-1}\mathbf{P}' = \begin{bmatrix} \frac{p_{11}}{p_{\cdot 1}} & \frac{p_{21}}{p_{\cdot 1}} & \cdots & \frac{p_{a1}}{p_{\cdot 1}} \\ \frac{p_{12}}{p_{\cdot 2}} & \frac{p_{22}}{p_{\cdot 2}} & \cdots & \frac{p_{a2}}{p_{\cdot 2}} \\ \vdots & \vdots & \cdots & \vdots \\ \frac{p_{1b}}{p_{\cdot b}} & \frac{p_{2b}}{p_{\cdot b}} & \cdots & \frac{p_{ab}}{p_{\cdot b}} \end{bmatrix}$$

are called *column profiles* in the a-dimensional space. Note that the sum of the elements of any column profile is one as well. If we denote the jth column profile by $\mathbf{c}'_j$, then

$$\mathbf{c}_j = (\frac{p_{1j}}{p_{\cdot j}}, \frac{p_{2j}}{p_{\cdot j}}, \ldots, \frac{p_{aj}}{p_{\cdot j}})'.$$

For example, the second column profile for our numerical example is calculated as

$$\mathbf{c}_2 = (0.05/0.36, 0.30/0.36, 0.00/0.36, 0.01/0.36)' = (0.14, 0.83, 0.00, 0.03)'.$$

The matrix $\mathbf{C}$, consisting of three column profiles for the matrix in Display 7.1, is

$$\mathbf{C} = \begin{bmatrix} .14 & .27 & .32 & .27 \\ .14 & .83 & .00 & .03 \\ .19 & .37 & .44 & .00 \end{bmatrix}.$$

Note, as was the case with row profiles, the sum of the elements of each column profile, as expected, is one, except for the round-off error.

The $b \times 1$ vector $\mathbf{c} = (p_{\cdot 1}, \ldots, p_{\cdot b})'$ is called the *average row profile* or *row centroid* or the vector of *row masses* and $a \times 1$ vector $\mathbf{r} = (p_{1\cdot}, \ldots, p_{a\cdot})'$ is called the *average column profile* or *column centroid* or the vector of *column masses*. Thus, the average row and column profiles are appropriately weighted averages, that is, they are weighted averages of the row and column profiles. More specifically, the average row profile is $\mathbf{c} = \sum_{i=1}^{a} p_{i\cdot}\mathbf{r}'_i$, where $\mathbf{r}'_i$ is the ith row profile. Similarly, the average column profile is $\mathbf{r} = \sum_{i=1}^{b} p_{\cdot j}\mathbf{c}'_j$. For our numerical example, the average row and column profiles respectively are $\mathbf{c} = (0.37, 0.36, 0.27)'$ and $\mathbf{r} = (0.15, 0.50, 0.24, 0.11)'$. As a linear combination of $\mathbf{r}'_i$, the average row profile $\mathbf{c}$ is

$$.15\begin{bmatrix} .33 \\ .33 \\ .33 \end{bmatrix} + .50\begin{bmatrix} .20 \\ .60 \\ .20 \end{bmatrix} + .24\begin{bmatrix} .50 \\ .00 \\ .50 \end{bmatrix} + .11\begin{bmatrix} .91 \\ .09 \\ .00 \end{bmatrix} = \begin{bmatrix} .37 \\ .36 \\ .27 \end{bmatrix}. \tag{7.4}$$

Consider the problem of displaying the row profiles in a lower dimensional Euclidean space. The problem of displaying the column profiles can be handled in a similar way. In our numerical example, in Equation 7.3 there are four rows, each of dimension three, representing the four row profiles. Because the elements of each row add up to one, each row of $\mathbf{R}$ can be represented as a point in a two-dimensional regular simplex, that is, the triangle in the three-dimensional space formed by the vertices $(1, 0, 0)$, $(0, 1, 0)$, and $(0, 0, 1)$. As described in Greenacre and Hastie (1987) each of these vertices can be thought of as row profiles with all the mass concentrated at one position. For example, the vertex (0, 1, 0) can be thought of as a row profile for which all the mass (all the frequency in that row) is at the second column. For this reason, this vertex in the plot is considered as a representative of the second column. Hence in our discussion, the projections of vertices are sometimes called as column points.

The natural question before we present the contingency table data as a graphical display is this: what is the appropriate distance (metric) to be used? Under the methodology discussed here, the appropriate distance turns out to be the chi-square distance (to be defined). To elaborate, consider Pearson's chi-square test statistic for testing the independence between X and Y,

$$\chi^2 = \sum_{i=1}^{a}\sum_{j=1}^{b} \frac{(n_{ij} - \frac{n_{i\cdot}n_{\cdot j}}{n})^2}{\frac{n_{i\cdot}n_{\cdot j}}{n}} = n\sum_{i=1}^{a}\sum_{j=1}^{b} \frac{(p_{ij} - p_{i\cdot}p_{\cdot j})^2}{p_{i\cdot}p_{\cdot j}} = n\,tr(\mathbf{E}) = n\sum_{i=1}^{m}\lambda_i^2,$$

where

$$\mathbf{E} = \mathbf{D_r}^{-1}(\mathbf{P} - \mathbf{rc}')\mathbf{D_c}^{-1}(\mathbf{P} - \mathbf{rc}')', \tag{7.5}$$

$\lambda_1^2 \geq \cdots \geq \lambda_m^2$ are the nonzero eigenvalues of $\mathbf{E}$ and

$$m = Rank(\mathbf{E}) = Rank(\mathbf{P} - \mathbf{rc}') = Rank(\mathbf{P}) = min(a, b) - 1.$$

We can also express the χ^2 as

$$\chi^2 = n\sum_i p_{i\cdot}[\sum_j (\frac{p_{ij}}{p_{i\cdot}} - p_{\cdot j})^2/p_{\cdot j}] = \sum_i np_{i\cdot}[(\mathbf{r}_i - \mathbf{c})'\mathbf{D}_c^{-1}(\mathbf{r}_i - \mathbf{c})]$$

$$= n\sum_i p_{i\cdot}d_i^2, \tag{7.6}$$

where

$$d_i^2 = (\mathbf{r}_i - \mathbf{c})'\mathbf{D}_c^{-1}(\mathbf{r}_i - \mathbf{c}).$$

The quantity d_i^2 represents the squared distance between the ith row profile and its average. This distance is called the *chi-square distance*. In fact, d_i^2 is similar to the usual squared Euclidean distance $(\mathbf{r}_i - \mathbf{c})'(\mathbf{r}_i - \mathbf{c})$ between the vectors $\mathbf{r}_i$ and $\mathbf{c}$, except that it has been weighted by the elements of vector $\mathbf{c}$, the average row profile. The squared chi-square distance d_2^2 between the second row profile and the average row profile in our example is

$$d_2^2 = (\mathbf{r}_2 - \mathbf{c})'\mathbf{D}_c^{-1}(\mathbf{r}_2 - \mathbf{c})$$

$$= \begin{bmatrix} .20 - .37 \\ .60 - .36 \\ .20 - .27 \end{bmatrix}' \begin{bmatrix} .37 & .00 & .00 \\ .00 & .36 & .00 \\ .00 & .00 & .27 \end{bmatrix}^{-1} \begin{bmatrix} .20 - .37 \\ .60 - .36 \\ .20 - .27 \end{bmatrix}$$

$$= \begin{bmatrix} -.17 & .24 & -.07 \end{bmatrix} \begin{bmatrix} 1/.37 & .00 & .00 \\ .00 & 1/.36 & .00 \\ .00 & .00 & 1/.27 \end{bmatrix} \begin{bmatrix} -.17 \\ .24 \\ -.07 \end{bmatrix}$$

$$= 0.2563.$$

However, the corresponding squared Euclidean distance $(\mathbf{r}_2 - \mathbf{c})'(\mathbf{r}_2 - \mathbf{c}) = 0.0914$. Since the weights in this case are comparable (around 0.33), the two distances, after appropriate scaling (that is, after dividing the Euclidean distance by 0.33) also are seen to be comparable.

The quantity χ^2/n is referred to as *total inertia*. Equation 7.6 shows that the total inertia can be expressed as the weighted average of the squared chi-square distances between the row profiles and their average.

The squared chi-square distance between two row profiles, say $\mathbf{r}_i$ and $\mathbf{r}_j$, is

$$d_{ij}^2 = (\mathbf{r}_i - \mathbf{r}_j)'\mathbf{D}_c^{-1}(\mathbf{r}_i - \mathbf{r}_j) \tag{7.7}$$

which is similar to the usual squared Euclidean distance $(\mathbf{r}_i - \mathbf{r}_j)'(\mathbf{r}_i - \mathbf{r}_j)$ between the two vectors $\mathbf{r}_i$ and $\mathbf{r}_j$, except that the reciprocals of the elements of the average profile have been used as weights in its computation. The chi-square distances between a column profile and its average, and that between two column profiles, are similarly defined.

We next wish to determine a two- or three-dimensional Euclidean subspace and project all the row profiles to this space such that the distances between them, measured in terms of corresponding chi-square distances, are preserved to the extent possible. Such a subspace is determined using generalized singular value decomposition (see Section 1.5.25) of the matrix $(\mathbf{P} - \mathbf{rc}')$ given by

$$(\mathbf{P} - \mathbf{rc}') = \mathbf{A}\mathbf{\Lambda}\mathbf{B}'. \tag{7.8}$$

Here the $a \times m$ matrix $\mathbf{A}$ and $b \times m$ matrix $\mathbf{B}$ are such that $\mathbf{A}'\mathbf{D}_r^{-1}\mathbf{A} = \mathbf{I}_m$ and $\mathbf{B}'\mathbf{D}_c^{-1}\mathbf{B} = \mathbf{I}_m$, and $\mathbf{\Lambda}$ is the diagonal matrix whose diagonal elements are the singular values $\lambda_1, \ldots, \lambda_m$ of $(\mathbf{P} - \mathbf{rc}')$. As observed in Chapter 1, the matrices $\mathbf{A}$ and $\mathbf{B}$ can be obtained from the usual singular value decomposition of $\mathbf{T} = \mathbf{D}_r^{-1/2}(\mathbf{P} - \mathbf{rc}')\mathbf{D}_c^{-1/2}$. Note that $\lambda_1^2, \ldots, \lambda_m^2$ are the eigenvalues of $\mathbf{TT}'$ as well as those of matrix $\mathbf{E}$ defined in Equation 7.5.

Suppose we want to represent the row and column profiles on a k ($\leq m$) dimensional space. Usually the value of k is taken to be 2 or 3. The coordinates for the a row profiles are the a rows of the matrix formed by taking the first k columns of

$$\mathbf{F} = \mathbf{D}_r^{-1}\mathbf{A}\mathbf{\Lambda}$$

and those for the b column profiles are the b rows of the matrix formed by taking the first k columns of

$$\mathbf{G} = \mathbf{D}_c^{-1}\mathbf{B}\mathbf{\Lambda}.$$

Since the total inertia (which represents total information in the whole space) is $\chi^2/n = tr(\mathbf{E}) = \sum_{i=1}^m \lambda_i^2$, we say the approximation of the entire m dimensional space by k dimensional subspace is good if $\sum_{i=1}^k \lambda_i^2$ is approximately equal to total inertia $\sum_{i=1}^m \lambda_i^2$, or alternatively if $\sum_{i=k+1}^m \lambda_i^2$ is close to zero. The quantities λ_1^2, λ_2^2, and so on can be interpreted as the amounts contributed to the total inertia by each of the first, second, and other dimensions. Thus we can take

$$\mathcal{L} = \frac{\sum_{i=k+1}^m \lambda_i^2}{\sum_{i=1}^m \lambda_i^2} = 1 - \frac{\sum_{i=1}^k \lambda_i^2}{\sum_{i=1}^m \lambda_i^2}$$

as a measure of relative loss of information (Rao, 1995).

Returning to our numerical example, m, the rank of $\mathbf{P}$, is 2 and hence if $k = 2$, there will be no loss of information. The two singular values are $\lambda_1 = 0.5522$ and $\lambda_2 = 0.3300$. The total inertia is 0.4138 and the parts accounted by the two dimensions respectively are $\lambda_1^2 = 0.3049$ and $\lambda_2^2 = 0.1089$. The matrices $\mathbf{F}$ and $\mathbf{G}$, respectively, are of dimensions 4 by 2 and 3 by 2 and, as defined above, are given by

$$\mathbf{F} = \begin{bmatrix} .0245 & .1409 \\ -.5050 & -.0353 \\ .6864 & .3613 \\ .7645 & -.8199 \end{bmatrix} \text{ and } \mathbf{G} = \begin{bmatrix} .5361 & -.2877 \\ -.7174 & -.0989 \\ .2219 & .5261 \end{bmatrix}.$$

In correspondence analysis, a rows of the matrix that is formed by the first two columns of $\mathbf{F}$ and b rows of the matrix that is formed by the first two columns of $\mathbf{G}$ are generally plotted on a plane on the same graph. Greenacre (1984) calls such a plot a symmetric display or *symmetric plot* of points corresponding to the row and column profiles. This actually amounts to placing two displays one on top of the other. In this plot the distance between the points corresponding to the row profiles or that between the points corresponding to the column profiles are the approximations to the corresponding chi-square distance between the respective profiles. However, no such interpretation can be assigned for the distance between two points, one corresponding a row profile and another corresponding to a column profile. Hence, in this display, only the distance between points corresponding to either two rows or two columns should be interpreted.

Greenacre (1984) suggests an *asymmetric plot* for the row and column profiles in which interpretation of distance between row and column points is possible. For example, suppose a points corresponding to a rows and the first 2 columns of $\mathbf{F}$ (representing a profiles) and b points corresponding to b rows and the first 2 columns of $\mathbf{D}_c^{-1}\mathbf{B}$ (representing columns) are plotted on the same plane. These points representing the columns in the plot, as discussed earlier, are the points corresponding to the vertices of a simplex. However, the squared distance between two vertices, say the ith and the jth, is not one but $1/p_{\cdot i} + 1/p_{\cdot j}$. The rows of $\mathbf{F}$ are called *principal coordinates* for rows and those of $\mathbf{D}_c^{-1}\mathbf{B}$ are called *standard coordinates* for columns (Greenacre, 1984). In this display, the distance between two row points has the same interpretation as before. But the Euclidean distance between two column points now has no interpretation. However, the distance between a row point, say corresponding to the ith row profile, and the column point, say corresponding to the jth column, is now approximated by $p_{i\cdot}/p_{ij}$, which is the inverse of the jth element $p_{ij}/p_{i\cdot}$ of $\mathbf{r}_i$, the ith row profile. The corresponding principal coordinates for columns and the standard coordinates of rows can be similarly defined.

For our example, the principal coordinates for rows are already given in $\mathbf{F}$ above and the standard coordinates for the $b = 3$ columns are

$$\mathbf{D}_c^{-1}\mathbf{B} = \begin{bmatrix} 0.9708 & -0.8719 \\ -1.2992 & -0.2997 \\ 0.4019 & 1.5944 \end{bmatrix}.$$

Greenacre (1993b) has shown that an asymmetric plot in correspondence analysis can be interpreted as a biplot. For details about this and for further in-depth geometric and other interpretations of these plots see Greenacre and Hastie (1987) and Greenacre (1993a, 1993b).

In SAS, the CORRESP procedure performs the computations needed for correspondence analysis and subsequent interpretation. SAS uses the correspondence matrix $\mathbf{P}$, instead of $\mathbf{P} - \mathbf{rc}'$, for computing the generalized singular value decomposition. However, the results are identical to those corresponding to $\mathbf{P} - \mathbf{rc}'$ if the first singular value (which is one) in $\mathbf{\Lambda}$ and first (trivial) columns corresponding to this singular value, of the matrices $\mathbf{A}$ and $\mathbf{B}$ are discarded from the calculations done in SAS. Options for computing the coordinates for the symmetric as well as the asymmetric plots are available. The coordinates for rows and columns denoted by the SAS variables DIM1, DIM2,. . . , are stored in a SAS data set using the OUT = option in the PROC CORRESP statement. The default value of k is taken as 2. These points can be plotted using the GPLOT procedure or the PLOTIT macro. We have used PROC GPLOT to plot these plots. The default coordinates are principal coordinates for both rows and columns that otherwise can be explicitly specified in the PROC CORRESP statement using the PROFILE= option. If we want the principal coordinates for row profiles and the standard coordinates for the columns, then the PROFILE=ROW option should be used. Similarly, PROFILE=COLUMN will give the other set of coordinates for column profiles and rows.

In the following we illustrate the correspondence analysis of a data set from medical research by using PROC CORRESP.

EXAMPLE 1 ***Analysis of Headache Types by Age Data*** The data set considered here is a two-way contingency table taken from Diehr et al. (1981) providing the frequencies of 681 headache sufferers classified according to their age group and headache categories. There are three headache categories, namely, no diagnosis (NDIAGNO), tension headache (TENSION), and migraine headache (MIGRAINE). The five age groups considered are 0-19, 20-29, 30-39, 40-49, and 50-59 years. These age groups are shown in Program 7.1 as Age:0-19, ..., Age:50-59. The data are provided in Program 7.1. These data have been previously analyzed in Greenacre (1992).

The problem of interest is to study the association (if any) between the two variables, namely, headache type and age group. Here a, the number of rows representing the number of age groups, is 5 and b, the number of columns representing the three types of headaches, is 3. Hence m, the rank of $\mathbf{P}$, is $min(5, 3) - 1 = 2$. Thus, a two-dimensional representation of the data will actually be a representation in the full space with no loss of information.

In Program 7.1, we used the OBSERVED option in the PROC CORRESP statement to print the observed frequencies along with their row and column sums as shown in Output 7.1. From this the average row and column profiles can be easily calculated. For example, the average row profile is $\mathbf{c} = (221/681, 279/681, 181/681)' = (0.3245, 0.4097, 0.2658)'$. The RP and CP options are included to print the row and column profiles. Output 7.1 shows that the first row profile corresponding to the age group Age:0-19 is the row vector $(0.5168, 0.3490, 0.1342)$. Similarly, the third column profile corresponding to the headache category Migraine is the column vector $(0.1105, 0.3757, 0.2044, 0.1713, 0.1381)'$. The SHORT option is used to suppress various extensive details in the output. The output also presents the nonzero singular values of $\mathbf{P}$, and these are $\lambda_1 = 0.23613$ and $\lambda_2 = 0.08205$. Therefore, the total inertia is

$$\lambda_1^2 + \lambda_2^2 = 0.23613^2 + 0.08205^2 = 0.05576 + 0.00673 = 0.06249.$$

The value of the chi-square statistic for the null hypothesis of no association between the two variables is $\chi^2 = n \times (total\ inertia) = 681 \times 0.06249 = 42.5557$. Under the null hypothesis, χ^2 has a chi-square distribution with $(a-1)(b-1) = (5-1)(3-1) = 8$ degrees of freedom. Since the p value of the test (not calculated in the SAS output) is negligible (< 0.0001), the null hypothesis is rejected. This suggests that there possibly is an association between the two categorical variables, namely, headache type and age group.

We explore this relationship graphically using correspondence analysis. The coordinates for the categories for a two-dimensional asymmetric plot are reported in Output 7.1 and are given below.

```
Category               (Dim1,        Dim2)

Age:0-19             (-0.43347,    -0.01049)
Age:20-29            ( 0.06953,     0.08674)
Age:30-39            ( 0.14150,     0.02630)
Age:40-49            ( 0.24405,    -0.03533)
Age:50-59            ( 0.09450,    -0.16244)

No Diagnosis         (-1.37711,     0.43014)
Tension Headache     ( 0.36942,    -1.14210)
Migraine             ( 1.11200,     1.23527)
```

These coordinates and other results are stored in a SAS data set by using the OUT= option in the PROC CORRESP statement and then later printed using PROC PRINT (the data set was revised to eliminate the coordinates with missing values).

A plot of these points is shown in Output 7.1. The actual plot of a point is denoted by a star (*) at the top of the corresponding label, which is achieved by using the POSITION='8' option in the LEGEND statement and SYMBOL V=STAR global option in the PLOT statement. The plane for this two-dimensional plot is that generated by the three-dimensional points $(1, 0, 0)$, $(0, 1, 0)$, and $(0, 0, 1)$ representing the categories No Diagnosis, Tension Headache, and Migraine (Greenacre, 1992). They are often used as the reference points. All five points representing all row profiles corresponding to different ages must fall within the triangle. The point corresponding to a row profile that has its elements closer to, say for example, $(1, 0, 0)$ will plot near that point. In the plot the point corresponding to the row profile $(0.5168, 0.3490, 0.1342)$ of the age group category Age:0-19, plots closer to No Diagnosis than any of the other two headache categories. Similarly the point corresponding to the last row profile $(0.2596, 0.5000, 0.2404)$ for the age category Age:50-59 plots near the point corresponding to Tension Headache.

We can also follow the movement of the points corresponding to the row profiles with respect to one of the reference points. For example, suppose we fix Migraine as our reference point and follow the points corresponding to the age categories. Initially for age category Age:0-19, the point plots far away from Migraine. Then the points plot nearer to Migraine as we move to higher age categories, before the last point corresponding to Age: 50-59 moves away. This is also evident from the last elements (corresponding to Migraine), namely $(0.1342, 0.3148, 0.3162, 0.3263, 0.2404)$ of the respective row profiles for the five age groups.

Examining the points corresponding to the age categories, we find that Age:20-29 and Age:30-39 can be considered to be close to each other with the squared chi-square distance between them equal to

$$d_{23}^2 = \frac{(.3056 - .2650)^2}{.3245} + \frac{(.3796 - .4188)^2}{.4097} + \frac{(.3148 - .3162)^2}{.2658} = .0088.$$

Similarly, the points corresponding to Age:30-39 and Age:40-49 are also close to each other, but the other row points are not very close to each other.

A symmetric plot of the row and column profiles is also given in Output 7.1. An examination of this plot reveals similar results about the row profiles. Unlike in the asymmetric plot, here we can also discuss the closeness of points corresponding to column profiles. It is observed that the three points representing the three headache categories are far apart.

```
/* Program 7.1 */

options ls=64 ps=45 nodate nonumber;
title1 'Output 7.1';
title2 'Correspondence Analysis of Headache Type by Age';
data headache;
length age_gp$ 10;
input age_gp$ NDiagno Tension Migraine;
cards;
 Age:0-19         77         52         20
Age:20-29         66         82         68
Age:30-39         31         49         37
Age:40-49         20         44         31
Age:50-59         27         52         25
run;
/* Source: Diehr et~al. (1981). Reprinted by permission of
the Pergamon Press. */

proc corresp data=headache observed rp cp short;
var _numeric_;
id age_gp;
run;
```

```
*Asymmetric plot of the age groups;
proc corresp data=headache out=plot profile=row noprint;
var _numeric_;
id age_gp;
label ndiagno='No Diagnosis' tension='Tension Headache'
migraine='Migraine';
run;

data plotprint;
set plot;
if dim1=. then delete;
run;
proc print data=plotprint label noobs;
var age_gp dim1 dim2;
label age_gp='Category';
title3 'Results for Asymmetric Display';
run;

data labels;
set plotprint;
x=dim1;
y=dim2;
retain xsys '2' ysys '2';
function='LABEL';
text=age_gp;
position='8';
run;
filename gsasfile "prog71a.graph";
goptions reset=all gaccess=gsasfile autofeed dev=psl;
goptions horigin=1in vorigin=2in;
goptions hsize=6in vsize=8in;
proc gplot data=plotprint;
plot dim2 * dim1/anno=labels frame
        href=0 vref=0 lvref=3 lhref=3 vaxis=axis2 haxis=axis1
        vminor=1 hminor=1;
axis1 length=5 in order=(-1.5 to 1.5 by .5)  offset=(2)
        label = (h=1.2 'Dimension 1');
axis2 length=5 in order =(-1.5 to 1.5 by .5) offset=(2)
label=(h=1.2 a=90 r=0  'Dimension 2');
symbol v=star;
title1 h=1.2 'Correspondence Analysis of Headache Type by Age';
title2 j=l 'Output 7.1';
title3 f=duplex 'Asymmetric Plot of Age Groups';
run;

*Symmetric plot of the age groups;
proc corresp data=headache out=plot2 profile=both noprint;
var _numeric_;
id age_gp;
label ndiagno='No Diagnosis' tension='Tension Headache'
migraine='Migraine';
run;
```

```
data plotprint2;
set plot2;
if dim1=. then delete;
run;
proc print data=plotprint2 label noobs;
var age_gp dim1 dim2;
label age_gp='Category';
title3 'Results for Symmetric Display';
run;

data label;
set plotprint2;
x=dim1;
y=dim2;
retain xsys '2' ysys '2';
function='LABEL';
text=age_gp;
position='8';
run;
filename gsasfile "prog71b.graph";
goptions reset=all gaccess=gsasfile autofeed dev=psl;
goptions horigin=1in vorigin=2in;
goptions hsize=6in vsize=8in;
proc gplot data=plotprint2;
plot dim2 * dim1/anno=label frame
        href=0 vref=0 lvref=3 lhref=3 vaxis=axis2 haxis=axis1
        vminor=1 hminor=1;
axis1 length=5 in order=(-.5 to .5 by .1)  offset=(2)
        label = (h=1.2 'Dimension 1');
axis2 length=5 in order =(-.2 to .2 by .05) offset=(2)
label=(h=1.2 a=90 r=0  'Dimension 2');
symbol v=star;
title1 h=1.2 'Correspondence Analysis of Headache Type by Age';
title2 j=l 'Output 7.1';
title3 f=duplex 'Symmetric Plot of Age Groups';
run;
```

Output 7.1

Output 7.1
Correspondence Analysis of Headache Type by Age

The CORRESP Procedure

Contingency Table

	NDiagno	Tension	Migraine	Sum
Age:0-19	77	52	20	149
Age:20-29	66	82	68	216
Age:30-39	31	49	37	117
Age:40-49	20	44	31	95
Age:50-59	27	52	25	104
Sum	221	279	181	681

Output 7.1
(continued)

Row Profiles

	NDiagno	Tension	Migraine
Age:0-19	0.516779	0.348993	0.134228
Age:20-29	0.305556	0.379630	0.314815
Age:30-39	0.264957	0.418803	0.316239
Age:40-49	0.210526	0.463158	0.326316
Age:50-59	0.259615	0.500000	0.240385

Column Profiles

	NDiagno	Tension	Migraine
Age:0-19	0.348416	0.186380	0.110497
Age:20-29	0.298643	0.293907	0.375691
Age:30-39	0.140271	0.175627	0.204420
Age:40-49	0.090498	0.157706	0.171271
Age:50-59	0.122172	0.186380	0.138122

Inertia and Chi-Square Decomposition

Singular Value	Principal Inertia	Chi-Square	Percent	Cumulative Percent
0.23613	0.05576	37.9706	89.23	89.23
0.08205	0.00673	4.5851	10.77	100.00
Total	0.06249	42.5557	100.00	

Degrees of Freedom = 8

```
    18   36   54   72   90
----+----+----+----+----+---
*************************
***
```

Results for Asymmetric Display

Category	Dim1	Dim2
Age:0-19	-0.43347	-0.01049
Age:20-29	0.06953	0.08674
Age:30-39	0.14150	0.02630
Age:40-49	0.24405	-0.03533
Age:50-59	0.09450	-0.16244
No Diagnosis	-1.37711	0.43014
Tension Headache	0.36942	-1.14210
Migraine	1.11200	1.23527

Output 7.1
(continued)

Results for Symmetric Display

Category	Dim1	Dim2
Age:0-19	-0.43347	-0.01049
Age:20-29	0.06953	0.08674
Age:30-39	0.14150	0.02630
Age:40-49	0.24405	-0.03533
Age:50-59	0.09450	-0.16244
No Diagnosis	-0.32518	0.03530
Tension Headache	0.08723	-0.09371
Migraine	0.26258	0.10136

Output 7.1

Correspondence Analysis of Headache Type by Age

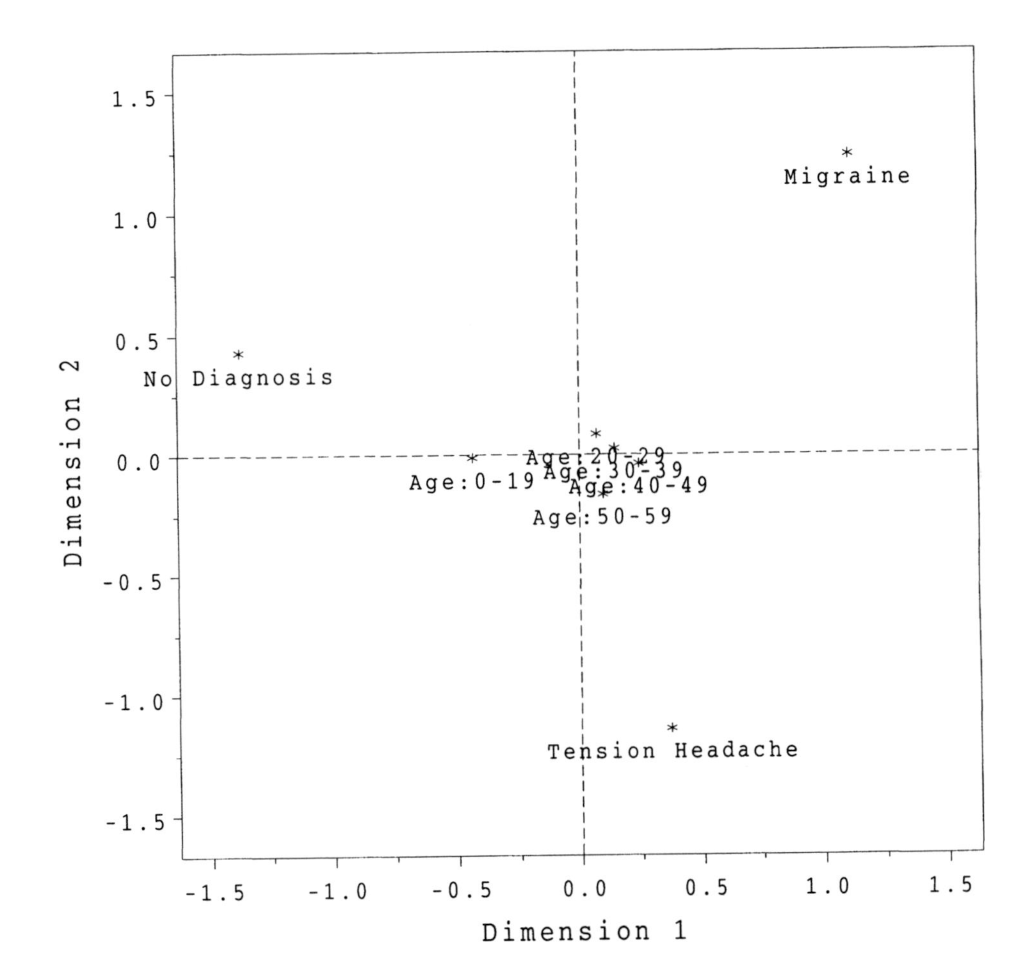

Output 7.1

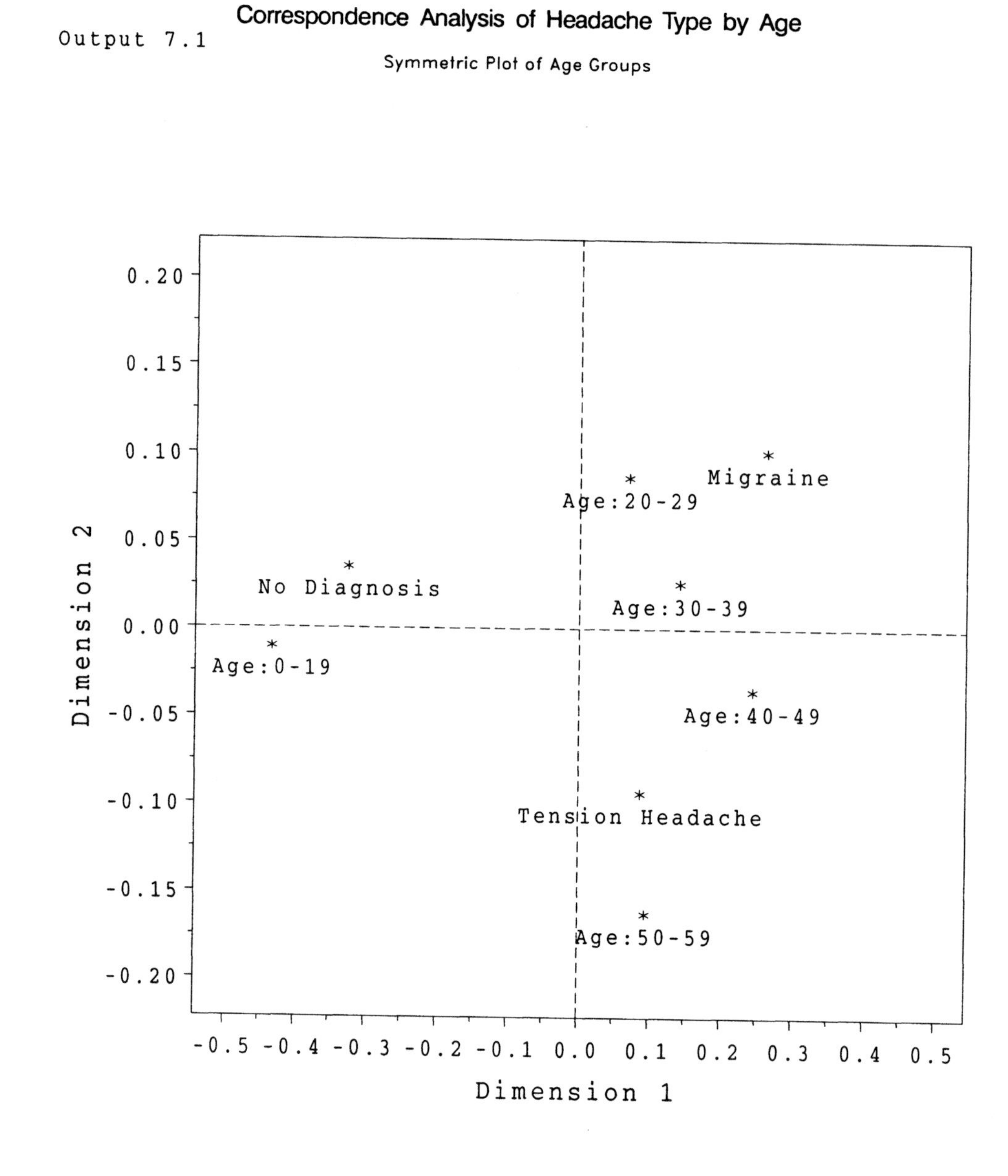

EXAMPLE 2 ***Correspondence Analysis of Socioeconomic Status by Mental Health of Children Data***
Since the number of columns in Example 1 was 3, the five row profiles were actually points on a two-dimensional space. Thus in that example, there was no loss of information in representing the profiles as points. In the example considered here the original profiles are in three-dimensional space but the projected points are in a two-dimensional plane.

The data considered here are from Srole et al. (1978) and are given in Table 7.1. We have already considered these data in Chapter 3. The objective of the study is to examine the relationship, if any, between children's mental impairment and parents' socioeconomic status. Six levels of socioeconomic status, 1 (high) to 6 (low), and four levels of mental health status, well (WELL), mild symptom formation (MILD), moderate symptom formation (MODERATE), and impairment (IMPAIRED) are considered. Data obtained in the form of a 6 by 4 contingency table are based on a sample of 1,660 residents of Manhattan. The data and the SAS code for performing correspondence analysis are provided in Program 7.2. As in Example 1, the program computes row and column profiles, the chi-square statistics for the independence of the two variables, the singular values corresponding to each dimension explaining the loss of information, and the coordinates for plotting. Output 7.2 lists all these values.

TABLE 7.1 Cross-Classification of Parents' Socioeconomic Status and Mental Health Status

Parents' Socioeconomic Status	Mental Health Status			
	(Well) A	(Mild Symptom Formation) B	(Moderate Symptom Formation) C	(Impaired) D
1 (high)	64	94	58	46
2	57	94	54	40
3	57	105	65	60
4	72	141	77	94
5	36	97	54	78
6 (low)	21	71	54	71

The chi-square test statistic for testing the null hypothesis of no association between the two variables is 45.9853 on $(a-1)(b-1) = (6-1)(4-1) = 15$ degrees of freedom. In view of the correspondingly small p value (<0.0000) we reject the null hypothesis and conclude that the two variables are not independent. The nonzero singular values of the matrix $\mathbf{P}$ are (see Output 7.2) $\lambda_1 = 0.1613$, $\lambda_2 = 0.0371$, and $\lambda_3 = 0.0173$. The total inertia is $\chi^2/n = 45.9853/1660 = 0.0277$, which can be viewed as the sum of the principal inertia corresponding to various dimensions. The principal inertia for the ith dimension is defined as λ_i^2. For example, for the first dimension the value of the principal inertia is $\lambda_1^2 = (0.1613)^2 = 0.0260$. The percentage of the total inertia explained by the one-dimensional approximation is already $(0.0260/0.0277)100 \approx 94\%$. This percentage explained by the two-dimensional approximation is close to 99%. Since the whole space here is three-dimensional ($min(a-1, b-1) = min(5, 3) = 3$) all 100% of the inertia is explained by all three dimensions. Since the corresponding loss of information here is only about 1%, we can be confident that the two-dimensional representation of the row profiles will be a reasonably good approximation to the whole space.

As in the previous example, we interpret asymmetric as well as symmetric plots. In the asymmetric plot provided in Output 7.2 the four points corresponding to the column categories are the projections of the four vertices of the triangle in the three-dimensional space. These points act as the reference points. The point corresponding to the status category 6 (low status) is closest to IMPAIRED, followed by other categories in decreasing order. At the same time, the point corresponding to the status 1 (high) is closest to the point corresponding to WELL, followed by other status categories in the increasing order.

For the symmetric plot of these data, the distance between row-row points and column-column points can be interpreted. However, the distance between row-column points has no interpretation. The four points representing the column profiles are quite far apart indicating that the four column categories are very different. However, the points representing the row profiles for rows 1 and 2 form a cluster. Hence, if we want, these status categories may be clubbed together to form one group.

If the categories are ordered (in some way) as in this case, it is also of interest to check whether the order is maintained along the first principal axis (horizontal axis in our case). The second graph in Output 7.2 shows that the status categories are found to be in an order (6, 5,... from left to right) along the horizontal axis. As we already noted above, categories 1 and 2 cannot be clearly distinguished and this is again reconfirmed. Hence these two status categories perhaps can be combined. There appears to be more or less equal distance between the other status categories. The categories corresponding to the mental status of children also follow an order from IMPAIRED to WELL along the horizontal axis. The two middle categories are quite close to each other, but there is a clear distinction between the other categories.

```
/* Program 7.2 */

options ls=64 ps=45 nodate nonumber;
title1 'Output 7.2';
title2 'Correspondence Analysis of Socioeconomic Status';
title3  'by Mental Health of Children';
data status;
length sestatus$8;
input sestatus$ Well Mild Moderate Impaired ;
lines;
1 64 94  58 46
2 57 94  54 40
3 57 105 65 60
4 72 141 77 94
5 36 97  54 78
6 21 71  54 71
;
/* Source: Srole et~al. (1978). Reprinted by permission of
the New York University Press. */

* Asymmetric plot of socioeconomic status;
proc corresp data=status out=plot short rp cp profile=row;
var well mild moderate impaired;
id sestatus;
run;

data plotprint;
set plot;
if dim1=. then delete;
run;
proc print data=plotprint label noobs;
var sestatus dim1 dim2;
label sestatus='Category';
title4 'Results for Asymmetric Display';
run;

data labels;
set plotprint;
x=dim1;
y=dim2;
retain xsys '2' ysys '2';
function='LABEL';
text=sestatus;
position='8';
run;

filename gsasfile "prog72a.graph";
goptions reset=all gaccess=gsasfile autofeed dev=psl;
goptions horigin=1in vorigin=2in;
goptions hsize=6in vsize=8in;
proc gplot data=plotprint;
plot dim2 * dim1/anno=labels frame href=0 vref=0
lvref=3 lhref=3 vaxis=axis2 haxis=axis1 vminor=1 hminor=1;
axis1 length=5 in order=(-2 to 2 by .5)  offset=(2)
        label = (h=1.2 'Dimension 1');
axis2 length=5 in order =(-.7 to 2 by .5) offset=(2)
label=(h=1.2 a=90 r=0  'Dimension 2');
```

```
symbol v=star;
title1 h=1.2 'Correspondence Analysis of Socioeconomic
Status';
title2 'by Mental Health of Children';
title3 j=l 'Output 7.2';
title4 f=duplex 'Asymmetric Display of the Categories';
run;

*Symmetric plot of socioeconomic status data;
proc corresp data=status out=plot2 profile=both noprint;
var well mild moderate impaired;
id sestatus;
run;

data plotprint2;
set plot2;
if dim1=. then delete;
run;
proc print data=plotprint2 label noobs;
var sestatus dim1 dim2;
label sestatus='Category';
title4 'Results for Symmetric Display';
run;

data label;
set plotprint2;
x=dim1;
y=dim2;
retain xsys '2' ysys '2';
function='LABEL';
text=sestatus;
position='8';
run;

filename gsasfile "prog72b.graph";
goptions reset=all gaccess=gsasfile autofeed dev=psl;
goptions horigin=1in vorigin=2in;
goptions hsize=6in vsize=8in;
proc gplot data=plotprint2;
plot dim2 * dim1/anno=label frame href=0 vref=0
lvref=3 lhref=3 vaxis=axis2 haxis=axis1 vminor=1 hminor=1;
axis1 length=5 in order=(-.3 to .3 by .1)  offset=(2)
        label = (h=1.2 'Dimension 1');
axis2 length=5 in order =(-.05 to .07 by .02) offset=(2)
label=(h=1.2 a=90 r=0  'Dimension 2');
symbol v=star;
title1 h=1.2 'Correspondence Analysis of Socioeconomic
Status';
title2 'by Mental Health of Children';
title3 j=l 'Output 7.2';
title4 f=duplex 'Symmetric Plot of Age Groups';
run;
```

Output 7.2

```
                    Output 7.2
   Correspondence Analysis of Socioeconomic Status
            by Mental Health of Children

               The CORRESP Procedure

                   Row Profiles

              Well          Mild      Moderate      Impaired

1         0.244275      0.358779      0.221374      0.175573
2         0.232653      0.383673      0.220408      0.163265
3         0.198606      0.365854      0.226481      0.209059
4         0.187500      0.367188      0.200521      0.244792
5         0.135849      0.366038      0.203774      0.294340
6         0.096774      0.327189      0.248848      0.327189


                  Column Profiles

              Well          Mild      Moderate      Impaired

1         0.208469      0.156146      0.160221      0.118252
2         0.185668      0.156146      0.149171      0.102828
3         0.185668      0.174419      0.179558      0.154242
4         0.234528      0.234219      0.212707      0.241645
5         0.117264      0.161130      0.149171      0.200514
6         0.068404      0.117940      0.149171      0.182519


          Inertia and Chi-Square Decomposition

  Singular    Principal      Chi-                 Cumulative
     Value      Inertia    Square     Percent        Percent

   0.16132      0.02602   43.2013       93.95          93.95
   0.03714      0.00138    2.2894        4.98          98.92
   0.01726      0.00030    0.4946        1.08         100.00

     Total      0.02770   45.9853      100.00

  Degrees of Freedom = 15

                 19   38   57   76   95
             ----+----+----+----+----+---
             ************************
             *
```

Output 7.2 *(continued)*

Results for Asymmetric Display

Category	Dim1	Dim2
1	0.18093	0.01925
2	0.18500	0.01163
3	0.05903	0.02220
4	-0.00889	-0.04208
5	-0.16539	-0.04361
6	-0.28769	0.06199
Well	1.60880	-0.32587
Mild	0.18341	-0.63685
Moderate	-0.08808	1.88225
Impaired	-1.47154	-0.50886

Results for Symmetric Display

Category	Dim1	Dim2
1	0.18093	0.019248
2	0.18500	0.011625
3	0.05903	0.022197
4	-0.00889	-0.042080
5	-0.16539	-0.043606
6	-0.28769	0.061994
Well	0.25954	-0.012102
Mild	0.02959	-0.023651
Moderate	-0.01421	0.069901
Impaired	-0.23739	-0.018897

Output 7.2

Correspondence Analysis of Socioeconomic Status
by Mental Health of Children
Output 7.2
Asymmetric Display of the Categories

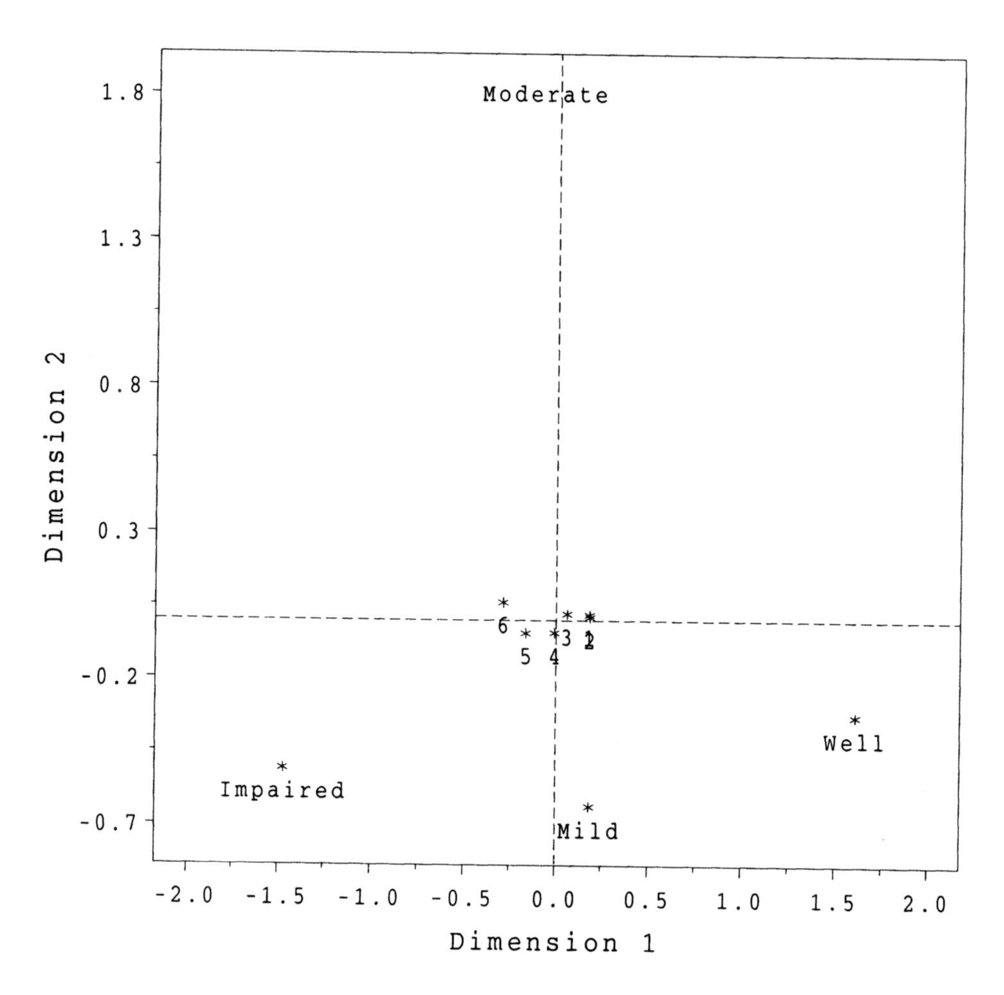
Moderate
Impaired
Mild
Well
6
5
4
3
2
1
1.8
1.3
0.8
0.3
-0.2
-0.7
Dimension 2
-2.0
-1.5
-1.0
-0.5
0.0
0.5
1.0
1.5
2.0
Dimension 1

Output 7.2

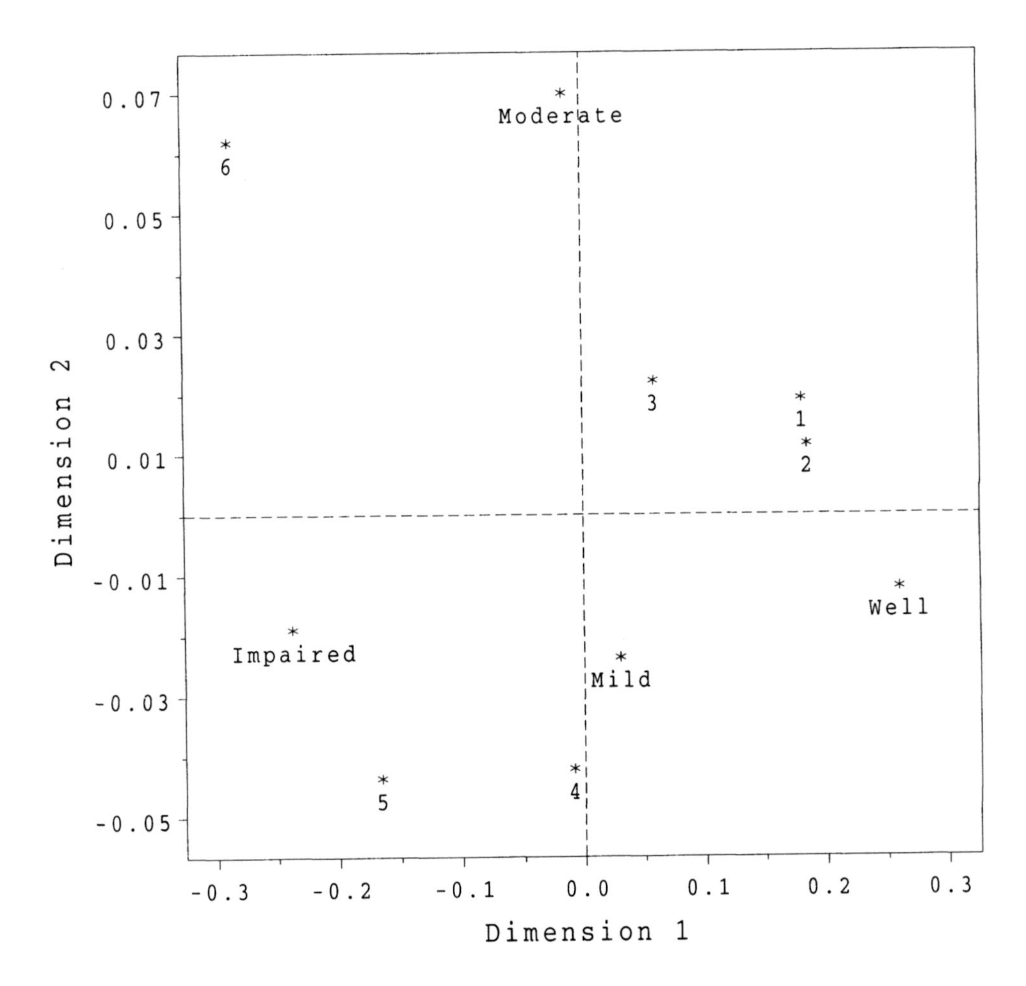

7.3 Multiple Correspondence Analysis

In the previous section we discussed the correspondence analysis of a two-way contingency table. The natural next step is analysis of a multi-way table. Suppose we have a multi-way contingency table and as before, the interest is in a graphical display of the information in this table as a two-dimensional display. This can be achieved by a graphical analysis known as *multiple correspondence analysis.* Multiple correspondence analysis (MCA) is a correspondence analysis applied to a particular indicator matrix (say **Z**) of data. The matrix has the number of rows equal to the total number of individuals in the sample and the number of columns equal to the sum of all categories corresponding to all the variables. Each element in a row of **Z** is one if the individual belongs to that particular category, and it is zero otherwise. This method can be best described by using examples. Therefore, here we take two examples—one for a two-way and one for a four-way contingency table data. First, we will consider the example of a two-way contingency table data, which are presented in Table 7.2. The basic problem is to study and compare the efficacies of several drugs. Thus the two variables are Drug with four types and Efficacy Ratings with five categories.

TABLE 7.2 Results of a Survey on Analgesic Efficacy of Drugs

	Ratings of Drug's Efficacy					
Drug	Poor	Fair	Good	Very Good	Excellent	Total
Z100	5	1	10	8	6	30
EC4	5	3	3	8	12	31
C60	10	6	12	3	0	31
C15	7	12	8	1	1	29
Total	27	22	33	20	19	121

There are a total of 121 individuals in this study and a total of $4 + 5 = 9$ categories. To define column order, suppose these nine categories are written in the order Z100, EC4, C60, C15, Poor, Fair, Good, Very Good, Excellent. Hence the matrix $\mathbf{Z}$ is a 121 by 9 matrix of zeros and ones. For example, the first five rows of this matrix representing the cell frequency of five for the first cell (Drug: Z100 and Efficacy Ratings: Poor) are all the same as $(1, 0, 0, 0, 1, 0, 0, 0, 0)$. Similarly, for example, in the ninth column corresponding to the category excellent, a "1" will be found only in the rows corresponding to those individuals who found the efficacy of (any of) the drug to be excellent. Thus a total of nineteen 1's will be found in that column. Another example where such a matrix was created can be found in Section 3.7.

Treating this matrix as the data in the form of a 121 by 9 contingency table, we perform a correspondence analysis on these data. Performing a correspondence analysis on $\mathbf{Z}$ is also equivalent to performing a correspondence analysis on the Burt matrix $\mathbf{Z}'\mathbf{Z}$ (see Gower and Hand, 1996, p. 57). For the data in Table 7.2, the Burt matrix $\mathbf{Z}'\mathbf{Z}$ is given in Table 7.3 below.

TABLE 7.3 Burt Matrix of Ratings of Efficacy of Drugs

	Z100	EC4	C60	C15	Poor	Fair	Good	V_Good	Excellent
Z100	30	0	0	0	5	1	10	8	6
EC4	0	31	0	0	5	3	3	8	12
C60	0	0	31	0	10	6	12	3	0
C15	0	0	0	29	7	12	8	1	1
Poor	5	5	10	7	27	0	0	0	0
Fair	1	3	6	12	0	22	0	0	0
Good	10	3	12	8	0	0	33	0	0
V_Good	8	8	3	1	0	0	0	20	0
Excellnt	6	12	0	1	0	0	0	0	19

The Burt matrix is essentially a block matrix

$$\mathbf{Z}'\mathbf{Z} = n \begin{bmatrix} \mathbf{P}_r & \mathbf{P} \\ \mathbf{P}' & \mathbf{P}_c \end{bmatrix}, \tag{7.9}$$

where $\mathbf{P}$ is the correspondence matrix that is obtained by dividing the frequencies of the original contingency table by the total frequency. The matrices $\mathbf{P}_r$ and $\mathbf{P}_c$ are both diagonal matrices with marginal column and marginal row totals, respectively, of $\mathbf{P}$ as the corresponding diagonal elements. For an r-way contingency table, $\mathbf{Z}'\mathbf{Z}$ will be a matrix with r diagonal blocks. Also, the $(i, j)^{th}$ nondiagonal block with, say, $i < j$ will be a two-way contingency table between the ith and j^{th} categorical variables (while ignoring all other $r-2$ categorical variables). The matrix $\frac{\mathbf{Z}'\mathbf{Z}}{n}$ is often referred to as the *standardized Burt matrix*.

Generally, the points on the plot produced by MCA do not have a simple geometric interpretation as in correspondence analysis (Greenacre and Hastie, 1987; Greenacre, 1988). But the points in the same vicinity (for example, the same quadrant and clusters) may be used to determine the association between the categories.

For illustration, we first take the two-way contingency table data given in Table 7.2 and perform the simple and multiple correspondence analysis and contrast the corresponding results.

EXAMPLE 3 ***Multiple Correspondence Analysis Rating of Drug's Efficacy: Analgesic Drugs Data*** The data reported in Table 7.2 are from Çalimlim et al. (1982). These have also been previously analyzed by Cox and Chuang (1984), Chuang and Agresti (1986), and Greenacre (1992). The objective of the study was to examine the ratings of the drugs by the hospital patients in order to determine whether a particular drug or a group of two or more drugs are favored by the patients. Four drugs, Z100, EC4, C60, and C15 and five ratings, poor (POOR), fair (FAIR), good (GOOD), very good (V_GOOD), and excellent (EXCELLNT) are considered. Data obtained in the form of a four by five contingency table are based on the ratings of the four drugs by 121 patients. We first perform the usual correspondence analysis discussed in the previous section and then perform multiple correspondence analysis based on the Burt matrix presented in Table 7.3.

The SAS code for performing these analyses is provided in Program 7.3. The first part of the program performs correspondence analysis and produces the asymmetric plot of the points of row profiles representing the profiles of the four drugs and the five points representing the five ratings. A part of the output and the plot are presented in Output 7.3.

The second part of the program performs the multiple correspondence analysis on these data. In the program we have read in the Burt matrix given in Table 7.2 as the data in a contingency table, and we use the MCA option in PROC CORRESP to get the results of multiple correspondence analysis. Another option that we used is NVARS=2, indicating that the Burt matrix provided is based on only two variables. In the VAR statement we list the names of the categories of all (two in this example) the variables. The program also performs the calculations needed to produce a graphical display of MCA results. A part of the MCA results and the plot are included in the later part of Output 7.3.

Examining the results in Output 7.3 corresponding to the correspondence analysis, we see that the total inertia is 0.3890 and the chi-square value is 47.0718 with 12 degrees of freedom. This high chi-square value indicates that there is a strong association between the two variables. More than 98% of the total inertia is explained by the first two dimensions. That is, the loss of information from approximating the present three-dimensional space by a two-dimensional Euclidean space is less than 2%. Hence, we examine the plots in a two-dimensional space and try to interpret the variables and then possibly determine the relationship between the variables.

From the asymmetric plot in Output 7.3 corresponding to the correspondence analysis, we conclude that the drug EC4 is rated very high, as the point representing the profile for this drug is placed near V_GOOD and EXCELLNT in the plot. The point representing the profile for Z100 is placed near V_GOOD. It is also the second point that is closest to EXCELLNT. However, these points are also close to the point POOR. If we examine the data in Table 7.1, we note that EC4 and Z100 are rated very high by most patients, but are also rated poor by many (the frequency is 5). The point corresponding to the drug C60 is placed very near to the point POOR, which itself is near the origin, although this drug has a higher frequency for a good rating. The location of the point POOR is justified by the fact that frequencies in the column corresponding to POOR in the contingency table are not small for any of the four drugs. At the same time, the drug C60 has a very large frequency corresponding to POOR, and hence the point C60 has been pulled near it. From the asymmetric display it is clear that EC4 being close to EXCELLNT and V_GOOD can be viewed as the best, followed by Z100, which is clearly close to V_GOOD. Since the drugs are represented by the row profiles, the distances between the points representing various drugs are relevant in asymmetric plot. The small distance between EC4 and Z100 reaffirms our conviction that these two are superior in comparison to the other two drugs.

We now examine the results of multiple correspondence analysis. The total inertia is 3.5000. The first two dimensions explain only about 40% of this total inertia. In the display, the points corresponding to EC4 and EXCELLNT plot in the first quadrant and are close to each other. The points Z100 and V_GOOD are together in the fourth quadrant. Thus, it reconfirms our earlier analysis of the same data. Similar conclusions can be made about the other two drugs with a similar type of ambiguity present when trying to interpret the category POOR.

```
/* Program 7.3 */

options ls=64 ps=45 nodate nonumber;

title1 'Output 7.3';
title2 'Correspondence Analysis Results';
title3 "Rating of Drug's Efficacy: Analgesic Drugs Data";
data rate;
input Drug$     Poor   Fair   Good   V_good   Excellnt;
cards;
Z100              5      1     10       8          6
EC4               5      3      3       8         12
C60              10      6     12       3          0
C15               7     12      8       1          1
;
/* Source: Calimlim et al. (1982).  Reprinted by permission of
the National Heart and Lung Institute, London, England. */

*Results for Asymmetric Plot;
proc corresp data=rate out=plot short profile=row;
var _numeric_;
id drug;
run;

data plotprint;
set plot;
if dim1=. then delete;
run;
proc print data=plotprint noobs label;
var drug dim1 dim2;
label drug='Category';
title4 'Results for Asymmetric Display';
run;

data labels;
set plotprint;
x=dim1;
y=dim2;
retain xsys '2' ysys '2';
function='LABEL';
text=drug;
position='8';
run;
```

```
filename gsasfile "prog73a.graph";
goptions reset=all gaccess=gsasfile autofeed dev=psl;
goptions horigin=1in vorigin=2in;
goptions hsize=6in vsize=8in;
proc gplot data=plotprint;
plot dim2 * dim1/anno=labels frame href=0 vref=0
lvref=3 lhref=3 vaxis=axis2 haxis=axis1 vminor=1 hminor=1;
axis1 length=5 in order=(-1.75 to 1.25 by .5)  offset=(2)
        label = (h=1.2 'Dimension 1');
axis2 length=5 in order =(-1.25 to 1.75 by .5) offset=(2)
label=(h=1.2 a=90 r=0  'Dimension 2');
symbol v=star;
title1 h=1.2 'Correspondence Analysis of Drug Efficacy Ratings Data';
title2 j=l 'Output 7.3';
title3 f=duplex 'Asymmetric Plot of Analgesic Drugs';
run;

title1 'Output 7.3';
title2 'Multiple Correspondence Analysis Results';
title3 "Rating of Drug's Efficacy: Analgesic Drugs Data";
data burt;
input row$ Z100  EC4  C60  C15  POOR  FAIR  GOOD  V_GOOD  EXCELLNT;
cards;
 Z100      30   0   0   0    5    1   10    8    6
 EC4        0  31   0   0    5    3    3    8   12
 C60        0   0  31   0   10    6   12    3    0
 C15        0   0   0  29    7   12    8    1    1
 POOR       5   5  10   7   27    0    0    0    0
 FAIR       1   3   6  12    0   22    0    0    0
 GOOD      10   3  12   8    0    0   33    0    0
 V_GOOD     8   8   3   1    0    0    0   20    0
 EXCELLNT   6  12   0   1    0    0    0    0   19
run;

proc corresp data=burt nvars=2 mca out=mplot;
var z100 ec4 c60 c15 poor fair good v_good excellnt;
run;

data mplotprint;
set mplot;
if dim1=. then delete;
run;
proc print data=mplotprint noobs label;
var _name_ dim1 dim2;
label _name_='Category';
title4 'Display of MCA Results';
run;

data mlabel;
set mplotprint;
x=dim1;
y=dim2;
retain xsys '2' ysys '2';
function='LABEL';
text=_name_;
position='8';
run;
```

```
filename gsasfile "prog73b.graph";
goptions reset=all gaccess=gsasfile autofeed dev=psl;
goptions horigin=1in vorigin=2in;
goptions hsize=6in vsize=8in;
proc gplot data=mplotprint;
plot dim2 * dim1/anno=mlabel frame
       href=0 vref=0 lvref=3 lhref=3 vaxis=axis2 haxis=axis1
       vminor=1 hminor=1;
axis1 length=5 in order=(-1. to 2 by .5)  offset=(2)
       label = (h=1.2 'Dimension 1');
axis2 length=5 in order =(-1.5 to 1.5 by .5) offset=(2)
label=(h=1.2 a=90 r=0  'Dimension 2');
symbol v=star;
title1 h=1.2 'MCA of Drug Efficacy Ratings Data';
title2 j=l 'Output 7.3';
title3 f=duplex 'Plot of Drugs and Ratings';
run;
```

Output 7.3

```
                        Output 7.3
              Correspondence Analysis Results
       Rating of Drug's Efficacy: Analgesic Drugs Data

                   The CORRESP Procedure

            Inertia and Chi-Square Decomposition

 Singular   Principal      Chi-                 Cumulative
    Value     Inertia    Square     Percent        Percent

  0.55197     0.30467   36.8647       78.32          78.32
  0.27810     0.07734    9.3583       19.88          98.20
  0.08376     0.00701    0.8488        1.80         100.00

    Total     0.38902   47.0718      100.00

 Degrees of Freedom = 12

                   16   32   48   64   80
               ----+----+----+----+----+---
               ************************
               ******
               *


               Results for Asymmetric Display

               Category        Dim1          Dim2

               Z100         -0.34931      -0.30117
               EC4          -0.70401       0.24600
               C60           0.45488      -0.24803
               C15           0.62767       0.31372
               Poor          0.44686      -0.26822
               Fair          1.16371       1.59462
               Good          0.48486      -1.07366
               V_good       -1.05595      -0.56352
               Excellnt     -1.71306       0.99271
```

Output 7.3 *(continued)*

Inertia and Chi-Square Decomposition

Singular Value	Principal Inertia	Chi-Square	Percent	Cumulative Percent
0.88090	0.77598	208.660	22.17	22.17
0.79941	0.63905	171.840	18.26	40.43
0.73612	0.54188	145.710	15.48	55.91
0.70711	0.50000	134.449	14.29	70.20
0.67685	0.45812	123.188	13.09	83.29
0.60079	0.36095	97.058	10.31	93.60
0.47330	0.22402	60.238	6.40	100.00
Total	3.50000	941.144	100.00	

Degrees of Freedom = 64

```
    4    8   12   16   20
----+----+----+----+----+---
****************************
***********************
*******************
******************
****************
*************
********
```

Display of MCA Results

Category	Dim1	Dim2
Z100	0.55748	-0.86570
EC4	1.12355	0.70714
C60	-0.72596	-0.71296
C15	-1.00172	0.90178
POOR	-0.39364	-0.21441
FAIR	-1.02511	1.27475
GOOD	-0.42711	-0.85829
V_GOOD	0.93019	-0.45048
EXCELLNT	1.50903	0.79358

Output 7.3
Correspondence Analysis of Drug Efficacy Ratings Data
Output 7.3
Asymmetric Plot of Analgesic Drugs

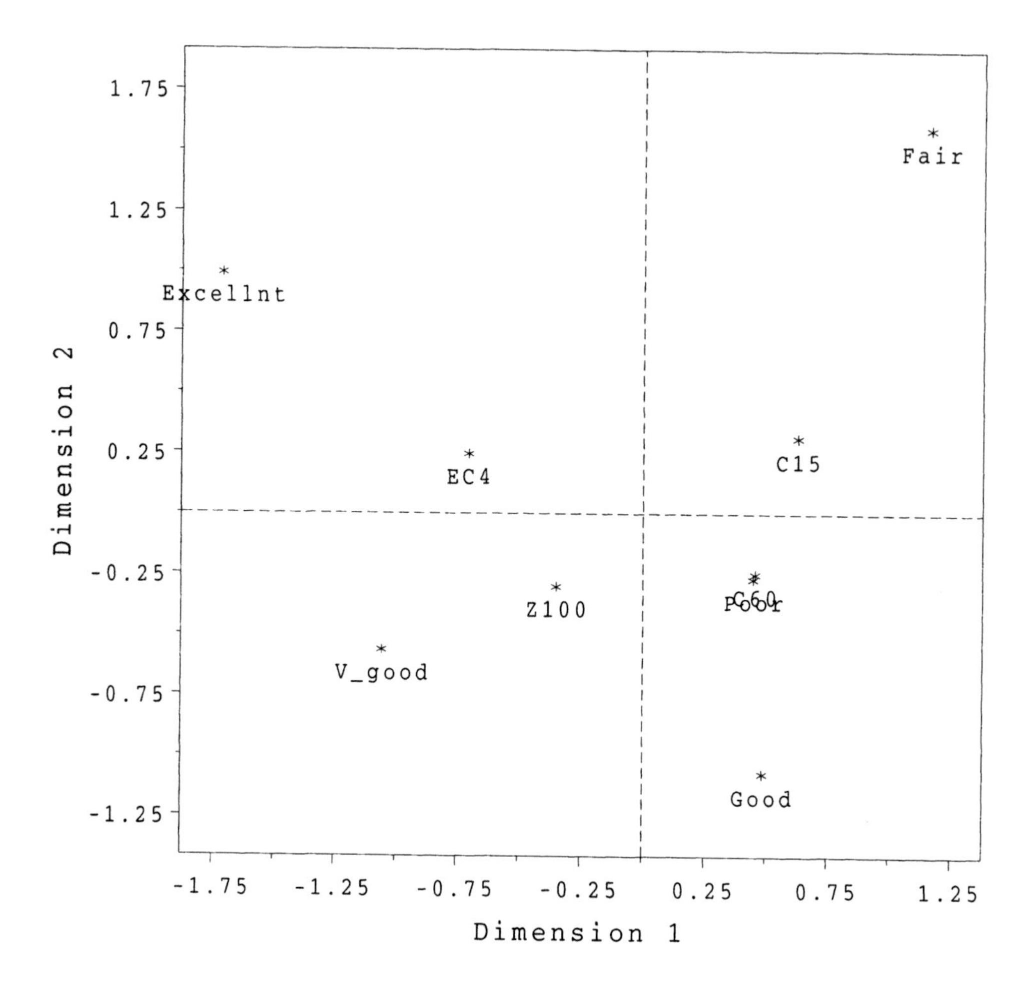
Fair
Excellnt
EC4
C15
Z100
V_good
Good
Dimension 2
Dimension 1
1.75
1.25
0.75
0.25
-0.25
-0.75
-1.25
-1.75
0.25
0.75
1.25

Output 7.3

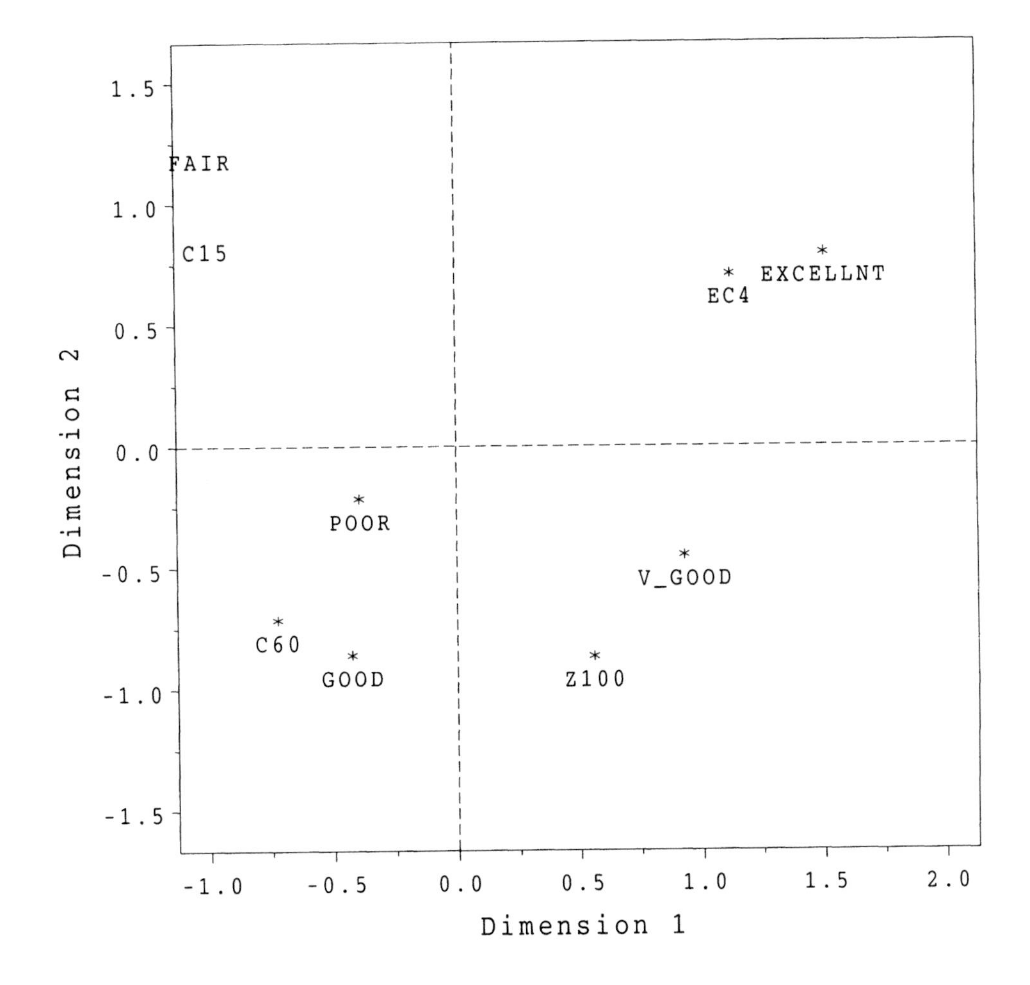

EXAMPLE 4 ***Multiple Correspondence Analysis of Environmental Factors and Farm Management Data*** The multiple correspondence analysis is more relevant in the present example. The data considered are a part of a large data set from Jongman et al. (1997) part of which are also reported in Gower and Hand (1996, p.75). The data on four categorical variables are available for twenty farms from the Dutch island of Terschelling. The objective was to study the effects of various environmental factors on different forms of farm management. The variables considered are *moisture* with five levels, *grassland management type* with four categories, *grassland use* with three categories, and *manure* with five levels. We label the five levels of moisture as MOIST1, MOIST2, MOIST3, MOIST4, and MOIST5. But MOIST3 does not appear in this data set. The four grassland management categories are: standard farming (SF), biological farming (BF), hobby farming (HF), and nature conservation management (NM). The three grassland use categories are: hay production (GU1), intermediate (GU2), and grazing (GU3). Finally, we label the five levels of manure considered by MANU0, MANU1, MANU2, MANU3, and MANU4. The level MANU0 represents that no manure was used on the farm.

In Program 7.4 we have read the raw data corresponding to the 20 farms and created a SAS data set named FARM. These data are then used to create a Burt matrix for the four variables in PROC CORRESP. We have achieved this by including MCA option in PROC CORRESP statement and the option statement TABLES with the list of variable names. The Burt matrix given in the form of a contingency table of all the variables is calculated using OBSERVED option in PROC CORRESP statement. The results are given in Output 7.4. A two dimensional display of the results is also provided in Output 7.4.

From the output observe that about 40% percent of the total inertia is explained by the first two dimensions. The two dimensional display of all the points corresponding to all seventeen categories indicates few interesting relationships. The third quadrant of the plot clearly indicates that the land with high level of moisture (MOIST5) and nature conservation management (NM) are associated. The point representing zero levels of manure (MANU0), rightly so, also falls in the same quadrant as presumably the nature conservation management would call for no external intervention and hence the use of no extra fertilizer. Moving anti-clockwise, the second level of moisture (MOIST2), standard farming (SF), fourth level of manure (MANU4), and the second category of grassland (GU2) are associated. In the next quadrant we see association between variety of categories. Since the HF and BF representing hobby farming and biological farming are included in this group, it makes sense to have an association of these with different (perhaps experimental) levels of manure in different levels of moisture. In the second quadrant both points correspond to the levels of the same variable, namely, grassland use. Since no other points corresponding to any other variable is present, no association is depicted in this quadrant.

```
/* Program 7.4 */

options ls=64 ps=45 nodate nonumber;
title1 'Output 7.4';
title2 'Multiple Correspondence Analysis';
title3 "Environmental Factors and Farm Management Data";

data farm;
input moisture$ managmnt$ grassuse$ manuare$;
cards;
Moist1 Sf Gu2 Manu4
Moist1 Bf Gu2 Manu2
Moist2 Sf Gu2 Manu4
Moist2 Sf Gu2 Manu4
Moist1 Hf Gu1 Manu2
Moist1 Hf Gu2 Manu2
Moist1 Hf Gu3 Manu3
Moist5 Hf Gu3 Manu3
Moist4 Hf Gu1 Manu1
Moist2 Bf Gu1 Manu1
Moist1 Bf Gu3 Manu1
Moist4 Sf Gu2 Manu2
Moist5 Sf Gu2 Manu3
Moist5 Nm Gu3 Manu0
Moist5 Nm Gu2 Manu0
Moist5 Sf Gu3 Manu3
Moist2 Nm Gu1 Manu0
Moist1 Nm Gu1 Manu0
Moist5 Nm Gu1 Manu0
Moist5 Nm Gu1 Manu0
run;
/* Source: Jongman, Ter Braak and Van Tongeren (1997).
Reprinted by permission of Cambridge Universtiy Press.*/
```

```
proc corresp data=farm mca short observed out=plot;
tables moisture managmnt grassuse manuare;
run;

data plotprint;
set plot;
if dim1=. then delete;
run;
proc print data=plotprint noobs label;
var _name_ dim1 dim2;
label _name_='Category';
title4 'Display of MCA Results';
run;

data labels;
set plotprint;
x=dim1;
y=dim2;
retain xsys '2' ysys '2';
function='LABEL';
text=_name_;
position='8';
run;
filename gsasfile "prog74.graph";
goptions reset=all gaccess=gsasfile autofeed dev=psl;
goptions horigin=1in vorigin=2in;
goptions hsize=6in vsize=8in;
proc gplot data=plotprint;
plot dim2 * dim1/anno=labels frame
        href=0 vref=0 lvref=3 lhref=3 vaxis=axis2 haxis=axis1
        vminor=1 hminor=1;
axis1 length=5 in order=(-1.5 to 1.5 by .5)  offset=(2)
        label = (h=1.2 'Dimension 1');
axis2 length=5 in order =(-2 to 1.5 by .5) offset=(2)
label=(h=1.2 a=90 r=0  'Dimension 2');
symbol v=star;
title1 h=1.2 'Multiple Correspondence Analysis: Farm Data';
title2 j=l 'Output 7.4';
title3 f=duplex "Plot of the Variables' Categories";
run;
```

Output 7.4

```
                         Output 7.4
               Multiple Correspondence Analysis
          Environmental Factors and Farm Management Data

                     The CORRESP Procedure

                          Burt Table
```

	Moist1	Moist2	Moist4	Moist5	Bf	Hf
Moist1	7	0	0	0	2	3
Moist2	0	4	0	0	1	0
Moist4	0	0	2	0	0	1
Moist5	0	0	0	7	0	1
Bf	2	1	0	0	3	0
Hf	3	0	1	1	0	5

Output 7.4 *(continued)*

Nm	1	1	0	4	0	0
Sf	1	2	1	2	0	0
Gu1	2	2	1	2	1	2
Gu2	3	2	1	2	1	1
Gu3	2	0	0	3	1	2
Manu0	1	1	0	4	0	0
Manu1	1	1	1	0	2	1
Manu2	3	0	1	0	1	2
Manu3	1	0	0	3	0	2
Manu4	1	2	0	0	0	0

Burt Table

	Nm	Sf	Gu1	Gu2	Gu3	Manu0
Moist1	1	1	2	3	2	1
Moist2	1	2	2	2	0	1
Moist4	0	1	1	1	0	0
Moist5	4	2	2	2	3	4
Bf	0	0	1	1	1	0
Hf	0	0	2	1	2	0
Nm	6	0	4	1	1	6
Sf	0	6	0	5	1	0
Gu1	4	0	7	0	0	4
Gu2	1	5	0	8	0	1
Gu3	1	1	0	0	5	1
Manu0	6	0	4	1	1	6
Manu1	0	0	2	0	1	0
Manu2	0	1	1	3	0	0
Manu3	0	2	0	1	3	0
Manu4	0	3	0	3	0	0

Burt Table

	Manu1	Manu2	Manu3	Manu4
Moist1	1	3	1	1
Moist2	1	0	0	2
Moist4	1	1	0	0
Moist5	0	0	3	0
Bf	2	1	0	0
Hf	1	2	2	0
Nm	0	0	0	0
Sf	0	1	2	3
Gu1	2	1	0	0
Gu2	0	3	1	3
Gu3	1	0	3	0
Manu0	0	0	0	0
Manu1	3	0	0	0
Manu2	0	4	0	0
Manu3	0	0	4	0
Manu4	0	0	0	3

Output 7.4
(continued)

```
                 The CORRESP Procedure

           Inertia and Chi-Square Decomposition

Singular    Principal     Chi-                  Cumulative
   Value      Inertia   Square    Percent         Percent

0.80617     0.64992    91.725      21.66           21.66
0.74511     0.55520    78.356      18.51           40.17
0.71899     0.51694    72.958      17.23           57.40
0.61806     0.38200    53.912      12.73           70.14
0.55704     0.31029    43.793      10.34           80.48
0.46999     0.22089    31.175       7.36           87.84
0.36506     0.13327    18.809       4.44           92.28
0.29847     0.08909    12.573       2.97           95.25
0.27829     0.07745    10.930       2.58           97.83
0.21800     0.04752     6.707       1.58           99.42
0.13202     0.01743     2.460       0.58          100.00

  Total     3.00000   423.399     100.00

 Degrees of Freedom = 225

                      4    8   12   16   20
                  ----+----+----+----+----+---
                  ***************************
                  ***********************
                  **********************
                  ****************
                  *************
                  *********
                  ******
                  ****
                  ***
                  **
                  *


                  Display of MCA Results

              Category        Dim1         Dim2

               Moist1      0.38514      0.63719
               Moist2      0.41615     -0.82695
               Moist4      0.71068      0.76325
               Moist5     -0.82599     -0.38272
               Bf          0.46611      1.11843
               Hf          0.22203      1.04503
               Nm         -1.33579     -0.47372
               Sf          0.91771     -0.95635
               Gu1        -0.70275      0.28718
               Gu2         0.79397     -0.57740
               Gu3        -0.28651      0.52178
               Manu0      -1.33579     -0.47372
               Manu1       0.20300      1.35201
               Manu2       0.83973      0.71305
               Manu3       0.01892      0.13400
               Manu4       1.32371     -1.53397
```

Output 7.4

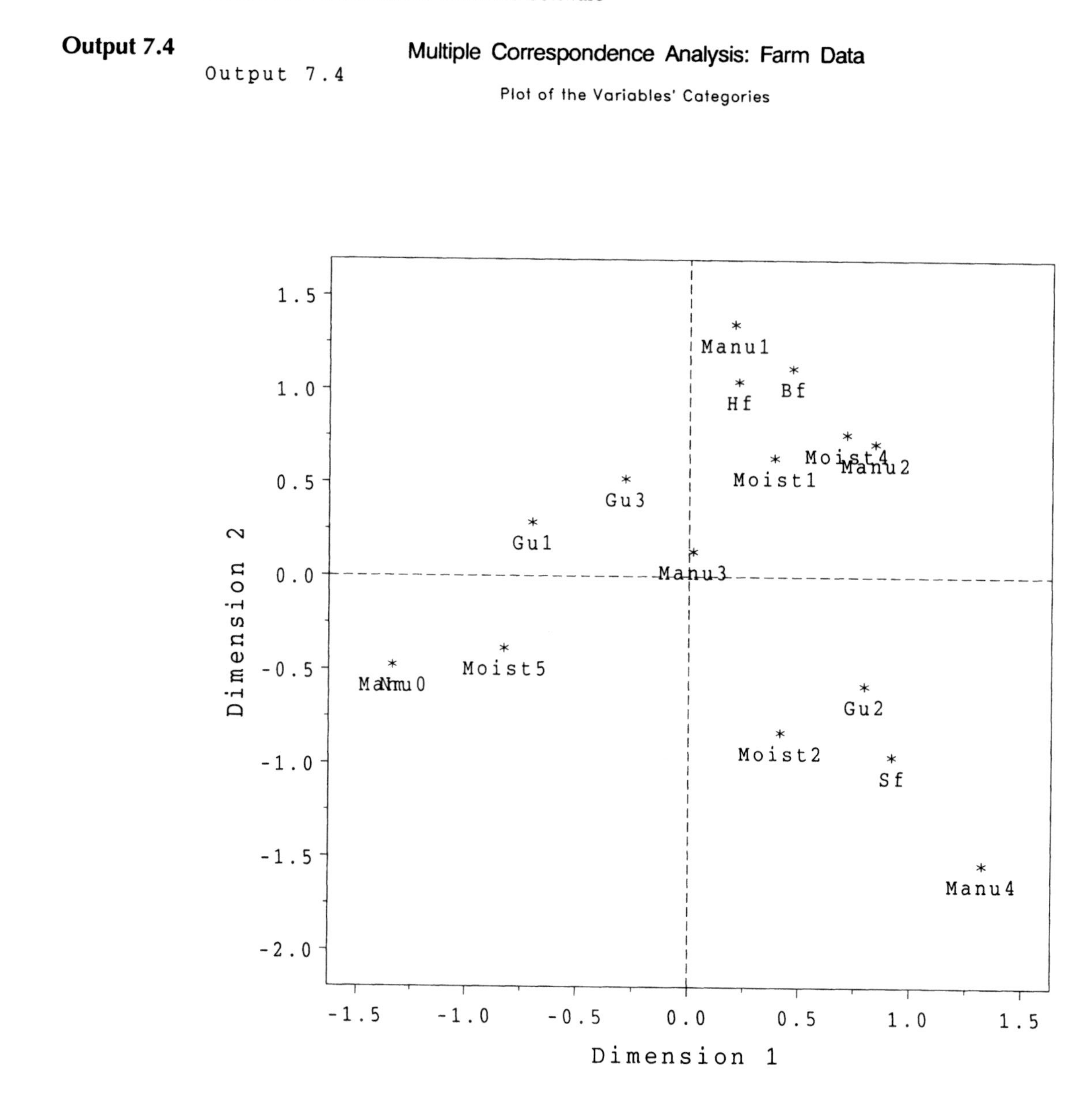

7.4 CA as a Canonical Correlation Analysis

Canonical correlation analysis (CCA) of two sets of variables has been discussed in Chapter 3. It is possible to apply the techniques developed there to determine the coordinates for plotting in correspondence analysis. The connection between CA and CCA is helpful in providing insight into the techniques. At the same time this connection can add to the interpretation by borrowing the similar (but relevant) interpretation from CCA.

Suppose as before we have two categorical variables X and Y with a and b categories respectively. Let $n \times (a+b)$ matrix $\mathbf{Z} = (\mathbf{X} : \mathbf{Y})$ be the data matrix in the form of indicators (0 or 1), indicating in which category of X and Y each of the n individuals belong. Then treating these as data on two multivariate random vector variables and using the matrix (matrix similar to the variance covariance matrix)

$$\mathbf{Z}'\mathbf{Z} = n \begin{bmatrix} \mathbf{P}_r & \mathbf{P} \\ \mathbf{P}' & \mathbf{P}_c \end{bmatrix}$$

as the basis, we can perform a canonical correlation analysis. This $(a+b) \times (a+b)$ matrix is the Burt matrix. Suppose $RV1$ is the raw canonical coefficient vector (see Chapter 3) for the first canonical variable, that is, the canonical variable corresponding to the first canonical correlation coefficient for variable X and suppose F_1 is the first column of the matrix of principal coordinates for row profiles. Then F_1 can be obtained in terms of $RV1$ as $F_1 = \sqrt{n}\rho_1 RV1$, where ρ_1 is the first canonical correlation coefficient. In the next example we will illustrate this relationship.

EXAMPLE 5 ***Correspondence Analysis as a Canonical Correlation Analysis Rating of Drug's Efficacy: Analgesic Drugs Data*** We consider the drug efficacy data used in Example 3. In Program 7.5 we first compute the principal coordinates for row and column profiles using PROC CORRESP and then also perform a canonical correlation analysis using the Burt matrix as the variance-covariance matrix of two sets of variables. Since the main objective of our analysis here is to get the coordinates for (symmetric) plot of the row profiles using the canonical correlation analysis, without going into much detail, we will present the relevant outputs from the two analyses in Output 7.5. Before we interpret the output we must realize that the first canonical correlation coefficient (= 1) and the first columns in each of the raw canonical coefficients should be discarded as trivial columns before any further analysis is performed.

We first note that the singular values, 0.5520, 0.2781, and 0.0838, from the correspondence analysis are the same as the canonical correlations from the canonical correlation analysis. Further (and apart from a sign) the Dim1 of row coordinates for the symmetric plot in the correspondence analysis can be obtained as $\sqrt{121}\rho_1 V2$. That is, it is the first canonical correlation times the V2 vector in Raw Canonical Coefficients for the 'VAR' Variables. Similarly, column coordinates for the symmetric plot can be obtained from the corresponding variables in the 'WITH' Variables section of the output.

```
/* Program 7.5 */

options ls=64 ps=45 nodate nonumber;

title1 'Output 7.5';
title2 'Correspondence Analysis Results';
title3 "Rating of Drug's Efficacy: Analgesic Drugs Data";
data rate;
input Drug$     Poor   Fair   Good   V_good   Excellnt;
cards;
Z100              5      1     10       8          6
EC4               5      3      3       8         12
C60              10      6     12       3          0
C15               7     12      8       1          1
;

proc corresp data=rate short;
var _numeric_;
id drug;
run;
```

```
title2 'Canonical Correlation Analysis Results';
title3 "Rating of Drug's Efficacy: Analgesic Drugs Data";
data cancor (type=cov);
_type_='cov';
input _name_$ Z100  EC4  C60  C15  POOR  FAIR  GOOD  V_GOOD  EXCELLNT;
cards;
 Z100       30    0    0    0    5    1   10    8    6
 EC4         0   31    0    0    5    3    3    8   12
 C60         0    0   31    0   10    6   12    3    0
 C15         0    0    0   29    7   12    8    1    1
 POOR        5    5   10    7   27    0    0    0    0
 FAIR        1    3    6   12    0   22    0    0    0
 GOOD       10    3   12    8    0    0   33    0    0
 V_GOOD      8    8    3    1    0    0    0   20    0
 EXCELLNT    6   12    0    1    0    0    0    0   19
run;

proc cancorr data=cancor(type=cov);
var z100 ec4 c60 c15;
with poor fair good v_good excellnt;
run;
```

Output 7.5

```
                        Output 7.5
               Correspondence Analysis Results
        Rating of Drug's Efficacy: Analgesic Drugs Data

              The Correspondence Analysis Procedure


               Inertia and Chi-Square Decomposition

Singular  Principal Chi-
Values    Inertias  Squares Percents   16   32   48   64   80
                                     ----+----+----+----+----+---
0.55197   0.30467   36.8647  78.32% ***********************
0.27810   0.07734    9.3583  19.88% ******
0.08376   0.00701    0.8488   1.80% *
          -------   -------
          0.38902   47.0718 (Degrees of Freedom = 12)


                       Row Coordinates

                           Dim1          Dim2

              Z100       -.349313      -.301166
              EC4        -.704011      0.246005
              C60        0.454881      -.248031
              C15        0.627670      0.313717


                     Column Coordinates

                           Dim1          Dim2

            POOR         0.246652      -.074592
            FAIR         0.642330      0.443469
            GOOD         0.267625      -.298590
            V_GOOD       -.582851      -.156716
            EXCELLNT     -.945552      0.276076
```

Output 7.5
(continued)

Canonical Correlation Analysis

	Canonical Correlation	Adjusted Canonical Correlation	Approx Standard Error	Squared Canonical Correlation
1	1.000000	1.000000	0.000000	1.000000
2	0.551966	0.551639	0.006954	0.304667
3	0.278104	0.277667	0.009227	0.077342
4	0.083755	0.083295	0.009930	0.007015

Raw Canonical Coefficients for the 'VAR' Variables

	V1	V2	V3	V4
Z100	0.0909090909	0.0575320165	-0.098447995	0.1098495785
EC4	0.0909090909	0.1159509004	0.0804162817	-0.063892191
C60	0.0909090909	-0.074919115	-0.081078661	-0.108659922
C15	0.0909090909	-0.103377443	0.1025508139	0.0708147636

Raw Canonical Coefficients for the 'WITH' Variables

	W1	W2	W3	W4
POOR	0.0909090909	-0.040623699	-0.024383329	-0.159684539
FAIR	0.0909090909	-0.10579197	0.1449653723	0.0629489733
GOOD	0.0909090909	-0.044078011	-0.097605712	0.0612963573
V_GOOD	0.0909090909	0.0959957628	-0.051228906	0.0671565134
EXCELLNT	0.0909090909	0.1557327551	0.0902462264	-0.023121311

7.5 Correspondence Analysis Using Andrews Plots

Andrews plots and modified Andrews plots have been found to be very useful tools for graphically representing multivariate data. A detailed discussion of these plots along with the SAS code for plotting them can be found in Khattree and Naik (1999). Some other applications of these plots in cluster analysis have been discussed in Chapter 6. For a p–dimensional multivariate observation $\mathbf{y} = (y_1, y_2, \ldots, y_p)'$ the Andrews function (Andrews, 1982) is defined as a function of t as

$$f_{\mathbf{y}}(t) = \frac{y_1}{\sqrt{2}} + y_2 sin(t) + y_3 cos(t) + y_4 sin(2t) + y_5 cos(2t) + \cdots, \quad -\pi \le t \le \pi. \tag{7.10}$$

For a given $t = t_0$, $(t_0, f_{\mathbf{y}}(t_0))$ represents a point on the curve, and by varying t between $-\pi$ and π, an Andrews curve is obtained as a collection of all such points. Suppose we have a sample of n multivariate observations. Corresponding to these n multivariate observations, there will be n different Andrews curves. A plot consisting of such curves is called an Andrews plot. The Andrews function given in Equation 7.10 has some very useful properties. Most noted among them is the distance property, which states that, apart from a constant, the Euclidean distance between the multi-dimensional points $\mathbf{x} = (x_1, x_2, \ldots, x_p)'$ and $\mathbf{y} = (y_1, y_2, \ldots, y_p)'$ is preserved as the L_2 distance between two curves $f_{\mathbf{x}}(t)$ and $f_{\mathbf{y}}(t)$, viz., $\int_{-\pi}^{\pi}(f_{\mathbf{x}}(t) - f_{\mathbf{y}}(t))^2 dt$. Specifically,

$$\int_{-\pi}^{\pi}(f_{\mathbf{x}}(t) - f_{\mathbf{y}}(t))^2 dt = \pi \sum_{i=1}^{p}(x_i - y_i)^2.$$

Thus, the points that are closer to each other in the p-dimensional space are represented as curves that are nearer to each other and vice versa.

The modified Andrews function proposed by Khattree and Naik (1998) is defined as

$$g_{\mathbf{y}}(t) = \frac{1}{\sqrt{2}}\{y_1 + y_2(sin(t) + cos(t)) + y_3(sin(t) - cos(t)) + y_4(sin(2t) + cos(2t)) + y_5(sin(2t) - cos(2t)) + \cdots\}, -\pi \le t \le \pi. \quad (7.11)$$

In practice this plot is found to produce a better display of the data in the form of curves. Khattree and Naik (1998) indicate that both Andrews plots or modified Andrews plots can be used in correspondence analysis for displaying the row and column profiles.

As described in Section 7.2 and using the same notations, in correspondence analysis the coordinates (principal coordinates) for representing the a row profiles on a plane are the points formed by a rows and the first two columns of $\mathbf{F} = \mathbf{D}_r^{-1}\mathbf{A}\mathbf{\Lambda}$, and those for representing b column profiles are the points formed by b rows and the first two columns of $\mathbf{G} = \mathbf{D}_c^{-1}\mathbf{B}\mathbf{\Lambda}$. These coordinates correspond to a symmetric plot in correspondence analysis. The plot is considered to be a good representation of the whole data set if $\lambda_1^2 + \lambda_2^2$ accounts for most of the total inertia. The graphical representation at this stage can be handled more effectively using an Andrews or modified Andrews plot. Instead of considering only the first two columns of **F** and **G**, we suggest transforming the rows of **F** and **G** (which represent the respective coordinates in $min(a-1, b-1)$−dimensional space) to an Andrews or modified Andrews curves and then display these curves as an Andrews or modified Andrews plot. Since no dimensionality reduction takes place and we do not discard the terms corresponding to the three and higher dimensions, there is no loss of information. This is an immediate advantage of using an Andrews or modified Andrews plot for correspondence analysis. In the following examples we restrict ourselves to using modified Andrews plots.

Modified Andrews plots corresponding to asymmetric plots in correspondence analysis can also be constructed by taking the appropriate choice of coordinates as discussed earlier. But we will present only the symmetric version of modified Andrews plots. If we want, by changing the PROFILE=BOTH option to PROFILE=ROW in the PROC CORRESP statement in Program 7.6, we can get the asymmetric version of the plot.

EXAMPLE 6 ***Modified Andrews Plot for Socioeconomic Status by Mental Health Status of Children Data*** To illustrate a modified Andrews plot, we reconsider cross-classified data on the socioeconomic status and mental status of children. The objective of the study was to examine the relationship, if any, between the mental impairment and the parents' socioeconomic status. The matrices **F** and **G** containing the principal coordinates for the row and column points collected from Output 7.2 are listed in Table 7.4. SAS code for plotting the modified Andrews plot is given in Program 7.6, and the plots are presented in Output 7.6.

The following observations, which are similar to what we have already observed in Example 7.2, are made from the first plot, where y_1 has a coefficient $1/\sqrt{2}$ (see the footnote in the plot).

(a) The two variables are ordered categorical in nature, and this fact clearly shows up in the plot. Also, it can be observed that there is possibly further subgrouping within the socioeconomic status. Status 1 and 2 are similar to each other; 3 and 4 are similar, and to some extent, there is some similarity between status 5 and 6. However, as subgroups, these three subgroups clearly separate themselves out in these plots.

(b) With respect to mental health status, the health status of Mild and Moderate appear to be similar, but are clearly different from Well and Impaired. Further, the ordinal nature of these health statuses is self evident in these plots.

(c) Although this cannot be said for sure in a symmetric version of this plot, there appears to be positive association between the mental health status of children and the parents' socioeconomic status, with mental health of children to be generally better at the higher levels of parents' socioeconomic status. Further, the upper two levels of parents' socioeconomic status groups generally correspond to a Well mental health status; the next two levels are closely related to Mild or Moderate and the lowest two levels of parents' socioeconomic status generally severely affect the mental health status of children toward impairment.

TABLE 7.4 Parents' Socioeconomic and Mental Health Status: Row and Column Points

Points		Coordinates		
Row		Dim 1	Dim 2	Dim 3
(Socioeconomic)	1	0.180932	0.019248	0.027525
	2	0.184996	0.011625	−.027386
	3	0.059031	0.022197	−.010575
	4	−.008887	−.042080	0.011025
	5	−.165392	−.043606	−.010368
	6	−.287690	0.061994	0.004824
Column	Well	0.259536	−.012102	0.022589
(Mental Health)	Mild	0.029588	−.023651	−.019818
	Moderate	−.014210	0.069901	−.003230
	Impaired	−.237392	−.018897	0.015848

The second plot in Output 7.6 corresponds to the case in which y_2 has been assigned to the constant $\frac{1}{\sqrt{2}}$ in the function $g_y(t)$. This plot reveals a few additional interesting observations.

(a) With respect to parents' socioeconomic status, curves corresponding to groups 1 and 2 show the patterns just opposite to those corresponding to groups 5 and 6.

(b) With respect to mental health status, the Well and Impaired groups follow patterns opposite to each other.

(c) There, possibly, is very little difference between groups 3 and 4 of parents' socioeconomic status and between groups Mild and Moderate with respect to mental health status. Conceivably, in each case, the corresponding two groups can be combined.

```
/* Program 7.6 */

options ls=64 ps=45 nodate nonumber;

title1 'Output 7.6';
*title2 'Modified Andrews Plots for Socioeconomic Status';
*title3  'by Mental Status of Children Data';
data status;
length sestatus$8;
input sestatus$ Well Mild Moderate Impaired ;
```

```
lines;
1 64 94  58 46
2 57 94  54 40
3 57 105 65 60
4 72 141 77 94
5 36 97  54 78
6 21 71  54 71
;

proc corresp data=status out=plot short rp cp profile=both dim=3;
var well mild moderate impaired;
id sestatus;
run;

data andrew;
set plot;
y1=dim1; y2=dim2; y3=dim3;
if y1=. then delete;
keep sestatus y1 y2 y3;
run;

filename gsasfile "prog76a.graph";
goptions reset=all gaccess=gsasfile autofeed dev=pslmono;
goptions horigin=1in vorigin=2in;
goptions hsize=6in vsize=8in;

data andrews;
set andrew;
Row_Col=_n_;
pi=3.14159265;
s=1/sqrt(2);
inc=2*pi/100;
do t=-pi to pi by inc;
mz1=s*(y1+(sin(t)+cos(t))*y2+(sin(t)-cos(t))*y3);
output;
end;

symbol1 i=join v=none l=1;
symbol2 i=join v=none l=1;
symbol3 i=join v=none l=1;
symbol4 i=join v=none l=1;
symbol5 i=join v=none l=1;
symbol6 i=join v=none l=1;
symbol7 i=join v=none l=20;
symbol8 i=join v=none l=20;
symbol9 i=join v=none l=20;
symbol10 i=join v=none l=20;

title1 h=1.2 'Modified Andrews Plot';
title2 j=l 'Output 7.6';
title3 'Mental Health vs. Socioeconomic Status';
footnote1 j=l 'Solid line: Rows; Broken line: Columns';
footnote2 j=l ' ';
footnote3 j=l 'Coeff. of y1 is=1/sqrt(2)';
```

```
data labels;
set andrews;
retain xsys '2' ysys '2';
x=3.25;
y=s*(y1+(sin(x)+cos(x))*y2+(sin(x)-cos(x))*y3);
function='LABEL';
text=sestatus;
proc gplot data=andrews;
plot mz1*t=Row_Col/vaxis=axis1 haxis=axis2 nolegend anno=labels;
axis1 label=(a=90 h=1.2 f=duplex 'Modified Function: g(t)');
axis2 label=(h=1.2 f=duplex 't') offset=(2);
run;

data andrews2;
set andrew;
Row_Col=_n_;
pi=3.14159265;
s=1/sqrt(2);
inc=2*pi/100;
do t=-pi to pi by inc;
mz2=s*(y2+(sin(t)+cos(t))*y1+(sin(t)-cos(t))*y3);
output;
end;

filename gsasfile "prog76b.graph";
goptions reset=all gaccess=gsasfile autofeed dev=pslmono;
goptions horigin=1in vorigin=2in;
goptions hsize=6in vsize=8in;
symbol1 i=join v=none l=1;
symbol2 i=join v=none l=1;
symbol3 i=join v=none l=1;
symbol4 i=join v=none l=1;
symbol5 i=join v=none l=1;
symbol6 i=join v=none l=1;
symbol7 i=join v=none l=20;
symbol8 i=join v=none l=20;
symbol9 i=join v=none l=20;
symbol10 i=join v=none l=20;
title1 h=1.2 'Modified Andrews Plot';
title2 j=l 'Output 7.6';
title3 'Mental Health vs. Socioeconomic Status';
footnote1 j=l 'Solid line: Rows; Broken line: Columns';
footnote2 j=l ' ';
footnote3 j=l 'Coeff. of y2 is=1/sqrt(2)';
data label;
set andrews2;
retain xsys '2' ysys '2';
x=3.25;
y=s*(y2+(sin(x)+cos(x))*y1+(sin(x)-cos(x))*y3);
function='LABEL';
text=sestatus;
proc gplot data=andrews2;
plot mz2*t=Row_Col/vaxis=axis1 haxis=axis2 nolegend anno=label;
 axis1 label=(a=90 h=1.2 f=duplex 'Modified Function: g(t)');
 axis2 label=(h=1.2 f=duplex 't') offset=(2);
run;
```

Output 7.6

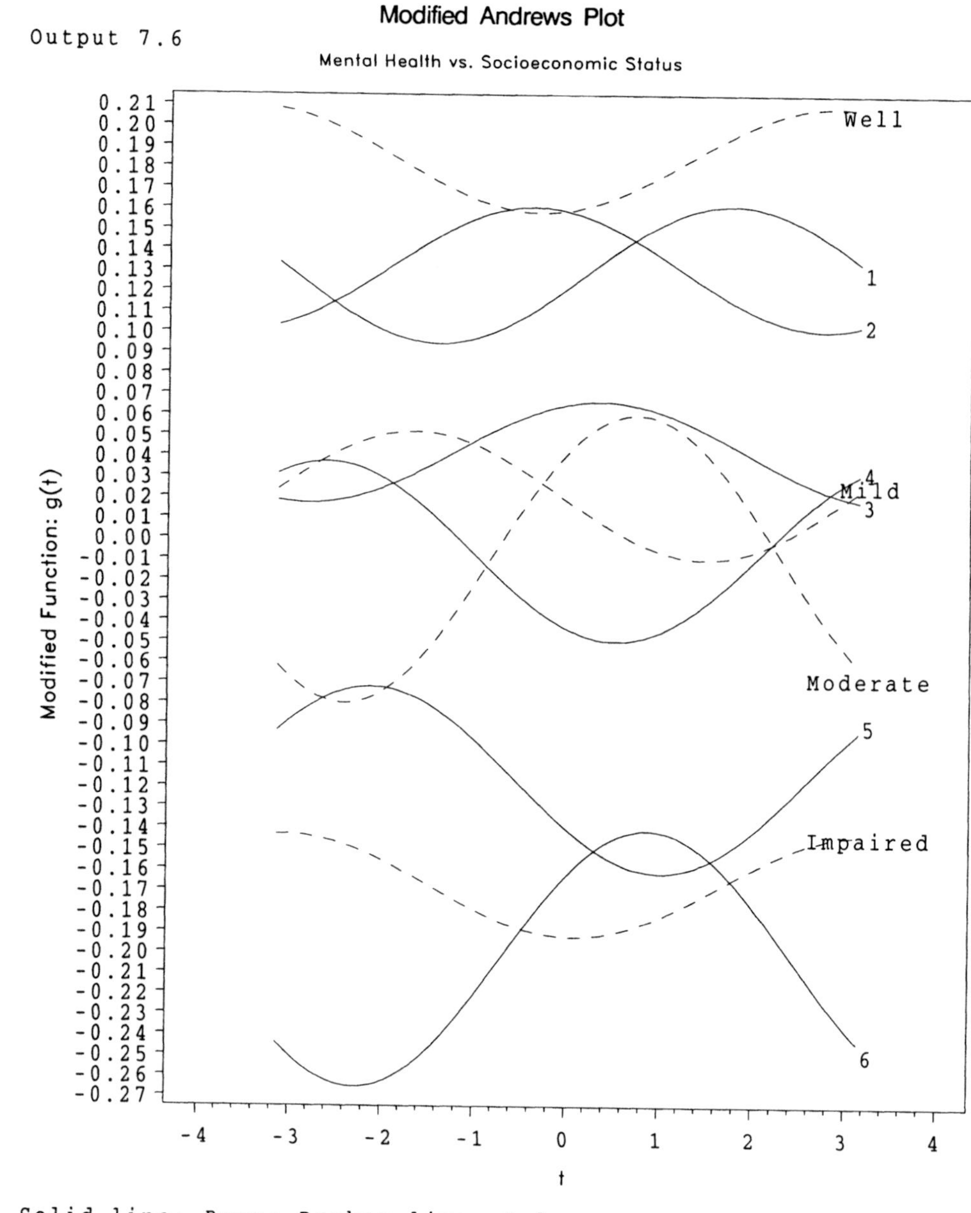
Modified Andrews Plot
Output 7.6
Mental Health vs. Socioeconomic Status
Modified Function: g(t)
Well
Mild
Moderate
Impaired
1
2
3
4
5
6
t
Solid line: Rows; Broken line: Columns
Coeff. of y1 is=1/sqrt(2)

Output 7.6

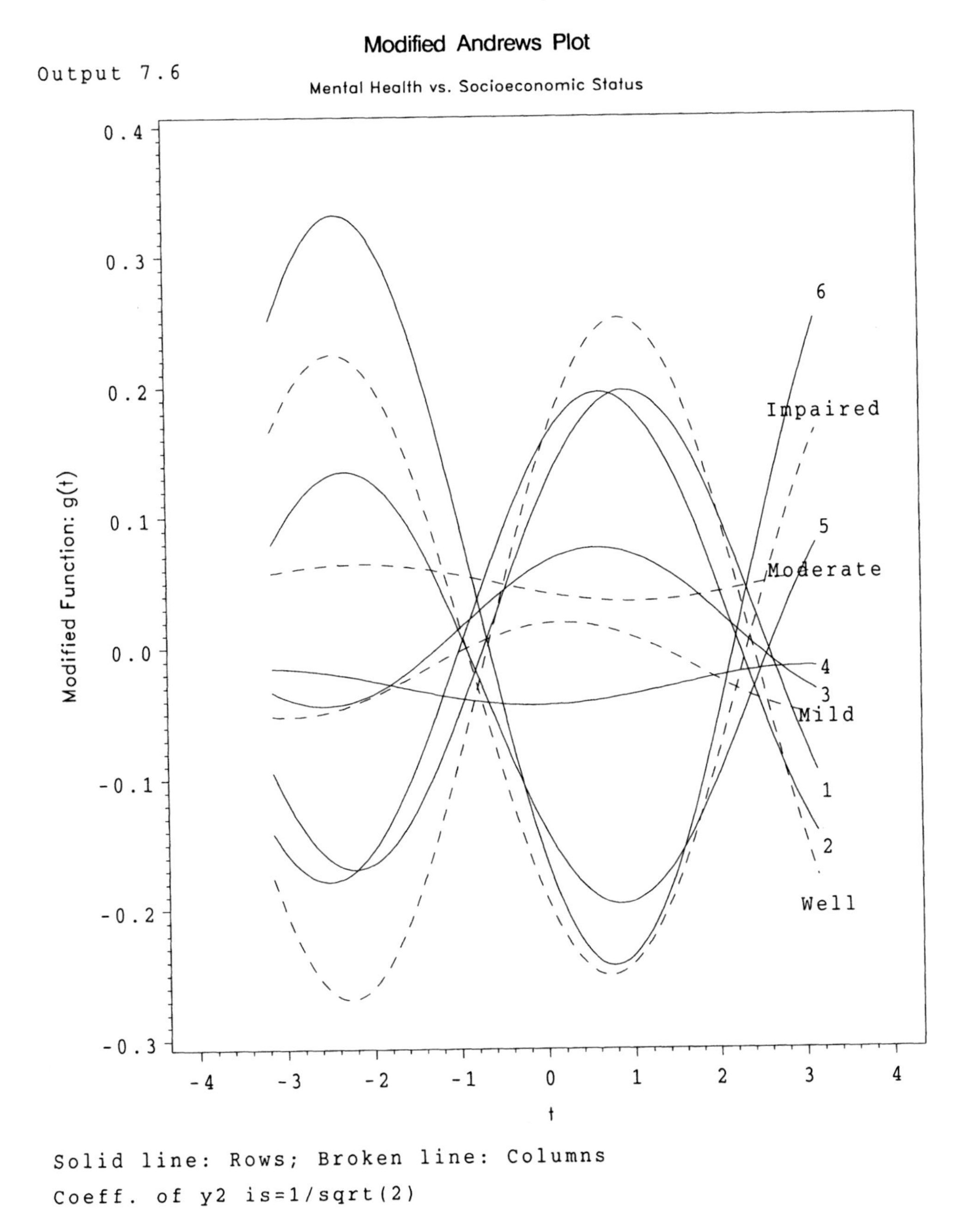

Solid line: Rows; Broken line: Columns

Coeff. of y2 is=1/sqrt(2)

EXAMPLE 7 ***Modified Andrews Plot for Catheterization and Cardiovascular Diagnosis Data*** For further illustration of a modified Andrews plot we consider a data set of Fisher et al. (1982). The data are from a study of reproducibility of a coronary arteriography diagnostic procedure that plays an important role in deciding if performing bypass surgery on a patient may be appropriate. It is important to monitor the quality of arteriography and ascertain the amount of agreement between the quality control site readings and the clinical site readings. The problem is essentially that of interlaboratory testing in a categorical data context. Table 7.5 presents the results for 870 randomly selected patients with two readings. For these readings the amount of disease was classified as none, zero vessel disease but some disease, one-, two-, and three-vessel disease.

For identification in plots, these groups for the quality control site are denoted by 1, 2, 3, 4 and 5, respectively. For the clinical site, the notations used are A, B, C, D, and E, respectively.

TABLE 7.5 Agreement with Respect to Number of Diseased Vessels

Quality Control Site Reading	Clinical Site Reading				
	Normal (A)	Some (B)	One (C)	Two (D)	Three (E)
Normal (1)	13	8	1	0	0
Some (2)	6	43	19	4	5
One (3)	1	9	155	54	24
Two (4)	0	2	18	162	68
Three (5)	0	0	11	27	240

In Program 7.7, using the CORRESP procedure we obtain the breakdown of the inertia among different dimensions. Output is shown in Output 7.7. The percentages of total inertia explained by each of the four dimensions are 43.80, 28.59, 16.31, and 11.30%. In this case, none of the four dimensions can be deemed redundant in that each explains a significant percent of total inertia. The loss $\mathcal{L}$ incurred in using only two dimensions here is about 28%. Whenever this happens, the use of only the first two dimensions to obtain a plot of points will not be very effective compared to the use of the modified Andrews plot, which uses all dimensions. This observation reinforces the usefulness of (modified) Andrews plots for graphical display instead of the usual two-dimensional plots.

The row and column points $\mathbf{y} = (y_1, y_2, y_3, y_4)'$ for the data are obtained using Program 7.7 and presented in Table 7.6. In Program 7.7 we also obtain the function $g_{\mathbf{y}}(t)$ for each of the five rows and columns. The plot is given in Output 7.7. In the case of perfect agreement between the clinical site and quality control site readings, all row points should be the same as the corresponding column points, and thus the corresponding (modified) Andrews curves should be identical also. This argument justifies looking at the closeness of the curves in this example.

As seen in the plot, the agreement between the two readings appears to be quite good for each of the five classification groups, and the (modified) Andrews curves of the corresponding points follow each other very closely.

TABLE 7.6 Clinical and QC Site Evaluations: Row and Column Points

Points		Coordinates			
		Dim 1	Dim 2	Dim 3	Dim 4
Row (QC Site)	Normal	3.14513	1.72789	0.91817	1.52739
	Some	1.67179	0.25096	−0.32230	−0.97903
	One	0.24290	−0.84605	−0.32512	0.27160
	Two	−0.34541	−0.16330	0.72270	−0.15729
	Three	−0.61364	0.68014	−0.34912	0.05434
Column (Clinical)	Normal	3.12841	1.74992	0.96954	1.71600
	Some	1.94382	0.40705	−0.25836	−1.07768
	One	0.35725	−0.89110	−0.45605	0.26900
	Two	−0.26104	−0.32340	0.72042	−0.12930
	Three	−0.56821	0.59772	−0.26196	0.02836

```
/* Program 7.7 */

options ls=64 ps=45 nodate nonumber;

title1 'Output 7.7';
title2 'Modified Andrews Plot';
title3 'Agreement w.r.t. No. of Diseased Vessels';

data cass;
input ser$ a b c d e;
lines;
1 13  8   1   0   0
2  6 43  19   4   5
3  1  9 155  54  24
4  0  2  18 162  68
5  0  0  11  27 240
;
/* Source: Fisher et al. (1982).  Courtesy: Wiley-Liss, Inc. */

proc corresp data=cass out=plot rp cp short dim=4;
var a b c d e;
id ser;
run;

data andrew;
set plot;
y1=dim1; y2=dim2; y3=dim3; y4=dim4;
if y1=. then delete;
keep ser y1 y2 y3 y4;
run;

data andrews;
set andrew;
Row_Col=_n_;
pi=3.14159265;
s=1/sqrt(2);
inc=2*pi/100;
do t=-pi to pi by inc;
mz1=s*(y1+(sin(t)+cos(t))*y2+(sin(t)-cos(t))*y3+(sin(2*t)+cos(2*t))*y4);
output;
end;
filename gsasfile 'prog77.graph';
goptions reset=all gaccess=gsasfile autofeed dev=pslmono;
goptions horigin=1in vorigin=2in;
goptions hsize=6in vsize=8in;
symbol1 i=join v=none l=1;
symbol2 i=join v=none l=1;
symbol3 i=join v=none l=1;
symbol4 i=join v=none l=1;
symbol5 i=join v=none l=1;
symbol6 i=join v=none l=20;
symbol7 i=join v=none l=20;
symbol8 i=join v=none l=20;
symbol9 i=join v=none l=20;
symbol10 i=join v=none l=20;
title1 h=1.2 'Modified Andrews Plot';
title2 j=l 'Output 7.7';
title3 f=duplex 'Agreement w.r.t. No. of Diseased Vessels';
footnote1 j=l 'Solid line: Rows; Broken line: Columns';
footnote2 j=l ' ';
```

```
data label;
set andrews;
retain xsys '2' ysys '2';
x=3.25;
y=s*(y1+(sin(x)+cos(x))*y2+(sin(x)-cos(x))*y3+(sin(2*x)+cos(2*x))*y4);
function='LABEL';
text=ser;
proc gplot data=andrews;
plot mz1*t=Row_Col/vaxis=axis1 haxis=axis2 nolegend anno=label;
axis1 label=(a=90 h=1.5 f=duplex 'Modified Function: g(t)');
axis2 label=(h=1.5 f=duplex 't') offset=(2);
footnote3 j=l 'Coeff. of y1 is=1/sqrt(2)';
run;
```

Output 7.7

```
                          Output 7.7
                     Modified Andrews Plot
              Agreement w.r.t. No. of Diseased Vessels

               The Correspondence Analysis Procedure
                Inertia and Chi-Square Decomposition

Singular  Principal Chi-
 Values   Inertias  Squares Percents    9   18   27   36   45
                                    ----+----+----+----+----+---
0.81767   0.66859   581.673  43.80% ************************
0.66067   0.43648   379.738  28.59% ****************
0.49907   0.24907   216.689  16.31% *********
0.41531   0.17248   150.061  11.30% ******
          -------   -------
          1.52662   1328.16 (Degrees of Freedom = 16)
```

Output 7.7

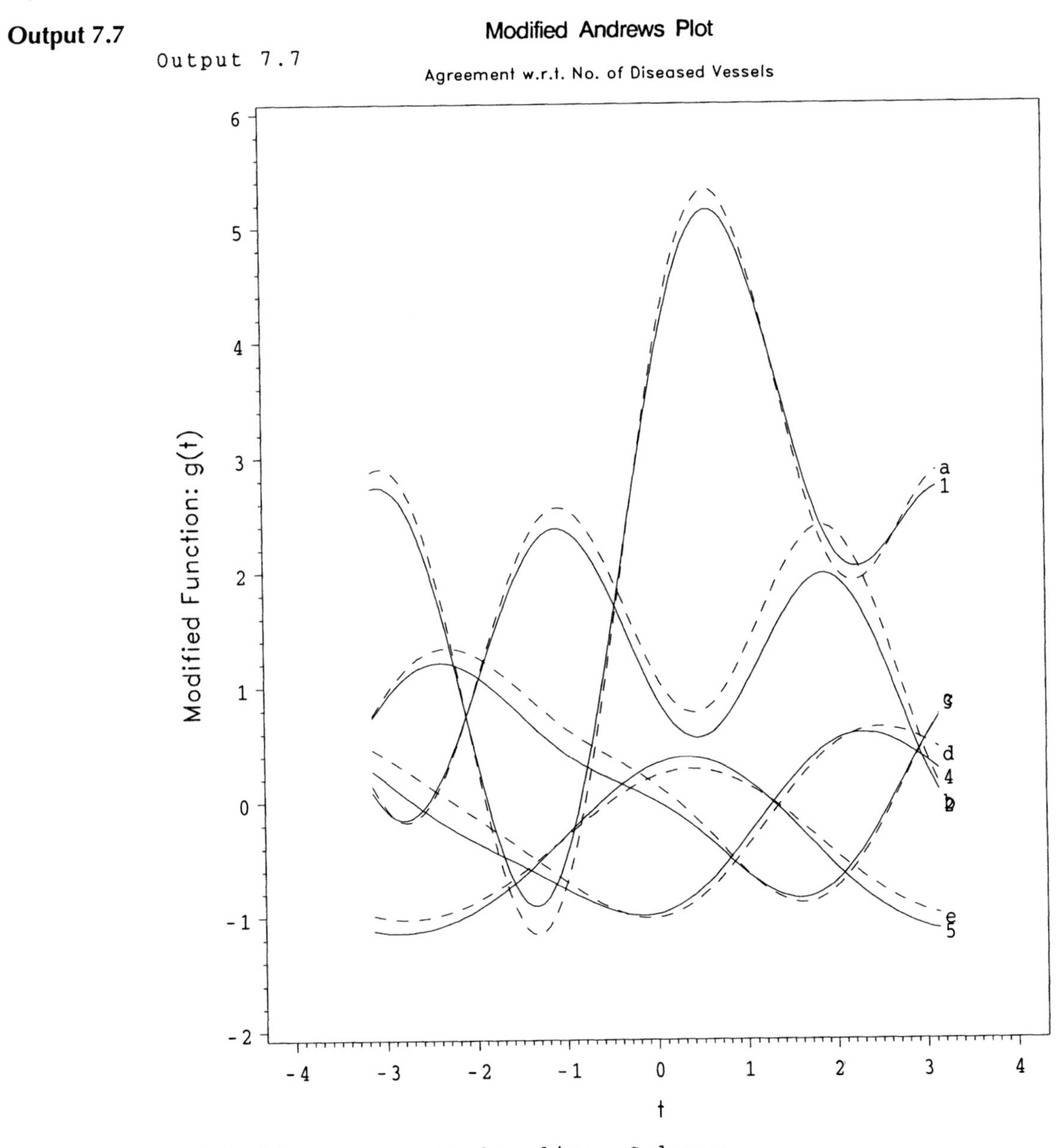
Output 7.7
Modified Andrews Plot
Agreement w.r.t. No. of Diseased Vessels
Modified Function: g(t)
6
5
4
3
2
1
0
-1
-2
-4
-3
-2
-1
0
1
2
3
4
t
a
1
c
d
4
b
2
e
5
Solid line: Rows; Broken line: Columns
Coeff. of y1 is=1/sqrt(2)

7.6 Correspondence Analysis Using Hellinger Distance

In correspondence analysis the distance of choice for measuring the distance between two profiles is the chi-square distance. The distance between the ith and j^{th} row profiles is measured by the chi-square distance given in Equation 7.7. Rao (1995) pointed out several drawbacks in using the chi-square distance as an obvious choice. Noted among them are

(a) the chi-square distance between the ith and j^{th} row profiles is not just a function of the these profiles, but also involves **c**, the average row profile and

(b) since the chi-square distance uses the elements of the average row profile in the denominator, undue emphasis is given to the categories with low frequencies in measuring affinities between profiles (Rao, 1995).

To overcome some of these drawbacks, he suggests using Hellinger distance for performing an analysis similar to the correspondence analysis.

Let $\mathbf{R} = (r_{ij})$ be the $a \times b$ matrix of row profiles as defined in Section 7.2, Equation 7.2. Define

$$\mathbf{H}_{a\times b} = (h_{ij}) = (\sqrt{r_{ij}}) = \begin{bmatrix} \mathbf{h}_1' \\ \mathbf{h}_2' \\ \vdots \\ \mathbf{h}_a' \end{bmatrix},$$

where $\mathbf{h}_i = (\sqrt{r_{i1}}, \sqrt{r_{i2}}, \ldots, \sqrt{r_{ib}})'$, $i = 1, \ldots, a$.

Then the Hellinger distance between the ith and j^{th} row profiles is defined as

$$\delta_{ij}^2 = (\mathbf{h}_i - \mathbf{h}_j)'(\mathbf{h}_i - \mathbf{h}_j). \tag{7.12}$$

The above distance δ_{ij}^2 depends only on the ith and j^{th} row profiles. Thus, this distance does not suffer from the drawbacks associated with the chi-square distance.

Rao's canonical coordinates for representing row profiles in a lower dimensional Euclidean space are determined as follows. Consider the singular value decomposition of the $a \times b$ matrix $\mathbf{D}_r^{1/2}(\mathbf{H} - \mathbf{1}_a\boldsymbol{\eta}')$ say,

$$\lambda_1\mathbf{u}_1\mathbf{v}_1' + \lambda_2\mathbf{u}_2\mathbf{v}_2' + \cdots + \lambda_m\mathbf{u}_m\mathbf{v}_m',$$

where m is the rank of $\mathbf{D}_r^{1/2}(\mathbf{H} - \mathbf{1}_a\boldsymbol{\eta}')$, and $\mathbf{D}_r$ is the $a \times a$ diagonal matrix (defined in Section 7.2) with the diagonal entries, $(n_{1\cdot}/n, \ldots, n_{a\cdot}/n)$. Rao (1995) suggested using $\boldsymbol{\eta} = \sqrt{\mathbf{c}} = (\sqrt{\frac{n_{\cdot 1}}{n}}, \sqrt{\frac{n_{\cdot 2}}{n}}, \ldots, \sqrt{\frac{n_{\cdot b}}{n}})'$ or $\boldsymbol{\eta} = \mathbf{H}'\mathbf{r}$, where $\mathbf{r} = (n_{1\cdot}/n, \ldots, n_{a\cdot}/n)'$ is the $a \times 1$ column vector.

Then, Rao's canonical coordinates for the row profiles are

$$\lambda_1\mathbf{D}_r^{-1/2}\mathbf{u}_1,\ \lambda_2\mathbf{D}_r^{-1/2}\mathbf{u}_2, \ldots, \lambda_m\mathbf{D}_r^{-1/2}\mathbf{u}_m.$$

Rao (1995) also suggests plotting the standard coordinates given by

$$\lambda_1\Delta_c^{-1}\mathbf{v}_1,\ \lambda_2\Delta_c^{-1}\mathbf{v}_2,\ \ldots, \lambda_m\Delta_c^{-1}\mathbf{v}_m$$

for representing the columns on the same plot. Here $\mathbf{\Delta}_c$ is the diagonal matrix with the ith diagonal element equal to the square root of the ith diagonal element of

$$\lambda_1^2 \mathbf{v}_1 \mathbf{v}_1' + \lambda_2^2 \mathbf{v}_2 \mathbf{v}_2' + \cdots + \lambda_m^2 \mathbf{v}_m \mathbf{v}_m' = (\mathbf{H} - \mathbf{1}\boldsymbol{\eta}')'(\mathbf{H} - \mathbf{1}\boldsymbol{\eta}').$$

The canonical coordinates for representing the column profiles and the standard coordinates for representing rows can be derived similarly. For example, by taking $\mathbf{H} = (h_{ij}) = (\sqrt{c_{ij}})$, where $\mathbf{C} = (c_{ij})$ is the matrix of column profiles as defined in Section 7.2, and $\boldsymbol{\eta} = \sqrt{\mathbf{r}} = \left(\sqrt{\frac{n_{1\cdot}}{n}}, \sqrt{\frac{n_{2\cdot}}{n}}, \ldots, \sqrt{\frac{n_{a\cdot}}{n}}\right)'$ or $\boldsymbol{\eta} = \mathbf{H}'\mathbf{c}$, where $\mathbf{c} = (n_{\cdot 1}/n, \ldots, n_{\cdot b}/n)'$ is the $b \times 1$ column vector, and by determining the singular value decomposition of $\mathbf{D}_c^{1/2}(\mathbf{H} - \mathbf{1}\boldsymbol{\eta}')$, these quantities can be easily determined.

Rao (1995) also pointed out that the statistic

$$4n(\lambda_1^2 + \cdots + \lambda_m^2)$$

can be used for testing the independence of the variables in the two-way contingency table, and this statistic under the null hypothesis of independence follows a chi-square distribution with $(a-1)(b-1)$ degrees of freedom. Further the quantity

$$\mathcal{L} = \frac{\sum_{i=k+1}^{m} \lambda_i^2}{\sum_{i=1}^{m} \lambda_i^2} = 1 - \frac{\sum_{i=1}^{k} \lambda_i^2}{\sum_{i=1}^{m} \lambda_i^2}$$

can be used as a measure of relative loss of information in representing the profiles in a k-dimensional space (Rao, 1995).

EXAMPLE 8 ***Correspondence Analysis Using Hellinger Distance, Rating of Drug's Efficacy: Analgesic Drugs Data*** To illustrate the use of Hellinger distance for performing correspondence analysis, we will reconsider the rating of drug's efficacy data of (Calimlim et al., 1982) presented earlier in Example 3. We compute the canonical coordinates for row profiles representing various drugs and standard coordinates for columns representing ratings of these drugs using the IML procedure. The code is provided in Program 7.8, and the plot is provided in Output 7.8.

```
/* Program 7.8 */

options ls=64 ps=45 nodate nonumber;

title1 'Output 7.8';
title2 'Correspondence Analysis using Hellinger Distance';
title3 "Rating of Drug's Efficacy: Analgesic Drugs Data";

data d1;
input Drug$     Poor   Fair   Good   V_good   Excellnt;
cards;
Z100               5      1     10        8          6
EC4                5      3      3        8         12
C60               10      6     12        3          0
C15                7     12      8        1          1
;
```

```
proc iml;
use d1;
read all var{Poor,Fair,Good,V_good,Excellnt} into Y;
read all var{drug} into id;
close d1;
vars={"Poor", "Fair", "Good", "V_good", "Excellnt"};
id=id//vars;
a=4;
b=5;
one=j(a,1,1);
n=y[+,+];
P=Y/n;
r_sum=j(a,1,0);
do i=1 to a;
r_sum[i,1]=y[i,+]/n;
end;
c_sum=j(1,b,0);
do i=1 to b;
c_sum[1,i]=y[+,i]/n;
end;

D_r=diag(r_sum);
R=inv(D_r)*P;
Drhalf=sqrt(D_r);
eta=sqrt(t(R))*r_sum;
*svd_mat=Drhalf*(sqrt(R)-j(a,1,1)*t(eta));
svd_mat=Drhalf*(sqrt(R)-j(a,1,1)*sqrt(c_sum));
call svd(u_rmat,lambda_r,v_rmat,svd_mat);

lsquare=lambda_r#lambda_r;
sumlsq=lsquare[+];
ratio=lsquare/sumlsq;

print lsquare, sumlsq, ratio;

m=min(a,b);
D_l=diag(lambda_r[1:m]);
row_prof=inv(Drhalf)*(u_rmat[,1:m])*D_l;
delta=sqrt(diag((t(svd_mat)*svd_mat)));
std_cpt=inv(delta)*v_rmat[,1:m]*d_l;
dim=2;
rplot=row_prof[,1:dim];
cpoint=std_cpt[,1:dim];
out=rplot//cpoint;
cvar = concat(shape({"DIM"},1,dim),char(1:dim,1.));
create plot from out[rowname=id colname=cvar];
append from out[rowname=id];
close plot;
run;
data labels;
set plot;
x=dim1;
y=dim2;
retain xsys '2' ysys '2';
function='LABEL';
text=id;
position='8';
```

```
filename gsasfile "prog78.graph";
goptions reset=all gaccess=gsasfile autofeed dev=pslmono;
goptions horigin=1in vorigin=2in;
goptions hsize=6in vsize=8in;
    proc gplot data=plot;
    plot dim2 * dim1/anno=labels frame
        href=0 vref=0 lvref=3 lhref=3 vaxis=axis2 haxis=axis1
        vminor=1 hminor=1;
    axis1 length=5 in order=(-1 to 1 by .2)  offset=(2)
        label = (h=1.3 'Dimension 1');
    axis2 length=5 in order =(-.8 to .8 by .2) offset=(2)
    label=(h=1.3 a=90 r=0  'Dimension 2');
    symbol v=star;
title1 h=1.2 'Correspondence Analysis Using Hellinger Distance';
title2 j=l 'Output 7.8';
title3 f=duplex 'Plot of Different Drugs';
    run;
```

Output 7.8

```
                    Output 7.8
Correspondence Analysis using Hellinger Distance
Rating of Drug's Efficacy: Analgesic Drugs Data

                     LSQUARE

                    0.1014853
                    0.0247938
                    0.0053433
                    0.0015377
                    2.878E-42
```

Output 7.8

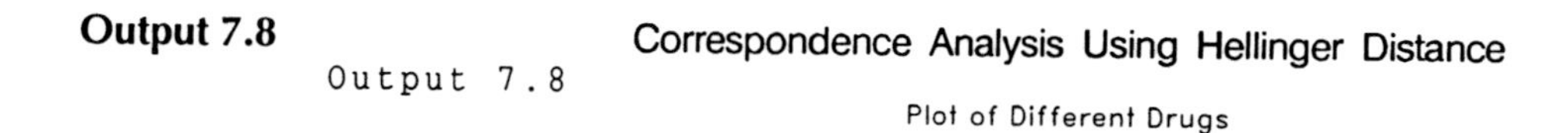

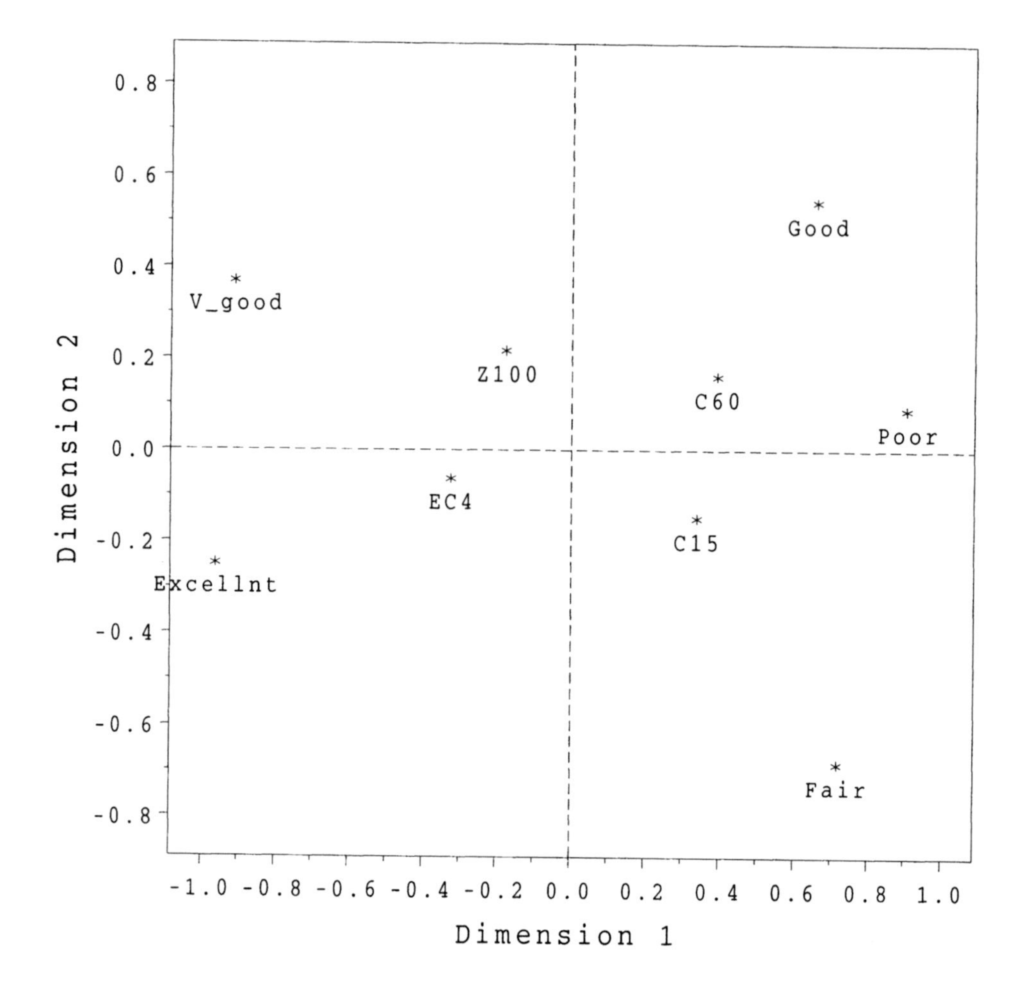

In the program we have used $\boldsymbol{\eta} = \sqrt{\mathbf{c}} = (\sqrt{\frac{n_{\cdot 1}}{n}}, \sqrt{\frac{n_{\cdot 2}}{n}}, \ldots, \sqrt{\frac{n_{\cdot b}}{n}})'$, and the statement that mentions the other choice, that is, $\boldsymbol{\eta} = \mathbf{H}'\mathbf{r}$, has been commented out. The values of the λ_i^2's are 0.1015, 0.0248, 0.0053, and 0.0015, respectively, with a total of 0.1332. Hence the loss incurred due to representing the row profiles in to a two-dimensional space is only about 5%. That is,

$$\mathcal{L} = 1 - \frac{.1015 + .0248}{.1332} = 1 - \frac{.1263}{.1332} \approx 1 - .95 = .05.$$

The four row profiles representing the four drugs, namely, EC4, Z100, C60, and C15 plot in an order similar to what we have observed before in Example 3. The points representing the columns graph consistently with what was observed before. The point representing the profile corresponding to C60 is placed more closely to the point Poor than to the point Good. This is consistent with the observed frequencies of 7 and 8 for those two categories, respectively. Thus we have an alternative distance to use for performing the correspondence analysis.

EXAMPLE 9 ***Hellinger Distance Plot for Catheterization and Cardiovascular Diagnosis Data*** For further illustration of using Hellinger distance for correspondence analysis we reconsider the catheterization and cardiovascular data of Fisher et al. (1982) considered in Example 7. Program 7.8 has been adapted to perform this analysis, and the code is provided in Program 7.9. The plot is provided in Output 7.9.

```
/* Program 7.9 */

options ls=64 ps=45 nodate nonumber;
title1 'Output 7.9';
title2 'Plot Based on Hellinger Distance';
title3 'Agreement w.r.t. No. of Diseased Vessels';

data d1;
input ser$ a b c d e;
lines;
1 13  8   1   0   0
2  6 43  19   4   5
3  1  9 155  54  24
4  0  2  18 162  68
5  0  0  11  27 240
;

proc iml;
use d1;
read all var{a,b,c,d,e} into Y;
read all var{ser} into id;
close d1;
vars={"a", "b", "c", "d", "e"};
id=id//vars;

a=5;
b=5;
one=j(a,1,1);
n=y[+,+];
P=Y/n;
r_sum=j(a,1,0);
do i=1 to a;
r_sum[i,1]=y[i,+]/n;
end;
c_sum=j(1,b,0);
do i=1 to b;
c_sum[1,i]=y[+,i]/n;
end;

D_r=diag(r_sum);
R=inv(D_r)*P;
Drhalf=sqrt(D_r);
eta=sqrt(t(R))*r_sum;
*svd_mat=Drhalf*(sqrt(R)-j(a,1,1)*t(eta));
svd_mat=Drhalf*(sqrt(R)-j(a,1,1)*sqrt(c_sum));
call svd(u_rmat,lambda_r,v_rmat,svd_mat);

lsquare=lambda_r#lambda_r;
sumlsq=lsquare[+];
ratio=lsquare/sumlsq;

print lsquare, sumlsq, ratio;
```

```
m=min(a,b);
D_l=diag(lambda_r[1:m]);
row_prof=inv(Drhalf)*(u_rmat[,1:m])*D_l;
delta=sqrt(diag((t(svd_mat)*svd_mat)));
std_cpt=inv(delta)*v_rmat[,1:m]*d_l;
dim=2;
rplot=row_prof[,1:dim];
cpoint=std_cpt[,1:dim];
out=rplot//cpoint;
cvar = concat(shape({"DIM"},1,dim),char(1:dim,1.));
create plot from out[rowname=id colname=cvar];
append from out[rowname=id];
close plot;
run;
data labels;
set plot;
x=dim1;
y=dim2;
retain xsys '2' ysys '2';
function='LABEL';
text=id;
position='5';

filename gsasfile "prog79.graph";
goptions reset=all gaccess=gsasfile autofeed dev=pslmono;
goptions horigin=1in vorigin=2in;
goptions hsize=6in vsize=8in;
    proc gplot data=plot;
    plot dim2 * dim1/anno=labels frame
        href=0 vref=0 lvref=3 lhref=3 vaxis=axis2 haxis=axis1
        vminor=1 hminor=1;
    axis1 length=5 in order=(-1 to 1 by .2)  offset=(2)
        label = (h=1.3 'Dimension 1');
    axis2 length=5 in order =(-.8 to .8 by .2) offset=(2)
        label=(h=1.3 a=90 r=0  'Dimension 2');
    symbol v=none;
title1 h=1.2 'Correspondence Analysis Using Hellinger Distance';
title2 j=l 'Output 7.9';
title3 f=duplex 'Agreement w.r.t. No. of Diseased Vessels';
    run;
```

Output 7.9

```
                 Output 7.9
       Plot Based on Hellinger Distance
    Agreement w.r.t. No. of Diseased Vessels

                  LSQUARE

                 0.1525156
                 0.0666546
                  0.036485
                 0.0268175
                 0.0048689

                  SUMLSQ

                 0.2873416
```

Output 7.9

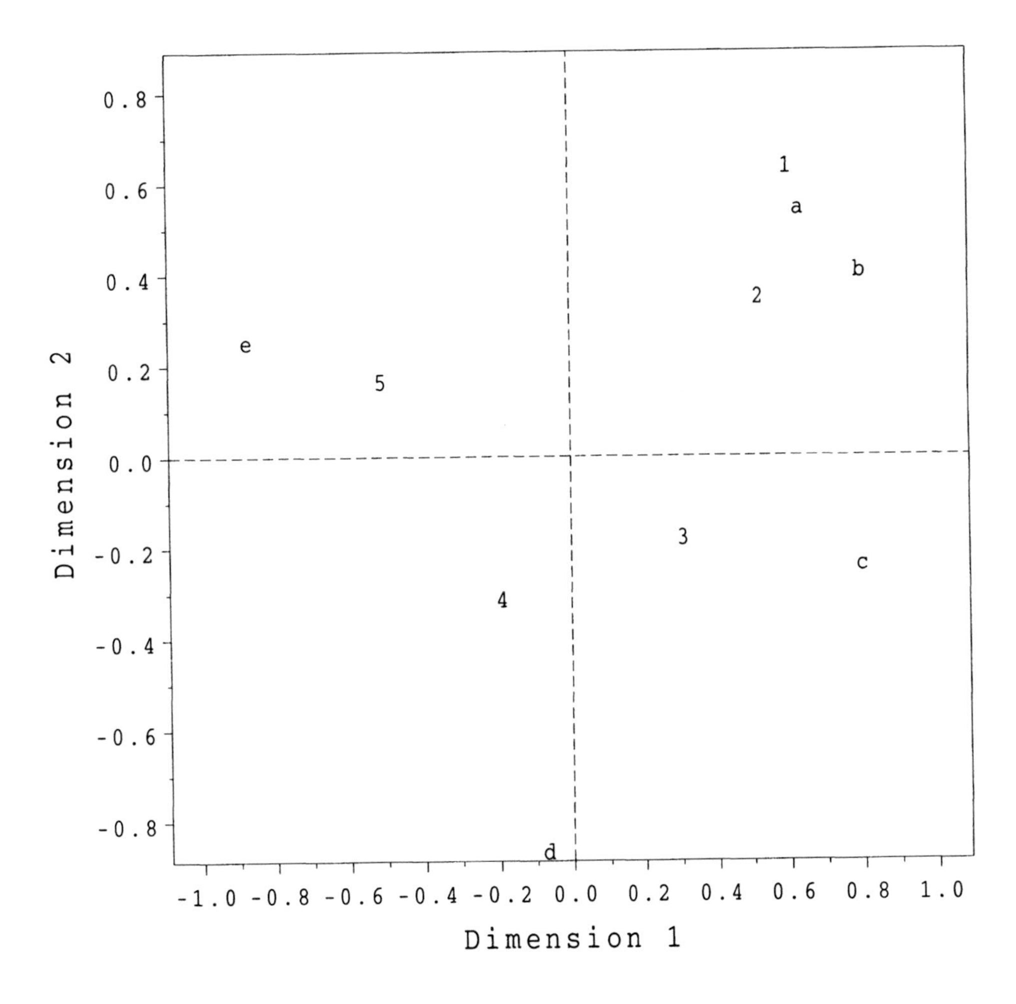

The quantities λ_i^2 are given in the output. The loss in using only two dimensions here is $\mathcal{L} \approx 23\%$, which is slightly less than the 28% that we incurred using chi-square distance. The plot presented in Output 7.9 reveals the similar conclusions that were made using the Andrews type of plots. The points representing the row profiles, namely 1, 2, 3, 4, and 5 follow an order in the graph. This is a feature that we look for in the plot, when the categories are ordinal, as is the case here. The points denoted by 1 and A plot very close to each other indicating that there is a close relationship between these two categories. This fact was observed in Example 7. Other similar features that we have observed in Example 7 are also observed here.

7.7 Canonical Correspondence Analysis

Ecologists analyze species-environment relations from data on biological communities and their environment. Generally, the data occurrences or abundance of each species of a taxonomic group are collected at several sites. Also, data on a set of important environmental variables useful in explaining the variation in the occurrences of species are collected. The sampling units are often defined as sites separated in space or time from other sites. By treating the species-abundance data over different sites as a contingency table, we can perform a correspondence analysis to graphically represent these data. Such graphical displays can be helpful in identifying the sites that have a maximum abundance of a certain species. Ecologists are also interested in determining the relationships between the environmental conditions (variables) favorable for the growth of certain species (measured in terms of abundance data). Essentially the idea is to depict the relationship between the environmental variables and species, using some graphical means. This can be done by using the biplots on the matrix of environmental scores obtained in a certain way and the scores representing the species abundance. This approach is called *canonical correspondence analysis*. These two scores can be obtained in many ways, depending on the particular problem at hand. We here present a scheme laid out by Ter Braak (1986).

Suppose y_{il} represents the abundance of the lth species at the ith site, where $i = 1, \ldots, n$ and $l = 1, \ldots, m$ and $z_{ij},\ j = 1, \ldots, q$ is the value of the jth environmental variable at the ith site. Also suppose x_i is the unknown site score (which is usually an unknown linear combination of the environmental variables) at site i. Let us denote the n by m matrix of species abundance by $\mathbf{Y} = (y_{il})$.

It is assumed that $y_{il},\ i = 1, \ldots, n;\ l = 1, \ldots, m$ have independent Poisson distributions with respective means $m_{il}'s$. Species typically follow unimodal relationships with respect to environmental variables $z_{ij}'s$. For instance, a plant species may grow well in a particular moisture content and not grow in other extreme moisture conditions. Thus, each species is largely confined to a specific interval along an environmental variable (Whittaker, 1956, 1967; Ter Braak, 1985). In view of this, a suggested model (Gauch and Whittaker, 1972) for m_{il}, relating it to $z_{ij}'s$ through the site score (which is a linear combination of environmental variables), is the Gaussian curve given as

$$m_{il} = c_l e^{-\frac{(x_i - \mu_l)^2}{2\sigma_l^2}}. \tag{7.13}$$

Thus μ_l can be interpreted as the value of x_i resulting in maximum abundance for the lth species, σ_l as an index of the tolerance and c_l as the value of maximum mean abundance (that is, the value of m_{il} when $x_i = \mu_l$).

Ter Braak (1986) makes the following additional assumptions:

- The species' tolerances σ_l are all equal to σ.
- The species' maxima c_l are all equal to c.
- The species' optima μ_l are homogeneously distributed over an interval I_1 that is large compared to σ.
- The site scores (x_i) are homogeneously distributed over a large interval I_2 that contains I_1.

We standardize each of the environmental variables z_{ij} such that their weighted means over all sites are all zeros, and the corresponding weighted standard deviations are all ones. That is,

$$\sum_i w_i z_{ij} = 0$$

and

$$\sum_i w_i z_{ij}^2 = 1,$$

where $w_i = \frac{y_{i\cdot}}{y_{\cdot\cdot}}$, $y_{i\cdot} = \sum_l y_{il}$ (total of the species at the ith site), and $y_{\cdot\cdot} = \sum_i \sum_l y_{il}$ (total species from all the sites). The n by q matrix of these standardized environmental variables will be denoted by $\mathbf{Z} = (z_{ij})$. The site score x_i, being a linear combination of the standardized $z'_{ij}s$, can be written as

$$x_i = \beta_1 z_{i1} + \cdots + \beta_q z_{iq} = \mathbf{z}'_i \boldsymbol{\beta},$$

where $\mathbf{z}_i = (z_{i1}, \ldots, z_{iq})'$, and $\boldsymbol{\beta} = (\beta_1, \ldots, \beta_q)'$.

The basic problem of interest in canonical correspondence analysis is to estimate the vectors of unknown parameters, namely $\boldsymbol{\mu} = (\mu_1, \ldots, \mu_m)'$ and $\boldsymbol{\beta}$.

Under the assumptions stated above, the maximum likelihood equations in matrix form for $\boldsymbol{\beta}$ and $\boldsymbol{\mu}$ are, respectively,

$$(\mathbf{S}_{22}^{-1}\mathbf{S}_{21}\mathbf{S}_{11}^{-1}\mathbf{S}_{12} - \lambda\mathbf{I})\boldsymbol{\beta} = \mathbf{0} \tag{7.14}$$

and

$$(\mathbf{S}_{11}^{-1}\mathbf{S}_{12}\mathbf{S}_{22}^{-1}\mathbf{S}_{21} - \lambda\mathbf{I})\boldsymbol{\mu} = \mathbf{0}, \tag{7.15}$$

where $\mathbf{S}_{21} = \mathbf{Z}'\mathbf{Y}$, $\mathbf{S}_{12} = \mathbf{Y}'\mathbf{Z}$, $\mathbf{S}_{11} = diag\,(y_{\cdot 1}, \ldots, y_{\cdot m})$, $y_{\cdot l} = \sum_i y_{il}$, being the total abundance of the lth species from all sites, and $\mathbf{S}_{22} = \mathbf{Z}'\mathbf{D}\mathbf{Z}$, with $\mathbf{D} = diag\,(y_{1\cdot}, \ldots, y_{n\cdot})$.

Equations 7.14 and 7.15 are familiar equations from the canonical correlation analysis (see Chapter 3), and hence the mathematical tools used to solve that problem can be used here. However, unlike the matrices in the canonical correlation analysis, the matrices here do not have the interpretation of the usual variance-covariance matrices of the variables involved. Hence, the CANCORR procedure cannot be used to solve this problem.

The solutions to Equations 7.14 and 7.15 are obtained using the singular value decomposition (see Chapter 1) of the m by q matrix $\mathbf{W} = \mathbf{S}_{11}^{-1/2}\mathbf{S}_{12}\mathbf{S}_{22}^{-1/2}$. If the rank of $\mathbf{W}$ is $r\ (\leq q)$, then let

$$\mathbf{W} = \mathbf{U}\boldsymbol{\Lambda}\mathbf{V}'. \tag{7.16}$$

Let the r solutions to Equations 7.14 and 7.15 respectively be given as the matrices

$$\hat{\mathbf{B}} = (\hat{\boldsymbol{\beta}}_1 : \cdots : \hat{\boldsymbol{\beta}}_r) = \mathbf{S}_{22}^{-1/2}\mathbf{V}$$

and

$$\hat{\mathbf{M}} = (\hat{\boldsymbol{\mu}}_1 : \cdots : \hat{\boldsymbol{\mu}}_r) = \mathbf{S}_{11}^{-1/2}\mathbf{U}.$$

The matrix $\hat{\mathbf{M}}$ is known as the species score matrix.

Using $\hat{\boldsymbol{\beta}}_1, \ldots, \hat{\boldsymbol{\beta}}_r$ and the relationship $x_i = \mathbf{Z}'_i\boldsymbol{\beta},\ i = 1, \ldots, n$, the r scores for the ith site, $i = 1, \ldots, n$ can be approximately determined as

$$x_i^{(k)} = \mathbf{Z}'_i\hat{\boldsymbol{\beta}}_k,\ k = 1, \ldots, r.$$

Let

$$\mathbf{x}'_i = (x_i^{(1)}, \ldots, x_i^{(r)}).$$

Writing these score vectors as rows, one on top of the other, we have

$$\begin{bmatrix} \mathbf{x}'_1 \\ \vdots \\ \mathbf{x}'_n \end{bmatrix} = \begin{bmatrix} \mathbf{Z}'_1 \\ \vdots \\ \mathbf{Z}'_n \end{bmatrix} \begin{bmatrix} \hat{\boldsymbol{\beta}}_1, \cdots, \hat{\boldsymbol{\beta}}_r \end{bmatrix}$$

or with self-explaining matrix notations, the site score matrix $\mathbf{X}$ is given by

$$\mathbf{X} = \mathbf{Z}\hat{\mathbf{B}}.$$

A bit of algebra also yields the following expression for the matrix $\hat{\mathbf{M}}$ of the species scores,

$$\hat{\mathbf{M}} = \mathbf{S}_{11}^{-1}\mathbf{Y}'\mathbf{X}\mathbf{\Lambda}^{-1}.$$

In practice, Ter Braak (1986) suggested that the matrix $\mathbf{X}$ be standardized before the species scores are computed. The suggested standardization is such that $\mathbf{X}'\mathbf{D}^*\mathbf{1} = \mathbf{0}$ and that the diagonal elements of $\mathbf{X}'\mathbf{D}^*\mathbf{X}$ are ones, where $\mathbf{D}^* = \frac{1}{y_{..}}\mathbf{D}$.

Suppose the matrix $\mathbf{X}$ has been standardized as suggested. Then the columns of $\mathbf{X}$ are called the *sample scores as linear combinations of environmental variables.* The elements of $\mathbf{S}_{11}^{-1}\mathbf{Y}'\mathbf{X}\mathbf{\Lambda}^{-1}$ are called *species scores.* Suppose $\mathbf{\Lambda}^2$ is the diagonal matrix with the eigenvalues of $\mathbf{W}\mathbf{W}'$ or $\mathbf{W}'\mathbf{W}$ at the diagonal entries. Two more differently scaled species scores can be obtained as $\mathbf{S}_{11}^{-1}\mathbf{Y}'\mathbf{X}\mathbf{\Lambda}^{-2\alpha}$, for $\alpha = 0,\ 0.5,\ 1$, where $\alpha = 0.5$ corresponds to the species scores defined above.

The elements of the matrix $\mathbf{X}^* = \mathbf{D}^{-1}\mathbf{Y}\mathbf{U}\mathbf{\Lambda}^{(2(\alpha-1)}$ are called the *sample scores* in Ter Braak (1988). A few other quantities of interest for the ecologists are

- the matrix $(\mathbf{Z}'\mathbf{Z})^{-1}\mathbf{Z}'\mathbf{X}$, which is called the matrix of canonical coefficients, corresponding to the r eigen axes,
- the *species-environment correlation matrix* $\mathbf{X}'\mathbf{D}^*\mathbf{X}^*$, and
- the correlation of an environmental variable with the sample scores given by $\mathbf{Z}'\mathbf{D}^*\mathbf{X}^*$.

For graphically examining various aspects of the species-environmental relationship using the biplots, one of the three matrices, namely $\mathbf{S}_{12}\mathbf{S}_{22}^{-1}$, $\mathbf{S}_{11}^{-1}\mathbf{S}_{12}$ or simply $\mathbf{S}_{12}$, can be used.

We will illustrate only the biplot corresponding to the matrix $\mathbf{S}_{11}^{-1}\mathbf{S}_{12}$ in the example that follows.

EXAMPLE 10 ***Canonical Correspondence Analysis of Hunting Spider Abundance Data*** The data considered here are from Ter Braak (1986, Table 3) and were originally adapted from Van der Aart and Smeek-Enserink (1975) after certain transformations. They consist of an abundance of 12 species of hunting spiders at 28 sites (representing pitfall traps) caught in pitfall traps over a period of 60 weeks, along with measurements on six environmental variables, namely percentage of soil dry mass (WATER_CONTENT), percentage cover of bare sand (BARE_SAND), percentage cover of fallen leaves and twigs (FALLEN_TWIGS), percentage cover of the moss layer (COVER_MOSS), percentage cover of the herb layer (COVER_HERBS), and reflection of the soil surface with cloudless sky (LIGHT_REFL). The objective of the study was to study the distribution of these 12 species of hunting spiders in a Dutch dune area in relation to environmental variables considered above. The data given in Ter Braak (1986) that we provide in Program 7.10 are such that the species abundance, that is, the number of individuals corresponding to each species, were transformed by taking the square root, and the environmental variables were transformed by taking the logarithms. Only the integer part of the square root transformed abundance was considered. A value of 9 for species abundance indicates the number of individuals of the species found is greater than or equal to 81. Further, the range of each transformed environmental variable was divided into ten equal categories denoted by 0-9, and these numbers were used as the data corresponding to the environmental variables. In Program 7.10 we use only abbreviated names of the 12 species.

In Program 7.10, adapting the SAS/IML code originally available in Hegde and Naik (1999), we read the data corresponding to species abundance as Y1-Y12 and that corresponding to the six environmental variables as Z1-Z6 in two separate files that are later combined into a single data set. All the above matrices have been computed but not presented in Output 7.10. However, the final biplot for the choice $\mathbf{S}_{11}^{-1}\mathbf{S}_{12}$ is presented here.

```
/* Program 7.10 */

options ls=64 ps=45 nodate nonumber;

title1 'Output 7.10';
title2 'Canonical Correspondence Analysis';
title3 'Hunting Spider Abundance Data';

data c1;
*Species abundance (y1 to y12) at 28 sites;
input y1-y12;
cards;
0 2 1 0 0 0 5 0 0 0 0 0
0 3 1 1 0 0 4 1 0 0 0 0
0 3 1 0 0 0 4 1 0 0 0 0
0 2 2 1 0 0 5 1 0 0 0 0
0 1 1 0 0 0 4 0 0 0 0 0
0 2 0 0 0 0 5 1 0 0 0 0
0 1 3 3 6 5 8 1 1 0 0 0
0 7 1 1 1 2 5 3 1 0 0 0
0 4 1 0 1 0 4 1 1 0 0 0
1 1 4 9 8 3 9 4 1 1 0 0
2 0 5 5 4 2 7 2 3 0 0 0
1 1 5 3 8 2 9 1 3 0 0 0
1 1 5 5 9 4 9 2 2 1 0 0
3 1 4 9 9 4 9 2 5 1 0 0
1 1 4 7 8 4 9 6 4 1 1 0
1 1 1 4 6 3 8 4 5 3 1 0
0 0 2 3 6 2 7 3 7 5 0 0
0 0 0 1 1 0 1 1 5 1 0 0
0 0 0 1 2 0 3 3 9 4 0 0
0 1 2 2 0 1 4 1 3 3 3 0
0 0 0 0 1 1 2 1 9 3 1 0
0 0 0 0 0 0 1 0 4 1 1 0
0 0 0 0 0 0 1 0 2 3 3 1
0 1 0 0 0 0 1 0 2 4 3 2
0 0 0 0 0 0 1 0 1 2 4 1
0 0 0 0 0 0 0 0 1 5 3 2
0 0 0 0 0 0 0 0 1 3 4 2
0 0 0 0 0 0 1 0 0 1 2 4
run;
data c2;
*Data on Environmental Variables (z1 to z6) on 28 sites;
input z1-z6;
cards;
9 0 1 1 9 5
7 0 3 0 9 2
8 0 1 0 9 0
8 0 1 0 9 0
9 0 1 2 9 5
8 0 0 2 9 5
8 0 2 3 3 9
6 0 2 1 9 6
7 0 1 0 9 2
8 0 0 5 0 9
9 5 5 1 7 6
8 0 4 2 0 9
6 0 5 6 0 9
8 0 1 5 0 9
```

```
9 3 1 7 3 9
6 0 5 8 0 9
5 0 7 8 0 9
5 0 9 7 0 6
6 0 8 8 0 8
3 7 2 5 0 8
4 0 9 8 0 7
4 8 7 8 0 5
0 7 8 8 0 6
0 6 9 9 0 6
1 7 9 8 0 0
0 5 8 8 0 6
2 7 9 9 0 5
0 9 4 9 0 2
run;
/* Source for both of these data sets: Van der Aart
and Smeek-Enserink (1975).  Reprinted by permission
of Brill Academic Publishers, Inc.*/

data a;
merge c1;
merge c2;

proc print data=a noobs;
run;

* PROC IML is used to do all the calculations;

*Create Data and Other Matrices;
proc iml;
use a;
read all var{y1,y2,y3,y4,y5,y6,y7,y8,y9,y10,y11,y12} into Y;
read all var{z1,z2,z3,z4,z5,z6} into Z;
close a;

m=12; /*Number of species */
q=6; /*Number of environmental variables */
n=28; /*Number of sites */

one=j(n,1,1);

sum1=j(m,1,0);
do i=1 to m;
sum1[i,1]=y[+,i];
end;
S11=diag(sum1);

r1=j(n,1,0);
do i=1 to n;
r1[i,1]=y[i,+];
end;
R=diag(r1);
R_star=R/y[+,+];
```

```
*Standardize (Weighted Mean=0, SD=1) environmental
variables;
Z=Z-j(n,n,1)*R_star*Z;
temp1=Z'*R_star*Z;
temp2=diag(temp1);
temp2=sqrt(temp2);
scalem=inv(temp2);
Z=Z*scalem;

*Create W, the fundamental matrix for the analysis;
S12=Y'*Z;
S22=Z'*R*Z;
* Find S11^(1/2) and S11^(-1/2);
s11_hf=j(m,m,0);
s11_nhf=j(m,m,0);
do i=1 to m;
s11_hf[i,i]=sqrt(S11[i,i]);
s11_nhf[i,i]=1/sqrt(S11[i,i]);
end;

* Find S22^(1/2) and S22^(-1/2);
call eigen(lambd,ev,S22);
lambda=diag(lambd);
lmda_hf=j(q,q,0);
lmda_nhf=j(q,q,0);
do i=1 to q;
lmda_hf[i,i]=sqrt(lambda[i,i]);
lmda_nhf[i,i]=1/sqrt(lambda[i,i]);
end;
s22_hf=ev*lmda_hf*ev';
s22_nhf=ev*lmda_nhf*ev';
W=s11_nhf*S12*s22_nhf;
print 'The Fundamental Matrix W';
print W;

*The SVD of the fundamental matrix W ;
call svd(P_mat,Lambda,Q_mat,W);
D=diag(Lambda);
*The diagonal elements of D matrix are the eigenvalues;
D=D*D;
print 'Eigenvalues';
print D;
print 'Eigenvectors';
print P_mat Q_mat;
print ' ';

*Solutions to Canonical Correspondence Analysis;
u_mat=s11_nhf*P_mat;
beta_mat=s22_nhf*Q_mat;
print 'Solutions to Canonical Correspondence Analysis';
print u_mat beta_mat;

*Sample Scores: Linear combinations of environmental variables;
X=Z*beta_mat;

*Species scores;
U_hat=inv(s11)*Y'*X*diag(1/lambda); * Assuming alpha=0;
print 'Species scores';
print U_hat;
```

```
*Standardize (Weighted Mean=0, SD=1) the X matrix;
X=X-j(n,n,1)*R_star*X;
temp1=X'*R_star*X;
temp2=diag(temp1);
temp2=sqrt(temp2);
scalem=inv(temp2);
X=X*scalem;
print 'Sample Scores: Linear combinations of environmental
variables';
print X;

*Canonical Coefficients corresponding to eigenvectors;
B_hat=inv(Z'*Z)*Z'*X;
print 'Canonical Coefficients corresponding to eigenvectors';
print B_hat;

* Biplot scores of environmental variables;
*COEVO=Z'*R_star*X;
*print COEVO;

*Species scores;
U_hat=inv(s11)*Y'*X; * Assuming alpha=0;
print 'Species scores';
print U_hat;

*Sample Scores;
X_star=inv(R)*Y*U_hat*inv(D);
print 'Sample Scores';
print X_star;

*Standardize (Weighted Mean=0, SD=1) the X_star matrix;
X_star=X_star-j(n,n,1)*R_star*X_star;
temp1=X_star'*R_star*X_star;
temp2=diag(temp1);
temp2=sqrt(temp2);
scalem=inv(temp2);
X_star=X_star*scalem;

*Correlation of an Environmental variable with an
ordination axis;
EOCORR=Z'*R_star*X_star;
print 'Correlation of an Environmental variable with
an ordination axis';
print 'OR Inter set Correlations';
print EOCORR;

*Species-Environment Correlations;
SECORR=X'*R_star*X_star;
SECORR=diag(secorr);
print 'Species-Environment Correlations';
print SECORR;

*Biplots;
print ' ';
print 'Biplot Information';

id={'Arct_lute', 'Pard_lugu', 'Zora_spin', 'Pard_nigr',
    'Pard_pull', 'Aulo_albi', 'Troc_terr', 'Alop_cune',
    'Pard_mont', 'Alop_acce', 'Alop_fabr',  'Arct_peri'};
```

```
vars={"WATER_CONTENT" "BARE_SAND" "COVER_MOSS"
"LIGHT_REFL" "FALLEN_TWIGS" "COVER_HERBS"};

reset fw=8 noname;
percent = 100*lambda##2 / lambda[##];
*Cumulate by multiplying by lower triangular matrix of 1's;
j = nrow(lambda);
tri = (1:j)` * repeat(1,1,j) >= repeat(1,j,1)*(1:j);
cum = tri*percent;
Print "Singular values and variance accounted for",,
    Lambda [colname={'Singular Values'} format=9.4]
    percent [colname={'Percent'} format=8.2]
    cum [colname={'cum % '} format = 8.2];

dim=2;
power=0;
scale=0.01;
U=s11_nhf*P_mat;
V=s22_hf*Q_mat;
U=U[,1:dim];
V=V[,1:dim];
Lambda=Lambda[1:dim];

*Scale the vectors by DL ,DR;
DL= diag(Lambda ## power);
DR= diag(Lambda ## (1-power));
A = U * DL;
B = V * DR # scale;

OUT=A // B;
*Create observation labels;
id = id // vars`;

type = repeat({"OBS "},m,1) // repeat({"VAR "},q,1);
       id  = concat(type,id);
   cvar = concat(shape({"DIM"},1,dim),char(1:dim,1.));
    * Create sas data set BIPLOT;
create plot from out[rowname=id colname=cvar];
append from out[rowname=id];
close plot;
*proc print;
run;

*Split id into _type_ and _Name_;
    data plot;
    set plot;
    drop id;
    length _type_  $3 _name_ $16;
    _type_ = scan(id,1);
    _name_ = scan(id,2);
    run;
```

```
*Annotate observation labels and variable vectors;
    data label;
    set plot;
    length text $16;
    xsys='2'; ysys='2';
    text=_name_;
    if _type_='OBS' then do;
    x = dim1;
    y = dim2;
    position='5';
    function='LABEL';
    output;
    end;
* Draw line from the origin to the variable point;
    if _type_ ='VAR' then do;
    x=0; y=0;
    function ='MOVE';
    output;
    x=dim1;
    y=dim2;
    function ='DRAW';
    output;
    if dim1>=0 then position ='6';  /*left justify*/
    else position ='2';             /*right justify*/
    function='LABEL';               /*variable name*/
    output;
    end;
    run;

* Plot the biplot using proc gplot;
    filename gsasfile "prog710.graph";
    goptions reset=all gaccess=gsasfile autofeed dev=pslmono;
    goptions horigin=1in vorigin=2in;
    goptions hsize=6in vsize=8in;

    proc gplot data=plot;
    plot dim2*dim1/anno=label frame href=0 vref=0
    lvref=3 lhref=3 vaxis=axis2 haxis=axis1 vminor=1 hminor=1;
    axis1 length=5 in order=(-.20 to .20 by .05)  offset=(2)
        label = (h=1.2 'Dimension 1');
    axis2 length=5 in order =(-.15 to .15 by .05) offset=(2)
    label=(h=1.2 a=90 r=0  'Dimension 2');
    symbol v=none;
    title1 h=1.2 'Biplot of Hunting Spider Data ';
    title2 j=l 'Output 7.10';
    title3 f=duplex 'Observations are points,
                           Variables are vectors';
    run;
```

Output 7.10

```
                    Output 7.10
         Canonical Correspondence Analysis
           Hunting Spider Abundance Data

                 Biplot Information

     Singular values and variance accounted for

         Singular Values  Percent   cum %

                  0.7342    61.88    61.88
                  0.4737    25.77    87.65
                  0.2698     8.36    96.01
                  0.1408     2.28    98.28
                  0.1063     1.30    99.58
                  0.0603     0.42   100.00


Obs         ID                  DIM1         DIM2

  1    OBS Arct_lute         -0.02892      0.04434
  2    OBS Pard_lugu         -0.02800     -0.13207
  3    OBS Zora_spin         -0.02649     -0.00242
  4    OBS Pard_nigr         -0.02235      0.02745
  5    OBS Pard_pull         -0.02358      0.04039
  6    OBS Aulo_albi         -0.02334      0.02390
  7    OBS Troc_terr         -0.01863     -0.02451
  8    OBS Alop_cune         -0.01611     -0.00299
  9    OBS Pard_mont          0.02169      0.03810
 10    OBS Alop_acce          0.06693      0.01389
 11    OBS Alop_fabr          0.11273     -0.01969
 12    OBS Arct_peri          0.14314     -0.05339
 13    VAR WATER_CONTENT     -0.16260      0.00885
 14    VAR BARE_SAND          0.12992     -0.00642
 15    VAR COVER_MOSS         0.11677      0.03552
 16    VAR LIGHT_REFL         0.10991      0.06382
 17    VAR FALLEN_TWIGS      -0.06920     -0.08843
 18    VAR COVER_HERBS       -0.05807      0.08681
```

Output 7.10

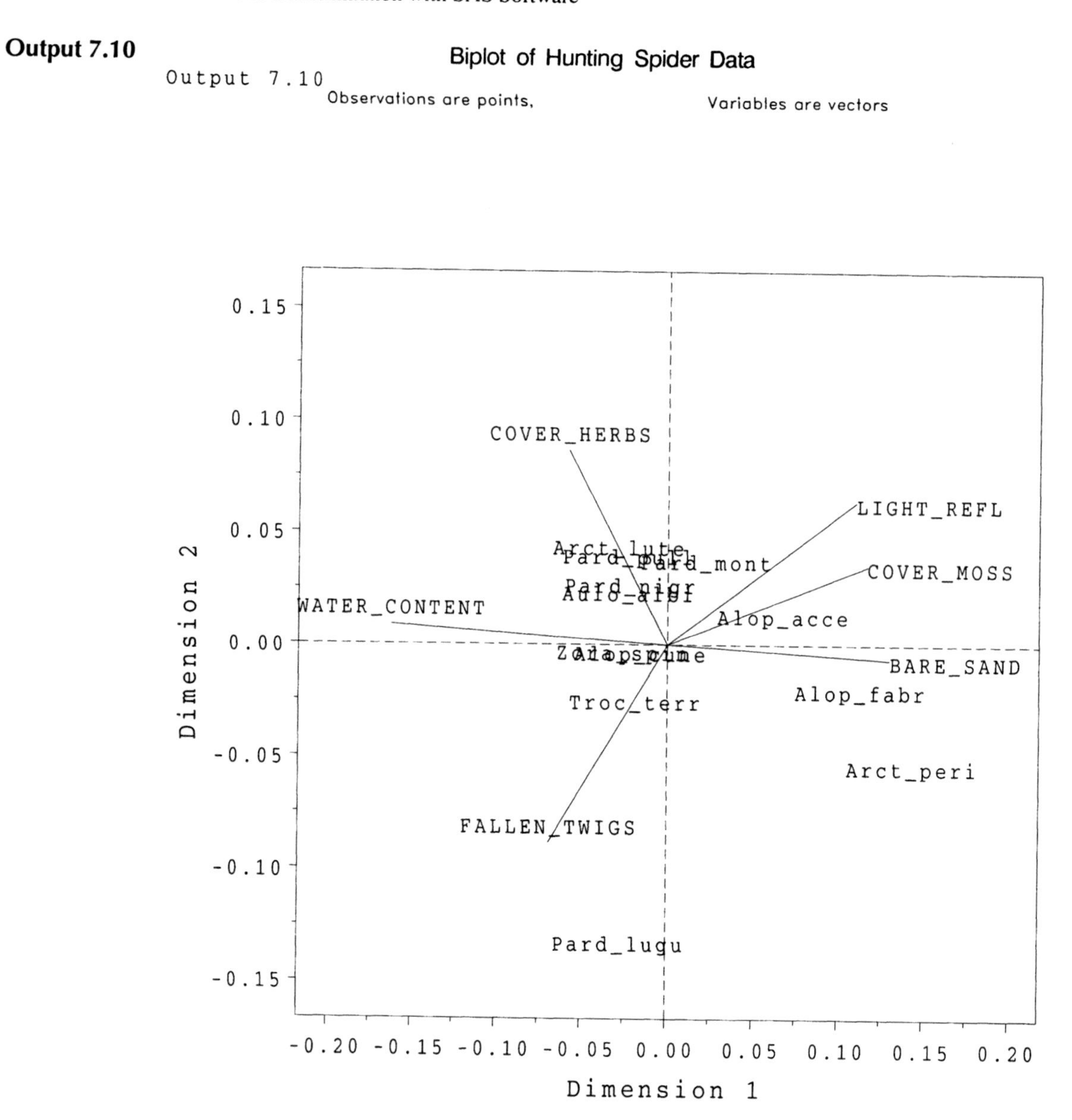

From the biplot presented in Output 7.10, we can identify certain habitats in which particular species are found. For example, the species Alop_fabr and Arct_peri were mainly found in habitats with higher percentages of sand (BARE_SAND). The species Arct_lute, Pard_pull, Pard_mont, Pard_nigr, and Aulo_albi are found in habitats with well developed herb layers (COVER_HERBS). Only the species Pard_lugu is found in the habitats with fallen twigs and leaves represented by the variable FALLEN_TWIGS in the graph.

7.8 Concluding Remarks

In this chapter, we have presented what we feel to be a brief account of correspondence analysis as a tool to graphically display information in a contingency table to a Euclidean plane. We have explained and illustrated both simple and multiple correspondence analysis using several practical examples.

The geometric basis behind the simple correspondence analysis allows us to determine the similarity among the categories of a variable and to some extent the association between two variables. Unfortunately, such a geometry for multiple correspondence analysis is not available. To overcome this problem with multiple correspondence analysis, Greenacre (1993a) suggested *joint correspondence analysis* in which all two-way associations are displayed in a single plot. However, we have not discussed this procedure here. See Greenacre (1993a) and Gower and Hand (1996) for details. A relationship between correspondence analysis and biplot display exists and such a relationship is described in detail in Greenacre (1993b) and Gower and Hand (1996).

While we have given only a brief account of correspondence analysis, we have discussed the use of Hellinger distance as an alternative distance to the usual chi-square distance for performing correspondence analysis (Rao, 1995). Further, we have provided SAS/IML code for performing an analysis called canonical correspondence analysis, which is a popular method among ecologists.

In conclusion, although correspondence analysis is one of the useful statistical tools, we must remember that it is only an exploratory data analysis technique and not a substitute for such statistical inference procedures as hypothesis testing for association and other analyses of contingency tables. The usual statistical inference techniques are still needed to complete the analysis of these data.

Appendix

Data Sets

Chapter 2 (Principal Component Analysis) Data Sets:

```
/*WOMEN TRACK DATA SET: womentrack.dat*/

11.61   22.94   54.50   2.15   4.43    9.79   178.52   Argentina
11.20   22.35   51.08   1.98   4.13    9.08   152.37   Australia
11.43   23.09   50.62   1.99   4.22    9.34   159.37   Austria
11.41   23.04   52.00   2.00   4.14    8.88   157.85   Belgium
11.46   23.05   53.30   2.16   4.58    9.81   169.98   Bermuda
11.31   23.17   52.80   2.10   4.49    9.77   168.75   Brazil
12.14   24.47   55.00   2.18   4.45    9.51   191.02   Burma
11.00   22.25   50.06   2.00   4.06    8.81   149.45   Canada
12.00   24.52   54.90   2.05   4.23    9.37   171.38   Chile
11.95   24.41   54.97   2.08   4.33    9.31   168.48   China
11.60   24.00   53.26   2.11   4.35    9.46   165.42   Columbia
12.90   27.10   60.40   2.30   4.84   11.10   233.22   Cookis
11.96   24.60   58.25   2.21   4.68   10.43   171.80   Costa
11.09   21.97   47.99   1.89   4.14    8.92   158.85   Czech
11.42   23.52   53.60   2.03   4.18    8.71   151.75   Denmark
11.79   24.05   56.05   2.24   4.74    9.89   203.88   Dominican
11.13   22.39   50.14   2.03   4.10    8.92   154.23   Finland
11.15   22.59   51.73   2.00   4.14    8.98   155.27   France
10.81   21.71   48.16   1.93   3.96    8.75   157.68   GDR
11.01   22.39   49.75   1.95   4.03    8.59   148.53   FRG
11.00   22.13   50.46   1.98   4.03    8.62   149.72   GB&NI
11.79   24.08   54.93   2.07   4.35    9.87   182.20   Greece
11.84   24.54   56.09   2.28   4.86   10.54   215.08   Guatemal
11.45   23.06   51.50   2.01   4.14    8.98   156.37   Hungary
11.95   24.28   53.60   2.10   4.32    9.98   188.03   India
11.85   24.24   55.34   2.22   4.61   10.02   201.28   Indonesia
11.43   23.51   53.24   2.05   4.11    8.89   149.38   Ireland
```

```
11.45   23.57   54.90   2.10   4.25    9.37   160.48   Israel
11.29   23.00   52.01   1.96   3.98    8.63   151.82   Italy
11.73   24.00   53.73   2.09   4.35    9.20   150.50   Japan
11.73   23.88   52.70   2.00   4.15    9.20   181.05   Kenya
11.96   24.49   55.70   2.15   4.42    9.62   164.65   Korea
12.25   25.78   51.20   1.97   4.25    9.35   179.17   DPRKorea
12.03   24.96   56.10   2.07   4.38    9.64   174.68   Luxembou
12.23   24.21   55.09   2.19   4.69   10.46   182.17   Malasiya
11.76   25.08   58.10   2.27   4.79   10.90   261.13   Mauritius
11.89   23.62   53.76   2.04   4.25    9.59   158.53   Mexico
11.25   22.81   52.38   1.99   4.06    9.01   152.48   Netherlands
11.55   23.13   51.60   2.02   4.18    8.76   145.48   NZealand
11.58   23.31   53.12   2.03   4.01    8.53   145.48   Norway
12.25   25.07   56.96   2.24   4.84   10.69   233.00   Guinea
11.76   23.54   54.60   2.19   4.60   10.16   200.37   Philippi
11.13   22.21   49.29   1.95   3.99    8.97   160.82   Poland
11.81   24.22   54.30   2.09   4.16    8.84   151.20   Portugal
11.44   23.46   51.20   1.92   3.96    8.53   165.45   Rumania
12.30   25.00   55.08   2.12   4.52    9.94   182.77   Singapore
11.80   23.98   53.59   2.05   4.14    9.02   162.60   Spain
11.16   22.82   51.79   2.02   4.12    8.84   154.48   Sweden
11.45   23.31   53.11   2.02   4.07    8.77   153.42   Switzerl
11.22   22.62   52.50   2.10   4.38    9.63   177.87   Taipei
11.75   24.46   55.80   2.20   4.72   10.28   168.45   Thailand
11.98   24.44   56.45   2.15   4.37    9.38   201.08   Turkey
10.79   21.83   50.62   1.96   3.95    8.50   142.72   USA
11.06   22.19   49.19   1.89   3.87    8.45   151.22   USSR
12.74   25.85   58.73   2.33   5.81   13.04   306.00   WSamoa

/* Source: IAAF/ATFS Track and Field Statistics Handbook for the 1984
   Los Angeles Olympics. Reproduced by permission of the International
   Amateur Athletics Federation. */
```

```
/*WOMEN TRACK DATA SET (SPEEDS): wtrack.dat*/

Argentin   8.61  8.72  7.34  6.20  5.64  5.11  3.94
Australi   8.93  8.95  7.83  6.73  6.05  5.51  4.62
Austria    8.75  8.66  7.90  6.70  5.92  5.35  4.41
Belgium    8.76  8.68  7.69  6.67  6.04  5.63  4.46
Bermuda    8.73  8.68  7.50  6.17  5.46  5.10  4.14
Brazil     8.84  8.63  7.58  6.35  5.57  5.12  4.17
Burma      8.24  8.17  7.27  6.12  5.62  5.26  3.68
Canada     9.09  8.99  7.99  6.67  6.16  5.68  4.71
Chile      8.33  8.16  7.29  6.50  5.91  5.34  4.10
China      8.37  8.19  7.28  6.41  5.77  5.37  4.17
Columbia   8.62  8.33  7.51  6.32  5.75  5.29  4.25
Cookis     7.75  7.38  6.62  5.80  5.17  4.50  3.02
Costa      8.36  8.13  6.87  6.03  5.34  4.79  4.09
Czech      9.02  9.10  8.34  7.05  6.04  5.61  4.43
Denmark    8.76  8.50  7.46  6.57  5.98  5.74  4.63
Dominica   8.48  8.32  7.14  5.95  5.27  5.06  3.45
Finland    8.98  8.93  7.98  6.57  6.10  5.61  4.56
France     8.97  8.85  7.73  6.67  6.04  5.57  4.53
GDR        9.25  9.21  8.31  6.91  6.31  5.71  4.46
FRG        9.08  8.93  8.04  6.84  6.20  5.82  4.73
GB&NI      9.09  9.04  7.93  6.73  6.20  5.80  4.70
Greece     8.48  8.31  7.28  6.44  5.75  5.07  3.86
Guatemal   8.45  8.15  7.13  5.85  5.14  4.74  3.27
Hungary    8.73  8.67  7.77  6.63  6.04  5.57  4.50
India      8.37  8.24  7.46  6.35  5.79  5.01  3.74
Indonesi   8.44  8.25  7.23  6.01  5.42  4.99  3.49
Ireland    8.75  8.51  7.51  6.50  6.08  5.62  4.71
Israel     8.73  8.49  7.29  6.35  5.88  5.34  4.38
Italy      8.86  8.70  7.69  6.80  6.28  5.79  4.63
Japan      8.53  8.33  7.44  6.38  5.75  5.43  4.67
Kenya      8.53  8.38  7.59  6.67  6.02  5.43  3.88
Korea      8.36  8.17  7.18  6.20  5.66  5.20  4.27
DPRKorea   8.16  7.76  7.81  6.77  5.88  5.35  3.93
Luxembou   8.31  8.01  7.13  6.44  5.71  5.19  4.03
Malasiya   8.18  8.26  7.26  6.09  5.33  4.78  3.86
Mauritiu   8.50  7.97  6.88  5.87  5.22  4.59  2.69
Mexico     8.41  8.47  7.44  6.54  5.88  5.21  4.44
Netherla   8.89  8.77  7.64  6.70  6.16  5.55  4.61
NZealand   8.66  8.65  7.75  6.60  5.98  5.71  4.83
Norway     8.64  8.58  7.53  6.57  6.23  5.86  4.83
Guinea     8.16  7.98  7.02  5.95  5.17  4.68  3.02
Philippi   8.50  8.50  7.33  6.09  5.43  4.92  3.51
Poland     8.98  9.00  8.12  6.84  6.27  5.57  4.37
Portugal   8.47  8.26  7.37  6.38  6.01  5.66  4.65
Rumania    8.74  8.53  7.81  6.94  6.31  5.86  4.25
Singapor   8.13  8.00  7.26  6.29  5.53  5.03  3.85
Spain      8.47  8.34  7.46  6.50  6.04  5.54  4.33
Sweden     8.96  8.76  7.72  6.60  6.07  5.66  4.55
Switzerl   8.73  8.58  7.53  6.60  6.14  5.70  4.58
Taipei     8.91  8.84  7.62  6.35  5.71  5.19  3.95
Thailand   8.51  8.18  7.17  6.06  5.30  4.86  4.17
Turkey     8.35  8.18  7.09  6.20  5.72  5.33  3.50
USA        9.27  9.16  7.90  6.80  6.33  5.88  4.93
USSR       9.04  9.01  8.13  7.05  6.46  5.92  4.65
WSamoa     7.85  7.74  6.81  5.72  4.30  3.83  2.30
```

```
/* SKULL DATA SET: skull.dat */

2  1512.5  181.5  187    186.5  145    96     137    136.5  117    103
3  1619.5  192.5  197    195    145    99.5   137    135.5  110.5  108.5
4  1595.5  199    201    202    149.5  97     133    132    111.4  104.5
5  1602    188    195    193    142.5  97.5   137.5  137.5  117.5  104
6  1497.5  194.5  197    196    141.5  96     136    135    112    101.5
7  1492.5  190.5  192    191.5  144    105.5  137    134    118.5  102.5
8  1605.5  194    199    197    154    101.5  133.5  134    121    110
9  1697    191.5  191    192    148    101.5  128    128    110    97.5
10 1579    187    187    187.5  147    97     126.5  127    109.3  97.5
11 1500    176    177    176.5  140    101.5  129    127.5  111.5  91
13 1481    191.5  196.5  196.5  149    106.5  128.5  126    114    103
14 1432.5  181    183.5  183.5  141    95.5   140    139    114.5  104.5
15 1670.5  186.5  183    186.5  154.5  106    140.5  138.5  113.8  104.5
16 1476    187    187    188    142    94     132.5  132    110.5  97
17 1590.5  190.5  195    192.5  145    95     139.5  139.5  116.3  105.5
18 1725    194    199    198    146.5  100    139.5  137    116    104.5
19 1613    189    188.5  189.5  142    99     134    134    111    103
20 1524    193    196    195.5  146    101    123.5  122    111    98
21 1345    179    179    180    144    102    127    125.5  115.5  99.5
22 1557.5  195.5  188.5  189    146.5  102    128    127    115.7  99.5
23 1632.5  184    187    185    149    104    130.5  130    113    97.5
24 1406    185    188    187.5  141.5  95     128    127.5  110.4  98
25 1726.5  191    193    191.5  145.5  94     136.5  136    120.5  102.5
27 1477    198.5  202.5  201.5  136    97     131.5  130.5  106.3  103.5
28 1296.5  182.5  191    190    134.5  85.5   128.5  127.5  107.2  102
30 1385.5  182    192    190.5  138    93     129.5  127.5  103.5  104
31 1359    180.5  181    181    140    100.5  129    127.5  106    90.5
32 1600.5  190    193    193.5  143.5  96.5   129    129    111.5  97.5
33 1521.5  182    186.5  185    144.5  99.5   132    131    110.3  101.5
34 1411    186.5  193    192.5  135.5  90.5   129.5  130.5  112.4  100
35 1502.5  185    193.5  193    140.5  95     128.5  128.5  107    99
36 1314    184    189.5  188.5  130.5  93.5   134    132    112.4  103.5
37 1309    182.5  186    185    137.5  100.5  130.5  131    109.5  105
38 1627    194    200    197.5  142.5  103    138.5  138    115.3  103
40 1478.5  185.5  184.5  185.5  142.5  102    133    133.5  114.4  104
41 1470.5  184    187.5  187    146    105    126    124.5  108.3  104.5
42 1519    185.5  184.5  184    148    97     136    137    114.7  97
43 1398    179.5  183    183.5  148    96.5   132.5  131.5  109.5  100.5
44 1292.5  180.5  188    187    137.5  87.5   122.5  122    105.3  104.5
47 1524    185.5  186    186.5  144    87.5   128.5  128.5  107.5  100
48 1600.5  192.5  199    198    148    97     122.5  121.5  102    98
49 1501.5  180    185    183    139    90.5   138    138    111.5  101
52 1554.5  187    186.5  187.5  149    92.5   135    133.5  110.5  97.5
53 1476    185    189    190    143    98.5   125    126    106    99.5
58 1229    177.5  182    182    136.5  93.5   122    122.5  102.5  95.5
63 1425    186.5  184.5  186    137    95     132    132    108.5  93
64 1241.5  171.5  175    174.5  141.5  95     123.5  122.5  104.4  101.5
67 1561    187.5  190    190.5  146    94.5   128.5  127.5  114.4  102.5
69 1549.5  187    190.5  191    148.5  101.5  128.5  127.5  97.6   95.5
70 1376.5  189    189    190.5  131    101    122.5  123.5  107    98.5
71 1248    187    188    188.5  129    98     122.5  122.5  101.3  101.5
74 1418.5  178    181    181    140    92     130.5  130    107.5  100
77 1476    190    189    191    138.5  97     127.5  127.5  112.7  96
83 1387    179.5  185    183    134    91.5   132    131    107    97.5
84 1568.5  180.5  183    182    147    96     138.5  138.5  115.2  94.5
85 1458    182.5  184    185.5  135    90.5   132.5  131.5  106.3  102
87 1283.5  178.5  183    182    130    89.5   123.5  123    99.5   96.5
```

```
88 1403.5   181.5  183    183     135    101     128    127.5 111.5  97.5
91 1359     184.5  189    188     141.5  93      126    126   104.7  93
92 1395.5   171    175    176     138    85      125    123.5 104.3  89.5
93 1581.5   192    197    196.5   137.5  101     134    133.5 111.5  107.5
94 1565     189.5  191    190.5   146    98      134    133.5 111.5  102.5
95 1212.5   178    184    185.5   133.5  94      116.5 118.5 93.5    100

/* Source: Hooke (1926).  Reprinted with permission from Biometrika.*/


/* HAEMATO DATA SET: haemato.dat */

13.4 39 4100 14 25 17 14.6 46 5000 15 30 20 13.5 42 4500 19 21 18
15.0 46 4600 23 16 18 14.6 44 5100 17 31 19 14.0 44 4900 20 24 19
16.4 49 4300 21 17 18 14.8 44 4400 16 26 29 15.2 46 4100 27 13 27
15.5 48 8400 34 42 36 15.2 47 5600 26 27 22 16.9 50 5100 28 17 23
14.8 44 4700 24 20 23 16.2 45 5600 26 25 19 14.7 43 4000 23 13 17
14.7 42 3400  9 22 13 16.5 45 5400 18 32 17 15.4 45 6900 28 36 24
15.1 45 4600 17 29 17 14.2 46 4200 14 25 28 15.9 46 5200  8 34 16
16.0 47 4700 25 14 18 17.4 50 8600 37 39 17 14.3 43 5500 20 31 19
14.8 44 4200 15 24 19 14.9 43 4300  9 32 17 15.5 45 5200 16 30 20
14.5 43 3900 18 18 25 14.4 45 6000 17 37 23 14.6 44 4700 23 21 27
15.3 45 7900 43 23 23 14.9 45 3400 17 15 24 15.8 47 6000 23 32 21
14.4 44 7700 31 39 23 14.7 46 3700 11 23 23 14.8 43 5200 25 19 22
15.4 45 6000 30 25 18 16.2 50 8100 32 38 18 15.0 45 4900 17 26 24
15.1 47 6000 22 33 16 16.0 46 4600 20 22 22 15.3 48 5500 20 23 23
14.5 41 6200 20 36 21 14.2 41 4900 26 20 20 15.0 45 7200 40 25 25
14.2 46 5800 22 31 22 14.9 45 8400 61 17 17 16.2 48 3100 12 15 18
14.5 45 4000 20 18 20 16.4 49 6900 35 22 24 14.7 44 7800 38 34 16
17.0 52 6300 19 21 16 15.4 47 3400 12 19 18 13.8 40 4500 19 23 21
16.1 47 4600 17 28 20 14.6 45 4700 23 22 27 15.0 44 5800 14 39 21
16.2 47 4100 16 24 18 17.0 51 5700 26 29 20 14.0 44 4100 16 24 18
15.4 46 6200 32 25 16 15.6 46 4700 28 16 16 15.8 48 4500 24 20 23
13.2 38 5300 16 26 20 14.9 47 5000 22 25 15 14.9 47 3900 15 19 16
14.0 45 5200 23 25 17 16.1 47 4300 19 22 22 14.7 46 6800 35 25 18
14.8 45 8900 47 36 17 17.0 51 6300 42 19 15 15.2 45 4600 21 22 18
15.2 43 5600 25 28 17 13.8 41 6300 25 27 15 14.8 43 6400 36 24 18
16.1 47 5200 18 28 25 15.0 43 6300 22 34 17 16.2 46 6000 25 25 24
14.8 44 3900  9 25 14 17.2 44 4100 12 27 18 17.2 48 5000 25 19 25
14.6 43 5500 22 31 19 14.4 44 4300 20 20 15 15.4 48 5700 29 26 24
16.0 52 4100 21 15 22 15.0 45 5000 27 18 20 14.8 44 5700 29 23 23
15.4 43 3300 10 20 19 16.0 47 6100 32 23 26 14.8 43 5100 18 31 19
13.8 41 8100 52 24 17 14.7 43 5200 24 24 17 14.6 44 9899 69 28 18
13.6 42 6100 24 30 15 14.5 44 4800 14 29 15 14.3 39 5000 25 20 19
15.3 45 4000 19 19 16 16.4 49 6000 34 22 17 14.8 44 4500 22 18 25
16.6 48 4700 17 27 20 16.0 49 7000 36 28 18 15.5 46 6600 30 33 13
14.3 46 5700 26 20 21

/* Source: Royston (1983) (Health Survey on 103 Paint Sprayers:
variables are haemoglobin concentration, packed cell volume,
white blood cell count, lymphocyte count, neutrophil count,
and serum lead concentration).  Reprinted by permission of the
Royal Statistical Society.*/
```

Chapter 3 (Canonical Correlation Analysis) Data Sets:

```
/* LIVESTOCK DATA SET: livestock.dat */

1921 8.41 8.73 55.7 65.0
1922 8.68 9.26 59.2 65.9
1923 8.22 6.60 59.8 74.5
1924  8.24 7.23 59.9 74.7
1925  8.64 10.04 60.0 67.3
1926 7.63 9.95 60.7 64.6
1927  9.50 8.32 54.9 68.2
1928 11.41 7.56 49.0 71.3
1929 10.49 7.93 49.8 69.8
1930 9.84 8.51 48.9 67.0
1931 9.20 7.03 48.5 68.3
1932 10.58 6.05 46.7 70.6
1933 8.92 6.48 51.4 69.9
1934 9.52 6.55 55.8 64.2
1935 12.76 11.53 53.5 48.6
1936 9.47 10.62 58.8 55.6
1937 11.37 9.93 55.3 55.9
1938 10.10 8.70 54.5 58.3
1939 9.88 6.66 54.5 64.4
1940 9.91 5.42 55.2 72.5

/* Source: F. W. Waugh (1942) (Livestock Prices and Meat
Consumption: variables are year priceb priceh consumpb
consumph). Reprinted by permission of the Econometric
Society.*/


/* URINE DATA SET: urine.dat */

1 5.7  4.67  17.6  1.50  .104  1.50  1.88  5.15  8.40   7.5  .14  205  24
1 5.5  4.67  13.4  1.65  .245  1.32  2.24  5.75  4.50   7.1  .11  160  32
1 6.6  2.70  20.3   .90  .097   .89  1.28  4.35  1.20   2.3  .10  480  17
1 5.7  3.49  22.3  1.75  .174  1.50  2.24  7.55  2.75   4.0  .12  230  30
1 5.6  3.49  20.5  1.40  .210  1.19  2.00  8.50  3.30   2.0  .12  235  30
1 6.0  3.49  18.5  1.20  .275  1.03  1.84 10.25  2.00   2.0  .12  215  27
1 5.3  4.84  12.1  1.90  .170  1.87  2.40  5.95  2.60  16.8  .14  215  25
1 5.4  4.84  12.0  1.65  .164  1.68  3.00  6.30  2.72  14.5  .14  190  30
1 5.4  4.84  10.1  2.30  .275  2.08  2.68  5.45  2.40    .9  .20  190  28
1 5.6  4.48  14.7  2.35  .210  2.55  3.00  3.75  7.00   2.0  .21  175  24
1 5.6  4.48  14.8  2.35  .050  1.32  2.84  5.10  4.00    .4  .12  145  26
1 5.6  4.48  14.4  2.50  .143  2.38  2.84  4.05  8.00   3.8  .18  155  27
2 5.2  3.48  18.1  1.50  .153  1.20  2.60  9.00  2.35  14.5  .13  220  31
2 5.2  3.48  19.7  1.65  .203  1.73  1.88  5.30  2.52  12.5  .20  300  23
2 5.6  3.48  16.9  1.40  .074  1.15  1.72  9.85  2.45   8.0  .07  305  32
2 5.8  2.63  23.7  1.65  .155  1.58  1.60  3.60  3.75   4.9  .10  275  20
2 6.0  2.63  19.2   .90  .155   .96  1.20  4.05  3.30    .2  .10  405  18
2 5.3  2.63  18.0  1.60  .129  1.68  2.00  4.40  3.00   3.6  .18  210  23
2 5.4  4.46  14.8  2.45  .245  2.15  3.12  7.15  1.81  12.0  .13  170  31
2 5.6  4.46  15.6  1.65  .422  1.42  2.56  7.25  1.92   5.2  .15  235  28
2 5.3  2.80  16.2  1.65  .063  1.62  2.04  5.30  3.90  10.2  .12  185  21
2 5.4  2.80  14.1  1.25  .042  1.62  1.84  3.10  4.10   8.5  .30  255  20
2 5.5  2.80  17.5  1.05  .030  1.56  1.48  2.40  2.10   9.6  .20  265  15
2 5.4  2.57  14.1  2.70  .194  2.77  2.56  4.25  2.60   6.9  .17  305  26
2 5.4  2.57  19.1  1.60  .139  1.59  1.88  5.80  2.30   4.7  .16  440  24
2 5.2  2.57  22.5   .85  .046  1.65  1.20  1.55  1.50   3.5  .21  430  16
```

```
3 5.5  1.26  17.0  .70   .094   .97  1.24  4.55   2.90   1.9  .12  350  18
3 5.9  1.26  12.5  .80   .039   .80   .64  2.65   0.72    .7  .13  475  10
3 5.6  2.52  21.5  1.80  .142  1.77  2.60  6.50   2.48   8.3  .17  195  33
3 5.6  2.52  22.2  1.05  .080  1.17  1.48  4.85   2.20   9.3  .14  375  25
3 5.3  2.52  13.0  2.20  .215  1.85  3.84  8.75   2.40  13.0  .11  160  35
3 5.6  3.24  13.0  3.55  .166  3.18  3.48  5.20   3.50  18.3  .22  240  33
3 5.5  3.24  10.9  3.30  .111  2.79  3.04  4.75   2.52  10.5  .21  205  31
3 5.6  3.24  12.0  3.65  .180  2.40  3.00  5.85   3.00  14.5  .21  270  34
3 5.4  1.56  22.8  .55   .069  1.00  1.14  2.85   2.90   3.3  .15  475  16
4 5.3  1.56  16.5  2.05  .222  1.49  2.40  6.55   3.90   6.3  .11  430  31
4 5.2  1.56  18.4  1.05  .267  1.17  1.36  6.60   2.00   4.9  .11  490  28
4 5.8  4.12  12.5  5.90  .093  3.80  3.84  2.90   3.00  22.5  .24  105  32
4 5.7  4.12  8.7   4.25  .147  3.62  5.32  3.00   3.55  19.5  .20  115  25
4 5.5  4.12  9.4   3.85  .217  3.36  5.52  3.40   5.20   1.3  .31   97  28
4 5.4  2.14  15.0  2.45  .418  2.38  2.40  5.40   1.81  20.0  .17  325  27
4 5.4  2.14  12.9  1.70  .323  1.74  2.48  4.45   1.88   1.0  .15  310  23
4 4.9  2.03  12.1  1.80  .205  2.00  2.24  4.30   3.70   5.0  .19  245  25
4 5.0  2.03  13.2  3.65  .348  1.95  2.12  5.00   1.80   3.0  .15  170  26
4 4.9  2.03  11.5  2.25  .320  2.25  3.12  3.40   2.50   5.1  .18  220  34

/* Source: Smith, Gnanadesikan, and Hughes (1962).
(Data on urine samples from four groups of men:
y1-y11: response variables z1-z2: covariates
group: class variable.  Reproduced by permission of the
International Biometric Society.*/
```

Chapter 4 (Factor Analysis) Data Sets:

```
/* MICE DATA SET: mice.dat */

.120 .138 .258 .408 .549 .659 .642
.150 .287 .475 .628 .800 .894 .941
.151 .270 .452 .600 .663 .800 .954
.127 .260 .459 .648 .802 .902 1.000
.152 .242 .410 .570 .668 .768 .855
.149 .240 .402 .573 .742 .870 .855
.141 .260 .472 .662 .760 .885 .878
.154 .271 .430 .593 .689 .775 .920
.170 .300 .499 .519 .599 .600 .750
.160 .283 .365 .430 .618 .719 .813
.164 .240 .397 .547 .702 .802 .865


.180 .329 .516 .521 .710 .780 .850
.169 .268 .429 .548 .689 .750 .844
.162 .300 .512 .682 .753 .845 .831
.180 .292 .454 .616 .716 .730 .880
.175 .327 .513 .661 .765 .795 .870
.180 .349 .504 .600 .660 .700 .762
.163 .327 .530 .730 .779 .801 .877
.181 .350 .571 .719 .812 .827 .872
.192 .330 .520 .703 .796 .860 .938
.180 .328 .493 .631 .774 .821 .968
```

```
0.190 0.388 0.621 0.823 1.078 1.132 1.191
0.218 0.393 0.568 0.729 0.839 0.852 1.004
0.141 0.260 0.472 0.662 0.760 0.885 0.878
0.211 0.394 0.549 0.700 0.783 0.870 0.925
0.209 0.419 0.645 0.850 1.001 1.026 1.069
0.193 0.362 0.520 0.530 0.641 0.640 0.751
0.201 0.361 0.502 0.530 0.657 0.762 0.888
0.202 0.370 0.498 0.650 0.795 0.858 0.910
0.190 0.350 0.510 0.666 0.819 0.879 0.929
0.219 0.399 0.578 0.699 0.709 0.822 0.953
0.225 0.400 0.545 0.690 0.796 0.825 0.836
0.224 0.381 0.577 0.756 0.869 0.929 0.999

/* Source: A.J.Izenman and J.S.Williams (1989).  Reproduced
by permission of the International Biometric Society.
(Ref: Biometrics, 45,831-849, "A class of linear spectral
models and analyses for the study of longitudinal data."
Data are weights of 33 male mice measured over the 21 days
(in three groups) from birth to weaning.  Data in this first
block represent weights of 11 mice at ages 0, 3, 6, 9, 12, 15,
and 18 days following birth. This is Group 1.  Data in the
second block  represent weights of 10 mice at ages 1, 4, 7,
10, 13, 16, and 19 days following birth. This is Group 2.
Finally data in the third block represent weights of 12 mice
at ages 2, 5, 8, 11, 14, 17, and 20 days following birth.
This is Group 3. */


/* DIABETICS DATA SET: diabetic.dat */

188 5 1 5 1 2 4 2 5 2 5 4 5 4 4 2 5 2 4 3 4 2 5 1 5 5
200 5 1 4 1 4 5 2 5 1 4 1 1 4 5 4 4 5 4 1 5 1 4 2 5 4
202 5 1 1 1 5 3 1 3 4 1 1 4 2 4 2 1 1 2 1 5 4 1 5 5 1
204 5 1 4 1 4 4 1 5 4 1 4 1 2 5 1 5 4 3 4 5 1 4 2 5 3
213 5 1 2 1 3 4 1 5 1 3 3 1 3 5 4 5 4 3 3 3 1 3 3 5 2
124 5 2 5 2 2 4 2 5 3 3 5 5 2 4 2 2 5 4 3 3 1 5 2 5 4
126 5 1 5 1 4 4 1 5 2 5 2 1 5 5 4 1 5 5 5 5 1 3 4 5 5
129 5 4 5 2 2 4 1 5 3 5 5 5 1 4 1 1 5 5 5 4 1 5 2 5 5
150 5 2 5 2 1 2 5 2 2 4 4 4 1 5 2 1 5 5 3 5 4 5 2 5 2
152 5 1 2 1 5 4 1 4 2 3 4 1 2 5 2 2 1 2 1 3 2 4 3 5 4
121 5 1 5 1 1 5 2 5 5 2 2 2 4 5 5 5 3 4 4 4 1 5 1 5 5
122 4 1 1 1 5 5 3 1 1 1 1 1 1 5 1 1 5 3 1 5 1 5 4 5 2
123 5 1 1 1 1 1 1 5 5 1 4 1 2 5 4 5 5 5 3 2 4 4 3 5 5
102 5 1 5 1 3 5 3 5 2 1 5 1 4 5 4 3 2 4 2 5 1 1 1 5 2
104 5 1 5 1 2 5 1 5 1 1 4 5 4 4 4 1 4 4 5 1 4 4 2 5 4
106 5 1 3 1 3 5 4 5 2 4 5 1 1 5 5 5 5 5 5 5 1 4 2 5 4
108 5 2 4 1 4 5 1 5 1 3 2 5 5 5 5 2 4 2 5 3 1 5 2 5 5
214 5 2 5 1 2 5 3 5 1 2 2 4 2 5 2 1 5 2 2 2 2 3 2 5 2
215 5 5 4 1 2 5 1 4 4 5 5 2 1 5 4 5 5 4 5 5 1 5 1 5 4
217 5 1 5 1 1 5 2 5 4 5 5 4 2 5 5 5 5 5 3 5 1 5 2 5 5
154 5 1 5 4 1 4 1 4 3 4 5 4 1 5 1 1 5 4 4 3 3 4 2 5 4
157 5 2 5 2 4 4 4 4 4 3 5 5 5 5 4 5 4 4 3 5 2 3 2 5 3
165 5 1 5 2 1 2 2 5 2 5 3 2 2 5 3 1 3 5 3 4 3 4 2 5 5
168 5 1 5 1 1 3 5 5 1 5 3 5 1 3 3 3 4 5 4 1 2 3 2 5 1
169 5 1 5 1 2 2 1 3 1 2 2 1 3 5 1 3 4 4 1 5 5 1 5 5 1
170 5 1 5 2 2 2 3 2 2 5 4 5 4 2 2 1 5 5 1 2 4 3 2 4 2
171 3 1 1 1 5 5 5 1 2 . 1 1 5 5 5 3 5 1 1 1 5 2 1 5 5
172 5 2 5 4 2 2 3 4 3 5 5 5 3 3 3 2 2 1 3 3 2 3 2 5 2
178 5 3 4 1 4 4 5 4 4 5 5 4 5 5 3 5 5 5 5 5 4 5 2 5 5
```

```
187 5 3 3 1 4 4 5 2 4 1 3 4 1 5 5 3 4 5 3 2 4 2 4 5 4
219 5 1 4 1 4 4 2 4 3 3 4 2 2 4 2 2 2 4 1 2 3 4 2 5 3
221 5 1 1 1 5 4 1 4 4 2 2 2 1 5 5 1 5 4 3 4 3 5 2 5 3
222 5 3 4 1 2 3 3 3 5 3 4 4 3 4 3 4 3 5 3 4 3 2 3 5 3
109 5 1 3 1 5 5 2 5 1 4 5 2 2 4 3 3 5 1 2 2 1 4 2 5 4
110 5 1 5 1 1 5 1 5 4 5 5 2 3 4 4 5 5 2 3 2 2 4 3 5 4
111 5 3 3 1 5 5 1 5 1 5 3 1 4 5 4 5 5 4 3 5 2 5 2 5 4
112 5 1 5 1 2 3 5 4 2 1 4 1 2 5 4 1 2 3 4 2 2 4 2 5 3
113 5 1 3 1 5 4 2 4 1 3 1 1 1 5 1 1 5 1 1 4 3 3 5 5 3
114 5 4 5 1 2 5 1 5 1 5 5 1 1 5 5 2 5 4 3 2 1 5 1 5 4
116 5 2 4 1 3 5 1 5 1 3 3 1 2 5 4 5 5 3 3 4 2 5 3 5 4
151 5 2 4 1 3 5 2 4 2 1 4 1 3 5 4 4 5 5 3 3 5 4 1 5 5
153 5 1 3 1 2 4 1 5 4 1 3 1 5 5 5 2 4 5 1 5 2 4 3 5 5
216 5 5 4 1 4 4 4 4 4 5 5 5 5 4 4 3 5 5 4 3 4 4 2 5 3
131 5 3 3 1 1 2 3 3 3 3 4 1 4 5 1 3 4 4 . 3 3 3 4 5 3
132 5 3 5 1 1 5 1 5 1 5 2 1 1 5 5 5 5 5 . 3 2 5 1 5 5
133 5 5 5 1 1 5 1 5 1 4 5 5 5 5 4 5 5 5 3 4 4 5 2 4 5
135 5 4 3 1 4 3 5 2 1 3 2 4 3 4 3 1 1 3 3 3 4 2 4 5 1
136 5 1 5 1 2 4 3 1 3 4 2 1 3 3 4 1 1 5 3 4 2 5 2 5 1
137 5 2 1 1 4 5 1 1 1 1 1 1 1 5 5 1 5 1 5 1 1 5 3 5 5
103 5 1 5 1 1 4 5 4 5 4 4 1 4 5 3 5 5 5 3 5 4 5 5 5 3
101 5 2 1 1 4 5 1 4 4 3 2 2 4 4 5 5 5 4 3 5 1 5 1 5 5
105 5 1 5 1 1 4 3 3 3 5 5 3 5 4 3 4 4 5 3 4 3 3 3 5 4
117 5 1 1 1 1 1 2 2 1 1 1 1 1 5 3 5 4 1 1 2 1 2 2 1 4
118 5 3 5 1 2 3 5 5 5 4 5 1 4 5 4 1 4 5 1 3 1 5 1 5 5
119 5 1 3 1 3 5 5 3 1 1 3 2 4 5 1 4 3 3 5 3 2 2 5 5 2
120 5 1 5 1 4 5 1 5 4 2 4 1 3 5 3 2 3 4 1 3 4 5 2 5 5
127 5 2 4 1 3 5 4 4 4 3 5 1 4 5 5 3 3 5 3 3 1 4 1 5 5
125 5 1 5 1 4 5 4 5 4 5 5 4 5 5 5 2 5 5 3 1 4 5 1 5 4
212 5 1 5 1 1 4 5 4 3 1 3 1 4 . 4 3 5 4 4 4 3 3 3 5 3
185 5 3 4 1 2 4 3 4 4 3 4 1 2 5 3 2 4 5 3 5 2 3 2 5 4
186 5 1 5 1 2 3 4 4 2 1 2 4 1 4 4 3 5 4 3 4 1 4 2 4 4
189 5 2 5 1 4 4 3 5 4 3 5 1 3 5 4 3 5 5 5 4 1 5 1 5 5
191 5 3 3 1 1 3 3 5 5 1 5 1 3 5 3 1 1 5 3 5 5 4 5 5 3
193 5 1 5 1 1 4 4 4 1 1 4 4 2 2 1 2 5 5 3 4 4 2 4 5 1
180 5 3 2 1 4 4 3 3 4 3 3 1 4 3 1 4 4 5 3 5 3 2 3 5 2
181 5 1 4 1 2 4 1 5 2 3 2 2 1 5 3 3 4 5 3 4 1 5 1 5 4
182 5 1 4 1 4 5 1 5 1 4 4 1 1 5 5 1 5 1 4 3 2 4 1 5 5
183 5 1 5 1 1 5 5 5 2 3 5 2 3 5 4 2 4 5 2 4 1 5 2 5 5
184 5 2 3 1 2 3 3 3 3 1 4 1 4 5 4 3 4 1 3 5 3 2 3 5 4
138 5 1 5 1 2 5 1 5 3 4 4 1 5 4 5 5 5 5 3 5 1 5 1 5 5
139 5 4 2 5 5 3 4 5 4 2 2 5 5 3 5 2 5 2 . 5 2 4 2 5 4
140 5 1 2 3 3 4 1 5 5 5 3 2 2 4 4 5 5 5 4 5 4 5 1 5 5
146 5 1 1 1 1 5 1 4 1 1 1 1 1 5 1 1 5 1 4 5 5 4 2 5 4
130 5 1 5 1 1 5 1 5 1 1 1 2 1 5 1 5 5 4 5 5 2 5 2 5 5
145 5 1 3 2 1 3 2 4 4 5 4 3 3 3 4 3 4 5 3 5 3 3 3 5 3
211 5 4 4 2 1 5 4 2 1 2 3 1 3 5 3 3 5 4 3 5 3 3 4 5 3
161 5 1 3 1 5 1 3 3 1 1 1 1 4 3 1 1 2 1 1 1 1 1 1 5 4
162 5 2 3 1 5 5 1 5 1 1 4 1 4 4 2 5 5 4 3 5 1 5 1 5 5
164 5 5 3 3 2 2 3 3 4 4 4 2 4 2 4 5 4 4 1 4 3 3 3 4 3
173 5 1 1 1 4 2 5 5 3 1 1 4 1 5 1 1 1 3 3 1 1 1 3 5 1
218 4 1 5 1 1 3 3 5 5 5 5 1 3 5 1 3 5 5 3 5 5 5 5 5 5
220 4 3 4 2 2 3 4 2 2 1 2 1 3 3 2 2 3 1 3 3 4 2 4 4 2
107 4 5 5 2 1 4 4 1 3 1 1 3 4 3 1 1 1 1 1 3 1 1 5 4 1
115 5 3 5 1 3 3 1 5 3 5 3 1 3 3 3 5 3 3 3 5 3 5 1 5 5
128 5 1 1 1 2 5 1 5 2 1 2 1 1 5 1 1 5 1 5 4 1 5 1 5 4
134 5 4 5 1 2 5 1 4 4 5 5 5 5 5 5 5 5 5 5 4 4 5 4 5 5
141 5 1 5 1 1 5 1 5 1 3 5 1 5 5 5 1 5 5 5 5 1 5 1 5 5
142 5 3 4 1 3 4 2 4 1 3 4 1 4 5 5 3 5 1 3 3 1 3 1 5 4
```

```
143 5 2 5 1 3 5 5 5 1 5 5 5 2 5 5 5 5 2 2 1 1 4 4 5 5
144 5 1 3 1 5 3 2 4 4 4 4 1 4 3 3 1 5 1 3 1 4 3 2 5 3
210 5 2 3 1 3 1 1 3 2 1 1 1 1 5 3 1 2 1 1 1 5 1 5 5 3
179 5 1 2 1 5 4 2 4 1 5 2 1 1 5 2 4 4 5 . 5 4 4 2 5 5
190 5 1 5 1 3 4 4 3 1 3 3 2 3 5 3 5 5 3 3 5 1 5 3 5 4
192 5 1 5 1 3 4 3 5 2 3 5 5 4 4 2 2 5 4 4 4 1 4 2 5 5
194 5 1 1 1 4 4 1 1 1 1 4 1 1 5 4 1 5 5 . 4 3 5 3 5 5
195 5 1 3 1 4 5 1 2 1 1 1 1 1 5 1 1 . 2 . 1 5 5 1 5 3
196 5 4 5 4 4 5 1 5 4 4 4 5 5 4 3 5 5 4 5 5 1 5 4 5 5
197 5 1 1 1 4 5 5 1 1 4 1 5 1 5 1 4 3 5 . 3 1 1 5 5 4
198 4 2 5 1 3 4 3 4 4 4 5 1 5 5 4 4 5 4 5 3 1 5 2 5 5
199 5 1 1 1 5 4 1 5 2 2 3 1 3 5 3 5 5 3 5 5 1 4 2 5 4
201 5 1 5 1 2 4 1 5 1 5 4 1 4 5 1 4 5 4 4 5 1 5 2 5 4
203 5 1 4 1 4 5 1 5 4 1 4 1 4 5 1 4 2 4 4 5 1 3 2 5 4
205 5 1 3 1 4 4 3 4 3 3 4 1 3 4 5 5 3 1 . 5 1 3 2 5 5
206 5 2 4 4 2 4 4 4 2 3 4 4 5 4 3 5 5 3 . 5 3 2 3 5 5
207 5 1 5 1 3 5 2 4 1 1 4 1 5 4 3 5 3 5 3 5 2 4 2 5 4
208 5 1 4 1 1 5 1 5 3 1 1 1 1 4 5 5 5 4 4 3 1 4 1 5 5
209 5 1 3 1 3 5 2 2 3 2 3 1 3 5 3 3 5 5 5 3 3 3 3 5 3
177 5 3 2 1 5 5 1 2 1 2 1 1 2 5 2 1 5 3 4 2 5 3 4 5 3
147 5 1 2 1 5 2 2 3 2 1 1 1 2 4 4 1 1 2 2 2 3 3 5 5 3
148 5 1 5 1 4 5 4 5 1 1 5 4 4 5 4 4 1 5 4 2 5 5 5 5 5
149 5 2 5 1 2 5 3 5 2 5 5 4 3 5 4 5 3 5 3 3 2 5 2 5 5
155 5 1 5 1 3 3 2 5 4 1 5 1 4 4 4 2 4 4 1 5 1 5 1 5 5
156 5 3 5 1 3 5 3 4 1 5 5 4 5 5 5 5 3 5 . 1 1 4 1 5 5
158 5 1 5 1 3 5 3 3 1 3 3 3 3 5 3 3 3 1 5 3 1 3 3 5 3
159 5 1 5 1 2 3 1 5 4 5 5 1 2 1 4 1 5 5 2 2 2 5 1 5 2
160 5 1 2 1 3 4 3 4 2 1 4 1 2 5 4 1 2 4 3 3 2 5 2 5 5
163 5 4 5 2 1 3 5 5 5 4 5 4 4 4 4 3 5 5 4 3 4 4 3 5 5
166 5 1 3 1 2 5 1 5 2 2 4 2 1 5 2 4 5 4 4 4 2 5 2 5 5
167 5 1 5 1 3 3 4 4 2 4 3 1 4 4 3 2 3 3 3 2 4 3 4 5 2
174 5 3 4 1 2 4 2 4 1 3 2 3 1 5 5 3 5 3 1 5 1 3 4 5 3
175 5 1 5 4 3 5 2 5 3 4 2 1 3 5 5 2 4 4 2 5 2 4 2 5 4
176 3 1 4 1 2 5 2 3 1 2 4 1 1 5 3 4 4 4 4 4 4 2 . 5 5

/* Data courtesy of Dr. G. R. Patwardhan, Strategic
Marketing Corporation.  These data are on diabetic
patients' attitudes toward the disease.  There are 122
patients and their opinions on 25 questions are recorded
in a five-point scale.  Scales: 1 (Strongly disagree),
2 (Somewhat disagree), 3 (Neither disagree nor agree)
4 (Somewhat agree), 5 (Strongly agree). */
```

```
/* SEDIMENTOLOGY DATA SET: sedimnt.dat */

 1 Apr_83 high  1  19.00  6.26 12.74 20.92 32.24  8.84 0.00
 2 Apr_83 high  2   0.00  0.42  6.74 42.63 42.24  7.97 0.00
 3 Apr_83 high  3   0.00  0.18 16.52 51.49 28.36  3.45 0.00
 4 Apr_83 high  4   0.00 16.47 45.37 29.65  7.18  1.33 0.00
 5 Apr_83 high  5  11.13 29.28 43.36 13.87  2.20  0.16 0.00
 6 Apr_83 high  6   0.04  3.58 36.83 38.37 21.18  0.00 0.00
 7 Apr_83 high  7   0.00  0.00 25.74 25.95 43.13  4.84 0.34
 8 Apr_83 high  8   0.00  0.00  5.64  6.71 61.77 23.67 2.21
 9 Apr_83 high  9   0.00  0.00  1.00  5.98 72.50 19.65 0.87
10 Apr_83 high 10   0.00  0.00  5.62 13.76 62.93 16.89 0.80
11 Apr_83 med   1  18.60  6.23 12.17 26.80 24.66 11.54 0.00
12 Apr_83 med   2   0.00  0.28  2.27 10.82 44.82 41.81 0.00
13 Apr_83 med   3   0.00  2.14 19.66 42.23 30.96  5.01 0.00
14 Apr_83 med   4   0.00  6.09 24.26 45.82 21.18  2.65 0.00
15 Apr_83 med   5  11.18 22.04 36.52 30.26  0.00  0.00 0.00
16 Apr_83 med   6   3.88 20.20 28.30 16.77 30.85  0.00 0.00
17 Apr_83 med   7   0.00  0.00 16.43 16.52 54.21 11.99 0.85
18 Apr_83 med   8   0.00  0.00 16.46  8.05 52.59 21.63 1.27
19 Apr_83 med   9   0.00  0.00  3.21  8.12 62.01 25.31 1.35
20 Apr_83 med  10   0.00  0.00  1.73  2.97 54.15 37.64 3.60
21 Apr_83 low   1  50.60  2.07  3.71 13.85 18.61 11.16 0.00
22 Apr_83 low   2   0.00  6.68 14.02 20.38 33.17 25.75 0.00
23 Apr_83 low   3   0.00  6.85 27.37 34.38 25.78  5.62 0.00
24 Apr_83 low   4   0.00 12.96 23.33 31.53 23.93  8.25 0.00
25 Apr_83 low   5  48.89 21.18 14.60  8.84  5.59  0.90 0.00
26 Apr_83 low   6   2.28 16.10 28.50 21.86 31.26  0.00 0.00
27 Apr_83 low   7   0.00  0.00  7.41 13.74 61.81 15.53 1.51
28 Apr_83 low   8   0.00  0.00  0.67  1.69 69.05 26.60 1.99
29 Apr_83 low   9   0.00  0.00  0.16  2.51 67.24 28.25 1.84
30 Apr_83 low  10   0.00  0.00  6.96  4.33 63.98 23.37 1.36
31 Jul_83 high  1   0.00  6.07 25.05 30.62 22.50 15.76 0.00
32 Jul_83 high  2   0.00  0.00  3.07 30.26 50.76 14.78 1.13
33 Jul_83 high  3   0.00  9.15 49.60 33.54  6.14  1.57 0.00
34 Jul_83 high  4   0.00  1.41 28.16 47.89 15.89  6.67 0.00
35 Jul_83 high  5   4.73 24.22 52.31 16.95  1.79  0.00 0.00
36 Jul_83 high  6   0.00  0.95 21.63 43.32 34.10  0.00 0.00
37 Jul_83 high  7   0.00  0.00 16.28 38.86 41.86  2.81 0.19
38 Jul_83 high  8   0.00  0.00 23.60 18.64 41.40 14.84 1.52
39 Jul_83 high  9   0.00  0.00  0.55  3.13 40.65 49.83 5.84
40 Jul_83 high 10   0.00  0.00  4.29 17.86 49.08 22.88 5.89
41 Jul_83 med   1   0.00 13.43 37.74 24.28 16.06  8.49 0.00
42 Jul_83 med   2   0.00  0.00  6.44 41.36 42.00  9.06 1.14
43 Jul_83 med   3   0.00 26.69 48.82 20.74  3.25  0.50 0.00
44 Jul_83 med   4   0.00  8.67 39.18 40.61  9.94  1.60 0.00
45 Jul_83 med   5   0.11  4.19 37.63 44.87 13.20  0.00 0.00
46 Jul_83 med   6   0.00  0.84 27.12 45.58 23.17  3.29 0.00
47 Jul_83 med   7   0.00  0.00  9.61 39.08 49.33  1.24 0.74
48 Jul_83 med   8   0.00  0.00 51.00  7.30 27.23 12.91 1.56
49 Jul_83 med   9   0.00  0.00  1.37  1.59 30.88 58.22 7.95
50 Jul_83 med  10   0.00  0.00 20.86 26.60 38.90 13.28 0.36
51 Jul_83 low   1   0.00 20.23 48.92 22.19  6.14  2.52 0.00
52 Jul_83 low   2   0.00  0.00  7.42 36.27 46.76  6.60 2.95
53 Jul_83 low   3   0.00  4.21 25.35 49.01 18.98  2.45 0.00
54 Jul_83 low   4   9.38 18.10 37.76 25.72  8.19  0.85 0.00
55 Jul_83 low   5  45.12 24.36 21.41  6.98  2.13  0.00 0.00
56 Jul_83 low   6   0.00  1.90 20.53 35.97 34.79  6.81 0.00
```

```
57  Jul_83  low   7    0.00   0.00   0.77  17.18  72.24   9.76  0.05
58  Jul_83  low   8    0.00   0.00  45.11   3.96  27.38  20.45  3.10
59  Jul_83  low   9    0.00   0.00   3.25   5.91  47.85  39.96  3.03
60  Jul_83  low   10   0.00   0.00  38.44  18.00  29.33  13.03  1.20
61  Sep_83  high  1    0.00   3.59  21.76  43.03  25.80   5.82  0.00
62  Sep_83  high  2    0.00   0.00   1.41  30.37  51.83  14.69  1.70
63  Sep_83  high  3    0.00   2.13  43.29  46.32   7.81   0.45  0.00
64  Sep_83  high  4    0.00   0.03  30.33  45.95  17.82   5.87  0.00
65  Sep_83  high  5    3.78  17.27  48.65  25.68   4.62   0.00  0.00
66  Sep_83  high  6    0.00   0.16   0.73  17.27  60.01  21.83  0.00
67  Sep_83  high  7    0.00   0.00   0.17  21.50  61.20  16.95  0.18
68  Sep_83  high  8    0.00   0.00   7.68   9.76  62.36  18.55  1.65
69  Sep_83  high  9    0.00   0.00   0.38   0.79  29.99  60.09  8.75
70  Sep_83  high  10   0.00   0.00   9.75  26.73  48.52  12.26  2.74
71  Sep_83  med   1    0.00   9.96  27.07  40.04  17.88   5.05  0.00
72  Sep_83  med   2    0.00   0.00   4.63  32.35  41.95  18.59  2.48
73  Sep_83  med   3    0.00   8.16  43.41  36.72  10.63   1.08  0.00
74  Sep_83  med   4    0.00  20.14  42.98  28.35   8.48   0.05  0.00
75  Sep_83  med   5    3.40  11.67  33.41  29.65  16.93   4.94  0.00
76  Sep_83  med   6    0.00   0.00   0.00   0.00   0.00   0.00  0.00
77  Sep_83  med   7    0.00   0.00  16.02  28.99  47.60   6.95  0.44
78  Sep_83  med   8    0.00   0.00  12.71   2.03  37.32  42.56  5.38
79  Sep_83  med   9    0.00   0.00   9.95   7.54  40.85  37.67  3.99
80  Sep_83  med   10   0.00   0.00  72.70   7.21  11.37   7.98  0.74
81  Sep_83  low   1   41.10  11.19  13.21  12.61  19.92   1.97  0.00
82  Sep_83  low   2    0.00   0.00   2.98  21.56  40.99  29.04  5.43
83  Sep_83  low   3   16.07  39.34  30.76  10.10   3.22   0.51  0.00
84  Sep_83  low   4   36.86  15.31  18.45  18.00   9.94   1.44  0.00
85  Sep_83  low   5    6.23  11.96  31.35  32.73  17.73   0.00  0.00
86  Sep_83  low   6    0.34   3.74  23.84  27.23  34.37  10.48  0.00
87  Sep_83  low   7    0.00   0.00   4.69  16.99  58.91  17.79  1.62
88  Sep_83  low   8    0.00   0.00   1.84   1.80  41.00  49.45  5.91
89  Sep_83  low   9    0.00   0.00   2.18   3.60  38.32  49.39  6.51
90  Sep_83  low   10   0.00   0.00  16.85   6.26  40.08  33.29  3.52

/* Data courtesy of Dr. G. N. Nayak, Gao University, India.  The variables
are: id, season, tide, station, x1-x7. */
```

Chapter 5 (Discriminant Analysis) Data Sets:

```
/* BEETLE DATA SET: beetle.dat */

1 150 15
1 147 13
1 144 14
1 144 16
1 153 13
1 140 15
1 151 14
1 143 14
1 144 14
1 142 15
1 141 13
1 150 15
1 148 13
1 154 15
1 147 14
1 137 14
1 134 15
1 157 14
1 149 13
1 147 13
1 148 14

2 120 14
2 123 16
2 130 14
2 131 16
2 116 16
2 122 15
2 127 15
2 132 16
2 125 14
2 119 13
2 122 13
2 120 15
2 119 14
2 123 15
2 125 15
2 125 14
2 129 14
2 130 13
2 129 13
2 122 12
2 129 15
2 124 15
2 120 13
2 119 16
2 119 14
2 133 13
2 121 15
2 128 14
2 129 14
2 124 13
2 129 14
```

```
3 145 8
3 140 11
3 140 11
3 131 10
3 139 11
3 139 10
3 136 12
3 129 11
3 140 10
3 137 9
3 141 11
3 138 9
3 143 9
3 142 11
3 144 10
3 138 10
3 140 10
3 130 9
3 137 11
3 137 10
3 136 9
3 140 10

/* Source: Lubischew (1962, Biometrics).
(Variables: species, x1, x2). Reproduced by
permission of the International Biometric Society.*/


/* CRYSTALS IN URINE DATA: urine.dat */

no 1.021 4.91 725 . 443 2.45
no 1.017 5.74 577 20 296 4.49
no 1.008 7.2 321 14.9 101 2.36
no 1.011 5.51 408 12.6 224 2.15
no 1.005 6.52 187 7.5 91 1.16
no 1.020 5.27 668 25.3 252 3.34
no 1.012 5.62 461 17.4 195 1.40
no 1.029 5.67 1107 35.9 550 8.48
no 1.015 5.41 543 21.9 170 1.16
no 1.021 6.13 779 25.7 382 2.21
no 1.011 6.19 345 11.5 152 1.93
no 1.025 5.53 907 28.4 448 1.27
no 1.006 7.12 242 11.3 64 1.03
no 1.007 5.35 283 9.9 147 1.47
no 1.011 5.21 450 17.9 161 1.53
no 1.018 4.90 684 26.1 284 5.09
no 1.007 6.63 253 8.4 133 1.05
no 1.025 6.81 947 32.6 395 2.03
no 1.008 6.88 395 26.1 95 7.68
no 1.014 6.14 565 23.6 214 1.45
no 1.024 6.30 874 29.9 380 5.16
no 1.019 5.47 760 33.8 199 0.81
no 1.014 7.38 577 30.1 87 1.32
no 1.020 5.96 631 11.2 422 1.55
no 1.023 5.68 749 29.0 239 1.52
no 1.017 6.76 455 8.8 270 0.77
no 1.017 7.61 527 25.8 75 2.17
no 1.010 6.61 225 9.8 72 0.17
no 1.008 5.87 241 5.1 159 0.83
```

```
no 1.020 5.44 781 29.0 349 3.04
no 1.017 7.92 680 25.3 282 1.06
no 1.019 5.98 579 15.5 297 3.93
no 1.017 6.56 559 15.8 317 5.38
no 1.008 5.94 256 8.1 130 3.53
no 1.023 5.85 970 38.0 362 4.54
no 1.020 5.66 702 23.6 330 3.98
no 1.008 6.40 341 14.6 125 1.02
no 1.020 6.35 704 24.5 260 3.46
no 1.009 6.37 325 12.2 97 1.19
no 1.018 6.18 694 23.3 311 5.64
no 1.021 5.33 815 26.0 385 2.66
no 1.009 5.64 386 17.7 104 1.22
no 1.015 6.79 541 20.9 187 2.64
no 1.010 5.97 343 13.4 126 2.31
no 1.020 5.68 876 35.8 308 4.49
yes 1.021 5.94 774 27.9 325 6.96
yes 1.024 5.77 698 19.5 354 13.00
yes 1.024 5.60 866 29.5 360 5.54
yes 1.021 5.53 775 31.2 302 6.19
yes 1.024 5.36 853 27.6 364 7.31
yes 1.026 5.16 822 26.0 301 14.34
yes 1.013 5.86 531 21.4 197 4.74
yes 1.010 6.27 371 11.2 188 2.50
yes 1.011 7.01 443 21.4 124 1.27
yes 1.022 6.21 . 20.6 398 4.18
yes 1.011 6.13 364 10.9 159 3.10
yes 1.031 5.73 874 17.4 516 3.01
yes 1.020 7.94 567 19.7 212 6.81
yes 1.040 6.28 838 14.3 486 8.28
yes 1.021 5.56 658 23.6 224 2.33
yes 1.025 5.71 854 27.0 385 7.18
yes 1.026 6.19 956 27.6 473 5.67
yes 1.034 5.24 1236 27.3 620 12.68
yes 1.033 5.58 1032 29.1 430 8.94
yes 1.015 5.98 487 14.8 198 3.16
yes 1.013 5.58 516 20.8 184 3.30
yes 1.014 5.90 456 17.8 164 6.99
yes 1.012 6.75 251 5.1 141 .65
yes 1.025 6.90 945 33.6 396 4.18
yes 1.026 6.29 833 22.2 457 4.45
yes 1.028 4.76 312 12.4 10 .27
yes 1.027 5.40 840 24.5 395 7.64
yes 1.018 5.14 703 29.0 272 6.63
yes 1.022 5.09 736 19.8 418 8.53
yes 1.025 7.90 721 23.6 301 9.04
yes 1.017 4.81 410 13.3 195 0.58
yes 1.024 5.40 803 21.8 394 7.82
yes 1.016 6.81 594 21.4 255 12.20
yes 1.015 6.03 416 12.8 178 9.39

/* Data courtesy of Dr. D. P. Byar formerly of National
Cancer Institute.  Also available in Andrews and
Herzberg's Data (1985).  (Physical characteristics of
urine with and without crystals;  Variables: crystals
sg ph mosm mmho urea calcium). */
```

```
/* ADMISSION DATA SET: admission.dat */

yes 296 596
yes 314 473
yes 322 482
yes 329 527
yes 369 505
yes 346 693
yes 303 626
yes 319 663
yes 363 447
yes 359 588
yes 330 563
yes 340 553
yes 350 572
yes 378 591
yes 344 692
yes 348 528
yes 347 552
yes 335 520
yes 339 543
yes 328 523
yes 321 530
yes 358 564
yes 333 565
yes 340 431
yes 338 605
yes 326 664
yes 360 609
yes 337 559
yes 380 521
yes 376 646
yes 324 467

no 254 446
no 243 425
no 220 474
no 236 531
no 257 542
no 235 406
no 251 412
no 251 458
no 236 399
no 236 482
no 266 420
no 268 414
no 248 533
no 246 509
no 263 504
no 244 336
no 213 408
no 241 469
no 255 538
no 231 505
no 241 489
no 219 411
no 235 321
no 260 394
no 255 528
```

```
no 272 399
no 285 381
no 290 384

border 286 494
border 285 496
border 314 419
border 328 371
border 289 447
border 315 313
border 350 402
border 289 485
border 280 444
border 313 416
border 301 471
border 279 490
border 289 431
border 291 446
border 275 546
border 273 467
border 312 463
border 308 440
border 303 419
border 300 509
border 303 438
border 305 399
border 285 483
border 301 453
border 303 414
border 304 446

/* Source: Johnson and Wichern (1998).
(Variables are Admission status, GPA and GMAT
scores of students).  Reproduced by permission
of Prentice Hall International, Inc. */


/* CUSHING SYNDROME DATA SET: cushing.dat */

a 3.1 11.7
a 3.0 1.3
a 1.9 .1
a 3.8 .04
a 4.1 1.1
a 1.9 .4
b  8.3 1
b  3.8 .2
b  3.9 .6
b  7.8 1.2
b  9.1 .6
b  15.4 3.6
b  7.7 1.6
b  6.5 .4
b  5.7 .4
b  13.6 1.6
c 10.2 6.4
c 9.2 7.9
c 9.6 3.1
c 53.8 2.5
c 15.8 7.6
```

```
/* Source: Aitchison and Dunsmore (1975).
(Variables are type of syndromes and two measurements on
urinary extraction rates namely, Tetrahydrocortisone
(tetra) and Pregnanetriol ( preg)). Reprinted by
permission of Cambridge University Press. */


/* EARTHQUAKE DATA: earthquake.dat */

equake 5.60 4.25
equake 5.18 3.93
equake 6.31 6.30
equake 5.36 4.49
equake 5.96 6.39
equake 5.26 4.42
equake 5.17 5.10
equake 4.75 4.40
equake 5.35 5.49
equake 5.01 4.48
equake 5.27 4.41
equake 5.27 4.69
equake 4.98 3.66
equake 5.22 3.99
equake 5.06 4.58
equake 5.09 4.90
equake 5.15 4.82
equake 4.56 4.08
equake 5.00 4.94
equake 5.43 5.48
explosn 6.04 4.33
explosn 5.97 4.39
explosn 5.84 4.35
explosn 5.79 4.14
explosn 5.87 3.90
explosn 6.51 4.49
explosn 5.74 4.22
explosn 5.98 4.08
explosn 6.07 4.30

/* Source: Shumway (1988).
(Variables are the population (Nuclear explosion vs.
Earthquake) and two seismological features namely body
wave magnitude and surface wave magnitude). Reproduced
by permission of the Prentice Hall International, Inc. */
```

```
/* LANDSAT SATELLITE DATA: satellite.dat */

limeston 158 82 114 108 250 139
limeston 159 81 114 109 255 148
limeston 136 64 84 74 174 102
limeston 189 100 140 130 254 131
limeston 177 91 121 110 249 143
limeston 190 104 139 120 248 141
limeston 184 99 133 116 241 134
limeston 213 119 158 136 254 136
limeston 225 122 159 136 253 132
limeston 182 97 141 129 255 152
limeston 184 99 142 128 238 138
limeston 183 96 141 125 216 128

gypsum 223 119 160 141 208 79
gypsum 228 125 172 152 234 98
gypsum 223 119 161 141 223 92
gypsum 221 118 162 142 223 96
gypsum 234 124 165 144 238 93
gypsum 216 115 158 138 228 102
gypsum 220 117 156 137 238 104
gypsum 212 111 148 129 207 85

sandston 144 73 108 95 171 100
sandston 152 77 111 100 198 117
sandston 153 77 105 102 245 149
sandston 142 72 99 88 155 96
sandston 142 71 97 85 158 98
sandston 139 80 140 131 233 136
sandston 141 79 132 123 235 138
carbonat 251 144 190 165 251 162
carbonat 242 133 180 153 255 162
carbonat 198 116 166 150 255 168
carbonat 240 136 187 163 255 171
carbonat 246 141 188 164 255 164
carbonat 235 131 180 157 255 166
carbonat 194 110 161 147 254 166

m_basalt 121 54 64 53 134 93
m_basalt 117 51 61 50 75 41
m_basalt 111 49 65 59 127 81
m_basalt 111 46 55 44 72 39
m_basalt 114 48 58 47 77 43
m_basalt 119 53 66 54 84 49
m_basalt 127 60 76 62 91 53
m_basalt 123 55 69 56 90 52

i_dacite 132 62 83 78 215 131
i_dacite 121 56 75 70 178 114
i_dacite 152 72 95 84 201 124
i_dacite 123 56 76 72 168 108
i_dacite 165 81 98 80 132 60
i_dacite 167 87 108 91 157 73
i_dacite 123 58  75 65 138 88
i_dacite 177 91 119 105 170 81
i_dacite 125 58 74 62 129 80

Source: Fox (1993, Landsat Thematic Mapper satellite data).
(Variables are type of rock/mineral and six geological
measurements).
```

```
/* SOIL DATA: soil.dat */

1 5.40 .188 .92 215 16.35 7.65 .72 1.14 1.09
1 5.65 .165 1.04 208 12.25 5.15 .71 .94 1.35
1 5.14 .260 .95 300 13.02 5.68 .68 .60 1.41
1 5.14 .169 1.10 248 11.92 7.88 1.09 1.01 1.64
2 5.14 .164 1.12 174 14.17 8.12 .70 2.17 1.85
2 5.10 .094 1.22 129 8.55 6.92 .81 2.67 3.18
2 4.70 .100 1.52 117 8.74 8.16 .39 3.32 4.16
2 4.46 .112 1.47 170 9.49 9.16 .70 3.76 5.14
3 4.37 .112 1.07 121 8.85 10.35 .74 5.74 5.73
3 4.39 .058 1.54 115 4.73 6.91 .77 5.85 6.45
3 4.17 .078 1.26 112 6.29 7.95 .26 5.30 8.37
3 3.89 .070 1.42 117 6.61 9.76 .41 8.30 9.21
4 3.88 .077 1.25 127 6.41 10.96 .56 9.67 10.64
4 4.07 .046 1.54 91 3.82 6.61 .50 7.67 10.07
4 3.88 .055 1.53 91 4.98 8.00 .23 8.78 11.26
4 3.74 .053 1.40 79 5.86 10.14 .41 11.04 12.15
5  5.11 .247 .94 261 13.25 7.55 .61 1.86 2.61
5  5.46 .298 .96 300 12.30 7.50 .68 2.00 1.98
5  5.61 .145 1.10 242 9.66 6.76 .63 1.01 .76
5  5.85 .186 1.20 229 13.78 7.12 .62 3.09 2.85
6  4.57 .102 1.37 156 8.58 9.92 .63 3.67 3.24
6  5.11 .097 1.30 139 8.58 8.69 .42 4.70 4.63
6  4.78 .122 1.30 214 8.22 7.75 .32 3.07 3.67
6  6.67 .083 1.42 132 12.68 9.56 .55 8.30 8.10
7  3.96 .059 1.53 98 4.80 10.00 .36 6.52 7.72
7  4.00 .050 1.50 115 5.06 8.91 .28 7.91 9.78
7  4.12 .086 1.55 148 6.16 7.58 .16 6.39 9.07
7  4.99 .048 1.46 97 7.49 9.38 .40 9.70 9.13
8  3.80 .049 1.48 108 3.82 8.80 .24 9.57 11.57
8  3.96 .036 1.28 103 4.78 7.29 .24 9.67 11.42
8  3.93 .048 1.42 109 4.93 7.47 .14 9.65 13.32
8  4.02 .039 1.51 100 5.66 8.84 .37 10.54 11.57
9  5.24 .194 1.00 445 12.27 6.27 .72 1.02 .75
9  5.20 .256 .78 380 11.39 7.55 .78 1.63 2.20
9  5.30 .136 1.00 259 9.96 8.08 .45 1.97 2.27
9  5.67 .127 1.13 248 9.12 7.04 .55 1.43 .67
10  4.46 .087 1.24 276 7.24 9.40 .43 4.17 5.08
10  4.91 .092 1.47 158 7.37 10.57 .59 5.07 6.37
10  4.79 .047 1.46 121 6.99 9.91 .30 5.15 6.82
10  5.36 .095 1.26 195 8.59 8.66 .48 4.17 3.65
11  3.94 .054 1.60 148 4.85 9.62 .18 7.20 10.14
11  4.52 .051 1.53 115 6.34 9.78 .34 8.52 9.74
11  4.35 .032 1.55 82 5.99 9.73 .22 7.02 8.60
11  4.64 .065 1.46 152 4.43 10.54 .22 7.61 9.09
12 3.82 .038 1.40 105 4.65 9.85 .18 10.15 12.26
12 4.24 .035 1.47 100 4.56 8.95 .33 10.51 11.29
12 4.22 .030 1.56 97 5.29 8.37 .14 8.27 9.51
12 4.41 .058 1.58 130 4.58 9.46 .14 9.28 12.69

/* Source: Horton, Russell, and Moore (1968).
(Variables are soil group, pH value, Nitrogen percentage,
Bulk density, Phosphorus in ppm, calcium, magnesium, potassium,
sodium, and conductivity).  Reproduced by permission of the
International Biometric Society. */
```

Chapter 6 (Cluster Analysis) Data Sets:

```
/* AIDS DATA SET: aids.dat */

a 8 8 20 10 9 5 3 1 4 1 1 3 0
a 2 11 10 12 5 3 2 2 2 3 0 0 0
a 18 12 10 18 14 10 8 5 1 3 2 2 0
a 19 12 7 11 9 6 3 1 3 1 2 1 0
a 14 17 8 8 2 4 3 3 1 3 1 2 0
a 10 16 9 4 6 4 0 1 2 0 0 0 0
a 3 1 4 3 3 2 2 0 2 0 2 1 1
a 27 22 16 7 6 5 2 2 4 2 0 0 0
a 12 16 16 17 9 7 4 5 1 2 0 0 1
a 10 14 10 10 7 3 1 0 0 2 1 1 0
a 2 3 12 7 4 9 6 2 3 3 2 3 1
o 21 19 9 4 3 2 1 0 0 0 0 0 0
o 20 17 11 7 2 6 1 0 1 0 0 0 0
o 16 27 22 12 9 2 3 3 2 4 2 0 0
o 11 15 13 14 9 9 5 2 2 0 0 1 0
o 10 13 5 12 5 9 5 1 1 2 3 3 0
o 23 13 14 8 4 6 2 3 1 1 2 1 2
o 15 14 15 10 7 12 5 4 0 1 1 2 0
o 14 15 13 11 7 4 1 2 1 0 0 0 0
o 10 19 16 19 8 3 4 1 1 1 1 1 0

/* Source: (Everitt, 1993). Variables: group s1- s13.
Reprinted by permission of Edward Arnold. */


/* HOSPITAL DATA SET: hospital.dat */

 1  Sloan-Ketterning        100.0  73.0  1.05  Yes  6.0  3544  1.85
 2  M. D. Anderson           99.7  67.5  0.78  Yes  6.0  3683  1.87
 3  Johns Hopkins            65.6  33.4  0.70  Yes  7.0  1278  1.33
 4  Mayo Clinic              60.4  26.0  0.57  Yes  7.0  2589  1.10
 5  U. Wash. Medical         39.0   9.1  0.67  Yes  6.0   606  2.03
 6  Duke U Medical           38.6  10.7  0.82  Yes  7.0  2523  1.63
 7  University of Chicago    37.2   6.6  0.68  Yes  7.0  1116  1.26
 8  Fox Chase Cancer         35.5   5.7  0.54  Yes  4.0   872  1.88
 9  U. Michigan Medical      35.2   1.3  0.56  Yes  7.0  1221  1.46
10  Roswell Park Cancer      34.9   8.4  0.81  Yes  5.5  1481  2.34
11  U. Pittsburgh Medical    33.3   3.0  0.69  Yes  7.0  1730  1.34
12  U. Virginia Health       32.9   0.0  0.54  Yes  6.0  1026  1.91
13  UCLA Medical Center      32.2   8.9  0.80  Yes  6.0   825  0.91
14  Clarian Health           32.1   4.2  0.78  Yes  7.0  1139  1.48
15  Barnes-Jewish            31.2   3.3  0.77  Yes  6.5  1872  1.52
16  Vanderbilt U.            31.2   1.4  0.71  Yes  7.0   779  1.88
17  U. Kentucky              31.1   0.0  0.63  Yes  6.0   698  1.85
18  Massachusetts Gen.       30.6   6.3  0.94  Yes  7.0  1547  1.35
19  Arthur James Cancer      30.0   0.4  0.50  Yes  5.0  1611  1.10
20  Allegheny U-Hos.         29.9   0.3  0.64  Yes  5.0   885  1.89
21  Loyola U.                29.9   0.5  0.63  Yes  6.0   683  1.39
22  U. Pennsylvania          29.4   5.5  0.96  Yes  6.0  1431  1.98
23  U. Florida               29.3   0.5  0.46  Yes  7.0   561  0.60
24  Cleveland Clinic         29.0   2.4  0.86  Yes  7.0  1686  1.69
25  Stanford U. Hos.         28.5   9.5  0.83  Yes  5.3   869  1.10
26  Lutheran Gen.            28.5   0.4  0.60  Yes  5.0   935  0.92
```

```
27  Emory University          28.5  0.0  0.61  Yes  5.5   762 0.98
28  Luke's Medical            28.5  0.0  0.67  Yes  7.0   820 1.01
29  U. Hos. Columbia MO       28.4  0.0  0.66  Yes  6.0   791 1.26
30  Kettering, Ohio           28.3  0.0  0.57  Yes  6.0   512 0.94
31  Baltimore Medical         28.3  0.0  0.54  Yes  4.0   903 1.47
32  U. Iowa Hos.              28.1  0.9  0.75  Yes  7.0  1138 1.12
33  Harper Hos. Detroit       28.0  0.4  0.68  Yes  5.5  1348 1.10
34  H. Lee Moffitt Cancer     27.8  0.8  0.61  Yes  3.0   980 1.43
35  Carolina Hospitals        27.7  0.0  0.69  Yes  6.0   751 1.49
36  Evanston Hospital         27.7  0.0  0.61  Yes  5.0   914 0.81
37  Baylor U.                 27.6  1.9  0.82  Yes  6.0  1392 1.53
38  U. Hos. Portland OR       27.4  0.0  0.63  Yes  6.0   569 0.90
39  Brigham and Women's       27.3  2.9  0.80  Yes  6.0   939 1.30
40  U. Wis. Madison           27.2  1.2  0.77  Yes  7.0   759 1.19
41  Nebraska Health           26.6  0.0  0.74  Yes  7.0   648 1.28
42  U. Hos. Cleveland         26.4  0.0  0.81  Yes  7.0  1414 1.26
43  Yale Hospital             26.3  1.0  0.77  Yes  6.5  1296 0.84
44  Meritcare Health Fargo    26.2  0.4  0.64  Yes  6.0   627 0.41
45  NY Presbyterian           25.9  3.9  1.01  Yes  7.0  1923 1.13
46  Loma Linda                25.6  0.0  0.64  Yes  4.8   768 1.79
47  Jefferson U.              25.4  0.0  0.81  Yes  6.0  1310 1.47
48  UC-Davis                  25.3  0.0  0.84  Yes  7.0   543 1.89
49  NC Baptist Hospital       25.2  0.5  0.83  Yes  6.0  1364 1.34
50  SW Hos. Temple TX         25.1  0.0  0.60  Yes  3.5   716 0.92

/* Source: US News and World Report 7/19/99 (best hospitals for cancer).
Variables are: Rank, Hospital (read in 6-31 columns in the data set),
index, Reput, Mortalit, COTH (yes), techscor, discharg, and  rn_bed.
Reproduced by permission of US News and World Report, Inc. */
```

```
/* MUTUAL FUNDS DATA: m_funds.dat */

Amer Century Ultra                 36.91 35.66 3.50 1.60 -3.68 0.90 10.48 -10.46
Babson Enterprise II               23.79 22.35 6.46 -2.02 -4.00 2.54 4.71 -9.32
Baron Asset                        54.50 54.52 -0.04 -3.08 -8.57 -6.98 7.84 -12.91
Berger Small Compan Growth          4.65 4.01 16.00 3.10 1.31 12.86 23.02 -12.72
Dreyfus New Leaders                47.45 44.14 7.50 0.02 -2.81 0.49 15.19 -10.09
Fidelity Capital Appreciation      25.53 24.00 6.38 0.47 -2.85 2.16 15.83 -9.54
Fidelity Fifty                     19.49 18.38 6.05 1.36 -2.90 0.59 24.65 -11.97
Fidlty Low-Prcd Stck               23.50 22.86 2.81 -3.17 -4.55 -0.51 2.84 -9.86
Founders Discovery                 31.80 26.71 19.08 -1.70 1.63 13.90 30.49 -10.55
Invesco Dynamics                   19.54 17.51 11.62 0.77 -2.06 4.83 23.67 -10.67
Invesco Small Company              13.52 11.98 12.83 -0.66 0.30 9.74 16.75 -11.62
Janus Olympus                      35.05 32.11 9.14 2.25 -1.74 4.35 27.08 -10.16
Kaufmann Fund                       5.40 5.37 0.60 -2.17 -4.76 2.27 -4.93 -11.05
Managers Special Equity            68.67 61.65 11.38 0.38 0.13 8.48 12.17 -10.30
Oakmark Small Cap                  14.36 14.49 -0.88 -5.90 -7.30 -4.14 -2.78 -11.26
Emerging Growth                    23.03 21.91 5.11 1.05 -1.92 6.52 -3.84 -10.65
Price Small Cap Stock              21.46 20.31 5.69 -1.24 -2.63 3.62 3.22 -9.55
Profund Ultra OTC (sold 2/8/99)    44.35 36.11 22.82 10.90 2.02 25.21 48.78 -14.87
Rob Stephens Em Growth             35.60 29.15 22.13 5.42 1.19 14.25 55.12 -14.32
Rob Stephens Value + Gwth          28.76 27.62 4.14 0.77 -2.57 5.74 10.96 -8.84
Rydex OTC                          54.16 47.17 14.83 6.01 2.36 14.05 30.95 -7.46
Safeco Growth                      20.32 21.11 -3.74 -1.45 -5.00 -2.68 -10.48 -11.68
Schwab Small Cap Index             17.13 16.26 5.35 -1.78 -3.44 2.15 5.42 -10.67
Scudder Development                39.41 39.15 0.67 -2.23 -7.75 -3.17 4.65 -10.00
SteinRoe Capital Opportunities     28.72 28.73 -0.03 -1.71 -7.06 -2.35 -2.18 -13.46
US Global Bonnel Growth            22.90 20.54 11.48 4.28 0.97 10.20 17.08 -8.54
```

Value Line Leveraged Growth 54.14 51.64 4.84 1.61 -2.73 3.64 11.81 -11.50
Warburg Pincus Emerging Grth 41.12 39.23 4.82 -1.08 -3.86 3.94 2.88 -10.55
Wasatch Micro-cap 4.38 3.98 10.12 -1.35 0.92 12.02 9.77 -10.09
Amer Cent Bnhm Eq Gwth 24.23 23.06 5.08 1.25 -1.42 4.13 7.77 -9.31
Baron Growth 29.44 27.52 6.99 -2.03 -4.57 -1.44 18.38 -13.63
Bramwell Growth 25.25 24.63 2.50 0.56 -4.32 1.24 4.68 -8.11
Columbia Growth 46.78 45.30 3.26 -0.47 -5.02 0.75 10.04 -10.13
Dreyfus Appreciation 44.92 43.41 3.47 2.96 -1.06 2.53 6.77 -8.19
Dreyfus Core Value 32.38 31.24 3.64 -0.03 -5.98 -2.88 12.25 -8.91
Dreyfus Third Century 13.58 12.94 4.97 1.34 -2.79 3.82 10.76 -10.55
Fidelity Blue Chip Growth 54.28 52.33 3.72 2.03 -2.02 2.92 7.83 -8.14
Fidelity Contrafund (clsd) 61.13 59.39 2.93 -1.18 -3.97 0.28 8.85 -8.73
Fidelity Dividend Growth 31.14 30.18 3.18 0.00 -4.33 0.32 8.39 -7.47
Fidelity Aggrssv Gw 44.01 36.94 19.14 6.92 3.43 14.28 42.00 -9.31
Fidelity Growth Company 64.80 55.71 16.32 6.00 5.47 13.82 28.13 -9.65
Fidelity Magellan (clsd) 126.31 121.12 4.29 0.73 -3.84 1.50 11.07 -9.23
Fidelity New Millenium (clsd) 35.97 32.72 9.95 0.22 -3.67 4.20 37.25 -9.32
Fidelity OTC 52.22 47.88 9.07 1.34 -1.21 5.54 19.69 -8.97
Fidelity Spartan Market Index 93.27 89.29 4.46 1.58 -2.92 1.79 10.44 -8.33
Fidelity Stock Selector 32.54 30.32 7.31 0.96 -0.46 5.07 13.34 -8.84
Fidelity Value 53.37 50.31 6.09 -1.98 -6.65 -3.63 15.15 -8.09
Founders Growth 22.34 21.61 3.38 2.06 -3.16 2.34 9.46 -10.27
Fremont Growth 16.63 16.03 3.75 1.53 -4.04 2.40 8.34 -9.95
Invesco Blue Chip Gw 7.07 6.74 4.89 4.74 -1.26 4.28 9.44 -7.99
Janus Enterprise 49.02 41.83 17.18 5.46 3.00 9.44 35.34 -8.81
Janus Fund 39.30 36.92 6.44 0.43 -3.53 3.20 16.79 -8.32
Janus Mercury 32.05 28.67 11.80 4.19 -1.75 6.55 32.93 -9.78
Janus Twenty 62.22 59.56 4.46 3.00 -2.87 2.44 16.74 -9.15
Longleaf Partners 26.99 27.04 -0.17 -1.85 -9.25 -8.20 10.66 -8.50
Montgomery Growth 22.99 22.02 4.41 -2.09 -6.28 -1.75 11.49 -9.36
Neuberger & Berman Manhattan 12.24 11.74 4.28 1.41 -3.47 4.79 2.43 -11.70
Neuberger & Berman Partners 26.94 26.85 0.34 -2.32 -7.83 -4.97 5.65 -9.81
Oakmark Fund 37.05 37.15 -0.26 -3.21 -7.00 -5.61 3.43 -9.54
Price Growth Stock 33.98 32.97 3.07 0.65 -3.66 1.04 5.96 -10.26
Price Midcap Growth Stock 36.84 34.98 5.30 -3.10 -4.21 0.99 8.10 -10.42
Rydex Nova (sold 8/5/99) 36.91 35.30 4.55 2.13 -5.12 1.54 11.07 -12.75
Schwab 1000 36.50 35.23 3.62 1.05 -3.49 1.00 8.92 -8.57
Schwab S&P 500 20.92 20.05 4.32 1.55 -2.92 1.75 10.34 -8.35
Scudder Large Co Value 29.98 29.19 2.70 -0.23 -5.40 -1.19 9.06 -9.02
SIT Large Cap Growth 52.26 50.64 3.20 2.39 -2.93 3.24 7.44 -9.94
Strong Opportunity 43.63 40.74 7.08 -2.44 -3.35 0.53 16.53 -9.36
Transamerica Equity Growth 26.48 26.64 -0.59 -0.71 -6.30 -0.79 6.86 -10.27
Tweedy Browne American Value 24.22 23.34 3.78 -0.90 -5.09 0.00 6.51 -8.68
Vanguard Index Trust 500 124.82 119.50 4.45 1.60 -2.90 1.82 10.56 -8.33
Vanguard/Primecap (clsd) 57.18 51.48 11.08 1.85 0.95 8.13 21.50 -7.59
Vanguard US Growth 39.55 38.36 3.10 2.41 -2.80 2.94 5.49 -7.84
Warburg Pincus Capital App 24.46 23.26 5.18 0.87 -1.53 4.04 11.38 -9.62
Amer Cent Eq Income 6.60 6.30 4.77 -0.90 -2.51 0.91 6.44 -2.61
American Century Value 6.60 6.26 5.39 -1.35 -4.07 -2.08 9.79 -5.35
Amer Cent Benhm Incm & Gwth 31.50 30.13 4.54 1.29 -2.33 2.98 8.56 -8.61
Babson Value 49.42 48.12 2.71 -0.62 -5.31 -4.13 8.37 -10.90
Berger Growth and Income 15.67 14.66 6.87 0.19 -2.43 4.82 17.37 -8.22
Columbia Common Stock 26.92 25.89 3.97 0.19 -4.06 1.09 10.45 -9.56
Fidelity Equity-Income 59.51 57.54 3.43 -0.58 -4.62 -0.95 9.44 -9.40
Fidelity Equity-Income II 31.03 30.57 1.52 0.10 -4.04 -0.47 4.92 -8.80
Fidelity Fund 36.91 36.08 2.31 0.13 -4.20 -0.07 7.33 -8.02
Fidelity Growth & Income (clsd) 47.54 46.68 1.85 0.57 -3.59 0.41 4.18 -8.58
Founders Grwth & Incm 7.38 7.31 0.95 1.23 -3.40 0.96 0.82 -7.83
Invesco Industrial Income 16.24 15.69 3.50 -0.43 -3.62 1.37 8.67 -6.75

```
Janus Equity Income               21.42 20.52 4.39 -1.02 -3.82 0.51 14.80 -7.63
Janus Growth & Income             33.43 31.74 5.33 0.91 -3.30 2.74 14.95 -8.93
Lexington Corporate Leaders       17.58 16.59 5.94 1.91 -1.79 0.57 12.74 -8.19
Lexington Gwth & Incm             22.38 22.26 0.52 -0.58 -5.09 -0.84 2.15 -10.03
Price Equity-Income               28.14 27.14 3.67 -1.12 -4.19 -1.81 8.91 -6.00
Price Growth & Income             28.06 27.25 2.99 -1.27 -4.78 -1.48 8.22 -7.75
Rob Stephs Mid Cap Opprtnties     16.72 15.52 7.75 -1.01 -4.95 1.39 19.00 -11.49
Safeco Equity                     25.32 24.30 4.21 1.28 -1.78 0.83 9.39 -8.25
Safeco Income                     22.10 22.87 -3.37 -2.86 -6.79 -4.75 -4.60 -9.85
Scudder Growth & Income           27.54 27.09 1.68 -1.64 -5.78 -3.13 5.95 -9.24
Selected American Shares          34.61 33.09 4.58 0.29 -5.26 0.99 12.01 -9.97
Vanguard Equity Income            25.65 25.13 2.06 -0.77 -4.11 -1.53 4.97 -6.24
Vanguard Growth and Income        34.49 32.72 5.41 1.74 -2.76 3.08 12.59 -9.06
Vanguard Windsor II               30.96 31.00 -0.14 -1.09 -6.78 -4.55 4.69 -8.06
Weiss Peck & Greer Gwth & Incm    40.09 39.55 1.37 1.96 -1.47 1.77 -1.13 -10.30

/* Source:  Fabian Premium Investment Resource (1999).
Variables: Fundname read in columns 1-33, x1-x8.
Reproduced by permission of Fabian Premium Investment Resource. */


/* MEANS FOR SIX GROUPS IN SATELITE DATA: meanfox.dat */

carbonat 229.4285714 130.1428571 178.8571429 157.0000000 254.2857143
165.5714286
gypsum  222.1250000 118.5000000 160.2500000 140.5000000 224.8750000
93.6250000
i_dacite 142.7777778 69.0000000 89.2222222 78.5555556 165.3333333
95.4444444
limeston 181.6666667 96.1666667 32.1666667 118.4166667 240.5833333
135.3333333
m_basalt 117.8750000 52.0000000 64.2500000 53.1250000 93.7500000
56.3750000
sandston 144.7142857 75.5714286 113.1428571 103.4285714 199.2857143
119.1428571

/* Raw data from Fox (1993). Means on six variables for six
Rock/Mineral groups.
Variables: group, tm1, tm2, tm3, tm4, tm5,
tm7. */
```

References

Acito, F. and Anderson, R. D. (1980), "A Monte Carlo Comparison of Factor Analytic Methods," *Journal of Marketing Research*, 17, 228-236.

Agresti, A. (1990), *Categorical Data Analysis*, New York: John Wiley & Sons, Inc.

Agresti, A. (1996), *An Introduction to Categorical Data Analysis*, New York: John Wiley & Sons, Inc.

Ahamad, B. (1967), "An Analysis of Crimes by the Method of Principal Components," *Applied Statistics*, 16, 17-35.

Aitchison, J. (1983), *The Statistical Analysis of Compositional Data*, London: Chapman & Hall.

Aitchison, J. and Dunsmore, I. R. (1975), *Statistical Prediction Analysis*, Cambridge: Cambridge University Press.

Akaike, H. (1971), "Autoregressive Model Fitting for Control," *Annals of Institute of Statistical Mathematics*, 23, 163-180.

Akaike, H. (1973), "Information Theory and the Extension of the Maximum Likelihood Principle," in *Second International Symposium on Information Theory*, edited by V. N. Petrov and F. Csaki, 267–281, Budapest: Akailseoniai-Kiudo.

Akaike, H. (1987), "Factor Analysis and AIC," *Psychometrika*, 52, 317-332.

Allison, T. and Cicchetti, D. V. (1976), "Sleep in Mammals: Ecological and Constitutional Correlates," *Science*, 194, 732-734.

Anderberg, M. R. (1973), *Cluster Analysis for Applications*, New York: Academic Press, Inc.

Anderson, J. C. and Gerbing, D. W. (1984), "The Effect of Sampling Error on Convergence, Improper Solutions, and Goodness-of-fit Indices for Maximum Likelihood Confirmatory Factor Analysis," *Psychometrika*, 49, 155-173.

Anderson, T. W. (1963), "A Test for Equality of Means When Covariance Matrices Are Unequal," *Annals of Mathematical Statistics*, 34, 671-672.

Anderson, T. W. (1984), *An Introduction to Multivariate Statistical Analysis*, New York: John Wiley & Sons, Inc.

Andrews, D. F. (1972), "Plots of High-Dimensional Data," *Biometrics*, 28, 125-136.

Andrews, D. F. and Herzberg, A. M. (1985), *Data: A Collection of Problems from Many Fields for the Student and Research Worker*, New York: Springer Verlag.

Armitage, P. (1971), *Statistical Methods in Medical Research*, Oxford: Blackwell Scientific.

Art, D., Gnanadesikan, R., and Kettenring, R. (1982), "Data-based Metrics for Cluster Analysis," *Utilitas Mathematica*, 21A, 75-99.

Arthanari, T. S. and Dodge, Y. (1993), *Mathematical Programming in Statistics*, New York: Wiley.

Bartholomew, D. J. (1981), "Posterior Analysis of the Factor Model," *British Journal of Mathematical and Statistical Psychology*, 34, 93-99.

Bartlett, M. S. (1937), "The Statistical Conception of Mental Factors," *British Journal of Psychology*, 28, 97-104.

Bartlett, M. S. and Please, N. W. (1963), "Discrimination in the Case of Zero Mean Differences," *Biometrika*, 50, 17–21.

Basilevsky, A. (1994), *Statistical Factor Analysis and Related Methods*, New York: John Wiley & Sons.

Belcham, P. and Hymans, R., eds. (1984), *IAAF/ATFS Track and Field Statistics Handbook for the 1984 Los Angeles Olympic Games*, London: IAAF.

Belsley, D. A. (1991), *Conditioning Diagnostics: Collinearity and Weak data in Regression*, John Wiley & Sons, Inc.

Belsley, D. A., Kuh, E., and Welsch, R. E. (1980), *Regression Diagnostics: Identifying Influential Data and Source of Collinearity*, New York: John Wiley & Sons, Inc.

Benzecri, J. P. (1992), *Correspondence Analysis Handbook*, New York: Marcel Dekker.

Blashfield, R. K. (1976), "Mixture Model Tests of Cluster Analysis: Accuracy of Few Agglomerative Methods," *Psychological Bulletin*, 83, 377-388.

Bollen, K. A. (1987), "Outliers and Improper Solutions: A Confirmatory Factor Analysis Example," *Sociological Methods and Research*, 15, 375-384.

Bolton, R. J. and Krzanowski, W. J. (1999), "A Characterization of Principal Components for Projection Pursuit," *The American Statistician*, 53, 108-109.

Bookstein, F. L. (1989), "Size and Shape: A Comment on Semantics," *Systematic Zoology*, 38, 173-180.

Boomsma, A. (1985), "Nonconvergence, Improper Solutions, and Starting Values in LISREL Maximum Likelihood Estimation," *Psychometrika*, 50, 229-242.

Browne, M. W. (1968), "A Comparison of Factor Analytic Techniques," *Psychometrika*, 33, 267-334.

Burt, C. (1950), "Factor Analysis of Qualitative Data," *British Journal of Statistical Psychology*, 3, 166-185.

Byth, K. and McLachlan, G. J. (1980), "Logistic Regression Compared to Normal Discrimination for Nonnormal Population," *Australian Journal of Statistics*, 22, 188-196.

Calimlim, J. F., Wardell, W. M., Davis, H. T., Lasagna, L., and Gillies, A. J. (1982), "Analgesic Efficacy of an Orally Administered Combination of Pentazocine and Aspirin: with Observations on the Use and Statistical Efficacy of Global Subjective Efficacy Ratings," *Clinical Pharmacology and Therapeutics*, 21, 34-43.

Chatterjee, S. and Hadi A. S. (1988), *Sensitivity Analysis in Linear Regression*, New York: John Wiley & Sons.

Chuang, C. and Agresti, A. (1986), "A New Model for Ordinal Pain Data from a Pharmaceutical Study," *Statistics in Medicine*, 5, 15-20.

Cook, R. D. and Weisberg, S. (1982), *Residuals and Influence in Regression*, London: Chapman & Hall.

Cooley, W. W. and Lohnes, P. R. (1971), *Multivariate Procedures for the Behavioral Sciences*, New York: John Wiley & Sons.

Costanza, M. C. and Afifi, A. A. (1979), "Comparison of Stopping Rules in Forward Stepwise Discriminant Analysis," *Journal of American Statistical Association*, 74, 777-785.

Cox, C. and Chuang, C. (1984), "A Comparison of Chi-square Partitioning and Two Logit Analyses of Ordinal Pain Data from a Pharmaceutical Study," *Statistics in Medicine*, 3 273-285.

Crawford, C. (1967), "A General Method of Rotation for Factor Analyis," Paper presented at the spring meeting of the Psychometric Society, Madison, Wis.

Crawley, D. R. (1979), "Logistic Discrimination as an Alternative to Fisher's Linear Discrimination Function," *New Zealand Statistician*, 14, 21-25.

Cronbach, L. J. (1951), "Coefficient of Alpha and the Internal Structure of Tests," *Psychometrika*, 16, 297-334.

Cureton, E. E. and D'Agostino, R. B. (1983), *Factor Analysis: An Applied Approach*, New Jersey: Lawrence Erlbaum Associates, Inc.

Dawkins, B. (1989), "Multivariate Analysis of National Track Records," *The American Statistician*, 43, 110-115.

Diehr, P., Wood, R. W., Barr, V., Wolcott, B., Slay, L., and Tompkins, R. K. (1981), "Acute Headaches: Presenting Symptoms and Diagnostic Rules to Identify Patients wth Tension and Migraine Headache," *Journal of Chronic Diseases*, 34, 147-158.

Draper, N. R. and Smith, H. (1998), *Applied Regression Analysis*, 3d ed., New York: John Wiley & Sons, Inc.

Eckart, C. and Young, G. (1936), "The Approximation of One Matrix by Another of Lower Rank," *Psychometrika*, 1, 211-308.

Efron, B. (1975), "The Efficiency of Logistic Regression Compared to Normal Discriminant Analysis," *Journal of American Statistical Association*, 70, 892-898.

El Oksh, H. A., Sutherland, T. M., and Williams, J. S. (1967), " Prenatal and Postnatal Influence on Growth of Mice," *Genetics*, 57, 79-94.

Everitt, B. S. (1989), *Statistical Methods in Medical Investigation*, London: Edward Arnold.

Everitt, B. S. (1993), *Cluster Analysis*, 3d ed., London: Edward Arnold.

Fabian Premium Investment Resource (1999), Volume 22, No. 9, 8-9.

Fisher, L. D., Judkins, M. P., Lesperance, J., Cameron, A., Swaye, P., Ryan, T. J., Maynard, C., Bourassa, M., Kennedy, J. W., Gosselin, A., Kemp, H., Faxon, D., Wexler, L., and Davis, K. (1982), "Reproducibility of coronary arteriographic reading in the coronary artery urgery study (CASS)," *Catheterization and Cardiovascular Diagnosis*, 8, 565-575.

Fisher, R. A. (1936), "The Use of Multiple Measurements in Taxonomic Problems," *Annals of Eugenics*, 7, 179-188.

Fisher, R. A. (1940), "The Precision of Discriminant Functions," *Annals of Eugenics*, 10, 422-429.

Flury, B. (1988), *Common Principal Components and Related Multivariate Models*, New York: John Wiley & Sons.

Flury, B. and Riedwyl, H. (1985), "T^2 tests, the Linear Two Group Discriminant Function and their Computation by Linear Regression, *The American Statistician*, 39, 20-25.

Fox, C. W. (1993), *Lithologic discrimination using Landsat Thematic Mapper data and canonical discriminant analysis*, unpublished M.S. Thesis, Wayne State University.

Friendly, M. (1991), *SAS System for Statistical Graphics, First Edition*, Cary, NC: SAS Institute Inc.

Fuller, W. A. (1987), *Measurement Error Models*, New York: John Wiley & Sons, Inc.

Gabriel, K. R. (1971), "The Biplot Graphic Display of Matrices with Application to Principal Component Analysis," *Biometrika*, 58, 453-467.

Gauch, H. G. (1982), *Multivariate Analysis in Community Ecology*, Cambridge: Cambridge University Press.

Gauch, H. G., and Whittaker, R. H., (1972), "Coenocline Simulation," *Ecology*, 53, 446-451.

Gifi, A. (1990), *Nonlinear Multivariate Analysis*, 2d ed., New York: John Wiley & Sons, Inc.

Gittins, R. (1985), *Canonical Analysis: A Review with Applications in Ecology*, Berlin: Springer-Verlag.

Gnanadesikan, R. (1997), *Methods for Statistical Data Analysis of Multivariate Observations, 2d ed.*, New York: John Wiley & Sons.

Gordon, A. D. (1981), *Classification*, London: Chapman and Hall.

Gower, J. C. (1967), "A Comparison of Some Methods of Cluster Analysis," *Biometrics*, 23, 623-628.

Gower, J. C. (1971), "A general coefficient of similarity and some of its properties," *Biometrics*, 27, 857-874.

Gower, J. C. (1993), "Recent Advances in Biplot Methodology," in *Multivariate Analysis: Future Directions 2*, edited by C. M. Cuadras and C. R. Rao, 295-325, North Holland.

Gower, J. C. and Hand, D. J. (1996), *Biplots*, London: Chapman & Hall.

Greenacre, M. J. (1984), *Theory and Applications of Correspondence Analysis*, London: Academic Press.

Greenacre, M. J. (1988), "Correspondence Analysis of Multivariate Categorical Data by Weighted Least Squares," *Biometrika*, 75, 457-467.

Greenacre, M. J. (1992), "Correspondence Analysis in Medical Research," *Statistical Methods in Medical Research*, 1, 97-117.

Greenacre, M. J. (1993a), *Correspondence Analysis in Practice*, London: Academic Press.

Greenacre, M. J. (1993b), "Biplots in Correspondence Analysis," *Journal of Applied Statistics*, 20, 251-269.

Greenacre, M. J. and Hastie, T. J. (1987), "The Geometric Interpretation of Correspondence Analysis," *Journal of American Statistical Association*, 82, 437-447.

Guttman, L. (1941), "The Quantification of a Class of Attributes: A Theory and Method of Scale Construction," in *The Prediction of Personal Adjustment*, edited by P. Horst, et al., 319-348, New York: Social Science Research Council.

Guttman, L. (1953), "Image Theory for the Structure of Quantitative Variates," *Psychometrika*, 18, 277-296.

Guttman, L. (1956), " 'Best Possible' Systematic Estimates of Communalities," *Psychometrika*, 21, 273-285.

Hadi, A. S. and Ling, R. F. (1998), "Some Cautionary Notes on the Use of Principal Components Regression," *The American Statistician*, 52, 15-19.

Hand, D. J. (1981), *Discrimination and Classification*, New York: John Wiley & Sons.

Harman, H.H. (1976), *Modern Factor Analysis*, 3d ed., Chicago: University of Chicago Press.

Harris, C. W. (1962), "Some Rao-Guttman Relationships," *Psychometrika*, 27, 247-263.

Harris, C. W. and Kaiser, H. F. (1964), "Oblique Factor Analytic Solutions by Orthogonal Transformations," *Psychometrika*, 29, 347-362.

Hartigan J. A. (1975), *Clustering Algorithms*, New York: John Wiley & Sons, Inc.

Harville D. A. (1997), *Matrix Algebra from a Statistician's Perspective*, New York: Springer-Verlag.

Hatcher, L. (1994), *A Step-by-Step Approach to Using the SAS System for Factor Analysis and Structural Equation Modeling*, Cary, NC: SAS Institute Inc.

Hawkins, D. M. (1974), "The Detection of Errors in Multivariate Data Using Principal Components," *Journal of American Statistical Association*, 69, 340-344.

Hayashi, C. (1950), "On the Quantification of Qualitative Data From the Mathematico-Statistical Point of View," *Annals of Institute of Statistical Mathematics*, 2, 35-47.

Hegde, L. M. and Naik, D. N. (1999), "Canonical Correspondence Analysis in SAS Software," in *Proceedings of the Twenty-Fourth Annual SAS Users Group International Conference*, 1607-1613.

Hendrickson, A. E. and White, P. O. (1964), "PROMAX: A Quick Method for Rotation to Oblique Simple Structure," *British Journal of Statistical Psychology*, 17, 65-70.

Heywood, H. B. (1931), "On Finite Sequences of Real Numbers," *Proceedings of the Royal Society, Series A*, 134, 486-501.

Hirschfeld, H. O. (1935), "A Connection Between Correlation and Contingency," *Proceedings Cambridge Philosophical Society*, 31, 520-524.

Hooke, B. G. E. (1926), "A Third Study of the English Skull with Special Reference to the Farringdon Street Crania," *Biometrika*, 18, 1-55.

Horst, P. (1935), "Measuring Complex Attitudes," *Journal of Social Psychology*, 6, 369-374.

Horton, I. F., Russell, J. S., and Moore, A. W. (1968), "Multivariate-Covariance and Canonical Analysis: A method for selecting the most effective Discriminators in a multivariate situation," *Biometrics*, 24, 845–858.

Hotelling, H. (1933), "Analysis of a Complex of Statistical Variables into Principal Components," *Journal of Educational Psychology*, 24, 417-441, 498-520.

Hotelling, H. (1936), "Simplified Computation of Principal Components," *Psychometrika*, 1, 27-35.

Huba, G. J., Wingard, J. A., and Bentler, P. M. (1981), "A Comparison of Two Latent Variable Causal Models for Adolescent Drug Use," *Journal of Personality and Social Psychology*, 40, 180-193.

Izenman, A. J. and Williams, J. S. (1989), "A Class of Linear Spectral Models and Analyses for the Study of Longitudinal Data," *Biometrics*, 45, 831-849.

Jackson, J. E. (1991), *A User's Guide to Principal Component Analysis*, New York: John Wiley & Sons, Inc.

Jackson, J. E. and Bradley, R. A. (1966), "Sequential Multivariate Procedures for Means with Quality Control Applications," in *Multivariate Analysis*, edited by P. R. Krishnaiah, 507-519, New York: Academic Press.

Jardine, N. and Sibson, R. (1968), "The Construction of Hierarchic and Nonhierarchic Classifications," *Computer Journal*, 11, 177-184.

Jeffers, J. N. R. (1967), "Two Case Studies in the Application of Principal Components Analysis," *Applied Statistics*, 16, 225-236.

Jobson, J. D. (1992), *Applied Multivariate Data Analysis, Vol. 2*, New York: Springer-Verlag.

Johnson, R. A. and Wichern, D. W. (1998), *Applied Multivariate Statistical Analysis, 4th ed.*, Englewood Cliffs, NJ: Prentice Hall.

Jolicoeur, P. and Mosimann, J. (1960), "Size and Shape Variation in the Painted Turtle: A Principal Component Analysis," *Growth*, 24, 339-354.

Jolliffe, I. T. (1972), "Discarding Variables in a Principal Component Analysis, I: Artificial Data," *Applied Statistics*, 21, 160-173.

Jolliffe, I. T. (1973), "Discarding Variables in a Principal Component Analysis, II: Real Data," *Applied Statistics*, 22, 21-31.

Jolliffe, I. T. (1982), "A Note on the Use of Principal Components in Regression," *Applied Statistics*, 31, 300-303.

Jolliffe, I. T. (1986), *Principal Component Analysis*, New York: Springer-Verlag.

Jongman, R. H. G., Ter Braak, C. J. F., and van Tongeren, O. F. R. (1997), *Data Analysis in Community and Landscape Ecology*, New York: Cambridge University Press.

Jöreskog, K. G. (1967), "Some Contributions to Maximum Likelihood Factor Analysis," *Psychometrika*, 32, 443-482.

Jöreskog, K. G. (1977), "Factor Analysis by Least Squares and Maximum-Likelihood Methods," in *Statistical Methods for Digital Computers*, Vol. III of Mathematical Methods for Digital Computers, edited by K. Enslein, A. Ralston, and H. S. Wilf, New York: John Wiley & Sons.

Kaiser, H. F. (1958), "The Varimax Criterion for Analytic Rotation in Factor Analysis," *Psychometrika*, 23, 187-200.

Kaiser, H. F. (1960), "The Application of Electronic Computers to Factor Analysis," *Educational and Psychological Measurement*, 20, 141-151.

Kaiser, H. F. (1970), "A Second Generation Little Jiffy," *Psychometrika*, 35, 401-415.

Kaiser, H. F. and Caffrey, J. (1965), "Alpha Factor Analysis," *Psychometrika*, 30, 1-14.

Kaiser, H. F. and Derflinger, G. (1990), "Some Contrasts Between Maximum Likelihood Factor Analysis and Alpha Factor Analysis," *Applied Psychological Measurement*, 14, 29-32.

Khattree, R. and Naik, D. N. (1998), "Andrews Plots for Multivariate Data: Some New Suggestions and Applications," *To appear in Journal of Statistical Planning and Inference*.

Khattree, R. and Naik, D. N. (1999), *Applied Multivariate Statistics with SAS Software, Second Edition*, a copublication of Cary, NC: SAS Institute Inc. and New York: John Wiley & Sons.

Krzanowski, W. J. and Marriott, F. H. C. (1994), *Multivariate Analysis, Part 1, Distributions, Ordination and Inference*, London: Edward Arnold.

Kshirsager, A. M. (1972), *Multivariate Analysis*, New York: Marcel Dekker.

Kshirsagar, A. M., Kocherlakota, S., and Kocherlakota, K. (1990), "Classification Procedures Using Principal Component Analysis and Stepwise Discriminant Function," *Communications in Statistics, Theory and Methods*, 19, 91–109.

Lachenbruch, P. A. (1975), *Discriminant Analysis*, New York: Hafner.

Lance, G. N. and Williams, W. T. (1967), "A General Theory of Classificatory sorting Strategies, I, Hierarchical Systems," *Computer Journal*, 1, 15-20.

Lawley, D. N. (1940), "The Estimation of Factor Loadings by the Method of Maximum Likelihood," *Proceedings of Royal Society Edinburgh (A)*, 60, 64-82.

Lawley, D. N. (1941), "Further Investigation in Factor Estimation," *Proceedings of Royal Society Edinburgh (A)*, 61, 176-185.

Lawley, D. N. (1959), "Tests of Significance in Canonical Analysis," *Biometrika*, 46, 59-66.

Lawley, D. N. and Maxwell, A. E. (1971), *Factor Analysis as a Statistical Method*, London: Butterworth & Co.

Lembart, L., Morineau, A., and Warwick, K. M. (1984), *Multivariate Descriptive Statistical Analysis: Correspondence Analysis and Related Techniques for Large Matrices*, New York: John Wiley & Sons, Inc.

Linn, R. L. (1968), "A Monte Carlo Approach to the Number of Factors Problem," *Psychometrika*, 33, 37-71.

Lubischew, A. A. (1962), "On the Use of Discriminant Functions in Taxonomy," *Biometrics*, 18, 455-477.

Mardia, K. V. (1970), "Measures of Multivariate Skewness and Kurtosis with Applications," *Biometrika*, 519-530.

Mardia, K. V., Kent, J. J., and Bibby, J. M. (1979), *Multivariate Analysis*, New York: Academic Press.

Marshall, A. W. and Olkin, I. (1979), *Inequalities: Theory of Majorization and Its Applications*, San Diego: Academic Press.

McDonald, R. P. (1970), "Three Common Factor Models for Groups of Variables," *Psychometrika*, 35, 111-128.

Miesch, A. T. (1976), "Q-Mode Factor Analysis of Geochemical and Petrologic Data Matrices with Constant Row-Sums," *Geological Survey Professional Paper*, 574-G, U.S. Dept. of the Interior.

Miller, J. K. (1969), "The Development and Application of Bi-Multivariate Correlation: a Measure of Statistical Association Between Multivariate Measurement Sets," Ed. D. Dissertation, Faculty of Educational Studies, State University of New York at Buffalo.

Milligan, G. and Cooper, M. C. (1985), "An Examination of Procedures for Determining the Number of Clusters in a Data Set," *Psychometrika*, 50, 159-179.

Mizoguchi, R. and Shimura, M. (1980), "An Algorithm for Detecting Clusters Using Hierarchical Structure," *IEEE Transactions on Pattern Analysis and Machine Intelligence*, PAMI-2, 292-300.

Morrison, D. F. (1976), *Multivariate Statistical Methods*, New York: McGraw Hill.

Muirhead, R. J. (1982), *Aspects of Multivariate Statistical Theory*, New York: John Wiley & Sons, Inc.

Mulaik, S. A. (1972), *The Foundations of Factor Analysis*, New York: McGraw-Hill.

Naik, D. N. and Khattree, R. (1996), "Revisiting Olympic Track Records: Some Practical Considerations in the Principal Component Analysis," *The American Statistician*, 50, 140-144.

Nayak, G. N. (1993), *Beaches of Karwar: Morphology, Texture and Mineralogy*, Panaji, Goa: P. Bhinde Publishers.

Nayak, G. N. and Chavadi, V. C. (1988), "Studies on Sediment Size Distribution of North Karnatak Beaches, West Coast of India, Using Empirical Orthogonal Function Analysis," *Indian Journal of Marine Sciences*, 17, 63-66.

Neuhaus, J. O. and Wrigley, C. (1954), "The Quartimax Method: An Analytical Approach to Orthogonal Simple Structure," *British Journal of Statistical Psychology*, 7, 81-91.

Nishisato, S. (1980), *Analysis of Categorical Data: Dual Scaling and Its Applications*, Toronto: University of Toronto Press.

O'Hara, T. F., Hosmer, D. W., Lemeshow, S., and Hartz, S. C. (1982), "A Comparison of Discriminant Function and Maximum Likelihood Estimates of Logistic Coefficients for Categorical-scaled Data," *Journal of Statistical Computation and Simulation*, 14, 169-178.

Okamoto, M. (1961), "Discrimination for Variance Matrices," *Osaka Mathematical Journal*, 13, 1-39.

Pearson, K. (1901), "On Lines Planes of Closest Fit to a System of Points in Space," *Philosophical Magazine*, 2, 557-572.

Pregibon, D. (1981), "Logistic regression diagnostics," *Annals of Statistics*, 9, 705–724.

Press, S. J. and Wilson, S. (1978), "Choosing Between Logistic Regression and Discriminant Analysis," *Journal of American Statistical Association*, 73, 699-705.

Rao, C. R. (1948), "The Utilization of Multiple Measurements in Problems of Biological Classification (with discussion)," *Journal of Royal Statistical Society*, Series B, 10, 159-193.

Rao, C. R. (1955), "Estimation and Tests of Significance in Factor Analysis," *Psychometrika*, 20, 93-111.

Rao, C. R. (1964), "The Use and Interpretation of Principal Components in Applied Research," *Sankhyā*, A, 26, 329-358.

Rao, C. R. (1970), "Inference on Discriminant Function Coefficients," in *Essays in Probability and Statistics*, edited by R. C. Bose et al., 587-602, Chapel Hill, NC: University of North Carolina Press.

Rao, C. R. (1973), *Linear Statistical Inference and Its Applications*, New York: John Wiley & Sons, Inc.

Rao, C. R. (1995), "A Review of Canonical Coordinates and an Alternative to Correspondence Analysis using Hellinger Distance," *QUWSTIIO*, 19, 23-63.

Rao, C. R. and Shaw, D. C. (1948), "On a Formula for the Prediction of Cranial Capacity," *Biometrics*, 4, 247-253.

Rao, C. R. and Rao, M. B. (1998), *Matrix Algebra and its Applications to Statistics and Economics*, Singapore: World Scientific.

Remme, J., Habbema, J. D. F., and Hermans, J. (1980), "A Simulative Comparison of Linear, Quadratic and Kernel Discrimination," *Journal of Statistical Computation and Simulation*, 11, 87-106.

Rencher, A. C. (1995), *Methods of Multivariate Analysis*, New York: John Wiley & Sons, Inc.

Richardson, M. and Kuder, G. F. (1933), "Making a Rating Scale that Measures," *Personal Journal*, 12, 36-40.

Rosenblatt, M. (1956), "Remarks on Some Nonparametric Estimates of a Density Function," *Annals of Mathematical Statistics*, 27, 832-837.

Royston, J. P. (1983), "Some Techniques for Assessing Multivariate Normality Based on the Shapiro-Wilk W," *Applied Statistics*, 32, 121-133.

Sarle, W. S. (1983), "Cubic Clustering Criterion," *SAS Technical Report A-108*, Cary, NC: SAS Institute Inc.

SAS Institute Inc. (1989), *SAS/STAT User's Guide, Version 6, Fourth Edition, Volume 1*, Cary, NC: SAS Institute Inc.

SAS Institute Inc. (1989), *SAS/STAT User's Guide, Version 6, Fourth Edition, Volume 2*, Cary, NC: SAS Institute Inc.

SAS Institute Inc. (1989), *SAS/IML Software: Usage and Reference, Version 6, First Edition*, Cary, NC: SAS Institute Inc.

SAS Institute Inc. (1993), *SAS/INSIGHT User's Guide, Version 6, Second Edition*, Cary, NC: SAS Institute Inc.

SAS Institute Inc. (1993), SAS Technical Report P-256: *SAS/STAT Software: The MODECLUS Procedure*, Cary, NC: SAS Institute Inc.

SAS Institute Inc. (1996), *SAS/STAT Software: Changes and Enhancements, through Release 6.11 and 6.12*, Cary, NC: SAS Institute Inc.

SAS Institute Inc., *Multivariate Statistical Methods: Practical Applications Course Notes*, Cary, NC: SAS Institute Inc.

Saunders, D. R. (1962), "Trans-varimax: Some Properties of the Ratiomax and Equamax Criteria for Blind Orthogonal Rotation," Paper presented at the meeting of the American Psychological Association, St. Louis.

Scheuchenpflug, T. and Blettner, M. (1996), "Coding Confusion using PROC LOGISTIC in SAS," *Computational Statistics and Data Analysis*, 21, 111-115.

Schott, J. R. (1997), *Matrix Analysis for Statistics*, New York: John Wiley & Sons, Inc.

Scott, D. W. (1992), *Multivariate Density Estimation: Theory, Practice and Visualization*, New York: John Wiley & Sons.

Searle, S. R. (1971), *Linear Models*, New York: John Wiley & Sons, Inc.

Seber, G. A. F. (1984), *Multivariate Observations*, New York: John Wiley & Sons, Inc.

Sharma, S. (1996), *Applied Multivariate Techniques*, New York: John Wiley & Sons, Inc.

Shumway, R. H. (1988), *Applied Statistical Time Series Analysis*, Englewood Cliffs, NJ: Prentice Hall.

Silverman, B. W. (1986), *Density Estimation for Statistics and Data Analysis*, London: Chapman and Hall.

Simonds, J. L. (1963), "Applications of Characteristic Vector Analysis to Photographic and Optical Response Data," *Journal of Optical Society of America*, 53, 968-974.

Sinha, R. N. and Lee, P. J. (1970), "Maximum Likelihood, Factor Analysis of Natural Arthropod Infestations in Stored Grain Bulks," *Research in Population Ecology*, 12, 51-60.

Smith, H. Gnanadesikan, R., and Hughes, J. B. (1962), "Multivariate Analysis of Variance (MANOVA)," *Biometrics*, 18, 22-41.

Sneath, P. H. A. and Sokal, R. R. (1973), *Numerical Taxonomy: The Principles and Practice of Numerical Classifications*, San Francisco: Freeman.

Spearman, C. (1904), "General Intelligence, Objectively Determined and Measured," *American Journal of Psychology*, 15, 201-293.

Srole, L., Langner, T. S., Michael, S. T., Kirkpatrick, P., Opler, M. K. and Rennie, T. A. C. (1978), *Mental Health in the Metropolis: The Midtown Manhattan Study*, Revised ed., New York: New York University Press.

Stewart, D. K. and Love, W. A. (1968), "A General Canonical Correlation Index," *Psychological Bulletin*, 70, 160-163.

Stokes, M. E., Davis, C. S., and Koch, G. G. (1995), *Categorical Data Analysis Using the SAS System*, Cary, NC: SAS Institute, Inc.

Ter Braak, C.J.F. (1985), "Correspondence Analysis of Incidence and Abundance Data: Properties in Terms of a Unimodal Response Model," *Biometrics*, 41, 859-873.

Ter Braak, C.J.F. (1986), "Canonical Correspondence Analysis: A New Eigenvector Technique for Multivariate Direct Gradient Analysis," *Ecology*, 67, 1167-1179.

Ter Braak, C.J.F. (1988), *CANOCO: A Fortran Program for Canonical Community Ordination by Partial Detrended Canonical Correlation Analysis, Principal Components Analysis and Redundancy Analysis (Version 2.1)*, Ithaca, NY: Microcomputer Power, USA.

The World Bank World Development Report, 1988, Washington, D.C.: The World Bank.

Thomson, G. H. (1951), *The Factorial Analysis of Human Ability*, London: London University Press.

Thurstone, L. L. (1947), *Multiple Factor Analysis*, Chicago: University of Chicago Press.

Tong, Y. L. (1990), *Multivariate Normal Distribution*, New York: Springer Verlag.

U. S. News & World Report, July 19, 1999, 77.

Vance, L. C. (1983), "Statistical Determination of Numerical Color Tolerances," *Modern Paint and Coatings*, 73, 49-51.

Vance, L. C. (1989), "Statistical Determination of Numerical Color Tolerances," in *Statistics, A Guide to the Unknown*, edited by Tanur et al. 3rd ed., 170-177.

Van der Aart, P. J. M. and Smeek-Enserink, N. (1975), "Correlations Between Distributions of Hunting Spiders and Environmental Characteristics in a Dune Area," *Netherlands Journal of Zoology*, 25, 1-45.

Van Driel, O. P. (1978), "On Various Causes of Improper Solutions in Maximum Likelihood Factor Analysis," *Psychometrika*, 43, 225-243.

Velicer, W. F. (1977), "An Empirical Comparison of the Similarity of Principal Component, Image and Factor Patterns," *Multivariate Behavioral Research*, 12, 3-22.

Velu, R. P. and Wichern, D. W. (1985), "A Note on Computing Wilks' Lambda for the Multivariate General Linear Test," *Journal of Statistical Computing and Simulation*, 20, 333-340.

Waugh, F. W. (1942), "Regression Between Sets of Variates," *Econometrica*, 10, 290-310.

Whittaker, R. H., (1956), "Vegetation of the Great Smoky Mountains," *Ecological Monographs*, 26, 1-80.

Whittaker, R. H., (1967) "Gradient Analysis of Vegetation," *Biological Reviews of the Cambridge Philosophical Society*, 49, 207-264.

Williams, J. S. and Izenman, A. J. (1981), "A Class of Linear Spectral Models and Analyses for the Study of Longitudinal Data." Technical Report, Dept. of Statistics, Colorado State University.

Williams, W. T. and Lance, G. N. (1977), "Hierarchical Classificatory Methods," in *Statistical Methods for Digital Computers*, edited by K. Enslein et al., 269-295, Volume 3, New York: John Wiley & Sons, Inc.

Williams, W. T., Lance, G. N., Dale, M. B, and Clifford, H. T. (1971), "Controversy Concerning the Criteria for Taxonometric Strategies," *Computer Journal*, 14, 162-165.

Wright, S. (1954), "The Interpretation of Multivariate Systems," in *Statistics and Mathematics in Biology*, edited by O. Kempthorne, T. A. Bancroft, J. W. Gowen, and J. L. Lush, 11-33, Ames: Iowa State University Press.

Index

A

B

C

D

E

F

G

H

I

J

K

L

M

N

O

P

Q

R

S

W

Special Characters

Books and Tapes from SAS® Institute's Books by Users℠ program:

An Array of Challenges — Test Your SAS® Skills
by **Robert Virgile**

Annotate: Simply the Basics
by **Art Carpenter**

Applied Multivariate Statistics with SAS® Software, Second Edition
by **Ravindra Khattree**
and **Dayanand N. Naik**

Applied Statistics and the SAS® Programming Language, Fourth Edition
by **Ronald P. Cody**
and **Jeffrey K. Smith**

Beyond the Obvious with SAS® Screen Control Language
by **Don Stanley**

Carpenter's Complete Guide to the SAS® Macro Language
by **Art Carpenter**

The Cartoon Guide to Statistics
by **Larry Gonick**
and **Woollcott Smith**

Categorical Data Analysis Using the SAS® System
by **Maura E. Stokes, Charles S. Davis,**
and **Gary G. Koch**

Cody's Data Cleaning Techniques Using SAS® Software
by **Ron Cody**

Common Statistical Methods for Clinical Research with SAS® Examples
by **Glenn A. Walker**

Concepts and Case Studies in Data Management
by **William S. Calvert**
and **J. Meimei Ma**

Efficiency: Improving the Performance of Your SAS® Applications
by **Robert Virgile**

Essential Client/Server Survival Guide, Second Edition
by **Robert Orfali, Dan Harkey,**
and **Jeri Edwards**

Extending SAS® Survival Analysis Techniques for Medical Research
by **Alan Cantor**

A Handbook of Statistical Analyses Using SAS®
by **B.S. Everitt**
and **G. Der**

The How-To Book for SAS/GRAPH® Software
by **Thomas Miron**

In the Know ... SAS® Tips and Techniques From Around the Globe
by **Phil Mason**

Integrating Results through Meta-Analytic Review Using SAS® Software
by **Morgan C. Wang**
and **Brad J. Bushman**

Learning SAS® in the Computer Lab
by **Rebecca J. Elliott**

The Little SAS® Book: A Primer
by **Lora D. Delwiche**
and **Susan J. Slaughter**

The Little SAS® Book: A Primer, Second Edition
by **Lora D. Delwiche**
and **Susan J. Slaughter**
(updated to include Version 7 features)

Logistic Regression Using the SAS System: Theory and Application
by **Paul D. Allison**

Mastering the SAS® System, Second Edition
by **Jay A. Jaffe**

Multiple Comparisons and Multiple Tests Using the SAS® System
by **Peter H. Westfall, Randall D. Tobias,**
Dror Rom, Russell D. Wolfinger,
and **Yosef Hochberg**

The Next Step: Integrating the Software Life Cycle with SAS® Programming
by **Paul Gill**

Multivariate Data Reduction and Discrimination with SAS® Software
by **Ravindra Khattree**
and **Dayanand N. Naik**

Painless Windows 3.1: A Beginner's Handbook for SAS® Users
by **Jodie Gilmore**

Painless Windows: A Handbook for SAS® Users
by **Jodie Gilmore**
(for Windows NT and Windows 95)

Painless Windows: A Handbook for SAS® Users, Second Edition
by **Jodie Gilmore**
(updated to include Version 7 features)

PROC TABULATE by Example
by **Lauren E. Haworth**

Professional SAS® Programmers Pocket Reference, Second Edition
by **Rick Aster**

Professional SAS® Programming Secrets, Second Edition
by **Rick Aster**
and **Rhena Seidman**

Professional SAS® User Interfaces
by **Rick Aster**

Programming Techniques for Object-Based Statistical Analysis with SAS® Software
by **Tanya Kolosova**
and **Samuel Berestizhevsky**

Quick Results with SAS/GRAPH® Software
by **Arthur L. Carpenter**
and **Charles E. Shipp**

Quick Start to Data Analysis with SAS®
by **Frank C. Dilorio**
and **Kenneth A. Hardy**

Reporting from the Field: SAS® Software Experts Present Real-World Report-Writing Applications

SAS® Applications Programming: A Gentle Introduction
by **Frank C. Dilorio**

SAS® Foundations: From Installation to Operation
by **Rick Aster**

SAS® Macro Programming Made Easy
by **Michele M. Burlew**

SAS® Programming by Example
by **Ron Cody**
and **Ray Pass**

SAS® Programming for Researchers and Social Scientists
by **Paul E. Spector**

SAS® Software Roadmaps: Your Guide to Discovering the SAS® System
by **Laurie Burch**
and **SherriJoyce King**

SAS® Software Solutions: Basic Data Processing
by **Thomas Miron**

SAS® System for Elementary Statistical Analysis, Second Edition
by **Sandra D. Schlotzhauer**
and **Dr. Ramon C. Littell**

SAS® System for Forecasting Time Series, 1986 Edition
by **John C. Brocklebank**
and **David A. Dickey**

SAS® System for Linear Models, Third Edition
by **Ramon C. Littell, Rudolf J. Freund,**
and **Philip C. Spector**

SAS® System for Mixed Models
by **Ramon C. Littell, George A. Milliken, Walter W. Stroup,**
and **Russell D. Wolfinger**

SAS® System for Regression, Second Edition
by **Rudolf J. Freund**
and **Ramon C. Littell**

SAS® System for Statistical Graphics, First Edition
by **Michael Friendly**

SAS® Today! A Year of Terrific Tips
by **Helen Carey** *and* **Ginger Carey**

The SAS® Workbook and Solutions (books in this set also sold separately)
by **Ron Cody**

Selecting Statistical Techniques for Social Science Data: A Guide for SAS® Users
by **Frank M. Andrews, Laura Klem, Patrick M. O'Malley, Willard L. Rodgers, Kathleen B. Welch,**
and **Terrence N. Davidson**

Solutions for your GUI Applications Development Using SAS/AF® FRAME Technology
by **Don Stanley**

Statistical Quality Control Using the SAS® System
by **Dennis W. King, Ph.D.**

A Step-by-Step Approach to Using the SAS® System for Factor Analysis and Structural Equation Modeling
by **Larry Hatcher**

A Step-by-Step Approach to Using the SAS® System for Univariate and Multivariate Statistics
by **Larry Hatcher**
and **Edward Stepanski**

Strategic Data Warehousing Principles Using SAS® Software
by **Peter R. Welbrock**

Survival Analysis Using the SAS® System: A Practical Guide
by **Paul D. Allison**

Table-Driven Strategies for Rapid SAS® Applications Development
by **Tanya Kolosova**
and **Samuel Berestizhevsky**

Tuning SAS® Applications in the MVS Environment
by **Michael A. Raithel**

Univariate and Multivariate General Linear Models: Theory and Applications Using SAS® Software
by **Neil H. Timm**
and **Tammy A. Mieczkowski**

Visualizing Categorical Data
by **Michael Friendly**

Working with the SAS® System
by **Erik W. Tilanus**

Your Guide to Survey Research Using the SAS® System
by **Archer Gravely**

JMP® Books

Basic Business Statistics: A Casebook
by **Dean P. Foster, Robert A. Stine,**
and **Richard P. Waterman**

Business Analysis Using Regression: A Casebook
by **Dean P. Foster, Robert A. Stine,**
and **Richard P. Waterman**

JMP® Start Statistics, Version 3
by **John Sall** *and* **Ann Lehman**

WILEY SERIES IN PROBABILITY AND STATISTICS

ESTABLISHED BY WALTER A. SHEWHART AND SAMUEL S. WILKS

Probability and Statistics Section

*ANDERSON · The Statistical Analysis of Time Series
ARNOLD, BALAKRISHNAN, and NAGARAJA · A First Course in Order Statistics
ARNOLD, BALAKRISHNAN, and NAGARAJA · Records
BACCELLI, COHEN, OLSDER, and QUADRAT · Synchronization and Linearity: An Algebra for Discrete Event Systems
BARNETT · Comparative Statistical Inference, *Third Edition*
BASILEVSKY · Statistical Factor Analysis and Related Methods: Theory and Applications
BERNARDO and SMITH · Bayesian Statistical Concepts and Theory
BILLINGSLEY · Convergence of Probability Measures, *Second Edition*
BOROVKOV · Asymptotic Methods in Queuing Theory
BOROVKOV · Ergodicity and Stability of Stochastic Processes
BRANDT, FRANKEN, and LISEK · Stationary Stochastic Models
CAINES · Linear Stochastic Systems
CAIROLI and DALANG · Sequential Stochastic Optimization
CONSTANTINE · Combinatorial Theory and Statistical Design
COOK · Regression Graphics
COVER and THOMAS · Elements of Information Theory
CSÖRGŐ and HORVÁTH · Weighted Approximations in Probability Statistics
CSÖRGŐ and HORVÁTH · Limit Theorems in Change Point Analysis
*DANIEL · Fitting Equations to Data: Computer Analysis of Multifactor Data, *Second Edition*
DETTE and STUDDEN · The Theory of Canonical Moments with Applications in Statistics, Probability, and Analysis
DEY and MUKERJEE · Fractional Factorial Plans
*DOOB · Stochastic Processes
DRYDEN and MARDIA · Statistical Shape Analysis
DUPUIS and ELLIS · A Weak Convergence Approach to the Theory of Large Deviations
ETHIER and KURTZ · Markov Processes: Characterization and Convergence
FELLER · An Introduction to Probability Theory and Its Applications, Volume 1, *Third Edition,* Revised; Volume II, *Second Edition*
FULLER · Introduction to Statistical Time Series, *Second Edition*
FULLER · Measurement Error Models
GHOSH, MUKHOPADHYAY, and SEN · Sequential Estimation
GIFI · Nonlinear Multivariate Analysis
GUTTORP · Statistical Inference for Branching Processes
HALL · Introduction to the Theory of Coverage Processes
HAMPEL · Robust Statistics: The Approach Based on Influence Functions
HANNAN and DEISTLER · The Statistical Theory of Linear Systems
HUBER · Robust Statistics
HUSKOVA, BERAN, and DUPAC · Collected Works of Jaroslav Hajek—with Commentary
IMAN and CONOVER · A Modern Approach to Statistics
JUREK and MASON · Operator-Limit Distributions in Probability Theory
KASS and VOS · Geometrical Foundations of Asymptotic Inference
KAUFMAN and ROUSSEEUW · Finding Groups in Data: An Introduction to Cluster Analysis
KELLY · Probability, Statistics, and Optimization
KENDALL, BARDEN, CARNE, and LE · Shape and Shape Theory
LINDVALL · Lectures on the Coupling Method
McFADDEN · Management of Data in Clinical Trials
MANTON, WOODBURY, and TOLLEY · Statistical Applications Using Fuzzy Sets
MORGENTHALER and TUKEY · Configural Polysampling: A Route to Practical Robustness
MUIRHEAD · Aspects of Multivariate Statistical Theory
OLIVER and SMITH · Influence Diagrams, Belief Nets and Decision Analysis
*PARZEN · Modern Probability Theory and Its Applications
PRESS · Bayesian Statistics: Principles, Models, and Applications
PUKELSHEIM · Optimal Experimental Design
RAO · Asymptotic Theory of Statistical Inference
RAO · Linear Statistical Inference and Its Applications, *Second Edition*

*Now available in a lower priced paperback edition in the Wiley Classics Library.

Probability and Statistics (Continued)

RAO and SHANBHAG · Choquet-Deny Type Functional Equations with Applications to Stochastic Models
ROBERTSON, WRIGHT, and DYKSTRA · Order Restricted Statistical Inference
ROGERS and WILLIAMS · Diffusions, Markov Processes, and Martingales, Volume I: Foundations, *Second Edition;* Volume II: Îto Calculus
RUBINSTEIN and SHAPIRO · Discrete Event Systems: Sensitivity Analysis and Stochastic Optimization by the Score Function Method
RUZSA and SZEKELY · Algebraic Probability Theory
SCHEFFE · The Analysis of Variance
SEBER · Linear Regression Analysis
SEBER · Multivariate Observations
SEBER and WILD · Nonlinear Regression
SERFLING · Approximation Theorems of Mathematical Statistics
SHORACK and WELLNER · Empirical Processes with Applications to Statistics
SMALL and McLEISH · Hilbert Space Methods in Probability and Statistical Inference
STAPLETON · Linear Statistical Models
STAUDTE and SHEATHER · Robust Estimation and Testing
STOYANOV · Counterexamples in Probability
TANAKA · Time Series Analysis: Nonstationary and Noninvertible Distribution Theory
THOMPSON and SEBER · Adaptive Sampling
WELSH · Aspects of Statistical Inference
WHITTAKER · Graphical Models in Applied Multivariate Statistics
YANG · The Construction Theory of Denumerable Markov Processes

Applied Probability and Statistics Section

ABRAHAM and LEDOLTER · Statistical Methods for Forecasting
AGRESTI · Analysis of Ordinal Categorical Data
AGRESTI · Categorical Data Analysis
ANDERSON, AUQUIER, HAUCK, OAKES, VANDAELE, and WEISBERG · Statistical Methods for Comparative Studies
ARMITAGE and DAVID (editors) · Advances in Biometry
*ARTHANARI and DODGE · Mathematical Programming in Statistics
ASMUSSEN · Applied Probability and Queues
*BAILEY · The Elements of Stochastic Processes with Applications to the Natural Sciences
BARNETT and LEWIS · Outliers in Statistical Data, *Third Edition*
BARTHOLOMEW, FORBES, and McLEAN · Statistical Techniques for Manpower Planning, *Second Edition*
BASU and RIGDON · Statistical Methods for the Reliability of Repairable Systems
BATES and WATTS · Nonlinear Regression Analysis and Its Applications
BECHHOFER, SANTNER, and GOLDSMAN · Design and Analysis of Experiments for Statistical Selection, Screening, and Multiple Comparisons
BELSLEY · Conditioning Diagnostics: Collinearity and Weak Data in Regression
BELSLEY, KUH, and WELSCH · Regression Diagnostics: Identifying Influential Data and Sources of Collinearity
BHAT · Elements of Applied Stochastic Processes, *Second Edition*
BHATTACHARYA and WAYMIRE · Stochastic Processes with Applications
BIRKES and DODGE · Alternative Methods of Regression
BLISCHKE AND MURTHY · Reliability: Modeling, Prediction, and Optimization
BLOOMFIELD · Fourier Analysis of Time Series: An Introduction, *Second Edition*
BOLLEN · Structural Equations with Latent Variables
BOULEAU · Numerical Methods for Stochastic Processes
BOX · Bayesian Inference in Statistical Analysis
BOX and DRAPER · Empirical Model-Building and Response Surfaces
*BOX and DRAPER · Evolutionary Operation: A Statistical Method for Process Improvement
BUCKLEW · Large Deviation Techniques in Decision, Simulation, and Estimation
BUNKE and BUNKE · Nonlinear Regression, Functional Relations and Robust Methods: Statistical Methods of Model Building
CHATTERJEE and HADI · Sensitivity Analysis in Linear Regression
CHERNICK · Bootstrap Methods: A Practitioner's Guide
CHILÈS and DELFINER · Geostatistics: Modeling Spatial Uncertainty
CHOW and LIU · Design and Analysis of Clinical Trials: Concepts and Methodologies
CLARKE and DISNEY · Probability and Random Processes: A First Course with Applications, *Second Edition*
*COCHRAN and COX · Experimental Designs, *Second Edition*
CONOVER · Practical Nonparametric Statistics, *Second Edition*
CORNELL · Experiments with Mixtures, Designs, Models, and the Analysis of Mixture Data, *Second Edition*
*COX · Planning of Experiments
CRESSIE · Statistics for Spatial Data, *Revised Edition*

*Now available in a lower priced paperback edition in the Wiley Classics Library.

DANIEL · Applications of Statistics to Industrial Experimentation
DANIEL · Biostatistics: A Foundation for Analysis in the Health Sciences, *Sixth Edition*
DAVID · Order Statistics, *Second Edition*
*DEGROOT, FIENBERG, and KADANE · Statistics and the Law
DODGE · Alternative Methods of Regression
DOWDY and WEARDEN · Statistics for Research, *Second Edition*
DUNN and CLARK · Applied Statistics: Analysis of Variance and Regression, *Second Edition*
*ELANDT-JOHNSON and JOHNSON · Survival Models and Data Analysis
EVANS, PEACOCK, and HASTINGS · Statistical Distributions, *Second Edition*
*FLEISS · The Design and Analysis of Clinical Experiments
FLEISS · Statistical Methods for Rates and Proportions, *Second Edition*
FLEMING and HARRINGTON · Counting Processes and Survival Analysis
GALLANT · Nonlinear Statistical Models
GLASSERMAN and YAO · Monotone Structure in Discrete-Event Systems
GNANADESIKAN · Methods for Statistical Data Analysis of Multivariate Observations, *Second Edition*
GOLDSTEIN and LEWIS · Assessment: Problems, Development, and Statistical Issues
GREENWOOD and NIKULIN · A Guide to Chi-Squared Testing
*HAHN · Statistical Models in Engineering
HAHN and MEEKER · Statistical Intervals: A Guide for Practitioners
HAND · Construction and Assessment of Classification Rules
HAND · Discrimination and Classification
HEIBERGER · Computation for the Analysis of Designed Experiments
HEDAYAT and SINHA · Design and Inference in Finite Population Sampling
HINKELMAN and KEMPTHORNE: · Design and Analysis of Experiments, Volume 1: Introduction to Experimental Design
HOAGLIN, MOSTELLER, and TUKEY · Exploratory Approach to Analysis of Variance
HOAGLIN, MOSTELLER, and TUKEY · Exploring Data Tables, Trends and Shapes
HOAGLIN, MOSTELLER, and TUKEY · Understanding Robust and Exploratory Data Analysis
HOCHBERG and TAMHANE · Multiple Comparison Procedures
HOCKING · Methods and Applications of Linear Models: Regression and the Analysis of Variables
HOGG and KLUGMAN · Loss Distributions
HOSMER and LEMESHOW · Applied Logistic Regression
HØYLAND and RAUSAND · System Reliability Theory: Models and Statistical Methods
HUBERTY · Applied Discriminant Analysis
JACKSON · A User's Guide to Principle Components
JOHN · Statistical Methods in Engineering and Quality Assurance
JOHNSON · Multivariate Statistical Simulation
JOHNSON and KOTZ · Distributions in Statistics
JOHNSON, KOTZ, and BALAKRISHNAN · Continuous Univariate Distributions, Volume 1, *Second Edition*
JOHNSON, KOTZ, and BALAKRISHNAN · Continuous Univariate Distributions, Volume 2, *Second Edition*
JOHNSON, KOTZ, and BALAKRISHNAN · Discrete Multivariate Distributions
JOHNSON, KOTZ, and KEMP · Univariate Discrete Distributions, *Second Edition*
JUREČKOVÁ and SEN · Robust Statistical Procedures: Aymptotics and Interrelations
KADANE · Bayesian Methods and Ethics in a Clinical Trial Design
KADANE AND SCHUM · A Probabilistic Analysis of the Sacco and Vanzetti Evidence
KALBFLEISCH and PRENTICE · The Statistical Analysis of Failure Time Data
KELLY · Reversability and Stochastic Networks
KHURI, MATHEW, and SINHA · Statistical Tests for Mixed Linear Models
KLUGMAN, PANJER, and WILLMOT · Loss Models: From Data to Decisions
KLUGMAN, PANJER, and WILLMOT · Solutions Manual to Accompany Loss Models: From Data to Decisions
KOTZ, BALAKRISHNAN, and JOHNSON · Continuous Multivariate Distributions, Volume 1, *Second Edition*
KOVALENKO, KUZNETZOV, and PEGG · Mathematical Theory of Reliability of Time-Dependent Systems with Practical Applications
LACHIN · Biostatistical Methods: The Assessment of Relative Risks
LAD · Operational Subjective Statistical Methods: A Mathematical, Philosophical, and Historical Introduction
LANGE, RYAN, BILLARD, BRILLINGER, CONQUEST, and GREENHOUSE · Case Studies in Biometry
LAWLESS · Statistical Models and Methods for Lifetime Data
LEE · Statistical Methods for Survival Data Analysis, *Second Edition*
LEPAGE and BILLARD · Exploring the Limits of Bootstrap
LINHART and ZUCCHINI · Model Selection
LITTLE and RUBIN · Statistical Analysis with Missing Data
LLOYD · The Statistical Analysis of Categorical Data
MAGNUS and NEUDECKER · Matrix Differential Calculus with Applications in Statistics and Econometrics, *Revised Edition*
MALLER and ZHOU · Survival Analysis with Long Term Survivors
MANN, SCHAFER, and SINGPURWALLA · Methods for Statistical Analysis of Reliability and Life Data
McLACHLAN and KRISHNAN · The EM Algorithm and Extensions
McLACHLAN · Discriminant Analysis and Statistical Pattern Recognition
McNEIL · Epidemiological Research Methods

*Now available in a lower priced paperback edition in the Wiley Classics Library.

Applied Probability and Statistics (Continued)

MEEKER and ESCOBAR · Statistical Methods for Reliability Data
*MILLER · Survival Analysis, *Second Edition*
MONTGOMERY and PECK · Introduction to Linear Regression Analysis, *Second Edition*
MYERS and MONTGOMERY · Response Surface Methodology: Process and Product in Optimization Using Designed Experiments
NELSON · Accelerated Testing, Statistical Models, Test Plans, and Data Analyses
NELSON · Applied Life Data Analysis
OCHI · Applied Probability and Stochastic Processes in Engineering and Physical Sciences
OKABE, BOOTS, and SUGIHARA · Spatial Tesselations: Concepts and Applications of Voronoi Diagrams
PANKRATZ · Forecasting with Dynamic Regression Models
PANKRATZ · Forecasting with Univariate Box-Jenkins Models: Concepts and Cases
PIANTADOSI · Clinical Trials: A Methodologic Perspective
PORT · Theoretical Probability for Applications
PUTERMAN · Markov Decision Processes: Discrete Stochastic Dynamic Programming
RACHEV · Probability Metrics and the Stability of Stochastic Models
RÉNYI · A Diary on Information Theory
RIPLEY · Spatial Statistics
RIPLEY · Stochastic Simulation
ROLSKI, SCHMIDLI, SCHMIDT, and TEUGELS · Stochastic Processes for Insurance and Finance
ROUSSEEUW and LEROY · Robust Regression and Outlier Detection
RUBIN · Multiple Imputation for Nonresponse in Surveys
RUBINSTEIN · Simulation and the Monte Carlo Method
RUBINSTEIN and MELAMED · Modern Simulation and Modeling
RYAN · Statistical Methods for Quality Improvement, *Second Edition*
SCHUSS · Theory and Applications of Stochastic Differential Equations
SCOTT · Multivariate Density Estimation: Theory, Practice, and Visualization
*SEARLE · Linear Models
SEARLE · Linear Models for Unbalanced Data
SEARLE, CASELLA, and McCULLOCH · Variance Components
SENNOTT · Stochastic Dynamic Programming and the Control of Queueing Systems
STOYAN, KENDALL, and MECKE · Stochastic Geometry and Its Applications, *Second Edition*
STOYAN and STOYAN · Fractals, Random Shapes and Point Fields: Methods of Geometrical Statistics
THOMPSON · Empirical Model Building
THOMPSON · Sampling
THOMPSON · Simulation: A Modeler's Approach
TIJMS · Stochastic Modeling and Analysis: A Computational Approach
TIJMS · Stochastic Models: An Algorithmic Approach
TITTERINGTON, SMITH, and MAKOV · Statistical Analysis of Finite Mixture Distributions
UPTON and FINGLETON · Spatial Data Analysis by Example, Volume I: Point Pattern and Quantitative Data
UPTON and FINGLETON · Spatial Data Analysis by Example, Volume II: Categorical and Directional Data
VAN RIJCKEVORSEL and DE LEEUW · Component and Correspondence Analysis
VIDAKOVIC · Statistical Modeling by Wavelets
WEISBERG · Applied Linear Regression, *Second Edition*
WESTFALL and YOUNG · Resampling-Based Multiple Testing: Examples and Methods for p-Value Adjustment
WHITTLE · Systems in Stochastic Equilibrium
WOODING · Planning Pharmaceutical Clinical Trials: Basic Statistical Principles
WOOLSON · Statistical Methods for the Analysis of Biomedical Data
*ZELLNER · An Introduction to Bayesian Inference in Econometrics

Texts and References Section

AGRESTI · An Introduction to Categorical Data Analysis
ANDERSON · An Introduction to Multivariate Statistical Analysis, *Second Edition*
ANDERSON and LOYNES · The Teaching of Practical Statistics
ARMITAGE and COLTON · Encyclopedia of Biostatistics: Volumes 1 to 6 with Index
BARTOSZYNSKI and NIEWIADOMSKA-BUGAJ · Probability and Statistical Inference
BENDAT and PIERSOL · Random Data: Analysis and Measurement Procedures, *Third Edition*
BERRY, CHALONER, and GEWEKE · Bayesian Analysis in Statistics and Econometrics: Essays in Honor of Arnold Zellner
BHATTACHARYA and JOHNSON · Statistical Concepts and Methods
BILLINGSLEY · Probability and Measure, *Second Edition*
BOX · R. A. Fisher, the Life of a Scientist
BOX, HUNTER, and HUNTER · Statistics for Experimenters: An Introduction to Design, Data Analysis, and Model Building
BOX and LUCEÑO · Statistical Control by Monitoring and Feedback Adjustment
BROWN and HOLLANDER · Statistics: A Biomedical Introduction
CHATTERJEE and PRICE · Regression Analysis by Example, *Third Edition*

*Now available in a lower priced paperback edition in the Wiley Classics Library.

Texts and References (Continued)

COOK and WEISBERG · Applied Regression Including Computing and Graphics
COOK and WEISBERG · An Introduction to Regression Graphics
COX · A Handbook of Introductory Statistical Methods
DILLON and GOLDSTEIN · Multivariate Analysis: Methods and Applications
*DODGE and ROMIG · Sampling Inspection Tables, *Second Edition*
DRAPER and SMITH · Applied Regression Analysis, *Third Edition*
DUDEWICZ and MISHRA · Modern Mathematical Statistics
DUNN · Basic Statistics: A Primer for the Biomedical Sciences, *Second Edition*
FISHER and VAN BELLE · Biostatistics: A Methodology for the Health Sciences
FREEMAN and SMITH · Aspects of Uncertainty: A Tribute to D. V. Lindley
GROSS and HARRIS · Fundamentals of Queueing Theory, *Third Edition*
HALD · A History of Probability and Statistics and their Applications Before 1750
HALD · A History of Mathematical Statistics from 1750 to 1930
HELLER · MACSYMA for Statisticians
HOEL · Introduction to Mathematical Statistics, *Fifth Edition*
HOLLANDER and WOLFE · Nonparametric Statistical Methods, *Second Edition*
HOSMER and LEMESHOW · Applied Survival Analysis: Regression Modeling of Time to Event Data
JOHNSON and BALAKRISHNAN · Advances in the Theory and Practice of Statistics: A Volume in Honor of Samuel Kotz
JOHNSON and KOTZ (editors) · Leading Personalities in Statistical Sciences: From the Seventeenth Century to the Present
JUDGE, GRIFFITHS, HILL, LÜTKEPOHL, and LEE · The Theory and Practice of Econometrics, *Second Edition*
KHURI · Advanced Calculus with Applications in Statistics
KOTZ and JOHNSON (editors) · Encyclopedia of Statistical Sciences: Volumes 1 to 9 wtih Index
KOTZ and JOHNSON (editors) · Encyclopedia of Statistical Sciences: Supplement Volume
KOTZ, REED, and BANKS (editors) · Encyclopedia of Statistical Sciences: Update Volume 1
KOTZ, REED, and BANKS (editors) · Encyclopedia of Statistical Sciences: Update Volume 2
LAMPERTI · Probability: A Survey of the Mathematical Theory, *Second Edition*
LARSON · Introduction to Probability Theory and Statistical Inference, *Third Edition*
LE · Applied Categorical Data Analysis
LE · Applied Survival Analysis
MALLOWS · Design, Data, and Analysis by Some Friends of Cuthbert Daniel
MARDIA · The Art of Statistical Science: A Tribute to G. S. Watson
MASON, GUNST, and HESS · Statistical Design and Analysis of Experiments with Applications to Engineering and Science
MURRAY · X-STAT 2.0 Statistical Experimentation, Design Data Analysis, and Nonlinear Optimization
PURI, VILAPLANA, and WERTZ · New Perspectives in Theoretical and Applied Statistics
RENCHER · Linear Models in Statistics
RENCHER · Methods of Multivariate Analysis
RENCHER · Multivariate Statistical Inference with Applications
ROSS · Introduction to Probability and Statistics for Engineers and Scientists
ROHATGI · An Introduction to Probability Theory and Mathematical Statistics
RYAN · Modern Regression Methods
SCHOTT · Matrix Analysis for Statistics
SEARLE · Matrix Algebra Useful for Statistics
STYAN · The Collected Papers of T. W. Anderson: 1943–1985
TIERNEY · LISP-STAT: An Object-Oriented Environment for Statistical Computing and Dynamic Graphics
WONNACOTT and WONNACOTT · Econometrics, *Second Edition*
WU and HAMADA · Experiments: Planning, Analysis, and Parameter Design Optimization

WILEY SERIES IN PROBABILITY AND STATISTICS

ESTABLISHED BY WALTER A. SHEWHART AND SAMUEL S. WILKS

Editors
Robert M. Groves, Graham Kalton, J. N. K. Rao, Norbert Schwarz, Christopher Skinner

Survey Methodology Section

BIEMER, GROVES, LYBERG, MATHIOWETZ, and SUDMAN · Measurement Errors in Surveys
COCHRAN · Sampling Techniques, *Third Edition*
COUPER, BAKER, BETHLEHEM, CLARK, MARTIN, NICHOLLS, and O'REILLY (editors) · Computer Assisted Survey Information Collection
COX, BINDER, CHINNAPPA, CHRISTIANSON, COLLEDGE, and KOTT (editors) · Business Survey Methods
*DEMING · Sample Design in Business Research
DILLMAN · Mail and Telephone Surveys: The Total Design Method

*Now available in a lower priced paperback edition in the Wiley Classics Library.

Survey Methodology (Continued)

GROVES and COUPER · Nonresponse in Household Interview Surveys
GROVES · Survey Errors and Survey Costs
GROVES, BIEMER, LYBERG, MASSEY, NICHOLLS, and WAKSBERG · Telephone Survey Methodology
*HANSEN, HURWITZ, and MADOW · Sample Survey Methods and Theory, Volume 1: Methods and Applications
*HANSEN, HURWITZ, and MADOW · Sample Survey Methods and Theory, Volume II: Theory
KISH · Statistical Design for Research
*KISH · Survey Sampling
KORN and GRAUBARD · Analysis of Health Surveys
LESSLER and KALSBEEK · Nonsampling Error in Surveys
LEVY and LEMESHOW · Sampling of Populations: Methods and Applications, *Third Edition*
LYBERG, BIEMER, COLLINS, de LEEUW, DIPPO, SCHWARZ, TREWIN (editors) · Survey Measurement and Process Quality
SIRKEN, HERRMANN, SCHECHTER, SCHWARZ, TANUR, and TOURANGEAU (editors) · Cognition and Survey Research

*Now available in a lower priced paperback edition in the Wiley Classics Library.

Printed in the United States
89002LV00006B/152/A